Louis Leithold
Pepperdine University

College Algebra and Trigonometry

ADDISON-WESLEY PUBLISHING COMPANY

Reading, Massachusetts ■ Menlo Park, California ■ New York
Don Mills, Ontario ■ Wokingham, England ■ Amsterdam ■ Bonn
Sydney ■ Singapore ■ Tokyo ■ Madrid ■ San Juan

Sponsoring Editor: *David F. Pallai*
Production Manager: *Barbara Pendergast*
Production Supervisor: *Marion E. Howe*
Text Designer: *Piñeiro Design Associates*
Cover Drawing: *Allen Harrison, courtesy of Tortue Gallery, Santa Monica, California*
Cover Designer: *Marshall Henrichs*
Illustrator: *J & R Technical Services, Inc.*
Art Coordinator: *Loretta Bailey*
Copy Editor: *Emily Arulpragassam*
Manufacturing Supervisor: *Roy Logan*

Library of Congress Cataloging-in-Publication Data

Leithold, Louis.
    College algebra and trigonometry / by Louis Leithold.
      p. cm.
    ISBN 0-201-15730-6 :
    1. Algebra. 2. Trigonometry.   I. Title
QA154.2.L433 1989
512′.13—dc19   88-15422

To my son Marc
and his sons Justin and Matthew

THE COVER ARTIST

The drawing reproduced on the cover is by Allen Harrison, a painter living and working in Los Angeles. His art deals with multiple planes of color and surface considerations involving a reductive technique. Harrison builds up layers of paint and wet sands between each layer to leave smooth squares or color rectangles. As the planes overlap, edges emerge with varying degrees of visibility that divide the interiors of the planes. Often he will use transparent or translucent colors built up over a layer of opaque color to create a luminous surface. In a structural sense, he uses symmetrical compositions which are layered in such a way as to alter the appearance of symmetry.

Harrison is represented by Tortue Gallery, Santa Monica, and O.K. Harris Works of Art, New York City.

Preface

In a course titled "College Algebra and Trigonometry" a student should gain an appreciation of mathematics as a logical science, and the subject matter should be expounded in such a way that it conforms to the experience and maturity of the freshman mathematics student. Furthermore, the course is rarely a terminal one in mathematics, and thus another of its purposes is to present the means to develop skills that will enable a person to study effectively more-advanced courses. With these objectives in mind, I have attempted to reflect the consensus that mathematics should be meaningful, and I have made every effort to write a textbook that students can read advantageously on their own.

Chapter 1 I assume that the student has had courses in plane geometry and intermediate algebra, or the equivalent, and may need a review of some of the material but not a redevelopment. Consequently, the pace is quick for the algebra topics that appear in Chapter 1.

The sections in this chapter may be covered in detail, treated as a review, or omitted. If the chapter is omitted, I recommend that students read on their own Section 1.1 because the understanding of subsequent material requires a familiarity with facts about the set of real numbers and its subsets. The last section in the chapter introduces the set of complex numbers in preparation for their appearance later as solutions to some equations.

Chapter 2 The first four sections of Chapter 2, devoted to solving linear and quadratic equations in one variable, also constitute a review of Intermediate Algebra. In Section 2.2 a step-by-step method, with numerous examples, is given as a possible procedure for solving word problems. Because of the importance of inequalities in later work in mathematics, the last three sections of the chapter should be studied in detail. Polynomial and rational inequalities are solved by finding critical numbers and determining the sign of the polynomial or rational expression on intervals on the real number line. We use extensively tables summarizing the results and figures showing the intervals to obtain solution sets of inequalities.

v

Chapter 3 An introduction to analytic geometry is presented in Chapter 3. The topics appearing include the traditional material on lines, circles, and parabolas. Graphs of these curves are discussed here so that they are available prior to the treatment of functions in Chapter 4. Supplementary Section 2.6 gives a brief discussion of the ellipse, the hyperbola, and conic sections.

Chapter 4 Chapter 4 is especially significant because the notion of a function is crucial for the study of more-advanced mathematics and is used as a unifying concept throughout much of the remainder of the book. I introduce a function as a correspondence from one set of real numbers to another but I define it as a set of ordered pairs to make its meaning precise. This formal definition of a function leads to the definition of the graph of a function as a set of points in a plane. Because of the importance of these two definitions there are numerous examples, illustrations, and exercises in Section 4.1 pertaining to them.

In Section 4.3 the mathematical models of functions are those encountered in later work and are designed to prepare the student to express a practical situation in terms of a functional relationship. The discussion of quadratic functions in Section 4.4 utilizes graphs of parabola studied in Chapter 3. In Section 4.5 the treatment of graphs of polynomial functions is as complete as possible without using calculus. Just as thorough a coverage of graphs of rational functions occurs in Section 4.6. The inverse of a function is defined in Section 4.7 in preparation for its application in Chapter 5.

Chapter 5 The emphasis in Chapter 5 is on the properties of exponential and logarithmic functions. I introduce the number e in Section 5.1 by considering interest on an investment at a rate compounded continuously and follow this discussion by an intuitive demonstration of how e can be defined in calculus: the number that the expression $\left(1 + \dfrac{1}{x}\right)^x$ approaches as x increases without bound.

I discuss exponential functions in Section 5.2 and define a logarithmic function in Section 5.3 as the inverse of an exponential function. Properties of logarithmic functions appear in Section 5.4, and Section 5.5 is devoted to exponential and logarithmic equations. Throughout the chapter there are applications in all of the sciences. These applications involve exponential growth and decay as well as bounded growth.

Chapter 6 We begin our study of trigonometry in Chapter 6 with trigonometric functions of angles. Section 6.1 is devoted to angles and their measurement, and the six trigonometric functions of angles are defined in Section 6.2. I define the sine and cosine of a real number t in Section 6.4 as the corresponding function of an angle having radian measure t, and I introduce the tangent, cotangent, secant, and cosecant of real numbers in Section 6.6 as quotients involving the sine and cosine.

Graphs of the trigonometric functions and their properties are treated in Sections 6.5 and 6.6. Solutions of right triangles are presented in Section 6.7 as applications of trigonometric functions of angles. I discuss applications of trigonometric functions of real numbers to simple harmonic motion and alternating electrical current in Section 6.8. Other graphs involving sine and cosine functions are given in Supplementary Section 6.9. In that section we draw a sketch of the graph of the function defined by $\dfrac{\sin t}{t}$, which suggests that the quotient approaches 1 as t approaches zero, a fact with important consequences in later work in mathematics.

Chapter 7   Trignometric identities and equations, as well as inverse trigonometric functions, are the topics of analytic trigonometry in Chapter 7. The eight fundamental identities are introduced in Section 7.1. We use these identities to prove others in Section 7.2, which includes a step-by-step summary of suggestions for proving identities. I have included in the exercises of Section 7.2 some that require the student to determine whether a particular equation is an identity. The computational skills acquired by proving identities are beneficial in later work in mathematics when it is necessary to convert a trigonometric expression from one form to another.

Other important trigonometric identities are those in Sections 7.3, 7.4, and Supplementary Section 7.7. Identities are used in Section 7.6 to solve equations involving trigonometric functions. Because inverse trigonometric functions are applied to solve trigonometric equations, I have defined them and obtained their graphs in Section 7.5.

Chapter 8   Applications of trigonometry that involve the solutions of oblique triangles appear in the first two sections of Chapter 8. We solve them by the law of sines in Section 8.1 and by the law of cosines in Section 8.2. A brief treatment of vectors and their applications is presented in Section 8.3.

Polar coordinates, an important topic in analytic geometry, are introduced in Section 8.4. I apply the polar form of a complex number in Section 8.5 to obtain the product and quotient of complex numbers and in Section 8.6 to determine powers and roots of complex numbers by De Moivre's theorem. Supplementary Section 8.7 is devoted to graphs of equations in polar coordinates.

Chapter 9   A straightforward coverage of systems of equations and inequalities and of matrices appears in Chapter 9. In Section 9.1 systems of two linear equations are discussed, and systems involving three linear equations are treated in Section 9.2. I present systems involving quadratic equations in two variables in Section 9.3 and systems of linear inequalities along with their use in linear programming in Section 9.4.

We solve systems of linear equations by the Gaussian reduction method involving matrices in Section 9.5 and by determinants and Cramer's rule in Section 9.6. There are two supplementary sections, 9.7 and 9.8, involving properties of matrices and solutions of linear systems by matrix inverses.

Chapter 10   Chapter 10 pertains to polynomial functions and polynomial equations. Synthetic division, discussed in Section 10.1, is a computational tool used in Section 10.2 to find rational zeros of polynomial equations. The discussion of roots of polynomial equations, begun in Section 10.2, is continued in Section 10.3 where I develop a systematic method to find the exact or approximate value of all real roots. The treatment of zeros of polynomial functions is completed in Section 10.4 where we determine complex zeros.

In Section 10.4 I state and prove the theorem used in Supplementary Section 10.5 when working with partial fractions: a polynomial with real coefficients can be expressed as a product of linear or quadratic polynomials with real coefficients.

Chapter 11   There is a variety of algebra topics in Chapter 11. Sequences and series play an important part in later work in mathematics, and in Section 11.1 there is a brief introduction to them based on the function concept. The first section also includes a discussion of the sigma notation, which is applied to write a summation. Section 11.2 is devoted to mathematical induction, a technique used to prove certain theorems involving positive integers. I cover arithmetic sequences and series in Section 11.3, geometric sequences and series in Section 11.4, and infinite geometric series in Section 11.5.

Counting, permutations, and combinations are applied to everyday-life situations in Section 11.6, and in Section 11.7 counting forms a basis for an introduction to the theory of probability. In the discussion of the binomial theorem in Section 11.8, I treat the binomial coefficients as combinations and utilize this concept in the proof of the binomial theorem.

Flexibility   Because there is a diversity of opinion regarding the content of a text in college algebra and trigonometry, there are more topics in this book than can be covered in a course of three or four semester hours. The instructor should choose the material that is appropriate for his or her class. As mentioned above, Chapter 1 may be omitted. Chapters 2 through 9 should be covered in most courses, with the stipulation that any supplementary section can be included or omitted without affecting the understanding of subsequent subject matter. Chapters 10 and 11 are self-contained, and either or both can be deleted from a short course.

Examples and Illustrations

Examples and illustrations are included in each section. The examples were carefully selected to prepare students for the exercises and should be used as models for their solutions. An illustration serves to demonstrate a particular concept, definition, or theorem; it is a prototype of the idea being presented.

Exercises

There are over 5500 exercises, appearing either at the end of a section or as review exercises following the last section of a chapter. These exercises are varied in scope and have been graded in difficulty, ranging from elementary to very challenging. They stress the computational, theoretical, and applied aspects of the subject. The applications encompass a multitude of fields including physics, chemistry, engineering, astronomy, navigation, biology, medicine, business, economics, sociology, psychology, and statistics.

The answers to the odd-numbered exercises are given in the back of the book, and the answers to the even-numbered ones are available in a separate booklet.

Calculators and Tables

I recommend that students studying this text own a scientific calculator containing keys for exponential, logarithmic, and trigonometric functions. In Chapters 5 through 8 there are examples, illustrations, and exercises involving elementary computations with such a calculator. The speed and accuracy afforded by a calculator makes its use preferable to that of tables. Nevertheless there are tables in the back of the book for exponential functions, natural and common logarithms, and trigonometric functions of both angles and real numbers.

Pacific Palisades, California   L.L.

ACKNOWLEDGMENTS

Addison-Wesley and Louis Leithold are grateful for the contributions of those who reviewed the manuscript as it developed through its various stages:

Dean S. Burbank
Gulf Coast Community College

John F. Cavalier
West Virginia Institute of Technology

August Garver
University of Missouri at Rolla

Frank A. Gunnip
Oakland Community College

Mary Lou Hart
Brevard Community College

Shirley Huffman
Southwest Missouri State University

David Price
Tarrant County Junior College

Vincent P. Schielack, Jr.
Texas A&M University

Gordon Shilling
University of Texas at Arlington

Sandra Spears
Jefferson Community College

Addison-Wesley thanks those who responded to a survey in Precalculus Mathematics:

Anne Albert
Findlay College

John Annulis
University of Arkansas at Monticello

Bernard Avant
Atlanta Junior College

Leonard Baldwin
Southeastern Illinois College

Frank Baum
Pasadena City College

Edward Beckenstein
St. John's University

Philip Beckman
Black Hawk College

Marianne Bedee
Bowling Green State University
at Firelands

Bruce Bemis
Westminister College

F. B. Bennett
Sampson Community College

Donald Bigwood
Bismarck State College

Bill Bompart
Augusta College

Martha Bouknight
Meredith College

Joan Brenizer
Lamar University

F. Brunschuk
Empire State College

Catherine Bureiss
Olivet Nazarene University

Dennis Callas
SUNY Argicultural and Technical
College at Delhi

L. Carson
Hagerstown Junior College

E. Warren Chilton
Surry Community College

B. I. Chotlos
Butler University

Z. H. Chowdhury
Clinch Valley College

Deann Christianson
University of the Pacific

Rick Cleary
Saint Michael's College

P. Coffman
Spokane Falls Community College

Robert Cole
Davidson County Community College

G. S. Colonna
Park College

Daniel Comenetz
University of Massachusetts at Boston

Cecil Crawford
Howard University

Ginny Crowder
Brenau College

Joanne Crowley
Wayne General and Technical
College

John G. Cubbage
Central Arizona College at Aravaipa

N. Cuppy
Allen County Community College

Lloyd Davis
Montreat–Anderson College

Roger E. Davis
Williamsport Area Community College

William Lloyd Davis
College of San Mateo

Franklin Demana
Ohio State University

David DeVries
Georgia College

M. Dotson
Belmont Abbey College

Bob Dressel
Kent State University in Stark County

Lucy Duphouse
Trinity College

Paul Eenigenburg
Western Michigan University

Constance B. Elko
Marist College

Eric Ellis
Essex Community College

John Enopre
Cayuga County Community College

Lyle Espenochero
University of Wisconsin Center at
    Marinette

Alice W. Essary
University of Southern Mississippi

Charles Eylet
Johnson State College

Neale Fadden
Belleville Area College

Harold Farmer
Wallace State College

David A. Felland
Ellsworth Community College

Kathy Fenimore
Frederick Community College

Richard Finley
Northeast Louisiana University

Sr. Mary Fitzpatrick
Avila College

Elton Fors
Northern State College

Lenore Frank
SUNY at Stony Brook

D. S. Frederick
Saginaw Valley State University

John C. Friedell
Loras College

Marvin Gamble
Westmar College

Charles Gammill
University of Tennessee at Martin

Mary Ann Gay
Purdue University Southeast

Nina Girard
University of Pittsburgh at Johnstown

Mark Goldstein
West Virginia Northern Community
    College

Peter Grassi
Hofstra University

Allen Gray
Northern Arizona University

Evelyn Grensporn
Mass Bay Community College

Charlotte Grossbeck
SUNY at Cobleskill

Julie Guelich
Normandale Community College

William G. Hackman
Rappahannock Community College

Lewis Hall
Lees–McRae College

B. Hardiman
Cuyahoga Community College

A. Hemingway
Morgan Community College

B. Hindle
Seton Hall College

Bruce Hoelter
Raritan Valley Community College

Larry M. Hopkins
Gogebic Community College

V. Dwight House
Methodist College

Jack Howard
Cumberland University

Susan Indorf
Skagit Valley College

Mysore Jagadish
Barry University

A. Jha
Claflin College

D. A. Josephson
Wheaton College

Philip Keller
Lincoln Land Community College

Matthew Kennedy
Alabama State University

James Kirkpatrick
Alabama A. & M. University

B. A. Kirkwood
Malcolm X College

Elizabeth Klipsch
Broome Community College

L. L. Krajewski
Viterbo College

Richard Kratzer
University of Southern Maine

Helen Kriegsman
Pittsburgh State University

Anne Landry
Dutchess Community College

Richard Langlie
North Hennepin Community
    College

William La Roue
Mitchell College

D. Larson
Walker College

R. E. Lentz
Mankato State University

Robert Levnue
Community College of Allegheny
    County at Boyce

E. Llinas
University of Pittsburgh at
    Greensburg

ACKNOWLEDGEMENTS

K. Alan Loper
Hope College

Eric Lubot
Bergen Community College

Robert B. Ludwig
Buffalo State College

Peggy Lumpkin
Wallace State Community College

B. H. Maddox
Eckerd College

L. Michael Majeske
Greater Hartford Community College

Giles Maloof
Boise State University

T. H. Maney
Northern Virginia Community
College

Raj Markanda
Northern State College

Jim D. Martin
Navajo Community College

Ann Mason
Aquinas College

Richard Mason
Indian Hills Community College

Fred Massey
Georgia State University

J. E. Master
Jackson Community College

Saburo Matsumoto
Bob Jones University

Clarence McBride
Eastern Arizona College

Peter McCarthy
Northern Essex Community College

Otis B. McCowan
Belmont College

Ann B. Megan
University of Texas at Austin

L. H. Michal
El Paso Community College

Wayne Miller
Lee College

Lyman Milroy
Williamsport Area Community
College

Mary Moore
Thiel College

Shellia Morman
North Lake College

R. G. Morrison
Chapman College

Ronald Narode
University of Massachusetts at
Amherst

Thomas Nelson
Sonoma State University

Stanley Obermueller
Concordia College

Reed Parr
Salt Lake Community College

William Pero
University of Lowell

Anthony J. Perrotto
El Paso Community College

David Peterson
University of Central Arkansas

Delmas H. Petrea
Pfeiffer College

Charles G. Pickens
Kearney State College

Hadley A. Pobst
Virginia Highlands Community
College

Arthur Quickenton
Appalachian State University

R. Quint
Ventura College

Otho Rater
Grand View College

Marion Reed
Dominican College

Tom Rice
State Fair Community College

Brenda Roberts
University of Tulsa

Richard Robinson
Wofford College

Jim Roche
Loyola College

Jan Roy
Montcalm Community College

D. C. Royster
University of North Carolina at
Charlotte

A. J. Ruddel
City University of New York
Bernard Baruch College

John L. Savige
Saint Petersburg Junior College

Paul Schaeffer
SUNY College at Geneseo

Gordon Schlafmann
Western Wisconsin Technical
Institute

Eugene Schlereth
University of Tennessee at
Chattanooga

Margaret Schmid
Black Hawk College
Main Campus

Margaret Schoenfelt
County College of Morris

Alan Schuloff
Mesa Community College

J. A. Schumaker
Rockford College

Richard Semmler
Northern Virginia Community
College

Mark Serebranksy
Camden County College

Anne Sevin
Framingham State College

Roy Shortt
Keuka College

Jean Shutters
Harrisburg Area Community
College

Delores S. Smith
Coppin State College

Steve Smith
Harding University

Carole Sokolowski
Merrimack College

Carol Spiegel
Clarke College

S. Srivastava
Bowie State College

P. A. St. Ours
University of New England

Thomas Stark
Cincinnati Tech College

Sally Stevens
Midway College

Donald G. Stewart
Arizona State University

Bernard C. Swartz
Glen Oaks Community College

N. L. Taylor
Lake Michigan College

Mohinda Tewari
Virginia State University

Connie Ting
Brazosport College

William K. Tomhave
Concordia College

Ray Treadway
Bennett College

Dorothy Tuggle
Red Rocks Community College

Sherman B. Vanaman
Carson–Newman College

A. Van De Water
Winona State University

C. Ralph Verno
West Chester University

Barbara Victor
Illinois Benedictine College

Ron Virden
Lehigh County Community College

John G. Vonhold, Jr.
Genesee Community College

John Watkins
Colorado College

M. A. Watkins
Community College of Allegheny
County—North

Louis Weiner
Northeastern Illinois University

Arden Welsh
West Liberty State College

Larry Welsh
Lees McRae College

Karen Williams
University of Wisconsin at Stout

Roman Wong
Washington and Jefferson College

Frieda Zames
New Jersey Institute of Technology

Manuel Zax
Worcester State College

Natalia Zotov
Gannon University

SUPPLEMENTS

For the Instructor **Complete Solutions Manuel** Containing the worked-out solutions for all the exercises in the text.

Even Numbered Answer Book Containing the answers for all the even-numbered exercises in the text.

For the Student **Student's Solution Manual** Containing worked-out solutions for every third exercise in the text.

Computerized Testing Systems **AWTest** Based on the learning objectives of the text, this easy to use, algorithm-based system allows the instructor to generate tests, quizzes, or remedial exercises. AWTest is available for the IBM-PC*. Departmental software free upon adoption.

AWTest Edit This is a computerized test bank containing over 2500 multiple-choice test items for the IBM-PC*. The program allows the instructor to edit existing test items and/or enter new items. Also tests may be created with both multiple-choice and open-ended questions. Departmental software free upon adoption.

Printed Test Bank At least three alternate tests per chapter are included in this valuable supplement. Instructors can use these items as actual tests or as a reference for creating tests with or without the computer.

Transparencies This package of acetates includes a selection of key definitions, figures, proofs, formulas, tables, and applications that appear in the text.

Computer Supplements for the Instructor and the Student **Master Grapher and 3D Graphing Software** by Bert Waits and Frank Demana (Ohio State University). These two interactive, utilities packages are available for the IBM-PC*, MacIntosh*, and Apple II series computers.

Master Grapher enables the user to graph and manipulate functions in two dimensions. This program features the ability to change function parameters, rotate axes, overlay one graph with another, and perform a variety of transformations.

Contents

C H A P T E R

1

Review Topics in Algebra

Introduction

Even though you have had a course in Intermediate Algebra, or the equivalent, you may need a review of some of the material necessary for College Algebra. This chapter is designed to meet that need. It includes a discussion of the real number system, laws of exponents and radicals, polynomials, factoring, rational expressions, and an introduction to the set of complex numbers.

Depending on your background, some sections may need to be covered in detail, while others may be treated lightly or omitted. You may wish to postpone the study of a particular section until it is required for a later discussion. For instance, Sections 1.3 and 1.4 on polynomials and factoring may be taken up prior to their use in solving equations and inequalities in Chapter 2. Section 1.5 on rational expressions may be reviewed before reading Section 4.6 on graphs of rational functions. The treatment of exponents and radicals in Sections 1.2, 1.6, and 1.7 can be examined just before studying Chapter 5 on exponential and logarithmic functions. The introduction to complex numbers in Section 1.8 is needed to solve some quadratic equations in Section 2.3.

1.1 The Set of Real Numbers

Algebra, like arithmetic, involves numbers on which are performed operations such as addition, multiplication, subtraction, and division. While arithmetic is concerned with operations on specific numbers such as

$$2 + 5 \qquad 8 \cdot 9 \qquad 43 - 18 \qquad 36 \div 4$$

in algebra we deal with operations on unspecified or unknown numbers that are designated by symbols or letters such as x, y, z, a, b, and c. Thus in algebra we consider

$$x + y \qquad a \cdot b \qquad y - z \qquad b \div c$$

The word "algebra" originated from the Arabic word *al-jabr,* which appears in the title *ilm al-jabr w'al muqâbalah* (translated as "the science of reduction and cancellation"), an early ninth-century work. The algebraic symbolism used to generalize the operations of arithmetic was formulated in the sixteenth and seventeenth centuries.

From time to time we will use some **set** notation and terminology. The idea of set is used extensively in mathematics and is such a basic concept that we will not give a formal definition here. Instead, let us say simply that a set is a collection of objects, and the objects in a set are called the **elements** of the set. Each particular object must be either in the set or not in the set.

A pair of braces, { }, is used with words or symbols to describe a set. For example, if S is the set of natural numbers less than 6, we can write set

S as

$$\{1, 2, 3, 4, 5\}$$

or as

$$\{x, \text{ such that } x \text{ is a natural number less than } 6\}$$

In the first notation, we have listed all the elements in the set. In the second notation, called **set builder notation,** we have given the criteria for deciding whether an object belongs to the set. Here the symbol x is called a **variable** because it is used to represent any element of the given set. The given set is called the **domain** of the variable.

Another way of writing the set S in set-builder notation is to use a vertical bar in place of the words "such that":

$$\{x|x \text{ is a natural number less than } 6\}$$

which is read "the set of all x such that x is a natural number less than 6."

Two sets A and B are said to be **equal,** written $A = B$, if and only if A and B have identical elements. For example,

$$\{1, 2, 3\} = \{3, 1, 2\}$$

The **union** of two sets A and B, denoted by $A \cup B$ and read "A union B," is the set of all elements that are in A or in B or in both A and B. The **intersection** of A and B, denoted by $A \cap B$ and read "A intersection B," is the set of all elements that are in both A and B. The set that contains no elements is called the **empty set** and is denoted by $\varnothing$.

Illustration   1

Suppose $A = \{2, 4, 6, 8, 10, 12\}$, $B = \{1, 4, 9, 16\}$, and $C = \{2, 10\}$. Then

$$A \cup B = \{1, 2, 4, 6, 8, 9, 10, 12, 16\} \qquad A \cap B = \{4\}$$
$$B \cup C = \{1, 2, 4, 9, 10, 16\} \qquad B \cap C = \varnothing \qquad \blacksquare$$

Observe in Illustration 1 that the intersection of sets B and C is the empty set. These two sets have no elements in common and they are called **disjoint sets.**

The symbol $\in$ indicates that a specific element belongs to a set. Hence for the set C of Illustration 1 we may write $2 \in C$, which is read "2 is an element of C." The notation $a, b \in S$ indicates that both a and b are elements of S. The symbol $\notin$ is read "is not an element of." Therefore we read $5 \notin A$ as "5 is not an element of A."

If every element of a set S is also an element of a set T, then S is a **subset** of T, written $S \subseteq T$. In Illustration 1 every element of C is also an element of A; thus C is a subset of A, and we write $C \subseteq A$. The symbol $\nsubseteq$ is read "is not a subset of." Thus we may write $\{1, 2, 3, 4\} \nsubseteq \{1, 2, 3\}$.

We referred earlier to **natural numbers.** The set of natural numbers is also called the set of **positive integers.** Thus the set of natural numbers, which we denote by N, may be written

$$N = \{1, 2, 3, \ldots \}$$

where the three dots are used to indicate that the list goes on and on with no last number.

The number *zero*, denoted by the symbol 0, is the number having the property that if it is added to any number, the result is that number. The set of numbers whose elements are the natural numbers and zero is called the set of **whole numbers,** denoted by W.

Corresponding to each positive integer n there is a negative integer such that if it is added to n, the result is 0. For example, the negative integer -5, read "negative five," is the number which when added to 5 gives a result of 0. The set of **negative integers** can be written as follows: $\{-1, -2, -3, \ldots \}$. The set of numbers whose elements are the positive integers, the negative integers, and zero is called the set of **integers,** denoted by J; thus

$$J = \{\ldots, -3, -2, -1, 0, 1, 2, 3, \ldots \}$$

The set of integers then is the union of three disjoint subsets: the set of positive integers, the set of negative integers, and the set consisting of the single number 0. Note that the number 0 is an integer, but it is neither positive nor negative. Sometimes we refer to the set of **nonnegative integers,** which is the set consisting of the positive integers and 0 or, equivalently, the set of whole numbers. Similarly, the set of **nonpositive integers** is the set consisting of the negative integers and 0.

Consider now the set whose elements are those numbers that can be represented by the quotient of two integers p and q, where q is not 0, that is, the numbers that can be represented symbolically as

$$\frac{p}{q} \qquad \text{where } q \text{ is not } 0$$

This set of numbers is called the set of **rational numbers,** which is denoted by Q. Thus

$$Q = \left\{ x \middle| x \text{ can be represented by } \frac{p}{q},\ p \in J,\ q \in J,\ q \text{ is not } 0 \right\}$$

Some numbers in the set Q are $\frac{1}{2}$, $\frac{3}{4}$, $\frac{11}{5}$, $\frac{-2}{3}$, and $\frac{-31}{12}$. Every integer is a rational number because every integer can be represented as the quotient of itself and 1; that is, 8 can be represented by $\frac{8}{1}$, 0 by $\frac{0}{1}$, and -15 by $\frac{-15}{1}$. Hence $J \subseteq Q$.

Any rational number can be written as a decimal. You are familiar with the process of using long division to do this. For example, $\frac{3}{10}$ can be written 0.3, $\frac{9}{4}$ can be written 2.25, and $\frac{83}{16}$ can be written 5.1875. These decimals are called **terminating decimals.** There are rational numbers whose decimal representation is nonterminating and repeating; for example, $\frac{1}{3}$ has the decimal representation $0.333\ldots$, where the digit 3 is repeated, and $\frac{47}{11}$ can be represented as 4.272727 . . . , where the digits 2 and 7 are repeated in that order. It can be proved that the decimal representation of every rational number is either a terminating decimal or a nonterminating repeating decimal. We shall show in Section 11.5 that every nonterminating repeating decimal is a representation of a rational number.

The following question now arises: Are there numbers whose decimal representation is nonterminating and nonrepeating? The answer is yes. One example of such a number is the principal square root of 2, denoted symbolically by $\sqrt{2}$ and indicated by a nonterminating nonrepeating decimal as 1.41421 . . . . Another such number is π (pi), which is the ratio of the circumference of a circle to its diameter and indicated by a nonterminating nonrepeating decimal as 3.14159 . . . . The numbers whose decimal representations are nonterminating and nonrepeating cannot be expressed as the quotient of two integers and hence are not rational numbers. This set of numbers is called the set of **irrational numbers,** which we denote by H. It may be defined symbolically as

$H = \{x|$the decimal representation of x is nonterminating, nonrepeating$\}$

The union of the set of rational numbers and the set of irrational numbers is the set of **real numbers.** Denoting the set of real numbers by R, we define R symbolically by

$$R = Q \cup H$$

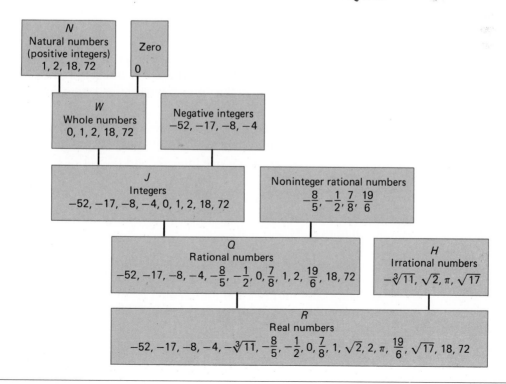

FIGURE 1

Figure 1 shows the relationships among the sets of numbers discussed above. Examples of each classification of numbers appear in the corresponding rectangle.

Example 1 The sets $N, J, Q, H,$ and R are the sets of numbers defined in this section. Insert either $\subseteq$ or $\nsubseteq$ to make the statement correct.

a) $N\underline{}J$　　　　　　b) $Q\underline{}J$　　　　　　c) $\{\sqrt{2}, \pi, 3.5\}\underline{}H$

d) $\{0\}\underline{}Q$　　　　　　e) $N\underline{}R$

Solution a) Because every natural number (or positive integer) is an integer, $N \subseteq J$.

b) Because there are rational numbers that are not integers, $Q \nsubseteq J$.

c) $\sqrt{2}$ and π are irrational numbers, but 3.5 is a rational number; therefore $\{\sqrt{2}, \pi, 3.5\} \nsubseteq H$.

d) Zero is a rational number, and thus $\{0\} \subseteq Q$.

e) Every positive integer is a real number; hence $N \subseteq R$. ∎

Example 2 In each of the following, determine which one of the sets N, $J, Q, H, R,$ and $\varnothing$ is equal to the given set.

a) $J \cup Q$　　　　　b) $J \cap Q$　　　　　c) $N \cap H$　　　　　d) $H \cup R$

Solution a) The union of J and Q is the set of numbers that are either integers or rational. Because the set of integers is a subset of the set of rational numbers, this union is the set of rational numbers. Hence $J \cup Q = Q$.

b) The intersection of J and Q is the set of numbers that are both integers and rational. This intersection is the set of integers, and thus $J \cap Q = J$.

c) Because the set of positive integers and the set of irrational numbers have no elements in common, $N \cap H = \varnothing$.

d) The union of H and R is the set of numbers that are either irrational or real. Because the set of irrational numbers is a subset of the set of real numbers, $H \cup R = R$. ∎

The **real number system** consists of the set of real numbers and two operations called **addition** and **multiplication.** Addition is denoted by the symbol $+$ and multiplication is denoted by the symbol $\cdot$ (or $\times$). If a and b are real numbers, $a + b$ denotes the **sum** of a and b, and $a \cdot b$ (or ab) denotes their **product.**

A discussion of the real number system appears in Section A.1 of the appendix, where it is described by a set of *axioms* (the word axiom is used to indicate a formal statement that is assumed to be true without proof).

Subtraction and *division* of real numbers are defined in terms of addition and multiplication, respectively. The definition of subtraction is as follows:

If a and b are real numbers, the operation of **subtraction** assigns to a and b a real number, denoted by $a - b$, called the **difference** of a and b, and

$$a - b = d \quad \text{if and only if} \quad a = b + d$$

Illustration 2

$$7 - 4 = 3 \quad \text{because} \quad 7 = 4 + 3 \qquad \blacksquare$$

In the above definition the "if and only if" qualification is used to combine two statements:

1. "$a - b = d$ if $a = b + d$."
2. "$a - b = d$ only if $a = b + d$," which is equivalent to the statement "$a = b + d$ if $a - b = d$."

We now define division.

If a and b are real numbers, and $b \neq 0$, the operation of **division** assigns to a and b a real number, denoted by $a \div b$, called the **quotient** of a and b, and

$$a \div b = q \quad \text{if and only if} \quad a = bq \tag{1}$$

Illustration 3

$$24 \div 6 = 4 \quad \text{because} \quad 24 = 6 \cdot 4 \qquad \blacksquare$$

Observe in the definition of division that $b \neq 0$. The reason for this restriction can be explained by allowing b to be 0 in statement (1). For instance, if in that statement $b = 0$ and $a = 3$ (any other nonzero value of a can be used instead of 3), the statement becomes

$$3 \div 0 = q \quad \text{if and only if} \quad 3 = 0 \cdot q$$

Of course, there is no value of q satisfying this statement because $0 \cdot q = 0$ and $3 \neq 0$. Furthermore, in statement (1) if $b = 0$ and $a = 0$, the statement becomes

$$0 \div 0 = q \quad \text{if and only if} \quad 0 = 0 \cdot q$$

Because $0 \cdot q = 0$ for any value of q, $0 \div 0$ could equal any real number; that is, $0 \div 0$ is indeterminate. Therefore, for every real number a, no meaning can be attached to $a \div 0$. Hence:

> Division by zero is not defined

An ordering of the set of real numbers can be accomplished by means of a relation denoted by the symbols $<$ (read "is less than") and $>$ (read "is greater than"). In the following definition of these symbols we use the concept of a positive number given by Axiom 8 in Section A.1 of the appendix.

Definition

> **The Symbols $<$ and $>$**
> If a and b are real numbers,
> (i) $a < b$ if and only if $b - a$ is positive;
> (ii) $a > b$ if and only if $a - b$ is positive.

Illustration 4

$3 < 5$ because $5 - 3 = 2$, and 2 is positive

$-10 < -6$ because $-6 - (-10) = 4$, and 4 is positive

$7 > 2$ because $7 - 2 = 5$, and 5 is positive

$-2 > -7$ because $-2 - (-7) = 5$, and 5 is positive

$\frac{3}{4} > \frac{2}{3}$ because $\frac{3}{4} - \frac{2}{3} = \frac{1}{12}$, and $\frac{1}{12}$ is positive   ■

Observe that

$$3 > 0 \quad \text{because} \quad 3 - 0 = 3, \text{ and } 3 \text{ is positive}$$
$$-4 < 0 \quad \text{because} \quad 0 - (-4) = 4, \text{ and } 4 \text{ is positive}$$

These statements are special cases of the following properties that are obtained from the definitions of $>$ and $<$:

> $a > 0$ if and only if a is positive
> $a < 0$ if and only if a is negative

If we write $a \le b$ (read "a is less than or equal to b") we mean that either $a < b$ or $a = b$. Similarly, $a \ge b$ (read "a is greater than or equal to b") indicates that either $a > b$ or $a = b$.

The statements $a < b$, $a > b$, $a \le b$, and $a \ge b$ are called **inequalities.** In particular, $a < b$ and $a > b$ are **strict** inequalities, whereas $a \le b$ and $a \ge b$ are **nonstrict** inequalities.

A number x is between a and b if $a < x$ and $x < b$. We can write this as a **continued inequality** as follows: $a < x < b$. Therefore

$$2 < 3 < 4 \qquad -5 < 1 < \tfrac{4}{3} \qquad \tfrac{1}{2} < \tfrac{2}{3} < \tfrac{3}{4}$$

Another continued inequality is $a \le x \le b$, which means that both $a \le x$ and $x \le b$. Other continued inequalities are $a \le x < b$ and $a < x \le b$.

Example   3   Use set notation and one or more of the symbols $<$, $>$, $\leq$, and $\geq$ to denote each of the following sets.

a) The set of all x such that x is between -2 and 2

b) The set of all t such that $4t - 1$ is nonnegative

c) The set of all y such that $y + 3$ is positive and less than or equal to 15

d) The set of all z such that $2z$ is greater than or equal to -5 and less than -1

Solution

a) $\{x \mid -2 < x < 2\}$ b) $\{t \mid 4t - 1 \geq 0\}$

c) $\{y \mid 0 < y + 3 \leq 15\}$ d) $\{z \mid -5 \leq 2z < -1\}$ ∎

We now give a geometric interpretation to the set R of real numbers by associating them with points on a horizontal line called an **axis.** A point, called the **origin,** is chosen to represent the number 0. A unit of distance is selected arbitrarily. Then each positive integer n is represented by the point at a distance of n units to the right of the origin, and each negative integer $-n$ is represented by the point at a distance of n units to the left of the origin. We call these points **unit points.** They are labeled with the numbers with which they are associated. For example, 4 is represented by the point 4 units to the right of the origin and -4 is represented by the point 4 units to the left of the origin. Figure 2 shows the points representing 0 and the first 12 positive integers and their corresponding negative integers.

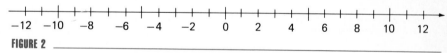

FIGURE 2

The rational numbers are associated with points on the axis of Figure 2 by dividing the segments between the points that represent successive integers. For instance, if the segment from 0 to 1 is divided into seven equal parts, the endpoint of the first such subdivision is associated with $\frac{1}{7}$, the endpoint of the second is associated with $\frac{2}{7}$, and so on. The point associated with the number $\frac{24}{7}$ is three-sevenths of the distance from point 3 to point 4. A negative rational number, in a similar manner, is associated with a point to the left of the origin. Figure 3 shows some of the points associated with rational numbers.

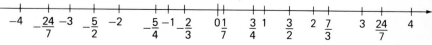

FIGURE 3

For certain irrational numbers, such as $\sqrt{2}$, $\sqrt{3}$, $\sqrt{5}$, and so on, geometrical constructions can be used to find the points corresponding to them. See Exercises 35 and 36. Points corresponding to other irrational numbers can be found by using decimal approximations of the numbers. For example, the point corresponding to π can be approximated by using some of the digits in the decimal representation 3.14159 . . . . Every irrational number can be associated with a unique point on the axis, and every point that does not correspond to a rational number can be associated with an irrational number. This indicates that a one-to-one correspondence between the set of real numbers and the points on the horizontal axis can be established. For this reason the horizontal axis is referred to as the **real number line.** Because the points on this line are identified with the numbers they represent, the same symbol is used for that number and the point.

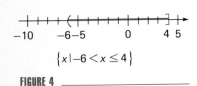

$\{x \mid -6 < x \leq 4\}$

FIGURE 4

Illustration 5

Consider the set $\{x \mid -6 < x \leq 4\}$. This set is represented on the real number line in Figure 4. The bracket at 4 indicates that 4 is in the set, and the parenthesis at -6 indicates that -6 is not in the set. ∎

The set of all numbers x satisfying the continued inequality $a < x < b$ is called an **open interval** and is denoted by (a, b). Therefore

$$(a, b) = \{x \mid a < x < b\}$$

The **closed interval** from a to b is the open interval (a, b) together with the two endpoints a and b and is denoted by $[a, b]$. Thus

$$[a, b] = \{x \mid a \leq x \leq b\}$$

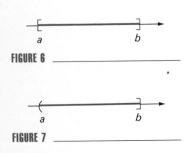

FIGURE 5

FIGURE 6

FIGURE 7

FIGURE 8

Figure 5 illustrates the open interval (a, b) and Figure 6 shows the closed interval $[a, b]$.

The **interval half-open on the left** is the open interval (a, b) together with the right endpoint b. It is denoted by $(a, b]$; so

$$(a, b] = \{x \mid a < x \leq b\}$$

We define an **interval half-open on the right** in a similar way and denote it by $[a, b)$. Thus

$$[a, b) = \{x \mid a \leq x < b\}$$

The interval $(a, b]$ appears in Figure 7, and the interval $[a, b)$ is shown in Figure 8.

We shall use the symbol $+\infty$ (positive infinity) and the symbol $-\infty$ (negative infinity); however, take care not to confuse these symbols with real numbers, for they do not obey the properties of the real numbers. We have the following intervals:

$$\begin{aligned}
(a, +\infty) &= \{x \mid x > a\} \\
(-\infty, b) &= \{x \mid x < b\} \\
[a, +\infty) &= \{x \mid x \geq a\} \\
(-\infty, b] &= \{x \mid x \leq b\} \\
(-\infty, +\infty) &= R
\end{aligned}$$

Figure 9 shows the interval $(a, +\infty)$, and Figure 10 illustrates the interval $(-\infty, b)$. Note that $(-\infty, +\infty)$ denotes the set of all real numbers.

FIGURE 9

FIGURE 10

Example   4 Show on the real number line each of the following sets and represent the set by interval notation.

a) $\{x \mid -7 \leq x < -2\}$

b) $\{x \mid x > 1 \text{ and } x < 10\}$

c) $\{x \mid x \leq -5 \text{ or } x \geq 5\}$

d) $\{x \mid x \geq 2\} \cap \{x \mid x < 9\}$

e) $\{x \mid x < 0\} \cup \{x \mid x \geq 3\}$

Solution The sets are shown on the real number line in Figure 11(a), (b), (c), (d), and (e), respectively. The sets are written in interval notation as follows.

a) $\{x \mid -7 \leq x < -2\} = [-7, -2)$

b) $\{x \mid x > 1 \text{ and } x < 10\} = (1, 10)$

c) $\{x \mid x \leq -5 \text{ or } x \geq 5\} = (-\infty, -5] \cup [5, +\infty)$

d) $\{x \mid x \geq 2\} \cap \{x \mid x < 9\} = [2, 9)$

e) $\{x \mid x < 0\} \cup \{x \mid x \geq 3\} = (-\infty, 0) \cup [3, +\infty)$

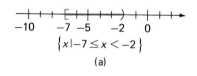

(a)

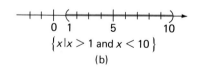

(b)

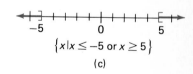

(c)

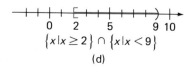

(d)

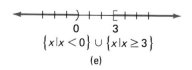

(e)

FIGURE 11

Example   5   Show on the real number line each of the following intervals and use set notation and inequality symbols to denote the interval.

a) $(-2, 4)$ b) $[3, 7]$ c) $[1, 6)$

d) $(-4, 0]$ e) $[0, +\infty)$ f) $(-\infty, 5)$

Solution   The intervals are shown on the real number line in Figure 12(a), (b), (c), (d), (e), and (f), respectively. They are written in set notation as follows.

a) $(-2, 4) = \{x | -2 < x < 4\}$ b) $[3, 7] = \{x | 3 \le x \le 7\}$

c) $[1, 6) = \{x | 1 \le x < 6\}$ d) $(-4, 0] = \{x | -4 < x \le 0\}$

e) $[0, +\infty) = \{x | x \ge 0\}$ f) $(-\infty, 5) = \{x | x < 5\}$

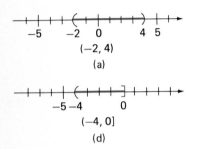

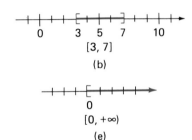

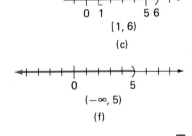

FIGURE 12

Associated with each real number is a nonnegative number, called its *absolute value*.

Definition

Absolute Value
If a is a real number, the **absolute value** of a, denoted by $|a|$, is a if a is nonnegative and is $-a$ if a is negative. With symbols, we write

$$|a| = \begin{cases} a & \text{if} & a \ge 0 \\ -a & \text{if} & a < 0 \end{cases}$$

Illustration   6

If in the above definition we take a as 6, 0, and -6, we have, respectively,

$$|6| = 6 \qquad |0| = 0 \qquad |-6| = -(-6)$$
$$= 6$$

The absolute value of a number can be considered as its distance (without regard to direction, left or right) from the origin. In particular, the points 6 and -6 are each six units from the origin.

From the definition of absolute value

$$|a - b| = \begin{cases} a - b & \text{if} \quad a - b \geq 0 \\ -(a - b) & \text{if} \quad a - b < 0 \end{cases}$$

or, equivalently,

$$|a - b| = \begin{cases} a - b & \text{if} \quad a \geq b \\ b - a & \text{if} \quad a < b \end{cases}$$

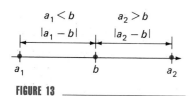

FIGURE 13

As a distance on the real number line, $|a - b|$ units can be interpreted as the distance between a and b without regard to direction. Refer to Figure 13.

Example 6 Show the points corresponding to the numbers -10, -7, -5, -3, 0, 3, 5, 7, and 10 on the real number line. Find the distance between u and v in each of the following cases.

a) $u = 10$, $v = 3$ b) $u = 3$, $v = 7$
c) $u = 5$, $v = -3$ d) $u = -7$, $v = 0$
e) $u = -3$, $v = -5$ f) $u = -10$, $v = -7$

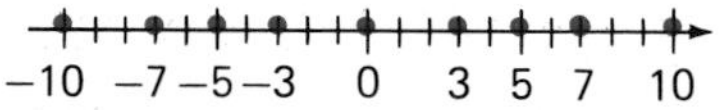

FIGURE 14

Solution Figure 14 shows the points corresponding to the given numbers on the real number line. In each part the distance between u and v is $|u - v|$ units.

a) $|u - v| = |10 - 3|$ b) $|u - v| = |3 - 7|$
$\quad = |7|$ $\quad = |-4|$
$\quad = 7$ $\quad = 4$

c) $|u - v| = |5 - (-3)|$ d) $|u - v| = |-7 - 0|$
$\quad = |8|$ $\quad = |-7|$
$\quad = 8$ $\quad = 7$

e) $|u - v| = |-3 - (-5)|$ f) $|u - v| = |-10 - (-7)|$
$\quad = |2|$ $\quad = |-3|$
$\quad = 2$ $\quad = 3$ ∎

Exercises 1.1

In Exercises 1 through 10, N is the set of natural numbers, J is the set of integers, Q is the set of rational numbers, H is the set of irrational numbers, and R is the set of real numbers.

In Exercises 1 and 2, insert $\in$ or $\notin$ in the blank to make the statement correct.

1. (a) 15 ____N; (b) 1.41421 ____H; (c) -3 ____J;
   (d) π ____Q

2. (a) 0 ____Q; (b) 2007 ____J; (c) $\frac{3}{7}$ ____R;
   (d) -5 ____H

In Exercises 3 and 4, use the symbol $\subseteq$ to give a correct statement involving the two sets.

3. (a) N and Q; (b) R and Q; (c) J and N; (d) J and R
4. (a) R and N; (b) J and Q; (c) H and R; (d) $\{0\}$ and J

In Exercises 5 and 6, insert either $\subseteq$ or $\nsubseteq$ to make the statement correct.

5. (a) J____R; (b) $\{0\}$____N; (c) H____R;
   (d) $\{0, \frac{1}{3}, 1.732\}$____$Q$
6. (a) N____Q; (b) Q____H; (c) Q____R;
   (d) $\{-\sqrt{3}, 0, \sqrt{3}\}$____$H$

In Exercises 7 and 8, determine which of the sets $N, J, Q,$ $H, R,$ and $\varnothing$ is equal to the given set.

7. (a) $Q \cap R$; (b) $Q \cup H$; (c) $J \cup N$; (d) $H \cap J$
8. (a) $H \cap R$; (b) $Q \cup R$; (c) $H \cap Q$; (d) $J \cap N$

In Exercises 9 and 10, for the set S define each of the following sets: (a) $S \cap N$; (b) $S \cap Q$; (c) $S \cap H$; (d) $S \cap J$.

9. $S = \{12, \frac{5}{3}, \sqrt{7}, 0, -38, -\sqrt{2}, 571, \pi, -\frac{1}{10}, 0.666$
   $\ldots, 16.34\}$

10. $S = \{-\frac{1}{4}, 26, \sqrt{3}, 1.23, -\sqrt{9}, -0.333$
    $\ldots, -6214, \frac{1}{2}\pi, \frac{4}{7}, 1\}$

In Exercises 11 and 12, arrange the elements of the given subset of R in the same order as their corresponding points from left to right on the real number line.

11. $\{-2, 3, 21, 5, -7, \frac{2}{3}, \sqrt{2}, -\frac{7}{4}, -\sqrt{5}, -10, 0, \frac{3}{4}, -\frac{5}{3}, -1\}$
12. $\{\frac{11}{3}, \pi, -8, -\sqrt{2}, 3, -\sqrt{3}, 4, \frac{21}{4}, -\frac{3}{2}, 1.26, \frac{1}{2}\pi\}$

In Exercises 13 through 16, use set notation and one or more of the symbols $<, >, \leq,$ and $\geq$ to denote the set.

13. (a) The set of all x such that x is greater than -9 and less than 8
    (b) The set of all y between -12 and -3
    (c) The set of all z such that $4z - 5$ is negative
14. (a) The set of all x between -5 and 3
    (b) The set of all y such that y is greater than or equal to -26 and less than -16
    (c) The set of all t such that $8t - 4$ is positive
15. (a) The set of all x such that $2x + 4$ is nonnegative
    (b) The set of all r such that r is greater than or equal to 2 and less than 8
    (c) The set of all a such that $a - 2$ is greater than -5 and less than or equal to 7
16. (a) The set of all s such that $2s + 3$ is nonpositive
    (b) The set of all x such that $3x$ is greater than 10 and less than or equal to 20

(c) The set of all z such that $2z + 5$ is between and including -1 and 15

In Exercises 17 through 24, show the set on the real number line and represent the set by interval notation.

17. (a) $\{x | x > 2\}$; (b) $\{x | -4 < x \leq 4\}$
18. (a) $\{x | x \leq 8\}$; (b) $\{x | 3 < x < 9\}$
19. (a) $\{x | x > 2$ and $x < 12\}$; (b) $\{x | x \leq -4$ or $x > 4\}$
20. (a) $\{x | x \geq -5$ and $x \leq 5\}$; (b) $\{x | x < 3$ or $x > 6\}$
21. (a) $\{x | x > 2\} \cap \{x | x < 12\}$; (b) $\{x | x \leq -4\} \cup \{x | x > 4\}$
22. (a) $\{x | x \geq -5\} \cap \{x | x \leq 5\}$; (b) $\{x | x < 3\} \cup \{x | x > 6\}$
23. (a) $\{x | x > -4\} \cap \{x | x \leq 0\}$; (b) $\{x | x \leq 0\} \cup \{x | x \leq 7\}$
24. (a) $\{x | x > -8\} \cap \{x | x \leq 0\}$; (b) $\{x | x > 2\} \cup \{x | x > 10\}$

In Exercises 25 through 28, show the interval on the real number line and use set notation and inequality symbols to denote the interval.

25. (a) $(2, 7)$; (b) $[-3, 6]$; (c) $(-5, 4]$; (d) $[-10, -2)$
26. (a) $(-5, 5)$; (b) $[1, 9]$; (c) $[-8, 3)$; (d) $(-7, 0]$
27. (a) $[3, +\infty)$; (b) $(-\infty, 0)$; (c) $(-4, +\infty)$; (d) $(-\infty, +\infty)$
28. (a) $(-\infty, -2]$; (b) $(-1, +\infty)$; (c) $(-\infty, 10)$; (d) $[0, +\infty)$

In Exercises 29 and 30, write the number without absolute value bars.

29. (a) $|7|$; (b) $|-\frac{3}{4}|$; (c) $|3 - \sqrt{3}|$; (d) $|\sqrt{3} - 3|$
30. (a) $|\frac{1}{3}|$; (b) $|-8|$; (c) $|\pi - 2|$; (d) $|3 - \pi|$

In Exercises 31 through 34, show the points corresponding to u and v on the real number line and then find the distance between u and v.

31. (a) $u = 8, v = 2$; (b) $u = -8, v = 2$;
    (c) $u = 8, v = -2$; (d) $u = -8, v = -2$
32. (a) $u = 6, v = 4$; (b) $u = -6, v = 4$;
    (c) $u = 6, v = -4$; (d) $u = -6, v = -4$
33. (a) $u = t, v = 2t,$ and $t > 0$;
    (b) $u = t, v = 2t,$ and $t < 0$
34. (a) $u = t, v = \frac{1}{2}t,$ and $t > 0$;
    (b) $u = t, v = \frac{1}{2}t,$ and $t < 0$
35. To determine the point on the real number line corresponding to the irrational number $\sqrt{2}$, we can use the construction indicated in the figure.

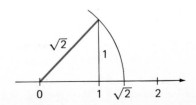

From point 1, a line segment of length one unit is drawn perpendicular to the axis. Then a right triangle is formed by connecting the endpoint of this segment with the origin. The length of the hypotenuse of this right triangle is $\sqrt{2}$ units. (This fact follows from the Pythagorean theorem, which states that c^2 has the same value as $a^2 + b^2$, where c units is the length of the hypotenuse and a units and b units are the lengths of the other two sides.) An arc of a circle having center at the origin and radius $\sqrt{2}$ is then drawn; the point where this arc intersects the axis is $\sqrt{2}$. Use this method to determine the point corresponding to $\sqrt{5}$.

36. Determine the point on the real number line corresponding to the irrational number $\sqrt{10}$. (*Hint:* See Exercise 35.)

1.2   Integer Exponents

To indicate a product, we use the centered dot, ·, or parentheses around one or more symbols. Sometimes we omit the symbol for multiplication. For example, the product of a and b can be written in the following ways:

$$a \cdot b \qquad (a)(b) \qquad a(b) \qquad (a)b \qquad ab$$

The numbers a and b are called **factors** of the product ab.

Suppose we have the product of two factors, each being x. We can use the notation x^2 to indicate this product, where the numeral 2 written to the upper right of the symbol x is called an *exponent;* thus

$$x^2 = x \cdot x$$

In particular, $3^2 = 3 \cdot 3$; that is, $3^2 = 9$.

In general, if a is a real number and n is a positive integer,

$$a^n = a \cdot a \cdot a \cdots a \qquad (n \text{ factors of } a)$$

where n is called the **exponent,** a is called the **base,** and a^n is called the **nth power of a.** For example, x^2 is the second power of x and y^5 is the fifth power of y, where

$$y^5 = y \cdot y \cdot y \cdot y \cdot y$$

When a symbol is written without an exponent, the exponent is understood to be 1. Hence $x = x^1$. It is customary to read x^2, the second power of x, as "x squared" and x^3, the third power of x, as "x cubed."

The representation of positive-integer powers by exponents was introduced by René Descartes (1596–1650) in 1637.

Illustration 1

a) $2^5 = 2 \cdot 2 \cdot 2 \cdot 2 \cdot 2$
      $= 32$

b) $\left(-\tfrac{1}{2}\right)^3 = \left(-\tfrac{1}{2}\right)\left(-\tfrac{1}{2}\right)\left(-\tfrac{1}{2}\right)$
      $= -\tfrac{1}{8}$

c) $(-3)^4 = (-3)(-3)(-3)(-3)$
   d) $-3^4 = -(3 \cdot 3 \cdot 3 \cdot 3)$
       $= 81$ $= -81$ ■

Consider now the product of two powers of the same base. For instance, the product of x^3 and x^4 can be written as

$$x^3 \cdot x^4 = (x \cdot x \cdot x)(x \cdot x \cdot x \cdot x)$$
$$= x^7$$

Thus the sum of the exponents of the two factors x^3 and x^4 is the exponent of the product. This fact is a special case of the following theorem.

Theorem 1

> If n and m are positive integers and a is a real number, then
> $$a^n \cdot a^m = a^{n+m}$$

The proof of this theorem requires mathematical induction, which is discussed in Section 11.2 and this theorem is proved in Example 5 of that section. However, we now give an informal argument for Theorem 1.

By the definition of a positive-integer exponent

$$a^n = a \cdot a \cdot a \cdots a \qquad (n \text{ factors of } a)$$
$$a^m = a \cdot a \cdot a \cdots a \qquad (m \text{ factors of } a)$$

Therefore the product of a^n and a^m gives $n + m$ factors of a; that is,

$$a^n \cdot a^m = a \cdot a \cdot a \cdots a \qquad (n + m \text{ factors of } a)$$
$$= a^{n+m}$$

Illustration 2

a) $2^3 \cdot 2^2 = 2^{3+2}$ b) $x^4 \cdot x = x^{4+1}$ c) $y^3 \cdot y^7 = y^{3+7}$
          $= 2^5$ $= x^5$ $= y^{10}$
          $= 32$ ■

Example   **1**   Find the following products. In part (b), n is a positive integer.

a) $(4x^4y^3)(-5x^5y^2)$ b) $(4nx^{5n})(5nx^{4n})$

Solution

a) $(4x^4y^3)(-5x^5y^2) = [4(-5)](x^4x^5)(y^3y^2)$
                    $= -20x^{4+5}y^{3+2}$
                    $= -20x^9y^5$

b) $(4nx^{5n})(5nx^{4n}) = (4 \cdot 5)(n \cdot n)(x^{5n} \cdot x^{4n})$
                     $= 20n^{1+1}x^{5n+4n}$
                     $= 20n^2x^{9n}$ ■

Suppose we have $(x^3)^4$, which is the fourth power of x cubed. By the definition of a positive-integer exponent,

$$\begin{aligned}(x^3)^4 &= x^3 \cdot x^3 \cdot x^3 \cdot x^3 \\ &= x^{3+3+3+3} \\ &= x^{3\cdot 4} \\ &= x^{12}\end{aligned}$$

Therefore the product of 3 and 4 is the exponent of x after simplification. In general we have the following theorem.

Theorem 2

> If n and m are positive integers and a is a real number, then
> $$(a^n)^m = a^{nm}$$

Because the proof of this theorem requires mathematical induction, we give an informal argument instead:

$$\begin{aligned}(a^n)^m &= a^n \cdot a^n \cdot a^n \cdots a^n &&(m \text{ factors of } a^n) \\ &= a^{n+n+n+\,\cdots\,+n} &&(m \text{ terms of } n \text{ in the exponent}) \\ &= a^{nm}\end{aligned}$$

Illustration 3

a) $\begin{aligned}(2^3)^2 &= 2^{3\cdot 2} \\ &= 2^6 \\ &= 64\end{aligned}$ b) $\begin{aligned}(x^2)^5 &= x^{2\cdot 5} \\ &= x^{10}\end{aligned}$ c) $\begin{aligned}(y^4)^6 &= y^{4\cdot 6} \\ &= y^{24}\end{aligned}$ ∎

Example   2   Find each of the following products, where n is any positive integer.

a) $(n^2)^2$ b) $(x^2)^n$ c) $(x^n)^n$

Solution

a) $\begin{aligned}(n^2)^2 &= n^{2\cdot 2} \\ &= n^4\end{aligned}$ b) $(x^2)^n = x^{2n}$ c) $\begin{aligned}(x^n)^n &= x^{n\cdot n} \\ &= x^{(n^2)}\end{aligned}$ ∎

The fourth power of the product of a and b is written as $(ab)^4$ and

$$\begin{aligned}(ab)^4 &= (ab)(ab)(ab)(ab) \\ &= (a \cdot a \cdot a \cdot a)(b \cdot b \cdot b \cdot b) \\ &= a^4 b^4\end{aligned}$$

This is a special case of the following theorem.

Theorem 3

If n is a positive integer and a and b are real numbers, then

$$(ab)^n = a^n b^n$$

Again we give an informal argument because the proof of Theorem 3 requires mathematical induction:

$$
\begin{aligned}
(ab)^n &= (ab)(ab)(ab)\cdots(ab) && \text{(n factors of ab)} \\
&= (a \cdot a \cdot a \cdots a)(b \cdot b \cdot b \cdots b) && \text{(n factors of a and n factors of b)} \\
&= a^n b^n
\end{aligned}
$$

Illustration 4

a) $(2 \cdot 5)^3 = 2^3 \cdot 5^3$ b) $(x^2 y^4)^5 = (x^2)^5 (y^4)^5$ c) $(3a^3)^4 = 3^4 (a^3)^4$

$\qquad\qquad = 8 \cdot 125 \qquad\qquad\qquad = x^{2 \cdot 5} y^{4 \cdot 5} \qquad\qquad\qquad = 81 a^{3 \cdot 4}$

$\qquad\qquad = 1000 \qquad\qquad\qquad\quad\; = x^{10} y^{20} \qquad\qquad\qquad\;\; = 81 a^{12}$ ■

By applying the associative law for multiplication, the formulas of Theorems 1–3 can be extended to involve any number of factors. For instance,

$$
\begin{aligned}
x^4 \cdot x^8 \cdot x \cdot x^6 &= x^{4+8+1+6} && (a^4 b^5 cd^2)^3 = (a^4)^3 (b^5)^3 c^3 (d^2)^3 \\
&= x^{19} && \qquad\qquad\quad\; = a^{12} b^{15} c^3 d^6
\end{aligned}
$$

Example 3   Find each of the following products.

a) $(-5r^3 s^4 t^2)(-6s^5 t^4)(-r^2 s)$ b) $(2x^3 y^2 z)^3 (-x^2 y^3 z^4)^4$

Solution

a) $(-5r^3 s^4 t^2)(-6s^5 t^4)(-r^2 s) = [(-5)(-6)(-1)]r^{3+2} s^{4+5+1} t^{2+4}$

$\qquad\qquad\qquad\qquad\qquad\qquad\quad = -30 r^5 s^{10} t^6$

b) $(2x^3 y^2 z)^3 (-x^2 y^3 z^4)^4 = [2^3 (x^3)^3 (y^2)^3 z^3][(-1)^4 (x^2)^4 (y^3)^4 (z^4)^4]$

$\qquad\qquad\qquad\qquad\qquad\quad = [8x^9 y^6 z^3][1x^8 y^{12} z^{16}]$

$\qquad\qquad\qquad\qquad\qquad\quad = 8x^{17} y^{18} z^{19}$ ■

The next theorem gives the formula for computing the quotient of two powers of the same base a, if $a \neq 0$. The stipulation that a is not 0 is necessary because division by 0 is not defined.

Theorem 4

If n and m are positive integers and a is a real number where $a \neq 0$, then

$$
\frac{a^n}{a^m} = \begin{cases} a^{n-m} & \text{if} \quad n > m \\[2mm] \dfrac{1}{a^{m-n}} & \text{if} \quad n < m \\[2mm] 1 & \text{if} \quad n = m \end{cases}
$$

Before proving this theorem, we give an illustration showing its application.

Illustration 5

a) $\dfrac{x^6}{x^2} = x^{6-2} = x^4$ b) $\dfrac{x^3}{x^7} = \dfrac{1}{x^{7-3}} = \dfrac{1}{x^4}$ c) $\dfrac{x^5}{x^5} = 1$ ■

Proof of Theorem 4 If $n = m$, the result follows immediately because

$$\frac{a^m}{a^m} = 1$$

If $n > m$, then the integer $n - m$ is positive. Therefore, from Theorem 1 $a^n = a^m \cdot a^{n-m}$. Thus

$$\begin{aligned}\frac{a^n}{a^m} &= \frac{a^m \cdot a^{n-m}}{a^m}\\ &= \frac{a^m}{a^m} \cdot a^{n-m}\\ &= 1 \cdot a^{n-m}\\ &= a^{n-m}\end{aligned}$$

If $n < m$, the integer $m - n$ is positive; thus $a^m = a^n \cdot a^{m-n}$. Hence

$$\begin{aligned}\frac{a^n}{a^m} &= \frac{a^n}{a^n \cdot a^{m-n}}\\ &= \frac{a^n}{a^n} \cdot \frac{1}{a^{m-n}}\\ &= 1 \cdot \frac{1}{a^{m-n}}\\ &= \frac{1}{a^{m-n}}\end{aligned}$$

■

Example 4 Find the following quotients. In part (b), n is a positive integer.

a) $\dfrac{8a^5b^2}{2a^3b^8}$

b) $\dfrac{6nx^{6n}}{3nx^{3n}}$

Solution

a) $\dfrac{8a^5b^2}{2a^3b^8} = \dfrac{8}{2} \cdot \dfrac{a^5}{a^3} \cdot \dfrac{b^2}{b^8}$

$\qquad = 4a^{5-3} \cdot \dfrac{1}{b^{8-2}}$

$\qquad = \dfrac{4a^2}{b^6}$

b) $\dfrac{6nx^{6n}}{3nx^{3n}} = \dfrac{6}{3} \cdot \dfrac{n}{n} \cdot \dfrac{x^{6n}}{x^{3n}}$

$\qquad = 2 \cdot 1 \cdot x^{6n-3n}$

$\qquad = 2x^{3n}$ ∎

The power of a quotient appears in the next theorem.

Theorem 5

If n is a positive integer and a and b are real numbers where $b \neq 0$, then

$$\left(\frac{a}{b}\right)^n = \frac{a^n}{b^n}$$

Informal argument

$$\left(\frac{a}{b}\right)^n = \frac{a}{b} \cdot \frac{a}{b} \cdot \frac{a}{b} \cdots \frac{a}{b} \quad \left(n \text{ factors of } \frac{a}{b}\right)$$

$$= \frac{a \cdot a \cdot a \cdots a}{b \cdot b \cdot b \cdots b} \quad (n \text{ factors of } a \text{ and } n \text{ factors of } b)$$

$$= \frac{a^n}{b^n}$$

Illustration 6

a) $\left(\dfrac{2}{3}\right)^5 = \dfrac{2^5}{3^5}$

$\qquad = \dfrac{32}{243}$

b) $\left(\dfrac{x^4z^2}{y^3}\right)^3 = \dfrac{(x^4z^2)^3}{(y^3)^3}$

$\qquad = \dfrac{(x^4)^3(z^2)^3}{y^9}$

$\qquad = \dfrac{x^{12}z^6}{y^9}$ ∎

Recall that the definition of a positive-integer exponent is as follows:

$$a^n = a \cdot a \cdot a \cdots a \quad (n \text{ factors of } a)$$

This definition has meaning only when the exponent n is a positive integer. Therefore, when the exponent is zero or a negative integer, a different definition must be given. We want these definitions to be such that the same laws that apply for positive-integer exponents also hold for zero and negative-integer exponents. In particular, if the formula of Theorem 1 is to hold for a zero exponent, then if $a \neq 0$

$$a^0 \cdot a^n = a^{0+n}$$
$$a^0 \cdot a^n = a^n$$

Because 1 is the number having the property that $1 \cdot a^n = a^n$, we must define a^0 as 1.

Now suppose that n is a positive integer, and thus $-n$ is a negative integer. If the formula of Theorem 1 is to hold for a negative-integer exponent, then

$$a^n \cdot a^{-n} = a^0$$
$$a^n \cdot a^{-n} = 1$$

Hence we must define a^{-n} as $\dfrac{1}{a^n}$. We state the definitions formally.

Definition

Zero and Negative-Integer Exponents
If n is a positive integer and a is a real number where $a \neq 0$, then

$$a^0 = 1$$

$$a^{-n} = \frac{1}{a^n}$$

Illustration 7

a) $6^0 = 1$ b) $4^{-2} = \dfrac{1}{4^2}$ c) $(-5)^0 (-2)^{-3} = 1 \cdot \dfrac{1}{(-2)^3}$

$$= \frac{1}{16} \qquad\qquad = \frac{1}{-8}$$

$$= -\frac{1}{8}$$ ∎

It can be shown that the laws for positive-integer exponents given by Theorems 1–5 also hold for zero and negative-integer exponents. Following is a summary of these laws. In each case n and m are any integers and a and b are real numbers.

1. $a^n \cdot a^m = a^{n+m}$ **2.** $(a^n)^m = a^{nm}$

3. $(ab)^n = a^n b^n$ **4.** $\dfrac{a^n}{a^m} = a^{n-m}$ $a \neq 0$

5. $\left(\dfrac{a}{b}\right)^n = \dfrac{a^n}{b^n}$ $b \neq 0$

Illustration 8

a) $(2^{-3} \cdot 3^{-2})^{-1} = 2^{-3(-1)} \cdot 3^{-2(-1)}$ b) $(2^{-3} + 3^{-2})^{-1} = \dfrac{1}{(2^{-3} + 3^{-2})^1}$

$\qquad\qquad\qquad = 2^3 \cdot 3^2$

$\qquad\qquad\qquad = 72$

$$= \dfrac{1}{\dfrac{1}{2^3} + \dfrac{1}{3^2}}$$

$$= \dfrac{1}{\dfrac{1}{8} + \dfrac{1}{9}}$$

$$= \dfrac{1}{\dfrac{17}{72}}$$

$$= \dfrac{72}{17} \qquad\blacksquare$$

Example 5 Write as a simple fraction with only positive exponents:

$$\left(\frac{x^{-3}y^4z^{-5}}{x^6y^{-2}z^{-4}} \right)^{-2}$$

Solution

$$\left(\frac{x^{-3}y^4z^{-5}}{x^6y^{-2}z^{-4}} \right)^{-2} = \frac{x^{-3(-2)}y^{4(-2)}z^{-5(-2)}}{x^{6(-2)}y^{-2(-2)}z^{-4(-2)}}$$

$$= \frac{x^6}{x^{-12}} \cdot \frac{y^{-8}}{y^4} \cdot \frac{z^{10}}{z^8}$$

$$= x^{6-(-12)}y^{-8-4}z^{10-8}$$

$$= x^{18}y^{-12}z^2$$

$$= x^{18} \cdot \frac{1}{y^{12}} \cdot z^2$$

$$= \frac{x^{18}z^2}{y^{12}} \qquad\blacksquare$$

Scientific notation involves powers of 10. It is especially important in physics, chemistry, astronomy, and computer science, because it is useful for computations with very large or very small numbers. We use the fact that any positive number x can be written as follows:

$$x = a \cdot 10^c \qquad \text{where } 1 \le a < 10 \text{ and } c \text{ is an integer}$$

When a number is expressed in this form, it is said to be written in **scientific notation**.

Illustration 9

Each of the following numbers is written in scientific notation:

$$582 = (5.82)10^2 \qquad 0.627 = (6.27)10^{-1}$$
$$97{,}136 = (9.7136)10^4 \qquad 0.00002381 = (2.381)10^{-5}$$
$$92{,}900{,}000{,}000 = (9.29)10^{10} \qquad 2.04 = (2.04)10^0 \qquad ■$$

To write a number in scientific notation, obtain the first factor by placing a decimal point after the first left-hand nonzero digit. Then write the second factor as a power of 10. To obtain the exponent, count the number of digits that must be passed over to move from the new position of the decimal point to the original position of the decimal point. If the movement is to the right, the exponent is positive; if the movement is to the left, the exponent is negative. You should verify this rule by applying it to the numbers in Illustration 9.

If a number is written in scientific notation, it can be converted to standard form by moving the decimal point in the first factor the number of places indicated by the exponent of the power of 10. The decimal point is moved to the right if the exponent is positive and to the left if the exponent is negative. This rule is applied in the next illustration.

Illustration 10

$$(3.659)10^4 = 36{,}590 \qquad (8.007)10^2 = 800.7$$
$$(9.46)10^{12} = 9{,}460{,}000{,}000{,}000 \qquad (3.92)10^{-3} = 0.00392$$
$$(4.018)10^{-2} = 0.04018 \qquad (7.2)10^{-16} = 0.0000000000000072 \quad ■$$

Scientific calculators display very large or very small numbers by using scientific notation. The manual for your calculator will tell you how to enter and read in the display a number written in scientific notation.

Scientific notation affords a convenient way of indicating the significant digits in a numeral like 83,200. For example, a measurement of 83,200 ft may be written as $(8.32)10^4$ ft to indicate there are three significant digits, meaning the measurement is accurate to the nearest hundred feet. If we write $(8.320)10^4$ ft, then there are four significant digits and the measurement is accurate to the nearest ten feet. Similarly, if we write $(8.3200)10^4$ ft, there are five significant digits and the measurement is accurate to the nearest foot.

A numeral having more than k significant digits is said to be **rounded off to k significant digits** if it is replaced by the number, having k significant digits, to which it is closest in value. For instance, the numeral 0.52368 is rounded off to four significant digits as 0.5237, while the numeral 78.142 is rounded off to four significant digits as 78.14. To round off to four significant digits a five-digit numeral whose fifth digit is 5, we adopt the following convention: if the fifth digit is 5 and the fourth digit is even, we round off to the fourth digit; if the fifth digit is 5 and the fourth digit is odd, we increase the fourth digit by one. Hence we round off 261.85 to 261.8 and 0.0039235 is rounded off to 0.003924. A similar convention is used to round off to any number of significant digits.

Example   6 Use scientific notation to perform the following computations.

a) $(0.00002350)(56,300)$, where 56,300 has four significant digits;

b) $\dfrac{(92,900,000,000)(0.00000262)}{(0.000310)(581)}$, where 92,900,000,000 has three significant digits.

Solution We first write each number in scientific notation.

a) $[(2.350)10^{-5}][(5.630)10^4] = [(2.350)(5.630)][10^{-5} \cdot 10^4]$
$$= (13.23)10^{-1}$$
$$= 1.323$$

b) $\dfrac{[(9.29)10^{10}][(2.62)10^{-6}]}{[(3.10)10^{-4}][(5.81)10^2]} = \dfrac{(9.29)(2.62)}{(3.10)(5.81)} \cdot \dfrac{10^{10} \cdot 10^{-6}}{10^{-4} \cdot 10^2}$

$$= (1.35)10^{10-6+4-2}$$
$$= (1.35)10^6 \qquad \blacksquare$$

Exercises   1.2

In Exercises 1 through 14, use laws of exponents to write the quantity as a rational number so that the exponent is 1.

1. (a) 5^3; (b) $(-2)^5$; (c) $(\frac{1}{3})^4$; (d) $(-\frac{1}{10})^6$; (e) 5^0
2. (a) 3^5; (b) $(-\frac{1}{4})^2$; (c) $(-10)^3$; (d) $(\frac{1}{5})^4$; (e) $(-2)^0$
3. (a) $(3^2)^4$; (b) $(2 \cdot 3)^5$; (c) $(\frac{7}{13})^2$; (d) $\left(\dfrac{2^3 \cdot 3^2}{5}\right)^2$
4. (a) $(2^2)^5$; (b) $(3 \cdot 5)^4$; (c) $(\frac{3}{11})^3$; (d) $\left(\dfrac{2^4 \cdot 5^2}{7}\right)^2$
5. (a) 5^{-4}; (b) $(-6)^{-2}$; (c) $(\frac{1}{2})^{-3}$; (d) $(-\frac{2}{3})^{-5}$; (e) $(-\frac{1}{4})^0$
6. (a) 3^{-2}; (b) $(-4)^{-3}$; (c) $(\frac{1}{5})^{-5}$; (d) $(-\frac{3}{4})^{-4}$; (e) $(\frac{9}{5})^0$
7. (a) $(2^5)(2^{-3})(2^{-4})$; (b) $\dfrac{(3^2)(3^{-1})}{3^{-4}}$; (c) $\dfrac{(10^{-2})(10^3)}{(10^7)(10^{-4})}$;
   (d) $\dfrac{(2^{-2})(3^4)}{(3^{-1})(2^{-3})}$
8. (a) $(5^2)(5^{-4})(5^{-1})$; (b) $\dfrac{(4^{-3})(4^2)}{4^{-4}}$; (c) $\dfrac{(3^{-5})(3^6)}{(3^7)(3^{-3})}$;
   (d) $\dfrac{(7^{-1})(2^{-3})}{(2^{-5})(7^2)}$

9. (a) $2^{-1} + 3^{-2}$; (b) $(2^{-1} \cdot 3^{-2})^2$; (c) $3^{-3} - 9^{-1}$;
   (d) $(2^{-2} + 4^{-3})^{-1}$
10. (a) $4^{-2} + 2^{-4}$; (b) $(4^{-1} \cdot 2^{-3})^2$; (c) $2^{-3} - 3^{-2}$;
    (d) $(3^{-3} + 3^{-1})^{-2}$
11. (a) $\dfrac{5^2}{2^{-3}}$; (b) $\left(\dfrac{2}{3}\right)^{-2}$; (c) $\dfrac{(-3)^{-4}}{(-4)^{-3}}$; (d) $\left(\dfrac{7^{-1}}{10^2}\right)^{-2}$
12. (a) $\dfrac{2^3}{7^{-2}}$; (b) $\left(\dfrac{3}{5}\right)^{-3}$; (c) $\dfrac{(-2)^{-5}}{(-5)^{-2}}$; (d) $\left(\dfrac{3^2}{4^{-3}}\right)^{-1}$
13. (a) $\dfrac{3^0 + 3^{-1}}{3^{-3} + 3^0}$; (b) $\dfrac{2^{-3} + 3^{-2}}{2^{-2} - 3^{-1}}$; (c) $\dfrac{5^2 + 10^{-1}}{5^{-2} + 2^2}$
14. (a) $\dfrac{4^{-1} + 2^{-3}}{5^0 - 4^{-2}}$; (b) $\dfrac{5^{-2} + 2^{-2}}{6^0 + 10^{-2}}$; (c) $\dfrac{3^2 + 9^{-1}}{2^4 - 3^{-1}}$

In Exercises 15 through 26, use laws of exponents to write the expression so that each variable occurs only once and all the exponents are positive. None of the denominators is zero.

15. (a) $x^2 \cdot x^5$; (b) $a^{-3} \cdot a^6$; (c) $\dfrac{y^5}{y^{-7}}$; (d) $\dfrac{x^{-3}}{x^2}$

16. (a) $b^5 \cdot b^3$; (b) $x^{-4} \cdot x^{-2}$; (c) $\dfrac{a^{-2}}{a^3}$; (d) $\dfrac{y^{-1}}{y^{-2}}$

17. (a) $(x^{-3})^4$; (b) $(x^3 y^2)^{-3}$; (c) $(s^5 t^{-1})^{-2}$; (d) $\dfrac{a^{-3} b^{-1}}{c^{-2}}$

18. (a) $(t^2)^{-5}$; (b) $(u^3 v^{-2})^3$; (c) $\left(\dfrac{x^{-1}}{y^3}\right)^{-4}$; (d) $\left(\dfrac{a^{-1} b^2}{c^{-4} d^0}\right)^2$

19. (a) $(5x^3)(4x^5)$; (b) $(x^3)^5$; (c) $(3ab^4)(-2a^3 b^2)$;
    (d) $-7(-x^6 y)(-x^6 y^5)$

20. (a) $(-6a^2)(3a^6)$; (b) $(a^2)^6$; (c) $(x^4 y^3)(2xy)$;
    (d) $5(-4r^2 s^2)(3r^2 s^3)$

21. (a) $(2x^2)^3 (3y^3)^2$; (b) $(x^2 y^3)^3$; (c) $\left(\dfrac{3u^4}{4v^3}\right)^4$;
    (d) $\left(-\dfrac{r^2 s^3 t^4}{2u^5}\right)^5$

22. (a) $(4s^4)^4 (2t^3)^3$; (b) $(s^4 t^3)^5$; (c) $\left(\dfrac{9w^2}{10z^3}\right)^4$;
    (d) $\left(\dfrac{7x^5}{8w^3 y^4 z^6}\right)^2$

23. (a) $\dfrac{(20r^2 s^3 t^4)(2r^2 s^2 t)}{(-4rst)(3rs^4 t^2)}$; (b) $\dfrac{(-3x^3 yz^2)(-2x^4 y^2 z^5)}{-3x^4 yz^4}$

24. (a) $\dfrac{(3a^4 b^3)(-10ab^6)}{(-2a^3 b^2)(15b^5)}$; (b) $\dfrac{3x^{20} y^{12} z^{16}}{(-x^9 y^3 z^3)(-9x^4 y^6 z^4)}$

25. (a) $\dfrac{3^{-2} x^{-4} y^0}{(3x^2 y^3)^{-4}}$; (b) $\left(\dfrac{2a^{-2} b^2 c^{-4}}{3a^{-3} b^{-1} c^2}\right)^{-2}$

26. (a) $\dfrac{8^{-1} s^{-3} t^0}{(2s^{-1} t)^{-5}}$; (b) $\left(\dfrac{3x^{-1} y^2 z^3}{2x^{-3} y^{-2} z^{-1}}\right)^{-1}$

In Exercises 27 through 30, use laws of exponents to simplify the expression and write it with only positive exponents. In each exercise, n is a positive integer.

27. (a) $(x^n)(x^{3n})$; (b) $(x^n)^{3n}$; (c) $\dfrac{x^{3n}}{x^n}$; (d) $(x^{3+n})(x^{3n-1})$;
    (e) $\dfrac{x^{3n+2}}{x^{2n+1}}$

28. (a) $(x^{2n})^n$; (b) $(x^{2n} \cdot x^n)^2$; (c) $\dfrac{x^{2n} \cdot x^2}{(x^n)^2}$;
    (d) $(x^{2n+2})(x^{2n-1})$; (e) $\dfrac{x^{5n-2}}{x^{n+1}}$

29. $\dfrac{(10^{1-n})^n}{(1000^n)^{n+1}} \cdot \dfrac{10{,}000^{n^2+2}}{100^{3-n}}$ 30. $\dfrac{9^{2n-1}}{(3^{n+1})^n} \cdot \dfrac{(81^{n-1})^{n+1}}{(27^{n+2})^{n-1}}$

In Exercises 31 through 34, use a calculator to compute the power to three significant figures. In Exercises 31 and 32, express the result in scientific notation.

31. (a) $(35.7)^3$; (b) $(3.78)^{-5}$; (c) $(0.261)^8$; (d) $(0.403)^{-3}$
32. (a) $(6.23)^4$; (b) $(15.7)^{-4}$; (c) $(0.362)^6$; (d) $(0.916)^{-2}$

33. (a) $(4.26)^{25}$; (b) $(0.0312)^5$; (c) $(0.00172)^{-12}$;
    (d) $(324)^{-10}$
34. (a) $(78.5)^{15}$; (b) $(0.00247)^8$; (c) $(0.0311)^{-7}$;
    (d) $(589)^{-12}$

In Exercises 35 through 38, write the number in scientific notation.

35. (a) 52.60; (b) 0.0061; (c) 172,000 (3 significant digits); (d) 172,000 (4 significant digits).

36. (a) 43,851; (b) 0.276;
    (c) 3400 (2 significant digits);
    (d) 3400 (3 significant digits).

37. (a) 0.03960; (b) 0.0000080022; (c) 1.723;
    (d) 426.0

38. (a) 0.00006405; (b) 0.0001030; (c) 98.0;
    (d) 7820.0

In Exercises 39 and 40, the number is written in scientific notation; write it in standard form.

39. (a) $(2.432)10^2$; (b) $(6.013)10^4$; (c) $(8.390)10^{-3}$;
    (d) $(5.08)10^{-5}$
40. (a) $(5.738)10^1$; (b) $(2.64)10^5$; (c) $(7.425)10^{-2}$;
    (d) $(9.620)10^{-7}$

In Exercises 41 and 42, use scientific notation and a calculator to perform the indicated operations. Assume three significant digits for each number.

41. (a) $(0.0470)(320{,}000)^2$;
    (b) $\dfrac{(180{,}000)^3 (0.0000450)^2}{(623{,}000)^2}$

42. (a) $\dfrac{(0.0000831)^2}{(140)^3}$;
    (b) $\dfrac{(256{,}000{,}000)^2 (0.0712)^3}{(0.000348)^3 (5100)^4}$

In Exercises 43 through 45, use scientific notation and a calculator.

43. Determine the number of meters in the distance from the earth to the sun, assuming that the sun is $(9.29)10^7$ mi from the earth and 1 mi is 1.61 km.

44. Determine the number of miles in the distance from the earth to a star that is 7.00 light-years away, assuming that the speed of light is $(1.86)10^5$ mi/sec and 1 year is 365 days.

45. Determine the number of tons in the mass of the earth, assuming that the number of grams in the earth's mass is $(5.97)10^{27}$ and 1 g is $(2.205)10^{-3}$ lb.

1.3 Polynomials

In Section 1.1 we stated that a variable is a symbol used to represent any element of a given domain. If the domain is R, then the variable represents a real number. A **constant** is a symbol whose domain contains only one element. For example, if we write the sum

$$6x^2 + 2x + 5$$

x is a symbol for a variable while 6, 2, and 5 are constants. Sometimes letters are used as symbols for constants if the letters designate fixed but unspecified numbers. For instance, we can write the sum

$$ax + b$$

where a and b are symbols for constants and x is a symbol for a variable. The sum $6x^2 + 2x + 5$ is a particular *algebraic expression*. The terminology **algebraic expression** is used to mean a constant, a variable, or a combination of variables and constants involving a finite number of indicated operations (addition, subtraction, multiplication, division, raising to a power, and extraction of a root) on them. Examples of algebraic expressions are

$$3x^2y^5 \qquad 5x^2 - 8x + 2 \qquad \frac{3x^2 - 6xy + y^2}{x + 2y} \qquad \frac{\sqrt{x+y} - 4}{(z+2)^3 - \sqrt[3]{x}}$$

An algebraic expression involving only nonnegative-integer powers of one or more variables and containing no variable in a denominator is called a **polynomial.** For example,

$$2x \qquad 5x^2 - 8x + 2 \qquad 4x^5 - 6x^3 + 3x^2 - 2x + 1$$

are polynomials in the variable x. Examples of polynomials in the variables x and y are

$$3x^2y^5 \qquad 6x^2 + 8y^2 \qquad 8xy - 7x + y - 3$$

A **term** of a polynomial is a constant or a constant multiplied by nonnegative-integer powers of variables. A polynomial can be considered as the sum of a finite number of terms. For example, $5x^2 - 8x + 2$ can be written as $5x^2 + (-8x) + 2$ and the terms of this polynomial are $5x^2$, $-8x$, and 2. The polynomial $8xy - 7x + y - 3$ can be written as $8xy + (-7x) + y + (-3)$ and the terms are $8xy$, $-7x$, y, and -3.

Any factor of a product is said to be the **coefficient** of the other factors. For instance, in the product $5xyz$ the coefficient of yz is $5x$, the coefficient of x is $5yz$, the coefficient of $5z$ is xy, and so on. If a coefficient is a constant, then it is called a **constant coefficient.** Hence in the product $5xyz$, 5 is the constant coefficient of xyz. Terms that may differ only in their constant coefficients are called **like terms.** For example, $6x^2$ and $3x^2$ are like terms, as are x and $-4x$. Like terms of a polynomial are combined algebraically by using the distributive law. In particular,

$$6x^2 + x + 7 + 3x^2 - 4x = (6x^2 + 3x^2) + (x - 4x) + 7$$
$$= (6 + 3)x^2 + (1 - 4)x + 7$$
$$= 9x^2 - 3x + 7$$

If after combining like terms a polynomial has one term, it is called a **monomial;** if it has two terms, it is called a **binomial;** and if it has three terms, it is called a **trinomial.** Polynomials $2x$ and $3x^2y^5$ are monomials, $6x^2 + 8y^2$ is a binomial, and $5x^2 - 8x + 2$ is a trinomial.

By the **degree** of a monomial in one variable, we mean the exponent of that variable. In particular, $5x^3$ has degree 3. If a monomial has more than one variable, its degree is the sum of the exponents of all the variables that appear. For example, the degree of $3x^2y^5$ is 7. The degree of $-2xyz$ is 3. The degree of a nonzero-constant monomial, such as 4, is zero. The constant 0 has no degree.

The **degree** of a polynomial is the same as the degree of the term with highest degree in the polynomial. Therefore $7x^2 - 4x + 2$ is a second-degree polynomial and $3x + 6$ is a first-degree polynomial. The degree of $6x^2y^2 - 4x^3 + 2y$ is 4 because $6x^2y^2$ is the term with highest degree, 4.

Because the variables used for polynomials represent real numbers, we can apply to polynomials the definitions, axioms, and theorems involving the real numbers. Some polynomials can be simplified by using distributive laws:

$$ax + bx = (a + b)x$$
$$ax - bx = (a - b)x$$
$$ax + bx + cx = (a + b + c)x$$

and so on.

Illustration   1

a) $8x^2 + 5x + 2x^2 - 4x = (8 + 2)x^2 + (5 - 4)x$
$$= 10x^2 + x$$

b) $7a^2b - 3a^2b + 8ab^2 - 5ab^2 - 6ab^2 = (7 - 3)a^2b + (8 - 5 - 6)ab^2$
$$= 4a^2b + (-3ab^2)$$
$$= 4a^2b - 3ab^2$$

c) $(4y^3 + 7y^2 + 3y - 8) + (6y^3 - 2y^2 + 4)$
$$= (4y^3 + 6y^3) + (7y^2 - 2y^2) + 3y + (-8 + 4)$$
$$= (4 + 6)y^3 + (7 - 2)y^2 + 3y + (-4)$$
$$= 10y^3 + 5y^2 + 3y - 4$$ ■

Sometimes the addition of polynomials is performed by placing them in separate rows with like terms in the same vertical column.

Illustration 2

The addition of the polynomials in part (c) of Illustration 1 can be performed as follows.

$$4y^3 + 7y^2 + 3y - 8$$
$$6y^3 - 2y^2 \qquad + 4$$
$$\overline{10y^3 + 5y^2 + 3y - 4}$$

∎

Subtraction of one polynomial from another can also be performed by arranging the two polynomials, one above the other, so that like terms are in the same column. Then, because $a - b = a + (-b)$, we mentally replace each term in the lower polynomial by its negative and add, as indicated in the next illustration.

Illustration 3

We wish to subtract $4a^3 - 8a^2 + 2a + 6$ from $2a^3 + a^2 + 4a - 3$. We arrange the two polynomials in columns of like terms.

$$2a^3 + \ a^2 + 4a - 3$$
$$\underline{4a^3 - 8a^2 + 2a + 6} \qquad \text{(subtract)}$$

Now think of each term in the lower polynomial replaced by its negative and then add.

$$2a^3 + \ a^2 + 4a - 3$$
$$\underline{-4a^3 + 8a^2 - 2a - 6} \qquad \text{(add)}$$
$$-2a^3 + 9a^2 + 2a - 9$$

∎

Distributive laws can be used to write the product of a monomial and a polynomial. For example,

$$a(x + y + z) = ax + ay + az$$

Illustration 4

a) $4t^3(5t^3 - 7t^2 - 3t) = 4t^3(5t^3) + 4t^3(-7t^2) + 4t^3(-3t)$
$$= 20t^6 - 28t^5 - 12t^4$$

b) If n is a positive integer,

$$3x^{3n}(4x^4 + 2x^{2n}) = 3x^{3n}(4x^4) + 3x^{3n}(2x^{2n})$$
$$= 12x^{3n+4} + 6x^{5n}$$

∎

Distributive laws can be applied repeatedly to obtain the product of two polynomials, as shown in the next illustration.

Illustration 5

$$(2x + 7)(3x - 4) = 2x(3x - 4) + 7(3x - 4)$$
$$= 6x^2 - 8x + 21x - 28$$
$$= 6x^2 + 13x - 28$$

∎

In the following example the multiplication of two polynomials is performed by first placing them in separate rows and then writing the various products of monomials so that like terms are in the same vertical column. When applying this procedure, each polynomial should be written in descending powers of the variable.

Example   1   Find the product $(3x^2 - 2x + 1)(3 + x - 4x^2)$.

Solution   When placing the two polynomials in separate rows, we rewrite the second polynomial so that it is in descending powers of x.

$$
\begin{array}{r}
3x^2 - 2x + 1 \\
-4x^2 + x + 3 \\
\hline
-12x^4 + 8x^3 - 4x^2 \\
3x^3 - 2x^2 + x \\
9x^2 - 6x + 3 \\
\hline
-12x^4 + 11x^3 + 3x^2 - 5x + 3
\end{array}
$$

(the product of $-4x^2$ and $3x^2 - 2x + 1$)
(the product of x and $3x^2 - 2x + 1$)
(the product of 3 and $3x^2 - 2x + 1$)   ∎

Example   2   Find the product $(2x - 3y)(5x^2 - xy + 4y^2)$.

Solution

$$
\begin{array}{r}
5x^2 - xy + 4y^2 \\
2x - 3y \\
\hline
10x^3 - 2x^2y + 8xy^2 \\
- 15x^2y + 3xy^2 - 12y^3 \\
\hline
10x^3 - 17x^2y + 11xy^2 - 12y^3
\end{array}
$$

(the product of $2x$ and $5x^2 - xy + 4y^2$)
(the product of $-3y$ and $5x^2 - xy + 4y^2$)   ∎

There are certain products of polynomials, called **special products,** that occur frequently and should be recognized. In these special products, x and y represent variables and a, b, and c represent constants. The products can be verified by performing the multiplications. You are asked to do this in Exercises 27 and 28.

Special product 1: $(x + a)(x + b) = x^2 + (a + b)x + ab$

Illustration   6

Applying special product 1, where a is 7 and b is 2, we obtain

$$(x + 7)(x + 2) = x^2 + (7 + 2)x + 7 \cdot 2$$
$$= x^2 + 9x + 14$$ ∎

Special product 2: $(x + y)^2 = x^2 + 2xy + y^2$

Illustration 7

From special product 2, where x is $2w$ and y is $5z$,

$$(2w + 5z)^2 = (2w)^2 + 2(2w)(5z) + (5z)^2$$
$$= 4w^2 + 20wz + 25z^2$$

■

$$\boxed{\textbf{\textit{Special product 3:}} \quad (x + y)(x - y) = x^2 - y^2}$$

Illustration 8

Applying special product 3, where x is $3u^3$ and y is $4v^2$, we have

$$(3u^3 + 4v^2)(3u^3 - 4v^2) = (3u^3)^2 - (4v^2)^2$$
$$= 9u^6 - 16v^4$$

■

$$\boxed{\textbf{\textit{Special product 4:}} \quad (ax + by)(cx + dy) = acx^2 + (ad + bc)xy + bdy^2}$$

Illustration 9

From special product 4, where a is 5, b is -2, c is 3, and d is 6,

$$(5x - 2y)(3x + 6y) = (5 \cdot 3)x^2 + [5 \cdot 6 + (-2) \cdot 3]xy + [(-2) \cdot 6]y^2$$
$$= 15x^2 + 24xy - 12y^2$$

■

The formulas giving the special products 1 through 4 are called *identities,* because a true statement is obtained if any real number is substituted for a variable. For example, in special product 2, if x is 3 and y is 4,

$$
\begin{array}{ll}
(x + y)^2 = (3 + 4)^2 & \quad x^2 + 2xy + y^2 = 3^2 + 2(3)(4) + 4^2 \\
\qquad\quad\; = 7^2 & \qquad\qquad\qquad\;\; = 9 + 24 + 16 \\
\qquad\quad\; = 49 & \qquad\qquad\qquad\;\; = 49
\end{array}
$$

To divide a polynomial by a monomial, each term of the polynomial is divided by the monomial. In particular, if $d \neq 0$ (because division by zero is not defined),

$$\frac{u + v + w}{d} = \frac{u}{d} + \frac{v}{d} + \frac{w}{d}$$

Illustration 10

If $x \neq 0$,

$$\frac{10x^5 - 12x^4 + 6x^3}{2x^2} = \frac{10x^5}{2x^2} + \frac{-12x^4}{2x^2} + \frac{6x^3}{2x^2}$$

$$= \frac{10}{2}x^{5-2} + \frac{-12}{2}x^{4-2} + \frac{6}{2}x^{3-2}$$

$$= 5x^3 - 6x^2 + 3x \qquad \blacksquare$$

Example   3 Find each of the following quotients, where none of the variables is zero. In part (b), n is a positive integer.

a) $\dfrac{30x^5y^4z^2 - 18x^2y^3z^4}{-6xy^2z}$ b) $\dfrac{8x^{8n} - 6x^{6n}}{2x^{2n}}$

Solution

a) $\dfrac{30x^5y^4z^2 - 18x^2y^3z^4}{-6xy^2z} = \dfrac{30x^5y^4z^2}{-6xy^2z} + \dfrac{-18x^2y^3z^4}{-6xy^2z}$

$$= -5x^4y^2z + 3xyz^3$$

b) $\dfrac{8x^{8n} - 6x^{6n}}{2x^{2n}} = \dfrac{8x^{8n}}{2x^{2n}} + \dfrac{-6x^{6n}}{2x^{2n}}$

$$= 4x^{6n} - 3x^{4n} \qquad \blacksquare$$

In Illustration 5 we showed that

$$(2x + 7)(3x - 4) = 6x^2 + 13x - 28$$

Therefore, if $x \neq \frac{4}{3}$, it follows that

$$\frac{6x^2 + 13x - 28}{3x - 4} = 2x + 7$$

We show now a formal method for finding the quotient $2x + 7$ if the dividend is $6x^2 + 13x - 28$ and the divisor is $3x - 4$. Such a systematic procedure is called an **algorithm.** The algorithm used here is similar to the algorithm of long division in arithmetic and it takes the following form:

$$\begin{array}{r} 2x + 7 \qquad \textit{quotient} \\ \textit{divisor} \quad 3x - 4 \overline{)\, 6x^2 + 13x - 28} \quad \textit{dividend} \\ \underline{6x^2 - 8x} \\ 21x - 28 \\ \underline{21x - 28} \\ 0 \end{array}$$

The explanation of the above procedure, in which we apply the division algorithm, is as follows:

1. We divide $6x^2$ (the first term of the dividend) by $3x$ (the first term of the divisor) to obtain $2x$ (the first term of the quotient).

2. We multiply $3x - 4$ (the divisor) by $2x$ to obtain the product $6x^2 - 8x$, which we write under the like terms of the dividend.

3. We subtract and obtain the remainder $21x - 28$, which we consider as a new dividend.

4. We divide $21x$ (the first term of the new dividend) by $3x$ (the first term of the divisor) and obtain 7 (the second term of the quotient).

5. We multiply $3x - 4$ by 7 to obtain the product $21x - 28$, which we write under the new dividend and subtract. We obtain the remainder 0 for this particular problem.

As we did when multiplying polynomials, in long division we write the dividend and divisor in descending powers of some variable. Furthermore, if any power of the variable is missing, we adjust for this by writing that term with a coefficient of zero, as in part (b) of the next example.

Example 4 Find the following quotients.

a) $\dfrac{3z^3 - 11z^2 - 18z - 6}{z - 5}, \ z \neq 5$ b) $\dfrac{x^3 + y^3}{x + y}, \ x + y \neq 0$

Solution

a)
$$
\begin{array}{r}
3z^2 + 4z + 2 \\
z - 5 \overline{)\ 3z^3 - 11z^2 - 18z - 6} \\
\underline{3z^3 - 15z^2} \\
4z^2 - 18z \\
\underline{4z^2 - 20z} \\
2z - 6 \\
\underline{2z - 10} \\
4
\end{array}
$$

The quotient is $3z^2 + 4z + 2$, and there is a remainder of 4. We write the result as

$$
\frac{3z^2 - 11z^2 - 18z - 6}{z - 5} = 3z^2 + 4z + 2 + \frac{4}{z - 5}
$$

b) We write the dividend in descending powers of x, where $0x^2$ and $0x$ appear for the terms involving x^2 and x, respectively.

$$
\begin{array}{r}
x^2 - xy + y^2 \\
x + y \overline{)\ x^3 + 0x^2 + 0x + y^3} \\
\underline{x^3 + x^2y} \\
-x^2y + 0x \\
\underline{-x^2y - xy^2} \\
xy^2 + y^3 \\
\underline{xy^2 + y^3} \\
0
\end{array}
$$

Hence

$$\frac{x^3 + y^3}{x + y} = x^2 - xy + y^2 \qquad \blacksquare$$

EXERCISES 1.3

In Exercises 1 and 2, indicate whether the polynomial is a monomial, a binomial, or a trinomial; for each polynomial, state its degree.

1. (a) $7x^2 - 2x + 5$; (b) $4u + 3v$; (c) $4xy^3z^2$;
   (d) $2rs - s^3 + 6r^2s^2$
2. (a) $8x^4 + 6x^2 - 1$; (b) $-8rst^2$; (c) $2u^3v^2w + 5uv^3$;
   (d) $xyz - 3x^2 + 2z^2$

In Exercises 3 through 6, simplify the algebraic expression.

3. (a) $4x^2 - 5x + 6x^2 - 2x$; (b) $-5(6y^3 - 4y^2 + y - 3)$
4. (a) $7y^2 - y^3 - 3y^2 - 8y^3$; (b) $2(3u - 4v) - (5u - 3v)$
5. (a) $3(-t^2 + 3st - 2s^2) - 2(7t^2 - st - s^2)$;
   (b) $-(x - 9) - 4[3x - 2(6 + x) - 5]$
6. (a) $4(x^3 + 3x^2y - 2xy^2 - y^3) - 2x(3x^2 + xy - 5y^2)$;
   (b) $3(2w - 3z) - [w - z - (w + z)]$

In Exercises 7 through 10, (a) add the two polynomials and (b) subtract the second polynomial from the first.

7. $4x^3 - 7x^2 + 2x - 4$; $3x^3 + 8x^2 + 3x - 7$
8. $5x^3 - 3x + 10$; $x^3 - 6x^2 + 4x - 2$
9. $6y^4 - 2y^2 + 3y - 1$; $8y^4 - 5y^3 - 2y^2 + 6$
10. $2z^5 + 7z^4 - z^2 + 4z + 1$; $4z^5 - 6z^3 + z - 8$

In Exercises 11 through 26, find the product. In Exercises 13, 14, 25, and 26, n is a positive integer.

11. (a) $-3x^2(4x^2 - 2x + 7)$; (b) $\frac{1}{2}a^3b^2(2a^2 + 5ab - b^2)$
12. (a) $5y^3(3 - 6y + 2y^2)$; (b) $-6ab^4(4a^2 - \frac{2}{3}ab - \frac{1}{2}b^2)$
13. (a) $2xyz^2(3xz - 6yz - xy - 1)$;
    (b) $3x^{2n}(x^{n+1} - 4x^n + 5)$
14. (a) $uvw(uv - 2uw + vw - 8)$;
    (b) $x^{2n-1}(2x^{2n+1} - x^{2n} + 3x - 1)$
15. (a) $(2x + 3)(x - 7)$; (b) $(5s - 6t)(2s + 4t)$
16. (a) $(y + 8)(4y - 3)$; (b) $(2x^2 - 5y^2)(-3x^2 + y^2)$
17. (a) $-4(9a - 5b)(a + 3b)$; (b) $2y(4y + 1)^2$
18. (a) $-x(x - 7)^2$; (b) $(a - 6)(3a^2 - 4a + 2)$
19. $(3x + 5y)(3x^2 - xy + 4y^2)$
20. $(2u - 5v)(4u^2 - uv - 3v^2)$
21. $(2x^2 + 4x - 3)(x^2 - 6x + 5)$
22. $(3t^2 + 2t + 1)(-2t^2 - 5t + 6)$
23. $(b - 3b^2 + 7)(5b^2 + 2 - 3b)$

24. $(2x - 5 + x^2)(8x^2 - 6 + x)$
25. $(3x^{2n} + y^n)(4x^{2n} - 5y^n)$ 26. $(3a^{3n} + b^{2n})(a^{2n} - b^n)$
27. Verify special products 1 and 2 by performing the multiplications.
28. Verify special products 3 and 4 by performing the multiplications.

In Exercises 29 through 36, apply special products 1 through 4 to find the indicated product.

29. (a) $(x + 4)(x + 5)$; (b) $(2x + 3y)^2$
30. (a) $(y - 2)(y + 1)$; (b) $(5t + 3)^2$
31. (a) $(w + 6)(w - 6)$; (b) $(3r - 2s)(4r + s)$
32. (a) $(x + 4)(x - 4)$; (b) $(6x - y)(3x + 2y)$
33. (a) $(t^2 - 5)(t^2 + 9)$; (b) $(4x^2 - 3y^2)^2$
34. (a) $(x^3 - 7y)(3x^3 - 8y)$; (b) $(6a + 5b)^2$
35. (a) $(11u + 8v)(11u - 8v)$; (b) $(7a^2 - 2b^2)(5a^2 + 3b^2)$
36. (a) $(3r - 10s)(3r + 10s)$; (b) $(2x + 9y)(7x - 4y)$

In Exercises 37 through 56, find the quotient. None of the divisors is zero. In Exercises 39 and 40, n is a positive integer.

37. (a) $\dfrac{8x^2 - 28x^4}{4x^2}$; (b) $\dfrac{35u^2v^3 - 20u^3v^2}{-5u^2v}$

38. (a) $\dfrac{a^2b - ab^3}{ab}$; (b) $\dfrac{-48y^3 + 30y^2 - 18y}{6y}$

39. (a) $\dfrac{-24a^3b^3c^4 + 32a^2b^4c^2 - 16a^5b^3c^3}{8a^2b^2c^2}$;

    (b) $\dfrac{16t^{4n} - 64t^{6n}}{2t^{2n}}$

40. (a) $\dfrac{28x^3y^6z^4 - 49x^2y^5z^3 - 21xy^3z^3}{-7xy^3z^2}$; (b) $\dfrac{4x^{4n} - 8x^{8n}}{2x^{2n}}$

41. $\dfrac{2x^2 - 5x + 2}{x - 2}$ 42. $\dfrac{3y^2 - 13y + 4}{y - 4}$

43. $\dfrac{6a^2 + 5a - 6}{3a - 2}$ 44. $\dfrac{12x^2 + 5x - 2}{3x + 2}$

45. $\dfrac{12 + 2x - 3x^2}{x + 1}$ 46. $\dfrac{t^3 - 7t - 6}{t + 2}$

47. $\dfrac{5b^3 + 7b^2 + 12}{b + 2}$ 48. $\dfrac{a^3 - 3a^2 - a + 3}{a - 2}$

49. $\dfrac{y^3 - 6y + 5}{y^2 + 3y - 2}$ 50. $\dfrac{2x^3 + 4x^2 - 5}{x^2 + 3}$ 53. $\dfrac{4x^2 - 2xy - 3y^2}{2x + 3y}$ 54. $\dfrac{a^4 + 16}{a + 2}$

51. $\dfrac{x^3 - 27}{x - 3}$ 52. $\dfrac{6t^2 - ts - s^2}{3t + s}$ 55. $\dfrac{a^6 - b^6}{a - b}$ 56. $\dfrac{x^5 + y^5}{x + y}$

1.4   Factoring Polynomials

If a polynomial is the product of other polynomials, then each of the latter polynomials is called a **factor** of the original one. Because

$$x^2 - 25 = (x - 5)(x + 5)$$

$x - 5$ and $x + 5$ are factors of $x^2 - 25$. The process of finding factors of a given polynomial is called **factoring** the polynomial. Factoring is important when working with fractions and solving equations.

    We begin by considering the *prime factors* of natural numbers. If a natural number greater than 1 has no natural number factors other than itself and 1, then the natural number is said to be a **prime number.** Therefore the prime numbers are 2, 3, 5, 7, 11, 13, 17, and so on. A natural number greater than 1 that is not a prime number is called a **composite number.** Hence 4, 6, 8, 9, 10, 12, 14, 15, 16, and so on are composite numbers. A composite number is said to be in **completely factored form** when it is written as a product of only prime numbers. To express a particular composite number in completely factored form, we first write it as a product of smaller factors (found by either observation or trial and error), and continue the process until each of the factors is a prime number. For example, $120 = 4 \cdot 30$; furthermore, $4 = 2^2$ and $30 = 6 \cdot 5$, and $6 = 2 \cdot 3$. Therefore

$$
\begin{aligned}
120 &= 4 \cdot 30 \\
&= (2^2)(6 \cdot 5) \\
&= (2^2)(2 \cdot 3)(5) \\
&= 2^3 \cdot 3 \cdot 5
\end{aligned}
$$

If we first express 120 as the product of 10 and 12 and then write $10 = 2 \cdot 5$ and $12 = 4 \cdot 3$ with $4 = 2^2$, we have

$$
\begin{aligned}
120 &= 10 \cdot 12 \\
&= (2 \cdot 5)(4 \cdot 3) \\
&= (2 \cdot 5)(2^2 \cdot 3) \\
&= 2^3 \cdot 5 \cdot 3
\end{aligned}
$$

Note that in each case, except for the order of the factors, we obtain the same prime factors of 120. In general, a composite number can be expressed in completely factored form in one and only one way. This fact is given in the following theorem, called *the fundamental theorem of arithmetic,* the proof of which is beyond the scope of this book.

The Fundamental Theorem of Arithmetic

The completely factored form of any composite number is unique except for the order of the factors.

Example   1 Express each of the following numbers in completely factored form.

   a) 252   b) 5005

Solution

a) $252 = 2 \cdot 126$
$ = 2 \cdot 6 \cdot 21$
$ = 2 \cdot 2 \cdot 3 \cdot 3 \cdot 7$
$ = 2^2 \cdot 3^2 \cdot 7$

b) $5005 = 5 \cdot 1001$
$ = 5 \cdot 11 \cdot 91$
$ = 5 \cdot 11 \cdot 7 \cdot 13$
$ = 5 \cdot 7 \cdot 11 \cdot 13$ ∎

A polynomial with integer coefficients is said to be **prime** if it has no monomial or polynomial factors with integer coefficients other than itself and one. Furthermore, a polynomial with integer coefficients is said to be in **completely factored form** when each of its polynomial factors is prime. We shall now study some methods for factoring polynomials. The discussion of each procedure is preceded by a heading describing either the given polynomial or the technique used to factor it.

Removing a Common Monomial Factor

If every term of a polynomial contains a common monomial factor, then, by the distributive law, the polynomial can be written as the product of the common monomial factor and the quotient obtained by dividing the original polynomial by the common factor. For instance, a is a common monomial factor of each term of the trinomial $ax + ay + az$; thus

$$ax + ay + az = a(x + y + z)$$

Illustration   1

To factor the polynomial $6x^3y^2 - 3x^2y + 9xy$, we first observe that $3xy$ is a common monomial factor of each term; hence

$$6x^3y^2 - 3x^2y + 9xy = 3xy(2x^2y - x + 3)$$

Note that the factor $2x^2y - x + 3$ is obtained by dividing the given polynomial by $3xy$.   ∎

Example 2 Factor each of the polynomials. In part (c), n is a positive integer.

a) $6y^6 - 4y^5 + 2y^4 - 8y^2$ b) $-10r^3s^2t^4 - 20r^3s^2t^3 + 5r^2s^4t^4$

c) $x^{2n} + x^{n+2}$

Solution

a) $6y^6 - 4y^5 + 2y^4 - 8y^2 = 2y^2(3y^4 - 2y^3 + y^2 - 4)$

b) $-10r^3s^2t^4 - 20r^3s^2t^3 + 5r^2s^4t^4 = -5r^2s^2t^3(2rt + 4r - s^2t)$

c) $x^{2n} + x^{n+2} = x^n(x^n + x^2)$ ■

Difference of Two Squares

From special product 3 of Section 1.3 we have the formula

$$x^2 - y^2 = (x + y)(x - y)$$

The left side of this formula is the difference of two squares, and the formula states that it can be written as the product of the sum and the difference of square roots of the two squares.

Illustration 2

a) The binomial $x^2 - 4$ is the difference of the two squares x^2 and 2^2. Therefore

$$x^2 - 4 = (x + 2)(x - 2)$$

b) The binomial $9u^2 - 49v^2$ is the difference of the squares of $3u$ and $7v$; that is, $9u^2 - 49v^2 = (3u)^2 - (7v)^2$; hence

$$9u^2 - 49v^2 = (3u + 7v)(3u - 7v)$$ ■

Example 3 Factor each of the polynomials. In part (c), n is a positive integer.

a) $4x^2 - 1$ b) $25a^4 - 36b^8$ c) $x^{2n} - y^{4n}$

Solution a) We have a binomial that is the difference of the two squares $(2x)^2$ and 1^2; thus

$$4x^2 - 1 = (2x + 1)(2x - 1)$$

b) Because $25a^4 - 36b^8 = (5a^2)^2 - (6b^4)^2$, we have

$$25a^4 - 36b^8 = (5a^2 + 6b^4)(5a^2 - 6b^4)$$

c) The binomial is the difference of the squares of x^n and y^{2n}. Therefore

$$x^{2n} - y^{4n} = (x^n)^2 - (y^{2n})^2$$
$$= (x^n + y^{2n})(x^n - y^{2n})$$ ∎

Factoring Trinomials

From special product 1 of Section 1.3 we have the formula

$$x^2 + (a + b)x + ab = (x + a)(x + b)$$

Illustration 3

The trinomial $x^2 + 3x - 28$ is of the type on the left side of the above formula. It can be factored into the product of two binomials $x + a$ and $x + b$ if there are two integers a and b such that $ab = -28$ and $a + b = 3$. The integers -4 and 7 satisfy these conditions; thus

$$x^2 + 3x - 28 = (x - 4)(x + 7)$$ ∎

Special product 2 of Section 1.3 gives the following formula:

$$x^2 + 2xy + y^2 = (x + y)^2$$

Note that whenever you are attempting to factor a trinomial that has two perfect-square terms, then this formula applies provided that the other term is twice the product of square roots of the perfect-square terms. Such a trinomial is called a **perfect-square trinomial.**

Illustration 4

The trinomial $16t^2 + 40t + 25$ has two perfect-square terms, namely $16t^2$, which is $(4t)^2$, and 25, which is 5^2; furthermore, the other term is $40t$, which is $2(4t)(5)$. Hence it is a perfect-square trinomial. Thus

$$16t^2 + 40t + 25 = (4t + 5)^2$$ ∎

The next formula is obtained from special product 4 of Section 1.3.

$$acx^2 + (ad + bc)xy + bdy^2 = (ax + by)(cx + dy)$$

The application of this formula requires a certain amount of trial and error.

Illustration 5

To factor the trinomial $15x^2 + 7xy - 2y^2$ as a product $(ax + by)(cx + dy)$, we wish to find two numbers a and c whose product is 15, and two numbers b and d whose product is -2 such that $ad + bc$ is 7. If a and c are to be positive, the possibilities for a and c are either 1 and 15 or 3 and 5. The possibilities for b and d are 1 and -2 and -1 and 2. By trial and error we see that we get the required middle term of $7xy$ if we write

$$15x^2 + 7xy - 2y^2 = (3x + 2y)(5x - y) \qquad \blacksquare$$

Example 4 Factor each of the polynomials.

a) $9x^2 + 24xy + 16y^2$ b) $9x^2 - 24xy + 16y^2$ c) $9x^2 + 25xy + 16y^2$

d) $9x^2 - 145xy + 16y^2$ e) $9x^2 - 16y^2$ f) $9x^2 + 16y^2$

Solution   a) The first and third terms are perfect squares, namely $(3x)^2$ and $(4y)^2$; and $24xy$ is $2(3x)(4y)$. Therefore we have a perfect-square trinomial.

$$9x^2 + 24xy + 16y^2 = (3x + 4y)^2$$

b) The first and third terms are the same as in part (a), but because the middle term is $-24xy$, we consider $16y^2$ as $(-4y)^2$. Because we have a perfect-square trinomial,

$$9x^2 - 24xy + 16y^2 = (3x - 4y)^2$$

c) We have a second-degree trinomial for which we use trial and error and obtain

$$9x^2 + 25xy + 16y^2 = (9x + 16y)(x + y)$$

d) Once more we use trial and error to obtain

$$9x^2 - 145xy + 16y^2 = (9x - y)(x - 16y)$$

e) We have a binomial that is the difference of two squares. Thus

$$9x^2 - 16y^2 = (3x + 4y)(3x - 4y)$$

f) The binomial $9x^2 + 16y^2$ is the sum of two squares, and it is prime. $\blacksquare$

Sum and Difference of Two Cubes

Computing the product of $x + y$ and $x^2 - xy + y^2$, we get

$$(x + y)(x^2 - xy + y^2) = x(x^2 - xy + y^2) + y(x^2 - xy + y^2)$$
$$= x^3 - x^2y + xy^2 + x^2y - xy^2 + y^3$$
$$= x^3 + y^3$$

Therefore we have

$$x^3 + y^3 = (x + y)(x^2 - xy + y^2)$$

This formula is used to factor the sum of two cubes. To factor the difference of two cubes we use the formula

$$x^3 - y^3 = (x - y)(x^2 + xy + y^2)$$

which can be verified by computing the product on the right side.

Illustration 6

The binomial $8 - b^3$ is the difference of the cubes of 2 and b. Thus

$$\begin{aligned}
8 - b^3 &= 2^3 - b^3 \\
&= (2 - b)(2^2 + 2b + b^2) \\
&= (2 - b)(4 + 2b + b^2)
\end{aligned}$$

■

Example   5   Factor the binomial

$$125x^3 + y^9$$

Solution   The given binomial is the sum of the cubes of $5x$ and y^3. Therefore

$$\begin{aligned}
125x^3 + y^9 &= (5x)^3 + (y^3)^3 \\
&= (5x + y^3)[(5x)^2 - (5x)(y^3) + (y^3)^2] \\
&= (5x + y^3)(25x^2 - 5xy^3 + y^6)
\end{aligned}$$

■

Factoring by Grouping

Sometimes, even though the terms of a polynomial do not have a common monomial factor, it may be possible to group terms in such a way that each group has a common factor.

Illustration 7

To factor the polynomial

$$3x^2 + 7x - 6xy - 14y$$

we group the first two terms and the last two terms, and we have

$$(3x^2 + 7x) + (-6xy - 14y)$$

The first two terms have a common monomial factor of x and the last two terms have a common monomial factor of $-2y$. Therefore the polynomial can be expressed as

$$x(3x + 7) - 2y(3x + 7)$$

We observe that there is a common binomial factor of $3x + 7$ in each term. Hence we have

$$(3x + 7)(x - 2y)$$

As an alternative solution, we can group the first and third terms and the second and fourth terms to get

$$(3x^2 - 6xy) + (7x - 14y) = 3x(x - 2y) + 7(x - 2y)$$
$$= (x - 2y)(3x + 7) \qquad ■$$

Example 6 Factor the polynomials.

a) $2xy + 8x + 3y + 12$ b) $5xz - 5yz - x + y$

Solution

a) $2xy + 8x + 3y + 12 = (2xy + 8x) + (3y + 12)$
$$= 2x(y + 4) + 3(y + 4)$$
$$= (y + 4)(2x + 3)$$

b) $5xz - 5yz - x + y = (5xz - 5yz) + (-x + y)$
$$= 5z(x - y) - 1(x - y)$$
$$= (x - y)(5z - 1) \qquad ■$$

The following illustration and example show some other ways that factoring can be done by grouping terms.

Illustration 8

a) If we have the polynomial $x^2 + 2xy + y^2 - 1$, we can group the first three terms to form a perfect-square trinomial, which we factor as the square of a binomial. Then we have the difference of two squares. The computation follows:

$$x^2 + 2xy + y^2 - 1 = (x^2 + 2xy + y^2) - 1$$
$$= (x + y)^2 - 1^2$$
$$= [(x + y) + 1][(x + y) - 1]$$
$$= (x + y + 1)(x + y - 1)$$

b) If we have the polynomial $x^2 - y^2 + x - y$, we can group the first two terms to form the difference of two squares, which we factor; then we have a common binomial factor of $x - y$ and we get

$$x^2 - y^2 + x - y = (x^2 - y^2) + (x - y)$$
$$= (x + y)(x - y) + (x - y)$$
$$= (x - y)[(x + y) + 1]$$
$$= (x - y)(x + y + 1) \qquad ■$$

Example 7 Factor the polynomials.

a) $16a^2 - 8ab + b^2 - c^2 + 6c - 9$ 　　b) $a^3 - b^3 - a + b$

Solution

a) $16a^2 - 8ab + b^2 - c^2 + 6c - 9 = (16a^2 - 8ab + b^2) - (c^2 - 6c + 9)$
$$= (4a - b)^2 - (c - 3)^2$$
$$= [(4a - b) + (c - 3)][(4a - b) - (c - 3)]$$
$$= (4a - b + c - 3)(4a - b - c + 3)$$

b) $a^3 - b^3 - a + b = (a^3 - b^3) - (a - b)$
$$= (a - b)(a^2 + ab + b^2) - (a - b)$$
$$= (a - b)[(a^2 + ab + b^2) - 1]$$
$$= (a - b)(a^2 + ab + b^2 - 1)$$ ■

Sometimes the completely factored form contains more than two factors, as happens in the following illustration and example.

Illustration 9

In the trinomial $2st^4 - 8st^2 - 90s$ there is a common monomial factor of $2s$. Hence we can write the trinomial as $2s(t^4 - 4t^2 - 45)$. The new trinomial can be factored and written as the product of two binomials, one of which is the difference of two squares. The computation follows:

$$2st^4 - 8st^2 - 90s = 2s(t^4 - 4t^2 - 45)$$
$$= 2s(t^2 + 5)(t^2 - 9)$$
$$= 2s(t^2 + 5)(t + 3)(t - 3)$$ ■

Example 8 Factor the polynomials.

a) $x^6 - 64$ 　　b) $x^6 - 16x^3 + 64$

Solution

a) $x^6 - 64 = (x^3)^2 - 8^2$
$$= (x^3 + 8)(x^3 - 8)$$
$$= (x + 2)(x^2 - 2x + 4)(x - 2)(x^2 + 2x + 4)$$

b) $x^6 - 16x^3 + 64 = (x^3)^2 + 2(-8)x^3 + (-8)^2$
$$= (x^3 - 8)^2$$
$$= [(x - 2)(x^2 + 2x + 4)]^2$$
$$= (x - 2)^2(x^2 + 2x + 4)^2$$ ■

EXERCISES 1.4

In Exercises 1 through 4, express the number in completely factored form.

1. (a) 50; (b) 72; (c) 750; (d) 1188

2. (a) 18; (b) 98; (c) 660; (d) 1911
3. (a) 300; (b) 3276; (c) 40,425
4. (a) 450; (b) 5544; (c) 255,255

In Exercises 5 through 92, factor the polynomial. In Exercises 15, 16, 25, 26, 47, 48, 91, and 92, n is a positive integer.

5. $8x^2 + 4x$

6. $9y^2 - 3y$

7. $a^5 - 3a^4 + a^3$

8. $x^4 + 2x^3 + 3x^2$

9. $36xy - 6y^2$

10. $12x^2y^3 - 4x^3y^2$

11. $a^4b^3 - a^3b^4 + a^2b^6$

12. $6u^2v^3 + 3u^3v^4 - 9u^5v^6$

13. $-12xy^3z^2 - 28y^3z - 20x^2y^2z^2$

14. $r^2s^4t^3 + r^3s^3t^2 + r^3s^2t^4$

15. $y^{3n} - y^{2n+1} + y^{2n}$

16. $a^{2n+1} + a^{n+2} + a^{n+1}$

17. $a^2 - 64$

18. $16 - y^2$

19. $x^2 - 49y^2$

20. $4u^2 - 9v^2$

21. $4s^2 - 25r^2$

22. $36x^2 - 81y^2$

23. $9x^2y^2 - 16w^6$

24. $49t^8 - 25x^{10}$

25. $x^{4n} - y^{6n}$

26. $b^{2n} - c^{8n}$

27. $x^2 + 7x + 10$

28. $x^2 - 9x + 18$

29. $a^2 - 10a + 24$

30. $y^2 + 13y + 42$

31. $t^2 - 4t - 32$

32. $x^2 + 5x - 24$

33. $21x^2 - 10xy + y^2$

34. $a^2 + 4ab - 21b^2$

35. $x^2 + 6x + 9$

36. $y^2 - 10y + 25$

37. $16x^2 - 8x + 1$

38. $9x^2 - 30xy + 25y^2$

39. $4a^2 - 12ab + 9b^2$

40. $25y^6 - 10y^3 + 1$

41. $5t^2 - 7t - 6$

42. $10y^2 - 11y - 6$

43. $18u^2 + 9uv - 20v^2$

44. $32a^2 + 12ab - 9b^2$

45. $18x^2 - 57xy + 35y^2$

46. $20x^2 + 43xy + 14y^2$

47. $x^{4n} - 4x^{2n} - 12$

48. $x^{6n} - 14x^{3n} + 49$

49. $t^3 + 8$

50. $27 - x^3$

51. $64x^3 - y^3$

52. $125a^3 + 64b^3$

53. $a^6b^3 - 27c^3$

54. $x^9y^3 - 8z^6$

55. $ac + ad + 2bc + 2bd$

56. $x^3 + 3x^2 + x + 3$

57. $4y^3 - y^2 + 4y - 1$

58. $a^2 + ab + ac + bc$

59. $10a^3 + 25a - 4a^2 - 10$

60. $4t^3 + 4t^2 - t - 1$

61. $3xy - yz + 3xw - zw$

62. $28 - 16x - 21x^2 + 12x^3$

63. $6st^2 - 9s^2t - 2t^3 + 27s^3$

64. $4 - (3a + 2b)^2$

65. $(2x - 3y)^2 - 16$

66. $(2x + y)^2 - (5z - 3w)^2$

67. $r^2 + 10rs + 25s^2 - 9$

68. $u^2 - v^2 + 4v - 4$

69. $x^2 - 8xy + 16y^2 - 36a^2 + 12ab - b^2$

70. $9t^2 - 24t + 16 - 4r^2 + 4rs - s^2$

71. $9a^2 - 16b^2 - 3a - 4b$

72. $16x^2 - y^2 - 4x + y$

73. $x - 4y - x^3 + 64y^3$

74. $a^3 + 27b^3 + a + 3b$

75. $(x + 2y)^3 - 1$

76. $(a - 3)^3 - 27b^3$

77. $y^4 - 16$

78. $81c^4 - d^4$

79. $x^8 - 1$

80. $t^6 - 1$

81. $a^6 + 2a^3 + 1$

82. $x^6 - 9x^3 + 8$

83. $8x^6 + 7x^3 - 1$

84. $y^8 - 5y^4 + 4$

85. $t^6 + 1 + t^2 + t^4$

86. $a^3 + 1 + a^2 + a$

87. $64x^6 - y^6$

88. $x^{12} - 2x^6 + 1$

89. $abx + acx - bcy - aby + bcx - acy$

90. $xy - vy + xz + wy + wz - vz$

91. $64a^{6n} - b^{6n}$

92. $x^{8n} - 16y^{4n}$

1.5 Rational Expressions

If a and b are real numbers and $b \neq 0$, there are three ways of denoting the operation of division of a by b:

$$a \div b \qquad \frac{a}{b} \qquad a/b$$

The numerals $\dfrac{a}{b}$ and a/b are **fractions,** in which a is the **numerator** and b is the **denominator.** If the numerator and denominator of a fraction are polynomials, then the fraction is called a **rational expression.** Examples of rational expressions are

$$\frac{5}{y - 7} \qquad \frac{3x + 2}{x^2 - 4} \qquad \frac{2}{5rs} \qquad \frac{3t^2 + t + 5}{t^4 + 1}$$

Since the denominator cannot be zero, it is understood that in the above rational expressions $y \neq 7$, $x \neq \pm 2$, $r \neq 0$, and $s \neq 0$. Because a rational expression denotes a quotient of real numbers, properties of fractions also hold for rational expressions.

A rational expression is said to be in **lowest terms** if the numerator and denominator have no common factor other than 1 and -1. If a given rational expression is not in lowest terms it can be replaced by an equivalent one in lowest terms by factoring the numerator and denominator and then dividing the numerator and denominator by the common factors. The procedure is called **reducing the rational expression to lowest terms** and is justified by the property of fractions that states

$$\frac{ak}{bk} = \frac{a}{b} \quad \text{if} \quad k \neq 0$$

Example   **1**   Reduce each of the following rational expressions to lowest terms.

a) $\dfrac{6x^8}{8x^6}$ b) $\dfrac{7a^5b^4}{21a^2b^5}$ c) $\dfrac{18x^2 + 9xy - 2y^2}{9x^2 - 4y^2}$

Solution

a) $\dfrac{6x^8}{8x^6} = \dfrac{3x^2(2x^6)}{4(2x^6)}$ b) $\dfrac{7a^5b^4}{21a^2b^5} = \dfrac{a^3(7a^2b^4)}{3b(7a^2b^4)}$

$\qquad = \dfrac{3x^2}{4}$ $= \dfrac{a^3}{3b}$

c) $\dfrac{18x^2 + 9xy - 2y^2}{9x^2 - 4y^2} = \dfrac{(6x - y)(3x + 2y)}{(3x - 2y)(3x + 2y)}$

$\qquad\qquad\qquad\qquad = \dfrac{6x - y}{3x - 2y}$

In part (a), we divided numerator and denominator by $2x^6$; in part (b), we divided numerator and denominator by $7a^2b^4$; and in part (c), we divided numerator and denominator by $3x + 2y$. ∎

To multiply rational expressions, we use the following property of fractions:

$$\frac{a}{b} \cdot \frac{c}{d} = \frac{ac}{bd}$$

Before multiplying two rational expressions, the numerators and denominators should be factored completely; this facilitates reducing to lowest terms the rational expression representing the product.

Example 2 Find each of the following products in lowest terms.

a) $\dfrac{5t}{8} \cdot \dfrac{4}{3t^2}$ b) $\dfrac{4x}{3y} \cdot \dfrac{3x^2y^2}{10}$ c) $\dfrac{x-5}{4x^2-9} \cdot \dfrac{4x^2+12x+9}{2x^2-11x+5}$

Solution

a) $\dfrac{5t}{8} \cdot \dfrac{4}{3t^2} = \dfrac{(5t)(2^2)}{(2^3)(3t^2)}$

$\qquad = \dfrac{5(2^2 t)}{(2 \cdot 3t)(2^2 t)}$

$\qquad = \dfrac{5}{6t}$

b) $\dfrac{4x}{3y} \cdot \dfrac{3x^2y^2}{10} = \dfrac{(2^2 x)(3x^2 y^2)}{(3y)(2 \cdot 5)}$

$\qquad = \dfrac{2x^3 y(2 \cdot 3y)}{5(2 \cdot 3y)}$

$\qquad = \dfrac{2x^3 y}{5}$

c) $\dfrac{x-5}{4x^2-9} \cdot \dfrac{4x^2+12x+9}{2x^2-11x+5} = \dfrac{x-5}{(2x+3)(2x-3)} \cdot \dfrac{(2x+3)^2}{(2x-1)(x-5)}$

$\qquad = \dfrac{(2x+3)[(2x+3)(x-5)]}{(2x-3)(2x-1)[(2x+3)(x-5)]}$

$\qquad = \dfrac{2x+3}{(2x-3)(2x-1)}$ ∎

The property of fractions used to divide rational expressions is

$$\frac{a}{b} \div \frac{c}{d} = \frac{a}{b} \cdot \frac{d}{c} \qquad \text{if} \qquad \frac{c}{d} \neq 0$$

Example 3 Find each of the following quotients in lowest terms.

a) $\dfrac{45a^3b^2}{28c^4d^3} \div \dfrac{-75a^4b}{8c^2d^4}$ b) $\dfrac{4x^2-9y^2}{xy+y^2} \div \dfrac{6x^2-xy-12y^2}{xy+x^2}$

Solution

a) $\dfrac{45a^3b^2}{28c^4d^3} \div \dfrac{-75a^4b}{8c^2d^4} = \dfrac{3^2 \cdot 5a^3b^2}{2^2 \cdot 7c^4d^3} \cdot \dfrac{2^3c^2d^4}{-3 \cdot 5^2a^4b}$

$\qquad\qquad = \dfrac{2^3 \cdot 3^2 \cdot 5a^3b^2c^2d^4}{-2^2 \cdot 3 \cdot 5^2 \cdot 7a^4bc^4d^3}$

$\qquad\qquad = \dfrac{2 \cdot 3bd(2^2 \cdot 3 \cdot 5a^3bc^2d^3)}{-5 \cdot 7ac^2(2^2 \cdot 3 \cdot 5a^3bc^2d^3)}$

$\qquad\qquad = -\dfrac{6bd}{35ac^2}$

b) $\dfrac{4x^2 - 9y^2}{xy + y^2} \div \dfrac{6x^2 - xy - 12y^2}{xy + x^2} = \dfrac{(2x - 3y)(2x + 3y)}{y(x + y)} \cdot \dfrac{x(y + x)}{(2x - 3y)(3x + 4y)}$

$\qquad\qquad = \dfrac{x(2x - 3y)(2x + 3y)(y + x)}{y(x + y)(2x - 3y)(3x + 4y)}$

$\qquad\qquad = \dfrac{x(2x + 3y)[(x + y)(2x - 3y)]}{y(3x + 4y)[(x + y)(2x - 3y)]}$

$\qquad\qquad = \dfrac{x(2x + 3y)}{y(3x + 4y)}$

$\qquad\qquad = \dfrac{2x^2 + 3xy}{4y^2 + 3xy}$ ■

The sum and difference of rational expressions are found by using the following properties of fractions:

$$\frac{a}{d} + \frac{b}{d} = \frac{a + b}{d} \qquad \frac{a}{d} - \frac{b}{d} = \frac{a - b}{d}$$

To apply these properties, it is necessary that the fractions have the same denominator. If we wish to add or subtract fractions that do not have the same denominator we replace them with equivalent fractions having the *least common denominator*. The **least common denominator** (LCD) of given rational expressions is the polynomial of smallest degree that has each of the given denominators as a factor. To determine this polynomial, we first obtain the completely factored form of the denominators. Then the LCD is the product of the different prime factors that occur in any of the denominators, where the power of each factor is the highest power to which it appears. The next two illustrations show the method.

Illustration 1

To perform the addition

$$\frac{5}{4x^2} + \frac{7}{6x}$$

we first write each denominator in completely factored form.

$$4x^2 = 2^2x^2 \qquad 6x = 2 \cdot 3x$$

The LCD is the product of the prime factors 2^2, 3, and x^2, which is $2^2 \cdot 3 \cdot x^2 = 12x^2$. We now write each of the given rational expressions as an equivalent one having this denominator. We multiply the numerator and denominator of the first fraction by 3 and of the second fraction by $2x$. Thus

$$\frac{5}{2^2x^2} + \frac{7}{2 \cdot 3x} = \frac{5 \cdot 3}{2^2x^2 \cdot 3} + \frac{7 \cdot 2x}{2 \cdot 3x \cdot 2x}$$

$$= \frac{15}{12x^2} + \frac{14x}{12x^2}$$

$$= \frac{15 + 14x}{12x^2}$$

∎

Illustration 2

To compute

$$\frac{5}{x^2 - 4} - \frac{4}{x^2 + 4x + 4}$$

we first write each denominator in completely factored form:

$$x^2 - 4 = (x + 2)(x - 2) \qquad x^2 + 4x + 4 = (x + 2)^2$$

The LCD is $(x + 2)^2(x - 2)$. Then each of the given rational expressions is replaced by an equivalent one having the LCD as its denominator. Therefore

$$\frac{5}{x^2 - 4} - \frac{4}{x^2 + 4x + 4} = \frac{5}{(x + 2)(x - 2)} - \frac{4}{(x + 2)^2}$$

$$= \frac{5(x + 2)}{(x + 2)^2(x - 2)} - \frac{4(x - 2)}{(x + 2)^2(x - 2)}$$

$$= \frac{5(x + 2) - 4(x - 2)}{(x + 2)^2(x - 2)}$$

$$= \frac{5x + 10 - 4x + 8}{(x + 2)^2(x - 2)}$$

$$= \frac{x + 18}{(x + 2)^2(x - 2)}$$

∎

Example **4** Express in lowest terms:

$$\frac{2y + 1}{2y^2 + y - 1} + \frac{3}{2 - 3y - 2y^2}$$

Solution

$$\frac{2y + 1}{2y^2 + y - 1} + \frac{3}{2 - 3y - 2y^2} = \frac{2y + 1}{(2y - 1)(y + 1)} + \frac{3}{(1 - 2y)(2 + y)}$$

Because $-(1 - 2y) = 2y - 1$, we multiply the numerator and denominator of the second fraction by -1. The LCD is then $(2y - 1)(y + 1)(y + 2)$. We have

$$\frac{2y + 1}{2y^2 + y - 1} + \frac{3}{2 - 3y - 2y^2} = \frac{2y + 1}{(2y - 1)(y + 1)} + \frac{(-1)(3)}{(-1)(1 - 2y)(2 + y)}$$

$$= \frac{2y + 1}{(2y - 1)(y + 1)} + \frac{-3}{(2y - 1)(y + 2)}$$

$$= \frac{(2y + 1)(y + 2)}{(2y - 1)(y + 1)(y + 2)} + \frac{-3(y + 1)}{(2y - 1)(y + 1)(y + 2)}$$

$$= \frac{2y^2 + 5y + 2 - 3y - 3}{(2y - 1)(y + 1)(y + 2)}$$

$$= \frac{2y^2 + 2y - 1}{(2y - 1)(y + 1)(y + 2)} \quad ■$$

If a fraction contains a fraction in either the numerator or denominator, or both, it is called a **complex fraction.** In contrast, a fraction that is not complex is referred to as a **simple fraction.**

Illustration 3

The fraction

$$\frac{\dfrac{3}{5} - \dfrac{4}{7}}{\dfrac{2}{5} + \dfrac{1}{3}}$$

is a complex fraction. We show how to find an equivalent simple fraction by two methods. In part (a), we multiply the numerator and denominator by the LCD of each of the fractions appearing. The LCD is $5 \cdot 3 \cdot 7 = 105$. In part (b), we perform the indicated operations in the numerator and denominator of the given complex fraction and then divide the numerator by the denominator.

a) $\dfrac{\dfrac{3}{5} - \dfrac{4}{7}}{\dfrac{2}{5} + \dfrac{1}{3}} = \dfrac{105 \cdot \dfrac{3}{5} - 105 \cdot \dfrac{4}{7}}{105 \cdot \dfrac{2}{5} + 105 \cdot \dfrac{1}{3}}$

$= \dfrac{63 - 60}{42 + 35}$

$= \dfrac{3}{77}$

b) $\dfrac{\dfrac{3}{5} - \dfrac{4}{7}}{\dfrac{2}{5} + \dfrac{1}{3}} = \dfrac{\dfrac{21}{35} - \dfrac{20}{35}}{\dfrac{6}{15} + \dfrac{5}{15}} = \dfrac{\dfrac{1}{35}}{\dfrac{11}{15}}$

$= \dfrac{1}{5 \cdot 7} \cdot \dfrac{3 \cdot 5}{11}$

$= \dfrac{3(5)}{7 \cdot 11(5)}$

$= \dfrac{3}{77}$ ∎

The methods used for simplifying complex rational expressions are the same as those shown in Illustration 3. Multiplying the numerator and denominator by the LCD, as shown in part (a), is usually the easier method.

Example 5 For each of the following rational expressions, find a simple rational expression equivalent to it. In part (a), multiply the numerator and denominator by the LCD. In part (b), first perform the indicated operations in the numerator and denominator.

a) $\dfrac{\dfrac{x-1}{x-2} - \dfrac{x-2}{x-1}}{\dfrac{1}{x-1} - \dfrac{1}{x-2}}$

b) $\dfrac{1 - \dfrac{2}{a}}{a - 1 - \dfrac{1}{a-1}}$

Solution

a) The LCD is $(x-1)(x-2)$. We have

$$\dfrac{\dfrac{x-1}{x-2} - \dfrac{x-2}{x-1}}{\dfrac{1}{x-1} - \dfrac{1}{x-2}} = \dfrac{(x-1)(x-2)\dfrac{x-1}{x-2} - (x-1)(x-2)\dfrac{x-2}{x-1}}{(x-1)(x-2)\dfrac{1}{x-1} - (x-1)(x-2)\dfrac{1}{x-2}}$$

$$= \dfrac{(x-1)(x-1) - (x-2)(x-2)}{(x-2) - (x-1)}$$

$$= \dfrac{x^2 - 2x + 1 - x^2 + 4x - 4}{x - 2 - x + 1}$$

$$= \dfrac{2x - 3}{-1}$$

$$= 3 - 2x$$

b)
$$\frac{1 - \dfrac{2}{a}}{a - 1 - \dfrac{1}{a - 1}} = \frac{\dfrac{a}{a} - \dfrac{2}{a}}{\dfrac{a(a - 1)}{a - 1} - \dfrac{a - 1}{a - 1} - \dfrac{1}{a - 1}}$$

$$= \frac{\dfrac{a - 2}{a}}{\dfrac{a^2 - a - a + 1 - 1}{a - 1}} = \frac{\dfrac{a - 2}{a}}{\dfrac{a^2 - 2a}{a - 1}}$$

$$= \frac{a - 2}{a} \cdot \frac{a - 1}{a(a - 2)}$$

$$= \frac{(a - 1)(a - 2)}{a^2(a - 2)}$$

$$= \frac{a - 1}{a^2} \qquad\blacksquare$$

The complex rational expression in the next example is similar to a type occurring in calculus.

Example 6 Simplify

$$\frac{\dfrac{1}{(x + h)^2} - \dfrac{1}{x^2}}{h}$$

Solution   The LCD is $x^2(x + h)^2$.

$$\frac{\dfrac{1}{(x + h)^2} - \dfrac{1}{x^2}}{h} = \frac{x^2(x + h)^2 \dfrac{1}{(x + h)^2} - x^2(x + h)^2 \dfrac{1}{x^2}}{x^2(x + h)^2 h}$$

$$= \frac{x^2 - (x + h)^2}{hx^2(x + h)^2}$$

$$= \frac{x^2 - (x^2 + 2hx + h^2)}{hx^2(x + h)^2}$$

$$= \frac{x^2 - x^2 - 2hx - h^2}{hx^2(x + h)^2}$$

$$= \frac{-h(2x + h)}{hx^2(x + h)^2}$$

$$= -\frac{2x + h}{x^2(x + h)^2} \qquad\blacksquare$$

EXERCISES 1.5

In Exercises 1 through 8, reduce the rational expression to lowest terms.

1. (a) $\dfrac{12a^2}{3a^6}$; (b) $\dfrac{7x^2y^5}{21x^5y^3}$; (c) $\dfrac{-36rs^6t^7}{30rs^7t^6}$

2. (a) $\dfrac{9x^5}{36x^2}$; (b) $\dfrac{25a^2b^2}{10a^4b^2}$; (c) $\dfrac{20x^3y^2z^4}{15x^6y^6z}$

3. (a) $\dfrac{2x-4}{x^2-4}$; (b) $\dfrac{x^2-y^2}{x^2+xy}$

4. (a) $\dfrac{y^2-4y+4}{y^2-4}$; (b) $\dfrac{5c+5d}{c^2+cd}$

5. (a) $\dfrac{t^2-3t-28}{t^2-4t-21}$; (b) $\dfrac{6x^2-xy-y^2}{2x^2-9xy+4y^2}$

6. (a) $\dfrac{a^2+8a+15}{a^2-a-12}$; (b) $\dfrac{6x^2+12x+7}{12x^2+13x-35}$

7. (a) $\dfrac{x^2-16}{4-x}$; (b) $\dfrac{x^3-27}{x-3}$

8. (a) $\dfrac{5-a}{a^2-25}$; (b) $\dfrac{2x-y}{8x^3-y^3}$

In Exercises 9 through 20, find the product in lowest terms.

9. (a) $\dfrac{5x^3}{7y^4}\cdot\dfrac{21y^2}{10x^2}$; (b) $\dfrac{45ad}{56bc}\cdot\dfrac{32ab}{27cd}$

10. (a) $\dfrac{15a^2}{16b^3}\cdot\dfrac{8b}{3a^4}$; (b) $\dfrac{5z}{42x^2y}\cdot\dfrac{63xy^2}{10z^2}$

11. (a) $\dfrac{20r^3st^3}{33uv^3w^2}\cdot\dfrac{-11uv^2w^4}{15r^3s^2w}$; (b) $\dfrac{8x^2}{9y^2z}\cdot\dfrac{3xy}{10z^3}\cdot\dfrac{5yz}{12x}$

12. (a) $\dfrac{36xy^2}{25wz^3}\cdot\dfrac{20x^2z}{27y^2w^2}$; (b) $\dfrac{27c^2}{16a^2b}\cdot\dfrac{100ac}{63b^3}\cdot\dfrac{7ab}{15c}$

13. $\dfrac{x^2-x-12}{5x-5}\cdot\dfrac{3x-3}{x^2-9}$

14. $\dfrac{6c-9}{c^2-25}\cdot\dfrac{c^2-3c-10}{12-4c}$

15. $\dfrac{x^2-3xy-4y^2}{x^2-xy-2y^2}\cdot\dfrac{x^2-xy-6y^2}{x^2-xy-12y^2}$

16. $\dfrac{x-y}{x+3y}\cdot\dfrac{x^2-9y^2}{x^2-y^2}$

17. $\dfrac{4a^2-3ab-b^2}{16a^2b^2-b^4}\cdot\dfrac{b^2-4ab}{b-a}$

18. $\dfrac{t^2-2t-15}{t^2-9}\cdot\dfrac{t^2-6t+9}{12-4t}$

19. $\dfrac{9x^2-16y^2}{3x^2-5xy-12y^2}\cdot\dfrac{xy-2x^2}{3xy-4y^2}\cdot\dfrac{x-3y}{2x-y}$

20. $\dfrac{2ab-a^2}{6a^2+13ab+5b^2}\cdot\dfrac{3a^2-4ab-15b^2}{2a^2-7ab+3b^2}\cdot\dfrac{b^2-4a^2}{2a^2-4ab}$

In Exercises 21 through 30, find the quotient in lowest terms.

21. (a) $\dfrac{81xz^3}{36y}\div\dfrac{27x^2z^2}{12xy}$; (b) $\dfrac{24r^3s^4t}{15uvw^4}\div\dfrac{18rs^3t^2}{25uv^3w}$

22. (a) $\dfrac{15x^3y^2z}{2abc^3}\div\dfrac{5x^2y^3z^2}{6a^2bc^2}$; (b) $\dfrac{21r^2s^5}{55tu^6}\div\dfrac{18r^3s^3}{25t^6u^4}$

23. $\dfrac{x^2+9x+14}{x^2+4x-21}\div\dfrac{x^2-3x-10}{x^2+2x-35}$

24. $\dfrac{x^2-4y^2}{4y-2x}\div\dfrac{2x^2+xy-6y^2}{6x-9y}$

25. $\dfrac{x^2-y^2}{xy-2y^2}\div\dfrac{x^2-2xy+y^2}{2x^2-4xy}$

26. $\dfrac{10a^2-29a+10}{6a^2-29a+20}\div\dfrac{10a^2-19a+6}{12a^2-28a+15}$

27. $\dfrac{9y^3-18y^2-4y+8}{3y^2-4y-4}\div(8-12y)$

28. $\dfrac{x^6-7x^3-8}{4x^2-4x-8}\div(2x^2+4x+8)$

29. $\dfrac{a}{b-a}\cdot\dfrac{a^2+b^2}{a+b}\div\dfrac{a^2-b^2}{a^2-2ab+b^2}$

30. $\dfrac{c^4}{9a^2-b^2}\cdot\dfrac{27a^3-b^3}{ac+bc}\div\dfrac{ac^3-bc^3}{36a^2-2ab-b^2}$

In Exercises 31 through 40, find the sum or difference in lowest terms.

31. $\dfrac{a+1}{a^3}-\dfrac{a+2}{a^2}+\dfrac{a+3}{a}$

32. $\dfrac{x-y}{xy}+\dfrac{x-z}{xz}-\dfrac{z-y}{yz}$

33. $\dfrac{x-2y}{x+2y}-\dfrac{2x-y}{2x+y}$

34. $\dfrac{x+y}{x-y}-\dfrac{x-y}{x+y}$

35. $\dfrac{2t+1}{3t-3}+\dfrac{6-t}{t^2-5t+4}$

36. $\dfrac{a-1}{2a^2-18}+\dfrac{a+2}{9a-3a^2}$

37. $\dfrac{2x+4}{x^2-8x+15}-\dfrac{x-5}{x^2-x-6}-\dfrac{x+3}{x^2-3x-10}$

38. $\dfrac{2x + 5}{x^2 + 8x + 16} + \dfrac{3}{2x} - \dfrac{x - 2}{x^2 + 4x}$

39. $\dfrac{4t^2}{s^2 - t^2} - \dfrac{t - s}{t + s} + \dfrac{s + t}{t - s}$

40. $\dfrac{2a + b}{2a - b} + \dfrac{8ab}{4a^2 - b^2} - \dfrac{b - 2a}{b + 2a}$

In Exercises 41 through 46, find a simple rational expression in lowest terms equivalent to the rational expression.

41. $\dfrac{\dfrac{1}{x} - \dfrac{1}{y}}{\dfrac{1}{x} + \dfrac{1}{y}}$

42. $\dfrac{\dfrac{a}{b} - \dfrac{b}{a}}{\dfrac{a}{b} + \dfrac{b}{a}}$

43. $\dfrac{\dfrac{x^2}{y^2} - 1}{\dfrac{x^2}{y^2} - \dfrac{2x}{y} + 1}$

44. $\dfrac{t}{1 + \dfrac{t - 1}{t + 1}}$

45. $\dfrac{\dfrac{1 + x}{1 - x} - \dfrac{1 - x}{1 + x}}{\dfrac{1}{1 + x} - \dfrac{1}{1 - x}}$

46. $\dfrac{\dfrac{a - b}{a + b} - \dfrac{b}{a - b}}{1 + b\left(\dfrac{2}{a + b} - \dfrac{3}{b - a}\right)}$

In Exercises 47 through 50, the complex rational expression is similar to a type occurring in calculus. Find a simple rational expression equivalent to it.

47. $\dfrac{\dfrac{1}{x + h} - \dfrac{1}{x}}{h}$

48. $\dfrac{\dfrac{1}{(x + h)^3} - \dfrac{1}{x^3}}{h}$

49. $\dfrac{\dfrac{1}{3x + 3h + 2} - \dfrac{1}{3x + 2}}{h}$

50. $\dfrac{\dfrac{1}{2x + 2h - 5} - \dfrac{1}{2x - 5}}{h}$

In Exercises 51 through 56, write the expression as a simple rational expression with only positive exponents.

51. $\dfrac{x^{-1} - y^{-1}}{x^2 - y^2}$

52. $\dfrac{a^{-1} + b^{-1}}{a^{-2} - b^{-2}}$

53. $\dfrac{b^2 a^{-2} - a^2 b^{-2}}{ba^{-1} + ab^{-1}}$

54. $\dfrac{2x^{-1} + 3xy^{-2}}{4x^{-2} - 9x^2 y^{-4}}$

55. $(x^{-1} + y^{-1})^{-1}$

56. $\left(\dfrac{x^{-1} + y^{-1}}{x^{-1} - y^{-1}}\right)^{-1}$

1.6 **Rational Exponents**

In Section 1.2 we defined a^n if n is any integer (positive, negative, or zero). We now wish to define a power of a where the exponent is any rational number, not specifically an integer. That is, we wish to attach a meaning to $a^{1/n}$ and $a^{m/n}$, where the exponents are fractions. Before discussing fractional exponents, we give the following definition.

Definition

> **The nth Root of a Real Number**
> If n is a positive integer greater than 1 and a and b are real numbers such that
>
> $$b^n = a$$
>
> then b is an **nth root of a.**

Illustration 1

a) 2 is a square root of 4 because $2^2 = 4$; furthermore, -2 is also a square root of 4 because $(-2)^2 = 4$.

b) 3 is a fourth root of 81 because $3^4 = 81$. Also, -3 is a fourth root of 81 because $(-3)^4 = 81$.

c) 4 is a cube root of 64 because $4^3 = 64$.

d) -4 is a cube root of -64 because $(-4)^3 = -64$. ■

Observe in part (a) of Illustration 1 that there are two real square roots of 4 and in part (b) there are two real fourth roots of 81. To distinguish between the two roots in such cases, we introduce the concept of **principal nth root.**

Definition

> **The Principal nth Root of a Real Number**
> If n is a positive integer greater than 1, a is a real number, and $\sqrt[n]{a}$ denotes the **principal nth root of a,** then
> (i) if $a > 0$, $\sqrt[n]{a}$ is the positive nth root of a;
> (ii) if $a < 0$, and n is odd, $\sqrt[n]{a}$ is the negative nth root of a;
> (iii) $\sqrt[n]{0} = 0$.

In the above definition, the symbol $\sqrt{}$ is called a **radical sign.** The entire expression $\sqrt[n]{a}$ is called a **radical,** where the number a is the **radicand** and the number n is the **index** that indicates the **order** of the radical. If no index appears, the order is understood to be 2.

Illustration 2

In parts (a), (b), and (c), we use (i) of the above definition, and in part (d), we use (ii).

a) $\sqrt{4} = 2$ (read "the principal square root of 4 equals 2")

Note that -2 is also a square root of 4 but it is not the principal square root of 4. However, we can write

$$-\sqrt{4} = -2$$

b) $\sqrt[4]{81} = 3$ (read "the principal fourth root of 81 equals 3")

The number -3 is also a fourth root of 81, and we can write

$$-\sqrt[4]{81} = -3$$

c) $\sqrt[3]{64} = 4$ (read "the principal cube root of 64 equals 4")

d) $\sqrt[3]{-64} = -4$ (read "the principal cube root of -64 equals -4") ■

Observe that if $a < 0$, $\sqrt[n]{a}$ is defined only if n is odd. For instance $\sqrt{-16}$ is not defined as a real number because there is no real number whose square is -16. Complex numbers are needed to define an even-order root of a negative number, and these numbers are discussed in Section 1.8.

The principal nth root of a real number b is a rational number if and only if b is the nth power of a rational number. For instance,

$$\sqrt{9} = 3 \qquad \text{because} \qquad 3^2 = 9$$
$$\sqrt[3]{-\tfrac{1}{27}} = -\tfrac{1}{3} \qquad \text{because} \qquad (-\tfrac{1}{3})^3 = -\tfrac{1}{27}$$
$$\sqrt[4]{625} = 5 \qquad \text{because} \qquad 5^4 = 625$$

Recall from Section 1.1 that a real number that is not rational is called an irrational number, and an irrational number cannot be represented by a terminating decimal or a nonterminating repeating decimal. Because 3 is not the square of a rational number, $\sqrt{3}$ is an irrational number. Other examples of irrational numbers are

$$\sqrt{2} \qquad \sqrt[3]{4} \qquad \sqrt[3]{-5} \qquad \sqrt[4]{15}$$

We are now ready to define a rational exponent of the form $\dfrac{1}{n}$ where n is a positive integer. Theorem 2 of Section 1.2 states that when m and n are positive integers

$$(a^n)^m = a^{nm}$$

If this formula is to hold when the exponent is $\dfrac{1}{n}$, then we must have

$$(a^{1/n})^n = a^{n/n}$$
$$(a^{1/n})^n = a$$

This equality states that the nth power of $a^{1/n}$ equals a. Thus we define $a^{1/n}$ as the principal nth root of a.

Definition

> $a^{1/n}$
>
> If n is a positive integer greater than 1, and a is a real number, then if $\sqrt[n]{a}$ is a real number
>
> $$a^{1/n} = \sqrt[n]{a}$$

Illustration   3

a) $25^{1/2} = \sqrt{25}$ b) $(-8)^{1/3} = \sqrt[3]{-8}$ c) $(\tfrac{1}{81})^{1/4} = \sqrt[4]{\tfrac{1}{81}}$
$\qquad\qquad = 5 \qquad\qquad\qquad\qquad = -2 \qquad\qquad\qquad\qquad = \tfrac{1}{3}$ ■

Consider now how we should define expressions such as

$$9^{3/2} \qquad 8^{2/3} \qquad (-27)^{4/3} \qquad 7^{3/4}$$

If the formula $a^{pq} = (a^p)^q$ is to hold for rational exponents as well as for integer exponents, then $a^{m/n}$ must be defined in such a way that

$$a^{m/n} = (a^{1/n})^m$$

Therefore we have the following definition, where the double arrow $\Leftrightarrow$ is used to mean that the statement preceding it and the statement following it are equivalent.

Definition

$a^{m/n}$

If m and n are positive integers that are relatively prime, and a is a real number, then if $\sqrt[n]{a}$ is a real number

$$a^{m/n} = (\sqrt[n]{a})^m \qquad \Leftrightarrow \qquad a^{m/n} = (a^{1/n})^m$$

Illustration 4

In the following we apply the definition of $a^{m/n}$.

a) $9^{3/2} = (\sqrt{9})^3$ b) $8^{2/3} = (\sqrt[3]{8})^2$ c) $(-27)^{4/3} = (\sqrt[3]{-27})^4$

        $= 3^3$ $= 2^2$ $= (-3)^4$

        $= 27$ $= 4$ $= 81$ ■

It can be shown that the commutative law holds for rational exponents, and therefore

$$(a^m)^{1/n} = (a^{1/n})^m$$

from which it follows that

$$\sqrt[n]{a^m} = (\sqrt[n]{a})^m$$

The next theorem follows from this equality and the definition of $a^{m/n}$.

Theorem 1

If m and n are positive integers that are relatively prime, and a is a real number, then if $\sqrt[n]{a}$ is a real number

$$a^{m/n} = \sqrt[n]{a^m} \qquad \Leftrightarrow \qquad a^{m/n} = (a^m)^{1/n}$$

Illustration 5

Theorem 1 is applied in the following:

a) $9^{3/2} = \sqrt{9^3}$ b) $8^{2/3} = \sqrt[3]{8^2}$ c) $(-27)^{4/3} = \sqrt[3]{(-27)^4}$

        $= \sqrt{729}$ $= \sqrt[3]{64}$ $= \sqrt[3]{531,441}$

        $= 27$ $= 4$ $= 81$ ■

Observe that $a^{m/n}$ can be evaluated by finding either $(\sqrt[n]{a})^m$ or $\sqrt[n]{a^m}$. Compare Illustrations 4 and 5 and you will see that the computation of $(\sqrt[n]{a})^m$ in Illustration 4 is simpler than that for $\sqrt[n]{a^m}$ in Illustration 5.

The laws of positive-integer exponents are satisfied by positive-rational exponents with one exception: For certain values of p and q, $(a^p)^q \neq a^{pq}$ for $a < 0$. This situation arises in the following illustration.

Illustration   6

a) $[(-9)^2]^{1/2} = 81^{1/2}$ and $(-9)^{2(1/2)} = (-9)^1$
$\qquad\qquad\qquad = 9$ $\qquad\qquad\qquad\qquad\qquad = -9$

Therefore $[(-9)^2]^{1/2} \neq (-9)^{2(1/2)}$.

b) $[(-9)^2]^{1/4} = 81^{1/4}$ and $(-9)^{2(1/4)} = (-9)^{1/2}$ (not a real number)
$\qquad\qquad\qquad = 3$

Therefore $[(-9)^2]^{1/4} \neq (-9)^{2(1/4)}$.   ■

The problems that arise in Illustration 6 are avoided by adopting the following rule: If m and n are positive even integers and a is a real number, then

$$(a^m)^{1/n} = |a|^{m/n}$$

A particular case of this equality occurs when $m = n$. We then have

$$(a^n)^{1/n} = |a| \qquad \text{(if } n \text{ is a positive even integer)}$$

or, equivalently,

$$\sqrt[n]{a^n} = |a| \qquad \text{(if } n \text{ is even)}$$

If n is 2, we have

$$\sqrt{a^2} = |a|$$

Illustration   7

a) $[(-9)^2]^{1/2} = |-9|$ b) $[(-9)^2]^{1/4} = |-9|^{2/4}$
$\qquad\qquad\qquad = 9$ $\qquad\qquad\qquad\qquad\quad = 9^{1/2}$
$\qquad\qquad\qquad\qquad\qquad\qquad\qquad\qquad = 3$ ■

For the formula $a^{pq} = (a^p)^q$ to hold for negative rational exponents, we must have

$$a^{-m/n} = (a^{1/n})^{-m}$$

By the definition of a negative-integer exponent, if $a \neq 0$,

$$(a^{1/n})^{-m} = \frac{1}{(a^{1/n})^m}$$

Therefore we give the following definition.

Definition

> **Negative Rational Exponent**
> If m and n are positive integers that are relatively prime, and a is a real number and $a \neq 0$, then if $\sqrt[n]{a}$ is a real number,
>
> $$a^{-m/n} = \frac{1}{a^{m/n}}$$

The first complete explanation of fractional and negative exponents was given by John Wallis (1616–1703) in 1655. Sir Isaac Newton (1642–1727) also used such exponents in his work.

Rational exponents (positive, negative, and zero) satisfy the laws of positive-integer exponents, with the understanding that $(a^m)^{1/n} = |a|^{m/n}$ when m and n are even integers.

Illustration 8

We compute $8^{-2/3}$ by three different methods. In part (a), we first use the definition of a negative rational exponent. In parts (b) and (c), we apply laws of exponents to negative rational exponents.

a) $8^{-2/3} = \dfrac{1}{8^{2/3}}$

$= \dfrac{1}{(\sqrt[3]{8})^2}$

$= \dfrac{1}{2^2}$

$= \dfrac{1}{4}$

b) $8^{-2/3} = (8^{-1/3})^2$

$= \left(\dfrac{1}{8^{1/3}}\right)^2$

$= \left(\dfrac{1}{2}\right)^2$

$= \dfrac{1}{4}$

c) $8^{-2/3} = (8^{-2})^{1/3}$

$= \left(\dfrac{1}{8^2}\right)^{1/3}$

$= \left(\dfrac{1}{64}\right)^{1/3}$

$= \dfrac{1}{4}$ ∎

Computations with rational exponents appear in the following examples. In Examples 2 and 3, the restriction that the variables are positive allows us to eliminate the absolute value bars when applying the law $(a^p)^q = a^{pq}$.

Example   **1**   Express each of the following in terms of a power of x with a positive exponent. Assume that $x > 0$.

a) $\dfrac{x^{1/3}}{x^{1/4}}$ b) $\left(\dfrac{1}{x^{-2/3}}\right)^{-1/2}$

Solution

a) $\dfrac{x^{1/3}}{x^{1/4}} = x^{1/3} \cdot \dfrac{1}{x^{1/4}}$ b) $\left(\dfrac{1}{x^{-2/3}}\right)^{-1/2} = (x^{2/3})^{-1/2}$

$\qquad = x^{1/3} \cdot x^{-1/4}$ $\qquad\qquad = x^{(2/3)(-1/2)}$

$\qquad = x^{(1/3)-1/4}$ $\qquad\qquad = x^{-1/3}$

$\qquad = x^{1/12}$ $\qquad\qquad = \dfrac{1}{x^{1/3}}$ ■

Example   **2**   Simplify the expression so that each variable appears only once and all the exponents are positive. Assume that $x > 0$ and $y > 0$.

$$\left(\frac{4x^8}{y^2}\right)^{3/2} \left(\frac{y^3}{27x^6}\right)^{2/3}$$

Solution

$$\left(\frac{4x^8}{y^2}\right)^{3/2} \left(\frac{y^3}{27x^6}\right)^{2/3} = \frac{(4^{1/2})^3 (x^8)^{3/2}}{(y^2)^{3/2}} \cdot \frac{(y^3)^{2/3}}{(27^{1/3})^2 (x^6)^{2/3}}$$

$$= \frac{2^3 x^{12}}{y^3} \cdot \frac{y^2}{3^2 x^4}$$

$$= \frac{8x^{12}y^2}{9x^4 y^3}$$

$$= \frac{8x^{12-4}}{9y^{3-2}}$$

$$= \frac{8x^8}{9y}$$ ■

Example   **3**   Find each of the following products and express the result with positive exponents. Assume that $a > 0$ and $b > 0$.

a) $(a^{1/2} + b^{1/2})(a^{1/2} - b^{1/2})$ b) $(a^{1/2} - b^{1/2})(a^{-1/2} + b^{-1/2})$

Solution

a) $(a^{1/2} + b^{1/2})(a^{1/2} - b^{1/2}) = (a^{1/2})^2 - (b^{1/2})^2$

$\qquad\qquad\qquad\qquad\qquad = a - b$

b) $(a^{1/2} - b^{1/2})(a^{-1/2} + b^{-1/2}) = a^0 + a^{1/2}b^{-1/2} - a^{-1/2}b^{1/2} - b^0$

$$= 1 + a^{1/2} \cdot \frac{1}{b^{1/2}} - \frac{1}{a^{1/2}} \cdot b^{1/2} - 1$$

$$= \frac{a^{1/2}}{b^{1/2}} - \frac{b^{1/2}}{a^{1/2}}$$

$$= \frac{a^{1/2} \cdot a^{1/2}}{b^{1/2} \cdot a^{1/2}} - \frac{b^{1/2} \cdot b^{1/2}}{a^{1/2} \cdot b^{1/2}}$$

$$= \frac{a - b}{a^{1/2}b^{1/2}}$$ ■

Example   4 Simplify the following expressions. Each variable can be any real number.

a) $(u^2v^4)^{1/4}$ b) $[(-x)^2(y - 3)^2]^{1/2}$

Solution

a) $(u^2v^4)^{1/4} = (u^2)^{1/4}(v^4)^{1/4}$ b) $[(-x)^2(y - 3)^2]^{1/2} = [(-x)^2]^{1/2}[(y - 3)^2]^{1/2}$
  $= |u|^{2/4}|v|^{4/4}$ $= |-x||y - 3|$
  $= |u|^{1/2}|v|$ $= |x||y - 3|$ ■

The next example involves factoring with rational exponents.

Example   5 Factor each of the expressions by removing the factor containing the lowest power of the variable.

a) $8x^{2/3} + 5x^{5/3}$ b) $4y^{-1/4} + 7y^{7/4}$ c) $2(x - 1)^{-1/2} + (x - 1)^{1/2}$

Solution

a) $8x^{2/3} + 5x^{5/3} = x^{2/3}(8 + 5x)$ b) $4y^{-1/4} + 7y^{7/4} = y^{-1/4}(4 + 7y^2)$
c) $2(x - 1)^{-1/2} + (x - 1)^{1/2} = (x - 1)^{-1/2}[2 + (x - 1)]$
$$= (x - 1)^{-1/2}(x + 1)$$ ■

The expression involving rational exponents in the following example occurs in calculus.

Example   6 Simplify:

$$\frac{4x^3(x^2 + 1)^{1/2} - x^4[\frac{1}{2}(x^2 + 1)^{-1/2}(2x)]}{[(x^2 + 1)^{1/2}]^2}$$

Solution

$$\frac{4x^3(x^2+1)^{1/2} - x^4[\frac{1}{2}(x^2+1)^{-1/2}(2x)]}{[(x^2+1)^{1/2}]^2} = \frac{(x^2+1)^{-1/2}[4x^3(x^2+1) - x^5]}{x^2+1}$$

$$= \frac{4x^5 + 4x^3 - x^5}{(x^2+1)^{1/2}(x^2+1)}$$

$$= \frac{3x^5 + 4x^3}{(x^2+1)^{3/2}} \quad\blacksquare$$

EXERCISES 1.6

In Exercises 1 through 8, find the value.

1. (a) $81^{1/2}$; (b) $27^{1/3}$; (c) $625^{1/4}$; (d) $32^{1/5}$
2. (a) $16^{1/2}$; (b) $125^{1/3}$; (c) $16^{1/4}$; (d) $100,000^{1/5}$
3. (a) $16^{3/2}$; (b) $125^{2/3}$; (c) $(\frac{1}{36})^{1/2}$; (d) $(\frac{4}{49})^{3/2}$
4. (a) $25^{3/2}$; (b) $27^{2/3}$; (c) $(\frac{1}{121})^{1/2}$; (d) $(\frac{16}{9})^{3/2}$
5. (a) $144^{-1/2}$; (b) $(-1)^{-1/3}$; (c) $243^{-1/5}$; (d) $(\frac{1}{10,000})^{-1/4}$
6. (a) $(\frac{1}{100})^{-1/2}$; (b) $(-1)^{-1/5}$; (c) $(-64)^{-1/3}$; (d) $(\frac{1}{81})^{-1/4}$
7. (a) $(-\frac{1}{8})^{-2/3}$; (b) $(\frac{81}{16})^{-3/4}$; (c) $(-\frac{1000}{343})^{-5/3}$;
   (d) $-0.16^{3/2}$
8. (a) $(\frac{1}{16})^{-3/4}$; (b) $(\frac{8}{729})^{-4/3}$; (c) $(\frac{16}{625})^{-5/4}$;
   (d) $-0.0016^{-3/4}$

In Exercises 9 through 18, write the expression so that each variable occurs only once and the exponents are positive. Assume that all the variables are positive.

9. (a) $x^{-1/3} \cdot x^{1/2}$; (b) $\dfrac{a^{5/8}}{a^{1/4}}$; (c) $(y^6)^{4/3}$; (d) $(x^{-3/4})^{-1/3}$

10. (a) $b^{3/4} \cdot b^{-1/2}$; (b) $\dfrac{x^{2/3}}{x^{7/6}}$; (c) $(t^4)^{3/2}$; (d) $(y^{4/3})^{-1/2}$

11. (a) $x^{-3/4} \cdot x^{5/6} \cdot x^{-1/3}$; (b) $\left(\dfrac{y^{-3/4}}{y^{3/2}}\right)^{-1/9}$

12. (a) $y^{1/4} \cdot y^{-3/2} \cdot y^{-5/8}$; (b) $\left(\dfrac{x^{-3/5}}{x^{-7/10}}\right)^{-1/4}$

13. (a) $\dfrac{10a^{1/3}b^{-1/4}}{15a^{-1/2}b^{3/4}}$; (b) $\left(\dfrac{t^{-3}s^{1/2}}{t^{1/3}s^{-2}}\right)^{1/5}$

14. (a) $\dfrac{9x^{-3/5}y^{1/6}}{6x^{2/5}y^{-4/3}}$; (b) $\left(\dfrac{64w^{-1/4}z}{w^{1/2}z^{-1/5}}\right)^{-1/3}$

15. $\left(\dfrac{16x^{-4}}{81y^8}\right)^{3/4} \left(\dfrac{8y^{-6}}{27x^9}\right)^{-2/3}$

16. $\left(\dfrac{25a^8}{16b^{-4}}\right)^{-3/2} \left(\dfrac{125b^6}{64a^{-9}}\right)^{2/3}$

17. $\left(\dfrac{x^{7/2}y^{4/3}z^{-9}}{x^0y^{-1}z^{-2}}\right)^{-1/7}$

18. $\left(\dfrac{u^{-2/3}v^{-4/3}w^{-4}}{u^{-1/3}v^{2/3}w^{-7/3}}\right)^{-3}$

In Exercises 19 through 22, find the product and express the result with positive exponents. Assume that all the variables are positive.

19. (a) $x^{1/3}(x - x^{2/3})$; (b) $a^{-1/4}(a^{3/8} + a^{3/2})$
20. (a) $y^{1/2}(y^{3/2} + y^{1/2})$; (b) $b^{-2/3}(b^{-1/3} - b^{2/3})$
21. $(x^{1/3} - x^{-2/3})(x^{2/3} - x^{-1/3})$
22. $(a^{1/4} - a^{1/2})(a^{-1/4} + a^{-1/2})$

In Exercises 23 through 26, factor each of the expressions by removing the factor containing the lowest power of the variable.

23. (a) $2y^{3/2} - 3y^{5/2}$; (b) $5x^{-4/3} + 4x^{5/3}$
24. (a) $6t^{3/4} + t^{7/4}$; (b) $4w^{4/5} - 3w^{-6/5}$
25. (a) $b^{-3/5} - 7b^{-2/5} + 2b^{2/5}$;
    (b) $3(x+3)^{-1/3} + 2(x+3)^{4/3}$
26. (a) $x^{-3/2} - x^{-1/2} - 4x^{1/2}$;
    (b) $(2x-3)^{-3/4} - 2(2x-3)^{5/4}$

In Exercises 27 through 30, write the expression as a fraction in lowest terms with positive exponents only; n is a positive integer. Assume that all the variables are positive.

27. (a) $(a^3)^{n/3}(a^{3n})^{3/n}$; (b) $(x^{n/2})^{-1/2}(x^{-1/2})^{-n}$
28. (a) $(y^4)^{n/4}(y^{2n})^{2/n}$; (b) $(t^{n/3})^{-2/3}(t^{-1/3})^{-n}$
29. (a) $\left(\dfrac{y^{3n}y^{4n}}{y^{2n}}\right)^{-1/5}$; (b) $\dfrac{4^{n-1/2}16^{n^2+1/2}}{(2^{n-1})^n(8^n)^{n+1}}$
30. (a) $\left(\dfrac{x^{6n}x^{4n}}{x^{3n}}\right)^{-1/7}$; (b) $\dfrac{9^{n+1/2}27^{n^2+1/3}}{(81^n)^{n+1}(3^{n+2})^{-n}}$

In Exercises 31 through 36, the expression is similar to one occurring in calculus. Simplify it in a manner similar to that of Example 6.

31. $\dfrac{(x+1)^{1/2} - x[\frac{1}{2}(x+1)^{-1/2}]}{[(x+1)^{1/2}]^2}$

32. $\dfrac{(x-1)^{2/3} - x[\frac{2}{3}(x-1)^{-1/3}]}{[(x-1)^{2/3}]^2}$

33. $\dfrac{3x^2(2x+5)^{1/2} - x^3[\frac{1}{2}(2x+5)^{-1/2}(2)]}{[(2x+5)^{1/2}]^2}$

34. $\dfrac{5(x^2-1)^{1/2} - 5x[\frac{1}{2}(x^2-1)^{-1/2}(2x)]}{[(x^2-1)^{1/2}]^2}$

35. $\dfrac{4x^3(3x^2+1)^{1/2} - x^4[\frac{1}{2}(3x^2+1)^{-1/2}(6x)]}{[(3x^2+1)^{1/2}]^2}$

36. $\dfrac{\frac{1}{3}(3x+1)^{-2/3}(3)(2x-3)^{1/2} - (3x+1)^{1/3}[\frac{1}{2}(2x-3)^{-1/2}(2)]}{[(2x-3)^{1/2}]^2}$

In Exercises 37 through 42, simplify the expression. Each variable can be any real number.

37. (a) $(x^6y^8)^{1/4}$; (b) $(4s^4t^{10})^{1/2}$

38. (a) $(9x^2y^4)^{1/2}$; (b) $(a^4b^{12})^{1/4}$
39. (a) $[(-3y)^4(y-2)^2]^{1/2}$; (b) $[(-2)^8(x-2)^8(2-y)^4]^{1/4}$
40. (a) $[(-2a)^4(b+2)^8]^{1/4}$; (b) $[(-3)^6x^2(x^2+9)^2]^{1/2}$
41. $[(-4)^4(u+1)^8(u-4)^4]^{1/4}$
42. $\left[\dfrac{(-5)^6x^2}{(x^2+4)^2}\right]^{1/2}$

43. (a) Simplify the expression
$(x^2 + 6x + 9)^{1/2} - (x^2 - 6x + 9)^{1/2}$.
(b) For what values of x is the expression in part (a) equivalent to 6?

44. (a) Simplify the expression
$(x^2 - 8x + 16)^{1/2} + (x^2 + 8x + 16)^{1/2}$.
(b) For what values of x is the expression in part (a) equivalent to $2x$?

1.7 Properties of Radicals

We introduced radicals in Section 1.6 to define a rational exponent of the form $\dfrac{1}{n}$ where n is a positive integer; that is,

$$\sqrt[n]{a} = a^{1/n}$$

Because of this equality, properties of radicals can be proved by using laws of exponents. Two such properties are given in the following theorem.

Theorem 1

If a and b are real numbers,

(i) $\sqrt[n]{a} \cdot \sqrt[n]{b} = \sqrt[n]{ab}$

(ii) $\dfrac{\sqrt[n]{a}}{\sqrt[n]{b}} = \sqrt[n]{\dfrac{a}{b}}$ $\qquad (b \neq 0)$

where both $a \geq 0$ and $b \geq 0$ if n is even.

Proof of (i)

$$\sqrt[n]{a} \cdot \sqrt[n]{b} = a^{1/n} \cdot b^{1/n}$$
$$= (ab)^{1/n}$$
$$= \sqrt[n]{ab} \qquad \blacksquare$$

The proof of (ii) is similar.

In Theorem 1 a and b must be nonnegative when n is even. This restriction implies that $\sqrt[n]{a}$ and $\sqrt[n]{b}$ be real numbers. In Illustration 8 of Section 1.8 we show why Theorem 1 does not hold for $a < 0$ and $b < 0$ when n is even.

Illustration 1

The following results are obtained by applying Theorem 1.

a) $\sqrt{4} \cdot \sqrt{25} = \sqrt{4 \cdot 25}$
    b) $\sqrt[3]{-9} \cdot \sqrt[3]{-3} = \sqrt[3]{(-9)(-3)}$

$= \sqrt{100}$ $= \sqrt[3]{27}$

$= 10$ $= 3$

c) $\dfrac{\sqrt[4]{96}}{\sqrt[4]{6}} = \sqrt[4]{\dfrac{96}{6}}$
    d) $\dfrac{\sqrt[3]{-54}}{\sqrt[3]{2}} = \sqrt[3]{\dfrac{-54}{2}}$

$= \sqrt[4]{16}$ $= \sqrt[3]{-27}$

$= 2$ $= -3$ ■

Formula (i) of Theorem 1 can be written as

$$\sqrt[n]{ab} = \sqrt[n]{a} \cdot \sqrt[n]{b}$$

where both $a \geq 0$ and $b \geq 0$ if n is even. The next illustration shows how this equality can be used to simplify radicals.

Illustration 2

To simplify the radical $\sqrt{540}$, we first write 540 as the product of prime factors. Since $540 = 10 \cdot 54$ while $10 = 2 \cdot 5$ and $54 = 6 \cdot 9$, we have $540 = 2 \cdot 5 \cdot 2 \cdot 3 \cdot 3^2$. Therefore

$$\sqrt{540} = \sqrt{2^2 \cdot 3^3 \cdot 5}$$
$$= \sqrt{(2^2 \cdot 3^2) \cdot (3 \cdot 5)}$$
$$= \sqrt{(2 \cdot 3)^2} \sqrt{3 \cdot 5}$$
$$= (2 \cdot 3)\sqrt{15}$$
$$= 6\sqrt{15}$$

    ■

Example 1   Simplify each of the following radicals.

a) $\sqrt[3]{81x^4y^3}$ b) $\sqrt[4]{80x^8y^7}$, $x \geq 0$ and $y \geq 0$

Solution

a) $\sqrt[3]{81x^4y^3} = \sqrt[3]{3^4x^4y^3}$
           b) $\sqrt[4]{80x^8y^7} = \sqrt[4]{2^4 \cdot 5x^8y^7}$

$= \sqrt[3]{(3^3x^3y^3)(3x)}$ $= \sqrt[4]{(2^4x^8y^4)(5y^3)}$

$= \sqrt[3]{(3xy)^3(3x)}$ $= \sqrt[4]{(2x^2y)^4(5y^3)}$

$= \sqrt[3]{(3xy)^3} \sqrt[3]{3x}$ $= \sqrt[4]{(2x^2y)^4} \sqrt[4]{5y^3}$

$= 3xy \sqrt[3]{3x}$ $= 2x^2y \sqrt[4]{5y^3}$ ■

When finding the product of radicals having the same index, you should first express any constant in completely factored form.

Example 2 Find the product and simplify the result:

$$\sqrt[3]{126r^2s^2t} \cdot \sqrt[3]{36rs^2t^2}$$

Solution

$$
\begin{aligned}
\sqrt[3]{126r^2s^2t} \cdot \sqrt[3]{36rs^2t^2} &= \sqrt[3]{2 \cdot 3^2 \cdot 7r^2s^2t}\ \sqrt[3]{2^2 \cdot 3^2rs^2t^2} \\
&= \sqrt[3]{2^3 \cdot 3^4 \cdot 7r^3s^4t^3} \\
&= \sqrt[3]{(2^3 \cdot 3^3r^3s^3t^3)(3 \cdot 7s)} \\
&= \sqrt[3]{(2 \cdot 3rst)^3}\ \sqrt[3]{3 \cdot 7s} \\
&= 6rst\ \sqrt[3]{21s}
\end{aligned}
$$

$\blacksquare$

If the radicand of a radical is a fraction having a monomial in the denominator, we can replace the radical by an equivalent expression for which the radicand contains no fraction. This process is called **rationalizing the denominator.** For a radical of order n the procedure consists of first obtaining an equivalent fraction in which the denominator is an nth power of a monomial. Then we apply formula (ii) of Theorem 1, written as

$$\sqrt[n]{\frac{a}{b}} = \frac{\sqrt[n]{a}}{\sqrt[n]{b}}$$

where $a \geq 0$ and $b > 0$ if n is even. Illustration 3 shows the procedure.

Illustration 3

To rationalize the denominator of the radical

$$\sqrt{\frac{3}{5}}$$

we wish to build the fraction in the radicand to one in which the denominator is the square of an integer. Hence we first multiply the numerator and denominator by 5 and then we apply the above equality. Doing this, we have

$$
\begin{aligned}
\sqrt{\frac{3}{5}} &= \sqrt{\frac{3 \cdot 5}{5 \cdot 5}} \\
&= \frac{\sqrt{3 \cdot 5}}{\sqrt{5^2}} \\
&= \frac{\sqrt{15}}{5}
\end{aligned}
$$

The result can also be written as $\frac{1}{5}\sqrt{15}$.

$\blacksquare$

Example 3 Rationalize the denominator of each of the radicals.

a) $\sqrt{\dfrac{3}{2x}}, \; x > 0$ b) $\sqrt[3]{\dfrac{3x^2}{2y^2}}, \; y \neq 0$

Solution

a) $\sqrt{\dfrac{3}{2x}} = \sqrt{\dfrac{3(2x)}{(2x)(2x)}}$ b) $\sqrt[3]{\dfrac{3x^2}{2y^2}} = \sqrt[3]{\dfrac{(3x^2)(2^2 y)}{(2y^2)(2^2 y)}}$

$\quad\quad = \sqrt{\dfrac{6x}{(2x)^2}}$ $\quad\quad = \sqrt[3]{\dfrac{2^2 \cdot 3x^2 y}{(2y)^3}}$

$\quad\quad = \dfrac{\sqrt{6x}}{\sqrt{(2x)^2}}$ $\quad\quad = \dfrac{\sqrt[3]{2^2 \cdot 3x^2 y}}{\sqrt[3]{(2y)^3}}$

$\quad\quad = \dfrac{\sqrt{6x}}{2x}$ $\quad\quad = \dfrac{\sqrt[3]{12x^2 y}}{2y}$ ∎

Example 4 Express each of the quotients with a rationalized denominator.

a) $\dfrac{1}{\sqrt[3]{5y}}, \; y \neq 0$ b) $\dfrac{50ab^2}{\sqrt{20a^3 b}}, \; a > 0, \; b > 0$

Solution

a) $\dfrac{1}{\sqrt[3]{5y}} = \dfrac{1 \sqrt[3]{5^2 y^2}}{\sqrt[3]{5y} \, \sqrt[3]{5^2 y^2}}$ b) $\dfrac{50ab^2}{\sqrt{20a^3 b}} = \dfrac{2 \cdot 5^2 ab^2 \sqrt{5ab}}{\sqrt{2^2 \cdot 5a^3 b} \, \sqrt{5ab}}$

$\quad\quad = \dfrac{\sqrt[3]{5^2 y^2}}{\sqrt[3]{(5y)^3}}$ $\quad\quad = \dfrac{2 \cdot 5^2 ab^2 \sqrt{5ab}}{\sqrt{2^2 \cdot 5^2 a^4 b^2}}$

$\quad\quad = \dfrac{\sqrt[3]{25y^2}}{5y}$ $\quad\quad = \dfrac{2 \cdot 5^2 ab^2 \sqrt{5ab}}{\sqrt{(2 \cdot 5a^2 b)^2}}$

$\quad\quad\quad\quad\quad\quad\quad\quad = \dfrac{2 \cdot 5^2 ab^2 \sqrt{5ab}}{2 \cdot 5a^2 b}$

$\quad\quad\quad\quad\quad\quad\quad\quad = \dfrac{(2 \cdot 5ab)(5b \sqrt{5ab})}{(2 \cdot 5ab)a}$

$\quad\quad\quad\quad\quad\quad\quad\quad = \dfrac{5b \sqrt{5ab}}{a}$ ∎

Recall the special product

$$(a + b)(a - b) = a^2 - b^2$$

Each of the two factors is called the **conjugate** of the other. The concept of the conjugate is used to rationalize the denominator of a fraction when the denominator is a binomial containing a radical of order 2. The following illustration and example demonstrate the procedure.

Illustration 4

To rationalize the denominator of the fraction

$$\frac{\sqrt{5} - \sqrt{2}}{\sqrt{5} + \sqrt{2}}$$

we multiply the numerator and denominator by $\sqrt{5} - \sqrt{2}$, which is the conjugate of $\sqrt{5} + \sqrt{2}$, and we have

$$\frac{\sqrt{5} - \sqrt{2}}{\sqrt{5} + \sqrt{2}} = \frac{(\sqrt{5} - \sqrt{2})(\sqrt{5} - \sqrt{2})}{(\sqrt{5} + \sqrt{2})(\sqrt{5} - \sqrt{2})}$$

$$= \frac{\sqrt{5^2} - 2\sqrt{10} + \sqrt{2^2}}{\sqrt{5^2} - \sqrt{2^2}}$$

$$= \frac{5 - 2\sqrt{10} + 2}{5 - 2}$$

$$= \frac{7 - 2\sqrt{10}}{3}$$ ∎

Example   5   Rationalize the denominator of the fraction

$$\frac{2}{\sqrt{x} + 3\sqrt{y}}, \qquad x > 0, \, y > 0$$

Solution   We multiply the numerator and denominator by the conjugate of the denominator:

$$\frac{2}{\sqrt{x} + 3\sqrt{y}} = \frac{2(\sqrt{x} - 3\sqrt{y})}{(\sqrt{x} + 3\sqrt{y})(\sqrt{x} - 3\sqrt{y})}$$

$$= \frac{2\sqrt{x} - 6\sqrt{y}}{\sqrt{x^2} - 9\sqrt{y^2}}$$

$$= \frac{2\sqrt{x} - 6\sqrt{y}}{x - 9y}$$ ∎

In calculus it is sometimes necessary to **rationalize the numerator** of a fraction, which means that an equivalent fraction is written that contains no radical in the numerator. The computations involved in such situations are shown in the next example. The process is similar to that used to rationalize a denominator.

Example   6   Rationalize the numerators of the rational expressions.

a) $\dfrac{\sqrt{x} - 1}{x - 1}$, $x > 0$, $x \neq 1$ b) $\dfrac{\sqrt{x + h} - \sqrt{x}}{h}$, $x > 0$, $x + h > 0$, $h \neq 0$

Solution In each part we multiply the numerator and denominator by the conjugate of the numerator.

a) $\dfrac{\sqrt{x} - 1}{x - 1} = \dfrac{(\sqrt{x} - 1)(\sqrt{x} + 1)}{(x - 1)(\sqrt{x} + 1)}$

$= \dfrac{(\sqrt{x})^2 - 1}{(x - 1)(\sqrt{x} + 1)}$

$= \dfrac{x - 1}{(x - 1)(\sqrt{x} + 1)}$

$= \dfrac{1}{\sqrt{x} + 1}$

b) $\dfrac{\sqrt{x + h} - \sqrt{x}}{h} = \dfrac{(\sqrt{x + h} - \sqrt{x})(\sqrt{x + h} + \sqrt{x})}{h(\sqrt{x + h} + \sqrt{x})}$

$= \dfrac{(\sqrt{x + h})^2 - \sqrt{x}\sqrt{x + h} + \sqrt{x}\sqrt{x + h} - (\sqrt{x})^2}{h(\sqrt{x + h} + \sqrt{x})}$

$= \dfrac{(x + h) - x}{h(\sqrt{x + h} + \sqrt{x})}$

$= \dfrac{h}{h(\sqrt{x + h} + \sqrt{x})}$

$= \dfrac{1}{\sqrt{x + h} + \sqrt{x}}$

■

EXERCISES 1.7

In Exercises 1 through 4, find the root.

1. (a) $\sqrt{81}$; (b) $\sqrt[3]{-0.001}$; (c) $\sqrt{\frac{16}{625}}$; (d) $\sqrt[4]{\frac{16}{625}}$
2. (a) $\sqrt{64}$; (b) $\sqrt[3]{-64}$; (c) $\sqrt{\frac{81}{16}}$; (d) $\sqrt[4]{\frac{81}{16}}$
3. (a) $\sqrt[3]{\frac{216}{125}}$; (b) $\sqrt{(-5)^2}$; (c) $\sqrt[3]{(-4)^3}$; (d) $-\sqrt[5]{-243}$
4. (a) $\sqrt[5]{\frac{243}{100,000}}$; (b) $\sqrt{(-3)^4}$; (c) $\sqrt[3]{(-3)^3}$; (d) $-\sqrt[5]{-32}$

In Exercises 5 and 6, find the product.

5. (a) $\sqrt{16}\sqrt{36}$; (b) $\sqrt[3]{-4}\sqrt[3]{-2}$; (c) $\sqrt[4]{108}\sqrt[4]{12}$
6. (a) $\sqrt{24}\sqrt{6}$; (b) $\sqrt[3]{-5}\sqrt[3]{-25}$; (c) $\sqrt[4]{90}\sqrt[4]{9000}$

In Exercises 7 and 8, find the quotient.

7. (a) $\dfrac{\sqrt{45}}{\sqrt{5}}$; (b) $\dfrac{\sqrt[4]{324}}{\sqrt[4]{4}}$; (c) $\dfrac{\sqrt[3]{3}}{\sqrt[3]{24}}$
8. (a) $\dfrac{\sqrt[3]{72}}{\sqrt[3]{9}}$; (b) $\dfrac{\sqrt{3}}{\sqrt{48}}$; (c) $\dfrac{\sqrt[4]{0.007}}{\sqrt[4]{70}}$

In Exercises 9 through 12, simplify the radical. All the variables represent positive numbers.

9. (a) $\sqrt{48x^2}$; (b) $\sqrt[3]{54x^6}$; (c) $\sqrt[3]{8c^8}$
10. (a) $\sqrt[3]{-81y^3}$; (b) $\sqrt{16x^{16}}$; (c) $\sqrt[3]{-27x^{10}}$
11. (a) $\sqrt{b^2}$; (b) $\sqrt[5]{-96x^{25}y^{12}}$; (c) $\sqrt{40r^5t^3s^8}$
12. (a) $\sqrt[4]{x^4}$; (b) $\sqrt[3]{135a^9b^8}$; (c) $\sqrt[4]{16x^{16}y^4z^9}$

In Exercises 13 through 16, find the product and simplify the result. All the variables represent positive numbers.

13. (a) $\sqrt{10}\sqrt{30}$; (b) $(2\sqrt[3]{9})(4\sqrt[3]{-6})$;
    (c) $\sqrt[3]{-6s^2t^4}\sqrt[3]{9s^5t^2}$
14. (a) $\sqrt{18}\sqrt{12}$; (b) $\sqrt[4]{24x^3}\sqrt[4]{270x^2}$;
    (c) $\sqrt[4]{\frac{10}{3}a^3b^5}\sqrt[4]{24a^2b^3}$

15. (a) $\sqrt{10abc}\ \sqrt{15a^3c}\ \sqrt{12bc}$;
    (b) $\sqrt[3]{9xy^2}\ \sqrt[3]{6x^2y^4}\ \sqrt[3]{60x^5y}$

16. (a) $\sqrt[3]{4x^2y^2}\ \sqrt[3]{6x^2z^2}\ \sqrt[3]{45x^2y^2z}$;
    (b) $\sqrt{2uv}\ \sqrt{3v}\ \sqrt{6uw}\ \sqrt{12vw}$

In Exercises 17 through 32, rationalize the denominator. All the variables represent positive numbers.

17. (a) $\sqrt{\dfrac{7}{2}}$; (b) $\dfrac{6}{\sqrt[3]{2}}$ 18. (a) $\sqrt[3]{\dfrac{11}{3}}$; (b) $\dfrac{5}{\sqrt{10}}$

19. (a) $\dfrac{\sqrt{5}}{2\sqrt{3}}$; (b) $\sqrt[3]{\dfrac{25}{2}}$

20. (a) $\dfrac{-4\sqrt{3}}{5\sqrt{2}}$; (b) $\sqrt[3]{\dfrac{4}{49}}$

21. (a) $\dfrac{3s}{\sqrt{2t}}$; (b) $\dfrac{\sqrt[4]{x^3y^2}}{3\sqrt[4]{xy}}$

22. (a) $\sqrt{\dfrac{2x}{3y}}$; (b) $\dfrac{21}{\sqrt[3]{-98x}}$

23. (a) $\sqrt{\dfrac{8x^6y^3}{15z^7}}$; (b) $\dfrac{270u^2vw^3}{\sqrt[4]{288u^5v^6w^2}}$

24. (a) $\sqrt{\dfrac{27t^8s^3}{10r^5}}$; (b) $\dfrac{25x^2yz^2}{\sqrt[3]{100x^4y^5z^2}}$

25. $\dfrac{3}{2-\sqrt{3}}$ 26. $\dfrac{\sqrt{7}-3}{\sqrt{7}+2}$

27. $\dfrac{\sqrt{6}+1}{\sqrt{3}+\sqrt{2}}$ 28. $\dfrac{2\sqrt{3}}{2\sqrt{3}-3\sqrt{2}}$

29. $\dfrac{2\sqrt{2}-3\sqrt{7}}{3\sqrt{2}+2\sqrt{7}}$ 30. $\dfrac{5\sqrt{3}-4\sqrt{5}}{2\sqrt{5}+3\sqrt{3}}$

31. $\dfrac{1}{2\sqrt{x}+\sqrt{y}}$ 32. $\dfrac{\sqrt{a}+\sqrt{b}}{\sqrt{a}-\sqrt{b}}$

In Exercises 33 through 38, the expression is similar to one occurring in calculus. Rationalize the numerator. All the radicands and all the variables represent positive numbers; none of the denominators is zero.

33. $\dfrac{\sqrt{x}-3}{x-9}$ 34. $\dfrac{\sqrt{x}-2}{x-4}$

35. $\dfrac{\sqrt{x+4}-2}{x}$ 36. $\dfrac{\sqrt{x+3}-\sqrt{3}}{x}$

37. $\dfrac{\sqrt{2(x+h)+1}-\sqrt{2x+1}}{h}$

38. $\dfrac{\sqrt{3(x+h)-2}-\sqrt{3x-2}}{h}$

In Exercises 39 and 40 use the fact that $a^3 - b^3 = (a-b)(a^2 + ab + b^2)$.

39. Rationalize the denominator of $\dfrac{1}{\sqrt[3]{x}-\sqrt[3]{y}}$.

40. Rationalize the numerator of $\dfrac{\sqrt[3]{x+h}-\sqrt[3]{x}}{h}$.

41. Given that $r_1 = \dfrac{-b+\sqrt{b^2-4ac}}{2a}$ and
    $r_2 = \dfrac{-b-\sqrt{b^2-4ac}}{2a}$, find:
    (a) $r_1 + r_2$; (b) $r_1 \cdot r_2$.

42. For the values of r_1 and r_2 in Exercise 41, find:
    (a) $\dfrac{1}{r_1}$; (b) $r_1{}^2 - r_2{}^2$.

43. For the values of r_1 and r_2 in Exercise 41, find:
    (a) $\dfrac{1}{r_2}$; (b) $r_1{}^2 + r_2{}^2$.

1.8 The Set of Complex Numbers

In Section 1.6 we defined an nth root of a real number. This definition stated that if n is a positive integer greater than 1 and a and b are real numbers such that

$$b^n = a$$

then b is an nth root of a. We indicated that if $a < 0$ and n is an even positive integer, there is no real nth root of a because an even power of a real number is a nonnegative number. For instance, suppose we have the equation

$$b^2 = -25$$

There is no real number that can be substituted for b in this equation.

Therefore, there is no real square root of -25; in a similar manner it follows that there is no real square root of any negative number.

We see, then, that in order to consider square roots of negative numbers, we must deal with numbers other than real numbers. Thus we develop a set of numbers that contains the set R of real numbers as a subset and also contains square roots of negative numbers. We denote such a set of numbers by C and refer to it as the set of *complex numbers*. We first require that the set C is such that the real number -1 has a square root. Let i be a symbol for a number in C whose square is -1; that is, we define i as a number such that

$$i^2 = -1$$

Because every real number is to be an element of C, it follows that if $b \in R$, then $b \in C$. In order for the closure law of multiplication to hold, the number $b \cdot i$, abbreviated bi, must be an element of C. Furthermore, if $a \in R$, then $a \in C$, and if the closure law for addition is to hold, the number $a + bi$ must be an element of C. We now have a set C, which we define formally.

Definition

The Set of Complex Numbers

The set of all numbers of the form

$a + bi,$ where a and b are real numbers and $i^2 = -1$,

is called the **set of complex numbers** and is denoted by C; that is,

$$C = \{a + bi \mid a, b \in R, i^2 = -1\}$$

For the complex number $a + bi$ the number a is called the **real part** and the number b is called the **imaginary part.**

Illustration 1

a) The number $-3 + 6i$ is a complex number whose real part is -3 and whose imaginary part is 6.

b) The number $7 + (-4)i$ is a complex number whose real part is 7 and whose imaginary part is -4.   ■

If $-p$ is a negative number, then the complex number $a + (-p)i$ can be written as $a - pi$. Hence

$$7 + (-4)i = 7 - 4i$$

A real number is a complex number whose imaginary part is 0; that is, if a is a real number,

$$a = a + 0i$$

Therefore R is a subset of C. Another subset of C is the set I of **imaginary numbers,** defined by

$$I = \{a + bi \mid a, b \in R, i^2 = -1, b \neq 0\}$$

The number $0 + bi$ can be written more simply as bi; that is,

$$bi = 0 + bi$$

This number is called a **pure imaginary number.**

Illustration 2

a) The complex number $-5 + 2i$ is an imaginary number.

b) The complex number $8i$ is a pure imaginary number.

c) The real number -3 is a complex number, and it can be written as $-3 + 0i$.

d) The real number 0 is a complex number, and it can be written as $0 + 0i$. ∎

The terminology *imaginary number* is an unfortunate but historical choice, and arose from the fact that an equation such as $x^2 = -1$ has no solution in the set of real numbers. As far back as the fifteenth century, mathematicians found it desirable to treat an equation such as $x^2 = -1$ in the same manner as an equation like $x^2 = 4$, which has real solutions. From their viewpoint a number whose square is -1 was not real but something imaginary. René Descartes in 1637 introduced the words *real* (*vraye*) and *imaginary* (*imaginaire*) in connection with sets of numbers, and Leonhard Euler (1707–1783) in 1748 used the symbol i to represent a number whose square is -1. The terminology and symbolism have become standard even though imaginary numbers are just as "real things" as the numbers 7, -4, and $\sqrt{3}$.

We now state a definition that allows every real number (positive, negative, or zero) to have a square root.

Definition

> **Square Root of Any Real Number**
> A number s is said to be a **square root** of a real number r if and only if
> $$s^2 = r$$

You have learned that any positive number has two square roots, one positive and one negative, and the number 0 has only one square root, 0. What about a square root of a negative number? In particular, consider the number -2. Because

$$(i\sqrt{2})^2 = i^2(\sqrt{2})^2 \qquad \text{and} \qquad (-i\sqrt{2})^2 = (-1)^2 i^2 (\sqrt{2})^2$$
$$= (-1)(2) \qquad\qquad\qquad = (1)(-1)(2)$$
$$= -2 \qquad\qquad\qquad\qquad = -2$$

it follows from the definition that both $i\sqrt{2}$ and $-i\sqrt{2}$ are square roots of -2. More generally, if $-p$ is any negative number, then p is a positive number and both $i\sqrt{p}$ and $-i\sqrt{p}$ are square roots of $-p$. As we did with square roots of positive numbers, we distinguish between the two square roots by using the concept of *principal square root*.

Definition

The Principal Square Root of a Negative Number
If p is a positive number, then the **principal square root** of $-p$, denoted by $\sqrt{-p}$, is defined by
$$\sqrt{-p} = i\sqrt{p}$$

The two square roots of $-p$ are written as $\sqrt{-p}$ and $-\sqrt{-p}$, or as $i\sqrt{p}$ and $-i\sqrt{p}$.

Illustration 3

a) $\sqrt{-5} = i\sqrt{5}$

The two square roots of -5 are $i\sqrt{5}$ and $-i\sqrt{5}$.

b) $\sqrt{-16} = i\sqrt{16}$
$\qquad\quad = 4i$

The two square roots of -16 are $4i$ and $-4i$.

c) $\sqrt{-1} = i\sqrt{1}$
$\qquad\;\; = i$

The two square roots of -1 are i and $-i$.   ■

A complex number is said to be in **standard form** when it is written as $a + bi$, where a and b are real numbers.

Example   1 Write the complex number in the standard form $a + bi$.

a) $\sqrt{-9}$ b) $5 - 6\sqrt{-4}$ c) $-\sqrt{\frac{16}{49}} + \sqrt{-25}$ d) $\sqrt{24} + 5\sqrt{-27}$

Solution

a) $\sqrt{-9} = i\sqrt{9}$
$\qquad\quad = 3i$
$\qquad\quad = 0 + 3i$

b) $5 - 6\sqrt{-4} = 5 - 6(i\sqrt{4})$
$\qquad\qquad\quad = 5 - 6(2i)$
$\qquad\qquad\quad = 5 + (-12i)$

c) $-\sqrt{\frac{16}{49}} + \sqrt{-25} = -\frac{4}{7} + i\sqrt{25}$

$\qquad\qquad\qquad\quad = -\frac{4}{7} + 5i$

d) $\sqrt{24} + 5\sqrt{-27} = \sqrt{4}\sqrt{6} + 5(i\sqrt{9}\sqrt{3})$

$\qquad\qquad\qquad = 2\sqrt{6} + 5(i \cdot 3\sqrt{3})$

$\qquad\qquad\qquad = 2\sqrt{6} + 15\sqrt{3}i$ ∎

Definition

> ### Equality of Two Complex Numbers
> Two complex numbers $a + bi$ and $c + di$ are said to be **equal** if and only if $a = c$ and $b = d$.

Illustration 4

If

$$x + 4i = -6 + yi$$

then $x = -6$ and $y = 4$. ∎

We wish to define addition and multiplication of complex numbers so that the axioms for these operations on the set of real numbers are valid. To arrive at such definitions, we consider two complex numbers, $a + bi$ and $c + di$, as if they were polynomials in i and then simplify the result by letting $i^2 = -1$. Thus

$$(a + bi) + (c + di) = a + c + bi + di$$
$$= (a + c) + (b + d)i$$

and

$$(a + bi)(c + di) = ac + adi + bci + bdi^2$$
$$= ac + (ad + bc)i + bd(-1)$$
$$= (ac - bd) + (ad + bc)i$$

We have then the following definition.

Definition

> ### The Sum and Product of Two Complex Numbers
> If $a + bi$ and $c + di$ are complex numbers, then
> $$(a + bi) + (c + di) = (a + c) + (b + d)i$$
> $$(a + bi)(c + di) = (ac - bd) + (ad + bc)i$$

From the above definition, it can be shown that the set C is closed under the operations of addition and multiplication. It also can be proved that addition and multiplication on C are commutative and associative, and that multiplication is distributive over addition; that is, if $u, v, w \in C$, then

$$u + v = v + u \qquad\qquad uv = vu$$
$$(u + v) + w = u + (v + w) \qquad (uv)w = u(vw)$$
$$u(v + w) = uv + uw$$

You are advised to compute with complex numbers as in the following example rather than memorize the definitions.

Example   2   Find the sum and product of the complex numbers $5 - 4i$ and $-2 + 6i$.

Solution

$$(5 - 4i) + (-2 + 6i) = 5 - 2 - 4i + 6i$$
$$= 3 + 2i$$

$$(5 - 4i)(-2 + 6i) = -10 + 30i + 8i - 24i^2$$
$$= -10 + 38i - 24(-1)$$
$$= -10 + 38i + 24$$
$$= 14 + 38i \qquad ■$$

The additive identity element in the set of complex numbers is 0, which can be written as $0 + 0i$. The additive inverse of the complex number $a + bi$ is $-a - bi$ because

$$(a + bi) + (-a - bi) = [a + (-a)] + [b + (-b)]i$$
$$= 0 + 0i$$

Therefore

$$-(a + bi) = -a - bi$$

As with real numbers, subtraction of complex numbers is defined in terms of addition; that is,

$$(a + bi) - (c + di) = (a + bi) + [-(c + di)]$$
$$= (a + bi) + (-c - di)$$
$$= (a - c) + (b - d)i$$

Example   3   Find the difference of the complex numbers of Example 2.

Solution

$$(5 - 4i) - (-2 + 6i) = 5 - 4i + 2 - 6i$$
$$= 7 - 10i \qquad ■$$

Because

$$(a + bi)(1 + 0i) = a \cdot 1 + a \cdot 0i + 1 \cdot bi + bi \cdot 0i$$
$$= a + 0i + bi + 0i^2$$
$$= a + bi$$

it follows that the multiplicative identity in C is $1 + 0i$, which is the real number 1.

The **conjugate** of the complex number $a + bi$ is $a - bi$.

Illustration 5

a) The conjugate of $3 + 2i$ is $3 - 2i$.

b) The conjugate of $-4 - 5i$ is $-4 - (-5i)$ or, equivalently, $-4 + 5i$. ■

Let us compute the sum and product of a complex number and its conjugate:

$$(a + bi) + (a - bi) = 2a \qquad (a + bi)(a - bi) = a^2 - b^2 i^2$$
$$= a^2 - b^2(-1)$$
$$= a^2 + b^2$$

Observe that in each case we obtain a real number. The concept of the conjugate is useful in certain computations with complex numbers. For instance, to write the quotient

$$\frac{a + bi}{c + di}$$

in standard form $u + vi$, we multiply the numerator and denominator by the conjugate of the denominator. We do this in the following example.

Example 4 Find the quotient of the complex numbers of Example 2.

Solution

$$\frac{5 - 4i}{-2 + 6i} = \frac{(5 - 4i)(-2 - 6i)}{(-2 + 6i)(-2 - 6i)}$$
$$= \frac{-10 - 22i + 24i^2}{4 - 36i^2}$$
$$= \frac{-10 - 22i + 24(-1)}{4 - 36(-1)}$$
$$= \frac{-34 - 22i}{40}$$
$$= -\frac{17}{20} - \frac{11}{20}i$$

■

The **multiplicative inverse** (or **reciprocal**) of the complex number $a + bi$ is defined to be

$$\frac{1}{a + bi}$$

We use the method of Example 4 to write the multiplicative inverse in standard form, as shown in the following illustration.

Illustration 6

The multiplicative inverse of $4 - 3i$ is

$$\frac{1}{4 - 3i}$$

To write this complex number in the standard form $a + bi$, we multiply the numerator and denominator by the conjugate of $4 - 3i$. We have then

$$\frac{1}{4 - 3i} = \frac{1 \cdot (4 + 3i)}{(4 - 3i)(4 + 3i)}$$

$$= \frac{4 + 3i}{16 - 9i^2}$$

$$= \frac{4 + 3i}{16 - 9(-1)}$$

$$= \frac{4}{25} + \frac{3}{25}i$$

Thus the multiplicative inverse of $4 - 3i$ is $\dfrac{4}{25} + \dfrac{3}{25}i$. That is,

$$(4 - 3i)\left(\frac{4}{25} + \frac{3}{25}i\right) = \frac{16}{25} + \frac{12}{25}i - \frac{12}{25}i - \frac{9}{25}i^2$$

$$= \frac{16}{25} - \frac{9}{25}(-1)$$

$$= \frac{16 + 9}{25}$$

$$= 1$$ ∎

In summary, we have the following facts about the set C.

1. The set C is closed under the operations of addition and multiplication.
2. Addition and multiplication on C are commutative and associative; multiplication is distributive over addition.
3. There is an identity element for addition and an identity element for multiplication.
4. Every element in C has an additive inverse and every element in C, except $0 + 0i$, has a multiplicative inverse.

These facts are the field axioms discussed in Section A.1 of the appendix. Therefore, the set C is a field under the operations of addition and multiplication. Consequently, the laws of exponents apply to positive-integer powers of i.

Illustration 7

$$i^3 = i^2 i \qquad i^4 = i^2 i^2 \qquad i^5 = i^4 i \qquad i^6 = i^4 i^2$$
$$ = (-1)i \qquad = (-1)(-1) \qquad = (1)i \qquad = (1)(-1)$$
$$ = -i \qquad = 1 \qquad = i \qquad = -1 \qquad \blacksquare$$

In Illustration 7 we see that we obtain the results i, $-i$, 1, and -1. By noting that $i^4 = 1$, we can find any positive-integer power of i, and it will be one of these four numbers obtained in Illustration 7.

Example 5 Find each of the following.

a) i^9 b) i^{23} c) $\dfrac{1}{i^3}$

Solution

a) $i^9 = i^8 i$ 　　　b) $i^{23} = i^{20} i^2 i$ 　　　c) $\dfrac{1}{i^3} = \dfrac{1}{-i}$

$ = (i^4)^2 i$ 　　　　$\phantom{b) i^{23}} = (i^4)^5(-1)i$ 　　　　$\phantom{c) \dfrac{1}{i^3}} = \dfrac{1 \cdot i}{-i^2}$

$ = (1)^2 i$ 　　　　$\phantom{b) i^{23}} = (1)^5(-1)i$ 　　　　$\phantom{c) \dfrac{1}{i^3}} = \dfrac{i}{-(-1)}$

$ = i$ 　　　　　$\phantom{b) i^{23}} = -i$ 　　　　　$\phantom{c) \dfrac{1}{i^3}} = i$ 　　$\blacksquare$

Illustration 8

$$\sqrt{-4}\sqrt{-25} = (i\sqrt{4})(i\sqrt{25})$$
$$= (2i)(5i)$$
$$= 10i^2$$
$$= -10 \qquad \blacksquare$$

Observe in Illustration 8 that, before multiplying, we expressed $\sqrt{-4}$ and $\sqrt{-25}$ as $i\sqrt{4}$ and $i\sqrt{25}$, respectively. We did not apply Theorem 1(i) of Section 1.7, which states that if a, $b \in R$, then

$$\sqrt[n]{a}\sqrt[n]{b} = \sqrt[n]{ab}$$

because the theorem further states that $a \geq 0$ and $b \geq 0$ when n is even. As a matter of fact the equality is not valid if $a < 0$ and $b < 0$ when n is even. If, in the first step of Illustration 8, we use the equality, we obtain

$$\sqrt{(-4)(-25)} = \sqrt{100}$$
$$= 10$$

which is an incorrect result for the product $\sqrt{-4}\sqrt{-25}$. To avoid making such an error, you should replace the symbol $\sqrt{-p}$, when $p > 0$, by $i\sqrt{p}$ before performing any multiplication or division.

Example **6** Perform the indicated operations and express the result in the form $a + bi$.

a) $\sqrt{-5}(\sqrt{15} - \sqrt{-5})$ b) $(2 - \sqrt{-9}) \div (2 + \sqrt{-9})$

Solution

a) $\sqrt{-5}(\sqrt{15} - \sqrt{-5}) = i\sqrt{5}(\sqrt{3 \cdot 5} - i\sqrt{5})$

$$= i\sqrt{3 \cdot 5^2} - i^2\sqrt{5^2}$$
$$= i\sqrt{5^2}\sqrt{3} - (-1)\sqrt{5^2}$$
$$= 5 + 5i\sqrt{3}$$

b) $\dfrac{2 - \sqrt{-9}}{2 + \sqrt{-9}} = \dfrac{(2 - 3i)(2 - 3i)}{(2 + 3i)(2 - 3i)}$

$$= \frac{4 - 6i - 6i + 9i^2}{4 - 9i^2}$$

$$= \frac{4 - 12i + 9(-1)}{4 - 9(-1)}$$

$$= \frac{-5 - 12i}{13}$$

$$= -\frac{5}{13} - \frac{12}{13}i$$

■

EXERCISES **1.8**

In Exercises 1 through 4, write the complex number in the form $a + bi$.

1. (a) 5; (b) $\sqrt{-49}$; (c) $3 + \sqrt{-25}$; (d) $3 - \sqrt{-25}$
2. (a) -4; (b) $\sqrt{-36}$; (c) $-2 + \sqrt{-16}$;
   (d) $-2 - \sqrt{-16}$
3. (a) $8 - 5\sqrt{-1}$; (b) $-8 + 5\sqrt{-1}$;
   (c) $-\sqrt{36} + \sqrt{-36}$; (d) $\frac{1}{3} - \frac{1}{5}\sqrt{-45}$
4. (a) $-4 + \sqrt{-4}$; (b) $48 - \sqrt{-48}$; (c) $2 - \sqrt{-\frac{25}{16}}$;
   (d) $54 + \sqrt{-162}$

In Exercises 5 through 36, perform the indicated operations and express the result in the form $a + bi$.

5. (a) $(5 + 2i) + (7 + i)$; (b) $(3 - 6i) - (2 - 4i)$
6. (a) $(4 - 3i) + (-6 + 8i)$; (b) $(7 + 10i) - (1 - 5i)$
7. (a) $(-9 - 4i) + (3 + 4i)$; (b) $7i - (5 - i)$
8. (a) $(3 + 8i) + (-3 - 6i)$; (b) $9 - (2 - 4i)$
9. $(5 + 2\sqrt{-9}) + (3 + 4\sqrt{-25})$
10. $(4 - 3\sqrt{-16}) + (-1 - \sqrt{-4})$
11. $(-3 - \sqrt{-20}) - (6 - \sqrt{-45})$

12. $(4 - \sqrt{-18}) - (2 - \sqrt{-2})$

13. (a) $\sqrt{-9}\sqrt{-25}$; (b) $\sqrt{-2}\sqrt{-8}$

14. (a) $\sqrt{-4}\sqrt{-16}$; (b) $\sqrt{-5}\sqrt{-75}$

15. $\sqrt{-12}\sqrt{-16}\sqrt{-27}$ 16. $\sqrt{-27}\sqrt{-54}\sqrt{-162}$

17. $\sqrt{-8}(3\sqrt{-9} - \sqrt{-8})$

18. $\sqrt{-18}(\sqrt{-2} - 9\sqrt{-18})$

19. $(2 - 7i)(2 + 7i)$ 20. $(4 - 3i)(-1 + 2i)$

21. $(3 + 2\sqrt{-3})(-2 + 3\sqrt{-3})$

22. $(2 - \sqrt{-2})(2 - 3\sqrt{-2})$

23. $(-3 - 3\sqrt{-3})^2$ 24. $(\sqrt{-3} - \sqrt{-2})^2$

25. $-5 \div i$ 26. $7 \div 3i$

27. $1 \div (2i - 3)$ 28. $-4 \div (6 + i)$

29. $(3 + 2i) \div (2 - i)$ 30. $(2 - 5i) \div 3i$

31. $(3 + 2i) \div 4i$ 32. $(2 - 6i) \div (2i - 3)$

33. $1 \div (3 + \sqrt{-2})$

34. $(\sqrt{-5} - 3)(2\sqrt{-5} + 4)$

35. $1 \div (3 + 2i)^2$ 36. $1 \div (4 - 2i)^2$

In Exercises 37 through 44, simplify the expression.

37. (a) i^{11}; (b) i^{33}; (c) i^{26} 38. (a) i^{22}; (b) i^{37}; (c) i^{47}

39. (a) $\dfrac{1}{i^5}$; (b) $\dfrac{1}{i^{15}}$; (c) $\dfrac{1}{i^6}$ 40. (a) $\dfrac{1}{i^7}$; (b) $\dfrac{1}{i^9}$; (c) $\dfrac{1}{i^{14}}$

41. $(i^4 + i^3 - i^2 + 1)^2$ 42. $(2i + 3i^3 - 4i^5 - i^7)^2$

43. $(2i + 3i^2 + 4i^3 - i^6)^3$

44. $(i - 1)^2 - (-i - 1)^2 + i^4$

In Exercises 45 through 48, find the value of the expression for the indicated value of x.

45. $x^2 - 2x + 3$; $x = 1 - i\sqrt{2}$

46. $x^2 - 2x + 4$; $x = 1 - \sqrt{-3}$

47. $4x^2 + 4x + 3$; $x = \frac{1}{2}(-1 + \sqrt{-2})$

48. $3x^2 - 2x + 2$; $x = \frac{1}{3}(1 - \sqrt{-5})$

REVIEW EXERCISES FOR CHAPTER 1

In Exercises 1 through 3, N is the set of natural numbers, J is the set of integers, Q is the set of rational numbers, H is the set of irrational numbers, and R is the set of real numbers. In Exercises 1 and 2, determine which of these sets and $\varnothing$ is equal to the given set. In Exercise 3, list the elements of the given set where

$$S = \{-4, \sqrt{2}, 15, \tfrac{3}{4}, 7\pi, -5, 0, -\tfrac{1}{3}, 2, -\sqrt{3}, \tfrac{1}{2}\pi\}$$

1. (a) $N \cup Q$; (b) $Q \cap H$; (c) $J \cap N$; (d) $Q \cup J$

2. (a) $Q \cap J$; (b) $J \cup N$; (c) $J \cap H$; (d) $Q \cup H$

3. (a) $S \cap N$; (b) $S \cap J$; (c) $S \cap Q$; (d) $S \cap H$

In Exercises 4 and 5, show the set on the real number line and represent the set by interval notation.

4. (a) $\{x \mid x \le 4\}$; (b) $\{x \mid 2 < x < 8\}$;
    (c) $\{x \mid x \le 0\} \cup \{x \mid x > 2\}$;
    (d) $\{x \mid x > 1\} \cap \{x \mid x \le 5\}$

5. (a) $\{x \mid x > 5\}$; (b) $\{x \mid -2 < x \le 3\}$;
    (c) $\{x \mid x \le -4\} \cup \{x \mid x \ge 4\}$;
    (d) $\{x \mid x > -1\} \cap \{x \mid x < 6\}$

In Exercises 6 and 7, write the number without absolute value bars.

6. (a) $|7|$; (b) $|-\sqrt{5}|$; (c) $|2 - \sqrt{3}|$; (d) $|\sqrt{3} - 2|$

7. (a) $|\frac{1}{2}|$; (b) $|-\pi|$; (c) $|\sqrt{10} - 3|$; (d) $|2\sqrt{5} - 5|$

In Exercises 8 through 12, find the numerical value written without exponents.

8. (a) $(2^3)^2$; (b) $(\frac{1}{5})^{-3}$; (c) $(-64)^{2/3}$; (d) $(\frac{1}{16})^{-3/4}$

9. (a) $\left(\dfrac{5}{3^2}\right)^3$; (b) $(27)^{4/3}$; (c) $\left(\dfrac{4}{25}\right)^{-3/2}$; (d) $\dfrac{(-2)^{-3}}{(-7)^{-2}}$

10. (a) $(1^0 + 2^0 + 3^0 + 4^0)^{-5/2}$;
    (b) $2^0 + 2^{-1} + 2^{-2} + 2^{-3}$

11. (a) $\dfrac{2^{-5} + 4^{-2}}{2^{-3} + 4^0}$; (b) $(3^{-3} + 9^{-2})^{-1/2}$;
    (c) $\left(\dfrac{5^{-1}}{5^{-2} + 4 \cdot 5^{-3}}\right)^{-1/2}$

12. (a) $\dfrac{2^{-1} + 3^{-2}}{2^{-2}}$; (b) $(2^2 + 3^2 \cdot 2^{-2})^{1/2}$;
    (c) $\left(\dfrac{3^{-1} + 2^{-2}}{3^{-2} + 2^{-1}}\right)^{-1}$

In Exercises 13 through 20, use laws of exponents to write the expression so that each variable occurs only once and the exponents are positive. Assume that all the variables are positive.

13. (a) $x^8 \cdot x^4$; (b) $(x^8)^4$; (c) $\dfrac{8x^8}{4x^4}$; (d) $x^{1/4} \cdot x^{1/8}$

14. (a) $(y^{-6})^3$; (b) $y^{-6} \cdot y^3$; (c) $\dfrac{3y^3}{-6y^{-6}}$; (d) $y^{-1/6} \cdot y^{1/3}$

15. (a) $(5x^2y^5)(-4x^4y^3)$; (b) $\left(\dfrac{2a^3}{3b^2}\right)^4$; (c) $\dfrac{x^{-2}y^3}{x^4y^{-5}}$

16. (a) $(3x^2)^4(2y^3)^3$; (b) $\left(\dfrac{5t^3}{6r^4s^5u^6}\right)^2$; (c) $(u^{-3}v^{-4}w^2)^{-2}$

17. (a) $(4u^2v^2)(-3uv^2)(-5u^3v)$; (b) $\dfrac{(3x^3yz)^3(-4x^2y^2z^3)^2}{-2x^2y^3z^2}$

18. (a) $(-6xyz)(5x^3y^2z)^2$; (b) $\dfrac{(8r^4s^5t^3)^2}{(-2r^2s^2t)^3(-rs^2t^2)}$

19. (a) $\left(\dfrac{27b^6}{8a^{-9}}\right)^{2/3}$; (b) $\left(\dfrac{t^{4/3}u^{-3}}{t^{-1}u^{1/9}}\right)^{-3}$

20. (a) $\left(\dfrac{16x^{-8}}{81y^4}\right)^{-3/4}$; (b) $\left(\dfrac{a^{1/2}b^{3/8}c^{-1}}{a^{1/4}b^{1/8}c^{1/4}}\right)^4$

In Exercises 21 and 22, simplify the algebraic expression.

21. (a) $3(2x + y) - (5x - y)$;
    (b) $-[3y - 2(y^2 + y)] + 3[y - (2y^2 - 5y)]$
22. (a) $2(x^2 + x - 1) - 4(6 - x - 2x^2)$;
    (b) $-(3s + 8t) + 5[4s - 2(t + s)]$

In Exercises 23 through 26, (a) add the two polynomials and (b) subtract the second polynomial from the first.

23. $4x^3 + 7x^2 - 9x + 3$; $2x^3 - 5x^2 + 11x - 8$
24. $7y^4 - 8y^3 - 10y^2 + y - 4$; $9y^4 - 2y^3 + 3y^2 - 4y + 5$
25. $5y^4 - 8y^3 + 7y - 2$; $6y^4 + 5y^3 - 3y^2 + 9$
26. $6x^5 - x^4 + 8x^2 - 10x + 3$; $2x^5 + x^4 - 3x^3 + 11x^2 - 6$

In Exercises 27 through 30, find the product.

27. (a) $-2y^3(5y^2 - 3y + 8)$; (b) $(3x - 5)(7x + 4)$
28. (a) $5a^2b^3(3a^2 - ab + 4b^2)$; (b) $(2x + 5)^2$
29. (a) $(9t - 4s)(9t + 4s)$; (b) $(x^2 + 7x - 2)(3x^2 - x + 5)$
30. (a) $(6 - 5x)(3 + 7x)$; (b) $(4x + 5y)(6x^2 - 3xy + 2y^2)$

In Exercises 31 through 36, find the quotient. None of the divisors is zero.

31. (a) $\dfrac{45x^4 - 18x^2}{9x^2}$; (b) $\dfrac{y^2 - 2y - 35}{y - 7}$

32. (a) $\dfrac{-9x^4y^3 + 24x^2y^4 - 12x^3y^3}{6x^2y^2}$; (b) $\dfrac{x^2 + 3x - 40}{x + 8}$

33. (a) $\dfrac{w^3 - 3w + 18}{w + 3}$; (b) $\dfrac{10x^2 - 7xy - 15y^2}{2x - 3y}$

34. (a) $\dfrac{2x^3 - 9x^2 + 9x - 7}{2x - 7}$;
    (b) $\dfrac{2y^4 - 8y^3 + 17y^2 + 50y + 14}{y^2 + 7}$

35. $\dfrac{a^4 + b^4}{a + b}$ 36. $\dfrac{x^6 - y^6}{x - y}$

In Exercises 37 through 48, factor the polynomial completely.

37. (a) $x^2 - 16x + 63$; (b) $3w^2 + 13w + 14$
38. (a) $x^2 + 8x - 48$; (b) $7t^2 - 19t + 10$
39. (a) $x^2 - 64$; (b) $x^3 - 64$
40. (a) $x^2 - 81$; (b) $x^3 - 27$
41. (a) $x^4 - 64$; (b) $x^3 + 64$
42. (a) $x^4 - 81$; (b) $x^3 + 27$
43. (a) $6x^2 - 11xy - 35y^2$; (b) $9a^2c - 36b^2c$
44. (a) $25r^2 - 60rs + 36s^2$; (b) $32x^2 - 50y^2$
45. (a) $2x^2 + 3x - 2xy - 3y$; (b) $ab - cd + bc - ad$
46. (a) $4ac - 4bc - a + b$; (b) $3x - 4y + 4y^2 - 3xy$
47. (a) $x^4y^6 - 81y^2w^6$; (b) $x^8 - 2x^4 + 1$
48. (a) $81w^2 - 18w + 1 - y^2$; (b) $x^3 + x^2 - y^3 - y^2$

In Exercises 49 through 58, perform the indicated operations and express the result in lowest terms.

49. (a) $\dfrac{42a^2b}{5c} \cdot \dfrac{10c^2}{36ab^2}$; (b) $\dfrac{3y^2 - y - 4}{y^2 + 6y + 8} \cdot \dfrac{y^2 - 16}{3y^2 - 4y}$

50. (a) $\dfrac{11y^3}{5x^4} \cdot \dfrac{10x^2}{33y^4}$; (b) $\dfrac{s^3}{s^2 - st} \cdot \dfrac{t - s}{st}$

51. (a) $\dfrac{20r^2t}{27s^2u^2} \div \dfrac{25t^3u}{36rs^2}$; (b) $\dfrac{x^2 + x}{y} \div \dfrac{x^3 - x}{xy - y}$

52. (a) $\dfrac{15ab^2c}{120ac^3} \div \dfrac{9b^2c}{8a^2}$;

    (b) $\dfrac{x^2 + 3x + 9}{x - 2} \div \dfrac{x^3 - 27}{x^2 - 4}$

53. (a) $\dfrac{5}{2y - 3} + \dfrac{4}{y + 3}$; (b) $\dfrac{9}{x + 3} - \dfrac{9}{3 - x} + 2x$

54. (a) $\dfrac{6}{5x - 2} - \dfrac{3}{4x + 2}$;

    (b) $\dfrac{x}{x - y} + \dfrac{x}{x + y} - \dfrac{y}{x - y}$

55. (a) $\dfrac{1}{x^2 - 9x + 20} - \dfrac{1}{x - 5}$;

    (b) $\dfrac{5x - 1}{3x^2 - 2x - 8} - \dfrac{3x + 2}{2x^2 - 3x - 2}$

56. (a) $\dfrac{1}{x^2 + 15x + 54} - \dfrac{4}{36 - x^2}$;

    (b) $\dfrac{4}{w^2 + 6w + 8} - \dfrac{2}{w^2 + 7w + 12} + \dfrac{w}{w^2 + 5w + 6}$

57. $\left(\dfrac{r^3 - s^3}{r^3 + s^3} \cdot \dfrac{r^2 - rs + s^2}{r^2 + 2rs + s^2}\right) \div \left(1 - \dfrac{2s}{r + s}\right)$

58. $\left(\dfrac{3u^2 + 10uv + 3v^2}{2u^2 + 5uv - 3v^2} \cdot \dfrac{2u - v}{u + 3v}\right) \div \dfrac{6u^2 + 11uv + 3v^2}{4u^2 + 12uv + 9v^2}$

In Exercises 59 through 62, find a simple rational expression in lowest terms equivalent to the given rational expression.

59. $\dfrac{\dfrac{x}{x-1}+1}{\dfrac{3}{x-2}+2}$

60. $\dfrac{\dfrac{x-y}{x+y}-\dfrac{x+y}{x-y}}{\dfrac{x-y}{x+y}+\dfrac{x+y}{x-y}}$

61. $\dfrac{\dfrac{1}{2x+2h+1}-\dfrac{1}{2x+1}}{h}$

62. $\dfrac{\dfrac{x+h+1}{x+h-2}-\dfrac{x+1}{x-2}}{h}$

In Exercises 63 and 64, write the expression as a simple rational expression with only positive exponents.

63. $(x+y)^{-1}(x^{-2}-y^{-2})$

64. $\dfrac{25a^2b^{-4}-16a^{-2}}{5ab^{-2}+4a^{-1}}$

In Exercises 65 and 66, use laws of exponents to simplify the expression. Each variable can be any real number.

65. (a) $(4x^2y^4z^6)^{1/2}$; (b) $[(y+1)^4(x^2+4)^4]^{1/4}$
66. (a) $(81r^4s^8t^{12})^{1/4}$; (b) $[(u^4+81)^2(v-1)^2]^{1/2}$

In Exercises 67 and 68, find the numerical value written without radicals or exponents.

67. (a) $\sqrt{\dfrac{81}{16}}$; (b) $\sqrt[3]{-5}\sqrt[3]{-25}$; (c) $\dfrac{\sqrt[4]{729}}{\sqrt[4]{9}}$

68. (a) $\sqrt[3]{-\dfrac{216}{125}}$; (b) $\sqrt[4]{24}\sqrt[4]{54}$; (c) $\dfrac{\sqrt{5}}{\sqrt{80}}$

In Exercises 69 and 70, simplify the radical. All the variables represent positive numbers.

69. (a) $\sqrt{75y^2}$; (b) $\sqrt[3]{192x^4y^6}$; (c) $\sqrt[9]{125x^3y^3}$
70. (a) $\sqrt[3]{48x^6}$; (b) $\sqrt{98a^2b^3c^4}$; (c) $\sqrt[8]{16x^4y^4}$

In Exercises 71 through 74, rationalize the denominator. All the radicands and all the variables represent positive numbers.

71. (a) $\sqrt{\dfrac{2}{11}}$; (b) $\sqrt[3]{\dfrac{5}{4s^2t}}$; (c) $\dfrac{6xy}{\sqrt[4]{8x^3y^2}}$

72. (a) $\sqrt[3]{\dfrac{7}{9}}$; (b) $x\sqrt[4]{\dfrac{2y}{9x^3}}$; (c) $\dfrac{6a^2b}{\sqrt{12a^3b}}$

73. (a) $\dfrac{3+\sqrt{6}}{2+\sqrt{6}}$; (b) $\dfrac{1-\sqrt{x}}{1+\sqrt{x}}$

74. (a) $\dfrac{\sqrt{7}-3\sqrt{5}}{2\sqrt{5}-\sqrt{7}}$; (b) $\dfrac{1}{\sqrt{x}+\sqrt{x-1}}$

In Exercises 75 and 76, rationalize the numerator. All the radicands and all the variables represent positive numbers.

75. (a) $\dfrac{\sqrt{x}-4}{x-16}$; (b) $\dfrac{\sqrt{x+9}-3}{x}$

76. (a) $\dfrac{\sqrt{x+2}-\sqrt{2}}{x}$; (b) $\dfrac{\sqrt{2(x+h)-1}-\sqrt{2x-1}}{h}$

In Exercises 77 through 82, perform the indicated operations and express the result in the form $a+bi$.

77. (a) $(8+3i)+(10-2i)$; (b) $(\frac{1}{2}-i)-(\frac{1}{4}-\frac{1}{3}i)$
78. (a) $(11-2i)-(-5+6i)$; (b) $(3+\frac{2}{3}i)+(-1-i)$
79. (a) $\sqrt{-9}\sqrt{-49}$; (b) $(-4+2i)(-3+i)$
80. (a) $\sqrt{-8}\sqrt{-24}\sqrt{-48}$; (b) $(-5-i)^2$
81. $(5-2i)\div(-4-3i)$
82. $i\div(-6-i)$

In Exercises 83 and 84, simplify the expression.

83. (a) i^9; (b) i^{23}; (c) i^{-10}
84. (a) i^{19}; (b) i^{66}; (c) i^{-7}

In Exercises 85 and 86, in parts (a) and (b) write the number in scientific notation, and in parts (c) and (d) write the number in standard form.

85. (a) 452.3; (b) 0.003710; (c) $(8.620)10^3$; (d) $(5.09)10^{-4}$
86. (a) 3642.5; (b) 0.0973; (c) $(1.8040)10^4$; (d) $(2.614)10^{-5}$
87. Use a calculator and scientific notation to perform the indicated operations. Assume four significant digits for each number.
(a) $(425,000)^2\sqrt[3]{0.001763}$;
(b) $\dfrac{\sqrt[4]{51620}(0.07324)^5}{\sqrt[3]{0.009208}(32,500)^2}$

CHAPTER

2

Equations and Inequalities

Introduction

Many applications of mathematics require the solution of equations or inequalities. In this chapter we are concerned with equations and inequalities that involve one variable. Those containing more than one variable are dealt with in later chapters.

We discuss linear (first-degree) equations in Section 2.1 and apply them to word problems in Section 2.2. Quadratic (second-degree) equations are treated in Section 2.3. The other equations in one variable, discussed in Section 2.4, are equations involving radicals, equations quadratic in form, and particular polynomial equations of degree greater than 2.

We devote the last three sections to inequalities. First-degree inequalities are introduced in Section 2.5, and polynomial and rational inequalities are presented in Section 2.6. Section 2.7 pertains to both equations and inequalities involving absolute value.

2.1 Equations

An **algebraic equation** in the variable x is a statement that two algebraic expressions in x are equal. A variable in an equation is sometimes called an **unknown.** The **domain** of a variable in an equation is the set of numbers for which the algebraic expressions in the equation are defined.

Illustration 1

In the following algebraic equations let x be a real number:

$$x - 7 = 0 \tag{1}$$
$$x^2 + 12 = 7x \tag{2}$$
$$x + 5 = 5 + x \tag{3}$$
$$x + 2 = x + 3 \tag{4}$$
$$\frac{x}{x} = 1 \tag{5}$$
$$\frac{3}{x + 4} = \frac{2}{3x - 2} \tag{6}$$

For Equations (1) through (4), the domain is R. Because the left side of (5) is not defined if x is 0, the domain is the set of all real numbers except zero. The left side of (6) is not defined if x is -4 and the right side is not defined if x is $\frac{2}{3}$; therefore, the domain is the set of all real numbers except -4 and $\frac{2}{3}$. ∎

When the variable in an equation is replaced by a specific number, the resulting statement may be either true or false. If it is true, then that number is called a **solution** (or **root**) of the equation. The set of all solutions is called the **solution set** of the equation. A number that is a solution is said to **satisfy the equation.**

Illustration 2

a) In Equation (1), if x is replaced by 7, the resulting statement is true, but if x is replaced by a number other than 7, the resulting statement is false. Thus the only solution is 7, and the solution set is $\{7\}$.

b) In Equation (2), if x is replaced by either 3 or 4, we get a true statement, and if x is replaced by a number other than 3 or 4, we get a false statement. Therefore, the equation has two solutions, 3 and 4, and the solution set is $\{3, 4\}$.

c) In Equation (3), if any real number is substituted for x, we get a true statement. Hence the solution set is R.

d) In Equation (4), if any real number is substituted for x, we get a false statement. Thus the solution set of Equation (4) is $\varnothing$.

e) In Equation (5), if any number other than 0 is substituted for x, we get a true statement. Therefore the solution set is $\{x \,|\, x \in R, \ x \neq 0\}$.

f) The solution set of Equation (6) is not apparent by inspection; however, we show how to find it in Example 2.   ■

If the solution set of any equation in one variable is the same as the domain of the variable, the equation is called an **identity.** Equations (3) and (5) are identities. If there is at least one number in the domain of the variable that is not in the solution set, we have a **conditional equation.** Equations (1), (2), (4), and (6) are conditional equations.

An important type of equation is the polynomial equation in one variable, which can be written in the form $P = 0$, where P is a polynomial in one variable. The degree of the polynomial is the degree of the equation. Particular examples of polynomial equations in one variable are

$$
\begin{aligned}
7x - 21 &= 0 && \text{(first degree)} \\
2y^2 - 3y - 5 &= 0 && \text{(second degree)} \\
4z^3 - 8z^2 - z + 2 &= 0 && \text{(third degree)} \\
9w^4 - 13w^2 + 4 &= 0 && \text{(fourth degree)}
\end{aligned}
$$

Determining the solution set of a conditional equation is called **solving** an equation. As shown by Equation (1) in Illustration 2, it is sometimes possible to solve an equation by inspection. In general, however, we use the concept of **equivalent equations,** which are equations having the same solution set. For example, the following equations are equivalent:

$$
7x - 21 = 0 \qquad 7x = 21 \qquad x = 3
$$

We can often solve an equation by replacing it by a succession of equivalent equations, each in some way simpler than the preceding one, so that we eventually obtain an equation for which the solution set is apparent. We can apply the properties of real numbers to replace an equation by an equivalent one. For instance, the solution set of an equation is not changed if the same algebraic expression is added to or subtracted from both sides of an equation. Furthermore, the solution set is not changed if both sides are multiplied or divided by the same nonzero algebraic expression. The following illustration demonstrates the procedure.

Illustration 3

To solve the equation

$$5x - 5 = 2x + 7$$

we first add the algebraic expression $5 - 2x$ to both sides. In this way we obtain an equivalent equation whose left side contains only terms involving x and whose right side contains only constant terms. We have then the following equivalent equations, each one simpler than the one preceding it:

$$5x - 5 + (5 - 2x) = 2x + 7 + (5 - 2x)$$
$$5x - 2x = 7 + 5$$
$$3x = 12$$

We now divide both sides of the equation by 3 and obtain the equivalent equation

$$\frac{3x}{3} = \frac{12}{3}$$
$$x = 4$$

Because the solution of this last equation is 4, we conclude that the solution set of the original equation is {4}. ■

It is advisable to check your solution of an equation in case you have made an error in algebra or arithmetic. A solution is checked by substituting it into the original equation as shown in the following illustration.

Illustration 4

To check the solution of Illustration 3, we replace x by 4 on each side of the original equation.

$$\text{Does } 5(4) - 5 = 2(4) + 7?$$
$$5(4) - 5 = 20 - 5 \qquad 2(4) + 7 = 8 + 7$$
$$= 15 \qquad\qquad = 15$$

Therefore the solution checks. ■

Example 1 Find the solution set of the equation

$$1 - 3(2x - 4) = 4(6 - x) - 8$$

Solution We simplify the algebraic expression on each side and obtain a succession of equivalent equations.

$$1 - 3(2x - 4) = 4(6 - x) - 8$$
$$1 - 6x + 12 = 24 - 4x - 8$$
$$-6x + 13 = -4x + 16$$
$$-6x + 13 + (4x - 13) = -4x + 16 + (4x - 13)$$
$$-2x = 3$$
$$\frac{-2x}{-2} = \frac{3}{-2}$$
$$x = -\frac{3}{2}$$

Thus the solution set is $\{-\frac{3}{2}\}$.

We now check the solution by replacing x by $-\frac{3}{2}$ on each side of the original equation.

$$\text{Does } 1 - 3[2(-\tfrac{3}{2}) - 4] = 4[6 - (-\tfrac{3}{2})] - 8?$$

$$
\begin{aligned}
1 - 3[2(-\tfrac{3}{2}) - 4] &= 1 - 3[-3 - 4] & 4[6 - (-\tfrac{3}{2})] - 8 &= 4(6 + \tfrac{3}{2}) - 8 \\
&= 1 - 3(-7) & &= 4(\tfrac{15}{2}) - 8 \\
&= 1 + 21 & &= 30 - 8 \\
&= 22 & &= 22
\end{aligned}
$$

Thus the solution checks. ∎

The equations in Illustration 3 and Example 1 are first-degree polynomial equations in one variable. Such equations are called *linear equations* because, as you will learn in Section 3.3, their graphs are lines.

Definition

> **Linear Equation**
> An equation of the form
> $$ax + b = 0$$
> where a and b are real numbers and $a \neq 0$, or any equation equivalent to an equation of this form, is called a **linear equation.**

To solve the linear equation $ax + b = 0$ for x, we subtract b from both sides and then divide both sides by a, which we can do because $a \neq 0$. We then have the following equivalent equations:

$$ax + b = 0$$
$$ax + b - b = 0 - b$$
$$ax = -b$$
$$\frac{ax}{a} = \frac{-b}{a}$$
$$x = -\frac{b}{a}$$

We have proved the following theorem.

Theorem 1

> The linear equation $ax + b = 0$ has exactly one solution, $-\dfrac{b}{a}$.

To solve an equation involving rational expressions, we multiply both sides of the equation by the LCD of the fractions, as shown in the next example.

Example **2** Find the solution set of the equation

$$\frac{3}{x + 4} = \frac{2}{3x - 2}$$

Solution Observe that when x is -4 or $\frac{2}{3}$ we obtain 0 in the denominator of one of the fractions in the given equation. Therefore, -4 and $\frac{2}{3}$ are not in the domain of variable x.

The LCD is $(x + 4)(3x - 2)$. Multiplying both sides of the equation by the LCD, we get

$$(x + 4)(3x - 2)\,\frac{3}{x + 4} = (x + 4)(3x - 2)\,\frac{2}{3x - 2}$$
$$3(3x - 2) = 2(x + 4)$$
$$9x - 6 = 2x + 8$$
$$9x - 6 + (6 - 2x) = 2x + 8 + (6 - 2x)$$
$$9x - 2x = 8 + 6$$
$$7x = 14$$
$$x = 2$$

Therefore the solution set is $\{2\}$.
We check the solution.

$$\text{Does } \frac{3}{2 + 4} = \frac{2}{3(2) - 2}?$$

$$\frac{3}{2 + 4} = \frac{3}{6} \qquad \frac{2}{3(2) - 2} = \frac{2}{4}$$
$$= \frac{1}{2} \qquad\qquad = \frac{1}{2}$$

Thus the solution checks. ∎

It is possible that when multiplying both sides of an equation by the LCD the resulting equation will not be equivalent to the given one. When this happens you can obtain as a possible solution a number that does not satisfy the original equation because it is not in the domain of the variable. This situation occurs in the next example.

Example   3   Find the solution set of the equation

$$\frac{1}{2x + 5} - \frac{4}{2x - 1} = \frac{4x + 4}{4x^2 + 8x - 5}$$

Solution   We factor the denominator on the right side, and we have

$$\frac{1}{2x + 5} - \frac{4}{2x - 1} = \frac{4x + 4}{(2x + 5)(2x - 1)}$$

Because when x is $-\frac{5}{2}$ or $\frac{1}{2}$ we obtain 0 in one of the denominators, these numbers are not in the domain of x. The LCD is $(2x + 5)(2x - 1)$; we multiply each side of the equation by the LCD and obtain

$$2x - 1 - 4(2x + 5) = 4x + 4$$
$$2x - 1 - 8x - 20 = 4x + 4$$
$$-6x - 21 = 4x + 4$$
$$-10x = 25$$
$$x = -\tfrac{5}{2}$$

Thus the only possible solution of the given equation is $-\frac{5}{2}$; but $-\frac{5}{2}$ is not in the domain of x. Therefore $-\frac{5}{2}$ is not a solution. Hence the solution set is $\varnothing$.   ■

An equation may contain more than one variable or it may contain symbols, such as a and b, representing constants. An equation of this type is sometimes called a **literal equation,** and often we wish to solve for one of the variables in terms of the other variables or symbols. The method of solution consists of treating the variable for which we are solving as the unknown and the other variables and symbols as known.

Illustration   5

If F degrees is the Fahrenheit temperature reading and C degrees is the Celsius temperature reading, then

$$F = \tfrac{9}{5}C + 32$$

To solve this equation for C, we first multiply each side by 5 and obtain

$$5F = 9C + 160$$
$$5F - 160 = 9C$$
$$9C = 5(F - 32)$$
$$C = \tfrac{5}{9}(F - 32)$$

Example 4 If p dollars is invested at the rate of $100r$ percent at simple interest for t years, and A dollars is the amount of the investment at t years, then the formula for determining A is

$$A = p(1 + rt)$$

a) Solve for p.   b) Solve for t.

Solution a) We divide both sides of the given equation by $1 + rt$, and we get

$$p = \frac{A}{1 + rt}$$

b) We first use the distributive law on the right side of the given equation. We have

$$A = p + prt$$
$$A - p = prt$$
$$t = \frac{A - p}{pr}$$

∎

EXERCISES 2.1

In Exercises 1 through 24, find the solution set of the equation.

1. $7x + 4 = 25$

2. $7 = y + 10$

3. $4w - 3 = 11 - 3w$

4. $x - 9 = 3x + 3$

5. $2(t - 5) = 3 - (4 + t)$

6. $1 - 3(2x - 4) = 4(6 - x) - 8$

7. $x = x + 1$

8. $x + 3 = 1 + x + 2$

9. $3(4y + 9) = 7(2 - 5y) - 2y$

10. $-2[s - (5 - 4s)] + 4 = -3s$

11. $\dfrac{3x - 2}{3} + \dfrac{x - 3}{2} = \dfrac{5}{6}$

12. $\dfrac{3}{8} + \dfrac{1}{2x} = \dfrac{2}{x}$

13. $\dfrac{5}{2y} - \dfrac{1}{y} = \dfrac{3}{4}$

14. $\dfrac{1}{4 - t} + \dfrac{3}{6 + t} = 0$

15. $\dfrac{2}{3x - 4} = \dfrac{5}{6x - 7}$

16. $\dfrac{3}{x^2 - 9} - \dfrac{7}{x - 3} = -\dfrac{4}{x + 3}$

17. $\dfrac{4}{25w^2 - 1} + \dfrac{3}{5w - 1} = \dfrac{2}{5w - 1}$

18. $\dfrac{3}{x^2 - x - 6} = \dfrac{4}{2x^2 + x - 6}$

19. $\dfrac{5}{x^2 + 6x - 7} = \dfrac{2}{x^2 - 1}$

20. $\dfrac{2}{y + 1} - \dfrac{3}{1 - y} = \dfrac{5}{y}$

21. $\dfrac{t + 3}{t - 2} - \dfrac{t - 3}{t + 2} = \dfrac{5}{t^2 - 4}$

22. $\dfrac{x + 17}{x^2 - 6x + 8} + \dfrac{x - 2}{x - 4} = \dfrac{x - 4}{x - 2}$

23. $\dfrac{3x^2 + 4}{x^3 + 8} = \dfrac{3}{x + 2}$

24. $\dfrac{w}{3w^2 - 8w + 4} = \dfrac{w + 2}{3w^2 + w - 2}$

In Exercises 25 through 30, solve for x or y in terms of the other symbols.

25. $3ax + 6ab = 7ax + 3ab$

26. $\dfrac{a + 3x}{b} = \dfrac{c}{2}$

27. $a(y - a) - 2b(y - 3b) = ab$

28. $5a(5a + x) = 2a(2a - x)$

29. $\dfrac{x + b}{3a - 4b} = \dfrac{x - a}{2a - 5b}$

30. $\dfrac{1}{c - y} + \dfrac{2}{c + y} = \dfrac{1}{y}$

In Exercises 31 through 38, solve for the indicated quantity in the given formula.

31. $A = \frac{1}{2}(a + b)h$; for h

32. $A = \frac{1}{2}(a + b)h$; for b

33. $E = I(R + r)$; for r

34. $A = P\left(1 + \dfrac{r}{n}\right)$; for r

35. $\dfrac{1}{f} = \dfrac{1}{p} + \dfrac{1}{q}$; for p

36. $E = I\left(R + \dfrac{r}{n}\right)$; for n

37. $S = \dfrac{a - rl}{1 - r}$; for r

38. $S = \dfrac{a - rl}{1 - r}$; for l

39. Find the solution set of each of the following equations:

(a) $\dfrac{x}{x} = 1$;

(b) $\dfrac{x}{x} = 0$;

(c) $\dfrac{x}{x} = 2$

40. Find the solution set of each of the following equations:

(a) $\dfrac{x - 2}{x - 2} = 1$;

(b) $\dfrac{x - 2}{x - 2} = 0$;

(c) $\dfrac{x - 2}{x - 2} = 2$

2.2 Applications of Linear Equations

In many applications of algebra, the problems are stated in words. They are called *word problems,* and they give relationships between known numbers and unknown numbers to be determined. In this section we solve word problems by using linear equations. There is no specific method to use. However, here are some steps that give a possible procedure for you to follow. As you read through the examples, refer to these steps to see how they are applied.

1. Read the problem carefully so that you understand it. To gain understanding, it is often helpful to make up a specific example that involves a similar situation in which all the quantities are known. Another aid is to draw a diagram if it is feasible to do so, as shown in Examples 1 and 3.

2. Determine the quantities that are known and those that are unknown. Use a variable to represent one of the unknown quantities in the equation you will obtain. When employing only one equation, as we are in this section, any other unknown quantities should be expressed in terms of this one variable. Because the variable is a number, its definition should indicate this fact. For instance, if the unknown quantity is a length and lengths are measured in feet, then if x is the variable, x should be defined as the number of feet in the length or, equivalently, x feet is the length. If the unknown quantity is time, and time is measured in seconds, then if t is the variable, t should be defined as the number of seconds in the time or, equivalently, t seconds is the time.

3. Write down any numerical facts known about the variable. In many word problems, these facts can be incorporated in a table as indicated in Examples 2 through 6.

4. From the information in step 3, determine two algebraic expressions for the same number and form an equation from them. The use of a table as suggested in step 3 will help you to discover equal algebraic expressions.

5. Solve the equation you obtained in step 4. From the solution set, write a conclusion that answers the questions of the problem. For instance, suppose the problem requires you to find the length of a rectangle, and the variable x is defined so that x feet is the length. Then if the solution set of the equation is {5}, your conclusion should state that the length of the rectangle is 5 ft.

6. It is important to keep in mind that the variable represents a number and the equation involves numbers. The units of measurement do not appear in the equation or its solution set. However, these units are included in the definition of the variable as indicated in step 2. The units also appear in the answers to the questions of the problem as indicated in step 5.

7. Check your results by determining whether the conditions of the word problem are satisfied. This check is to verify the accuracy of the equation obtained in step 4 as well as the accuracy of its solution set.

Example 1 If a rectangle has a length that is 3 cm less than four times its width and its perimeter is 19 cm, what are the dimensions?

Solution We wish to determine the number of centimeters in each dimension of the rectangle. We now define these numbers.

w: the number of centimeters in the width of the rectangle
$4w - 3$: the number of centimeters in the length of the rectangle

Refer to Figure 1. The perimeter of the rectangle is the total distance around it. Therefore the number of centimeters in the perimeter can be represented by either $w + (4w - 3) + w + (4w - 3)$ or 19; thus we have the equation

$$w + (4w - 3) + w + (4w - 3) = 19$$
$$10w - 6 = 19$$
$$10w = 25$$
$$w = \tfrac{5}{2} \qquad 4w - 3 = 4(\tfrac{5}{2}) - 3$$
$$= 7$$

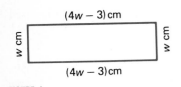

(4w − 3) cm

w cm   w cm

(4w − 3) cm

FIGURE 1

Hence the width of the rectangle is $\tfrac{5}{2}$ cm and the length is 7 cm.

Check The perimeter is $(\tfrac{5}{2} + 7 + \tfrac{5}{2} + 7)$ cm, which equals 19 cm. ∎

The word problem in the next example can be classified as an **investment problem** because it is one involving income from an investment. The income can be in the form of interest, and in that case we use the formula

$$I = P \cdot R$$

(1)

where I dollars is the annual interest earned when P dollars is invested at a rate R per year. The rate is usually given as a percent; thus if the rate is 8 percent, then $R = 0.08$.

Example   2 A man invested part of $15,000 at 12 percent and the remainder at 8 percent. If his annual income from the two investments is $1456, how much does he have invested at each rate?

Solution   Because we want to find the number of dollars invested at each rate we make the following definitions:

$$x:\quad \text{the number of dollars invested at 12 percent}$$
$$15{,}000 - x:\quad \text{the number of dollars invested at 8 percent}$$

We use Formula (1) and get Table 1.

TABLE 1

	Number of Dollars Invested	×	Rate	=	Number of Dollars in Interest
12 percent investment	x		0.12		$0.12x$
8 percent investment	$15{,}000 - x$		0.08		$0.08(15{,}000) - x$

Because the annual income from the two investments is $1456, the sum of the entries in the last column of the table is 1456; therefore we have the equation

$$0.12x + 0.08(15{,}000 - x) = 1456$$
$$0.12x + 1200 - 0.08x = 1456$$
$$0.04x = 256$$
$$x = 6400 \qquad 15{,}000 - x = 15{,}000 - 6400$$
$$= 8600$$

Thus the man has $6400 invested at 12 percent and $8600 at 8 percent.

Check   The number of dollars in the annual interest from the $6400 invested at 12 percent is $0.12(6400) = 768$ and from the $8600 invested at 8 percent is $0.08(8600) = 688$; and $768 + 688 = 1456$.   ∎

A **mixture problem** can involve mixing solutions containing different percents of a substance in order to obtain a solution containing a certain percent of the substance. For instance, in the next example a chemist wishes to obtain 6 liters of a 10 percent acid solution by mixing a 7 percent acid solution with a 12 percent acid solution. Another kind of mixture problem for which the method of solving is similar involves mixing commodities of different values to obtain a combination worth a specific sum of money. Problems of this type appear in Exercises 15 and 16.

Example   3   Determine how many liters of a 7 percent acid solution and how many liters of a 12 percent acid solution should be mixed by a chemist to obtain 6 liters of a 10 percent acid solution.

Solution   We need to determine the number of liters of each solution to be used. Therefore we make the following definitions:

x: the number of liters of the 7 percent acid solution
$6 - x$:   the number of liters of the 12 percent acid solution

The diagram shown in Figure 2 gives a visual interpretation of the problem. The information appearing there is incorporated in Table 2.

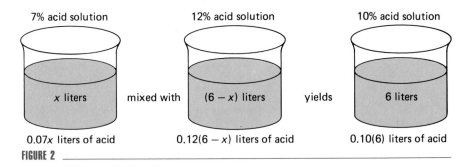

7% acid solution   12% acid solution   10% acid solution

x liters   mixed with $(6 - x)$ liters   yields   6 liters

0.07x liters of acid   0.12(6 − x) liters of acid   0.10(6) liters of acid

FIGURE 2

TABLE 2

	Percent of Acid	×	Number of Liters of Solution	=	Number of Liters of Acid
7% acid solution	7%		x		0.07x
12% acid solution	12%		$6 - x$		0.12(6 − x)
Mixture	10%		6		0.10(6)

From the last column in the table we see that the total number of liters of acid in the mixture can be represented by either 0.10(6) or $0.07x + 0.12(6 - x)$. Thus we have the equation

$$0.07x + 0.12(6 - x) = 0.10(6)$$
$$0.07x + 0.72 - 0.12x = 0.60$$
$$-0.05x = 0.60 - 0.72$$
$$-0.05x = -0.12$$
$$x = \frac{-0.12}{-0.05}$$
$$x = 2.4 \qquad 6 - x = 6 - 2.4$$
$$= 3.6$$

Therefore the chemist should use 2.4 liters of the 7 percent acid solution and 3.6 liters of the 12 percent acid solution.

Check   The 2.4 liters of the 7 percent acid solution gives 0.168 liter of acid, and 3.6 liters of the 12 percent acid solution gives 0.432 liter of acid; and $0.168 + 0.432 = 0.60$.   ∎

If an object travels at a uniform rate of r miles per hour for a time of t hours, then if d miles is the distance traveled,

$$r \cdot t = d \tag{2}$$

A problem involving the use of this formula is called a **uniform-motion problem.** In applying the formula, the units of measurement of the rate, time, and distance must be consistent. In Example 4 the rate is measured in meters per second; thus the time is measured in seconds and the distance is measured in meters. In Example 5, because the distance is measured in kilometers and the time is measured in hours, the rate is measured in kilometers per hour.

Example   4   One runner took 3 min 45 sec to complete a race and another runner required 4 min to run the same race. The rate of the faster runner is 0.4 m/sec more than the rate of the slower runner. Find their rates.

Solution   We choose seconds as the measurement of time. Because we want to determine the rates of the runners, we have the following definitions:

r: the number of meters per second in the rate of the slower runner

$r + 0.4$:   the number of meters per second in the rate of the faster runner

Because each runner is in the same race, they travel equal distances. This fact will be used to obtain an equation. We apply Formula (2) and make Table 3.

TABLE 3

	Number of Meters per Second in Rate	×	Number of Seconds in Time	=	Number of Meters in Distance
Slower runner	r		240		$240r$
Faster runner	$r + 0.4$		225		$225(r + 0.4)$

From the last column in the table, we observe that the number of meters in the distance can be represented by either $240r$ or $225(r + 0.4)$.

Hence we have the equation

$$240r = 225(r + 0.4)$$
$$240r = 225r + 90$$
$$15r = 90$$
$$r = 6 \qquad r + 0.4 = 6 + 0.4$$
$$= 6.4$$

Therefore the rate of the slower runner is 6 m/sec, and the faster runner's rate is 6.4 m/sec.

Check In 240 sec the slower runner travels 1440 m ($6 \cdot 240 = 1440$), and in 225 sec the faster runner travels 1440 m ($6.4 \cdot 225 = 1440$). ∎

If we solve Formula (2) for r, we obtain

$$r = \frac{d}{t} \tag{3}$$

and if we solve it for t, we get

$$t = \frac{d}{r} \tag{4}$$

The next example involves a uniform-motion problem that leads to an equation containing rational expressions.

Example 5 A father and daughter leave home at the same time in separate automobiles. The father drives to his office, a distance of 24 km, and the daughter drives to school, a distance of 28 km. They arrive at their destinations at the same time. What are their average rates, if the father's average rate is 12 km/hr less than his daughter's?

Solution We need to determine the average rates of the daughter and father; therefore, we make the following definitions:

r: the number of kilometers per hour in the daughter's average rate

$r - 12$:   the number of kilometers per hour in the father's average rate

We shall obtain an equation from the fact that the driving times of the daughter and the father are the same. We apply Formula (4) and make Table 4.

TABLE 4

	Number of Kilometers in Distance	÷	Number of Kilometers per Hour in Rate	=	Number of Hours in Time
Daughter	28		r		$\dfrac{28}{r}$
Father	24		$r - 12$		$\dfrac{24}{r - 12}$

The last column in the table indicates that the number of hours in the time can be represented by either $\dfrac{28}{r}$ or $\dfrac{24}{r - 12}$. Therefore, we have the following equation:

$$\frac{28}{r} = \frac{24}{r - 12}$$

We solve the equation by first multiplying on both sides by the LCD:

$$r(r - 12)\frac{28}{r} = r(r - 12)\frac{24}{r - 12}$$
$$(r - 12)28 = r(24)$$
$$(r - 12)7 = r(6)$$
$$7r - 84 = 6r$$
$$7r - 6r = 84$$
$$r = 84 \qquad r - 12 = 84 - 12$$
$$= 72$$

Therefore, the daughter's average rate is 84 km/hr and the father's average rate is 72 km/hr.

Check   The time for the daughter to travel 28 km is 20 min ($28 \div 84 = \frac{1}{3}$, and $\frac{1}{3}$ hour is 20 minutes). The time for the father to travel 24 km is 20 min ($24 \div 72 = \frac{1}{3}$). ■

A **work problem** is one in which a specific job is done in a certain length of time when a uniform rate of work is assumed. For instance, if it takes a man 10 hr to paint a room, then his rate of work is $\frac{1}{10}$ of the room per hour. To solve a work problem we multiply the rate of work by the time to obtain the fractional part of the work completed. In particular, if the painter works for 7 hr, then the fractional part of the work completed is $\frac{7}{10}$.

Example   6   One painter can paint a room in 12 hr and another can paint the same room in 10 hr. How long will it take to paint the room if they work together?

Solution   Since we wish to know how long it takes the two painters to paint the room together, we make the following definition:

> x: the number of hours in the time to paint the room when they are working together

Because the two painters complete the work together (they paint the room), the fractional part of the work done by the first painter plus the fractional part of the work done by the second painter equals 1. We make Table 5 to get expressions for these fractional parts of the work.

TABLE 5

	Fractional Part of Work Done Per Hour	×	Number of Hours Worked	=	Fractional Part of Work Done
First painter	$\dfrac{1}{12}$		x		$\dfrac{x}{12}$
Second painter	$\dfrac{1}{10}$		x		$\dfrac{x}{10}$

We obtain the following equation by setting the sum of the last entries in the table equal to 1:

$$\frac{x}{12} + \frac{x}{10} = 1$$

$$60 \cdot \frac{x}{12} + 60 \cdot \frac{x}{10} = 60 \cdot 1$$

$$5x + 6x = 60$$

$$11x = 60$$

$$x = \tfrac{60}{11}$$

Hence it takes the painters $\tfrac{60}{11}$ hr to paint the room together.

Check   The fractional part of the work done by the first painter is $\tfrac{60}{11} \cdot \tfrac{1}{12} = \tfrac{5}{11}$, and the fractional part of the work done by the second painter is $\tfrac{60}{11} \cdot \tfrac{1}{10} = \tfrac{6}{11}$; and $\tfrac{5}{11} + \tfrac{6}{11} = 1$.   ∎

Other types of work problems having solutions similar to that of Example 6 involve objects doing the work. For instance, in Exercise 30, pipes are working to fill a swimming pool, and in Exercise 31, printing presses are working to print a newspaper.

EXERCISES   2.2

1. The sum of two numbers is 9 and their difference is 6. What are the numbers?
2. Find two numbers whose sum is 7, given that one is 3 times the other.
3. The smaller of two numbers is 9 less than the larger, and their sum is 37. Find the numbers.
4. Find three consecutive even integers whose sum is 138.

5. If the width of a rectangle is 2 cm more than one-half its length and its perimeter is 40 cm, what are the dimensions?

6. The longest side of a triangle is twice as long as the shortest side and 2 cm longer than the third side. If the perimeter of the triangle is 33 cm, what is the length of each side?

7. Admission tickets to a motion picture theater were priced at $4 for adults and $3 for students. If 810 tickets were sold and the total receipts were $2853, how many of each type of ticket were sold?

8. A parking meter slot receives dimes and nickels. When the meter box was emptied there were 148 coins and a total of $10.65. How many dimes and how many nickels were there?

9. The profits of a business are shared among three stockholders. The first stockholder receives twice as much as the second and the second receives three times as much as the third. If the profits for last year were $26,400, how much did each stockholder receive?

10. A woman invested $25,000 in two business ventures. Last year she made a profit of 15 percent from the first venture but lost 5 percent from the second venture. If last year's income from the two investments was equivalent to a return of 8 percent on the entire amount invested, how much had she invested in each venture?

11. An investor wishes to realize a return of 12 percent on a total of two investments. If he has $10,000 invested at 10 percent, how much additional money should be invested at 16 percent?

12. A retail merchant invested $6500 in three kinds of cameras. The profit on the sales of camera A was 25 percent; on the sales of camera B the profit was 12 percent; and there was a loss of 1 percent on the sales of camera C. If the merchant invested an equal amount in cameras A and B, and the overall profit on the total investment was 14 percent, how much was invested in each kind of camera?

13. One alloy contains 80 percent gold (by weight) and another alloy contains 55 percent gold. How many grams of each alloy should be combined to make 40 g of an alloy that contains 70 percent gold?

14. How many pounds of a 35 percent salt solution and how many pounds of a 14 percent salt solution should be combined so that 50 lb of a 20 percent salt solution are obtained?

15. Tea worth $4.10 per pound is to be mixed with tea worth $4.90 per pound. How many pounds of each should be used to obtain 25 lb of a blend worth $4.40 per pound?

16. Perfume to sell for $80 an ounce is to be a blend of perfume selling for $104 an ounce and perfume selling for $50 an ounce. If 270 oz of the blend are desired, how much of each kind of perfume should be used?

17. A radiator contains 25 quarts of a water and antifreeze solution, of which 60 percent (by volume) is antifreeze. How much of this solution should be drained and replaced with water for the new solution to be 40 percent antifreeze?

18. Determine how much water is required to dilute 15 liters of a solution that is 12 percent dye so that a 5 percent dye solution is obtained.

19. How much water must be evaporated from the 15 liters of the 12 percent dye solution in Exercise 18 to obtain a solution that is 20 percent dye? Assume that the total amount of dye is not affected by the process of evaporation.

20. A gardener has 26 lb of a mixture of fertilizer and weed killer. If 1 lb of the mixture is replaced by weed killer the result is a mixture that is 5 percent weed killer. What percent of the original mixture was weed killer?

21. Two airplanes, traveling in opposite directions, leave an airport at the same time. If one plane averages 480 mi/hr and the other averages 520 mi/hr, how long will it take before they are 2000 mi apart?

22. A woman leaves home in an automobile on a business trip. Twenty minutes later, her husband discovers that she left her briefcase behind and decides to overtake her in another car. The husband knows that his wife averages 45 mi/hr and that he averages 60 mi/hr. How long will it be until he overtakes his wife?

23. One hour after a truck has left on an overnight haul, a messenger on a motorcycle leaves from the same starting point to overtake the truck. If the messenger travels at an average rate of 65 mi/hr and overtakes the truck in 4 hr, what is the average rate of the truck?

24. Two friends, living 39 mi apart, leave their homes at the same time on bicycles and travel toward

each other. If one person averages 2 mi/hr more than the other, and they meet in $1\frac{1}{2}$ hr, what is each person's average rate of cycling?

25. A sprinter who averaged 30 ft/sec completed a race 1 sec before another sprinter who averaged 28 ft/sec. What was the distance of the race?

26. In an automobile race, the rate of one car was 120 mi/hr and the rate of another was 105 mi/hr. If the faster car finished the race 20 min before the slower one, what was the distance of the race?

27. One runner ran 1000 yd in the same time that another ran 985 yd. If the average rate of the faster runner was $\frac{1}{3}$ ft/sec more than the average rate of the slower runner, what was the average rate of each runner?

28. A freight train, traveling at an average rate of 30 mi/hr, leaves San Francisco for Los Angeles. Five hours later, a passenger train, traveling at an average rate of 55 mi/hr, leaves San Francisco for Los Angeles. How far are the trains from San Francisco when the passenger train passes the freight train?

29. One secretary can do a particular typing job in 6 hr and another secretary can do the same job in 4 hr. How long will it take if the job is done by the two secretaries working together?

30. A pipe can fill a swimming pool in 10 hr. If a second pipe is open, the two pipes together can fill the pool in 4 hr. How long would it take the second pipe alone to fill the pool?

31. Two printing presses are available to print the daily college newspaper. If only one press is used, it takes the older press twice as long to print an edition as it takes the newer press. If the two presses together can print an edition in 3 hr, how long would it take each press alone to print an edition?

32. Each of two brothers can wash a car in 1 hr; however, their sister can wash the car in 45 min. If all three work together, how long will it take to wash the car?

33. A woman can do a certain job in 10 hr and her younger daughter can do it in 12 hr. After the woman and her younger daughter have been working for 1 hr, they are joined by the older daughter, and the three complete the job in 3 more hours. How long will it take the older daughter to do the job by herself?

34. One pipe can fill a tank in 45 min and another pipe can fill it in 30 min. If these two pipes are open and a third pipe is draining water from the tank, it takes 27 min to fill the tank. How long will it take the third pipe alone to empty a full tank?

35. Every freshman student at a particular college is required to take an English aptitude test. A student who passes the examination enrolls in English Composition, and a student who fails the test must enroll in English Fundamentals. In a freshman class of 1240 students there are more students enrolled in English Fundamentals than in English Composition. However, if 30 more students had passed the test, each course would have the same enrollment. How many students are taking each course?

36. The annual sophomore class picnic is planned by a committee consisting of 17 members. A vote to determine whether the picnic should be held at the beach or in the mountains resulted in a victory for the beach location. However, if two committee members had changed their vote from favoring the beach to favoring the mountains, the mountain site would have won by one vote. How many votes did each picnic location receive?

2.3 Quadratic Equations in One Variable

An equation that can be written as

$$ax^2 + bx + c = 0$$

where a, b, and c are real-number constants and $a \neq 0$, is called a second-degree polynomial equation, or **quadratic equation,** in the variable x. The word "quadratic" comes from quadrate, meaning square or rectangular. When a quadratic equation is written in the above manner (with all the nonzero terms on the left side and 0 on the right side), it is said to be in

standard form. It is convenient to have a standard form for a particular type of equation so that we have a way of referring to its properties when stating formulas and theorems.

Following are examples of quadratic equations in x, written in standard form, with the indicated values of a, b, and c.

$$6x^2 + 7x - 3 = 0 \qquad (a = 6, b = 7, c = -3)$$
$$x^2 - 7 = 0 \qquad (a = 1, b = 0, c = -7)$$
$$3x^2 - 4x = 0 \qquad (a = 3, b = -4, c = 0)$$

Note that it is possible for either b or c to be 0, as in the second and third of the above equations, respectively. However, the restriction that $a \neq 0$ is necessary in order to have a second-degree equation.

One method of finding the solution set of a quadratic equation involves the following theorem. Its proof is based on the properties of the set of real numbers given in Section A.1 of the appendix.

Theorem 1

If r and s are real numbers, then

$$rs = 0 \qquad \text{if and only if} \qquad r = 0 \quad \text{or} \quad s = 0$$

This theorem can be extended to a product of more than two factors. For instance, if $r, s, t, u \in R$, then $rstu = 0$ if and only if at least one of the numbers r, s, t, or u is 0.

Illustration 1

To find the solution set of the equation

$$x^2 + 3x - 10 = 0$$

we factor the left side and obtain

$$(x + 5)(x - 2) = 0$$

By applying Theorem 1, it follows that the equation gives a true statement if and only if

$$x + 5 = 0 \qquad \text{or} \qquad x - 2 = 0$$

The solution of the first of these equations is -5 and the solution of the second is 2. Therefore the solution set of the given equation is $\{-5, 2\}$.

The solutions can be checked by substituting -5 and 2 into the original equation as follows.

$$\text{Does } (-5)^2 + 3(-5) - 10 = 0? \qquad \text{Does } 2^2 + 3(2) - 10 = 0?$$
$$(-5)^2 + 3(-5) - 10 = 25 - 15 - 10 \qquad 2^2 + 3(2) - 10 = 4 + 6 - 10$$
$$= 0 \qquad\qquad\qquad = 0$$

Therefore both solutions check.

∎

The method used in Illustration 1 can be applied to any quadratic equation in standard form for which the left side can be factored. After the left side is factored, each factor is set equal to zero and the solutions of these linear equations are found. Then the solution set of the given quadratic equation is the union of the solution sets of the two linear equations.

Example   1 Find the solution set of each equation.

a) $1 + \dfrac{5x}{6} = \dfrac{2x^2}{3}$ b) $\dfrac{3x}{x + 2} = \dfrac{x + 4}{x}$

Solution We first write the given equation in standard form. Then we factor the left member, set each factor equal to zero, and solve the equations.

a)
$$1 + \frac{5x}{6} = \frac{2x^2}{3}$$
$$(6)(1) + (6)\frac{5x}{6} = (6)\frac{2x^2}{3}$$
$$6 + 5x = 4x^2$$
$$-4x^2 + 5x + 6 = 0$$
$$4x^2 - 5x - 6 = 0$$
$$(4x + 3)(x - 2) = 0$$
$$4x + 3 = 0 \qquad x - 2 = 0$$
$$4x = -3 \qquad\quad x = 2$$
$$x = -\frac{3}{4}$$

The solution set is $\{-\frac{3}{4}, 2\}$.

b)
$$\frac{3x}{x + 2} = \frac{x + 4}{x}$$
$$x(x + 2)\frac{3x}{x + 2} = x(x + 2)\frac{x + 4}{x}$$
$$3x^2 = (x + 2)(x + 4)$$
$$3x^2 = x^2 + 6x + 8$$
$$2x^2 - 6x - 8 = 0$$
$$x^2 - 3x - 4 = 0$$
$$(x - 4)(x + 1) = 0$$
$$x - 4 = 0 \quad x + 1 = 0$$
$$x = 4 \qquad\quad x = -1$$

The solution set is $\{-1, 4\}$. ∎

Suppose we have a quadratic equation of the form

$$x^2 = d$$

that is, there is no first-degree term. Then an equivalent equation is

$$x^2 - d = 0$$

and, factoring the left member, we obtain

$$(x - \sqrt{d})(x + \sqrt{d}) = 0$$

We set each factor equal to zero and solve the equations.

$$x - \sqrt{d} = 0 \qquad x + \sqrt{d} = 0$$
$$x = \sqrt{d} \qquad\qquad x = -\sqrt{d}$$

Therefore, the solution set of the equation $x^2 = d$ is $\{\sqrt{d}, -\sqrt{d}\}$. We can abbreviate this solution set as $\{\pm\sqrt{d}\}$. Thus

$$x^2 = d \qquad \text{if and only if} \qquad x = \pm\sqrt{d}$$

where $x = \pm\sqrt{d}$ stands for the two equations $x = \sqrt{d}$ and $x = -\sqrt{d}$.

You can think of solving the equation $x^2 = d$ as "taking the square root of both sides of the equation" and writing the symbol $\pm$ to indicate both square roots of d.

Example   2   Find the solution set of each of the following equations.

a) $x^2 = 25$ b) $x^2 = 11$ c) $x^2 = -9$

Solution   We take the square root of both sides of the equation.

a) $x^2 = 25$ b) $x^2 = 11$ c) $x^2 = -9$

$\quad x = \pm\sqrt{25}$ $x = \pm\sqrt{11}$ $x = \pm\sqrt{-9}$

$\quad x = \pm 5$ $x = \pm 3i$ ∎

Illustration   2

The solution set of the equation

$$(x - 4)^2 = 5$$

can be found by taking the square root of both sides. We then have

$$x - 4 = \pm\sqrt{5}$$
$$x = 4 \pm \sqrt{5}$$

Thus the two solutions are $4 + \sqrt{5}$ and $4 - \sqrt{5}$, and the solution set can be written as $\{4 \pm \sqrt{5}\}$.   ∎

We can apply the method of Illustration 2 to find the solution set of any quadratic equation. The first step is to write the equation in a form similar to the given equation in the illustration. That is, the left side will be the square of an algebraic expression containing the variable and the right side will be a constant. We use this procedure in the following illustration.

Illustration   3

To find the solution set of the equation

$$x^2 + 6x - 1 = 0$$

we first add 1 to each side and obtain

$$x^2 + 6x = 1 \qquad (1)$$

We now add to each side the square of one-half of the coefficient of x, or 3^2. We obtain

$$x^2 + 6x + 9 = 1 + 9 \qquad (2)$$

The left side is now the square of $x + 3$. Thus we have

$$(x + 3)^2 = 10 \qquad (3)$$

Taking the square root of both sides of the equation, we have

$$x + 3 = \pm\sqrt{10}$$
$$x = -3 \pm \sqrt{10}$$

Therefore the solution set of the given equation is $\{-3 \pm \sqrt{10}\}$. ■

The method used to get Equation (3) in Illustration 3 is called **completing the square.** The important step is obtaining Equation (3) equivalent to Equation (1). Note that in Equation (1) the left side is $x^2 + 6x$ (the coefficient of x^2 is 1 and the coefficient of x is 6). We added the square of one-half of 6 to each side of Equation (1), which gives Equation (2), in which the left side is $(x + 3)^2$. More generally:

> To complete the square of $x^2 + kx$, add the square of one-half the coefficient of x; that is, add $\left(\dfrac{k}{2}\right)^2$.

Observe that this rule for completing the square applies only to a quadratic expression of the form $x^2 + kx$, where the coefficient of the second-degree term is 1. Also note that when $\left(\dfrac{k}{2}\right)^2$ is added to $x^2 + kx$, we have the square of a binomial:

$$x^2 + kx + \left(\frac{k}{2}\right)^2 = \left(x + \frac{k}{2}\right)^2$$

If k is $2a$, this formula becomes

$$x^2 + 2ax + a^2 = (x + a)^2$$

Example 3 Add a term to each of the following algebraic expressions in order to make it the square of a binomial. Also write the resulting expression as the square of a binomial.

a) $x^2 + 12x$ b) $x^2 - 5x$ c) $x^2 + \frac{3}{4}x$

Solution   The coefficient of x^2 in each of the given expressions is 1. Therefore, to complete the square, we add the square of one-half of the coefficient of x.

a) $x^2 + 12x + 6^2 = x^2 + 12x + 36$
$$= (x + 6)^2$$

b) $x^2 - 5x + (-\frac{5}{2})^2 = x^2 - 5x + \frac{25}{4}$
$$= (x - \frac{5}{2})^2$$

c) $x^2 + \frac{3}{4}x + (\frac{3}{8})^2 = x^2 + \frac{3}{4}x + \frac{9}{64}$
$$= (x + \frac{3}{8})^2$$

         ■

Example   **4**   Find the solution set by completing the square.

$$3x^2 - 2x - 6 = 0$$

Solution   We first divide both sides of the equation by 3 so that the coefficient of the second-degree term is 1.

$$x^2 - \tfrac{2}{3}x - 2 = 0$$
$$x^2 - \tfrac{2}{3}x = 2$$

One-half the coefficient of x is $-\frac{1}{3}$ and $(-\frac{1}{3})^2 = \frac{1}{9}$. Therefore, we complete the square of the left side by adding $\frac{1}{9}$ to both sides.

$$x^2 - \tfrac{2}{3}x + \tfrac{1}{9} = 2 + \tfrac{1}{9}$$
$$(x - \tfrac{1}{3})^2 = \tfrac{19}{9}$$
$$x - \frac{1}{3} = \frac{\pm\sqrt{19}}{3}$$
$$x = \frac{1 \pm \sqrt{19}}{3}$$

The solution set is $\left\{ \dfrac{1 \pm \sqrt{19}}{3} \right\}$.

         ■

Consider now the general quadratic equation in standard form,

$$ax^2 + bx + c = 0$$

We solve this equation for x in terms of a, b, and c by completing the square. We first divide both sides of the equation by a (remember that $a \neq 0$), and then we add $-\dfrac{c}{a}$ to both sides and obtain

$$x^2 + \frac{b}{a}x = -\frac{c}{a}$$

We now add the square of one-half the coefficient of x to both sides.

$$x^2 + \frac{b}{a}x + \left(\frac{b}{2a}\right)^2 = -\frac{c}{a} + \left(\frac{b}{2a}\right)^2$$

$$\left(x + \frac{b}{2a}\right)^2 = \frac{b^2}{4a^2} - \frac{c}{a}$$

$$\left(x + \frac{b}{2a}\right)^2 = \frac{b^2 - 4ac}{4a^2}$$

$$x + \frac{b}{2a} = \pm\frac{\sqrt{b^2 - 4ac}}{2a}$$

$$x = -\frac{b}{2a} \pm \frac{\sqrt{b^2 - 4ac}}{2a}$$

$$x = \frac{-b \pm \sqrt{b^2 - 4ac}}{2a}$$

These two values of x are the solutions of the equation $ax^2 + bx + c = 0$. We have obtained the *quadratic formula*, which we now state formally.

Quadratic Formula

If $a \neq 0$, the solutions of the equation $ax^2 + bx + c = 0$ are given by

$$x = \frac{-b \pm \sqrt{b^2 - 4ac}}{2a}.$$

Simply substituting the values of a, b, and c into the quadratic formula allows us to solve any quadratic equation in standard form.

Example   5   Use the quadratic formula to find the solution set of the equation.

a) $6x^2 = 10 + 11x$ b) $x^2 + \frac{5}{3}x + 1 = 0$

Solution   a) We write the given equation in standard form as

$$6x^2 - 11x - 10 = 0$$

Using the quadratic formula, where a is 6, b is -11, and c is -10, we have

$$x = \frac{-b \pm \sqrt{b^2 - 4ac}}{2a}$$

$$= \frac{-(-11) \pm \sqrt{(-11)^2 - 4(6)(-10)}}{2(6)}$$

$$= \frac{11 \pm \sqrt{121 + 240}}{12}$$

$$= \frac{11 \pm \sqrt{361}}{12}$$

$$= \frac{11 \pm 19}{12}$$

$$x = \frac{11 + 19}{12} \qquad\qquad x = \frac{11 - 19}{12}$$

$$= \frac{30}{12} \qquad\qquad\qquad = \frac{-8}{12}$$

$$= \frac{5}{2} \qquad\qquad\qquad = -\frac{2}{3}$$

The solution set is $\{-\frac{2}{3}, \frac{5}{2}\}$.

b) Writing the given equation in standard form, we have

$$3x^2 + 5x + 3 = 0$$

We now use the quadratic formula, where a is 3, b is 5, and c is 3, and obtain

$$x = \frac{-b \pm \sqrt{b^2 - 4ac}}{2a}$$

$$= \frac{-5 \pm \sqrt{(5)^2 - 4(3)(3)}}{2(3)}$$

$$= \frac{-5 \pm \sqrt{25 - 36}}{6}$$

$$= \frac{-5 \pm \sqrt{-11}}{6}$$

$$= \frac{-5 \pm i\sqrt{11}}{6}$$

The solution set is $\left\{ \dfrac{-5 \pm i\sqrt{11}}{6} \right\}$.   ■

To summarize the procedures for solving quadratic equations, we have used the following methods:

1. Solution by factoring   (Illustration 1 and Example 1)
2. Solution by square root   (Example 2 and Illustration 2)
3. Solution by completing the square   (Illustration 3 and Example 4)
4. Solution by the quadratic formula   (Example 5)

Observe that method 4 is a generalization of method 3.

We now show how to obtain information about the **character of the roots** of a quadratic equation without actually solving the equation. Let r and s denote the roots of

$$ax^2 + bx + c = 0$$

where a, b, and c are real numbers, with

$$r = \frac{-b + \sqrt{b^2 - 4ac}}{2a} \quad \text{and} \quad s = \frac{-b - \sqrt{b^2 - 4ac}}{2a}$$

The number represented by $b^2 - 4ac$ is called the **discriminant** of the quadratic equation. The character of the roots can be determined by finding the value of the discriminant.

1. If $b^2 - 4ac = 0$, then

$$r = \frac{-b}{2a} \quad \text{and} \quad s = \frac{-b}{2a}$$

and therefore r and s are equal real numbers. In such a case the number $-\dfrac{b}{2a}$ is called a **double root** (or a **root of multiplicity two**).

2. If $b^2 - 4ac > 0$, then r and s are unequal real numbers.

3. If $b^2 - 4ac < 0$, then r and s are unequal imaginary numbers, each one being the complex conjugate of the other.

We summarize these results.

> $b^2 - 4ac = 0$: roots are real and equal; a double root
> $b^2 - 4ac > 0$: roots are real and unequal
> $b^2 - 4ac < 0$: roots are imaginary and unequal; they are complex conjugates of each other.

Example 6 Determine the character of the roots of each of the following equations.

a) $3x^2 - 2x - 6 = 0$　　b) $4x^2 - 12x + 9 = 0$　　c) $2x^2 + 6x + 7 = 0$

Solution

a) For the given equation, a is 3, b is -2, and c is -6. Thus

$$b^2 - 4ac = (-2)^2 - 4(3)(-6)$$
$$= 76$$

The discriminant is positive. Hence the roots are real and unequal.

b) Because a is 4, b is -12, and c is 9,

$$b^2 - 4ac = (-12)^2 - 4(4)(9)$$
$$= 0$$

The discriminant is zero; therefore the roots are equal real numbers. Observe that because $4x^2 - 12x + 9 = (2x - 3)^2$, the equation can be written as

$$(2x - 3)^2 = 0$$
$$(2x - 3)(2x - 3) = 0$$

and each factor in the equation gives the number $\frac{3}{2}$ as a solution.

c) For the given equation, a is 2, b is 6, and c is 7. Hence

$$b^2 - 4ac = 6^2 - 4(2)(7)$$
$$= -20$$

The discriminant is negative; therefore the roots are imaginary and unequal; they are complex conjugates of each other. ∎

Example 7 A park contains a flower garden, 50 m long and 30 m wide, and a path of uniform width around it. If the area of the path is 600 m², what is its width?

Solution Let w meters represent the width of the path. See Figure 1.

The area of the park minus the area of the garden is equal to the area of the path; thus we have the equation

$$(50 + 2w)(30 + 2w) - 50 \cdot 30 = 600$$
$$1500 + 160w + 4w^2 - 1500 = 600$$
$$4w^2 + 160w - 600 = 0$$
$$w^2 + 40w - 150 = 0$$

We solve this equation by using the quadratic formula where a is 1, b is 40 and c is -150.

$$w = \frac{-b \pm \sqrt{b^2 - 4ac}}{2a}$$
$$= \frac{-40 \pm \sqrt{(40)^2 - 4(1)(-150)}}{2(1)}$$
$$= \frac{-40 \pm \sqrt{2200}}{2}$$
$$= \frac{-40 \pm 10\sqrt{22}}{2}$$
$$= -20 \pm 5\sqrt{22}$$

Because w must be a positive number, we reject the negative root. To two

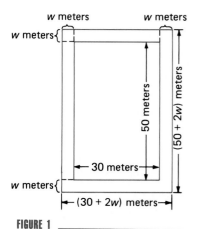

w meters $\qquad$ w meters

w meters

50 meters

$(50 + 2w)$ meters

30 meters

w meters

$(30 + 2w)$ meters

FIGURE 1

decimal places, $\sqrt{22} \approx 4.69$, where the symbol $\approx$ is read "approximately equals." Therefore

$$w \approx -20 + 5(4.69)$$
$$w \approx 3.45$$

Thus the width of the path is 3.45 m.

Check   The park is 56.90 m long and 36.90 m wide; hence the area of the park is 2100 m². The area of the garden is 1500 m² and the area of the path is 600 m²; and $2100 - 1500 = 600$. ∎

EXERCISES 2.3

In Exercises 1 through 14, find the solution set of the equation.

1. $x^2 = 49$
2. $25x^2 - 16 = 0$
3. $5t^2 - 12 = 0$
4. $3y^2 - 5 = 0$
5. $4x^2 = x$
6. $\frac{1}{6}x^2 + x = 0$
7. $x^2 = 8x - 15$
8. $y^2 - 11y + 28 = 0$
9. $8w^2 + 10w - 3 = 0$
10. $14x^2 - x - 3 = 0$
11. $49x^2 + 84x + 36 = 0$
12. $64y^2 - 80y + 25 = 0$
13. $\frac{3t}{3t + 4} + \frac{2}{5} = \frac{t}{3t - 4}$
14. $\frac{32}{x^2 + 3x + 2} - 3 = \frac{x - 3}{x + 1}$

In Exercises 15 through 18, add a term to the algebraic expression in order to make it a square of a binomial; also write the resulting expression as a square of a binomial.

15. (a) $x^2 + 6x$; (b) $x^2 - 5x$
16. (a) $x^2 - 4x$; (b) $x^2 + 7x$
17. (a) $x^2 - \frac{2}{3}x$; (b) $x^2 + \frac{3}{5}x$
18. (a) $x^2 + \frac{4}{3}x$; (b) $x^2 - \frac{5}{6}x$

In Exercises 19 through 26, find the solution set of the equation by completing the square.

19. $x^2 + 6x + 8 = 0$
20. $x^2 - 5x + 6 = 0$
21. $2x^2 - x - 6 = 0$
22. $4x^2 + 4x - 3 = 0$
23. $x^2 - x - 1 = 0$
24. $3x^2 + x - 1 = 0$
25. $3y^2 + 4y + 2 = 0$
26. $w = 2w^2 + 1$

In Exercises 27 through 36, find the solution set of the equation by using the quadratic formula.

27. $x^2 - 3x - 4 = 0$
28. $x^2 + 2x - 3 = 0$
29. $2x + 2 = x^2$
30. $x^2 + 1 = 6x$
31. $5y^2 - 4y - 2 = 0$
32. $4s^2 - 10s + 5 = 0$
33. $x^2 + \frac{1}{2} = x$
34. $\frac{x^2}{4} + 2 = x$
35. $t^2 - 4t + 7 = 0$
36. $25y^2 - 20y + 7 = 0$

In Exercises 37 through 40, find the discriminant and determine the character of the roots of the quadratic equation; do not solve the equation.

37. (a) $6x^2 - 11x - 10 = 0$; (b) $3x^2 - 4x = 3$
38. (a) $4x^2 + 12x + 9 = 0$; (b) $4t^2 + 2t = 1$
39. (a) $25x^2 - 40x + 16 = 0$; (b) $3y = 2y^2 + 5$
40. (a) $14y^2 = 11y - 15$; (b) $14y^2 + 11y - 15 = 0$

In Exercises 41 through 44, solve the formula for the indicated variable. All the variables represent positive numbers.

41. $V = \frac{1}{3}\pi r^2 h$; for r
42. $A = 2\pi r(r + h)$; for r
43. $F = \frac{kMv^2}{r}$; for v
44. $g = \frac{4\pi^2}{t^2}$; for t

In Exercises 45 through 48, solve for x in terms of the other symbols.

45. $5ax^2 - 3x - 2a = 0$
46. $6dx^2 - 3dx + 5 = 0$
47. $x^2 - 2xy - 4x - 3y^2 = 0$
48. $9x^2 - 6xy + y^2 - 3y = 0$
49. The sum of the reciprocals of two consecutive even integers is $\frac{9}{40}$. What are the integers?
50. Are there two consecutive even integers the sum of whose reciprocals is $\frac{8}{45}$? If your answer is yes, find them. If your answer is no, prove it.
51. It took a faster runner 10 sec longer to run a distance of 1500 ft than it took a slower runner to run a distance of 1000 ft. If the rate of the faster runner was 5 ft/sec more than the rate of the slower runner, what was the rate of each?
52. It takes a boy 15 min longer to mow the lawn than it takes his sister, and if they both work to-

gether it takes them 56 min. How long does it take the boy to mow the lawn by himself?

53. It is desired to have an open bin with a square bottom, rectangular sides, and a height of 3 m. If the material for the bottom costs $5.40 per square meter and the material for the sides costs $2.40 per square meter, what is the volume of the bin that can be constructed for $63 worth of material?

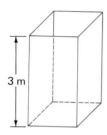

54. A page of print containing 144 cm² of printed region has a margin of $4\frac{1}{2}$ cm at the top and bottom and a margin of 2 cm at the sides. What are the dimensions of the page if the width across the page is four-ninths of the length?

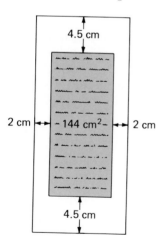

55. A park in the shape of a rectangle has dimensions 60 m by 100 m. If the park contains a rectangular garden enclosed by a concrete terrace, how wide is the terrace if the area of the garden is one-half the area of the park?

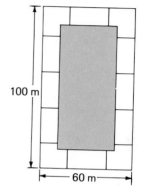

56. What is the width of a strip that must be plowed around a rectangular field 100 m long by 60 m wide so that the field will be two-thirds plowed?

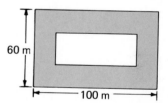

57. The standard form of an equation of a parabola having a vertical axis is $y = ax^2 + bx + c$. Solve for x in terms of y, a, b, and c.

58. If a regular polygon of ten sides is inscribed in a circle of radius r units, then, if s units is the length of a side,

$$\frac{r}{s} = \frac{s}{r - s}$$

Solve this formula for s in terms of r.

2.4   Other Equations in One Variable

If an algebraic equation contains radicals or rational exponents, we can solve it by raising both sides of the equation to the same integer power. When we do this we must apply the following theorem.

Theorem 1

> If E and F are algebraic expressions, then
> $$E = F$$
> is an algebraic equation, and its solution set is a subset of the solution set of the equation
> $$E^n = F^n$$
> where n is any positive integer.

The theorem follows immediately from the fact that if a and b are complex numbers and $a = b$, then $a^n = b^n$, where n is any positive integer.

Illustration 1

The equation
$$x = 5$$

has the solution set $\{5\}$. If we square each side we obtain the equation
$$x^2 = 25$$

which has the solution set $\{\pm 5\}$. The solution set of the first equation is a subset of the solution set of the second. This result agrees with Theorem 1. ∎

Theorem 1 is used to solve an equation having terms involving one or more radicals. The first step is to obtain an equivalent equation having the term involving the most complicated radical on one side and all the other terms on the other side. Then, applying Theorem 1, we raise both sides of the equation to the power corresponding to the index of the radical.

Illustration 2

To solve the equation
$$\sqrt{2x + 5} + x = 5$$

we first add $-x$ to both sides and get the equivalent equation
$$\sqrt{2x + 5} = 5 - x \tag{1}$$

Now, with the radical alone on one side of the equation, we apply Theorem 1 with $n = 2$ and square both sides to obtain

$$(\sqrt{2x + 5})^2 = (5 - x)^2$$
$$2x + 5 = 25 - 10x + x^2 \tag{2}$$
$$x^2 - 12x + 20 = 0$$
$$(x - 10)(x - 2) = 0$$
$$x - 10 = 0 \qquad x - 2 = 0$$
$$x = 10 \qquad \qquad x = 2$$

Thus the solution set of Equation (2) is $\{2, 10\}$.

According to Theorem 1 the solution set of Equation (1) is a subset of $\{2, 10\}$; that is, both of the numbers 2 and 10 may be solutions, only one may be a solution, or neither may be a solution. To determine which case applies, we substitute each number into Equation (1) to see whether the equation is satisfied.

$$\sqrt{2(2) + 5} \overset{?}{=} 5 - 2 \qquad \sqrt{2(10) + 5} \overset{?}{=} 5 - 10$$
$$\sqrt{9} \overset{?}{=} 3 \qquad\qquad \sqrt{25} \overset{?}{=} -5$$
$$3 = 3 \qquad\qquad 5 \neq -5$$

Hence 2 is a solution and 10 is not. Therefore the solution set of Equation (1) is $\{2\}$. ∎

In Illustration 2, the number 10 is called an extraneous solution of Equation (1); it was introduced when both sides were squared. The reason that this extraneous solution was introduced should be apparent after you read Illustration 3.

Illustration 3

If both sides of the equation

$$-\sqrt{2x + 5} = 5 - x \tag{3}$$

are squared, we obtain

$$(-\sqrt{2x + 5})^2 = (5 - x)^2$$
$$2x + 5 = 25 - 10x + x^2$$

which is Equation (2). In Illustration 2 we showed that the solution set of Equation (2) is $\{2, 10\}$. Hence, by Theorem 1, the solution set of Equation (3) is a subset of $\{2, 10\}$. Substituting each of these numbers into Equation (3), we have

$$-\sqrt{2(2) + 5} \overset{?}{=} 5 - 2 \qquad -\sqrt{2(10) + 5} \overset{?}{=} 5 - 10$$
$$-\sqrt{9} \overset{?}{=} 3 \qquad\qquad -\sqrt{25} \overset{?}{=} -5$$
$$-3 \neq 3 \qquad\qquad -5 = -5$$

Hence 10 is a solution and 2 is an extraneous solution. Therefore the solution set of Equation (3) is $\{10\}$. ∎

Observe in Illustration 2 that Equation (1) states that the principal square root of $2x + 5$ is $5 - x$, and in Illustration 3 that Equation (3) states that the negative square root of $2x + 5$ is $5 - x$. Thus, when squaring both sides of either equation, we obtain Equation (2); so in each case an extraneous solution is introduced. This discussion should convince you that when Theorem 1 is applied, all solutions obtained must be checked in the original equation. The check is for possible extraneous solutions, not just for computational errors.

Example 1 Find the solution set of the equation

$$\sqrt{x}\sqrt{x - 8} = 3$$

Solution   We square both sides of the equation.

$$(\sqrt{x}\sqrt{x-8})^2 = 3^2$$
$$x(x-8) = 9$$
$$x^2 - 8x = 9$$
$$x^2 - 8x - 9 = 0$$
$$(x-9)(x+1) = 0$$

$$x - 9 = 0 \qquad x + 1 = 0$$
$$x = 9 \qquad\qquad x = -1$$

Hence the solution set of the given equation is a subset of $\{-1, 9\}$. We substitute these numbers into the given equation.

$$\sqrt{9}\sqrt{9-8} \stackrel{?}{=} 3 \qquad \sqrt{-1}\sqrt{-1-8} \stackrel{?}{=} 3$$
$$3 \cdot 1 \stackrel{?}{=} 3 \qquad\qquad i \cdot 3i \stackrel{?}{=} 3$$
$$3 = 3 \qquad\qquad\qquad 3i^2 \stackrel{?}{=} 3$$
$$3(-1) \stackrel{?}{=} 3$$
$$-3 \neq 3$$

Therefore, 9 is a solution and -1 is an extraneous solution. Thus the solution set of the given equation is $\{9\}$.   ■

If an equation contains more than one radical, it is sometimes necessary to apply Theorem 1 more than once before obtaining an equation free of radicals.

Example **2**   Find the solution set of the equation.

a) $\sqrt{3 - 3x} - \sqrt{3x + 2} = 3$ b) $\sqrt{2x + 3} - \sqrt{x - 2} - 2 = 0$

Solution   In both parts (a) and (b), we first write an equivalent equation in which one radical is isolated on one side. Then, after applying Theorem 1, we obtain another equation involving a radical. So we use the same procedure for the new equation.

a) $\sqrt{3 - 3x} - \sqrt{3x + 2} = 3$
$$\sqrt{3 - 3x} = 3 + \sqrt{3x + 2}$$
$$(\sqrt{3 - 3x})^2 = (3 + \sqrt{3x + 2})^2$$
$$3 - 3x = 9 + 6\sqrt{3x + 2} + 3x + 2$$
$$-6\sqrt{3x + 2} = 6x + 8$$
$$-3\sqrt{3x + 2} = 3x + 4$$
$$9(\sqrt{3x + 2})^2 = (3x + 4)^2$$
$$9(3x + 2) = 9x^2 + 24x + 16$$
$$27x + 18 = 9x^2 + 24x + 16$$
$$-9x^2 + 3x + 2 = 0$$
$$9x^2 - 3x - 2 = 0$$
$$(3x - 2)(3x + 1) = 0$$

$$3x - 2 = 0 \qquad 3x + 1 = 0$$
$$3x = 2 \qquad\quad 3x = -1$$
$$x = \frac{2}{3} \qquad\qquad x = -\frac{1}{3}$$

Therefore the solution set of the given equation is a subset of $\{-\frac{1}{3}, \frac{2}{3}\}$. We substitute these numbers into the given equation.

$$\sqrt{3 - 3(-\tfrac{1}{3})} - \sqrt{3(-\tfrac{1}{3}) + 2} \overset{?}{=} 3 \qquad \sqrt{3 - 3(\tfrac{2}{3})} - \sqrt{3(\tfrac{2}{3}) + 2} \overset{?}{=} 3$$
$$\sqrt{4} - \sqrt{1} \overset{?}{=} 3 \qquad\qquad \sqrt{1} - \sqrt{4} \overset{?}{=} 3$$
$$2 - 1 \overset{?}{=} 3 \qquad\qquad 1 - 2 \overset{?}{=} 3$$
$$1 \neq 3 \qquad\qquad\qquad -1 \neq 3$$

Because each of the numbers is an extraneous solution, the solution set of the given equation is $\varnothing$.

b) $\sqrt{2x + 3} - \sqrt{x - 2} - 2 = 0$
$$\sqrt{2x + 3} = \sqrt{x - 2} + 2$$
$$(\sqrt{2x + 3})^2 = (\sqrt{x - 2} + 2)^2$$
$$2x + 3 = x - 2 + 4\sqrt{x - 2} + 4$$
$$x + 1 = 4\sqrt{x - 2}$$
$$(x + 1)^2 = 16(\sqrt{x - 2})^2$$
$$x^2 + 2x + 1 = 16(x - 2)$$
$$x^2 + 2x + 1 = 16x - 32$$
$$x^2 - 14x + 33 = 0$$
$$(x - 3)(x - 11) = 0$$
$$x - 3 = 0 \qquad x - 11 = 0$$
$$x = 3 \qquad\quad x = 11$$

Therefore the solution set of the given equation is a subset of $\{3, 11\}$. We substitute these numbers into the given equation.

$$\sqrt{2(3) + 3} - \sqrt{3 - 2} - 2 \overset{?}{=} 0 \qquad \sqrt{2(11) + 3} - \sqrt{11 - 2} - 2 \overset{?}{=} 0$$
$$\sqrt{9} - \sqrt{1} - 2 \overset{?}{=} 0 \qquad\qquad \sqrt{25} - \sqrt{9} - 2 \overset{?}{=} 0$$
$$3 - 1 - 2 \overset{?}{=} 0 \qquad\qquad\quad 5 - 3 - 2 \overset{?}{=} 0$$
$$0 = 0 \qquad\qquad\qquad\qquad 0 = 0$$

Each of the numbers, 3 and 11, is a solution of the given equation; so its solution set is $\{3, 11\}$. ∎

An equation in a single variable x is said to be **quadratic in form** if it can be written as

$$au^2 + bu + c = 0$$

where $a \neq 0$ and u is an algebraic expression in x.

Illustration 4

The following equations are quadratic in form.

a) $x^4 - 2x^2 - 15 = 0$ is quadratic in x^2 because if $u = x^2$, the equation becomes $u^2 - 2u - 15 = 0$.

b) The equation $2x^{2/3} - 5x^{1/3} - 3 = 0$ is quadratic in $x^{1/3}$ because if $u = x^{1/3}$ the equation becomes $2u^2 - 5u - 3 = 0$.

c) The equation $3\left(4x - \dfrac{1}{x}\right)^2 - 4\left(4x - \dfrac{1}{x}\right) - 15 = 0$ is quadratic in $4x - \dfrac{1}{x}$ because if $u = 4x - \dfrac{1}{x}$ the equation becomes $3u^2 - 4u - 15 = 0$. ■

In the next two examples, we solve the equations of Illustration 4.

Example 3 Find the solution set of the equation.

a) $x^4 - 2x^2 - 15 = 0$ b) $2x^{2/3} - 5x^{1/3} - 3 = 0$

Solution a) Let $u = x^2$. The equation then becomes

$$u^2 - 2u - 15 = 0$$

We solve this equation for u.

$$(u - 5)(u + 3) = 0$$
$$u - 5 = 0 \qquad u + 3 = 0$$
$$u = 5 \qquad u = -3$$

Now we replace u with x^2 and solve the resulting equations:

$$x^2 = 5 \qquad x^2 = -3$$
$$x = \pm\sqrt{5} \qquad x = \pm i\sqrt{3}$$

The solution set of the original equation is therefore $\{\pm\sqrt{5},\ \pm i\sqrt{3}\}$.

b) If $u = x^{1/3}$, we have

$$2u^2 - 5u - 3 = 0$$
$$(2u + 1)(u - 3) = 0$$
$$2u + 1 = 0 \qquad u - 3 = 0$$
$$u = -\tfrac{1}{2} \qquad u = 3$$

Replacing u by $x^{1/3}$, we get

$$x^{1/3} = -\tfrac{1}{2} \qquad x^{1/3} = 3$$
$$(x^{1/3})^3 = (-\tfrac{1}{2})^3 \qquad (x^{1/3})^3 = 3^3$$
$$x = -\tfrac{1}{8} \qquad x = 27$$

The solution set of the original equation is $\{-\tfrac{1}{8},\ 27\}$. ■

Example 4 Find the solution set of the equation

$$3\left(4x - \frac{1}{x}\right)^2 - 4\left(4x - \frac{1}{x}\right) - 15 = 0$$

Solution Let $u = 4x - \dfrac{1}{x}$. The equation then becomes

$$3u^2 - 4u - 15 = 0$$
$$(3u + 5)(u - 3) = 0$$
$$3u + 5 = 0 \qquad u - 3 = 0$$
$$u = -\tfrac{5}{3} \qquad u = 3$$

We replace u with $\left(4x - \dfrac{1}{x}\right)$ and solve for x.

$$4x - \frac{1}{x} = -\frac{5}{3} \qquad\qquad\qquad 4x - \frac{1}{x} = 3$$

$$3x(4x) - 3x\left(\frac{1}{x}\right) = 3x\left(-\frac{5}{3}\right) \qquad\qquad x(4x) - x\left(\frac{1}{x}\right) = x(3)$$

$$12x^2 - 3 = -5x \qquad\qquad\qquad 4x^2 - 1 = 3x$$
$$12x^2 + 5x - 3 = 0 \qquad\qquad\qquad 4x^2 - 3x - 1 = 0$$
$$(4x + 3)(3x - 1) = 0 \qquad\qquad\qquad (4x + 1)(x - 1) = 0$$

$$4x + 3 = 0 \qquad 3x - 1 = 0 \qquad\qquad 4x + 1 = 0 \qquad x - 1 = 0$$

$$x = -\frac{3}{4} \qquad x = \frac{1}{3} \qquad\qquad\qquad x = -\frac{1}{4} \qquad x = 1$$

Therefore the solution set is $\{-\tfrac{3}{4}, -\tfrac{1}{4}, \tfrac{1}{3}, 1\}$. ■

We have solved quadratic equations in standard form by factoring the left side and setting each factor equal to zero. This procedure can also be used to solve some higher-degree equations. For instance, in Example 3(a) we solved the equation

$$x^4 - 2x^2 - 15 = 0$$

by letting $u = x^2$. An alternative method is to factor the left side and get

$$(x^2 - 5)(x^2 + 3) = 0$$

and then set each factor equal to zero to obtain

$$x^2 - 5 = 0 \qquad x^2 + 3 = 0$$

The solutions of these equations give the solution set of the original equation. In the following example we apply the same method to a third-degree equation.

Example 5 Find the solution set of the equation

$$x^3 = 8$$

Solution The given equation is equivalent to

$$x^3 - 8 = 0$$

Recall that $a^3 - b^3 = (a - b)(a^2 + ab + b^2)$. Using this formula where a is x and b is 2, we can factor the left side to obtain

$$(x - 2)(x^2 + 2x + 4) = 0$$

Setting each factor equal to zero and solving the resulting equations, we have

$$x - 2 = 0 \qquad x^2 + 2x + 4 = 0$$

$$x = 2 \qquad\qquad x = \frac{-2 \pm \sqrt{4 - 16}}{2}$$

$$= \frac{-2 \pm \sqrt{-12}}{2}$$

$$= \frac{-2 \pm 2i\sqrt{3}}{2}$$

$$= -1 \pm i\sqrt{3}$$

Therefore the solution set of the original equation is $\{2, -1 \pm i\sqrt{3}\}$. ■

Observe in Example 5 that because $x^3 = 8$, x is a cube root of 8. Therefore, the solutions are cube roots of 8; one is real and two are imaginary.

EXERCISES 2.4

In Exercises 1 through 46, find the solution set of the equation.

1. $\sqrt{x} - 5 = 3$
2. $\sqrt{x} + 4 = 7$
3. $\sqrt{x + 5} = 3$
4. $\sqrt{x - 4} = 7$
5. $\sqrt{2x - 3} = 2$
6. $\sqrt{2x + 5} = 3$
7. $\sqrt{y} + y = 6$
8. $\sqrt{t + 6} = t$
9. $\sqrt{3x + 7} = x + 1$
10. $2\sqrt{x + 4} - 1 = x$
11. $\sqrt{x + 5} - \sqrt{x} = 1$
12. $\sqrt{2t + 1} - 2\sqrt{t - 1} = 0$
13. $\sqrt{3x - 4} + 8 = 0$
14. $\sqrt{5x + 1} + \sqrt{3x + 1} = 0$
15. $\sqrt{5w + 1} - \sqrt{3w - 1} = 0$
16. $\sqrt{2 + 4y} + \sqrt{3 - 4y} = 3$
17. $\sqrt{2x + 11} + 1 = \sqrt{5x + 1}$

18. $7 - \sqrt{8 - x} = \sqrt{2x + 25}$
19. $\sqrt{t}\sqrt{t - 6} + 4 = 0$
20. $\sqrt{x}\sqrt{x - 5} - 4 = 0$
21. $\sqrt{4 - 3x} + \sqrt{3x - 9} = \sqrt{3x - 14}$
22. $\sqrt{2x - 1} + \sqrt{x + 4} = 3\sqrt{x - 1}$
23. $\sqrt{y - 2}\sqrt{y + 3} + 6 = 0$
24. $\sqrt{w + 2}\sqrt{w - 1} + 2 = 0$
25. $\sqrt{\sqrt{5x - 1}} - x = 1$
26. $\sqrt{2x - 1} - \sqrt{x - 7} - 4 = 0$
27. $\sqrt[3]{3x + 1} = x + 1$
28. $\sqrt[3]{x^2 - 1} = x - 1$
29. $x^4 - 5x^2 + 4 = 0$
30. $9x^4 - 8x^2 - 1 = 0$
31. $t^4 - 5t^2 + 6 = 0$
32. $6w^4 - 17w^2 + 12 = 0$
33. $8x^4 - 6x^2 - 5 = 0$
34. $8x^4 + 6x^2 - 9 = 0$
35. $5t^{-2} - 9t^{-1} - 2 = 0$
36. $y^{-4} - 37y^{-2} + 9 = 0$

37. $y^6 - 35y^3 + 216 = 0$
38. $27z^6 - 35z^3 + 8 = 0$
39. $\sqrt[4]{x} + 2\sqrt{x} = 3$
40. $\sqrt{x} - 5\sqrt[4]{x} + 4 = 0$
41. $(x^2 + 2x)^2 - 14(x^2 + 2x) - 15 = 0$
42. $(2x^2 + 7x)^2 - 12(2x^2 + 7x) - 45 = 0$
43. $8\left(x + \dfrac{1}{2x}\right)^2 - 30\left(x + \dfrac{1}{2x}\right) + 27 = 0$
44. $\left(x - \dfrac{5}{x}\right)^2 - 2\left(x - \dfrac{5}{x}\right) = 8$
45. $\sqrt[4]{3x^2 - 3x + 1} = \sqrt{x - 2}$
46. $\sqrt[4]{2x^2 + 5} = \sqrt{x + 2}$
47. Find three cube roots of 1 by solving the equation $x^3 = 1$.
48. Find three cube roots of 27 by solving the equation $x^3 = 27$.
49. Find three cube roots of -8 by solving the equation $x^3 = -8$.
50. Find four fourth roots of 1 by solving the equation $x^4 = 1$.
51. Find four fourth roots of 81 by solving the equation $x^4 = 81$.
52. Find six sixth roots of 64 by solving the equation $x^6 = 64$.

53. If the length of the diagonal of a rectangle is 50 in. and the area is 1200 in^2, what are the dimensions?

54. What are the dimensions in Exercise 53 if the length of the diagonal is 13 in. and the area is 60 in^2?

55. A spherical balloon is being inflated so that its volume is increasing at the rate of $\frac{148}{3}\pi$ ft^3/min. If the radius is now 3 ft, what will the radius be in 1 min? (The formula for the volume of a sphere is $V = \frac{4}{3}\pi r^3$.)

56. The lateral surface area of a right-circular cone of base radius r units and height h units is S square units where $S = \pi r \sqrt{r^2 + h^2}$. Solve this formula for (a) h and (b) r.

2.5   Inequalities

In Section 1.1 we gave an ordering for the set R of real numbers by means of a relation denoted by the symbols $<$ and $>$. We introduced them to define intervals on the real number line. You may wish to review these concepts at this time.

In this section and the next two we are concerned with inequalities involving a variable. Preliminary to our discussion, we present the trichotomy and transitive properties of order.

Trichotomy Property of Order

If a and b are real numbers, exactly one of the following three statements is true:

$$a < b \qquad b < a \qquad a = b$$

The geometric interpretation of the trichotomy property is that either one point, a or b, on the real number line lies to the left of the other, or else they are the same point. Figure 1 shows the three possibilities.

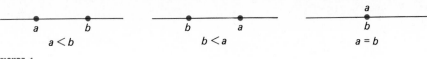

FIGURE 1

Transitive Property of Order

If a, b, and c are real numbers, and

$$\text{if} \quad a < b \quad \text{and} \quad b < c, \quad \text{then} \quad a < c$$

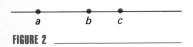

FIGURE 2

The geometric interpretation of the transitive property is shown in Figure 2; if point a is to the left of point b and point b is to the left of point c, then a is to the left of c.

Illustration   1

If $x < 5$ and $5 < y$, then by the transitive property it follows that $x < y$.   ■

The **domain** of a variable in an inequality is the set of real numbers for which the members of the inequality are defined. Examples of linear inequalities having the set R of real numbers as domain are

$$3x - 8 < 7 \qquad \frac{x - 7}{4} \leq x \qquad 2 < 4x + 6 \leq 14$$

An example of a quadratic inequality having domain R is

$$x^2 + 2x > 15$$

The inequality

$$\frac{3x}{x + 2} < 5$$

is rational. Because the left side is not defined when x is -2, its domain is the set of all real numbers except -2.

Any number in the domain for which the inequality is true is a **solution** of the inequality, and the set of all solutions is called the **solution set.** An **absolute inequality** is one that is true for every number in the domain. For instance, if x is a real number,

$$x + 1 < x + 2 \qquad \text{and} \qquad x^2 \geq 0$$

are absolute inequalities. A **conditional inequality** is one for which there is at least one number in the domain that is not in the solution set. To find the solution set of a conditional inequality we proceed in a manner similar to that used to solve an equation; that is, we obtain **equivalent inequalities** (those having the same solution set) until we have one whose solution set is apparent. The following properties are used to get equivalent inequalities.

Properties of $<$

> If a, b, and c are real numbers and
>
> (i) if $a < b$, then $a + c < b + c$　　　(Addition Property)
> (ii) if $a < b$, then $a - c < b - c$　　　(Subtraction Property)
> (iii) if $a < b$ and $c > 0$, then $ac < bc$　　(Multiplication Property)
> (iv) if $a < b$ and $c < 0$, then $ac > bc$　　(Multiplication Property)

These properties can be proved by using the definition of $<$ and the fact that the product of two positive numbers is positive.

Property (iii) states that if both sides of an inequality are multiplied by a positive number, the direction of the inequality remains unchanged, whereas property (iv) states that if both sides are multiplied by a negative number, the direction is reversed.

Illustration 2

a)　If $x < y$, then from property (iii), $4x < 4y$. For instance, because $-5 < 3$, then $4(-5) < 4(3)$ or, equivalently, $-20 < 12$.

b)　Because $-5 < 3$, then if $z < 0$, it follows from property (iv) that $-5z > 3z$. For instance, because $-5 < 3$, then $(-5)(-4) > 3(-4)$ or, equivalently, $20 > -12$.

■

In the following examples we apply properties (i)–(iv) to find the solution set of a linear inequality in one variable. Interval notation is used to represent the solution sets.

Example　1　Find and show on the real number line the solution set of the inequality

$$3x - 8 < 7$$

Solution　The following inequalities are equivalent:

$$3x - 8 < 7$$
$$3x - 8 + 8 < 7 + 8$$
$$3x < 15$$
$$\tfrac{1}{3}(3x) < \tfrac{1}{3}(15)$$
$$x < 5$$

FIGURE 3

Therefore the solution set of the given inequality is $\{x \mid x < 5\}$, which is the interval $(-\infty, 5)$. This interval appears in Figure 3.　■

Example　2　Find and show on the real number line the solution set of the inequality

$$\frac{x - 7}{4} \le x$$

Solution   The following inequalities are equivalent:

$$\frac{x-7}{4} \leq x$$

$$(4)\frac{x-7}{4} \leq (4)x$$

$$x - 7 \leq 4x$$

$$x - 7 - 4x + 7 \leq 4x - 4x + 7$$

$$-3x \leq 7$$

$$-\tfrac{1}{3}(-3x) \geq (-\tfrac{1}{3})7$$

$$x \geq -\tfrac{7}{3}$$

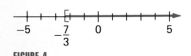

FIGURE 4

Thus the solution set of the given inequality is $\{x \mid x \geq -\tfrac{7}{3}\}$, which is the interval $[-\tfrac{7}{3}, +\infty)$, shown in Figure 4.   ■

Example 3   Find and show on the real number line the solution set of the inequality

$$3 < 4x + 7 \leq 15$$

Solution   A solution of the given inequality must be a solution of both of the inequalities

$$3 < 4x + 7 \qquad \text{and} \qquad 4x + 7 \leq 15$$

We solve each of these inequalities separately.

$$
\begin{array}{ll}
3 < 4x + 7 & 4x + 7 \leq 15 \\
3 - 7 < 4x + 7 - 7 & 4x + 7 - 7 \leq 15 - 7 \\
-4 < 4x & 4x \leq 8 \\
\tfrac{1}{4}(-4) < \tfrac{1}{4}(4x) & \tfrac{1}{4}(4x) \leq \tfrac{1}{4}(8) \\
-1 < x & x \leq 2
\end{array}
$$

A value of x will be a solution of the given inequality if and only if

$$-1 < x \qquad \text{and} \qquad x \leq 2$$

that is, if and only if

$$-1 < x \leq 2$$

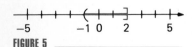

FIGURE 5

Therefore the solution set is the interval $(-1, 2]$, shown in Figure 5.

    The work can be shortened by performing the same computations with the given continued inequality as follows:

$$3 < 4x + 7 \leq 15$$

$$3 - 7 < 4x + 7 - 7 \leq 15 - 7$$

$$-4 < 4x \leq 8$$

$$\frac{-4}{4} < \frac{4x}{4} \leq \frac{8}{4}$$

$$-1 < x \leq 2 \qquad\qquad ■$$

Example 4 A company that builds and sells desks has a weekly over-head, including salaries and plant cost, of $3400. The cost of materials for each desk is $40 and the desk is sold for $200. How many desks must be built and sold each week so that the company is guaranteed a profit?

Solution   Let x be the number of desks built and sold each week. Then the number of dollars in the total revenue received each week is $200x$, and the number of dollars in the total cost each week is $3400 + 40x$. If P dollars is the weekly profit, then because profit equals revenue minus cost, we have

$$P = 200x - (3400 + 40x)$$
$$= 160x - 3400$$

For a profit we must have $P > 0$; that is,

$$160x - 3400 > 0$$
$$160x > 3400$$
$$x > 21\tfrac{1}{4}$$

Because x must be a positive integer, we conclude that the company must build and sell at least 22 desks each week to have a profit. ∎

In this section we have concentrated on linear inequalities. In Section 2.6 you will learn how to find the solution set of polynomial and rational inequalities.

EXERCISES   2.5

In Exercises 1 through 28, find the solution set of the inequality and write it with interval notation. Show the solution set on the real number line.

1. $20 \le 4x$

2. $3x - 5 < 7$

3. $2x - 1 < 6$

4. $3x + 1 \ge 4x - 3$

5. $2x + 1 > x - 4$

6. $5x + 6 \le x - 2$

7. $\dfrac{2x - 5}{4} \le 3$

8. $\dfrac{2x - 9}{7} > 0$

9. $-3 > \dfrac{3x + 5}{4}$

10. $\dfrac{x - 5}{4} \le x$

11. $\dfrac{2x - 5}{3} < x + 1$

12. $6x - 1 < \dfrac{5x - 1}{3}$

13. $\tfrac{1}{2}x + 2 \le \tfrac{1}{4}x + 3$

14. $5 - \tfrac{1}{3}x \ge \tfrac{1}{4}x - 2$

15. $10 - 3x > \dfrac{4x - 5}{-3}$

16. $\dfrac{3x + 8}{-5} < 4 - 2x$

17. $5 \le 2x - 3 < 13$

18. $11 \ge 3x - 5 > 2$

19. $-7 < 2x + 1 < 3$

20. $1 \le 3x - 2 \le 16$

21. $1 < \dfrac{4x - 1}{3} < 5$

22. $10 < 4 - 3x < 19$

23. $6 \le 2 - x \le 8$

24. $-1 < \dfrac{7 - 2x}{5} \le 5$

25. $2 > -3 - 3x \ge -7$

26. $-1 < 2 - 2x \le 3$

27. $-4 < \dfrac{5x + 3}{-6} \le 2$

28. $-4 \le \dfrac{5 + 3x}{-2} \le 5$

29. The perimeter of a rectangle must not be greater than 30 cm and the length must be 8 cm. What is the set of values for the width?

30. If the temperature on the Fahrenheit scale is F degrees and on the Celsius scale is C degrees, then

$$C = \tfrac{5}{9}(F - 32)$$

What is the set of values of F if C is between 10 and 20?

31. An investor has $8000 invested at 9 percent and wishes to invest some additional money at 16 percent in order to realize a return of at least 12 percent on the total investment. What is the least amount of money that should be invested?

32. Part of $20,000 is to be invested at 9 percent and the remainder is to be invested at 12 percent. What is the least amount of money that can be invested at 12 percent in order to have a yearly income of at least $2250 from the two investments?

33. A lamp manufacturer sells only to wholesalers through its showroom. The weekly overhead, including salaries, plant cost, and showroom rental, is $6000. If each lamp sells for $168 and the material used in its production costs $44, how many lamps must be made and sold each week so that the manufacturer realizes a profit?

34. If in a particular course a student has an average score of less than 90 and not below 80 on four examinations, the student will receive a grade of B in the course. If the student's scores on the first three examinations are 87, 94, and 73, what score on the fourth examination will result in a B grade?

35. A silversmith wishes to obtain an alloy containing at least 72 percent silver and at most 75 percent silver. Determine the greatest and least amounts of an 80 percent silver alloy that should be combined with a 65 percent silver alloy in order to have 30 g of the required alloy.

36. What is the minimum amount of pure alcohol that must be added to 24 liters of a 20 percent alcohol solution to obtain a mixture that is at least 30 percent alcohol?

37. Prove that if a, b, c, and d are real numbers, and

if $a < b$ and $c < d$, then $a + c < b + d$

(*Hint:* Use the definition of $<$ and the fact that the sum of two positive numbers is positive.)

2.6 Polynomial and Rational Inequalities

In the previous section we were concerned with finding solution sets of linear inequalities. We now discuss solving inequalities containing polynomials of degree greater than 1 as well as those containing rational expressions.

A quadratic inequality is one of the form

$$ax^2 + bx + c < 0 \qquad \text{(the symbol } < \text{ can be replaced by } >, \leq, \text{ or } \geq)$$

where a, b, and c are real numbers and $a \neq 0$. To solve a quadratic inequality we use the concepts of *critical number* and *test number*.

A **critical number** of the inequality above is a real root of the quadratic equation

$$ax^2 + bx + c = 0$$

Suppose r_1 and r_2 are critical numbers and $r_1 < r_2$. Then the polynomial $ax^2 + bx + c$ can change algebraic sign only at r_1 and r_2. Thus the sign (+ or −) of $ax^2 + bx + c$ will be constant on each of the intervals

$$(-\infty, r_1) \qquad (r_1, r_2) \qquad (r_2, +\infty)$$

To determine the sign on a particular one of these intervals we compute the value of $ax^2 + bx + c$ at an arbitrary **test number** in the interval. From the results we can obtain the solution set of the inequality. The procedure is shown in the following illustration and example.

Illustration 1

To solve the inequality

$$x^2 - 8 < 2x$$

we first write an equivalent inequality having all the nonzero terms on one side of the inequality sign. Thus we have

$$x^2 - 2x - 8 < 0$$
$$(x + 2)(x - 4) < 0$$

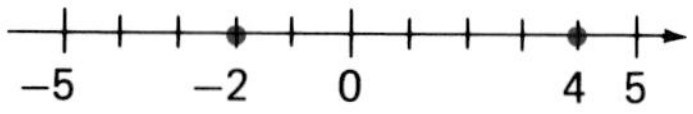

We observe from the factored form of the inequality that $x^2 - 2x - 8 = 0$ has the roots -2 and 4, which are the critical numbers of the inequality. We plot on the real number line the points corresponding to these numbers. See Figure 1. These points separate the line into the following three intervals:

$$(-\infty, -2) \qquad (-2, 4) \qquad (4, +\infty)$$

On each of these intervals the sign of $(x + 2)(x - 4)$ is constant. To determine the sign on an interval we choose an arbitrary test number in the interval and compute the sign of each of the factors $x + 2$ and $x - 4$ at this test number. We select -3 in $(-\infty, -2)$, 0 in $(-2, 4)$ and 5 in $(4, +\infty)$. The results are summarized in Table 1.

FIGURE 1 _____

TABLE 1

Interval	Test Number k	Sign of $x + 2$ at k	Sign of $x - 4$ at k	Sign of $(x + 2)(x - 4)$ on Interval
$(-\infty, -2)$	-3	$-$	$-$	$+$
$(-2, 4)$	0	$+$	$-$	$-$
$(4, +\infty)$	5	$+$	$+$	$+$

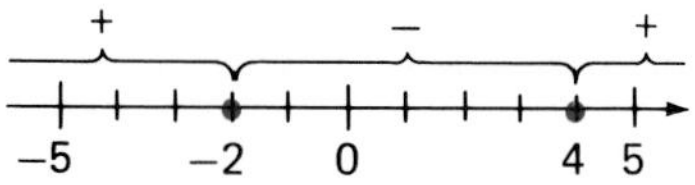

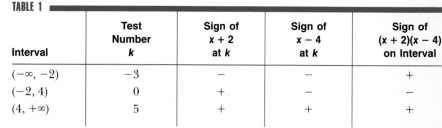

FIGURE 2 _____

FIGURE 3 _____

From the table we obtain Figure 2, which indicates on the real number line the points -2 and 4 and the intervals on which $(x + 2)(x - 4)$ is positive or negative. We conclude that the solution set of the inequality is the interval $(-2, 4)$, shown in Figure 3. ∎

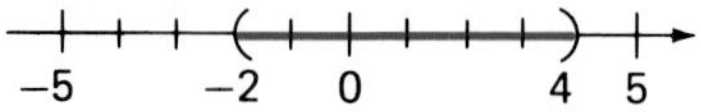

Example   1   Find and show on the real number line the solution set of the inequality

$$x^2 + 2x \geq 15$$

Solution   The given inequality is equivalent to

$$x^2 + 2x - 15 \geq 0$$
$$(x + 5)(x - 3) \geq 0$$

The critical numbers are -5 and 3. The points corresponding to these numbers are plotted in Figure 4 and the following intervals are determined:

$$(-\infty, -5) \qquad (-5, 3) \qquad (3, +\infty)$$

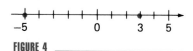

FIGURE 4 _____

Table 2 summarizes the results obtained by choosing a test number in each of these intervals and determining the sign of $(x + 5)(x - 3)$ on the intervals.

TABLE 2

Interval	Test Number k	Sign of $x + 5$ at k	Sign of $x - 3$ at k	Sign of $(x + 5)(x - 3)$ on Interval
$(-\infty, -5)$	-6	$-$	$-$	$+$
$(-5, 3)$	0	$+$	$-$	$-$
$(3, +\infty)$	4	$+$	$+$	$+$

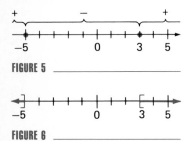

FIGURE 5

FIGURE 6

Figure 5 shows on the real number line the points -5 and 3 as well as the sign of $(x + 5)(x - 3)$ on the intervals $(-\infty, -5)$, $(-5, 3)$, and $(3, +\infty)$. Therefore $(x + 5)(x - 3) > 0$ if x is in either $(-\infty, -5)$ or $(3, +\infty)$. Furthermore, -5 and 3 are in the solution set because $(x + 5)(x - 3) = 0$ if x is either of these numbers. Thus the solution set of the given inequality is $(-\infty, -5] \cup [3, +\infty)$, appearing in Figure 6. ■

Example 2 Find the solution set of each of the following inequalities.

a) $5x^2 - 2x + 1 < x^2 + 2x$ b) $-6x^2 - 8x + 1 \le 3x^2 + 4x + 5$

Solution a) The given inequality is equivalent to

$$4x^2 - 4x + 1 < 0$$
$$(2x - 1)^2 < 0$$

Because there is no value of x for which $(2x - 1)^2$ is negative, there is no solution. Therefore, the solution set is $\varnothing$.

b) The given inequality is equivalent to

$$-9x^2 - 12x - 4 \le 0$$
$$9x^2 + 12x + 4 \ge 0$$
$$(3x + 2)^2 \ge 0$$

Because $(3x + 2)^2$ is nonnegative for all values of x, the solution set is the set R of all real numbers. ■

The method used in Illustration 1 and Example 1 to solve a quadratic inequality is applied in the next example to a polynomial inequality of the third degree.

Example 3 Find and show on the real number line the solution set of the inequality

$$(x + 1)(2x^2 - 5x + 2) > 0$$

Solution The given inequality is equivalent to

$$(x + 1)(2x - 1)(x - 2) > 0$$

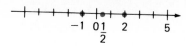

The critical numbers are -1, $\frac{1}{2}$, and 2, and the points corresponding to these numbers are plotted on the real number line in Figure 7. These points determine the following intervals:

$$(-\infty, -1) \qquad (-1, \tfrac{1}{2}) \qquad (\tfrac{1}{2}, 2) \qquad (2, +\infty)$$

We compute the sign of $(x + 1)(2x - 1)(x - 2)$ in each interval by selecting a test number there. The results are summarized in Table 3.

TABLE 3

Interval	Test Number k	Sign of $x + 1$ at k	Sign of $2x - 1$ at k	Sign of $x - 2$ at k	Sign of $(x + 1)(2x - 1)(x - 2)$ on Interval
$(-\infty, -1)$	-2	$-$	$-$	$-$	$-$
$(-1, \tfrac{1}{2})$	0	$+$	$-$	$-$	$+$
$(\tfrac{1}{2}, 2)$	1	$+$	$+$	$-$	$-$
$(2, +\infty)$	3	$+$	$+$	$+$	$+$

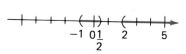

From the table we obtain Figure 8, showing on the real number line the points -1, $\frac{1}{2}$, and 2, as well as the sign of $(x + 1)(2x - 1)(x - 2)$ on the intervals $(-\infty, -1)$, $(-1, \tfrac{1}{2})$, $(\tfrac{1}{2}, 2)$, and $(2, +\infty)$. We conclude that $(x + 1)(2x - 1)(x - 2) > 0$ if x is in either $(-1, \frac{1}{2})$ or $(2, +\infty)$. The solution set of the given inequality is $(-1, \frac{1}{2}) \cup (2, +\infty)$, which appears in Figure 9. ∎

In the next illustration, we have a **rational inequality,** one that contains a rational expression involving the variable. The method of solution is similar to that used for polynomial inequalities.

Illustration 2

We wish to find the solution set of the inequality

$$\frac{5x}{x - 1} < 4$$

If we were to multiply both sides of the inequality by $x - 1$, we would have to consider the two possibilities: $x - 1$ is positive or negative. Alternatively, we write an equivalent inequality having zero on one side and nonzero terms on the other. Thus we have

$$\frac{5x}{x - 1} - 4 < 0$$

Combining terms on the left side gives

$$\frac{5x - 4(x - 1)}{x - 1} < 0$$

$$\frac{x + 4}{x - 1} < 0$$

A fraction can change sign only when either the numerator or denominator changes sign. Therefore the critical numbers of the inequality are those numbers for which either the numerator or denominator of the rational expression on the left is zero. Thus we solve the equations

$$x + 4 = 0 \quad \text{and} \quad x - 1 = 0$$

and obtain the critical numbers -4 and 1. Plotting the points corresponding to these numbers, as shown in Figure 10, we have the following intervals:

$$(-\infty, -4) \quad (-4, 1) \quad (1, +\infty)$$

We now choose a test number in each of these intervals and determine the sign of $\dfrac{x + 4}{x - 1}$ on the interval. Table 4 summarizes the results.

FIGURE 10

TABLE 4

Interval	Test Number k	Sign of $x + 4$ at k	Sign of $x - 1$ at k	Sign of $\dfrac{x + 4}{x - 1}$ on Interval
$(-\infty, -4)$	-5	$-$	$-$	$+$
$(-4, 1)$	0	$+$	$-$	$-$
$(1, +\infty)$	2	$+$	$+$	$+$

FIGURE 11

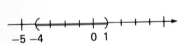

FIGURE 12

Figure 11 shows on the real number line the points -4 and 1 as well as the sign of $\dfrac{x + 4}{x - 1}$ on the intervals $(-\infty, -4)$, $(-4, 1)$, and $(1, +\infty)$. The solution set of the given inequality is then the interval $(-4, 1)$, appearing in Figure 12.　■

Example　4　Find and show on the real number line the solution set of the inequality

$$\frac{3x - 1}{x^2 - x - 6} \leq 1$$

Solution　We first write an equivalent inequality having zero on the right side and nonzero terms on the left. We then combine terms on the left and obtain a single rational expression. The computation is as follows:

$$\frac{3x - 1}{x^2 - x - 6} - 1 \leq 0$$

$$\frac{3x - 1 - (x^2 - x - 6)}{x^2 - x - 6} \leq 0$$

$$\frac{-x^2 + 4x + 5}{x^2 - x - 6} \leq 0$$

We now multiply on both sides of the inequality sign by -1, reverse the direction of the inequality, and factor the numerator and denominator. We have

$$\frac{x^2 - 4x - 5}{x^2 - x - 6} \geq 0$$

$$\frac{(x - 5)(x + 1)}{(x - 3)(x + 2)} \geq 0$$

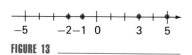

FIGURE 13 _____

The critical numbers are obtained by equating both the numerator and denominator to zero. This gives the numbers 5, -1, 3, and -2. Figure 13 shows that the points on the real number line corresponding to these numbers separate the line into five intervals:

$$(-\infty, -2) \qquad (-2, -1) \qquad (-1, 3) \qquad (3, 5) \qquad (5, +\infty)$$

We determine the sign of the rational expression on each of the five intervals by choosing a test number in the interval. Table 5 summarizes the results.

TABLE 5

Interval	Test Number k	Sign of $x + 2$ at k	Sign of $x + 1$ at k	Sign of $x - 3$ at k	Sign of $(x - 5)$ at k	Sign of $\dfrac{(x - 5)(x + 1)}{(x - 3)(x + 2)}$ on Interval
$(-\infty, -2)$	-3	$-$	$-$	$-$	$-$	$+$
$(-2, -1)$	$-\frac{3}{2}$	$+$	$-$	$-$	$-$	$-$
$(-1, 3)$	0	$+$	$+$	$-$	$-$	$+$
$(3, 5)$	4	$+$	$+$	$+$	$-$	$-$
$(5, +\infty)$	6	$+$	$+$	$+$	$+$	$+$

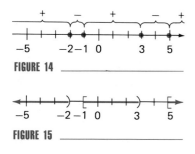

FIGURE 14 _____

FIGURE 15 _____

From the table we obtain Figure 14, showing on the real number line the points -2, -1, 3, and 5 as well as the sign of the rational expression on the intervals $(-\infty, -2)$, $(-2, -1)$, $(-1, 3)$, $(3, 5)$, and $(5, +\infty)$. The rational expression is positive if x is in one of the following intervals: $(-\infty, -2)$, $(-1, 3)$, and $(5, +\infty)$. Furthermore, this rational expression is zero if $x = -1$ or $x = 5$; thus -1 and 5 are in the solution set. Of course, the rational expression is not defined when $x = -2$ or $x = 3$ because then the denominator is zero. Therefore, the solution set is $(-\infty, -2) \cup [-1, 3) \cup [5, +\infty)$, appearing in Figure 15. ∎

Example 5 A decorator designs and sells wall fixtures and can sell at a price of \$75 each all the fixtures she produces. If x fixtures are manufactured each day, then the number of dollars in the daily total cost of production is $x^2 + 25x + 96$. How many fixtures should be produced each day so that the decorator is guaranteed a profit?

Solution The number of dollars in the total revenue received each day from the sale of x fixtures is $75x$. If P dollars is the daily profit from the sale of x fixtures, then because profit equals revenue minus cost, we have

$$P = 75x - (x^2 + 25x + 96)$$
$$= -x^2 + 50x - 96$$

For the decorator to be guaranteed a profit, we must have $P > 0$; that is,

$$-x^2 + 50x - 96 > 0$$
$$x^2 - 50x + 96 < 0$$
$$(x - 2)(x - 48) < 0$$

We wish to solve this inequality. Because x is the number of fixtures, the solution set is restricted to nonnegative values of x. The critical numbers are 2 and 48. The points corresponding to these two numbers separate the nonnegative side of the real number line into the following three intervals:

$$[0, 2) \quad (2, 48) \quad (48, +\infty)$$

Table 6 summarizes the results obtained by choosing a test number in each of these intervals to determine the sign of $(x - 2)(x - 48)$ on the interval.

TABLE 6

Interval	Test Number k	Sign of $x - 2$ at k	Sign of $x - 48$ at k	Sign of $(x - 2)(x - 48)$ on Interval
$[0, 2)$	1	$-$	$-$	$+$
$(2, 48)$	3	$+$	$-$	$-$
$(48, +\infty)$	49	$+$	$+$	$+$

From the table we conclude that the solution set of the inequality is the open interval $(2, 48)$. Thus for the decorator to be guaranteed a profit, the number of fixtures produced and sold each day must be more than 2 and less than 48. ■

EXERCISES 2.6

In Exercises 1 through 40, find the solution set of the inequality and show the solution set on the real number line.

1. $x^2 > 9$
2. $x^2 < 4$
3. $(x + 3)(x - 4) < 0$
4. $(x - 1)(x - 5) > 0$
5. $(2x + 1)(2x - 7) > 0$
6. $(3x + 5)(2x - 3) < 0$
7. $x^2 - 4x + 3 \le 0$
8. $x^2 + 6x + 8 \ge 0$
9. $4 - 3x - x^2 \ge 0$
10. $2x^2 + x - 1 \le 0$
11. $x^2 > 8 - 2x$
12. $x^2 < 15 + 2x$
13. $x \le 6 - 2x^2$
14. $4x^2 \ge 9 - 9x$
15. $16t^2 + 1 \ge 8t$
16. $9y^2 < 30y - 25$

17. $x(11 - 3x) < 10$

18. $x(6x + 1) \geq 15$

19. $(y + 3)(y - 1)(y - 4) > 0$

20. $(x + 4)(x^2 - 4) < 0$

21. $x^3 \leq 16x$

22. $6t^3 > 7t^2 + 3t$

23. $x^2 - x - 1 < 0$

24. $x^2 + 2x - 2 \geq 0$

25. $x^2 + x + 3 > 0$

26. $x^2 - 2x + 2 < 0$

27. $\dfrac{x - 1}{2x + 5} < 0$

28. $\dfrac{y + 6}{y - 4} \geq 0$

29. $\dfrac{5t}{t - 4} \geq 6$

30. $\dfrac{3x - 8}{2x + 3} < 4$

31. $\dfrac{3x - 11}{3x + 5} < 7$

32. $\dfrac{4x - 7}{5x + 1} \geq 2$

33. $\dfrac{6 - 2x}{4 + x} \leq 5$

34. $\dfrac{2 - 7t}{5 - 4t} > 4$

35. $\dfrac{2}{3 - y} > \dfrac{3}{2 + y}$

36. $\dfrac{4}{2x - 1} < \dfrac{1}{x + 1}$

37. $\dfrac{5}{2x - 1} < x - 2$

38. $\dfrac{6}{3w - 4} > w + 1$

39. $\dfrac{5x}{x^2 + 2x - 8} \geq 3$

40. $\dfrac{2x - 7}{x^2 - 6x + 8} \leq 1$

41. A firm can sell at a price of $100 per unit all of a particular commodity it produces. If x units are produced each day, the number of dollars in the total cost of each day's production is $x^2 + 20x + 700$. How many units should be produced each day so that the firm is guaranteed a profit?

42. A company that builds and sells desks can sell at a price of $400 per desk all the desks it produces. If x desks are built and sold each week, then the number of dollars in the total cost of the week's production is $2x^2 + 80x + 3000$. How many desks should be built each week in order that the manufacturer is guaranteed a profit?

43. A rectangular field is to be fenced off along the bank of a river; no fence is required along the river. The material for the fence costs $8 per running foot for the two ends and $16 per running foot for the side parallel to the river. If the area of the field is to be 12,000 ft² and the cost of the fence is not to exceed $3520, what are the restrictions on the dimensions of the field?

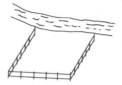

44. A rectangular plot of ground is to be enclosed by a fence and then divided down the middle by another fence. The fence down the middle costs $3 per running foot and the other fence costs $6 per running foot. If the area of the plot is to be 1800 ft² and the cost of the fence is not to exceed $2310, what are the restrictions on the dimensions of the plot?

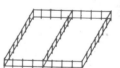

2.7 Equations and Inequalities Involving Absolute Value

According to the definition in Section 1.1, the absolute value of a real number a, denoted by $|a|$, is given by

$$|a| = \begin{cases} a, & \text{if} \quad a \geq 0 \\ -a, & \text{if} \quad a < 0 \end{cases}$$

Also recall from Section 1.1 that on the real number line $|a|$ is the distance (without regard to direction left or right) from the origin to the point a. See Figure 1, where $a > 0$, and Figure 2, where $a < 0$.

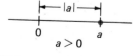

FIGURE 1

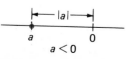

FIGURE 2

Example 1 Find the solution set of the equation $|3x + 5| = 9$.

Solution   The given equation will be satisfied if either

$$3x + 5 = 9 \quad \text{or} \quad -(3x + 5) = 9$$
$$3x = 4 \qquad\qquad -3x - 5 = 9$$
$$x = \tfrac{4}{3} \qquad\qquad -3x = 14$$
$$x = -\tfrac{14}{3}$$

The solution set is $\{\tfrac{4}{3}, -\tfrac{14}{3}\}$ ■

Example   2   Find the solution set of the equation

$$|2x - 3| = |7 - 3x|$$

Solution   The given equation will be satisfied if either

$$2x - 3 = 7 - 3x \quad \text{or} \quad 2x - 3 = -(7 - 3x)$$
$$2x + 3x = 7 + 3 \qquad\qquad 2x - 3 = -7 + 3x$$
$$5x = 10 \qquad\qquad 2x - 3x = -7 + 3$$
$$x = 2 \qquad\qquad -x = -4$$
$$x = 4$$

The solution set is $\{2, 4\}$.   ■

We now discuss inequalities involving absolute value.

Illustration   1

Suppose we have the inequality

$$|x| < 3$$

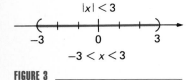

$-3 < x < 3$

FIGURE 3

This inequality states that on the real number line the distance from the origin to the point x is less than 3 units; that is, $-3 < x < 3$. Therefore, x is in the open interval $(-3, 3)$. See Figure 3. It appears then that the solution set of $|x| < 3$ is $\{x \mid -3 < x < 3\}$. That this is the case can be proved by using only properties of absolute value.   ■

In the above illustration, by comparing the given inequality and its solution set, we conclude that the inequality

$$|x| < 3 \quad \text{is equivalent to} \quad -3 < x < 3$$

More generally, if $b > 0$,

$$|x| < b \quad \text{is equivalent to} \quad -b < x < b$$

Furthermore, if instead of x we have an algebraic expression E and $b > 0$, then the inequality

$$|E| < b \quad \text{is equivalent to} \quad -b < E < b \tag{1}$$

Statement (1) is valid if the symbol $<$ is replaced by $\leq$.

Example 3 Find and show on the real number line the solution set of the inequality

$$|2x - 7| < 9$$

Solution The given inequality is equivalent to

$$-9 < 2x - 7 < 9$$
$$-9 + 7 < 2x < 9 + 7$$
$$-2 < 2x < 16$$
$$-1 < x < 8$$

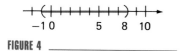

FIGURE 4

Therefore the solution set is the open interval $(-1, 8)$, shown in Figure 4. ∎

Illustration 2

Consider the inequality

$$|x| > 2$$

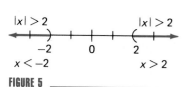

FIGURE 5

This inequality states that on the real number line the distance from the origin to the point x is greater than 2 units; that is, either $x > 2$ or $x < -2$. Therefore x is in $(-\infty, -2) \cup (2, +\infty)$. See Figure 5. Thus it appears that the solution set of $|x| > 2$ is $\{x | x > 2\} \cup \{x | x < -2\}$. Properties of absolute value can be used to prove that this is the situation. ∎

By comparing the given inequality and its solution set, we observe from Illustration 2 that the inequality

$$|x| > 2 \quad \text{is equivalent to} \quad x > 2 \quad \text{or} \quad x < -2$$

More generally, if $b > 0$,

$$|x| > b \quad \text{is equivalent to} \quad x > b \quad \text{or} \quad x < -b$$

If, instead of x, we have an algebraic expression E and $b > 0$, then the inequality

$$|E| > b \quad \text{is equivalent to} \quad E > b \ \text{ or } \ E < -b \qquad (2)$$

That is, the solution set of the inequality $|E| > b$ is the union of the solution sets of the inequalities $E > b$ and $E < -b$.

From Statement (1), $|E| < b$ is equivalent to the continued inequality $-b < E < b$; however, $|E| > b$ is not equivalent to any continued inequality.

Statement (2) is valid if the symbol $>$ is replaced by $\geq$ and the symbol $<$ is replaced by $\leq$.

Example   4   Find and show on the real number line the solution set of the inequality

$$\left|\tfrac{2}{3}x - 5\right| \geq 3$$

Solution   The solution set of the given inequality is the union of the solution sets of the inequalities

$$
\begin{array}{ll}
\tfrac{2}{3}x - 5 \geq 3 & \tfrac{2}{3}x - 5 \leq -3 \\
2x - 15 \geq 9 & 2x - 15 \leq -9 \\
2x \geq 24 & 2x \leq 6 \\
x \geq 12 & x \leq 3
\end{array}
$$

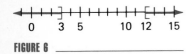

FIGURE 6

Thus the solution set is $\{x \mid x \leq 3\} \cup \{x \mid x \geq 12\}$ or, with interval notation, $(-\infty, 3] \cup [12, +\infty)$. The solution set is shown on the real number line in Figure 6. ■

The following theorems about absolute value are useful in later work.

Theorem 1

If a and b are real numbers, then

$$|ab| = |a| \cdot |b|$$

Proof   Recall from Section 1.6 that $|x| = \sqrt{x^2}$. Therefore

$$
\begin{aligned}
|ab| &= \sqrt{(ab)^2} \\
&= \sqrt{a^2 b^2} \\
&= \sqrt{a^2}\sqrt{b^2} \\
&= |a| \cdot |b|
\end{aligned}
$$

■

Theorem 2

If a and b are real numbers and $b \neq 0$, then

$$\left| \frac{a}{b} \right| = \frac{|a|}{|b|}$$

The proof is similar to that of Theorem 1 and is left as an exercise (see Exercise 49).

In calculus we often wish to show that one inequality is equivalent to a simpler one. The following example involves such a situation.

Example   **5**   Show that the inequality

$$|(3x + 2) - 8| < 1 \qquad \text{is equivalent to} \qquad |x - 2| < \tfrac{1}{3}$$

Solution   The following inequalities are equivalent:

$$|(3x + 2) - 8| < 1$$
$$|3x - 6| < 1$$
$$|3(x - 2)| < 1$$
$$|3||x - 2| < 1$$
$$3|x - 2| < 1$$
$$|x - 2| < \tfrac{1}{3}$$

   ■

The following theorem, called the *triangle inequality,* is used frequently in proving theorems in calculus.

The Triangle Inequality

If a and b are real numbers, then

$$|a + b| \leq |a| + |b|$$

In Exercise 50 you are asked to prove the triangle inequality. We demonstrate the content of the triangle inequality in the following illustration with four particular cases.

Illustration 3

If $a = 3$ and $b = 4$, then

$$|a + b| = |3 + 4| \qquad |a| + |b| = |3| + |4|$$
$$= |7| \qquad\qquad\qquad = 3 + 4$$
$$= 7 \qquad\qquad\qquad\quad = 7$$

If $a = -3$ and $b = 4$, then

$$|a + b| = |-3 + 4| \qquad |a| + |b| = |-3| + |4|$$
$$= |1| \qquad\qquad = 3 + 4$$
$$= 1 \qquad\qquad = 7$$

If $a = 3$ and $b = -4$, then

$$|a + b| = |3 + (-4)| \qquad |a| + |b| = |3| + |-4|$$
$$= |-1| \qquad\qquad = 3 + 4$$
$$= 1 \qquad\qquad = 7$$

If $a = -3$ and $b = -4$, then

$$|a + b| = |-3 + (-4)| \qquad |a| + |b| = |-3| + |-4|$$
$$= |-7| \qquad\qquad = 3 + 4$$
$$= 7 \qquad\qquad = 7$$

In each case $|a + b| \le |a| + |b|$.   ■

EXERCISES 2.7

In Exercises 1 through 14, find the solution set of the equation.

1. $|x - 5| = 4$
2. $|2x + 3| = 7$
3. $|3y - 8| = 4$
4. $|4t - 9| = 11$
5. $|4x + 5| = 15$
6. $|8 - x| = 4$
7. $|7 - 2w| = 9$
8. $|3x - 2| = |2x + 3|$
9. $|x - 4| = |5 - 2x|$
10. $|5y| = |6 - y|$
11. $\left|\dfrac{x + 3}{x - 3}\right| = 7$
12. $\left|\dfrac{2x + 1}{x - 1}\right| = 3$
13. $|3t - 2| = t^2$
14. $|x - 1| = \dfrac{x^2}{4}$

In Exercises 15 through 30, find the solution set of the inequality and write it with interval notation. Show the solution set on the real number line.

15. $|x| \le 5$
16. $|x| > 6$
17. $|x - 1| > 7$
18. $|x + 1| < 5$
19. $|x - 5| \le 3$
20. $|3x - 4| \le 2$
21. $|2x - 7| < 9$
22. $|3x - 4| \ge 2$
23. $|7 - 2x| > 9$
24. $6 < |4x + 7|$
25. $4 < |3x + 12|$
26. $|5 - 3x| \ge 10$
27. $|5x - 7| + 4 \le 6$
28. $|4x - 3| - 3 > 6$
29. $|\frac{3}{2} - 2x| > \frac{1}{2}$
30. $|2 - \frac{3}{4}x| \le \frac{1}{2}$

In Exercises 31 through 36, show that the two inequalities are equivalent.

31. $|(2x - 3) - 9| < 1$; $|x - 6| < \frac{1}{2}$
32. $|(2x + 3) - 1| < 1$; $|x + 1| < \frac{1}{2}$
33. $|(3x - 5) - 1| < \frac{1}{2}$; $|x - 2| < \frac{1}{6}$
34. $|(5x - 2) - 3| < \frac{1}{2}$; $|x - 1| < \frac{1}{10}$
35. $|(\frac{1}{2}x - 5) + 7| < \frac{1}{8}$; $|x + 4| < \frac{1}{4}$
36. $|(\frac{1}{4}x - 1) + 2| < \frac{1}{6}$; $|x + 4| < \frac{2}{3}$

In Exercises 37 through 46, find the solution set of the inequality and write it with interval notation. Show the solution set on the real number line.

37. $|x^2 - 5| < 4$
38. $|y^2 - 10| \le 6$
39. $|t^2 - 5t| \le 6$
40. $|x^2 - 3x - 1| < 3$
41. $|x^2 - 17| \ge 8$
42. $|x^2 + x - 4| > 2$
43. $\left|\dfrac{x + 4}{x}\right| < 3$
44. $\left|\dfrac{x + 4}{x - 4}\right| \ge 1$
45. $\left|\dfrac{r}{r + 1}\right| \ge \dfrac{1}{2}$
46. $\left|\dfrac{3t}{t + 1}\right| < 8$

In Exercises 47 and 48, verify the triangle inequality for the given values of a and b.

47. (a) $a = 5$ and $b = 7$; (b) $a = 5$ and $b = -7$;
 (c) $a = -5$ and $b = 7$; (d) $a = -5$ and $b = -7$
48. (a) $a = \frac{1}{2}$ and $b = \frac{1}{3}$; (b) $a = -\frac{1}{2}$ and $b = \frac{1}{3}$;
 (c) $a = \frac{1}{2}$ and $b = -\frac{1}{3}$; (d) $a = -\frac{1}{2}$ and $b = -\frac{1}{3}$
49. Prove Theorem 2.
50. Prove the triangle inequality. (*Hint:* First show that

$$-|a| \le a \le |a| \qquad \text{and} \qquad -|b| \le b \le |b|$$

Then apply the result of Exercise 37 of Exercises 2.5 and statement (1) of this section, where $<$ is replaced by $\le$.)

In Exercises 51 and 52, use the triangle inequality to prove the statement.

51. If $|x - 1| < \frac{1}{3}$ and $|y + 1| < \frac{1}{4}$, then $|x + y| < \frac{7}{12}$.
52. If $|x - 1| < \frac{1}{3}$ and $|y - 1| < \frac{1}{4}$, then $|x - y| < \frac{7}{12}$.

REVIEW EXERCISES FOR CHAPTER 2

In Exercises 1 through 42, find the solution set of the equation.

1. $6x + 5 = 23$
2. $5x - 6 = 2 + 9x$
3. $2(5x - 4) = 11 - (3 + 2x)$
4. $5(2t - 4) = 11 - (3 + 2t)$
5. $\dfrac{2}{5 - y} - \dfrac{1}{y - 2} = 0$
6. $\dfrac{x - 1}{x - 1} = 1$
7. $\dfrac{x - 1}{x - 1} = 0$
8. $\dfrac{x}{2x - 2} = \dfrac{x - 4}{2x - 4}$
9. $49x^2 - 64 = 0$
10. $5x^2 = x$
11. $t^2 - 3t - 10 = 0$
12. $5w^2 + 17w - 12 = 0$
13. $10x^2 + 7x - 6 = 0$
14. $y^2 + 2y - 1 = 0$
15. $2p^2 - 4p - 5 = 0$
16. $4x^2 + 2x + 1 = 0$
17. $3w^2 - 2w + 2 = 0$
18. $(3x + 10)(x - 3) = 2x + 14$
19. $(x - 4)(x + 2) = 7$
20. $\dfrac{x + 11}{x + 8} - \dfrac{3x - 2}{x - 2} = 0$
21. $\dfrac{y + 4}{y} = \dfrac{3y}{y + 2}$
22. $\dfrac{6}{t^2 - 1} + \dfrac{1}{t - 1} = \dfrac{1}{2}$
23. $\dfrac{70}{x^2 - 4x + 3} = \dfrac{23}{1 - x} - 3$
24. $\dfrac{w^2 - 3}{w^2 - 6w + 5} = \dfrac{2w + 3}{w - 5} - \dfrac{w + 3}{w - 1}$
25. $\sqrt{2x + 5} + x = 5$
26. $\sqrt{x + 2} + 2 + x = 0$
27. $\sqrt{3t + 7} + \sqrt{t + 6} = 3$
28. $\sqrt{y + 2} + \sqrt{y + 5} - \sqrt{8 - y} = 0$
29. $x^3 - 8 = 0$
30. $x^4 - 8x^2 = 9$
31. $36x^4 - 13x^2 + 1 = 0$
32. $8z^3 + 27 = 0$
33. $(x^2 + 3x)^2 - 3(x^2 + 3x) - 10 = 0$
34. $6\left(y + \dfrac{1}{y}\right)^2 - 35\left(y + \dfrac{1}{y}\right) + 50 = 0$
35. $2\sqrt{t - 5} + \sqrt[4]{t - 5} = 3$
36. $15t^{-2} - 14t^{-1} - 8 = 0$
37. $y^{-2/3} + y^{-1/3} - 6 = 0$
38. $3x^{1/3} + 5 = 2x^{-1/3}$
39. $|2x + 5| = 7$
40. $|3y - 4| = 8$
41. $\left|\dfrac{w + 1}{w - 1}\right| = 3$
42. $|4x - 1| = x + 5$

In Exercises 43 through 48, solve for x in terms of the other symbols.

43. $Ax + By + C = 0$
44. $\dfrac{d}{10x} - \dfrac{d}{5} = \dfrac{1}{x}$
45. $x^2 + b^2 = 2bx + a^2$
46. $x^2 + xy + 2x - 1 = 0$
47. $6x^2 - 2xy - x + y - 1 = 0$
48. $rsx^2 + s^2x + rtx + st = 0$

In Exercises 49 and 50, add a term to the algebraic expression to make it a square of a binomial; also write the resulting expression as a square of a binomial.

49. (a) $x^2 - 8x$; (b) $y^2 + 3y$; (c) $x^2 + \frac{5}{3}x$
50. (a) $w^2 + 10w$; (b) $x^2 - \frac{6}{5}x$; (c) $x^2 + 7bx$

In Exercises 51 and 52, find the discriminant and determine the character of the roots of the quadratic equation; do not solve the equation.

51. (a) $4x^2 + 20x + 25 = 0$; (b) $8x^2 - 10x = 3$; (c) $5t - 3 = 4t^2$
52. (a) $15x^2 - 19x = 10$; (b) $2y^2 = 9 - 4y$; (c) $9x^2 - 42x + 49 = 0$

In Exercises 53 through 84, find the solution set of the inequality and write it with interval notation. Show the solution set on the real number line.

53. $3x - 1 \le 11$
54. $5x + 7 \ge 2x - 2$
55. $3x - 2 > 7x + 3$
56. $4x < 8x - 7$
57. $\dfrac{x + 1}{2} - 3 \ge \dfrac{x + 2}{4}$
58. $\dfrac{2x + 1}{3} \le \dfrac{9 - 2x}{6}$
59. $-4 < 1 - 5x < 11$
60. $7 < 2x + 3 \le 15$
61. $-5 \le \dfrac{4x - 5}{-3} < 7$
62. $-7 \le \dfrac{7 - 3x}{-4} \le 8$
63. $|x + 1| \ge 2$
64. $|x - 4| \le 6$
65. $|2x - 5| < 7$
66. $|3x + 7| > 11$
67. $\frac{4}{3} < |\frac{3}{2} - x|$
68. $\left|\dfrac{3 - 2x}{5}\right| < \dfrac{1}{3}$
69. $x^2 < 25$
70. $x^2 + 3x - 10 \ge 0$
71. $2x^2 - 3x \ge 5$
72. $3t^2 < 4(t + 1)$
73. $4y^3 < 5y^2 - 6y$
74. $x^3 > x$
75. $\dfrac{2x + 1}{x - 5} > 1$
76. $\dfrac{3x - 4}{2x - 3} \le 2$

77. $\dfrac{3x - 1}{2x - 1} < \dfrac{3x + 4}{2}$

78. $2x - 3 \geq \dfrac{5}{x - 3}$

79. $\dfrac{8x + 6}{x^2 + x - 12} \leq 1$

80. $\dfrac{10 - 4x}{x^2 - 7x + 10} > 3$

81. $|x^2 - 3x - 7| < 3$

82. $|y^2 - 26| \geq 10$

83. $\left|\dfrac{2t + 5}{t + 1}\right| \geq 1$

84. $\left|\dfrac{4x + 1}{x + 2}\right| < 3$

In Exercises 85 and 86, show that the two inequalities are equivalent.

85. $|(4x + 3) - 11| < \frac{1}{2};\ |x - 2| < \frac{1}{8}$

86. $|(2x - 5) - 7| < \frac{4}{5};\ |x - 6| < \frac{2}{5}$

87. A company obtained two loans totaling $30,000. The interest rate for one loan was 16 percent, and for the other loan it was 12 percent. The total annual interest for the two loans was equal to the interest the company would have paid if the entire amount had been borrowed at 15 percent. What was the amount of each loan?

88. A group of four students decide to hire a tutor for a review session prior to an examination, and the tutor's fee is to be shared by each student. If two additional students join the group, the cost to each student is reduced by $6. What is the cost per student if four are in the group?

89. An automobile radiator contains 8 liters of a solution that is 10 percent antifreeze and 90 percent water. How much of the solution should be drained and replaced with pure antifreeze to obtain a solution that is 25 percent antifreeze?

90. How many liters of a solution that is 55 percent glycerine should be added to 25 liters of a solution that is 28 percent glycerine to give a solution that is 35 percent glycerine?

91. A man can paint a room by himself in 8 hr, and it takes his son 12 hr to do the same job. The man works by himself for 3 hr. Then his son joins him and the two together complete the job. How long does the son work?

92. A woman leaves home at 8 A.M. and walks to her office at the rate of 8 km/hr. At 8:15 A.M., the woman's daughter leaves home and rides her bicycle at a rate of 20 km/hr along the same route to school. At what time does she overtake her mother?

93. In a long-distance race around a 400-m track, the winner finished one lap ahead of the loser. If the average rate of the winner was 6 m/sec and the average rate of the loser was 5.75 m/sec, how soon after the start did the winner complete the race?

94. A small pipe takes 24 min longer to fill a tank than it takes a large pipe. The two pipes together can fill the tank in 9 min. How long does it take each pipe alone to fill the tank?

95. A farmer is plowing a rectangular field that is 400 rods long and 240 rods wide. What should the width of a strip the farmer must plow around it be so that the field will be one-half plowed?

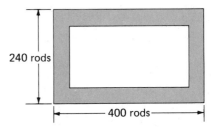

96. To form an open box, squares of side 4 cm are cut from each corner of a square piece of tin and then the sides of the tin are turned up. If the volume of the box is to be 400 cm^3, what should the area of the original piece of tin be?

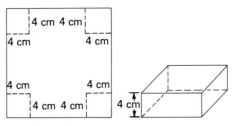

97. A train on its way east was delayed 1 hr when it was 560 km west of New York City. By increasing its normal rate 10 km/hr, the train arrived at New York on schedule. What is the train's normal rate?

98. In a long-distance race covering 42 km, one runner finished 12 min before another runner. If the faster runner's rate was 1 km/hr more than the slower runner's, what were their rates?

99. A student must receive an average score of at least 90 on five examinations to earn a grade of A in a particular course. If the student's scores on the first four examinations are 93, 95, 79, and 88, what must the score on the fifth examination be for the course grade to be A?

100. The perimeter of a rectangle must not be greater than 30 cm and the length must be 8 cm. What is the range of values for the width?

C H A P T E R

3

Graphs and Equations

Introduction

The union of geometry and algebra forms the basis of analytic geometry. The topics in this chapter will give you an introduction to that subject. In Section 3.1 we discuss the rectangular cartesian coordinate system in the plane and use it to obtain the distance between two points and the midpoint of a line segment. Graphs of equations and some properties of graphs are treated in Section 3.2. We obtain equations of circles in Section 3.2, equations of lines in Section 3.3, equations of parabolas in Sections 3.4 and 3.5, and equations of ellipses and hyperbolas in Supplementary Section 3.6. Translation of axes, presented in Section 3.5, is used to obtain more general equations of parabolas, ellipses, and hyperbolas. We include a brief discussion of conic sections in Section 3.6.

3.1 Points in a Plane

Ordered pairs of real numbers are important in our discussions. Any two real numbers form a pair. When the order of appearance of the numbers is significant, we call it an **ordered pair.** If x is the first real number and y is the second, this ordered pair is denoted by writing them in parentheses with a comma separating them as (x, y). Observe that the ordered pair $(5, 2)$ is different from the ordered pair $(2, 5)$.

The set of all ordered pairs of real numbers is called the **number plane,** denoted by R^2, and each ordered pair (x, y) is a **point** in the number plane. Just as we can identify R with points on an axis (a one-dimensional space), we can identify R^2 with points in a geometric plane (a two-dimensional space). The concept is attributed to the French mathematician René Descartes (1596–1650), who is credited with the origination of analytic geometry in 1637. A horizontal line, called the **x axis,** is chosen in the geometric plane. A vertical line is selected and is called the **y axis.** The point of intersection of the x axis and the y axis is called the **origin** and is denoted by the letter O. A unit of length, usually the same on each axis, is chosen. We establish the positive direction on the x axis to the right of the origin, and the positive direction on the y axis above the origin. See Figure 1.

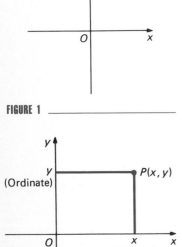

FIGURE 1

FIGURE 2

We now associate an ordered pair of real numbers (x, y) with a point in the geometric plane. At the point x on the horizontal axis and the point y on the vertical axis, line segments are drawn perpendicular to the respective axes. The intersection of these two perpendicular line segments is the point P associated with the ordered pair (x, y). Refer to Figure 2. The first number x of the pair is called the **abscissa** (or **x-coordinate**) of P, and the second number y is called the **ordinate** (or **y coordinate**) of P. If the abscissa is positive, P is to the right of the y axis; and if it is negative, P is to the left of the y axis. If the ordinate is positive, P is above the x axis; and if it is negative, P is below the x axis. The abscissa and ordinate of a point are called the **rectangular cartesian coordinates** of the point. There is a one-

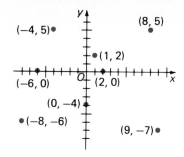

FIGURE 3 ▬▬▬▬▬▬▬▬▬▬▬

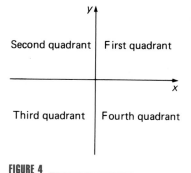

FIGURE 4 ▬▬▬▬▬▬▬▬▬▬▬

to-one correspondence between the points in a geometric plane and R^2; that is, with each point there corresponds a unique ordered pair (x, y), and with each ordered pair (x, y) there is associated only one point. This one-to-one correspondence is called a **rectangular cartesian coordinate system.**

Example 1 Plot each of the following points on a rectangular cartesian coordinate system: $(1, 2)$, $(2, 0)$, $(8, 5)$, $(9, -7)$, $(0, -4)$, $(-4, 5)$, $(-6, 0)$, and $(-8, -6)$.

Solution Figure 3 shows a rectangular cartesian coordinate system with the given points plotted. ∎

The x and y axes are called the **coordinate axes.** They divide the plane into four parts, called **quadrants.** The first quadrant is the one in which the abscissa and the ordinate are both positive, that is, the upper right quadrant. The other quadrants are numbered in the counterclockwise direction, with the fourth being the lower right quadrant. See Figure 4.

Because of the one-to-one correspondence, we identify R^2 with the geometric plane. For this reason we call an ordered pair (x, y) a point.

We now discuss the problem of finding the distance between two points in R^2. If A is the point (x_1, y_1) and B is the point (x_2, y_1) (that is, A and B have the same ordinate but different abscissas), then the **directed distance** from A to B is denoted by $\overline{AB}$ and we define

$$\overline{AB} = x_2 - x_1$$

Illustration 1

Refer to Figure 5(a), (b), and (c). If A is the point $(3, 4)$ and B is the point $(9, 4)$, then $\overline{AB} = 9 - 3$; that is, $\overline{AB} = 6$. If A is the point $(-8, 0)$ and B is the point $(6, 0)$, then $\overline{AB} = 6 - (-8)$; that is, $\overline{AB} = 14$. If A is the point $(4, 2)$ and B is the point $(1, 2)$, then $\overline{AB} = 1 - 4$; that is, $\overline{AB} = -3$. We see that $\overline{AB}$ is positive if B is to the right of A, and $\overline{AB}$ is negative if B is to the left of A. ∎

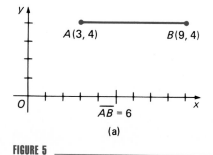
(a)

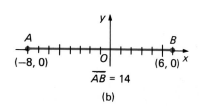
(b)

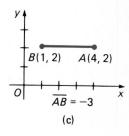
(c)

FIGURE 5 ▬▬▬▬▬▬▬▬▬▬▬▬▬▬▬▬▬▬▬▬▬▬▬▬▬▬▬▬▬▬▬▬▬▬▬

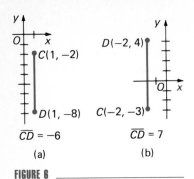

$\overline{CD} = -6$

(a)

$\overline{CD} = 7$

(b)

FIGURE 6 _____

If C is the point (x_1, y_1) and D is the point (x_1, y_2), then the **directed distance** from C to D, denoted by $\overline{CD}$, is defined by

$$\overline{CD} = y_2 - y_1$$

Illustration 2

Refer to Figure 6(a) and (b). If C is the point $(1, -2)$ and D is the point $(1, -8)$, then $\overline{CD} = -8 - (-2)$; that is, $\overline{CD} = -6$. If C is the point $(-2, -3)$ and D is the point $(-2, 4)$, then $\overline{CD} = 4 - (-3)$; that is, $\overline{CD} = 7$. The number $\overline{CD}$ is positive if D is above C, and $\overline{CD}$ is negative if D is below C. ∎

Observe that the terminology *directed distance* indicates both a distance and a direction (positive or negative). If we are concerned only with the length of the line segment between two points P_1 and P_2 (that is, the distance between the points P_1 and P_2 without regard to direction), then we use the terminology *undirected distance*. We denote the **undirected distance** from P_1 to P_2 by $|\overline{P_1P_2}|$, which is a nonnegative number. If we use the word "distance" without an adjective "directed" or "undirected," it is understood that we mean an undirected distance.

We now wish to obtain a formula for computing the distance $|\overline{P_1P_2}|$ if $P_1(x_1, y_1)$ and $P_2(x_2, y_2)$ are any two points in the plane. We use the Pythagorean theorem from plane geometry, which we now state. Refer to Figure 7.

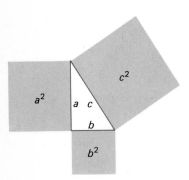

FIGURE 7 _____

The Pythagorean Theorem

In a right triangle, if a and b are the lengths of the perpendicular sides and c is the length of the hypotenuse, then

$$a^2 + b^2 = c^2$$

Figure 8 shows P_1 and P_2 in the first quadrant and the point $R(x_2, y_1)$. Note that $|\overline{P_1P_2}|$ is the length of the hypotenuse of right triangle P_1RP_2. Using the Pythagorean theorem, we have

$$|\overline{P_1P_2}|^2 = |\overline{P_1R}|^2 + |\overline{RP_2}|^2$$
$$|\overline{P_1P_2}| = \sqrt{|\overline{P_1R}|^2 + |\overline{RP_2}|^2}$$
$$|\overline{P_1P_2}| = \sqrt{(x_2 - x_1)^2 + (y_2 - y_1)^2}$$

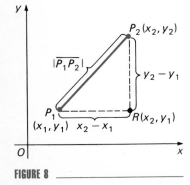

FIGURE 8 _____

Observe that in this formula we do not have a $\pm$ symbol in front of the radical because $|\overline{P_1P_2}|$ is a nonnegative number. The formula holds for all possible positions of P_1 and P_2 in all four quadrants. The length of the hypotenuse is always $|\overline{P_1P_2}|$, and the lengths of the legs are always $|\overline{P_1R}|$ and $|\overline{RP_2}|$. We have then the following theorem.

Distance Formula

> The distance between two points $P_1(x_1, y_1)$ and $P_2(x_2, y_2)$ is given by
> $$|\overline{P_1P_2}| = \sqrt{(x_2 - x_1)^2 + (y_2 - y_1)^2}$$

FIGURE 9

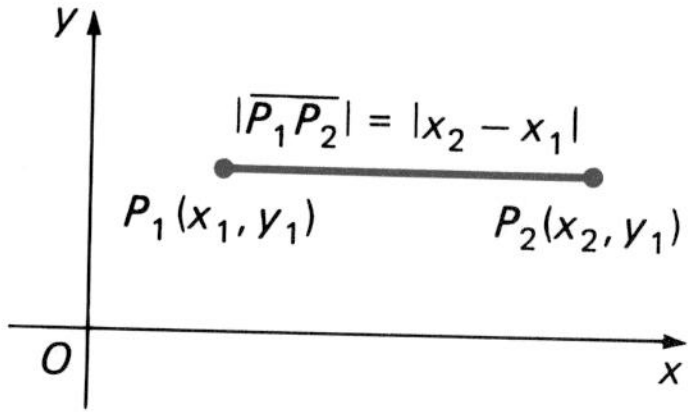

Observe that if P_1 and P_2 are on the same horizontal line, as in Figure 9, then $y_2 = y_1$ and

$$|\overline{P_1P_2}| = \sqrt{(x_2 - x_1)^2 + 0^2}$$
$$\Leftrightarrow \quad |\overline{P_1P_2}| = |x_2 - x_1| \qquad (\text{because } \sqrt{a^2} = |a|)$$

Furthermore, if P_1 and P_2 are on the same vertical line, as in Figure 10, then $x_2 = x_1$ and

$$|\overline{P_1P_2}| = \sqrt{0^2 + (y_2 - y_1)^2}$$
$$\Leftrightarrow \quad |\overline{P_1P_2}| = |y_2 - y_1|$$

In the following example, we use the converse of the Pythagorean theorem.

Converse of the Pythagorean Theorem

> If a, b, and c are the lengths of the sides of a triangle and $a^2 + b^2 = c^2$, then the triangle is a right triangle, and c is the length of the hypotenuse.

FIGURE 10

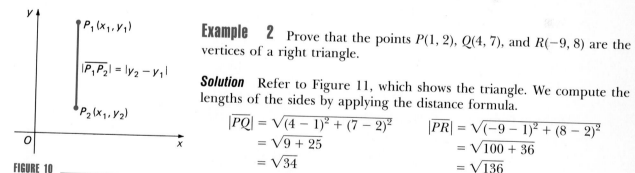

Example 2 Prove that the points $P(1, 2)$, $Q(4, 7)$, and $R(-9, 8)$ are the vertices of a right triangle.

Solution Refer to Figure 11, which shows the triangle. We compute the lengths of the sides by applying the distance formula.

$$\begin{aligned} |\overline{PQ}| &= \sqrt{(4-1)^2 + (7-2)^2} \\ &= \sqrt{9 + 25} \\ &= \sqrt{34} \end{aligned} \qquad \begin{aligned} |\overline{PR}| &= \sqrt{(-9-1)^2 + (8-2)^2} \\ &= \sqrt{100 + 36} \\ &= \sqrt{136} \end{aligned}$$

$$\begin{aligned} |\overline{QR}| &= \sqrt{(-9-4)^2 + (8-7)^2} \\ &= \sqrt{169 + 1} \\ &= \sqrt{170} \end{aligned}$$

FIGURE 11

Therefore

$$\begin{aligned} |\overline{PQ}|^2 + |\overline{PR}|^2 &= 34 + 36 \\ &= 170 \\ &= |\overline{QR}|^2 \end{aligned}$$

Thus the triangle is a right triangle and the hypotenuse is the side connecting the points Q and R. ■

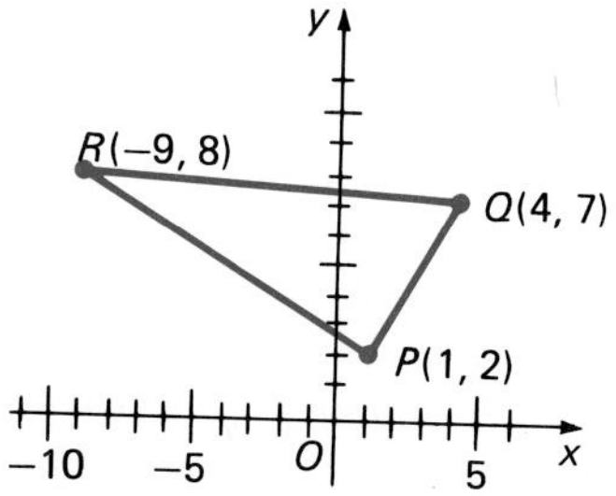

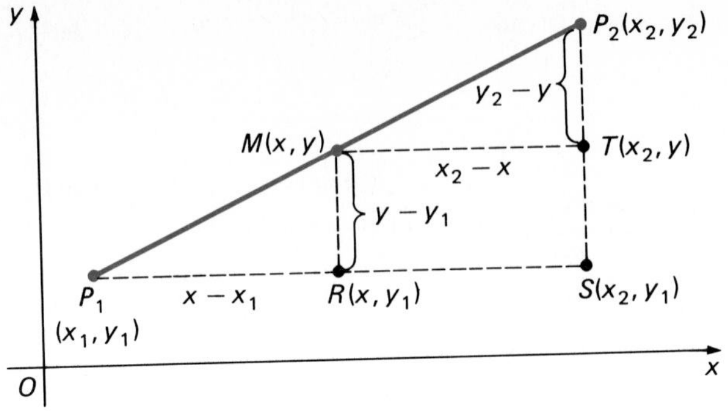

FIGURE 12

We now obtain the formulas for finding the midpoint of a line segment. Let $M(x, y)$ be the midpoint of the line segment from $P_1(x_1, y_1)$ to $P_2(x_2, y_2)$. Refer to Figure 12.

Because triangles P_1RM and MTP_2 are congruent,

$$|\overline{P_1R}| = |\overline{MT}| \qquad \text{and} \qquad |\overline{RM}| = |\overline{TP_2}|$$

Thus

$$x - x_1 = x_2 - x \qquad y - y_1 = y_2 - y$$
$$2x = x_1 + x_2 \qquad\quad 2y = y_1 + y_2$$
$$x = \frac{x_1 + x_2}{2} \qquad\quad y = \frac{y_1 + y_2}{2}$$

Midpoint Formulas

If $M(x, y)$ is the midpoint of the line segment from $P_1(x_1, y_1)$ to $P_2(x_2, y_2)$, then

$$x = \frac{x_1 + x_2}{2} \qquad y = \frac{y_1 + y_2}{2}$$

In the derivation of the midpoint formulas it was assumed that $x_2 > x_1$ and $y_2 > y_1$. The same formulas are obtained by using any orderings of these numbers.

Example 3

a) Determine the coordinates of the midpoint M of the line segment from $A(5, -3)$ to $B(-1, 6)$.

b) Plot the points A, M, and B, and show that $|\overline{AM}| = |\overline{MB}|$.

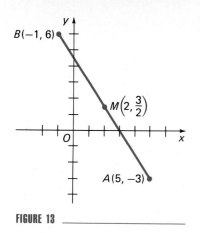

FIGURE 13

Solution   a) From the midpoint formulas, if M is the point (x, y), then

$$x = \frac{5 - 1}{2} \qquad y = \frac{-3 + 6}{2}$$

$$= 2 \qquad\qquad = \frac{3}{2}$$

Thus M is the point $(2, \frac{3}{2})$.

b) Figure 13 shows the points A, M, and B. From the distance formula,

$$|\overline{AM}| = \sqrt{(2 - 5)^2 + (\tfrac{3}{2} + 3)^2} \qquad |\overline{MB}| = \sqrt{(-1 - 2)^2 + (6 - \tfrac{3}{2})^2}$$

$$= \sqrt{9 + \tfrac{81}{4}} \qquad\qquad\qquad = \sqrt{9 + \tfrac{81}{4}}$$

$$= \tfrac{3}{2}\sqrt{13} \qquad\qquad\qquad\quad = \tfrac{3}{2}\sqrt{13}$$

Therefore $|\overline{AM}| = |\overline{MB}|$.   ■

EXERCISES 3.1

In Exercises 1 and 2, plot the point P on a rectangular cartesian coordinate system and state the quadrant in which it lies.

1. (a) $P(3, 7)$; (b) $P(-4, -6)$; (c) $P(2, -5)$;
    (d) $P(-1, 4)$
2. (a) $P(5, 6)$; (b) $P(8, -1)$; (c) $P(-7, -2)$;
    (d) $P(-9, 3)$

In Exercises 3 through 8, plot the point P and each of the following points as may apply: (a) The point Q such that the line through Q and P is perpendicular to the x axis and bisected by it. Give the coordinates of Q. (b) The point R such that the line through P and R is perpendicular to and bisected by the y axis. Give the coordinates of R. (c) The point S such that the line through P and S is bisected by the origin. Give the coordinates of S. (d) The point T such that the line through P and T is perpendicular to and bisected by the 45° line through the origin bisecting the first and third quadrants. Give the coordinates of T.

3. $P(1, -2)$ 4. $P(-2, 2)$
5. $P(2, 2)$ 6. $P(-2, -2)$
7. $P(-1, -3)$ 8. $P(0, -3)$

In Exercises 9 through 12, for the given points A and B, find the directed distances: (a) $\overline{AB}$; (b) $\overline{BA}$.

9. $A(-1, 7)$ and $B(6, 7)$ 10. $A(-2, 3)$ and $B(-4, 3)$
11. $A(3, -4)$ and $B(3, -8)$
12. $A(-4, -5)$ and $B(-4, 6)$

13. Given that A is the point $(-2, 3)$ and B is the point $(x, 3)$, find x such that (a) $\overline{AB} = -8$; (b) $\overline{BA} = -8$.
14. Given that A is the point $(-4, y)$ and B is the point $(-4, 3)$, find y such that (a) $\overline{AB} = -3$; (b) $\overline{BA} = -3$.

In Exercises 15 through 18, do the following: (a) plot the points A and B and draw the line segment between them; (b) find the distance between A and B; (c) find the midpoint of the line segment from A to B.

15. $A(1, 3)$ and $B(-2, 7)$ 16. $A(-4, -1)$ and $B(4, 5)$
17. $A(8, 5)$ and $B(3, -7)$ 18. $A(6, -5)$ and $B(2, -2)$

In Exercises 19 through 22, do the following: (a) determine the coordinates of the midpoint M of the line segment from A to B; (b) plot the points A, M, and B and show that $|\overline{AM}| = |\overline{MB}|$.

19. $A(-4, 7)$ and $B(1, -3)$ 20. $A(3, 4)$ and $B(4, -3)$
21. $A(1, 3)$ and $B(4, 0)$ 22. $A(0, -2)$ and $B(2, 0)$

In Exercises 23 and 24, draw the triangle having vertices at A, B, and C and find the lengths of the sides.

23. $A(4, -5)$, $B(-2, 3)$, $C(-1, 7)$
24. $A(2, 3)$, $B(3, -3)$, $C(-1, -1)$
25. A median of a triangle is a line segment from a vertex to the midpoint of the opposite side. Find the length of the medians of the triangle having vertices $A(2, 3)$, $B(3, -3)$, and $C(-1, -1)$.

26. Find the length of the medians of the triangle having vertices $A(-3, 5)$, $B(2, 4)$, and $C(-1, -4)$.

27. Prove that the triangle with vertices $A(3, -6)$, $B(8, -2)$, and $C(-1, -1)$ is a right triangle.

28. Find the midpoints of the diagonals of the quadrilateral whose vertices are $(0, 0)$, $(0, 4)$, $(3, 5)$, and $(3, 1)$.

29. Prove that the points $A(-7, 2)$, $B(3, -4)$, and $C(1, 4)$ are the vertices of an isosceles triangle.

30. Prove that the points $A(-4, -1)$, $B(-2, -3)$, $C(4, 3)$, and $D(2, 5)$ are the vertices of a rectangle.

31. By using the distance formula, prove that the points $(-3, 2)$, $(1, -2)$, and $(9, -10)$ lie on a line.

32. Determine whether the points $(14, 7)$, $(2, 2)$, and $(-4, -1)$ lie on a line by using the distance formula.

33. Prove that the points $A(6, -13)$, $B(-2, 2)$, $C(13, 10)$, and $D(21, -5)$ are the vertices of a square. Find the length of a diagonal.

34. One end of a line segment is the point $(-4, 2)$ and the midpoint is $(3, -1)$; find the coordinates of the other end of the line segment.

35. One end of a line segment is the point $(6, -2)$ and the midpoint is $(-1, 5)$; find the coordinates of the other end of the line segment.

36. By showing that the three sides have the same length, prove that the three points $A(-5, 0)$, $B(3, 0)$, and $C(-1, 4\sqrt{3})$ are the vertices of an equilateral triangle. Draw a sketch of the triangle.

37. The abscissa of a point is -6 and its distance from the point $(1, 3)$ is $\sqrt{74}$. Find the ordinate of the point.

38. Given the two points $A(-3, 4)$ and $B(2, 5)$, find the coordinates of a point P on the line through A and B such that P is (a) twice as far from A as from B and (b) twice as far from B as from A.

3.2 Graphs of Equations

In Section 3.1 we demonstrated how a rectangular cartesian coordinate system is used to obtain some geometrical facts by algebra. We now show how such a coordinate system enables us to associate a *graph* (a geometric concept) with an equation (an algebraic concept). Before defining the graph of an equation, we consider the particular equation

$$y = 3x - 2 \qquad (1)$$

where (x, y) is a point in R^2. A **solution** of this equation is an ordered pair of real numbers, one for x and one for y, that satisfies the equation. For example, if x is replaced by 2 in the equation, we see that $y = 4$; thus the ordered pair $(2, 4)$ constitutes a solution. If any number is substituted for x in the right side of Equation (1), a corresponding value for y is obtained. Therefore, Equation (1) has an unlimited number of solutions. Table 1 gives a few of them.

TABLE 1

x	-2	-1	0	1	2	3
$y = 3x - 2$	-8	-5	-2	1	4	7

Definition

The Graph of an Equation
The **graph of an equation** in R^2 is the set of all points in R^2 whose coordinates are numbers satisfying the equation.

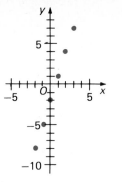

FIGURE 1

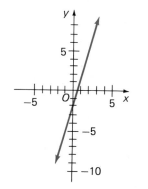

FIGURE 2

Because Equation (1) has an unlimited number of solutions, its graph consists of an unlimited number of points. The six points given by the solutions in Table 1 are plotted in Figure 1; they appear to lie on a line. In fact, it can be shown that every solution of Equation (1) corresponds to a point on the line and, conversely, the coordinates of each point on the line satisfy Equation (1). Hence the line is the graph of the equation. A sketch of this graph appears in Figure 2.

Unlike the sketch, the complete graph does not terminate but continues in both directions, as indicated by the arrow heads. Observe that the coordinates of any point (x, y) on the line satisfy Equation (1). The coordinates of any point not on the line do not satisfy the equation.

We now obtain a sketch of the graph of the equation

$$y = x^2 - 3$$

which is a second-degree (or quadratic) equation. Table 2 gives a few representative solutions for this equation. These solutions are obtained by substituting arbitrary numbers for x in the right side and computing the corresponding values for y.

TABLE 2

x	0	1	2	3	4	-1	-2	-3	-4
$y = x^2 - 3$	-3	-2	1	6	13	-2	1	6	13

In Figure 3 we have plotted the points having as coordinates the number pairs (x, y) in Table 2. We can get a better idea of the appearance of the graph by plotting additional points between these points. In particular, the points at which the graph intersects the x axis are found by substituting 0 for y and solving for x. Doing this, we have

$$x^2 - 3 = 0$$
$$x^2 = 3$$
$$x = \pm\sqrt{3}$$

Therefore the points $(\sqrt{3}, 0)$ and $(-\sqrt{3}, 0)$ are on the graph. The abscissas of these points, $\sqrt{3}$ and $-\sqrt{3}$, are called the x intercepts of the graph. Similarly, because the graph intersects the y axis at the point $(0, -3)$, -3 is called the y intercept of the graph. More points are obtained from the solutions that appear in Table 3.

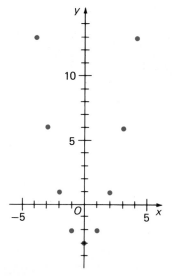

FIGURE 3

TABLE 3

x	$\frac{1}{2}$	$\frac{3}{2}$	$\frac{5}{2}$	$\frac{7}{2}$	$-\frac{1}{2}$	$-\frac{3}{2}$	$-\frac{5}{2}$	$-\frac{7}{2}$
$y = x^2 - 3$	$-\frac{11}{4}$	$-\frac{3}{4}$	$\frac{13}{4}$	$\frac{37}{4}$	$-\frac{11}{4}$	$-\frac{3}{4}$	$\frac{13}{4}$	$\frac{37}{4}$

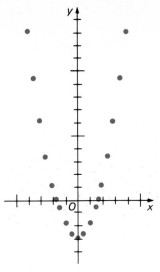

FIGURE 4 _____

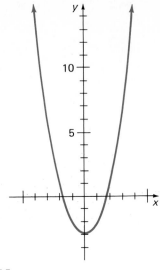

FIGURE 5 _____

Figure 4 shows the points obtained from the solutions in both Tables 2 and 3 as well as the points of intersection of the graph with the x axis. If we connect these points with a smooth curve, we get the sketch of the graph appearing in Figure 5.

Example **1** Draw a sketch of the graph of the equation $y^2 - 4x = 0$.

Solution Solving the equation for y, we have

$$y^2 = 4x$$
$$y = \pm 2\sqrt{x}$$

Thus the given equation is equivalent to the two equations

$$y = 2\sqrt{x} \quad \text{and} \quad y = -2\sqrt{x}$$

The coordinates of any point that satisfy either one of these two equations will satisfy the given equation. Conversely, the coordinates of any point satisfying the given equation will satisfy either $y = 2\sqrt{x}$ or $y = -2\sqrt{x}$. Table 4 gives some of these coordinates. Observe that for any value of $x < 0$, there is no real value for y. Also, for each value of $x > 0$ there are two values for y. Figure 6 shows a sketch of the graph.

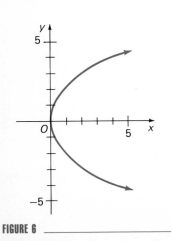

FIGURE 6 _____

TABLE 4

x	0	1	1	2	2	3	3	4	4
y	0	2	-2	$2\sqrt{2}$	$-2\sqrt{2}$	$2\sqrt{3}$	$-2\sqrt{3}$	4	-4

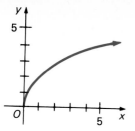

FIGURE 7

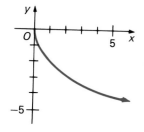

FIGURE 8

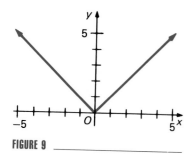

FIGURE 9

The graphs in Figures 5 and 6 are called *parabolas*. We discuss parabolas in Section 3.4.

Example   2   Draw sketches of the graphs of the equations

$$y = 2\sqrt{x} \quad \text{and} \quad y = -2\sqrt{x}$$

Solution   Recall from Example 1 that these two equations together are equivalent to the equation $y^2 - 4x = 0$. In the equation $y = 2\sqrt{x}$, the value of y is nonnegative. Hence the graph of this equation, which appears in Figure 7, is the upper half of the graph in Figure 6.

Similarly, the graph of the equation $y = -2\sqrt{x}$, a sketch of which appears in Figure 8, is the lower half of the graph of Figure 6.   ■

Example   3   Draw a sketch of the graph of the equation $y = |x|$.

Solution   From the definition of the absolute value of a number

$$y = \begin{cases} x & \text{if} \quad x \geq 0 \\ -x & \text{if} \quad x < 0 \end{cases}$$

Table 5 gives some values of x and y satisfying the given equation, and Figure 9 shows a sketch of the graph.

TABLE 5

x	0	1	2	3	4	-1	-2	-3	-4
y	0	1	2	3	4	1	2	3	4

■

When drawing sketches of graphs, it is often helpful to consider properties of *symmetry*.

Definition

> **Symmetry of Two Points**
> Two points P and Q are said to be **symmetric with respect to a line** if and only if the line is the perpendicular bisector of the line segment PQ. Two points P and Q are said to be **symmetric with respect to a third point** if and only if the third point is the midpoint of the line segment PQ.

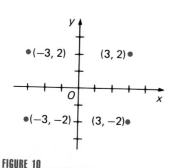

FIGURE 10

Illustration   1

The points $(3, 2)$ and $(3, -2)$ are symmetric with respect to the x axis, the points $(3, 2)$ and $(-3, 2)$ are symmetric with respect to the y axis, and the points $(3, 2)$ and $(-3, -2)$ are symmetric with respect to the origin. See Figure 10.   ■

In general, the points (x, y) and $(x, -y)$ are symmetric with respect to the x axis, the points (x, y) and $(-x, y)$ are symmetric with respect to the y axis, and the points (x, y) and $(-x, -y)$ are symmetric with respect to the origin.

Definition

> **Symmetry of a Graph**
> The graph of an equation is **symmetric with respect to a line** l if and only if for every point P on the graph there is a point Q, also on the graph, such that P and Q are symmetric with respect to l. The graph of an equation is **symmetric with respect to a point** R if and only if for every point P on the graph there is a point S, also on the graph, such that P and S are symmetric with respect to R.

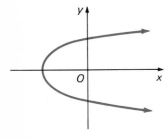

FIGURE 11

Figure 11 shows the sketch of a graph symmetric with respect to the x axis, Figure 12 shows one symmetric with respect to the y axis, and Figure 13 shows one symmetric with respect to the origin.

From the definition of symmetry of a graph, it follows that if a point (x, y) is on a graph symmetric with respect to the x axis, then the point $(x, -y)$ also must be on the graph. And if both the points (x, y) and $(x, -y)$ are on the graph, then the graph is symmetric with respect to the x axis. Therefore the coordinates of the point $(x, -y)$ as well as (x, y) must satisfy an equation of the graph. Hence the graph of an equation in x and y is symmetric with respect to the x axis if and only if an equivalent equation is obtained when y is replaced by $-y$ in the equation. We have thus proved part (i) in the following symmetry tests. The proofs of parts (ii) and (iii) are similar.

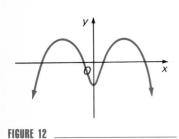

FIGURE 12

Symmetry Tests

The graph of an equation in x and y is

(i) symmetric with respect to the x axis if and only if an equivalent equation is obtained when y is replaced by $-y$ in the equation;

(ii) symmetric with respect to the y axis if and only if an equivalent equation is obtained when x is replaced by $-x$ in the equation;

(iii) symmetric with respect to the origin if and only if an equivalent equation is obtained when x is replaced by $-x$ and y is replaced by $-y$ in the equation.

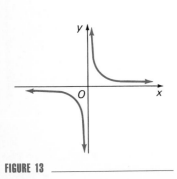

FIGURE 13

Refer back to the graph in Figure 5. It is symmetric with respect to the y axis and its equation is $y = x^2 - 3$. Observe that an equivalent equation is obtained when x is replaced by $-x$. In Example 1 we have the equation $y^2 - 4x = 0$, for which an equivalent equation is obtained when y is replaced by $-y$, and its graph, sketched in Figure 6, is symmetric with respect to the x

axis. The following example gives a graph that is symmetric with respect to the origin.

Example   4   Draw a sketch of the graph of the equation

$$2y = x^3$$

Solution   We test for symmetry. If x is replaced by $-x$ and y is replaced by $-y$ in the given equation, we have

$$2(-y) = (-x)^3$$
$$-2y = -x^3$$

which is equivalent to the given equation. Therefore, by symmetry test (iii), the graph is symmetric with respect to the origin. When $x = 0$, then $y = 0$; thus the origin is on the graph. We plot some points in the first quadrant by using the values of x and y in Table 6.

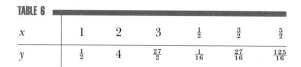

TABLE 6

x	1	2	3	$\frac{1}{2}$	$\frac{3}{2}$	$\frac{5}{2}$
y	$\frac{1}{2}$	4	$\frac{27}{2}$	$\frac{1}{16}$	$\frac{27}{16}$	$\frac{125}{16}$

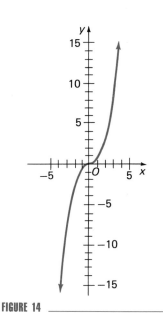

FIGURE 14

Then, from symmetry, we obtain the portion of the graph in the third quadrant. The required sketch appears in Figure 14.   ■

The first definition in this section pertains to the graph of an equation. We now define what is meant by an *equation of a graph*.

Definition

> **Equation of a Graph**
> An **equation of a graph** is an equation that is satisfied by the coordinates of those, and only those, points on the graph.

From this definition, it follows that an equation of a graph has the following properties:

1.   If a point $P(x, y)$ is on the graph, then its coordinates satisfy the equation.

2.   If a point $P(x, y)$ is not on the graph, then its coordinates do not satisfy the equation.

One of the simplest curves having a second-degree equation in two variables is the *circle*.

Definition

> **Circle**
> A **circle** is the set of all points in a plane equidistant from a fixed point. The fixed point is called the **center** of the circle and the constant equal distance is called the **radius** of the circle.

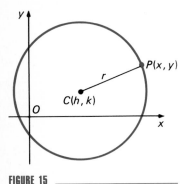

FIGURE 15

To obtain an equation of the circle having center at $C(h, k)$ and radius r, we use the distance formula. Refer to Figure 15. The point $P(x, y)$ is on the circle if and only if $|\overline{PC}| = r$; that is, if and only if

$$\sqrt{(x - h)^2 + (y - k)^2} = r$$

This equation is true if and only if

$$(x - h)^2 + (y - k)^2 = r^2 \qquad (r > 0)$$

This equation is satisfied by the coordinates of those and only those points that lie on the circle; and therefore it is an equation of the circle. We state this result formally.

Equation of a Circle

> An equation of the circle with center at the point (h, k) and radius r is
> $$(x - h)^2 + (y - k)^2 = r^2$$

Illustration 2

If the center of a circle is at $(3, -2)$ and its radius is 5, then we obtain an equation of the circle by using the above equation with $h = 3$, $k = -2$, and $r = 5$. We have

$$(x - 3)^2 + [y - (-2)]^2 = 5^2$$
$$(x - 3)^2 + (y + 2)^2 = 25$$
$$x^2 - 6x + 9 + y^2 + 4y + 4 = 0$$
$$x^2 + y^2 - 6x + 4y - 12 = 0 \qquad \blacksquare$$

If the center of a circle is at the origin, then $h = 0$ and $k = 0$; therefore an equation of the circle is

$$x^2 + y^2 = r^2$$

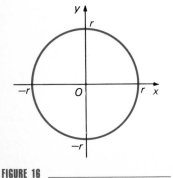

FIGURE 16

Such a circle appears in Figure 16. If a circle has radius 1, it is called a **unit circle.** If the center and radius of a circle are known, the circle can be drawn by using a compass.

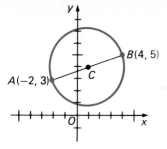

FIGURE 17

Example   5   Find an equation of the circle having a diameter with end-points at $A(-2, 3)$ and $B(4, 5)$.

Solution   The midpoint of the line segment from A to B is the center of the circle. See Figure 17. If $C(h, k)$ is the center of the circle, then

$$h = \frac{-2 + 4}{2} \qquad k = \frac{3 + 5}{2}$$

$$= 1 \qquad\qquad = 4$$

The center is at $C(1, 4)$. The radius of the circle can be computed as either $|\overline{CA}|$ or $|\overline{CB}|$. If $r = |\overline{CA}|$, then

$$r = \sqrt{(1 + 2)^2 + (4 - 3)^2}$$
$$= \sqrt{10}$$

An equation of the circle is therefore

$$(x - h)^2 + (y - k)^2 = r^2$$
$$(x - 1)^2 + (y - 4)^2 = (\sqrt{10})^2$$
$$x^2 - 2x + 1 + y^2 - 8y + 16 = 10$$
$$x^2 + y^2 - 2x - 8y + 7 = 0 \qquad\qquad \blacksquare$$

The equation $(x - h)^2 + (y - k)^2 = r^2$ is called the **center-radius form** of an equation of a circle. If we remove parentheses and combine like terms, we obtain

$$x^2 + y^2 - 2hx - 2ky + (h^2 + k^2 - r^2) = 0$$

By letting $D = -2h$, $E = -2k$, and $F = h^2 + k^2 - r^2$, this equation becomes

$$x^2 + y^2 + Dx + Ey + F = 0,$$

which is called the **general form** of an equation of a circle. Because every circle has a center and radius, its equation can be put in the center-radius form, and hence into the general form, as we did in Example 5. If we start with an equation of a circle in the general form, we can write it in the center-radius form by completing the squares. The next example shows the procedure.

Example   6   Find the center and radius of the circle having the equation

$$x^2 + y^2 + 6x - 2y - 15 = 0$$

Solution   The given equation may be written as

$$(x^2 + 6x) + (y^2 - 2y) = 15$$

Completing the squares of the terms in parentheses by adding 9 and 1 on both sides of the equation, we have

$$(x^2 + 6x + 9) + (y^2 - 2y + 1) = 15 + 9 + 1$$
$$(x + 3)^2 + (y - 1)^2 = 25$$

This equation is in the center-radius form with $h = -3$, $k = 1$, and $r = 5$. Thus it is an equation of a circle with its center at $(-3, 1)$ and radius 5. ∎

We now show that there are equations of the form

$$x^2 + y^2 + Dx + Ey + F = 0$$

whose graphs are not circles. Suppose when we complete the squares we obtain

$$(x - h)^2 + (y - k)^2 = d$$

If $d > 0$, we have a circle with center at (h, k) and radius $\sqrt{d}$. However, if $d < 0$, there are no real values of x and y that satisfy the equation; thus there is no graph. In such a case we can state that the graph is the empty set. Finally, if $d = 0$, we have

$$(x - h)^2 + (y - k)^2 = 0$$

The only real values of x and y satisfying this equation are $x = h$ and $y = k$. Thus the graph is the single point (h, k).

Illustration 3

Suppose we have the equation

$$x^2 + y^2 - 4x + 10y + 29 = 0$$

which can be written as

$$(x^2 - 4x) + (y^2 + 10y) = -29$$

Completing the squares of the terms in parentheses by adding 4 and 25 on both sides, we have

$$(x^2 - 4x + 4) + (y^2 + 10y + 25) = -29 + 4 + 25$$
$$(x - 2)^2 + (y + 5)^2 = 0$$

Because the only real values of x and y satisfying this equation are $x = 2$ and $y = -5$, the graph is the point $(2, -5)$. ∎

Illustration 4

The equation

$$x^2 + y^2 + 8x - 6y + 30 = 0$$

can be written as

$$(x^2 + 8x) + (y^2 - 6y) = -30$$

We complete the squares of the terms in parentheses by adding 16 and 9 on both sides of the equation to obtain

$$(x^2 + 8x + 16) + (y^2 - 6y + 9) = -30 + 16 + 9$$

$$(x + 4)^2 + (y - 3)^2 = -5$$

This equation has a negative number on the right side and the sum of the squares of two real numbers on the left. There are no real values of x and y that satisfy such an equation; thus the graph is the empty set. ■

EXERCISES 3.2

In Exercises 1 through 38, draw a sketch of the graph of the equation.

1. $y = x + 2$
2. $y = 5 - x$
3. $y = 5 - 2x$
4. $y = 4x - 3$
5. $2x + 5y + 10 = 0$
6. $3x - 2y - 6 = 0$
7. $y = 5$
8. $x = -3$
9. $y = x^2$
10. $y = -x^2$
11. $y = 4 - x^2$
12. $y = x^2 + 2$
13. $y^2 - 9x = 0$
14. $y^2 + 4x = 0$
15. $y = \sqrt{x + 5}$
16. $y = \sqrt{x - 3}$
17. $y = -\sqrt{x + 5}$
18. $y = -\sqrt{x - 3}$
19. $y^2 = x + 5$
20. $y^2 = x - 3$
21. $y = |x - 5|$
22. $y = |x + 2|$
23. $y = |x| - 5$
24. $y = |x| + 2$
25. $y = x^3$
26. $x^3 + 2y = 0$
27. $y = \sqrt{4 - x^2}$
28. $y = \sqrt{25 - x^2}$
29. $y = -\sqrt{25 - x^2}$
30. $x = -\sqrt{4 - y^2}$
31. $x = \sqrt{16 - y^2}$
32. $x^2 + y^2 = 16$
33. $x^2 + y^2 = 25$
34. $4x^2 + 4y^2 = 1$
35. $(x - 3)^2 + (y + 4)^2 = 4$
36. $(x + 1)^2 + (y - 5)^2 = 36$
37. $(x + 4)^2 + y^2 = 1$
38. $x^2 + (y - 2)^2 = 9$

In Exercises 39 through 44, find an equation of the circle with center at C and radius r. Write the equation in both the center-radius form and the general form. Draw a sketch of the circle.

39. $C(4, -3)$, $r = 5$
40. $C(0, 0)$, $r = 8$
41. $C(-5, -12)$, $r = 3$
42. $C(-1, 1)$, $r = 2$
43. $C(0, 7)$, $r = 1$
44. $C(-3, 0)$, $r = 4$

In Exercises 45 through 48, find an equation of the circle satisfying the given conditions.

45. Center is at $(1, 2)$ and through the point $(3, -1)$.
46. Center is at $(-3, 4)$ and through the point $(2, 0)$.
47. Diameter has endpoints at $(3, -4)$ and $(7, 2)$.
48. Diameter has endpoints at $(-1, -5)$ and $(4, -6)$.

In Exercises 49 through 54, find the center and radius of the circle and draw a sketch of the circle.

49. $x^2 + y^2 - 6x - 8y + 9 = 0$
50. $x^2 + y^2 - 10x - 10y + 25 = 0$
51. $x^2 + y^2 + 2x + 10y + 18 = 0$
52. $x^2 + y^2 + 6x - 1 = 0$
53. $3x^2 + 3y^2 + 4y - 7 = 0$
54. $2x^2 + 2y^2 - 2x + 2y + 7 = 0$

In Exercises 55 through 60, determine whether the graph is a circle, a point, or the empty set.

55. $x^2 + y^2 - 2x + 10y + 19 = 0$
56. $x^2 + y^2 + 2x - 4y + 5 = 0$
57. $x^2 + y^2 - 10x + 6y + 36 = 0$
58. $4x^2 + 4y^2 + 24x - 4y + 1 = 0$
59. $2x^2 + 2y^2 - 2x + 6y + 5 = 0$
60. $9x^2 + 9y^2 + 6x - 6y + 5 = 0$

3.3 Equations of a Line

Suppose it costs \$2 per unit to manufacture a particular commodity and there is a fixed daily overhead of \$400. Then if x units are produced per day and y dollars is the manufacturer's total daily cost,

$$y = 2x + 400 \tag{1}$$

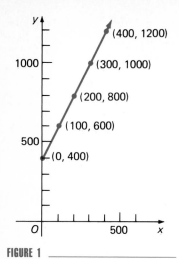

FIGURE 1

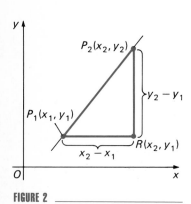

FIGURE 2

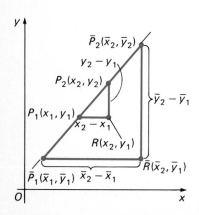

FIGURE 3

Some of the solutions of this equation are given in Table 1.

TABLE 1

x	0	100	200	300	400
$y = 2x + 400$	400	600	800	1000	1200

In Figure 1 we have plotted points whose coordinates are the number pairs in this table. They appear to lie on a line. That the graph of Equation (1) is indeed a line is established in Theorem 2 of this section. Observe that for each 100-unit increase in x, y increases by 200 units or, equivalently, for each 1-unit increase in x, y increases by 2 units. Thus the ratio of the change in y to the change in x is a constant 2. This constant ratio is called the *slope* of the line. We proceed now to arrive at a formal definition of *slope*.

Let l be a nonvertical line and $P_1(x_1, y_1)$ and $P_2(x_2, y_2)$ be any two distinct points on l. Figure 2 shows such a line. In the figure R is the point (x_2, y_1), and the points P_1, P_2, and R are vertices of a right triangle; furthermore, $\overline{P_1 R} = x_2 - x_1$ and $\overline{RP_2} = y_2 - y_1$. The number $y_2 - y_1$ gives the measure of the change in the ordinate from P_1 to P_2, and it may be positive, negative, or zero. The number $x_2 - x_1$ gives the measure of the change in the abscissa from P_1 to P_2, and it may be positive or negative. Because the line l is not vertical, $x_2 \neq x_1$, and therefore $x_2 - x_1$ is not zero. Let

$$m = \frac{y_2 - y_1}{x_2 - x_1} \tag{2}$$

The value of m computed from this equation is independent of the choice of the two points P_1 and P_2 on l. To show this, suppose we choose two different points $\overline{P}_1(\overline{x}_1, \overline{y}_1)$ and $\overline{P}_2(\overline{x}_2, \overline{y}_2)$ on line l, and compute a number $\overline{m}$ from (1):

$$\overline{m} = \frac{\overline{y}_2 - \overline{y}_1}{\overline{x}_2 - \overline{x}_1}$$

We shall show that $\overline{m} = m$. Refer to Figure 3. Triangles $\overline{P}_1 \overline{R} \overline{P}_2$ and $P_1 R P_2$ are similar; so the lengths of corresponding sides are proportional. Therefore

$$\frac{\overline{y}_2 - \overline{y}_1}{\overline{x}_2 - \overline{x}_1} = \frac{y_2 - y_1}{x_2 - x_1}$$

or

$$\overline{m} = m$$

Thus the value of m computed from Equation (2) is the same number no matter what two points on l are selected. This number m is called the *slope* of the line.

Definition

> **Slope**
>
> If $P_1(x_1, y_1)$ and $P_2(x_2, y_2)$ are any two distinct points on line l, which is not parallel to the y axis, then the slope of l, denoted by m, is given by
>
> $$m = \frac{y_2 - y_1}{x_2 - x_1}$$

Multiplying both sides of the above equation by $x_2 - x_1$, we obtain

$$y_2 - y_1 = m(x_2 - x_1)$$

It follows from this equation that if we consider a particle moving along a line, the change in the ordinate of the particle is equal to the product of the slope and the change in the abscissa.

Illustration 1

If l is the line through the points $P_1(2, 1)$ and $P_2(4, 7)$ and m is the slope of l, then

$$m = \frac{7 - 1}{4 - 2}$$
$$= 3$$

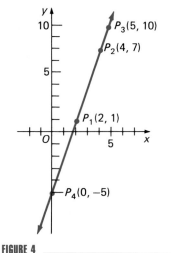

FIGURE 4

Refer to Figure 4. If a particle is moving along line l, the change in the ordinate is three times the change in the abscissa. That is, if the particle is at $P_2(4, 7)$ and the abscissa is increased by one unit, then the ordinate is increased by three units, and the particle is at the point $P_3(5, 10)$. Similarly, if the particle is at $P_1(2, 1)$ and the abscissa is decreased by two units, then the ordinate is decreased by six units, and the particle is at $P_4(0, -5)$. ■

If the slope of a line is positive, then as the abscissa of a point on the line increases, the ordinate increases. Such a line is shown in Figure 5. A line whose slope is negative appears in Figure 6. For this line, as the abscissa of a point on the line increases, the ordinate decreases. If a line is parallel to the x axis, then $y_2 = y_1$; so the slope of the line is zero. See Figure 7. If a line

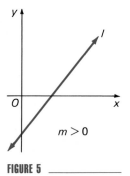

FIGURE 5

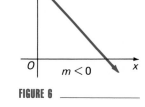

FIGURE 6

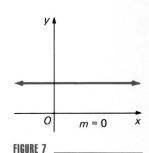

FIGURE 7

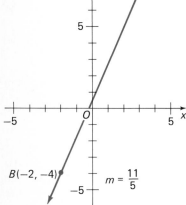

Slope is undefined

FIGURE 8

is parallel to the y axis, $x_2 = x_1$; thus the fraction $\dfrac{y_2 - y_1}{x_2 - x_1}$ is meaningless because we cannot divide by zero. For this reason, lines parallel to the y axis are excluded in the definition of slope. Thus the slope of a vertical line is not defined. See Figure 8.

Example **1** Draw a sketch of the line through each pair of points and determine the slope of the line.

a) $A(3, 7)$ and $B(-2, -4)$

b) $A(-2, 5)$ and $B(2, -3)$

c) $A(-3, 4)$ and $B(5, 4)$

d) $A(5, 3)$ and $B(5, -1)$

Solution Sketches of the lines appear in Figure 9(a)–(d). We compute the slope by applying the definition.

a) $m = \dfrac{-4 - 7}{-2 - 3}$ b) $m = \dfrac{-3 - 5}{2 - (-2)}$ c) $m = \dfrac{4 - 4}{5 - (-3)}$

$= \dfrac{-11}{-5}$ $= \dfrac{-8}{4}$ $= \dfrac{0}{8}$

$= \dfrac{11}{5}$ $= -2$ $= 0$

d) Because the line is vertical, the slope is not defined. If you attempt to use the definition to compute the slope, you obtain zero in the denominator.

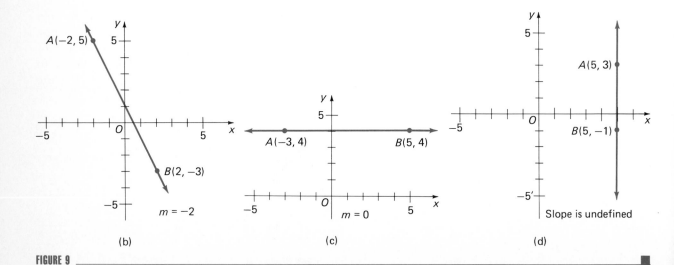

FIGURE 9

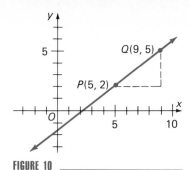

FIGURE 10

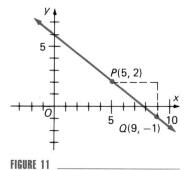

FIGURE 11

In the following illustration we show how to draw the line that passes through a particular point and has a given slope.

Illustration 2

a) Suppose a line contains the point $P(5, 2)$ and its slope is $\frac{3}{4}$. To determine another point on the line we start at P and, because the slope is $\frac{3}{4}$, we go 4 units to the right and then 3 units upward. We have then the point $Q(9, 5)$ also on the line. A sketch of the line is drawn through points P and Q as in Figure 10.

b) If a line through $P(5, 2)$ has the negative slope $-\frac{3}{4}$, we obtain another point on the line by starting at P and then going 4 units to the right and 3 units downward. This gives the point $Q(9, -1)$. Figure 11 shows a sketch of the line through these points P and Q. ∎

Because a point $P_1(x_1, y_1)$ and a slope m determine a unique line, we should be able to obtain an equation of this line. Let $P(x, y)$ be any point on the line except (x_1, y_1). Then since the slope of the line through P_1 and P is m, we have from the definition of slope

$$\frac{y - y_1}{x - x_1} = m$$

$$\boxed{y - y_1 = m(x - x_1)}$$

This equation is called the **point-slope form** of an equation of the line. It gives an equation of the line if a point on the line and the slope of the line are known.

Illustration 3

To find an equation of the line through the points $A(6, -3)$ and $B(-2, 3)$, we first compute m.

$$m = \frac{3 - (-3)}{-2 - 6}$$

$$= \frac{6}{-8}$$

$$= -\frac{3}{4}$$

Using the point-slope form of an equation of the line with A as P_1, we have

$$y - (-3) = -\tfrac{3}{4}(x - 6)$$
$$4y + 12 = -3x + 18$$
$$3x + 4y - 6 = 0$$

If B is taken as P_1 in the point-slope form, we have

$$y - 3 = -\tfrac{3}{4}[x - (-2)]$$
$$4y - 12 = -3x - 6$$
$$3x + 4y - 6 = 0$$

which of course is the same equation. ∎

If in the point-slope form we choose the particular point $(0, b)$ (that is, the point where the line intersects the y axis) for the point (x_1, y_1), we have

$$y - b = m(x - 0)$$

$\Leftrightarrow$

$$y = mx + b$$

The number b, the ordinate of the point where the line intersects the y axis, is the **y intercept** of the line. Consequently, the preceding equation is called the **slope-intercept form** of an equation of the line. This form is especially useful because it enables us to find the slope of a line from its equation. It is also important because it expresses the y coordinate of a point on the line explicitly in terms of its x coordinate.

Example 2 Find the slope of the line having the equation

$$6x + 5y - 7 = 0$$

Solution To obtain an equation of the line in slope-intercept form, we solve the given equation for y.

$$5y = -6x + 7$$
$$y = -\tfrac{6}{5}x + \tfrac{7}{5}$$

This equation is in the form $y = mx + b$, where $m = -\tfrac{6}{5}$ and $b = \tfrac{7}{5}$. Therefore the slope is $-\tfrac{6}{5}$. ∎

Because the slope of a vertical line is undefined, we cannot apply the point-slope form to obtain its equation. We use instead the following theorem, which also gives an equation of a horizontal line.

Theorem 1

(i) An equation of the vertical line having x intercept a is

$$x = a$$

(ii) An equation of the horizontal line having y intercept b is

$$y = b$$

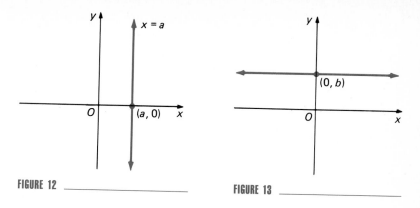

Proof

(i) Figure 12 shows the vertical line that intersects the x axis at the point $(a, 0)$. This line contains those and only those points on the line having the same abscissa. So $P(x, y)$ is any point on the line if and only if

$$x = a$$

(ii) The horizontal line that intersects the y axis at the point $(0, b)$ appears in Figure 13. For this line, $m = 0$. Therefore, from the slope-intercept form, an equation of this line is

$$y = b$$ ∎

We have shown that an equation of a nonvertical line is of the form $y = mx + b$ and an equation of a vertical line is of the form $x = a$. Because each of these equations is a special case of an equation of the form

$$Ax + By + C = 0 \tag{3}$$

where A, B, and C are constants, and A and B are not both zero, it follows that every line has an equation of the form of Equation (3). The converse of this fact is given by the next theorem.

Theorem 2

> The graph of the equation
>
> $$Ax + By + C = 0$$
>
> where A, B, and C are constants and where not both A and B are zero is a line.

The proof of this theorem is left as an exercise. See Exercise 51.
    Equation (3) is called a **linear equation** because its graph is a line. It is the general equation of the first degree in x and y.

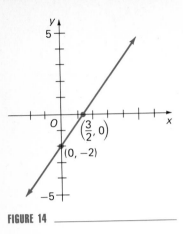

FIGURE 14

Since two points determine a line, to draw a sketch of its graph, we need only determine the coordinates of two points on the line, plot the points, and then draw the line. Any two points will suffice, but it is usually convenient to plot the two points where the line intersects the axes.

Illustration 4

To draw a sketch of the line having the equation

$$4x - 3y = 6$$

we first find the x intercept a and the y intercept b. In the equation we substitute 0 for y and get $a = \frac{3}{2}$. Substituting 0 for x, we obtain $b = -2$. Thus we have the line appearing in Figure 14. ■

An application of slopes is given by the following theorem.

Theorem 3

> If l_1 and l_2 are two distinct nonvertical lines having slopes m_1 and m_2, respectively, then l_1 and l_2 are parallel if and only if $m_1 = m_2$.

Proof Let equations of l_1 and l_2 be, respectively,

$$y = m_1x + b_1 \quad \text{and} \quad y = m_2x + b_2$$

See Figure 15, showing the two lines intersecting the y axis at the points $B_1(0, b_1)$ and $B_2(0, b_2)$. Let the vertical line $x = 1$ intersect l_1 at the point $A_1(1, m_1 + b_1)$ and l_2 at the point $A_2(1, m_2 + b_2)$. Then

$$|\overline{B_1B_2}| = b_2 - b_1 \quad \text{and} \quad |\overline{A_1A_2}| = (m_2 + b_2) - (m_1 + b_1)$$

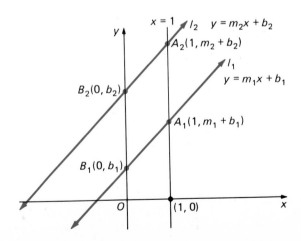

FIGURE 15

The two lines are parallel if and only if the vertical distances $|\overline{B_1B_2}|$ and $|\overline{A_1A_2}|$ are equal; that is, l_1 and l_2 are parallel if and only if

$$b_2 - b_1 = (m_2 + b_2) - (m_1 + b_1)$$
$$b_2 - b_1 = m_2 + b_2 - m_1 - b_1$$
$$m_1 = m_2$$

Thus l_1 and l_2 are parallel if and only if $m_1 = m_2$.   ■

Illustration 5

Let l_1 be the line through the points $A(1, 2)$ and $B(3, -6)$ and m_1 be the slope of l_1; and let l_2 be the line through the points $C(2, -5)$ and $D(-1, 7)$ and m_2 be the slope of l_2. Then

$$m_1 = \frac{-6 - 2}{3 - 1} \qquad m_2 = \frac{7 - (-5)}{-1 - 2}$$

$$= \frac{-8}{2} \qquad\qquad = \frac{12}{-3}$$

$$= -4 \qquad\qquad = -4$$

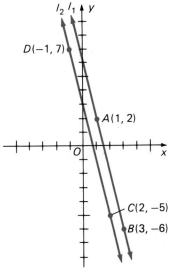

FIGURE 16

Because $m_1 = m_2$, it follows from Theorem 3 that l_1 and l_2 are parallel. See Figure 16.   ■

Any two distinct points determine a line. Three distinct points may or may not lie on the same line. If three or more points lie on the same line, they are said to be **collinear.** Hence three points A, B, and C are collinear if and only if the line through the points A and B is the same as the line through the points B and C. Because the line through A and B and the line through B and C both contain the point B, they are the same line if and only if their slopes are equal.

Example 3 Determine by means of slopes whether the points $A(-3, -4)$, $B(2, -1)$, and $C(7, 2)$ are collinear.

Solution If m_1 is the slope of the line through A and B, and m_2 is the slope of the line through B and C, then

$$m_1 = \frac{-1 - (-4)}{2 - (-3)} \qquad m_2 = \frac{2 - (-1)}{7 - 2}$$

$$= \frac{3}{5} \qquad\qquad = \frac{3}{5}$$

Hence $m_1 = m_2$. Therefore the line through A and B and the line through B and C have the same slope and contain the common point B. Thus they are the same line, and therefore A, B, and C are collinear.   ■

We now state and prove a theorem regarding the slopes of two perpendicular lines.

Theorem 4

> Two nonvertical lines l_1 and l_2, having slopes m_1 and m_2, respectively, are perpendicular if and only if $m_1 m_2 = -1$.

Proof Let us choose the coordinate axes so that the origin is at the point of intersection of l_1 and l_2. See Figure 17.

Because neither l_1 nor l_2 is vertical, these two lines intersect the line $x = 1$ at points P_1 and P_2, respectively. The abscissa of both P_1 and P_2 is 1. Let $\bar{y}$ be the ordinate of P_1. Since l_1 contains the points $(0, 0)$ and $(1, \bar{y})$ and its slope is m_1, then

$$m_1 = \frac{\bar{y} - 0}{1 - 0}$$

Thus $\bar{y} = m_1$. Similarly, the ordinate of P_2 is shown to be m_2. From the Pythagorean theorem and its converse, triangle $P_1 O P_2$ is a right triangle if and only if

$$|\overline{OP_1}|^2 + |\overline{OP_2}|^2 = |\overline{P_1 P_2}|^2 \qquad (4)$$

By applying the distance formula, we obtain

$$|\overline{OP_1}|^2 = (1 - 0)^2 + (m_1 - 0)^2 \qquad |\overline{OP_2}|^2 = (1 - 0)^2 + (m_2 - 0)^2$$
$$= 1 + m_1{}^2 \qquad\qquad = 1 + m_2{}^2$$

$$|\overline{P_1 P_2}|^2 = (1 - 1)^2 + (m_2 - m_1)^2$$
$$= m_2{}^2 - 2m_1 m_2 + m_1{}^2$$

Substituting into (4), we can conclude that $P_1 O P_2$ is a right triangle if and only if

$$(1 + m_1{}^2) + (1 + m_2{}^2) = m_2{}^2 - 2m_1 m_2 + m_1{}^2$$
$$2 = -2m_1 m_2$$
$$m_1 m_2 = -1 \qquad \blacksquare$$

Because $m_1 m_2 = -1$ is equivalent to

$$m_1 = -\frac{1}{m_2} \qquad \text{and} \qquad m_2 = -\frac{1}{m_1}$$

Theorem 4 states that two nonvertical lines are perpendicular if and only if the slope of one is the negative reciprocal of the slope of the other.

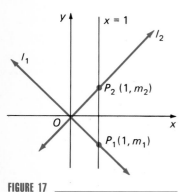

FIGURE 17

Example 4 Given the line l having the equation

$$5x + 4y - 20 = 0$$

find an equation of the line through the point $(2, -3)$ that is (a) parallel to l, and (b) perpendicular to l.

Solution We first determine the slope of l by writing its equation in the slope-intercept form. Solving the equation for y, we have

$$4y = -5x + 20$$
$$y = -\tfrac{5}{4}x + 5$$

The slope of l is the coefficient of x, which is $-\tfrac{5}{4}$.

a) The slope of a line parallel to l is also $-\tfrac{5}{4}$. Because the required line contains the point $(2, -3)$, we use the point-slope form, which gives

$$y - (-3) = -\tfrac{5}{4}(x - 2)$$
$$4y + 12 = -5x + 10$$
$$5x + 4y + 2 = 0$$

b) The slope of a line perpendicular to l is the negative reciprocal of $-\tfrac{5}{4}$, which is $\tfrac{4}{5}$. From the point-slope form, an equation of the line through $(2, -3)$ and having slope $\tfrac{4}{5}$ is

$$y - (-3) = \tfrac{4}{5}(x - 2)$$
$$5y + 15 = 4x - 8$$
$$4x - 5y - 23 = 0 \qquad \blacksquare$$

Example 5 Prove by means of slopes that the four points $A(6, 2)$, $B(8, 6)$, $C(4, 8)$, and $D(2, 4)$ are the vertices of a rectangle.

Solution See Figure 18, where l_1 is the line through A and B, l_2 is the line through B and C, l_3 is the line through D and C, and l_4 is the line through A and D; m_1, m_2, m_3, and m_4 are their respective slopes. Then

$$m_1 = \frac{6 - 2}{8 - 6} \qquad m_2 = \frac{8 - 6}{4 - 8} \qquad m_3 = \frac{8 - 4}{4 - 2} \qquad m_4 = \frac{4 - 2}{2 - 6}$$

$$= 2 \qquad\qquad = -\frac{1}{2} \qquad\qquad = 2 \qquad\qquad = -\frac{1}{2}$$

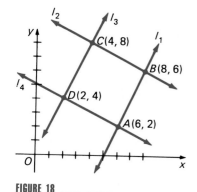

FIGURE 18

Because $m_1 = m_3$, l_1 is parallel to l_3; and because $m_2 = m_4$, l_2 is parallel to l_4. Because $m_1 m_2 = -1$, l_1 and l_2 are perpendicular. Therefore the quadrilateral has its opposite sides parallel, and a pair of adjacent sides are perpendicular. Thus the quadrilateral is a rectangle. $\qquad \blacksquare$

EXERCISES 3.3

In Exercises 1 through 6, draw a sketch of the line through the points A and B and determine the slope of the line.

1. (a) $A(1, 4)$, $B(6, 5)$; (b) $A(2, -3)$, $B(-4, 3)$
2. (a) $A(5, 2)$, $B(-2, -3)$; (b) $A(-4, 2)$, $B(8, 5)$

3. (a) $A(-4, 3)$, $B(0, 0)$; (b) $A(\tfrac{1}{3}, \tfrac{1}{2})$, $B(-\tfrac{5}{6}, \tfrac{2}{3})$
4. (a) $A(7, 0)$, $B(0, -6)$; (b) $A(-\tfrac{3}{4}, \tfrac{1}{8})$, $B(\tfrac{5}{4}, -\tfrac{1}{2})$
5. (a) $A(1, 5)$, $B(-2, 5)$; (b) $A(-2.1, 0.3)$, $B(2.3, 1.4)$
6. (a) $A(3, -5)$, $B(3, 4)$;
   (b) $A(5.2, -3.5)$, $B(-6.3, -1.4)$

In Exercises 7 and 8, draw a sketch of the line passing through the point P and having slope m.

7. (a) $P(3, 4)$, $m = \frac{2}{5}$; (b) $P(-1, 6)$, $m = -3$
8. (a) $P(4, 3)$, $m = 2$; (b) $P(2, -5)$, $m = -\frac{4}{3}$

In Exercises 9 through 20, find an equation of the line satisfying the given conditions.

9. (a) The slope is 4 and through the point $(2, -3)$; (b) through the two points $(-1, -5)$ and $(3, 6)$.
10. (a) The slope is -2 and through the point $(-4, 3)$; (b) through the two points $(3, 1)$ and $(-5, 4)$.
11. (a) The slope is $-\frac{2}{3}$ and the y intercept is 1; (b) the slope is 2 and the x intercept is $-\frac{1}{4}$.
12. (a) The slope is $\frac{3}{5}$ and the y intercept is -4; (b) the slope is -2 and the x intercept is 4.
13. (a) Through the point $(1, -7)$ and parallel to the x axis; (b) through the point $(2, 6)$ and parallel to the y axis.
14. (a) Through the point $(-5, 2)$ and parallel to the x axis; (b) through the point $(-3, -4)$ and parallel to the y axis.
15. (a) The x intercept is -3 and the y intercept is 4; (b) through the origin and bisecting the angle between the axes in the first and third quadrants.
16. (a) The x intercept is 5 and the y intercept is -6; (b) through the origin and bisecting the angle between the axes in the second and fourth quadrants.
17. Through the point $(-2, 3)$ and parallel to the line whose equation is $2x - y - 2 = 0$.
18. Through the point $(1, 4)$ and parallel to the line whose equation is $2x - 5y + 7 = 0$.
19. Through the point $(2, 4)$ and perpendicular to the line whose equation is $x - 5y + 10 = 0$.
20. Through the origin and perpendicular to the line whose equation is $2x - 5y + 6 = 0$.

In Exercises 21 through 24, find the slope and y intercept of the line having the given equation, and draw a sketch of the line.

21. (a) $x + 3y - 6 = 0$; (b) $4y - 9 = 0$
22. (a) $8x - 4y = 5$; (b) $3y - 5 = 0$
23. (a) $7x - 8y = 0$; (b) $x = 6 - 2y$
24. (a) $x = 4y - 2$; (b) $4x = 3y$

In Exercises 25 and 26, find an equation of the line through the two points and write the equation in slope-intercept form; draw a sketch of the line.

25. $(1, 3)$ and $(2, -2)$ 26. $(3, -5)$ and $(1, -2)$

In Exercises 27 and 28, draw a sketch of the line having the given equation by finding the x and y intercepts of the line.

27. (a) $3x - 2y + 6 = 0$; (b) $4x + 5y + 20 = 0$
28. (a) $5x + 2y - 10 = 0$; (b) $4x - 3y - 12 = 0$

In Exercises 29 and 30, draw a sketch of the line having the given equation.

29. (a) $x = 4$; (b) $y = -6$ 30. (a) $x = -2$; (b) $y = 7$
31. Show that the lines having the equations $3x + 5y + 7 = 0$ and $6x + 10y - 5 = 0$ are parallel, and draw sketches of their graphs.
32. Show that the lines having the equations $4x - 3y + 12 = 0$ and $8x - 6y + 15 = 0$ are parallel, and draw sketches of their graphs.
33. Show that the lines having the equations $2x - 3y + 6 = 0$ and $3x + 2y - 12 = 0$ are perpendicular, and draw sketches of their graphs.
34. Show that the lines having the equations $2y = 10 - 5x$ and $5y = 2x + 20$ are perpendicular, and draw sketches of their graphs.
35. Find the value of k such that the lines whose equations are $3x + 6ky = 7$ and $9kx + 8y = 15$ are parallel.
36. Find the value of k such that lines whose equations are $3kx + 8y = 5$ and $6y - 4kx = -1$ are perpendicular.

In Exercises 37 through 40, determine by means of slopes whether the points are collinear.

37. (a) $(2, 3)$, $(-4, -7)$, $(5, 8)$; (b) $(2, -1)$, $(1, 1)$, $(3, 4)$
38. (a) $(4, 6)$, $(1, 2)$, $(-5, -4)$; (b) $(-3, 6)$, $(3, 2)$, $(9, -2)$
39. (a) $(2, 5)$, $(-1, 4)$, $(3, -2)$; (b) $(0, 2)$, $(-3, -1)$, $(4, 6)$
40. (a) $(-1, 2)$, $(7, 4)$, $(2, -1)$; (b) $(4, -9)$, $(4, 1)$, $(4, 8)$
41. Show by means of slopes that the four points $(0, 0)$, $(-2, 1)$, $(3, 4)$, and $(5, 3)$ are vertices of a parallelogram (a quadrilateral with opposite sides parallel).
42. Show by means of slopes that the four points $(-4, -1)$, $(3, \frac{8}{3})$, $(8, -4)$, and $(2, -9)$ are vertices of a trapezoid (a quadrilateral with one pair of opposite sides parallel).
43. Show by means of slopes that the three points $(3, 1)$, $(6, 0)$, and $(4, 4)$ are the vertices of a right triangle, and find the area of the triangle.

44. Show by means of slopes that the points $(-6, 1)$, $(-4, 6)$, $(4, -3)$, and $(6, 2)$ are the vertices of a rectangle.
45. Find equations of the three medians of the triangle having vertices at $(3, -2)$, $(3, 4)$, and $(-1, 1)$.
46. Find the ordinate of the point whose abscissa is -3 and which is collinear with the points $(3, 2)$ and $(0, 5)$.
47. The producer of a particular commodity has a total cost consisting of a weekly overhead of $3000 and a manufacturing cost of $25 per unit.
    (a) Given that x units are produced per week and y dollars is the total weekly cost, write an equation involving x and y.
    (b) Draw a sketch of the graph of the equation in part (a).
48. A producer's total cost consists of a manufacturing cost of $20 per unit and a fixed daily overhead.
    (a) Given that the total cost of producing 200 units in 1 day is $4500, determine the fixed daily overhead.
    (b) Given that x units are produced per day and y dollars is the total daily cost, write an equation involving x and y.
    (c) Draw a sketch of the graph of the equation in part (b).

49. Do Exercise 48 with these changes: the producer's cost is $30 per unit and the total cost of producing 200 units in 1 day is $6600.
50. The graph of an equation relating the temperature reading in Celsius degrees and the temperature reading in Fahrenheit degrees is a line. Water freezes at $0°$ Celsius and $32°$ Fahrenheit, and water boils at $100°$ Celsius and $212°$ Fahrenheit.
    (a) Given that y degrees Fahrenheit corresponds to x degrees Celsius, write an equation involving x and y.
    (b) Draw a sketch of the graph of the equation in part (a).
    (c) What is the Fahrenheit temperature corresponding to $20°$ Celsius?
    (d) What is the Celsius temperature corresponding to $86°$ Fahrenheit?
51. Prove Theorem 2: The graph of the equation $Ax + By + C = 0$, where A, B, and C are constants and where not both A and B are zero, is a line. (*Hint:* Consider two cases $B \neq 0$ and $B = 0$. If $B \neq 0$, show that the equation is that of a line having slope $-A/B$ and y intercept $-C/B$. If $B = 0$, show that the equation is that of a vertical line.)

3.4   The Parabola

In Section 3.2 we stated that the graphs of Figures 5 and 6 in that section are called parabolas. There are many important applications of these curves. They are used in the design of parabolic mirrors, searchlights, and automobile headlights. The path of a projectile is a parabola if motion is considered to be in a plane and air resistance is neglected. Arches are sometimes parabolic in appearance, and the cable of a suspension bridge could hang in the form of a parabola. Dish antennas for receiving satellite television signals are also parabolic in shape.

A complete treatment of the parabola and its properties belongs to a course in analytic geometry. However, because of the importance of parabolas in our study of quadratic functions, we discuss them here.

In the definition of a parabola we refer to the distance from a point to a line. By such a distance we mean the length of the perpendicular line segment from the point to the line. See Figure 1, where $|\overline{PQ}|$ is the distance from point P to line l.

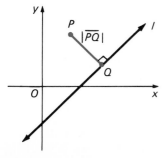

FIGURE 1

Definition

> **Parabola**
>
> A **parabola** is the set of all points in a plane equidistant from a fixed point and a fixed line. The fixed point is called the **focus** and the fixed line is called the **directrix.**

We now derive an equation of a parabola from the definition. For this equation to be as simple as possible, we choose the y axis as perpendicular to the directrix and containing the focus. The origin is taken as the point on the y axis midway between the focus and the directrix. Observe that we are choosing the axes (*not* the parabola) in a special way. See Figure 2.

Let p be the directed distance $\overline{OF}$. The focus is the point $F(0, p)$, and the directrix is the line having the equation $y = -p$. A point $P(x, y)$ is on the parabola if and only if P is equidistant from F and the directrix. That is, if $Q(x, -p)$ is the foot of the perpendicular line from P to the directrix, then P is on the parabola if and only if

$$|\overline{FP}| = |\overline{QP}|$$

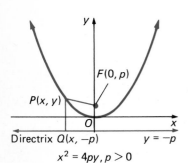

$x^2 = 4py, p > 0$

FIGURE 2 _____

Because

$$|\overline{FP}| = \sqrt{x^2 + (y - p)^2}$$

and

$$|\overline{QP}| = \sqrt{(x - x)^2 + (y + p)^2}$$

the point P is on the parabola if and only if

$$\sqrt{x^2 + (y - p)^2} = \sqrt{(y + p)^2}$$

By squaring on both sides of the equation, we obtain

$$x^2 + y^2 - 2py + p^2 = y^2 + 2py + p^2$$
$$x^2 = 4py$$

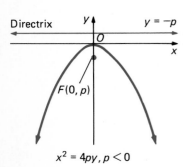

$x^2 = 4py, p < 0$

FIGURE 3 _____

We state this result formally.

Equation of a Parabola

> An equation of the parabola having its focus at $(0, p)$ and having as its directrix the line $y = -p$ is
>
> $$x^2 = 4py$$

In Figure 2, p is positive; p may be negative, however, because it is the directed distance $\overline{OF}$. Figure 3 shows a parabola for $p < 0$.

From Figures 2 and 3 we see that for the equation $x^2 = 4py$ the parabola opens upward if $p > 0$ and downward if $p < 0$. The line through the

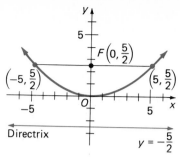

FIGURE 4

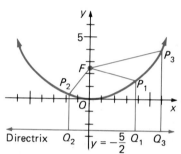

FIGURE 5

focus perpendicular to the directrix is called the **axis** of the parabola. The axis of the parabolas of Figures 2 and 3 is the y axis. The intersection of the parabola with its axis is called the **vertex,** which, of course, is midway between the focus and directrix. The vertex of the parabolas in Figures 2 and 3 is the origin.

Illustration 1

The graph of the equation

$$x^2 = 10y$$

is a parabola whose vertex is at the origin and whose axis is the y axis. Because $4p = 10$, $p = \frac{5}{2} > 0$, and therefore the parabola opens upward. The focus is at the point $F(0, \frac{5}{2})$ and an equation of the directrix is $y = -\frac{5}{2}$. Two points on the parabola are $(5, \frac{5}{2})$ and $(-5, \frac{5}{2})$. These points are the endpoints of a chord through the focus and perpendicular to the axis of the parabola. This chord is called the **latus rectum** of the parabola. The length of the latus rectum is $|4p|$; see Exercise 35. When drawing a sketch of a parabola it is helpful to plot the endpoints of the latus rectum. Figure 4 shows a sketch of the parabola, the focus, the directrix, and the latus rectum. ∎

Figure 5 shows the parabola of Figure 4 with three points P_1, P_2, and P_3 on it. Because the definition of a parabola states that any point on the parabola is equidistant from the focus and the directrix,

$$|\overline{FP_1}| = |\overline{Q_1P_1}| \qquad |\overline{FP_2}| = |\overline{Q_2P_2}| \qquad |\overline{FP_3}| = |\overline{Q_3P_3}|$$

Example 1 Draw a sketch of the parabola having the equation

$$x^2 = -8y$$

Find the focus, an equation of the directrix, and the endpoints of the latus rectum.

Solution The graph is a parabola whose vertex is at the origin and whose axis is the y axis. Because $4p = -8$, $p = -2$, and since $p < 0$, the parabola opens downward. The focus is at the point $F(0, -2)$, and an equation of the directrix is $y = 2$.

Because the latus rectum is the chord through the focus and perpendicular to the axis, we obtain its endpoints by substituting -2 for y in the equation of the parabola and solving for x. We have

$$x^2 = -8(-2)$$
$$x^2 = 16$$
$$x = \pm 4$$

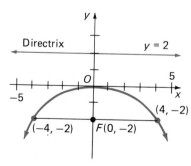

FIGURE 6

Therefore the endpoints of the latus rectum are $(4, -2)$ and $(-4, -2)$. Figure 6 shows a sketch of the parabola, the focus, and the directrix. ∎

Example 2 Find an equation of the parabola having its focus at $(0, 3)$ and having as its directrix the line $y = -3$. Draw a sketch of the graph.

Solution Because the focus is on the y axis and is also above the directrix, the parabola opens upward and $p = 3$. The vertex is at the origin. An equation of the parabola is of the form $x^2 = 4py$ with $4p = 12$; thus the required equation is

$$x^2 = 12y$$

Substituting 3 for y in this equation, we obtain

$$x^2 = 36$$
$$x = \pm 6$$

Thus the endpoints of the latus rectum are $(6, 3)$ and $(-6, 3)$. A sketch of the graph appears in Figure 7.

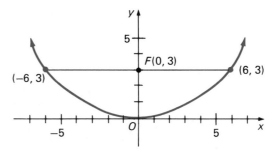

FIGURE 7

Example 3 A parabolic mirror has a depth of 12 cm at the center, and the distance across the top of the mirror is 32 cm. Find the distance from the vertex to the focus.

Solution See Figure 8. The coordinate axes are chosen so that the parabola has its vertex at the origin, has its axis along the y axis, and opens upward. Therefore an equation of the parabola is of the form

$$x^2 = 4py$$

where p centimeters is the distance from the vertex to the focus. Because the point $(16, 12)$ is on the parabola, its coordinates satisfy the equation, and we have

$$16^2 = 4p(12)$$
$$p = \tfrac{16}{3}$$

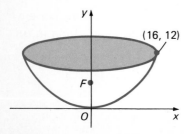

FIGURE 8

Therefore the distance from the vertex to the focus is $\tfrac{16}{3}$ cm.

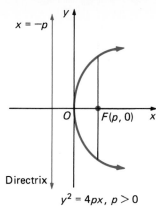

$x = -p$

$F(p, 0)$

Directrix

$y^2 = 4px, \ p > 0$

FIGURE 9

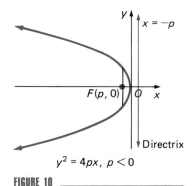

$x = -p$

$F(p, 0)$

Directrix

$y^2 = 4px, \ p < 0$

FIGURE 10

As we observed in Section 3.2, there are parabolas whose axes are horizontal. The graph having the equation

$$y^2 = 4x$$

and shown in Figure 6 of Section 3.2 is an example. This equation is of the form

$$y^2 = 4px$$

which can be obtained from the equation $x^2 = 4py$ by interchanging x and y. A parabola having the equation $y^2 = 4px$ has its vertex at the origin, the x axis as its axis, and its focus at the point $F(p, 0)$; an equation of its directrix is $x = -p$. If $p > 0$, the parabola opens to the right as in Figure 9, and if $p < 0$, the parabola opens to the left as in Figure 10.

Example   4 Draw a sketch of the parabola having the equation

$$y^2 = 7x$$

Find the focus, an equation of the directrix, and the endpoints of the latus rectum.

Solution   The given equation is of the form $y^2 = 4px$; therefore the vertex is at the origin and the x axis is its axis. Because $4p = 7$, $p = \frac{7}{4} > 0$; thus the parabola opens to the right. The focus is at the point $F(\frac{7}{4}, 0)$ and an equation of the directrix is $x = -\frac{7}{4}$. To obtain the endpoints of the latus rectum, let $x = \frac{7}{4}$ in the given equation and we have

$$y^2 = \tfrac{49}{4}$$
$$y = \pm \tfrac{7}{2}$$

Therefore the endpoints of the latus rectum are $(\frac{7}{4}, \frac{7}{2})$ and $(\frac{7}{4}, -\frac{7}{2})$. A sketch of the parabola, the focus, and the directrix are shown in Figure 11.

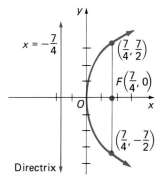

$x = -\dfrac{7}{4}$

$\left(\dfrac{7}{4}, \dfrac{7}{2}\right)$

$F\left(\dfrac{7}{4}, 0\right)$

$\left(\dfrac{7}{4}, -\dfrac{7}{2}\right)$

Directrix

FIGURE 11

EXERCISES 3.4

In Exercises 1 through 16, for the parabola having the given equation, find

(a) the vertex,

(b) the axis,

(c) the focus,

(d) an equation of the directrix, and

(e) the endpoints of the latus rectum.

(f) Draw a sketch of the parabola.

1. $x^2 = 4y$
2. $x^2 = 8y$
3. $x^2 = -16y$
4. $x^2 = -12y$
5. $x^2 - y = 0$
6. $x^2 - 2y = 0$
7. $y^2 = 12x$
8. $y^2 = -6x$
9. $y^2 = -8x$
10. $y^2 = x$
11. $y^2 - 5x = 0$
12. $y^2 + 3x = 0$
13. $3x^2 + 8y = 0$
14. $2x^2 + 5y = 0$
15. $2y^2 - 9x = 0$
16. $3y^2 - 4x = 0$

In Exercises 17 through 30, find an equation of the parabola having the given properties; draw a sketch of the parabola.

17. Focus, $(0, 4)$; directrix, $y = -4$
18. Focus, $(0, -2)$; directrix, $y = 2$
19. Focus, $(0, -5)$; directrix, $y - 5 = 0$
20. Focus, $(0, -\frac{1}{2})$; directrix, $2y - 1 = 0$
21. Focus, $(2, 0)$; directrix, $x = -2$
22. Focus, $(1, 0)$; directrix, $x = -1$
23. Focus, $(-\frac{5}{3}, 0)$; directrix, $5 - 3x = 0$
24. Focus, $(-\frac{3}{2}, 0)$; directrix, $2x - 3 = 0$
25. Vertex, the origin; opens upward; through the point $(6, 3)$.
26. Vertex, the origin; opens downward; through the point $(-4, -2)$.
27. Vertex, the origin; directrix, $3x = 2$.
28. Vertex, the origin; directrix, $2y + 5 = 0$.
29. Vertex, the origin; y axis is its axis; through the point $(-2, 4)$.
30. Vertex, the origin; x axis is its axis; through the point $(-3, 3)$.
31. A reflecting telescope has a parabolic mirror for which the distance from the vertex to the focus is 30 ft. If the distance across the top of the mirror is 64 in., how deep is the mirror at the center?
32. A parabolic arch has a height of 20 m and a width of 36 m at the base. If the vertex of the parabola is at the top of the arch, at which height above the base is it 18 m wide?

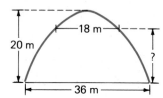

33. The cable of a suspension bridge hangs in the form of a parabola when the load is uniformly distributed horizontally. The distance between two towers is 150 m, the points of support of the cable on the towers are 22 m above the roadway, and the lowest point on the cable is 7 m above the roadway. Find the vertical distance to the cable from a point in the roadway 15 m from the foot of a tower.

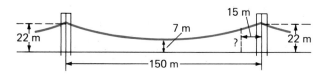

34. Assume that water issuing from the end of a horizontal pipe 25 ft above the ground describes a parabolic curve, the vertex of the parabola being at the end of the pipe. If at a point 8 ft below the line of the pipe the flow of water has curved outward 10 ft beyond a vertical line through the end of the pipe, how far beyond this vertical line will the water strike the ground?

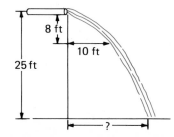

35. Prove that the length of the latus rectum of a parabola is $|4p|$.

3.5 | Translation of Axes

The shape of a graph is not affected by the position of the coordinate axes, but an equation of the graph is affected. For example, a circle with radius 3 and having its center at $(4, -1)$ has the equation

$$(x - 4)^2 + (y + 1)^2 = 9$$

However, if the coordinate axes are chosen so that the origin is at the center, the same circle has the simpler equation

$$x^2 + y^2 = 9$$

If we may select the coordinate axes as we please, we generally do so in such a way that the equations will be as simple as possible. If the axes are given, however, we may wish to find a simpler equation of a particular graph relative to a different set of axes. If these different axes are chosen parallel to the given ones, we say that there has been a **translation of axes.**

In particular, let the given x and y axes be translated to new axes x' and y', having origin (h, k) with respect to the given axes. Also assume that the positive numbers lie on the same side of the origin on the x' and y' axes as on the x and y axes. See Figure 1. A point P in the plane, having coordinates (x, y) with respect to the given coordinate axes, will have coordinates (x', y') with respect to the new axes. To obtain relationships between these two sets of coordinates, draw two lines through P, one parallel to the y and y' axes and one parallel to the x and x' axes. Let the first line intersect the x axis at point A and the x' axis at point A', and let the second line intersect the y axis at point B and the y' axis at point B', as shown in Figure 1.

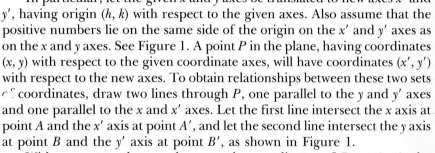

FIGURE 1 _____

With respect to the x and y axes, the coordinates of P are (x, y), the coordinates of A are $(x, 0)$, and the coordinates of A' are (x, k). Because $\overline{A'P} = \overline{AP} - \overline{AA'}$,

$$y' = y - k$$

With respect to the x and y axes, the coordinates of B are $(0, y)$, and the coordinates of B' are (h, y). Because $\overline{B'P} = \overline{BP} - \overline{BB'}$,

$$x' = x - h$$

We state these results formally.

Equations for Translating the Axes

> If (x, y) represents a point P with respect to a given set of axes, and (x', y') is a representation of P after the axes are translated to a new origin having coordinates (h, k) with respect to the given axes, then
>
> $$x' = x - h \quad \text{and} \quad y' = y - k$$

Example   **1**   Given the equation

$$x^2 + y^2 - 4x + 6y - 3 = 0$$

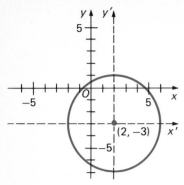

FIGURE 2

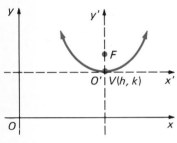

FIGURE 3

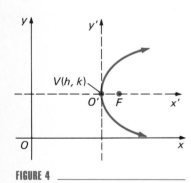

FIGURE 4

Translate the axes so that the equation of the graph with respect to the x' and y' axes contains no first-degree terms.

Solution   We rewrite the given equation as

$$(x^2 - 4x) + (y^2 + 6y) = 3$$

Completing the squares of the terms in parentheses by adding 4 and 9 on both sides of the equation, we have

$$(x^2 - 4x + 4) + (y^2 + 6y + 9) = 3 + 4 + 9$$
$$(x - 2)^2 + (y + 3)^2 = 16$$

If we let $x' = x - 2$ and $y' = y - 3$, we obtain

$$x'^2 + y'^2 = 16$$

The graph of this equation with respect to the x' and y' axes is a circle with its center at the origin and radius 4. Because the substitutions of $x' = x - 2$ and $y' = y + 3$ result in a translation of axes to a new origin of $(2, -3)$, the graph of the given equation with respect to the x and y axes is a circle with center at $(2, -3)$ and radius 4. This agrees with our discussion of the circle in Section 3.2. Figure 2 shows the circle with both sets of axes. ■

We now apply translation of axes to finding the general equation of a parabola having its vertex at (h, k) and either a vertical or horizontal axis. In particular, let the axis be vertical. Let the x' and y' axes be such that the origin is at $V(h, k)$. See Figure 3. With respect to the x' and y' axes, an equation of the parabola in the figure is

$$x'^2 = 4py'$$

To obtain an equation of this parabola with respect to the x and y axes, we let $x' = x - h$ and $y' = y - k$, which gives

$$(x - h)^2 = 4p(y - k)$$

In Figure 4, the axis of the parabola is horizontal and the vertex is at $V(h, k)$. By a similar argument its equation with respect to the x and y axes is

$$(y - k)^2 = 4p(x - h)$$

We have obtained the standard forms of an equation of a parabola.

Standard Forms of an Equation of a Parabola

> If p is the directed distance from the vertex to the focus, an equation of the parabola with its vertex at (h, k) and with its axis vertical is
>
> $$(x - h)^2 = 4p(y - k)$$
>
> A parabola with the same vertex and with its axis horizontal has the equation
>
> $$(y - k)^2 = 4p(x - h)$$

Example 2 Show that the graph of the equation

$$y = -\tfrac{1}{4}x^2 + x + 6$$

is a parabola. Find the vertex, an equation of the axis, the focus, and the endpoints of the latus rectum. Draw a sketch of the parabola.

Solution The given equation is equivalent to

$$4y = -x^2 + 4x + 24$$
$$x^2 - 4x = -4y + 24$$

Completing the square on the left by adding 4 to each side, we get

$$x^2 - 4x + 4 = -4y + 24 + 4$$
$$(x - 2)^2 = -4y + 28$$
$$(x - 2)^2 = -4(y - 7)$$

This equation is of the form

$$(x - h)^2 = 4p(y - k)$$

with $h = 2$, $k = 7$, and $p = -1$. Therefore its graph is a parabola with vertex at $(2, 7)$, and the axis is vertical. Thus the axis has the equation $x = 2$. Because $p < 0$, it opens downward. Furthermore, the focus is the point on the axis 1 unit below the vertex; thus the focus is at $(2, 6)$. Because the length of the latus rectum is $|4p| = 4$, its endpoints are 2 units to the right and left of the focus at $(4, 6)$ and $(0, 6)$. Figure 5 shows a sketch of the parabola.

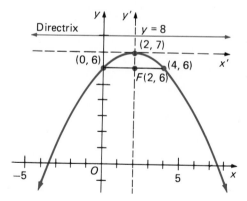

FIGURE 5

The graph of any quadratic equation of the form

$$y = ax^2 + bx + c \tag{1}$$

where a, b, and c are constants and $a \neq 0$, is a parabola whose axis is vertical. This statement can be proved by showing that Equation (1) is equivalent to an equation of the form

$$(x - h)^2 = 4p(y - k)$$

You are asked to do this in Exercise 43. Observe that the given equation in Example 2 is the special case of Equation (1) where $a = -\frac{1}{4}$, $b = 1$, and $c = 6$.

If x and y are interchanged in Equation (1), we have the equation

$$x = ay^2 + by + c \qquad (2)$$

The graph of any equation of this form is a parabola whose axis is horizontal. This fact can be verified by showing that Equation (2) is equivalent to an equation of the form

$$(y - k)^2 = 4p(x - h)$$

Example   3   Given the parabola having the equation

$$x = 2y^2 + 8y + 11$$

find the vertex, an equation of the axis, the focus, and the endpoints of the latus rectum. Draw a sketch of the parabola.

Solution   The given equation is equivalent to

$$2y^2 + 8y = x - 11$$
$$2(y^2 + 4y) = x - 11$$

To complete the square of the expression within the parentheses on the left, we add 4 to $y^2 + 4y$. We are actually adding $2 \cdot 4 = 8$ to the left side; so

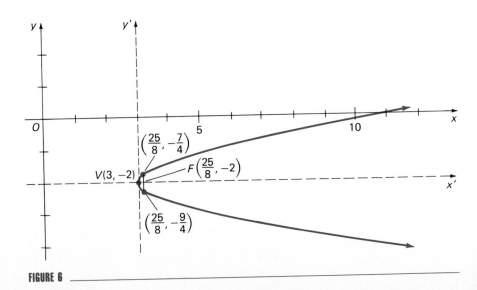

FIGURE 6

we also add 8 to the right side, and we have

$$2(y^2 + 4y + 4) = x - 11 + 8$$
$$2(y + 2)^2 = x - 3$$
$$(y + 2)^2 = \tfrac{1}{2}(x - 3)$$

This equation is of the form

$$(y - k)^2 = 4p(x - h)$$

with $h = 3$, $k = -2$, and $p = \tfrac{1}{8}$. Therefore the parabola has its vertex at $(3, -2)$, its axis is the horizontal line $y = -2$, and because $p > 0$, it opens to the right. Since the focus is $\tfrac{1}{8}$ unit to the right of the vertex, it is at the point $(\tfrac{25}{8}, -2)$. The length of the latus rectum is $4p = \tfrac{1}{2}$; thus the endpoints of the latus rectum are $\tfrac{1}{4}$ unit above and below the focus at $(\tfrac{25}{8}, -\tfrac{7}{4})$ and $(\tfrac{25}{8}, -\tfrac{9}{4})$. Figure 6 shows a sketch of the parabola. ∎

     In the next two examples, we apply translation of axes to other graphs.

Example   4   From the graph of $y = |x|$ and a suitable translation of axes, obtain the graph of $y = |x - 4| - 6$.

Solution   In Figure 9 of Section 3.2 we have the graph of $y = |x|$. It is reproduced here in Figure 7. To obtain the graph of

$$y = |x - 4| - 6$$
$$\Leftrightarrow \qquad\qquad y + 6 = |x - 4|$$

let

$$x' = x - 4 \qquad \text{and} \qquad y' = y + 6$$

We have translated the axes to the new origin $(4, -6)$, and the equation becomes $y' = |x'|$. The graph of this equation with respect to the x' and y' axes is the same as the graph in Figure 7 with respect to the x and y axes. Thus we obtain the graph shown in Figure 8.

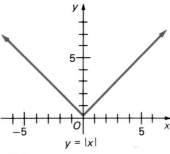

$$y = |x|$$

FIGURE 7 _____

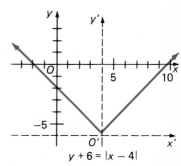

$$y + 6 = |x - 4|$$

FIGURE 8 _____ ∎

Example   **5**   Use the graph of $2y = x^3$ from Example 4 of Section 3.2 along with a suitable translation of axes to obtain the graph of the equation $2(y - 3) = (x + 5)^3$.

Solution   Figure 9 shows the graph of $2y = x^3$. To obtain the graph of

$$2(y - 3) = (x + 5)^3$$

let

$$x' = x + 5 \quad \text{and} \quad y' = y - 3$$

We have translated the axes to the new origin $(-5, 3)$, and the equation becomes $2y' = x'^3$. Figure 10 shows the graph of this equation with respect to the x' and y' axes. It is the same as the graph in Figure 9 with respect to the x and y axes. ∎

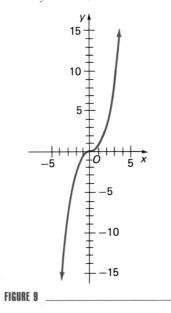

FIGURE 9 _____

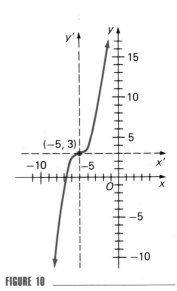

FIGURE 10 _____

EXERCISES   **3.5**

In Exercises 1 through 4, translate the axes so that an equation of the graph with respect to the new axes will contain no first-degree terms. Draw the original and the new axes and a sketch of the graph.

1. $x^2 + y^2 + 6x + 4y = 0$
2. $x^2 + y^2 - 2x - 8y + 1 = 0$
3. $x^2 + y^2 + x - 2y + 1 = 0$
4. $x^2 + y^2 - 10x + 4y + 13 = 0$

In Exercises 5 through 24, for the given parabola find (a) the vertex, (b) an equation of the axis, (c) the focus, and (d) the endpoints of the latus rectum. (e) Draw a sketch of the parabola.

5. $x^2 - 4x - 8y - 28 = 0$ 6. $x^2 + 4x + 2y = 0$
7. $y^2 + 6x + 6y + 39 = 0$ 8. $2y^2 - 2x - 4y + 3 = 0$
9. $x^2 - 6x - 4y + 13 = 0$ 10. $x^2 - 4x + 8x + 28 = 0$
11. $y^2 + 4x + 12y = 0$

12. $y^2 - 12x - 14y + 25 = 0$
13. $y = x^2 - 4$ 14. $x = -y^2 + 1$
15. $x = y^2 - 6y$ 16. $y = x^2 + 4x$
17. $y = -x^2 + 4x - 5$ 18. $y = x^2 + 6x - 2$
19. $x = -2y^2 - 8y - 5$ 20. $x = 2y^2 + 10y + 3$
21. $y = -\frac{1}{2}x^2 + 4x - 5$ 22. $y = \frac{1}{16}x^2 + \frac{1}{2}x$
23. $y = \frac{1}{8}x^2 - \frac{1}{2}x - \frac{3}{2}$ 24. $x = -\frac{1}{4}y^2 - \frac{3}{2}y - 2$

In Exercises 25 through 42, do the following: (a) draw a sketch of the graph of the first equation; (b) from the graph obtained in part (a) and a suitable translation of axes, obtain a sketch of the graph of the second equation.

25. $y = |x|$; $y = |x - 2|$ 26. $y = |x|$; $y = |x + 3|$
27. $y = |x|$; $y = |x| + 3$ 28. $y = |x|$; $y = |x| - 2$
29. $y = |x|$; $y = |x + 4| - 5$ 30. $y = |x|$; $y = |x - 1| + 6$
31. $y = x^3$; $y = (x - 4)^3$

32. $2y = -x^3$; $2y + 2 = -x^3$
33. $y = x^3$; $y = (x + 1)^3 + 1$
34. $2y = -x^3$; $2y = -(x - 4)^3 + 4$
35. $y = \sqrt{x}$; $y = \sqrt{x - 2} + 4$
36. $y = \sqrt{x}$; $y = \sqrt{x + 3} - 2$
37. $y = x^2$; $y = (x - 4)^2$ 38. $y = x^2$; $y = (x + 3)^2$
39. $y = x^2$; $y = x^2 + 3$ 40. $y = x^2$; $y = x^2 - 4$
41. $y = x^2$; $y = (x + 1)^2 - 5$
42. $y = x^2$; $y = (x - 2)^2 + 1$
43. Show that the equation

$$y = ax^2 + bx + c$$

is equivalent to an equation of the form

$$(x - h)^2 = 4p(y - k)$$

by solving the second equation for y.

3.6   The Ellipse, the Hyperbola, and Conic Sections (Supplementary)

Curves, other than the circle and the parabola, that are graphs of quadratic equations in two variables are the *ellipse* and the *hyperbola*. A complete treatment of these curves belongs to a course in analytic geometry. However, the curves occur so often that a brief treatment is given here. We begin with the ellipse.

Definition

> **Ellipse**
> An **ellipse** is the set of points in a plane, the sum of whose distances from two fixed points is a constant. Each fixed point is called a **focus.**

Let the undirected distance between the foci (the plural of focus) be $2c$, where $c > 0$. To obtain an equation of an ellipse we select the x axis as the line through the foci F and F' and we choose the origin as the midpoint of the segment between F and F'. See Figure 1. The foci F and F' have coordinates $(c, 0)$ and $(-c, 0)$, respectively. Let the constant sum referred to in the definition be $2a$. Then $a > c$, and the point $P(x, y)$ in Figure 1 is any point on the ellipse if and only if

$$|\overline{FP}| + |\overline{F'P}| = 2a$$

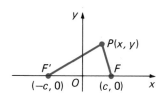

FIGURE 1

Because

$$|\overline{FP}| = \sqrt{(x - c)^2 + y^2} \quad \text{and} \quad |\overline{F'P}| = \sqrt{(x + c)^2 + y^2}$$

P is on the ellipse if and only if

$$\sqrt{(x - c)^2 + y^2} + \sqrt{(x + c)^2 + y^2} = 2a$$

To simplify this equation requires eliminating the radicals and performing some algebraic manipulations, which you are asked to do in Exercise 27. When this is done we obtain

$$\frac{x^2}{a^2} + \frac{y^2}{b^2} = 1$$

where $b^2 = a^2 - c^2$. We state this result formally.

Equation of an Ellipse

If $2a$ is the constant referred to in the definition of an ellipse, and if the foci are at $(c, 0)$ and $(-c, 0)$, then if $b^2 = a^2 - c^2$, an equation of the ellipse is

$$\frac{x^2}{a^2} + \frac{y^2}{b^2} = 1$$

To obtain a sketch of this ellipse, first observe from the equation that the graph is symmetric with respect to both the x and y axes. If we replace y by 0 in the equation we get $x = \pm a$, and if we replace x by 0 we get $y = \pm b$. Therefore the graph intersects the x axis at the points $(a, 0)$ and $(-a, 0)$ and it intersects the y axis at the points $(0, b)$ and $(0, -b)$. Because $b^2 = a^2 - c^2$, it follows that $a > b$. See Figure 2 and refer to it as you read the next paragraph.

The line through the foci is called the **principal axis** of the ellipse. For this ellipse, the x axis is the principal axis. The points of intersection of the ellipse and its principal axis are called the **vertices.** Thus for this ellipse the vertices are at $V(a, 0)$ and $V'(-a, 0)$. The point on the principal axis that lies halfway between the two vertices is called the **center.** The origin is the center of this ellipse. The segment of the principal axis between the two vertices is called the **major axis** and its length is $2a$ units. For this ellipse the segment of the y axis between the points $(0, b)$ and $(0, -b)$ is called the **minor axis.** Its length is $2b$ units.

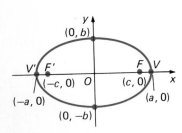

FIGURE 2

Example 1 Given the ellipse having the equation

$$\frac{x^2}{25} + \frac{y^2}{16} = 1$$

find the vertices, the endpoints of the minor axis, and the foci. Draw a sketch of the ellipse and show the foci.

Solution   Because the equation is of the form

$$\frac{x^2}{a^2} + \frac{y^2}{b^2} = 1$$

the center of the ellipse is at the origin and the principal axis is the x axis. Since $a^2 = 25$ and $b^2 = 16$, $a = 5$ and $b = 4$. Therefore the vertices are at $V(5, 0)$ and $V'(-5, 0)$ and the endpoints of the minor axis are at $B(0, 4)$ and $B'(0, -4)$.

To find the foci we solve for c from the equation $b^2 = a^2 - c^2$ with $a^2 = 25$ and $b^2 = 16$. Thus, because $c > 0$,

$$16 = 25 - c^2$$
$$c^2 = 9$$
$$c = 3$$

Therefore the foci are at $F(3, 0)$ and $F'(-3, 0)$.

As an aid in drawing a sketch of the ellipse we find a point on it in the first quadrant by substituting 3 for x in the equation and solving for y. Of course, any other value of x between 0 and 5 can be used. By symmetry we have corresponding points in the other three quadrants. Figure 3 shows a sketch of the ellipse and the foci.   ∎

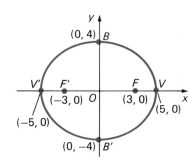

FIGURE 3

Observe from the definition of an ellipse that if P is any point on the ellipse of Example 1, $|\overline{FP}| + |\overline{F'P}| = 10$. See Figure 4, where we have taken P in the second quadrant.

There are applications of ellipses in astronomy because paths of many comets and orbits of planets and satellites are ellipses. Arches of bridges are sometimes elliptical in shape, and ellipses are used in making machine gears. There is a reflective property of the ellipse that is used in so-called whispering galleries, where the ceilings have cross sections that are arcs of ellipses with common foci. A person located at one focus F can hear another person whispering at the other focus F' because the sound waves originating from the whisperer at F' hit the ceiling and are reflected by the ceiling to the listener at F. A famous example of a whispering gallery is under the dome of the Capitol in Washington, D.C. Another is at the Mormon Tabernacle in Salt Lake City.

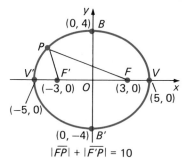

FIGURE 4

Example   2   An arch is in the form of a semiellipse. It is 48 ft wide at the base and has a height of 20 ft. How wide is the arch at a height of 10 ft above the base?

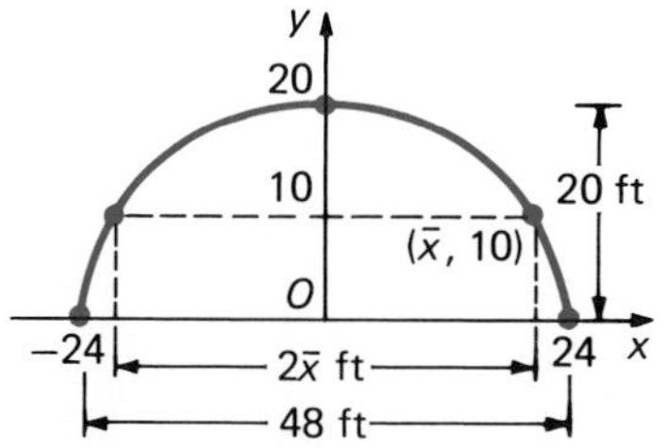

Solution   Figure 5 shows a sketch of the arch and the coordinate axes chosen so that the x axis is along the base and the origin is at the midpoint of the base. Then the ellipse has its principal axis on the x axis, its center at the

FIGURE 5

origin, $a = 24$, and $b = 20$. Thus an equation of the ellipse is

$$\frac{x^2}{576} + \frac{y^2}{400} = 1$$

Let $2\bar{x}$ be the number of feet in the width of the arch at a height of 10 ft above the base. Therefore the point $(\bar{x}, 10)$ is on the ellipse. Thus

$$\frac{\bar{x}^2}{576} + \frac{100}{400} = 1,$$
$$\bar{x}^2 = 432,$$
$$\bar{x} = 12\sqrt{3}$$

Hence at a height of 10 ft above the base, the width of the arch is $24\sqrt{3}$ ft. ■

If an ellipse has its center at the origin and principal axis on the y axis, then an equation of the ellipse is of the form

$$\frac{y^2}{a^2} + \frac{x^2}{b^2} = 1$$

This equation is obtained by interchanging x and y in the equation

$$\frac{x^2}{a^2} + \frac{y^2}{b^2} = 1$$

Illustration 1

Because for an ellipse $a > b$, it follows that the ellipse having the equation

$$\frac{x^2}{16} + \frac{y^2}{25} = 1$$

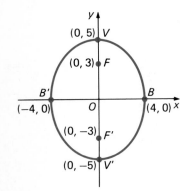

has its principal axis on the y axis. This ellipse has the same shape as the ellipse of Example 1. The vertices are at $(0, 5)$ and $(0, -5)$, the endpoints of the minor axis are at $(4, 0)$ and $(-4, 0)$, and the foci are at $(0, 3)$ and $(0, -3)$. A sketch of this ellipse appears in Figure 6. ■

FIGURE 6

Definition

> **Hyperbola**
> A **hyperbola** is the set of points in a plane, the absolute value of the difference of whose distances from two fixed points is a constant. The two fixed points are called the **foci.**

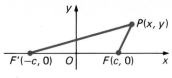

FIGURE 7 _____

To obtain an equation of a hyperbola we begin as we did with the ellipse by letting the undirected distance between the foci be $2c$, where $c > 0$. Then we choose the x axis as the line through the foci F and F', and we take the origin as the midpoint of the segment between F and F'. Refer to Figure 7.

The points $(c, 0)$ and $(-c, 0)$ are the foci F and F', respectively. Let $2a$ be the constant referred to in the definition. It can be shown that $c > a$. The point $P(x, y)$ in Figure 7 is any point on the hyperbola if and only if

$$\left| |\overline{FP}| - |\overline{F'P}| \right| = 2a$$

Because

$$|\overline{FP}| = \sqrt{(x - c)^2 + y^2} \qquad \text{and} \qquad |\overline{F'P}| = \sqrt{(x + c)^2 + y^2}$$

P is on the hyperbola if and only if

$$\left| \sqrt{(x - c)^2 + y^2} - \sqrt{(x + c)^2 + y^2} \right| = 2a$$

or, equivalently, without absolute value bars,

$$\sqrt{(x - c)^2 + y^2} - \sqrt{(x + c)^2 + y^2} = \pm 2a$$

This equation can be simplified by eliminating the radicals and performing some algebraic manipulations. You are asked to do this in Exercise 28. The resulting equation is

$$\frac{x^2}{a^2} - \frac{y^2}{b^2} = 1$$

where $b^2 = c^2 - a^2$. We have then the following statement.

Equation of a Hyperbola

If $2a$ is the constant referred to in the definition of a hyperbola, and if the foci are at $(c, 0)$ and $(-c, 0)$, then if $b^2 = c^2 - a^2$, an equation of the hyperbola is

$$\frac{x^2}{a^2} - \frac{y^2}{b^2} = 1$$

A sketch of the graph of this hyperbola appears in Figure 8. We now show how this graph is obtained. From the equation we observe that the graph is symmetric with respect to both the x and y axes. As with the ellipse, the line through the foci is called the **principal axis.** Thus for this hyperbola the x axis is the principal axis. The points where the hyperbola intersects the principal axis are called the **vertices,** and the point that is halfway between the vertices is called the **center.** For this hyperbola the vertices are at $V(a, 0)$ and $V'(-a, 0)$ and the center is at the origin. The segment $V'V$ of the principal axis is called the **transverse axis** and its length is $2a$ units. Substituting 0 for x in the equation of the hyperbola, we obtain $y^2 = -b^2$,

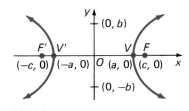

FIGURE 8 _____

which has no real solutions. Consequently, the hyperbola does not intersect the y axis. However, the line segment having endpoints at $(0, -b)$ and $(0, b)$ is called the **conjugate axis** of the hyperbola and its length is $2b$ units. If we solve the equation of the hyperbola for y in terms of x, we have

$$y = \pm \frac{b}{a}\sqrt{x^2 - a^2}$$

We conclude from this equation that if $|x| < a$, there is no real value of y. Therefore there are no points (x, y) on the hyperbola for which $-a < x < a$. We also observe that if $|x| > a$, then y has two real values. Thus the hyperbola has two *branches*. One branch contains the vertex $V(a, 0)$ and extends indefinitely to the right of V. The other branch contains the vertex $V'(-a, 0)$ and extends indefinitely to the left of V'.

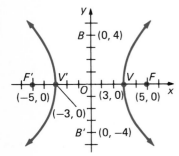

FIGURE 9

Example 3   Given the hyperbola having the equation

$$\frac{x^2}{9} - \frac{y^2}{16} = 1$$

find the vertices and foci. Draw a sketch of the hyperbola and show the foci.

Solution   Because the equation is of the form

$$\frac{x^2}{a^2} - \frac{y^2}{b^2} = 1$$

the center of the hyperbola is at the origin and the principal axis is the x axis. Since $a^2 = 9$ and $b^2 = 16$, $a = 3$ and $b = 4$. The vertices are therefore at $V(3, 0)$ and $V'(-3, 0)$. The number of units in the length of the transverse axis is $2a = 6$, and the number of units in the length of the conjugate axis is $2b = 8$. Because $b^2 = c^2 - a^2$, with $c > 0$, we have

$$16 = c^2 - 9$$
$$c^2 = 25$$
$$c = 5$$

Hence the foci are at $F(5, 0)$ and $F'(-5, 0)$. A sketch of the hyperbola with its foci appears in Figure 9.   ∎

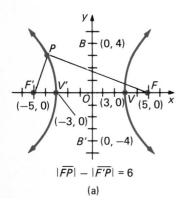

$$|\overline{FP}| - |\overline{F'P}| = 6$$

(a)

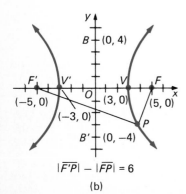

$$|\overline{F'P}| - |\overline{FP}| = 6$$

(b)

FIGURE 10

From the definition, if P is any point on the hyperbola of Example 3, $||\overline{FP}| - |\overline{F'P}|| = 6$. See Figures 10(a) and (b); in part (a) P is in the second quadrant and $|\overline{FP}| - |\overline{F'P}| = 6$; in (b) P is in the fourth quadrant and $|\overline{F'P}| - |\overline{FP}| = 6$.

If, in the equation

$$\frac{x^2}{a^2} - \frac{y^2}{b^2} = 1$$

x and y are interchanged, we obtain

$$\frac{y^2}{a^2} - \frac{x^2}{b^2} = 1$$

which is an equation of a hyperbola having its center at the origin and its principal axis on the y axis.

Illustration 2

The equation

$$\frac{y^2}{9} - \frac{x^2}{16} = 1$$

can be obtained from the one in Example 3 by interchanging x and y. The graph of this equation is a hyperbola with its center at the origin and its principal axis is the y axis. Its vertices are at $V(0, 3)$ and $V'(0, -3)$ and its foci are at $F(0, 5)$ and $F'(0, -5)$. A sketch of the hyperbola with its foci appears in Figure 11.   ∎

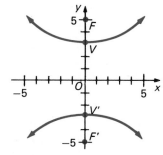

FIGURE 11

Refer now to Figure 12, showing the hyperbola having the equation

$$\frac{x^2}{a^2} - \frac{y^2}{b^2} = 1$$

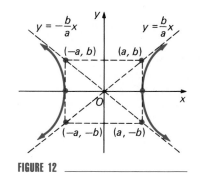

FIGURE 12

The dashed lines in the figure are called **asymptotes** of the hyperbola. These lines may be used as guides in drawing a sketch of the hyperbola. A rigorous definition of an asymptote of a graph requires the concept of *limit*, which is studied in a calculus course. However, the following statement can be considered as an intuitive idea of an asymptote: If the undirected distance between a graph and a line gets smaller and smaller (but not zero) as either $|x|$ or $|y|$ gets larger and larger, then the line is an asymptote of the graph.

Observe in Figure 12 that the diagonals of the rectangle having vertices at (a, b), $(a, -b)$, $(-a, b)$, and $(-a, -b)$ are on the asymptotes of the hyperbola. This rectangle is called the **auxiliary rectangle;** its sides have lengths $2a$ and $2b$. The vertices of the hyperbola are the points of intersection of the principal axis and the auxiliary rectangle. A fairly good sketch of a hyperbola can be made by first drawing the auxiliary rectangle. By extending the diagonals of the rectangle we have the asymptotes. Through each vertex we draw a branch of the hyperbola by using the asymptotes as guides.

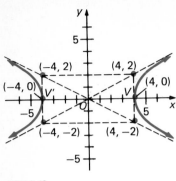

FIGURE 13

Example 4 Given the hyperbola having the equation

$$x^2 - 4y^2 = 16$$

find the vertices and draw a sketch of the graph. Show the auxiliary rectangle and the asymptotes.

Solution By dividing both sides of the given equation by 16, we obtain the following equivalent equation having 1 on the right side:

$$\frac{x^2}{16} - \frac{y^2}{4} = 1$$

Therefore the hyperbola has its center at the origin, and its principal axis is the x axis. Because $a^2 = 16$ and $b^2 = 4$, $a = 4$ and $b = 2$. The vertices are at $V(4, 0)$ and $V'(-4, 0)$ and the sides of the auxiliary rectangle have lengths $2a = 8$ and $2b = 4$. Figure 13 shows the auxiliary rectangle and the asymptotes. These asymptotes are used as guides to draw the sketch in the figure. ■

For a hyperbola there is no general inequality involving a and b corresponding to the inequality $a > b$ for an ellipse. That is, for a hyperbola it is possible to have $a < b$ as in Example 3, where $a = 3$ and $b = 4$; or it is possible to have $a > b$ as in Example 4, where $a = 4$ and $b = 2$.

The hyperbola has a reflective property that is used in the design of certain telescopes. Hyperbolas form the basis of several important navigational systems and they are also used in combat in sound ranging to locate the position of enemy guns by the sound of the firing of those guns. Some comets move in hyperbolic orbits.

A **conic section** (or **conic**) is a curve of intersection of a plane with a right-circular cone of two nappes. Three types of curves of intersection that occur are the parabola, the ellipse (including the circle as a special case), and the hyperbola. The Greek mathematician Apollonius studied conic sections, in terms of geometry, by using this concept.

In the geometry of conic sections, a cone is regarded as having two nappes, extending indefinitely far in both directions. A portion of a right-circular cone of two nappes is shown in Figure 14. A **generator** (or element) of the cone is a line lying in the cone, and all the generators of a cone contain the point V, called the **vertex** of the cone.

In Figure 15 we have a cone and a cutting plane that is parallel to one and only one generator of the cone. This conic is a parabola. If the cutting plane is parallel to two generators, it intersects both nappes of the cone and we have a hyperbola (see Figure 16). An ellipse is obtained if the cutting plane is parallel to no generator, in which case the cutting plane intersects each generator, as in Figure 17. A circle, a special case of an ellipse, is formed if the cutting plane that intersects each generator is also perpendicular to the axis of the cone. See Figure 18.

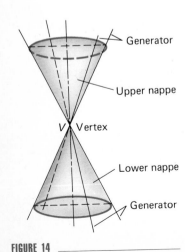

FIGURE 14

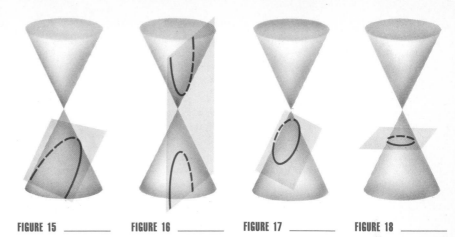

FIGURE 15 _____   FIGURE 16 _____   FIGURE 17 _____   FIGURE 18 _____

EXERCISES 3.6

In Exercises 1 through 8, for the ellipse having the given equation, find (a) the principal axis, (b) the vertices, (c) the endpoints of the minor axis, and (d) the foci. (e) Draw a sketch of the ellipse and show the foci.

1. $\dfrac{x^2}{25} + \dfrac{y^2}{9} = 1$ 2. $\dfrac{x^2}{100} + \dfrac{y^2}{64} = 1$

3. $\dfrac{x^2}{4} + \dfrac{y^2}{16} = 1$ 4. $\dfrac{x^2}{25} + \dfrac{y^2}{169} = 1$

5. $9x^2 + 25y^2 = 900$ 6. $4x^2 + 9y^2 = 36$

7. $9x^2 + y^2 = 9$ 8. $25x^2 + 4y^2 = 100$

In Exercises 9 through 14, for the hyperbola having the given equation, find (a) the principal axis, (b) the vertices, and (c) the foci. (d) Draw a sketch of the hyperbola and show the foci.

9. $\dfrac{x^2}{64} - \dfrac{y^2}{36} = 1$ 10. $\dfrac{x^2}{4} - \dfrac{y^2}{4} = 1$

11. $\dfrac{y^2}{25} - \dfrac{x^2}{144} = 1$ 12. $\dfrac{y^2}{16} - \dfrac{x^2}{9} = 1$

13. $9x^2 - 4y^2 = 36$ 14. $25y^2 - 4x^2 = 100$

In Exercises 15 through 20, for the hyperbola having the given equation, find (a) the principal axis and (b) the vertices. (c) Draw a sketch of the hyperbola and show the auxiliary rectangle and the asymptotes.

15. $\dfrac{x^2}{25} - \dfrac{y^2}{16} = 1$ 16. $\dfrac{x^2}{9} - \dfrac{y^2}{25} = 1$

17. $\dfrac{y^2}{4} - \dfrac{x^2}{16} = 1$ 18. $\dfrac{y^2}{100} - \dfrac{x^2}{49} = 1$

19. $25y^2 - 36x^2 = 900$ 20. $4x^2 - 9y^2 = 144$

21. The ceiling in a hallway 10 m wide is in the shape of a semiellipse and is 9 m high in the center and 6 m high at the side walls. Find the height of the ceiling 2 m from either wall.

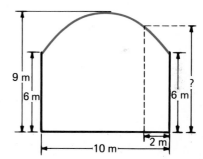

22. The arch of a bridge is in the shape of a semi-ellipse having a horizontal span of 40 m and a height of 16 m at its center. How high is the arch 9 m to the right or left of the center?

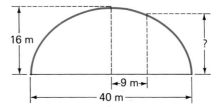

23. Suppose that the orbit of a planet is in the shape of an ellipse for which the distance between the vertices is 500 million km, and the distance between the foci is 400 million km. Find an equation of the orbit.

24. The definition of an ellipse gives a procedure for drawing the graph of an ellipse. To apply the method to the ellipse $4x^2 + 9y^2 = 36$, first determine the points of intersection with the coordinate axes. Obtain the foci on the x axis by using a compass with its center at one of the points of intersection with the y axis and with radius 3. Then fasten thumbtacks at each focus. Take a piece of string of length 6, which is $2a$, and attach an end at one thumbtack and an end at the other thumbtack. Place a pencil against the string and make it tight. Move the pencil against the string and trace a curve. This curve is an ellipse because the pencil traces a set of points the sum of whose distances from the two tacks is the constant 6.

25. The cost of production of a commodity is $12 less per unit at a point A than it is at a point B, and the distance between A and B is 100 km. Assuming that the route of delivery of the commodity is along a straight line, and that the delivery cost is 20 cents per unit per kilometer, find the curve at any point of which the commodity can be supplied from either A or B at the same total cost. (*Hint:* Take points A and B at $(-50, 0)$ and $(50, 0)$, respectively.)

26. Two LORAN (short for Long-Range Navigation) stations A and B lie on a line running east and west, and A is 80 mi due east of B. An airplane is traveling east on a straight line course that is 60 mi north of the line through A and B. Signals are sent at the same time from A and B, and the signal from A reaches the plane 350 μsec before the one from B. Assuming that the signals travel at the rate of 0.2 mi per microsecond, locate the position of the plane by the definition of a hyperbola.

27. Show that the equation

$$\sqrt{(x - c)^2 + y^2} + \sqrt{(x + c)^2 + y^2} = 2a$$

can be simplified to

$$\frac{x^2}{a^2} + \frac{y^2}{b^2} = 1 \qquad \text{where} \qquad b^2 = a^2 - c^2$$

28. Show that the equation

$$\sqrt{(x - c)^2 + y^2} - \sqrt{(x + c)^2 + y^2} = \pm 2a$$

can be simplified to

$$\frac{x^2}{a^2} - \frac{y^2}{b^2} = 1 \qquad \text{where} \qquad b^2 = c^2 - a^2$$

In Exercises 29 through 40, use the following information to draw a sketch of the graph of the given equation.

If the center of an ellipse or hyperbola is at the point (h, k), we can show by a translation of axes that an equation of the ellipse is of the form

$$\frac{(x - h)^2}{a^2} + \frac{(y - k)^2}{b^2} = 1 \qquad a > b$$

if the principal axis is horizontal, and of the form

$$\frac{(y - k)^2}{a^2} + \frac{(x - h)^2}{b^2} = 1 \qquad a > b$$

if the principal axis is vertical. By the same procedure we can show that an equation of the hyperbola is of the form

$$\frac{(x - h)^2}{a^2} - \frac{(y - k)^2}{b^2} = 1$$

if the principal axis is horizontal, and of the form

$$\frac{(y - k)^2}{a^2} - \frac{(x - h)^2}{b^2} = 1$$

if the principal axis is vertical. By expanding $(x - h)^2$ and $(y - k)^2$ and simplifying, each of these equations can be written in the form

$$Ax^2 + Cy^2 + Dx + Ey + F = 0$$

where A and C have the same sign for ellipses and A and C have opposite signs for hyperbolas. In the exercises, if the graph is an ellipse, find (a) the center, (b) an equation of the principal axis, (c) the vertices, and (d) the endpoints of the minor axis. If the graph is a hyperbola, find (a) the center, (b) an equation of the principal axis, and (c) the vertices. Also for a hyperbola show the auxiliary rectangle and the asymptotes.

29. $4x^2 + 9y^2 - 16x - 18y - 11 = 0$
30. $x^2 + 4y^2 - 6x + 8y - 3 = 0$
31. $x^2 - y^2 + 6x - 4y - 4 = 0$
32. $9y^2 - 4x^2 + 32x - 36y - 64 = 0$
33. $4x^2 + y^2 + 8x - 4y - 92 = 0$
34. $9x^2 - 16y^2 + 54x - 32y - 79 = 0$
35. $9y^2 - 25x^2 - 50x - 72y - 106 = 0$
36. $25x^2 + 16y^2 + 150x - 128y - 1119 = 0$
37. $9x^2 - 4y^2 - 18x - 16y + 29 = 0$
38. $25x^2 + y^2 - 4y - 21 = 0$
39. $x^2 + 3y^2 - 4x - 23 = 0$
40. $2x^2 - y^2 + 12x + 8y - 6 = 0$

REVIEW EXERCISES FOR CHAPTER 3

In Exercises 1 and 2, do the following: (a) plot the points A and B and draw the line segment between them; (b) find the distance between A and B; (c) find the midpoint of the line segment between A and B.

1. $A(7, -1)$ and $B(3, 2)$
2. $A(-5, 2)$ and $B(3, -4)$

In Exercises 3 and 4, do the following: (a) determine the coordinates of the midpoint M of the line segment between A and B; (b) plot the points A, M, and B and show that $|\overline{AM}| = |\overline{MB}|$.

3. $A(6, 5)$ and $B(-4, -9)$
4. $A(-1, -8)$ and $B(7, 6)$
5. Find the lengths of the sides of the triangle having vertices at $A(4, 1)$, $B(5, -5)$ and $C(1, -3)$.
6. Find the lengths of the medians of the triangle of Exercise 5.
7. Use the converse of the Pythagorean theorem to prove that the points $(2, 4)$, $(1, -4)$, and $(5, -2)$ are vertices of a right triangle.
8. Prove that the triangle with vertices at $(-8, 1)$, $(-1, -6)$, and $(2, 4)$ is isosceles.

In Exercises 9 through 18, draw a sketch of the graph of the equation.

9. $y = x - 4$
10. $3x + 2y - 6 = 0$
11. $y = x^2 + 3$
12. $y = 1 - x^2$
13. $y^2 - x + 1 = 0$
14. $y = |x - 2|$
15. $xy = 9$
16. $y = 2x^3$
17. $x^2 + y^2 = 9$
18. $(x + 2)^2 + (y - 4)^2 = 25$
19. Find an equation of the circle with center at $(3, -5)$ and radius 2. Write the equation in both the center-radius form and the general form. Draw a sketch of the circle.
20. Find an equation of the circle having a diameter whose endpoints are at $(2, -1)$ and $(6, -5)$.

In Exercises 21 and 22, find the center and radius of the circle and draw a sketch of the circle.

21. $x^2 + y^2 + 4x - 6y - 3 = 0$
22. $3x^2 + 3y^2 + 4x - 4 = 0$

In Exercises 23 and 24, do the following: (a) draw a sketch of the line through the two points; (b) determine the slope of the line; (c) find an equation of the line.

23. $(1, -3)$ and $(4, 5)$
24. $(-2, -5)$ and $(6, -7)$

In Exercises 25 and 26, do the following: (a) draw a sketch of the line through the point P and having slope m; (b) find an equation of the line.

25. $P(5, -2)$; $m = -\frac{3}{2}$
26. $P(-2, 1)$; $m = \frac{3}{4}$

In Exercises 27 and 28, do the following: (a) find the slope and y intercept of the line having the given equation; (b) draw a sketch of the line.

27. $2x - 5y - 10 = 0$
28. $2x + 3y + 12 = 0$
29. Find an equation of the line through the point $(-3, -2)$ and parallel to the line whose equation is $7x - 3y - 4 = 0$. Draw a sketch of each line on the same coordinate system.
30. Find an equation of the line through the point $(-1, 6)$ and perpendicular to the line whose equation is $4x + 2y - 5 = 0$. Draw a sketch of each line on the same coordinate system.
31. Prove that the lines having the equations $2x + 5y + 20 = 0$ and $5x - 2y - 10 = 0$ are perpendicular, and draw sketches of their graphs.
32. Prove that the lines having the equations $2x - 3y + 12 = 0$ and $4x - 6y - 3 = 0$ are parallel, and draw sketches of their graphs.
33. Find the abscissa of the point whose ordinate is -3 and for which the line through it and the point $(2, 7)$ is parallel to the line having the equation $3x - 4y = 12$.
34. Prove that the three points $(0, -3)$, $(1, 4)$, and $(2, 11)$ are collinear by two methods: (a) use slopes; (b) use the distance formula.
35. Prove that the three points $(2, 4)$, $(1, -4)$, and $(5, -2)$ are the vertices of a right triangle by two methods: (a) use slopes; (b) use the distance formula.
36. Prove that the quadrilateral having vertices at $(1, 2)$, $(5, -1)$, $(11, 7)$, and $(7, 10)$ is a rectangle.

In Exercises 37 through 40, for the parabola having the given equation, find (a) the vertex, (b) the axis, (c) the focus, (d) an equation of the directrix, and (e) the endpoints of the latus rectum. (f) Draw a sketch of the parabola and show the focus and directrix.

37. $x^2 = 16y$
38. $x^2 = -4y$
39. $y^2 = -10x$
40. $y^2 = 6x$

In Exercises 41 through 44, find an equation of the parabola having the given properties; draw a sketch of the parabola.

41. Focus, $(0, 2)$; directrix, $y = -2$
42. Focus, $(-4, 0)$; directrix, $x = 4$
43. Vertex, the origin; opens to the left; through the point $(-3, 6)$.
44. Vertex, the origin; opens upward; through the point $(-5, 2)$.

In Exercises 45 through 50, for the parabola having the given equation, find (a) the vertex, (b) an equation of the axis, (c) the focus, and (d) the endpoints of the latus rectum. (e) Draw a sketch of the parabola.

45. $y = x^2 - 3$
46. $x = y^2 + 6$
47. $x = y^2 - 8y$
48. $y = x^2 + 10x$
49. $y = -\frac{1}{8}x^2 + \frac{3}{4}x + \frac{7}{8}$
50. $x = -\frac{1}{6}y^2 - \frac{2}{3}y + \frac{7}{3}$

In Exercises 51 through 58, do the following: (a) draw a sketch of the graph of the first equation; (b) from the graph obtained in part (a) and a suitable translation of axes, obtain a sketch of the graph of the second equation.

51. $y = |x|$; $y = |x - 3| + 2$
52. $y = |x|$; $y = |x + 2| - 3$
53. $y = \sqrt{x}$; $y = \sqrt{x + 5}$
54. $y = x^2$; $y = (x - 5)^2$
55. $y = x^2$; $y = (x - 1)^2 - 1$
56. $x = \sqrt{y}$; $x = \sqrt{y + 4} + 3$
57. $2x = y^3$; $2x = (y - 3)^3 - 4$
58. $4y = x^3$; $4(y + 3) = (x + 1)^3$

In Exercises 59 through 62, the points A, B, C, and D are the vertices of a quadrilateral. Use slopes to determine whether the quadrilateral is a rectangle, a parallelogram, or a trapezoid. Draw a sketch of the quadrilateral.

59. $A(3, 1)$, $B(5, 2)$, $C(15, 5)$, $D(17, 6)$
60. $A(-8, 0)$, $B(-3, -5)$, $C(1, 4)$, $D(3, 2)$
61. $A(3, 1)$, $B(2, -2)$, $C(-1, -1)$, $D(0, 2)$
62. $A(2, 13)$, $B(-2, 5)$, $C(3, -1)$, $D(7, 7)$

In Exercises 63 through 74, draw a sketch of the graph of the given equation. If the graph is an ellipse, find (a) the center, (b) an equation of the principal axis, (c) the vertices, (d) the endpoints of the minor axis, and (e) the foci. If the graph is a hyperbola, find (a) the center, (b) an equation of the principal axis, and (c) the vertices. Also for a hyperbola show the auxiliary rectangle and the asymptotes. For Exercises 71 through 74, refer to the instructions for Exercises 29 through 40 in Exercises 3.6.

63. $\dfrac{x^2}{100} + \dfrac{y^2}{36} = 1$

64. $\dfrac{x^2}{16} + \dfrac{y^2}{64} = 1$

65. $\dfrac{y^2}{9} - \dfrac{x^2}{16} = 1$

66. $\dfrac{x^2}{36} - \dfrac{y^2}{64} = 1$

67. $9x^2 - 25y^2 = 225$
68. $4y^2 - 25x^2 = 400$
69. $25x^2 + 9y^2 = 225$
70. $4x^2 + 25y^2 = 400$
71. $25x^2 + y^2 - 100x + 8y + 91 = 0$
72. $16y^2 - 9x^2 - 64y - 80 = 0$
73. $4x^2 - 4y^2 - 32x + 16y + 39 = 0$
74. $x^2 + 2y^2 + 12x - 12y + 38 = 0$
75. The arch of a bridge is in the shape of a semi-ellipse having a horizontal span of 60 m and a height of 20 m at its center. How high is the arch 10 m to the right or left of the center?
76. Points A and B are 1000 m apart, and it is determined from the sound of an explosion heard at these points at different times that the location of the explosion is 600 m closer to A than to B. Show that the location of the explosion is restricted to a particular curve and find an equation of it.

C H A P T E R

4

Functions and Their Graphs

Introduction

The notion of a *function* is one of the most important concepts in mathematics. It is crucial for the study of calculus. A function is introduced in Section 4.1 as a correspondence between sets of real numbers. However, to be precise, we formally define a function as a set of ordered pairs. In Section 4.2 we discuss function notation and types of functions, and in Section 4.3 we give some examples showing how a practical situation can be expressed in terms of a functional relationship. The next three sections are devoted to the graphs of particular kinds of functions: quadratic, polynomial, and rational. The final section of the chapter pertains to inverse functions and their properties.

4.1 Functions

Often in practical applications the value of one quantity depends on the value of another. In particular, a person's salary may depend on the number of hours worked; the total production at a factory may depend on the number of machines used; the distance traveled by an object may depend on the time elapsed since it left a specific point; the volume of the space occupied by a gas having a constant pressure depends on the temperature of the gas; the resistance of an electrical cable of fixed length depends on its diameter; and so forth. A relationship between such quantities is often given by means of a *function*. For our purposes we confine the quantities in the relationship to real numbers. Then

> A function can be thought of as a correspondence from a set X of real numbers x to a set Y of real numbers y, where the number y is unique for a specific value of x.

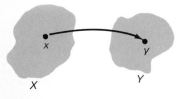

FIGURE 1

Figure 1 gives a visualization of such a correspondence where the sets X and Y consist of points in a plane region.

Stating the concept of a function another way, we intuitively consider the real number y in set Y to be a *function* of the real number x in set X if there is some rule by which a unique value of y is assigned to a value of x. This rule is often given by an equation. For example, the equation

$$y = x^2$$

defines a function for which X is the set of all real numbers and Y is the set of nonnegative numbers: The value of y in Y assigned to the value of x in X is obtained by multiplying x by itself. Table 1 gives the value of y assigned to some particular values of x and Figure 2 visualizes the correspondence for the numbers in the table.

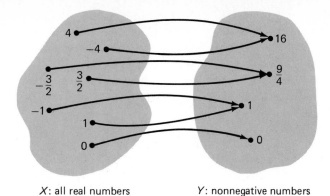

X: all real numbers   Y: nonnegative numbers

FIGURE 2

TABLE 1						
x	1	$\frac{3}{2}$	4	0	-1	$-\frac{3}{2}$	-4
$y = x^2$	1	$\frac{9}{4}$	16	0	1	$\frac{9}{4}$	16

We use symbols such as f, g, and h to denote a function. The set X of real numbers described above is the *domain* of the function and the set Y of real numbers assigned to the values of x in X is the *range* of the function.

Illustration 1

The equation

$$y = 2x^2 + 5$$

defines a function. Call this function f. The equation gives the rule by which a unique value of y can be determined whenever x is given; that is, multiply the number x by itself, then multiply that product by 2, and add 5. The domain of f is the set of all real numbers, and it can be denoted with interval notation as $(-\infty, +\infty)$. The smallest value that y can assume is 5 (when $x = 0$). The range of f is then the set of all positive numbers greater than or equal to 5, which is $[5, +\infty)$. ∎

Illustration 2

Let g be the function defined by the equation

$$y = \sqrt{x^2 - 9}$$

Because the numbers are confined to real numbers, y is a function of x only for $x \geq 3$ or $x \leq -3$ (or simply $|x| \geq 3$), because for any x satisfying either of these inequalities, a unique value of y is determined. However, if x is in the interval $(-3, 3)$, a square root of a negative number is obtained, and hence no real number y exists. Therefore we must restrict x so that $|x| \geq 3$. The domain of g is $(-\infty, -3] \cup [3, +\infty)$, and the range is $[0, +\infty)$. ∎

We can consider a function as a set of ordered pairs. For instance, the function defined by the equation $y = x^2$ consists of all the ordered pairs (x, y) satisfying the equation. The ordered pairs in this function given by Table 1 are $(1, 1)$, $(\frac{3}{2}, \frac{9}{4})$, $(4, 16)$, $(0, 0)$, $(-1, 1)$, $(-\frac{3}{2}, \frac{9}{4})$, and $(-4, 16)$. Of course, there is an unlimited number of ordered pairs in the function. Some others are $(2, 4)$, $(-2, 4)$, $(5, 25)$, $(-5, 25)$, $(\sqrt{3}, 3)$, and so on.

Illustration 3

The function f of Illustration 1 is the set of ordered pairs (x, y) for which $y = 2x^2 + 5$. With symbols we write

$$f = \{(x, y) | y = 2x^2 + 5\}$$

Some of the ordered pairs in f are $(0, 5)$, $(1, 7)$, $(\sqrt{2}, 9)$, $(2, 13)$, $(-1, 7)$, $(-\sqrt{5}, 15)$, and so on. ∎

Illustration 4

The function g of Illustration 2 is the set of ordered pairs (x, y) for which $y = \sqrt{x^2 - 9}$; that is,

$$g = \{(x, y) | y = \sqrt{x^2 - 9}\}$$

Some of the ordered pairs in g are $(3, 0)$, $(4, \sqrt{7})$, $(5, 4)$, $(-3, 0)$, $(-\sqrt{13}, 2)$, and so on. ∎

We now give the formal definition of a function. Defining a function as a set of ordered pairs rather than as a rule or correspondence makes its meaning precise.

Definition

> **Function**
> A **function** is a set of ordered pairs of real numbers (x, y) in which no two distinct ordered pairs have the same first number. The set of all admissible values of x is called the **domain** of the function, and the set of all resulting values of y is called the **range** of the function.

In the above definition, the restriction that no two distinct ordered pairs can have the same first number assures that y is unique for a specific value of x. The numbers x and y are **variables.** Because values are assigned to x, and because the value of y is dependent on the choice of x, x is the **independent variable** and y is the **dependent variable.**

The concept of a function as a set of ordered pairs permits us to give the following definition of the *graph of a function*.

Definition

> **Graph of a Function**
> If f is a function, then the **graph** of f is the set of all points (x, y) in R^2 for which (x, y) is an ordered pair in f.

Recall that to have a function there must be a unique value of the dependent variable for a value of the independent variable in the domain of the function. Thus we have the following geometric fact.

> The graph of a function can be intersected by a vertical line in at most one point.

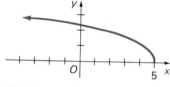

FIGURE 3

Illustration 5

Let $f = \{(x, y) \mid y = \sqrt{5 - x}\}$. A sketch of the graph of f appears in Figure 3. Observe that a vertical line having the equation $x = k$, where $k \leq 5$, intersects the graph in only one point. The domain of f is the set of all real numbers less than or equal to 5, which is the interval $(-\infty, 5]$, and the range is the set of all nonnegative real numbers, which is $[0, +\infty)$. ∎

In Illustration 6 a function is defined by more than one equation.

Illustration 6

Let g be the function which is the set of all ordered pairs (x, y) such that

$$y = \begin{cases} -3 & \text{if} & x \leq -1 \\ 1 & \text{if} & -1 < x \leq 2 \\ 4 & \text{if} & 2 < x \end{cases}$$

FIGURE 4

A sketch of the graph of g appears in Figure 4. The solid dot at $(-1, -3)$ and the open dot at $(-1, 1)$ indicate that the value of y is -3 when $x = -1$. Notice that every vertical line intersects the graph at only one point. The domain of g is $(-\infty, +\infty)$, while the range of g consists of the three numbers -3, 1, and 4. ∎

Illustration 7

Consider the set

$$\{(x, y) \mid x^2 + y^2 = 25\}$$

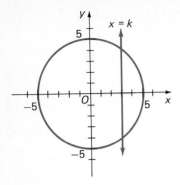

FIGURE 5

The graph of this set, shown in Figure 5, is the circle with center at the origin and radius 5. This set of ordered pairs is not a function, because for any x in the interval $(-5, 5)$ there are two ordered pairs having x as the first number. For example, both $(3, 4)$ and $(3, -4)$ are ordered pairs in the given set. Furthermore, observe from Figure 5 that a vertical line having the equation $x = k$, where $-5 < k < 5$, intersects the graph in two points. ∎

The domain of a function is usually apparent from the function's definition. Often the range can be determined from the graph of the function.

Example   1   The function h is defined by

$$h = \{(x, y)\ y = |x|\}$$

Draw a sketch of the graph of h and determine the domain and range.

Solution   Figure 6 shows the required sketch. From the definition of h, x can be any real number. Therefore the domain is $(-\infty, +\infty)$. Because we observe from Figure 6 that y can be any nonnegative number, the range is $[0, +\infty)$. ∎

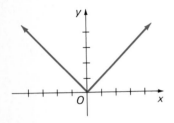

FIGURE 6

Example   2   Let F be the function which is the set of all ordered pairs (x, y) such that

$$y = \begin{cases} 3x - 2 & \text{if} & x < 1 \\ \frac{1}{2}(x^2 + 1) & \text{if} & 1 \le x \end{cases}$$

Draw a sketch of the graph of F and determine the domain and range.

Solution   The graph of $y = 3x - 2$ is a line, and the graph of $y = \frac{1}{2}(x^2 + 1)$ is a parabola. Thus we obtain the sketch of the graph of F appearing in Figure 7. Because x can be any real number, the domain of F is $(-\infty, +\infty)$. From Figure 7 we observe that y also can be any real number. Therefore the range is $(-\infty, +\infty)$. ∎

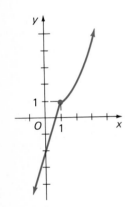

FIGURE 7

Example   3   Let G be the function which is the set of all ordered pairs (x, y) such that

$$y = \frac{x^2 - 9}{x - 3}$$

Determine the domain and range of G and draw a sketch of its graph.

Solution   Because a value for y is determined for each value of x except 3, the domain of G consists of all real numbers except 3. When $x = 3$, both the numerator and denominator are zero, and $\frac{0}{0}$ is undefined.

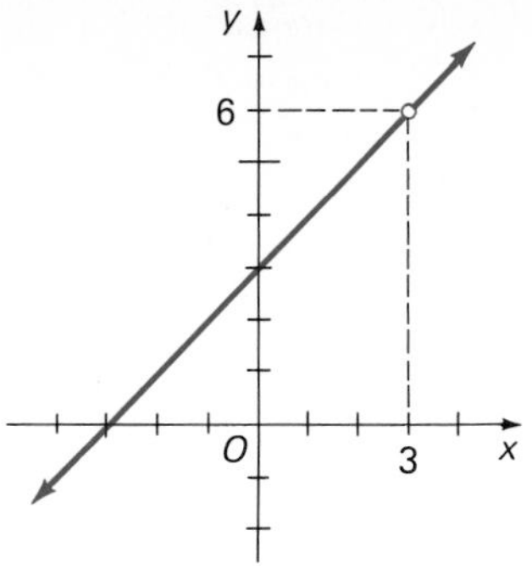

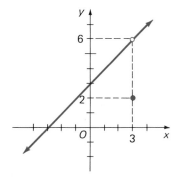

FIGURE 8

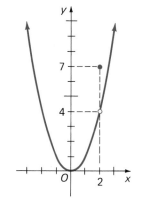

FIGURE 9

Factoring the numerator into $(x - 3)(x + 3)$, we obtain

$$y = \frac{(x - 3)(x + 3)}{x - 3}$$

or $y = x + 3$, provided that $x \neq 3$. In other words, the function G consists of all ordered pairs (x, y) such that

$$y = x + 3 \qquad \text{and} \qquad x \neq 3$$

From this definition of G it is apparent that the graph contains all points on the line $y = x + 3$ except the point $(3, 6)$. A sketch of the graph appears in Figure 8. The range of G is the set of all real numbers except 6. ■

Example   4   Let H be the function which is the set of all ordered pairs (x, y) such that

$$y = \begin{cases} x + 3 & \text{if} \quad x \neq 3 \\ 2 & \text{if} \quad x = 3 \end{cases}$$

Draw a sketch of the graph of H and determine the domain and range.

Solution   A sketch of the graph of H is shown in Figure 9. The graph contains the point $(3, 2)$ and all points on the line $y = x + 3$ except $(3, 6)$. Function H is defined for all values of x, and therefore the domain is $(-\infty, +\infty)$. The range is the set of all real numbers except 6. ■

Example   5   Let f be the function which is the set of all ordered pairs (x, y) such that

$$y = \begin{cases} x^2 & \text{if} \quad x \neq 2 \\ 7 & \text{if} \quad x = 2 \end{cases}$$

Draw a sketch of the graph of f and determine the domain and range.

Solution   A sketch of the graph of f appears in Figure 10. The graph consists of the point $(2, 7)$ and all points on the parabola $y = x^2$ except $(2, 4)$. Function f is defined for all real numbers; therefore the domain is $(-\infty, +\infty)$. As indicated in Figure 10, the range is the set of all nonnegative real numbers. ■

FIGURE 10

Example   6   Let h be the function which is the set of all ordered pairs (x, y) such that

$$y = \begin{cases} x + 5 & \text{if} \quad x < -3 \\ \sqrt{9 - x^2} & \text{if} \quad -3 \leq x \leq 3 \\ 5 - x & \text{if} \quad 3 < x \end{cases}$$

Determine the domain and range of h and draw a sketch of its graph.

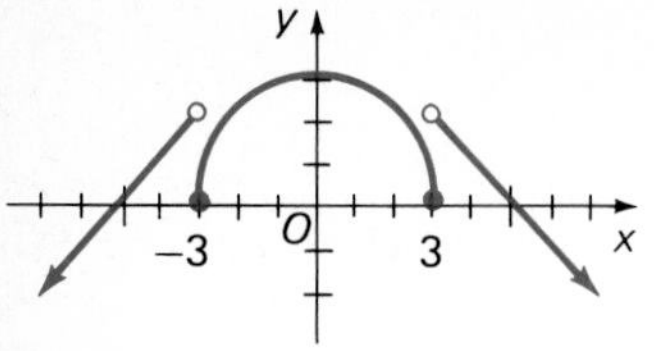

FIGURE 11

Solution   The part of the graph of h for $x < -3$ is a portion of the line $y = x + 5$. For $-3 \le x \le 3$, $y = \sqrt{9 - x^2}$. From our discussion of the circle in Section 3.2 we know that the graph of $x^2 + y^2 = 9$ is a circle having its center at the origin and radius 3. If we solve the equation $x^2 + y^2 = 9$ for y, we obtain $y = \pm\sqrt{9 - x^2}$. Thus the part of the graph of h for $-3 \le x \le 3$ is the upper half of the circle. The part of the graph for $3 < x$ is a portion of the line $y = 5 - x$. A sketch of the graph appears in Figure 11. The domain of h is $(-\infty, +\infty)$, and from Figure 11 the range is $(-\infty, 3]$. ∎

EXERCISES   4.1

In Exercises 1 through 20, draw a sketch of the graph of the function and determine its domain and range.

1. $f = \{(x, y) | y = 3x - 1\}$
2. $g = \{(x, y) | y = 4 - x\}$
3. $F = \{(x, y) | y = 2x^2\}$
4. $G = \{(x, y) | y = x^2 + 2\}$
5. $g = \{(x, y) | y = 5 - x^2\}$
6. $f = \{(x, y) | y = (x - 1)^2\}$
7. $G = \{(x, y) | y = \sqrt{x - 1}\}$
8. $F = \{(x, y) | y = \sqrt{9 - x}\}$
9. $f = \{(x, y) | y = \sqrt{x^2 - 4}\}$
10. $g = \{(x, y) | y = \sqrt{4 - x^2}\}$
11. $g = \{(x, y) | y = \sqrt{9 - x^2}\}$
12. $f = \{(x, y) | y = \sqrt{x^2 - 1}\}$
13. $H = \{(x, y) | y = |x - 3|\}$
14. $H = \{(x, y) | y = |5 - x|\}$
15. $F = \{(x, y) | y = |3x + 2|\}$

16. $G = \left\{(x, y) | y = \dfrac{x^2 - 4}{x - 2}\right\}$

17. $H = \left\{(x, y) | y = \dfrac{x^2 - 25}{x + 5}\right\}$

18. $f = \left\{(x, y) | y = \dfrac{2x^2 + 7x + 3}{x + 3}\right\}$

19. $f = \left\{(x, y) | y = \dfrac{x^2 - 4x + 3}{x - 1}\right\}$

20. $g = \left\{(x, y) | y = \dfrac{(x^2 - 4)(x - 3)}{x^2 - x - 6}\right\}$

In Exercises 21 through 36, the function is the set of all ordered pairs (x, y) satisfying the given equation. Draw a sketch of the graph of the function and determine its domain and range.

21. $f: y = \begin{cases} -2 & \text{if } x \le 3 \\ 2 & \text{if } 3 < x \end{cases}$

22. $g: y = \begin{cases} -4 & \text{if } x < -2 \\ -1 & \text{if } -2 \le x \le 2 \\ 3 & \text{if } 2 < x \end{cases}$

23. $g: y = \begin{cases} 2x - 1 & \text{if } x \ne 2 \\ 0 & \text{if } x = 2 \end{cases}$

24. $f: y = \begin{cases} 3x + 2 & \text{if } x \ne 1 \\ 8 & \text{if } x = 1 \end{cases}$

25. $F: y = \begin{cases} x^2 - 4 & \text{if } x \ne 3 \\ -2 & \text{if } x = 3 \end{cases}$

26. $G: y = \begin{cases} 9 - x^2 & \text{if } x \ne -3 \\ 4 & \text{if } x = -3 \end{cases}$

27. $G: y = \begin{cases} 1 - x^2 & \text{if } x < 0 \\ 3x + 1 & \text{if } 0 \le x \end{cases}$

28. $F: y = \begin{cases} x^2 - 4 & \text{if } x < 3 \\ 2x - 1 & \text{if } 3 \le x \end{cases}$

29. $g: y = \begin{cases} 6x + 7 & \text{if } x \le -2 \\ 4 - x & \text{if } -2 < x \end{cases}$

30. $f: y = \begin{cases} x - 2 & \text{if } x \le 0 \\ x^2 + 1 & \text{if } 0 < x \end{cases}$

31. $h: y = \begin{cases} x + 3 & \text{if } x < -5 \\ \sqrt{25 - x^2} & \text{if } -5 \le x \le 5 \\ 3 - x & \text{if } 5 < x \end{cases}$

32. $H: y = \begin{cases} x + 2 & \text{if } x \le -4 \\ \sqrt{16 - x^2} & \text{if } -4 < x < 4 \\ 2 - x & \text{if } 4 \le x \end{cases}$

33. $F: y = \dfrac{x^3 - 2x^2}{x - 2}$

34. $G: y = \dfrac{x^3 + 3x^2}{x + 3}$

35. $h: y = |x| + |x - 1|$

36. $H: y = |x^2 - 1|$

4.2   Function Notation, Operations on Functions, and Types of Functions

If f is the function having as its domain variable x and as its range variable y, the symbol $f(x)$ (read "f of x" or "f at x") denotes the particular value of y that corresponds to the value of x.

Illustration 1

In Illustration 5 of Section 4.1,

$$f = \{(x, y)|y = \sqrt{5 - x}\}$$

Thus

$$f(x) = \sqrt{5 - x}$$

Because when $x = 1$, $\sqrt{5 - x} = 2$, we have $f(1) = 2$. Also, $f(-6) = \sqrt{11}$, $f(0) = \sqrt{5}$, and so on. ∎

Illustration 2

In Example 1 of Section 4.1,

$$h = \{(x, y)|y = |x|\}$$

Therefore

$$h(x) = |x|$$

We now compute $h(x)$ for some specific values of x.

$$h(0) = |0| \qquad h(1) = |1| \qquad h(-1) = |-1| \qquad h(2) = |2| \qquad h(-2) = |-2|$$
$$\quad = 0 \qquad\qquad = 1 \qquad\qquad\quad = 1 \qquad\qquad = 2 \qquad\qquad\quad = 2$$

and so on. ∎

When defining a function, the domain of the function must be given either implicitly or explicitly. For instance, if f is defined by

$$f(x) = 3x^2 - 5x + 2$$

it is implied that x can be any real number. However, if f is defined by

$$f(x) = 3x^2 - 5x + 2 \qquad 1 \le x \le 10$$

then the domain of f consists of all real numbers between and including 1 and 10.

Similarly, if g is defined by the equation

$$g(x) = \frac{5x - 2}{x + 4}$$

it is implied that $x \neq -4$, because the quotient is undefined for $x = -4$: hence the domain of g is the set of all real numbers except -4.

If
$$h(x) = \sqrt{9 - x^2}$$

it is implied that x is in the closed interval $[-3, 3]$, because $\sqrt{9 - x^2}$ is not a real number for $x > 3$ or $x < -3$. Thus the domain of h is $[-3, 3]$ and the range is $[0, 3]$.

Example 1 Given that f is the function defined by
$$f(x) = x^2 + 3x - 4$$

find the following.

a) $f(0)$ b) $f(2)$ c) $f(h)$ d) $f(2h)$

e) $f(2x)$ f) $f(x + h)$ g) $f(x) + f(h)$

Solution

a) $f(0) = 0^2 + 3 \cdot 0 - 4$ b) $f(2) = 2^2 + 3 \cdot 2 - 4$ c) $f(h) = h^2 + 3h - 4$
$\quad = -4$ $\qquad\qquad\qquad = 6$

d) $f(2h) = (2h)^2 + 3(2h) - 4$ e) $f(2x) = (2x)^2 + 3(2x) - 4$
$\qquad\quad = 4h^2 + 6h - 4$ $\qquad\qquad = 4x^2 + 6x - 4$

f) $f(x + h) = (x + h)^2 + 3(x + h) - 4$
$\qquad\qquad = x^2 + 2hx + h^2 + 3x + 3h - 4$
$\qquad\qquad = x^2 + (2h + 3)x + (h^2 + 3h - 4)$

g) $f(x) + f(h) = (x^2 + 3x - 4) + (h^2 + 3h - 4)$
$\qquad\qquad\quad = x^2 + 3x + (h^2 + 3h - 8)$ ∎

Compare the computations in parts (f) and (g) of Example 1. In part (f) the computation is for $f(x + h)$, which is the function value at the sum of x and h. In part (g), where $f(x) + f(h)$ is computed, we obtain the sum of the two function values $f(x)$ and $f(h)$.

Example 2 Given that $f(x) = 3x^2 - 2x + 4$, find
$$\frac{f(x + h) - f(x)}{h}$$

where $h \neq 0$. Such a quotient often occurs in calculus.

Solution

$$\frac{f(x + h) - f(x)}{h} = \frac{3(x + h)^2 - 2(x + h) + 4 - (3x^2 - 2x + 4)}{h}$$

$$= \frac{3x^2 + 6hx + 3h^2 - 2x - 2h + 4 - 3x^2 + 2x - 4}{h}$$

$$= \frac{6hx - 2h + 3h^2}{h}$$

$$= 6x - 2 + 3h \qquad \blacksquare$$

We now define some operations on functions. In the definition new functions are formed from given functions by adding, subtracting, multiplying, and dividing function values. Accordingly, these new functions are known as the *sum, difference, product,* and *quotient* of the original functions.

Definition

Sum, Difference, Product, and Quotient of Two Functions

Given the two functions f and g:

(i) their **sum,** denoted by $f + g$, is the function defined by

$$(f + g)(x) = f(x) + g(x)$$

(ii) their **difference,** denoted by $f - g$, is the function defined by

$$(f - g)(x) = f(x) - g(x)$$

(iii) their **product,** denoted by $f \cdot g$, is the function defined by

$$(f \cdot g)(x) = f(x) \cdot g(x)$$

(iv) their **quotient,** denoted by f/g, is the function defined by

$$(f/g)(x) = f(x)/g(x)$$

In each case the *domain* of the resulting function consists of those values of x common to the domains of f and g, with the additional requirement in case (iv) that the values of x for which $g(x) = 0$ are excluded.

Example 3 Given that f and g are the functions defined by

$$f(x) = \sqrt{x + 1} \qquad \text{and} \qquad g(x) = \sqrt{x - 4}$$

find the following: (a) $(f + g)(x)$; (b) $(f - g)(x)$; (c) $(f \cdot g)(x)$; (d) $(f/g)(x)$. In each case determine the domain of the resulting function.

Solution

a) $(f + g)(x) = \sqrt{x + 1} + \sqrt{x - 4}$ b) $(f - g)(x) = \sqrt{x + 1} - \sqrt{x - 4}$

c) $(f \cdot g)(x) = \sqrt{x + 1} \cdot \sqrt{x - 4}$ d) $(f/g)(x) = \dfrac{\sqrt{x + 1}}{\sqrt{x - 4}}$

The domain of f is $[-1, +\infty)$ and the domain of g is $[4, +\infty)$. Hence in parts (a), (b), and (c), the domain of the resulting function is $[4, +\infty)$. In part (d), the denominator is zero when $x = 4$; thus 4 is excluded from the domain, and therefore the domain is $(4, +\infty)$. $\blacksquare$

Obtaining the *composite function* of two given functions is another operation on functions.

Definition

> **Composite Function**
> Given the two functions f and g, the **composite function,** denoted by $f \circ g$, is defined by
> $$(f \circ g)(x) = f(g(x))$$
> and the domain of $f \circ g$ is the set of all numbers x in the domain of g such that $g(x)$ is in the domain of f.

The definition indicates that when computing $(f \circ g)(x)$ we first apply function g to x and then function f to $g(x)$. The procedure is demonstrated in the following illustration and example.

Illustration 3

If f and g are defined by
$$f(x) = \sqrt{x} \quad \text{and} \quad g(x) = 2x - 3$$
then
$$\begin{aligned}(f \circ g)(x) &= f(g(x)) \\ &= f(2x - 3) \\ &= \sqrt{2x - 3}\end{aligned}$$

The domain of g is $(-\infty, +\infty)$, and the domain of f is $[0, +\infty)$. Therefore the domain of $f \circ g$ is the set of real numbers for which $2x - 3 \geq 0$ or, equivalently, $[\frac{3}{2}, +\infty)$. ∎

Example 4   Given that f and g are defined by
$$f(x) = \sqrt{x} \quad \text{and} \quad g(x) = x^2 - 1$$
find the following: (a) $f \circ f$; (b) $g \circ g$; (c) $f \circ g$; (d) $g \circ f$. Also determine the domain of each composite function.

Solution   The domain of f is $[0, +\infty)$, and the domain of g is $(-\infty, +\infty)$.

a) $\begin{aligned}(f \circ f)(x) &= f(f(x)) \\ &= f(\sqrt{x}) \\ &= \sqrt{\sqrt{x}} \\ &= \sqrt[4]{x}\end{aligned}$

The domain is $[0, +\infty)$.

b) $\begin{aligned}(g \circ g)(x) &= g(g(x)) \\ &= g(x^2 - 1) \\ &= (x^2 - 1)^2 - 1 \\ &= x^4 - 2x^2\end{aligned}$

The domain is $(-\infty, +\infty)$.

c) $(f \circ g)(x) = f(g(x))$
$= f(x^2 - 1)$
$= \sqrt{x^2 - 1}$

The domain is $(-\infty, -1] \cup [1, +\infty)$.

d) $(g \circ f)(x) = g(f(x))$
$= g(\sqrt{x})$
$= (\sqrt{x})^2 - 1$
$= x - 1$

The domain is $[0, +\infty)$.

In part (d), note that even though $x - 1$ is defined for all values of x, the domain of $g \circ f$, by the definition of a composite function, is the set of all numbers x in the domain of f such that $f(x)$ is in the domain of g. Thus the domain of $g \circ f$ must be a subset of the domain of f. ■

Observe, from the results of parts (c) and (d) of the above example, that $(f \circ g)(x)$ and $(g \circ f)(x)$ are not necessarily equal.

Definition

Even Function and Odd Function
(i) A function f is said to be an **even** function if for every x in the domain of f, $f(-x) = f(x)$.
(ii) A function f is said to be an **odd** function if for every x in the domain of f, $f(-x) = -f(x)$.
In both parts (i) and (ii) it is understood that $-x$ is the domain of f whenever x is.

Illustration 4

a) If $f(x) = 3x^4 - 2x^2 + 7$, then

$$f(-x) = 3(-x)^4 - 2(-x)^2 + 7$$
$$= 3x^4 - 2x^2 + 7$$
$$= f(x)$$

Therefore f is an even function.

b) If $g(x) = 3x^5 - 4x^3 - 9x$, then

$$g(-x) = 3(-x)^5 - 4(-x)^3 - 9(-x)$$
$$= -3x^5 + 4x^3 + 9x$$
$$= -(3x^5 - 4x^3 - 9x)$$
$$= -g(x)$$

Therefore g is an odd function.

c) If $h(x) = 2x^4 + 7x^3 - x^2 + 9$, then

$$h(-x) = 2(-x)^4 + 7(-x)^3 - (-x)^2 + 9$$
$$= 2x^4 - 7x^3 - x^2 + 9$$

Because $h(-x) \neq h(x)$ and $h(-x) \neq -h(x)$, h is neither even nor odd. ■

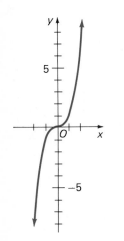

FIGURE 1

From the symmetry tests given in Section 3.2 it follows that the graph of an even function is symmetric with respect to the y axis, and the graph of an odd function is symmetric with respect to the origin.

Illustration 5

a) If $f(x) = x^2$, f is an even function, and its graph is a parabola that is symmetric with respect to the y axis. See Figure 1.

b) If $g(x) = x^3$, g is an odd function. The graph of g, shown in Figure 2, is symmetric with respect to the origin. ■

A function whose range consists of only one number is called a **constant function.** Thus if $f(x) = c$, and if c is any real number, then f is a constant function and its graph is a horizontal line at a directed distance of c units from the x axis.

Illustration 6

a) The function defined by $f(x) = 5$ is a constant function and its graph, shown in Figure 3, is a horizontal line 5 units above the x axis.

b) The function defined by $g(x) = -4$ is a constant function whose graph is a horizontal line 4 units below the x axis, as shown in Figure 4.

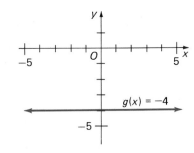

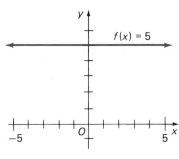

FIGURE 3

FIGURE 4 ■

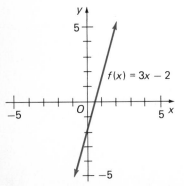

FIGURE 2

FIGURE 5

A **linear function** is defined by

$$f(x) = mx + b$$

where m and b are constants and $m \neq 0$. Its graph is a line having slope m and y intercept b.

Illustration 7

The function defined by

$$f(x) = 3x - 2$$

is linear. Its graph is the line appearing in Figure 5. ■

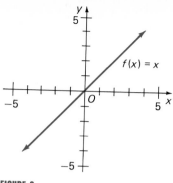

FIGURE 6

The particular linear function defined by

$$f(x) = x$$

is called the **identity function.** Its graph, shown in Figure 6, is the line bisecting the first and third quadrants.

If a function f is defined by

$$f(x) = a_n x^n + a_{n-1} x^{n-1} + a_{n-2} x^{n-2} + \cdots + a_1 x + a_0$$

where $a_0, a_1, \ldots a_n$ are real numbers $(a_n \neq 0)$ and n is a nonnegative integer, then f is called a **polynomial function** of degree n. Thus the function defined by

$$f(x) = 3x^5 - x^2 + 7x - 1$$

is a polynomial function of degree 5.

A linear function is a polynomial function of degree 1. If the degree of a polynomial function is 2, it is called a **quadratic function,** and if the degree is 3, it is called a **cubic function.**

Quadratic functions are discussed in Section 4.4, and graphs of polynomial functions are treated in Section 4.5.

If a function can be expressed as the quotient of two polynomial functions, it is called a **rational function.** Graphs of rational functions are considered in Section 4.6.

An **algebraic function** is a function formed by a finite number of algebraic operations on the identity function and the constant function. These algebraic operations include addition, subtraction, multiplication, division, raising to powers, and extracting roots. Polynomial and rational functions are particular kinds of algebraic functions. A complicated example of an algebraic function is the one defined by

$$f(x) = \frac{(x^2 - 3x + 1)^3}{\sqrt{x^4 + 1}}$$

Transcendental functions are also discussed in this text. Examples of transcendental functions are the exponential and logarithmic functions presented in Chapter 5.

EXERCISES   4.2

1. Given $f(x) = 2x - 1$, find: (a) $f(3)$; (b) $f(-2)$;
   (c) $f(0)$; (d) $f(a + 1)$; (e) $f(x + 1)$; (f) $f(2x)$;
   (g) $2f(x)$; (h) $f(x + h)$; (i) $f(x) + f(h)$;
   (j) $\dfrac{f(x + h) - f(x)}{h}$, $h \neq 0$.

2. Given $f(x) = \dfrac{3}{x}$, find: (a) $f(1)$; (b) $f(-3)$; (c) $f(6)$;
   (d) $f\left(\dfrac{1}{3}\right)$; (e) $f\left(\dfrac{3}{a}\right)$; (f) $f\left(\dfrac{3}{x}\right)$; (g) $\dfrac{f(3)}{f(x)}$; (h) $f(x - 3)$;

   (i) $f(x) - f(3)$; (j) $\dfrac{f(x + h) - f(x)}{h}$, $h \neq 0$.

3. Given $f(x) = 2x^2 + 5x - 3$, find: (a) $f(-2)$;
   (b) $f(-1)$; (c) $f(0)$; (d) $f(3)$; (e) $f(h + 1)$; (f) $f(2x^2)$;
   (g) $f(x^2 - 3)$; (h) $f(x + h)$; (i) $f(x) + f(h)$;
   (j) $\dfrac{f(x + h) - f(x)}{h}$, $h \neq 0$.

4. Given $g(x) = 3x^2 - 4$, find: (a) $g(-4)$; (b) $g(\tfrac{1}{2})$;
   (c) $g(x^2)$; (d) $g(3x^2 - 4)$; (e) $g(x - h)$;

(f) $g(x) - g(h)$; (g) $\dfrac{g(x + h) - g(x)}{h}$, $h \neq 0$.

5. Given $F(x) = \sqrt{2x + 3}$, find: (a) $F(-1)$; (b) $F(4)$; (c) $F(\tfrac{1}{2})$; (d) $F(11)$; (e) $F(2x + 3)$.

6. Given $G(x) = \sqrt{2x^2 + 1}$, find: (a) $G(-2)$; (b) $G(0)$; (c) $G(1)$; (d) $G(\tfrac{4}{7})$; (e) $G(2x^2 - 1)$.

In Exercises 7 through 16, the functions f and g are defined. In each exercise, define the following functions and determine the domain of the resulting function: (a) $f + g$: (b) $f - g$; (c) $f \cdot g$; (d) f/g; (e) g/f.

7. $f(x) = x - 5$; $g(x) = x^2 - 1$
8. $f(x) = \sqrt{x}$; $g(x) = x^2 + 1$
9. $f(x) = \dfrac{x + 1}{x - 1}$; $g(x) = \dfrac{1}{x}$
10. $f(x) = \sqrt{x}$; $g(x) = 4 - x^2$
11. $f(x) = \sqrt{x}$; $g(x) = x^2 - 1$
12. $f(x) = |x|$; $g(x) = |x - 3|$
13. $f(x) = x^2 + 1$; $g(x) = 3x - 2$
14. $f(x) = \sqrt{x + 4}$; $g(x) = x^2 - 4$
15. $f(x) = \dfrac{1}{x + 1}$; $g(x) = \dfrac{x}{x - 2}$
16. $f(x) = x^2$; $g(x) = \dfrac{1}{\sqrt{x}}$

In Exercises 17 through 26, the functions f and g are defined. In each exercise, define the following functions and determine the domain of the composite function: (a) $f \circ g$; (b) $g \circ f$; (c) $f \circ f$; (d) $g \circ g$.

17. $f(x) = x - 2$; $g(x) = x + 7$
18. $f(x) = 3 - 2x$; $g(x) = 6 - 3x$
19. The functions of Exercise 7.
20. The functions of Exercise 8.
21. $f(x) = \sqrt{x - 2}$; $g(x) = x^2 - 2$
22. $f(x) = x^2 - 1$; $g(x) = \dfrac{1}{x}$
23. $f(x) = \dfrac{1}{x}$; $g(x) = \sqrt{x}$ 24. $f(x) = \sqrt{x}$; $g(x) = -\dfrac{1}{x}$
25. $f(x) = |x|$; $g(x) = |x + 2|$
26. $f(x) = \sqrt{x^2 - 1}$; $g(x) = \sqrt{x - 1}$

In Exercises 27 and 28, a function f is defined. In each exercise, define the following functions, and determine the domain of the resulting function: (a) $f(x^2)$; (b) $f^2(x)$; (c) $(f \circ f)(x)$.

27. $f(x) = 2x - 3$ 28. $f(x) = \dfrac{2}{x - 1}$

In Exercises 29 through 32, show that $(f \circ g)(x) = x$ and $(g \circ f)(x) = x$.

29. $f(x) = 2x - 3$ and $g(x) = \dfrac{x + 3}{2}$

30. $f(x) = \dfrac{1}{x + 1}$ and $g(x) = \dfrac{1 - x}{x}$

31. $f(x) = x^2$, $x \geq 0$ and $g(x) = \sqrt{x}$
32. $f(x) = (x - 1)^3$ and $g(x) = 1 + \sqrt[3]{x}$
33. Given $G(x) = |x - 2| - |x| + 2$, express $G(x)$ without absolute-value bars if x is in the given interval: (a) $[2, +\infty)$; (b) $(-\infty, 0)$; (c) $[0, 2)$.

34. Given $f(t) = \dfrac{|3 + t| - |t| - 3}{t}$, express $f(t)$ without absolute-value bars if t is in the given interval: (a) $(0, +\infty)$; (b) $[-3, 0)$; (c) $(-\infty, -3)$.

35. Given

$$f(x) = \begin{cases} \dfrac{|x|}{x} & \text{if } x \neq 0 \\ 1 & \text{if } x = 0 \end{cases}$$

find: (a) $f(1)$; (b) $f(-1)$; (c) $f(4)$; (d) $f(-4)$; (e) $f(-x)$; (f) $f(x + 1)$; (g) $f(x^2)$; (h) $f(-x^2)$.

In Exercises 36 and 37, determine whether the function is even, odd, or neither.

36. (a) $f(x) = 2x^4 - 3x^2 + 1$ (b) $f(x) = 5x^3 - 7x$
   (c) $f(s) = s^2 + 2s + 2$ (d) $g(x) = x^6 - 1$
   (e) $h(t) = 5t^7 + 1$ (f) $f(x) = |x|$
   (g) $f(y) = \dfrac{y^3 - y}{y^2 + 1}$ (h) $g(z) = \dfrac{z - 1}{z + 1}$

37. (a) $g(x) = 5x^2 - 4$ (b) $f(x) = x^3 + 1$
   (c) $f(t) = 4t^5 + 3t^3 - 2t$ (d) $g(r) = \dfrac{r^2 - 1}{r^2 + 1}$
   (e) $f(x) = \begin{cases} 1 & \text{if } x \geq 0 \\ -1 & \text{if } x < 0 \end{cases}$ (f) $h(x) = \dfrac{4x^2 - 5}{2x^3 + x}$
   (g) $f(z) = (z - 1)^2$ (h) $g(x) = \dfrac{|x|}{x^2 + 1}$
   (i) $g(x) = \sqrt{x^2 - 1}$ (j) $f(x) = \sqrt[3]{x}$

38. There is one function that is both even and odd. What is it?

39. Determine whether the composite function $f \circ g$ is odd or even in each of the following cases: (a) f and g are both even; (b) f and g are both odd; (c) f is even and g is odd; (d) f is odd and g is even.

4.3 Functions As Mathematical Models

In applications it is often necessary to express a practical situation in terms of a functional relationship. The function obtained gives a mathematical **model** of the situation. In this section we give examples showing the procedure involved in obtaining some mathematical models.

Example 1 A clock manufacturer can produce a particular clock at a cost of $15 per clock. It is estimated that if the selling price of the clock is x dollars, then the number of clocks sold per week is $125 - x$.

a) Express the number of dollars in the manufacturer's weekly profit as a function of x.

b) Use the result of part (a) to determine the weekly profit if the selling price is $45 per clock.

Solution a) The profit can be obtained by subtracting the total cost from the total revenue. Let $R(x)$ dollars be the weekly revenue. Because the revenue is the product of the selling price of each clock and the number of clocks sold,

$$R(x) = x(125 - x)$$

Let $C(x)$ dollars be the total cost of the clocks that are sold per week. Because the total cost is the product of the cost of each clock and the number of clocks sold,

$$C(x) = 15(125 - x)$$

If $P(x)$ dollars is the weekly profit, then

$$P(x) = R(x) - C(x)$$
$$= x(125 - x) - 15(125 - x)$$
$$= (125 - x)(x - 15)$$

b) If the selling price is $45, the number of dollars in the weekly profit is $P(45)$. From the expression for $P(x)$ in part (a),

$$P(45) = (125 - 45)(45 - 15)$$
$$= 80 \cdot 30$$
$$= 2400$$

Therefore the weekly profit is $2400 when the clocks are sold at $45 each. ∎

The function of Example 1 is discussed again in Section 4.4, where we determine the selling price of a clock in order for a manufacturer's weekly profit to be a maximum.

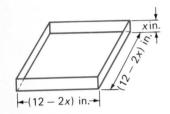

FIGURE 1 _____

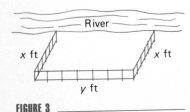

FIGURE 2 _____

Example **2**   A cardboard box manufacturer wishes to make open boxes from square pieces of cardboard of side 12 in. by cutting equal squares from the four corners and turning up the sides.

a) Let x inches be the length of the side of the square to be cut out; express the number of cubic inches in the volume of the box as a function of x.

b) What is the domain of the resulting function?

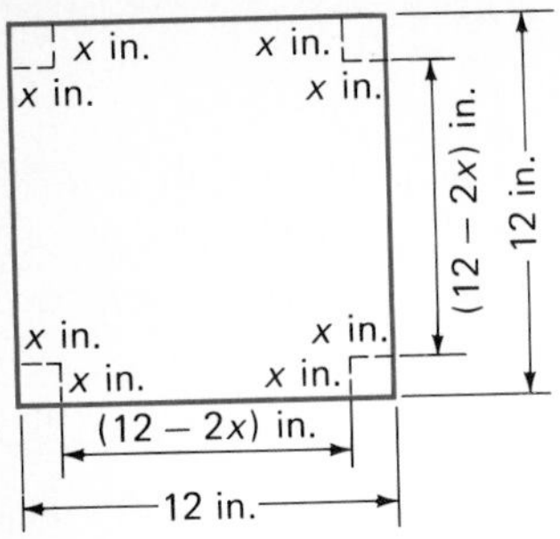

Solution a) Figure 1 represents a given piece of cardboard and Figure 2 represents the box obtained from the cardboard. The numbers of inches in the dimensions of the box are x, $12 - 2x$, and $12 - 2x$. The volume of the box is the product of the three dimensions. Therefore, if $V(x)$ cubic inches is the volume,

$$V(x) = x(12 - 2x)(12 - 2x)$$
$$= 144x - 48x^2 + 4x^3$$

b) From the expression for $V(x)$ in part (a), we observe that $V(0) = 0$ and $V(6) = 0$. From conditions of the problem, we know that x can be neither negative nor greater than 6. Thus the domain of V is the closed interval $[0, 6]$. ■

In Example 2, to determine the value of x that will give a box having the largest possible volume, we must find the maximum value of the cubic function V. The procedure for doing this requires techniques of calculus.

In the next example we first obtain two equations involving a dependent variable and two independent variables. We then express the dependent variable as a function of a single independent variable by eliminating the other independent variable from the pair of equations.

Example **3**   A rectangular field is to be fenced off along the bank of a river, and no fence is required along the river. The material for the fence costs $8 per running foot for the two ends and $12 per running foot for the side parallel to the river; $3600 worth of fence is to be used.

a) Let x feet be the length of an end; express the number of square feet in the area of the field as a function of x.

b) What is the domain of the resulting function?

Solution a) Let y feet be the length of the side of the field parallel to the river and A square feet be the area of the field. See Figure 3. Then

$$A = xy$$

Because the cost of the material for each end is $8 per running foot and the length of an end is x feet, the total cost of the fence for each end is $8x$ dollars. Similarly, the total cost of the fence for the third side

FIGURE 3 _____

is $12y$ dollars. We have then

$$8x + 8x + 12y = 3600 \qquad (1)$$

To express A in terms of a single variable, we first solve Equation (1) for y in terms of x.

$$12y = 3600 - 16x$$
$$y = 300 - \tfrac{4}{3}x$$

We substitute this value of y into the equation $A = xy$, yielding A as a function of x, and

$$A(x) = x(300 - \tfrac{4}{3}x)$$

b) Both x and y must be nonnegative. The smallest value that x can assume is 0. The smallest value that y can assume is 0, and when $y = 0$ we obtain, from Equation (1), $x = 225$. Thus 225 is the largest value that x can assume. Hence x must be in the closed interval $[0, 225]$, and this closed interval is the domain of A.   ∎

In Section 4.4 we return to the function of Example 3 and learn how to determine the dimensions of the field of largest possible area that can be enclosed by the $3600 worth of fence.

Applications involving the dependence of one variable on another occur in the physical, life, and social sciences. The formulas used in these applications often determine functions. For instance, if y dollars is the simple interest for one year earned by a principal of x dollars at the rate of 12 percent per year, then

$$y = 0.12x$$

For a given nonnegative value of x there corresponds a unique value of y; thus the value of y depends on the value of x. If f is the function defined by $f(x) = 0.12x$, and the domain of f is the set of nonnegative real numbers, then the equation $y = 0.12x$ can be written as $y = f(x)$. The equation $y = 0.12x$ is an example of *direct proportion*, and y is said to be *directly proportional* to x.

Definition

> **Directly Proportional**
> A variable y is said to be **directly proportional** to a variable x if
>
> $$y = kx$$
>
> where k is a nonzero constant. More generally, a variable y is said to be **directly proportional** to the nth power of x $(n > 0)$ if
>
> $$y = kx^n$$
>
> The constant k is called the **constant of proportionality**.

Example **4** A person's approximate brain weight is directly proportional to his or her body weight, and a person weighing 150 lb has an approximate brain weight of 4 lb.

a) Express the number of pounds in the approximate brain weight of a person as a function of the person's body weight.

b) Find the approximate brain weight of a person whose body weight is 176 lb.

Solution a) Let $f(x)$ pounds be the approximate brain weight of a person having a body weight of x pounds. Then

$$f(x) = kx \tag{2}$$

Because a person of body weight 150 lb has a brain weighing approximately 4 lb, we substitute 150 for x and 4 for $f(x)$ in (2), and we have

$$4 = k(150)$$
$$k = \tfrac{2}{75}$$

We replace k in (2) by this value and obtain

$$f(x) = \tfrac{2}{75}x$$

b) Because $f(x) = \tfrac{2}{75}x$,

$$f(176) = \frac{2}{75}(176) = 4.7$$

The approximate brain weight of a person weighing 176 lb is 4.7 lb. ∎

Definition

> **Inversely Proportional**
> A variable y is said to be **inversely proportional** to a variable x if
>
> $$y = \frac{k}{x}$$
>
> where k is a nonzero constant. More generally, a variable y is said to be **inversely proportional** to the nth power of $x(n > 0)$ if
>
> $$y = \frac{k}{x^n}$$

Example **5** The intensity of light from a given source is inversely proportional to the square of the distance from it.

a) Express the number of candlepower (CP) in the intensity of light as a function of the number of meters in the distance of the light from the source if the intensity is 225 CP at a distance of 5 m from the source.

b) Find the intensity at a point 15 m from the source.

Solution   a) Let $f(x)$ candlepower be the intensity of light from a source that is x meters from it. Then

$$f(x) = \frac{k}{x^2} \tag{3}$$

Because the intensity is 225 CP at a distance of 5 m from the source, we replace x by 5 and $f(x)$ by 225 in (3) and obtain

$$225 = \frac{k}{5^2}$$

$$k = 5625$$

Substituting this value of k in (3), we have

$$f(x) = \frac{5625}{x^2}$$

b) From the above expression for $f(x)$, we get

$$f(15) = \frac{5625}{15^2}$$

$$= \frac{5625}{225}$$

$$= 25$$

Therefore the intensity at a point 15 m from the source is 25 CP. ∎

Definition

> **Jointly Proportional**
> A variable z is said to be **jointly proportional** to variables x and y if
>
> $$z = kxy$$
>
> where k is a nonzero constant. More generally, a variable z is said to be **jointly proportional** to the nth power of x and the mth power of y ($n > 0$ and $m > 0$) if
>
> $$z = kx^n y^m$$

Example   6   In a limited environment where A is the maximum number of bacteria supportable by the environment, the rate of bacterial growth is jointly proportional to the number present and the difference between A and the number present. Suppose 1 million bacteria is the maximum number supportable by the environment and the rate of growth is 60 bacteria per minute when there are 1000 bacteria present.

a) Express the rate of bacterial growth as a function of the number of bacteria present.

b) Find the rate of growth when there are 100,000 bacteria present.

Solution   a) Let $f(x)$ bacteria per minute be the rate of growth when there are x bacteria present. Then

$$f(x) = kx(1,000,000 - x) \tag{4}$$

Because the rate of growth is 60 bacteria per minute when there are 1000 bacteria present, we replace x by 1000 and $f(x)$ by 60 in (4), and we have

$$60 = k(1000)(1,000,000 - 1000)$$

$$k = \frac{60}{999,000,000}$$

$$= \frac{1}{16,650,000}$$

Replacing k in (4) by this value, we obtain

$$f(x) = \frac{x(1,000,000 - x)}{16,650,000}$$

b) From the above expression for $f(x)$, we have

$$f(100,000) = \frac{100,000(1,000,000 - 100,000)}{16,650,000}$$

$$= \frac{100,000(900,000)}{16,650,000}$$

$$= 5405$$

Therefore the rate of growth is 5405 bacteria per minute when there are 100,000 bacteria present.   ■

EXERCISES   4.3

1. A carpenter can construct bookcases at a cost of $40 each. If the carpenter sells the bookcases for x dollars each, it is estimated that $300 - 2x$ bookcases will be sold per month.
   (a) Express the number of dollars in the carpenter's monthly profit as a function of x.
   (b) Use the result of part (a) to determine the monthly profit, given that the selling price is $110 per bookcase.

2. A toy manufacturer can produce a particular toy at a cost of $10 per toy. It is estimated that if the selling price of the toy is x dollars, then the number of toys that are sold each day is $45 - x$.
   (a) Express the number of dollars in the manufacturer's daily profit as a function of x.
   (b) Use the result of part (a) to determine the daily profit, given that the selling price is $30 per toy.

3. A manufacturer of open tin boxes wishes to make use of pieces of tin with dimensions 8 in. by 15 in. by cutting equal squares from the four corners and turning up the sides.

(a) Let x inches be the length of the side of the square to be cut out; express the number of cubic inches in the volume of the box as a function of x.

(b) What is the domain of the resulting function?

4. Suppose the manufacturer of Exercise 3 makes the open boxes from square pieces of tin that measure k centimeters on a side.

(a) Let x centimeters be the length of the side of the square cut out; express the number of cubic centimeters in the volume of the box as a function of x. Remember that k is a constant.

(b) What is the domain of the resulting function?

5. A rectangular field is to be enclosed with 240 m of fence.

(a) Let x meters be the length of the field; express the number of square meters in the area of the field as a function of x.

(b) What is the domain of the resulting function?

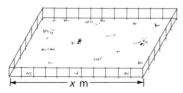

6. A rectangular garden is to be fenced off with 100 ft of fencing material.

(a) Let x feet be the length of the garden; express the number of square feet in the area of the garden as a function of x.

(b) What is the domain of the resulting function?

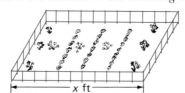

7. Do Exercise 5 with this change: One side of the field is to have a river as a natural boundary, and the fencing material is to be used for the other three sides. Let x meters be the length of the side of the field that is parallel to the river.

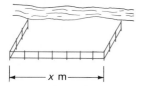

8. Do Exercise 6 with this change: The garden is to be placed so that a side of a house serves as a boundary, and the fencing material is to be used for the other three sides. Let x feet be the length of the side of the garden that is parallel to the house.

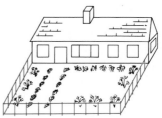

9. A rectangular plot of ground is to be enclosed by a fence and then divided down the middle by another fence. The fence down the middle costs \$2 per running foot and the other fence costs \$5 per running foot, and \$960 worth of fencing material is to be used.

(a) Let x feet be the length of the fence down the middle; express the number of square feet in the area of the plot as a function of x.

(b) What is the domain of the resulting function?

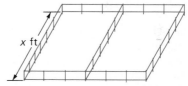

10. A package in the shape of a rectangular box with a square cross section is to have the sum of its length and girth (the perimeter of a cross section) equal to 100 in.

(a) Let x inches be the length of the package; express the volume of the box as a function of x.

(b) What is the domain of the resulting function?

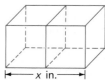

11. The graph of an equation relating the temperature reading in Celsius degrees and the temperature reading in Fahrenheit degrees is a straight line. Water freezes at 0° Celsius and 32° Fahrenheit and water boils at 100° Celsius and 212° Fahrenheit.

(a) Let x degrees be the Fahrenheit temperature; express the number of degrees in the Celsius temperature as a function of x.

(b) Find the Celsius temperature when the Fahrenheit temperature is 95°.

12. The approximate weight of a person's muscles is directly proportional to his or her body weight.

(a) Express the number of pounds in the approximate muscle weight of a person as a function of the person's body weight, given that a person weighing 150 lb has muscles weighing approximately 60 lb.

(b) Find the approximate muscle weight of a person weighing 130 lb.

13. The daily payroll for a work crew is directly proportional to the number of workers, and a crew of 12 workers earns a payroll of $540.

(a) Express the number of dollars in the daily payroll as a function of the number of workers.

(b) What is the daily payroll for a crew of 15 workers?

14. For a gas having a constant pressure, its volume is directly proportional to the absolute temperature, and at a temperature of 180° the gas occupies 100 m³.

(a) Express the number of cubic meters in the volume of the gas as a function of the number of degrees in the absolute temperature.

(b) What is the volume of the gas at a temperature of 150°?

15. The period (the time for one complete oscillation) of a pendulum is directly proportional to the square root of the length of the pendulum, and a pendulum of length 8 ft has a period of 2 sec.

(a) Express the number of seconds in the period of a pendulum as a function of the number of feet in its length.

(b) Find the period of a pendulum of length 2 ft.

16. For a vibrating string, the rate of vibrations is directly proportional to the square root of the tension on the string.

(a) Given that a particular string vibrates 864 times per second under a tension of 24 kg, express the number of vibrations per second as a function of the number of kilograms in the tension.

(b) Find the number of vibrations per second under a tension of 6 kg.

17. The weight of a body is inversely proportional to the square of its distance from the center of the earth.

(a) Given that a body weighs 200 lb on the earth's surface, express the number of pounds in its weight as a function of the number of miles from the center of the earth. Assume that the radius of the earth is 4000 mi.

(b) How much does the body weigh at a distance of 400 mi above the earth's surface?

18. For an electric cable of fixed length, the resistance is inversely proportional to the square of the diameter of the cable.

(a) Given that a cable having the fixed length is $\frac{1}{2}$ cm in diameter and has a resistance of 0.1 ohm, express the number of ohms in the resistance as a function of the number of centimeters in the diameter.

(b) What is the resistance of a cable having the fixed length and a diameter of $\frac{2}{3}$ cm?

19. In a small town of population 5000 the rate of growth of an epidemic (the rate of change of the number of infected persons) is jointly proportional to the number of people infected and the number of people who are not infected.

(a) Given that the epidemic is growing at the rate of 9 people per day when there are 100 people infected, express the rate of growth of the epidemic as a function of the number of infected people.

(b) How fast is the epidemic growing when 200 people are infected?

20. In a community of 8000 people the rate at which a rumor spreads is jointly proportional to the number of people who have heard the rumor and the number of people who have not heard it.

(a) Given that the rumor is spreading at the rate of 20 people per hour when 200 people have heard it, express the rate at which the rumor is spreading as a function of the number of people who have heard it.

(b) How fast is the rumor spreading when 500 people have heard it?

21. The maximum number of bacteria supportable by a particular environment is 900,000, and the rate of bacterial growth is jointly proportional to the number present and the difference between 900,000 and the number present.

(a) Let $f(x)$ bacteria per minute be the rate of

growth when there are x bacteria present; write an equation defining $f(x)$.

(b) What is the domain of the function f in part (a)?

22. A particular lake can support a maximum of 14,000 fish and the rate of growth of the fish population is jointly proportional to the number of fish present and the difference between 14,000 and the number present.

(a) Let $f(x)$ fish per day be the rate of growth when there are x fish present; write an equation defining $f(x)$.

(b) What is the domain of the function f in part (a)?

23. A page of print is to contain 24 in.^{2} of printed

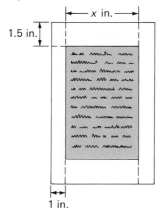

region, a margin of $1\frac{1}{2}$ in. at the top and bottom, and a margin of 1 in. at the sides.

(a) Let $A(x)$ in.^{2} be the total area of the page when x in. is the width of the printed portion; write an equation defining $A(x)$.

(b) What is the domain of the function A in part (a)?

24. A one-story building having a rectangular floor space of 13,200 ft^2 is to be constructed, with a walkway 22 ft wide in the front and back and a walkway 15 ft wide on each side.

(a) Let $A(x)$ ft^2 be the total area of the lot on which the building and walkways will be located, where x ft is the length of the front and back of the building; write an equation defining $A(x)$.

(b) What is the domain of the function A in part (a)?

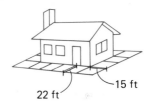

4.4 Quadratic Functions

The general quadratic function is defined by

$$f(x) = ax^2 + bx + c$$

where a, b, and c are constants representing real numbers and $a \neq 0$. The graph of f is the same as the graph of the equation

$$y = ax^2 + bx + c$$

In Section 3.5 we learned that the graph of such an equation is a parabola whose axis is vertical. Properties of the parabola discussed in Sections 3.4 and 3.5 are used as an aid in drawing a sketch of the graph.

Illustration   1

If the function f is defined by

$$f(x) = -2x^2 + 8x - 5$$

the graph of f is the same as the graph of the equation

$$y = -2x^2 + 8x - 5$$

This equation is equivalent to

$$2(x^2 - 4x) = -y - 5$$

To complete the square of the binomial within the parentheses, we add $2(4)$ to both sides of the equation and we have

$$2(x^2 - 4x + 4) = -y - 5 + 8$$
$$(x - 2)^2 = -\tfrac{1}{2}(y - 3)$$

This equation is of the form

$$(x - h)^2 = 4p(y - k)$$

where (h, k) is $(2, 3)$ and $p = -\tfrac{1}{8}$. Therefore the vertex of the parabola is at $(2, 3)$ and the axis is the line $x = 2$. Because $p < 0$, the parabola opens downward. We find a few more points on the parabola and draw the sketch shown in Figure 1. ■

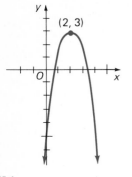

FIGURE 1

The **zeros** of a function f are the values of x for which $f(x) = 0$.

Illustration 2

The function of Illustration 1 is defined by

$$f(x) = -2x^2 + 8x - 5$$

To find the zeros of this function we substitute 0 for (x) and get

$$-2x^2 + 8x - 5 = 0$$
$$2x^2 - 8x + 5 = 0$$

Solving this equation by the quadratic formula, where a is 2, b is -8, and c is 5, we have

$$x = \frac{-b \pm \sqrt{b^2 - 4ac}}{2a}$$

$$= \frac{8 \pm \sqrt{64 - 40}}{4}$$

$$= \frac{8 \pm \sqrt{24}}{4}$$

$$= \frac{8 \pm 2\sqrt{6}}{4}$$

$$= \frac{4 \pm \sqrt{6}}{2}$$

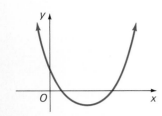

FIGURE 2

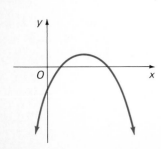

FIGURE 3

Hence the zeros of f are

$$\frac{4 + \sqrt{6}}{2} \approx 3.22 \quad \text{and} \quad \frac{4 - \sqrt{6}}{2} \approx 0.78$$

These numbers are also the x intercepts of the parabola in Figure 1 that is the graph of the given function. ■

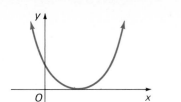

FIGURE 4

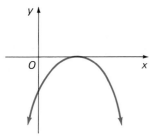

FIGURE 5

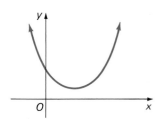

FIGURE 6

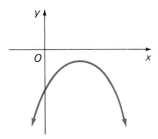

FIGURE 7

In general, if

$$f(x) = ax^2 + bx + c$$

then the zeros of f are the roots of the equation

$$ax^2 + bx + c = 0$$

If the zeros are real they are the x intercepts of the graph of f.

In Section 2.3 we learned that a quadratic equation in one variable can have two real roots, one real root (of multiplicity two), or two imaginary roots. If a quadratic equation has two real roots, the corresponding quadratic function has two real zeros and its graph intersects the x axis at two distinct points. This situation is shown in Figure 2, where the parabola opens upward ($a > 0$), and in Figure 3, where the parabola opens downward ($a < 0$).

If the function has one real zero, the graph intersects the x axis at a single point as shown in Figures 4 ($a > 0$) and 5 ($a < 0$). If the function has two imaginary zeros, the graph does not intersect the x axis; this situation occurs in Figures 6 ($a > 0$) and 7 ($a < 0$).

When the graph of a quadratic function opens upward, the function has a **minimum value,** which occurs at the vertex of the parabola. There is no maximum value for such a function. When the parabola opens downward, the function has a **maximum value** occurring at the vertex; it has no minimum value.

Illustration 3

The function of Illustrations 1 and 2 is defined by

$$f(x) = -2x^2 + 8x - 5$$

The graph of f is the parabola of Figure 1. The parabola opens downward and has its vertex at (2, 3). Therefore this function has a maximum value of 3 and it occurs when $x = 2$. ∎

Example 1 Find either a maximum or minimum value of the function defined by

$$f(x) = 3x^2 + 3x + 2$$

Solution The graph of f is the parabola having the equation

$$y = 3x^2 + 3x + 2$$

We write this equation in the form $(x - h)^2 = 4p(y - k)$. The equation is equivalent to

$$3(x^2 + x) = y - 2$$

Completing the square of the binomial in parentheses by adding $\frac{3}{4}$ to both sides of the equation, we have

$$3(x^2 + x + \tfrac{1}{4}) = y - 2 + \tfrac{3}{4}$$
$$3(x + \tfrac{1}{2})^2 = y - \tfrac{5}{4}$$
$$(x + \tfrac{1}{2})^2 = \tfrac{1}{3}(y - \tfrac{5}{4})$$

The parabola opens upward and its vertex is at $(-\tfrac{1}{2}, \tfrac{5}{4})$. Therefore the minimum value of f is $\tfrac{5}{4}$, occurring at $x = -\tfrac{1}{2}$.   ∎

We now apply the method used in the solution of Example 1 to the general quadratic function defined by

$$f(x) = ax^2 + bx + c$$

In this equation we replace $f(x)$ by y and obtain

$$y = ax^2 + bx + c$$

which is equivalent to

$$ax^2 + bx = y - c$$
$$a\left(x^2 + \frac{b}{a}x\right) = y - c$$

We complete the square of the binomial in parentheses.

$$a\left(x^2 + \frac{b}{a}x + \frac{b^2}{4a^2}\right) = y - c + \frac{b^2}{4a}$$
$$\left(x + \frac{b}{2a}\right)^2 = \frac{1}{a}\left(y + \frac{b^2 - 4ac}{4a}\right)$$

The graph of this equation is a parabola having its vertex at the point where $x = -\dfrac{b}{2a}$. If $a > 0$, the parabola opens upward and so f has a minimum value at the point where $x = -\dfrac{b}{2a}$. If $a < 0$, the parabola opens downward and so f has a maximum value at the point where $x = -\dfrac{b}{2a}$.

These results are given in the following theorem. In the statement of the theorem **extreme value** means either a maximum value or a minimum value.

Theorem 1

> The quadratic function defined by $f(x) = ax^2 + bx + c$, where $a \neq 0$, has an extreme value at the point where $x = -\dfrac{b}{2a}$. If $a > 0$, the extreme value is a minimum value, and if $a < 0$, the extreme value is a maximum value.

Example   2   Use Theorem 1 to find either a maximum or minimum value of the function g if

$$g(x) = -\tfrac{3}{2}x^2 + 6x - 10$$

Solution   For the given quadratic function, $a = -\tfrac{3}{2}$ and $b = 6$. Because $a < 0$, g has a maximum value at the point where

$$x = -\frac{b}{2a}$$

$$= -\frac{6}{2(-\tfrac{3}{2})}$$

$$= 2$$

The maximum value is

$$g(2) = -\tfrac{3}{2}(2)^2 + 6(2) - 10$$
$$= -4$$ ∎

Example   3   In Example 1 of Section 4.3 we had the following situation: A clock manufacturer can produce a particular clock at a cost of $15 per clock. It is estimated that if the selling price of the clock is x dollars, then the number of clocks sold per week is $125 - x$. Determine what the selling price should be in order for the manufacturer's weekly profit to be a maximum.

Solution   In Example 1 of Section 4.3 we showed that if $P(x)$ dollars is the manufacturer's weekly profit, then

$$P(x) = (125 - x)(x - 15)$$
$$P(x) = -x^2 + 140x - 1875$$

Function P is quadratic with $a = -1$ and $b = 140$. Because $a < 0$, P has a maximum value at the point where

$$x = -\frac{b}{2a}$$
$$= 70$$

Thus the manufacturer's weekly profit will be a maximum when the selling price of the clock is $70.   ∎

Example   4   In Example 3 of Section 4.3 we had the following situation: A rectangular field is to be fenced off along the bank of a river, and no fence is required along the river. The material for the fence costs $8 per running foot for the two ends and $12 per running foot for the side parallel to the river; $3600 worth of fence is to be used. Find the dimensions of the field of largest possible area that can be enclosed with the $3600 worth of fence. What is the largest area?

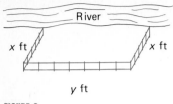

FIGURE 8

Solution Let x feet be the length of an end of the field. See Figure 8. In Example 3 of Section 4.3 we showed that the number of feet in the length of the side parallel to the river is $300 - \frac{4}{3}x$, and if $A(x)$ square feet is the area of the field,

$$A(x) = x(300 - \tfrac{4}{3}x)$$
$$A(x) = -\tfrac{4}{3}x^2 + 300x$$

Function A is quadratic with $a = -\frac{4}{3}$ and $b = 300$. Because $a < 0$, A has a maximum value at the point where

$$x = -\frac{b}{2a}$$
$$= -\frac{300}{2(-\frac{4}{3})}$$
$$= \frac{225}{2}$$

When $x = \frac{225}{2}$, $300 - \frac{4}{3}x = 150$. Furthermore,

$$A(\tfrac{225}{2}) = \tfrac{225}{2}(150)$$
$$= 16{,}875$$

Therefore the largest possible area that can be enclosed for $3600 is $16{,}875 \text{ ft}^2$, and this is obtained when the side parallel to the river is 150 ft long and the ends are each $\frac{225}{2}$ ft long.   ■

Example 5 The financial manager of a college newsletter determines that 1000 copies of the newsletter will be sold if the price is 50 cents and that the number of copies sold decreases by 10 for each 1 cent added to the price. What price will yield the largest gross income from sales and what is the largest gross income?

Solution The number of cents in the gross income depends on the price per copy. Let $f(x)$ cents be the gross income when x cents is the price per copy.

The amount by which x exceeds 50 is $x - 50$. To determine the number of copies sold when x cents is the price per copy, we must subtract from 1000 the product of 10 and this excess. Hence, when x cents is the price per copy, the number of copies sold is $1000 - 10(x - 50)$.

We obtain an expression for the gross income by multiplying the number of copies sold by the price per copy. Therefore

$$f(x) = [1000 - 10(x - 50)]\,x$$
$$f(x) = (1500 - 10x)x$$
$$f(x) = -10x^2 + 1500x$$

For this quadratic function $a = -10$, $b = 1500$, and $c = 0$. Because $a < 0$,

f has a maximum value at the point where

$$x = -\frac{b}{2a}$$

$$= -\frac{1500}{2(-10)}$$

$$= 75$$

The maximum value is

$$f(75) = -10(75)^2 + 1500(75)$$
$$= -10(5625) + 112,500$$
$$= 56,250$$

Thus the price of 75 cents per copy will yield the largest gross income from sales, $562.50. ■

EXERCISES 4.4

In Exercises 1 through 4, find the zeros of the function.

1. $f(x) = x^2 - 2x - 3$
2. $f(x) = x^2 - 3x + 1$
3. $f(x) = 2x^2 - 2x - 1$
4. $f(x) = 6x^2 - 7x - 5$

In Exercises 5 through 14, draw a sketch of the graph of the function and determine from the graph which of the following statements characterizes the zeros of the function: (a) two real zeros; (b) one real zero of multiplicity two; or (c) two imaginary zeros.

5. $f(x) = x^2 - 4x$
6. $f(x) = x^2 - 3$
7. $f(x) = -x^2 + 4$
8. $g(x) = x^2 - 6x + 11$
9. $g(x) = -4x^2 + 8x - 8$
10. $g(x) = 2x^2 + 4x + 1$
11. $h(x) = 9 - 6x + x^2$
12. $h(x) = 1 - 4x - x^2$
13. $f(x) = \frac{1}{8}(4x^2 + 20x + 49)$
14. $f(x) = -4x^2 + 12x - 9$

In Exercises 15 through 18, use the method of Example 1 to find either a maximum or minimum value of the function.

15. $f(x) = 4x^2 + 8x + 7$
16. $f(x) = 2 + 6x - x^2$
17. $g(x) = -\frac{1}{2}(x^2 + 6x + 5)$
18. $G(x) = \frac{1}{8}(x^2 - 4x - 4)$

In Exercises 19 through 22, use Theorem 1 to find either a maximum or minimum value of the function.

19. $f(x) = 2 + 4x - 3x^2$
20. $g(x) = 3x^2 + 6x + 9$
21. $G(x) = \frac{1}{8}(4x^2 + 12x - 9)$
22. $F(x) = -\frac{1}{2}(x^2 + 8x + 8)$
23. Find two numbers whose sum is 10 and whose product is a maximum.
24. Find two numbers whose difference is 14 and whose product is a minimum.

25. In Exercise 1 of Exercises 4.3 we had the following situation: A carpenter can construct bookcases at a cost of $40 each. If the carpenter sells the bookcases for x dollars each, it is estimated that $300 - 2x$ bookcases will be sold per month. Determine the selling price of each bookcase that will give the carpenter the greatest monthly profit.
26. In Exercise 2 of Exercises 4.3 we had the following situation: A toy manufacturer can produce a particular toy at a cost of $10 per toy. It is estimated that if the selling price of the toy is x dollars, then the number of toys that are sold each day is $45 - x$. Determine what the selling price of each toy should be in order for the manufacturer to realize the maximum daily profit.
27. Refer to Exercise 5 in Exercises 4.3. Find the dimensions of the largest rectangular field that can be enclosed with 240 m of fence.
28. Refer to Exercise 6 in Exercises 4.3. Find the dimensions of the largest rectangular garden that can be fenced off with 100 ft of fencing material.
29. Refer to Exercise 7 in Exercises 4.3. If one side of a rectangular field is to have a river as a natural boundary, what are the dimensions of the largest rectangular field that can be enclosed by using 240 m of fence for the other three sides?
30. Refer to Exercise 8 in Exercises 4.3. Find the dimensions of the largest rectangular garden that can be placed so that a side of a house serves as a boundary and 100 ft of fencing material is to be used for the other three sides.

31. Refer to Exercise 19 in Exercises 4.3. In a small town of population 5000 the rate of growth of an epidemic is jointly proportional to the number of people infected and the number of people who are not infected. Determine the number of people infected when the rate of growth is a maximum.

32. Refer to Exercise 20 in Exercises 4.3. In a community of 8000 people the rate at which a rumor spreads is jointly proportional to the number of people who have heard the rumor and the number of people who have not heard it. Determine how many people have heard the rumor when it is being spread at the greatest rate.

33. Refer to Exercise 21 in Exercises 4.3. The maximum number of bacteria supportable by a particular environment is 900,000, and the rate of bacterial growth is jointly proportional to the number present and the difference between 900,000 and the number present. Determine the number of bacteria present when the rate of growth is a maximum.

34. Refer to Exercise 22 in Exercises 4.3. A particular lake can support a maximum of 14,000 fish and the rate of growth of the fish population is jointly proportional to the number of fish present and the difference between 14,000 and the number present. What should the size of the fish population be in order for the growth rate to be a maximum?

35. Find two positive numbers whose sum is 50 and such that the sum of their squares is a minimum.

36. An object is thrown straight upward from the ground with an initial velocity of 96 ft/sec. If the height of the object is s feet after t seconds, and if air resistance is neglected, $s = 96t - 16t^2$. What is the maximum height reached by the object and how many seconds after it is thrown does it reach its maximum height?

37. A projectile is shot straight upward from a point 15 ft above the ground with an initial velocity of 176 ft/sec. If the height of the projectile is s feet after t seconds and if air resistance is neglected, $s = 15 + 176t - 16t^2$. How long does it take the projectile to reach its maximum height and what is the maximum height?

38. A travel agency offers an organization an all-inclusive tour for $800 per person if not more than 100 people take the tour. However the cost per person will be reduced $5 for each person in excess of 100. How many people should take the tour in order for the travel agency to receive the largest gross revenue, and what is this largest gross revenue?

39. A student club on a college campus charges annual membership dues of $10, less 5 cents for each member over 60. How many members would give the club the most revenue from annual dues?

4.5 Graphs of Polynomial Functions

Recall from Section 4.2 that a polynomial function f of degree n is defined by

$$f(x) = a_n x^n + a_{n-1}x^{n-1} + a_{n-2}x^{n-2} + \cdots + a_1 x + a_0$$

where $a_0, a_1, \ldots, a_n$ are real numbers ($a_n \neq 0$) and n is a nonnegative integer. We have already discussed the graph of three types of polynomial functions: constant functions, linear functions, and quadratic functions. Consider now the power function defined by

$$f(x) = ax^n$$

where n is a positive integer. In Figure 1, we have sketches of the graphs of the power function for $a = 1$ and n having values 1 through 6. Sketches of the graphs of the power function for $a = -1$ and n having values 1 through 6 are shown in Figure 2. They are mirror images in the x axis of the corre-

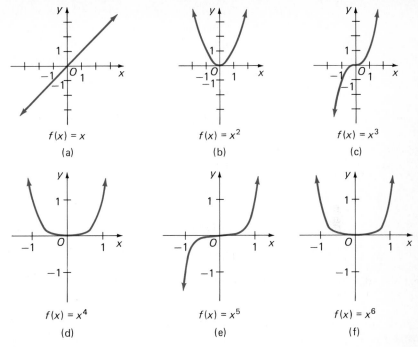

$f(x) = x$ (a) $f(x) = x^2$ (b) $f(x) = x^3$ (c)

$f(x) = x^4$ (d) $f(x) = x^5$ (e) $f(x) = x^6$ (f)

FIGURE 1

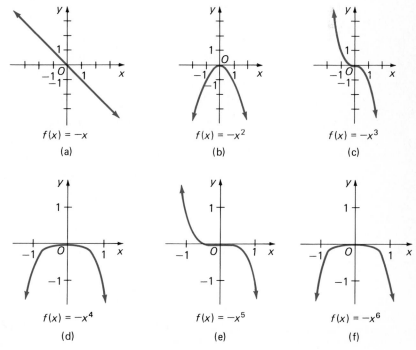

$f(x) = -x$ (a) $f(x) = -x^2$ (b) $f(x) = -x^3$ (c)

$f(x) = -x^4$ (d) $f(x) = -x^5$ (e) $f(x) = -x^6$ (f)

FIGURE 2

sponding graphs in Figure 1. All of the graphs contain the origin, and this is the only intersection of the curve with either axis. If $n > 1$, the x axis is tangent to the graph at the origin. If n is a positive even integer, the graph is in the first and second quadrants and is symmetric with respect to the y axis. If n is a positive odd integer, the graph is in the first and third quadrants and is symmetric with respect to the origin. As $|x|$ increases without bound, so does $|f(x)|$.

Graphs of the functions defined by equations of the form

$$f(x) = ax^n + c \qquad \text{and} \qquad f(x) = a(x + c)^n$$

can be obtained from those of the form $f(x) = ax^n$ by a translation of the y and x axes, respectively.

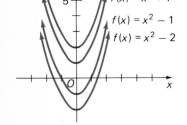

FIGURE 3

Illustration 1

Figure 3 shows the graphs of the functions defined by

$$f(x) = x^2 + c$$

where c takes on the values -2, -1, 1, and 2. Observe that the graphs are obtained from the graph of $f(x) = x^2$ (Figure 1b) by a vertical translation (or shift) of c units in the upward direction if $c > 0$ and $|c|$ units in the downward direction if $c < 0$. Furthermore, c is the y intercept of the graph. ■

Example 1 Draw a sketch of the graph of each of the following functions. (a) $f(x) = x^3 + 2$; (b) $f(x) = x^3 - 3$.

Solution a) The graph of $f(x) = x^3 + 2$ is obtained from the graph of $f(x) = x^3$ (Figure 1c) by a vertical translation of 2 units upward. See Figure 4.

b) The graph of $f(x) = x^3 - 3$ is shown in Figure 5. It is obtained from the graph of $f(x) = x^3$ by a vertical translation of 3 units downward.

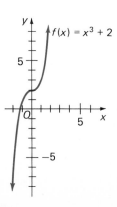

FIGURE 4

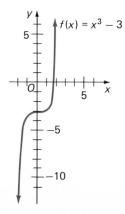

FIGURE 5 ■

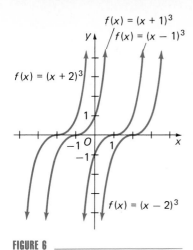
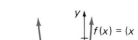
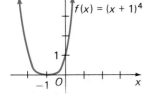
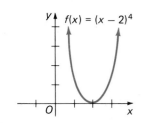

FIGURE 6

FIGURE 7

FIGURE 8

Illustration 2

In Figure 6 there are graphs of the functions defined by

$$f(x) = (x + c)^3$$

for c having the values $-2, -1, 1,$ and 2. These graphs are obtained from the graph of $f(x) = x^3$ by a horizontal translation of c units to the left if $c > 0$ and $|c|$ units to the right if $c < 0$. ∎

Example 2　Draw a sketch of the graph of each of the following functions. (a) $f(x) = (x + 1)^4$; (b) $f(x) = (x - 2)^4$.

Solution　a) The graph of $f(x) = (x + 1)^4$ is obtained from the graph of $f(x) = x^4$ (Figure 1d) by a horizontal translation of 1 unit to the left. The graph appears in Figure 7.

b) Figure 8 shows the graph of $f(x) = (x - 2)^4$. It is obtained from the graph of $f(x) = x^4$ by a horizontal translation of 2 units to the right. ∎

For graphs of more general polynomial functions, first recall that a polynomial function of the nth degree is defined by

$$P(x) = a_n x^n + a_{n-1} x^{n-1} + a_{n-2} x^{n-2} + \cdots + a_1 x + a_0$$

where $a_0, a_1, \ldots, a_n$ are real numbers with $a_n \neq 0$, and n is a nonnegative integer. When $n = 1$, we have

$$P(x) = a_1 x + a_0$$

which defines a linear function, and its graph is a line. When $n = 2$, we have

$$P(x) = a_2 x^2 + a_1 x + a_0$$

which defines a quadratic function, and its graph is a parabola.

We now consider graphs of polynomial functions for which $n \geq 3$. The first term of the polynomial is $a_n x^n$. As $|x|$ increases without bound, $|a_n x^n|$ increases without bound and will become larger than the sum of all the other terms in the polynomial. Therefore the form of the graph for large values of $|x|$ will be affected by the values of the term $a_n x^n$. We can conclude that the shape of the graph for large values of $|x|$ will be similar to that of the graph of the power function of degree n. Let us consider as separate cases a_n positive and a_n negative.

Case 1:　$a_n > 0$. The function values will be increasing for large values of x; so the graph will be going up to the right as in Figure 9(a) and (b). If n is even, the graph comes down from the left as in Figure 9(a), and if n is odd, the graph comes up from the left as in Figure 9(b).

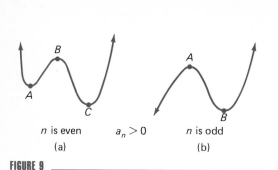

n is even $a_n > 0$ n is odd

(a)   (b)

FIGURE 9

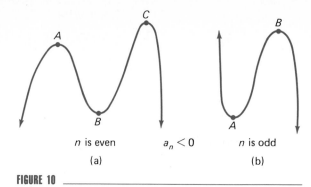

n is even $a_n < 0$ n is odd

(a)   (b)

FIGURE 10

Case 2: $a_n < 0$. The function values will be decreasing for large values of x; thus the graph will be going down to the right as in Figure 10(a) and (b). If n is even, the graph comes up from the left as in Figure 10(a), and if n is odd, the graph comes down from the left as in Figure 10(b).

These facts are summarized in Table 1.

TABLE 1   **Graphs of Polynomial Functions**

a_n	n	Properties of the graph	Example
Positive	Even	Comes down from the left, goes up to the right	Figure 9(a)
Positive	Odd	Comes up from the left, goes up to the right	Figure 9(b)
Negative	Even	Comes up from the left, goes down to the right	Figure 10(a)
Negative	Odd	Comes down from the left, goes down to the right	Figure 10(b)

When obtaining graphs of polynomial functions, we assume that the graphs are unbroken curves. This property follows from the fact that polynomial functions are *continuous* (a concept studied in calculus) which implies that a small change in x results in a small change in $f(x)$.

An important aid in drawing a sketch of the graph of a polynomial function is to determine the number of **turning points** of the graph. These turning points are either **high points** or **low points.** In Figure 9(a) the graph has one high point at B and two low points at A and C. In Figure 9(b) there are two turning points, a high point at A and a low point at B. There are also three turning points in Figure 10(a) and two turning points in Figure 10(b). We state without proof a theorem giving the number of turning points for a polynomial function.

Theorem 1	The graph of a polynomial function of the nth degree has at most $n - 1$ turning points.

From this theorem it follows that the graph of a polynomial function of the fourth degree has at most three turning points. Therefore the graph in Figure 9(a) could be that of a fourth-degree polynomial. The graph in Figure 9(b) could be that of a third-degree polynomial because such a graph has at most two turning points.

Example   3 Draw a sketch of the graph of the function P defined by

$$P(x) = x^3 - 6x^2 + 9x - 4$$

Solution   Because $P(x)$ is a third-degree polynomial, the graph has at most two turning points. Furthermore, the coefficient of x^3 is positive and the degree of the polynomial is odd. We refer to Table 1 and note that the graph comes up from the left and goes up to the right. Therefore the graph is probably similar to that shown in Figure 9(b). Table 2 gives the coordinates of some points on the graph.

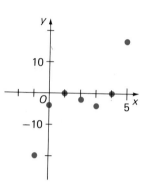

FIGURE 11

TABLE 2

x	-2	-1	0	1	2	3	4	5
$P(x)$	-54	-20	-4	0	-2	-4	0	16

In Figure 11 we have plotted the points $(-1, -20)$, $(0, -4)$, $(1, 0)$, $(2, -2)$, $(3, -4)$, $(4, 0)$, and $(5, 16)$ given from Table 2. We have chosen different-size units on the y axis than on the x axis. With these points and the previous information, we draw the sketch shown in Figure 12. ■

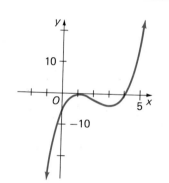

FIGURE 12

In Example 3, without using calculus, we cannot be absolutely sure that the turning points occur at $(1, 0)$ and $(3, -4)$, but we can assume that the graph has the general appearance given in Figure 12. A similar comment holds for the graph in the next example.

Example   4 Draw a sketch of the graph of the function P defined by

$$P(x) = 3x^4 - 4x^3 - 12x^2 + 12$$

Solution   The graph has at most three turning points because $P(x)$ is a fourth-degree polynomial. The coefficient of x^4 is positive and the degree of the polynomial is even. From Table 1 we observe that the graph comes down from the left and goes up to the right. Because $P(0) = 12$, the graph intersects the y axis at $(0, 12)$. Values of $P(x)$ are computed for some values of x and these are given in Table 3.

TABLE 3

x	-2	-1	0	1	2	3
$P(x)$	44	7	12	-1	-20	39

In Figure 13 we have plotted the points $(-2, 44)$, $(-1, 7)$, $(0, 12)$, $(1, -1)$, $(2, -20)$, and $(3, 39)$ given from Table 3. Observe that we have chosen different-size units on the y axis than on the x axis. With these points and the information found above, we draw the sketch of the graph appearing in Figure 14.

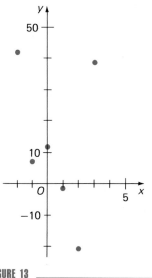

FIGURE 13

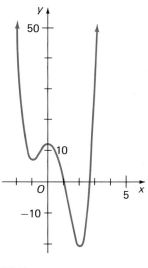

FIGURE 14

EXERCISES   4.5

In each exercise, draw a sketch of the graph of the polynomial function.

1. $f(x) = 2x^2$
2. $g(x) = -2x^2$
3. $g(x) = (x + 2)^2$
4. $f(x) = (x - 1)^2$
5. $f(x) = -4(x - 2)^2$
6. $g(x) = 3(x + 1)^2$
7. $P(x) = 3x^2 + 2$
8. $P(x) = 2x^2 - 3$
9. $F(x) = 2x^3$
10. $G(x) = -2x^3$
11. $P(x) = (x + 2)^3$
12. $P(x) = (x - 1)^3$
13. $f(x) = (x - 1)^4$
14. $g(x) = -(x + 2)^4$
15. $g(x) = x^3 - 2$
16. $f(x) = x^4 - 1$
17. $h(x) = (x + 1)^6$
18. $h(x) = -(x - 1)^5$
19. $f(x) = (x - 2)^3 + 3$
20. $g(x) = (x + 3)^4 - 2$

21. $P(x) = x^3 - 2x^2 - 5x + 6$
22. $P(x) = x^3 - 3x^2 - 9x + 9$
23. $F(x) = x^3 - 3x^2 + 3$ 24. $G(x) = x^3 + 4x^2 + 4x$
25. $g(x) = 3x^3 - 4x^2 - 5x + 2$
26. $f(x) = 6x^3 + 29x^2 + x - 6$
27. $f(x) = x^4 - 5x^3 + 2x^2 + 8x$
28. $g(x) = x^4 - 5x^2 + 4$
29. $P(x) = x^4 + x^3 - 7x^2 - x + 6$
30. $P(x) = x^4 - 6x^3 + 11x^2 - 6x$
31. $G(x) = 3x^4 + 5x^3 - 5x^2 - 5x + 2$
32. $h(x) = 2x^4 - x^3 - 6x^2 - x + 2$
33. $f(x) = -x^4 + x^2$ 34. $f(x) = x^5 - 4x^3$

4.6 Graphs of Rational Functions

In Section 4.2 we defined a rational function as one that can be expressed as the quotient of two polynomial functions. Thus, if P and Q are polynomial functions and f is the function defined by

$$f(x) = \frac{P(x)}{Q(x)}$$

then f is a rational function. The domain of f is the set of all real numbers except the zeros of Q. We shall assume that $P(x)$ and $Q(x)$ are polynomials having no common factor.

Knowing the behavior of $f(x)$ when x is close to a zero of Q is helpful when obtaining the graph of f. Consider, for example, the function defined by

$$f(x) = \frac{x + 2}{x - 3}$$

The domain of f is the set of all real numbers except 3. We shall investigate the function values when x is close to 3 but not equal to 3. First let x take on the values $4, \frac{7}{2}, \frac{10}{3}, \frac{13}{4}, \frac{31}{10}, \frac{301}{100}, \frac{3001}{1000}$, and so on. We are taking values of x closer and closer to 3 but greater than 3; in other words, the variable x is approaching 3 from the right. We illustrate this in Table 1.

TABLE 1

x	4	$\frac{7}{2}$	$\frac{10}{3}$	$\frac{13}{4}$	$\frac{31}{10}$	$\frac{301}{100}$	$\frac{3001}{1000}$
$f(x) = \dfrac{x + 2}{x - 3}$	6	11	16	21	51	501	5001

From the table we see intuitively that as x gets closer and closer to 3 from the right, $f(x)$ increases without bound. In other words, we can make $f(x)$ greater than any preassigned positive number by taking x close enough to 3 and greater than 3. To indicate that $f(x)$ increases without bound as x approaches 3 from the right, we use the symbolism

$$f(x) \to +\infty \qquad \text{as} \qquad x \to 3^+$$

The symbol $+\infty$ (positive infinity) is not a real number; it is used to indicate the behavior of the function values $f(x)$ as x gets closer and closer to 3. The $+$ symbol as a superscript after the 3 indicates that x is approaching 3 from the right.

TABLE 2

x	2	$\frac{5}{2}$	$\frac{8}{3}$	$\frac{11}{4}$	$\frac{29}{10}$	$\frac{299}{100}$	$\frac{2999}{1000}$
$f(x) = \dfrac{x + 2}{x - 3}$	-4	-9	-14	-19	-49	-499	-4999

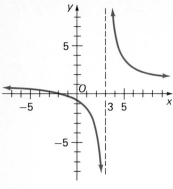

FIGURE 1 _____

Now let the variable x approach 3 through values less than 3; that is, let x take on the values $2, \frac{5}{2}, \frac{8}{3}, \frac{11}{4}, \frac{29}{10}, \frac{299}{100}, \frac{2999}{1000}$, and so on. Refer to Table 2. Notice that as x gets closer and closer to 3 from the left, the values of $f(x)$ decrease without bound (the values of $f(x)$ are negative numbers whose absolute values increase without bound); that is, we can make $f(x)$ less than any preassigned negative number by taking x close enough to 3 and x less than 3. We use the following notation to indicate that $f(x)$ decreases without bound as x approaches 3 from the left:

$$f(x) \to -\infty \qquad \text{as} \qquad x \to 3^-$$

In Figure 1 there is a sketch of the graph of f that shows the behavior of $f(x)$ near $x = 3$. As x gets closer and closer to 3 from either the right or left, the absolute value of $f(x)$ gets larger and larger. Observe that the graph does not intersect the line $x = 3$, which is shown as dashed in the figure. The line $x = 3$ is called a *vertical asymptote* of the graph of f.

Definition

> **Vertical Asymptote**
> The line $x = a$ is said to be a **vertical asymptote** of the graph of the function f if at least one of the following statements is true:
> (i) $f(x) \to +\infty$ as $x \to a^+$
> (ii) $f(x) \to -\infty$ as $x \to a^+$
> (iii) $f(x) \to +\infty$ as $x \to a^-$
> (iv) $f(x) \to -\infty$ as $x \to a^-$

In Figure 1 both statements (i) and (iv) of the above definition are true for the function f when a is 3.

Illustration 1

In Figure 2 there is a sketch of the graph of a function for which statements (i) and (iii) of the above definition are true. For the function whose graph appears in Figure 3, statements (ii) and (iii) are true.

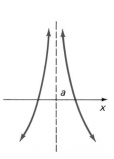

FIGURE 2 _____

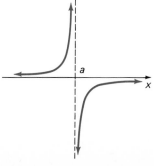

FIGURE 3 _____

The following theorem can be proved from the definition of a vertical asymptote.

Theorem 1

> The graph of a rational function of the form $P(x)/Q(x)$, where $P(x)$ and $Q(x)$ have no common factors, has the line $x = a$ as a vertical asymptote if $Q(a) = 0$.

We apply Theorem 1 in the next example.

Example   1 Draw a sketch of the graph of the function f defined by

$$f(x) = \frac{3}{x^2}$$

Solution The domain of f is the set of all real numbers except 0. Thus the graph has no y intercept. Because $f(x)$ is never 0, the graph has no x intercept. Also observe from the equation that $f(x)$ is never negative. Therefore the graph is confined to the first and second quadrants. Because $f(-x) = f(x)$, f is an even function, and the graph is symmetric with respect to the y axis.

To obtain any vertical asymptotes we use Theorem 1. We set the denominator equal to zero and get $x = 0$. Thus the y axis is a vertical asymptote.

A few points on the graph in the first quadrant are given in Table 3. We plot these points and use the symmetry property to obtain corresponding points in the second quadrant. We complete the sketch shown in Figure 4 by connecting the points in each quadrant with an unbroken curve and using the above information.

TABLE 3

x	$\frac{1}{2}$	1	2	3	4
$f(x) = \dfrac{3}{x^2}$	12	3	$\dfrac{3}{4}$	$\dfrac{1}{3}$	$\dfrac{3}{16}$

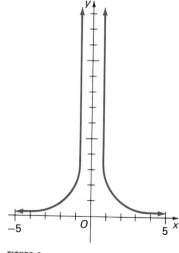

FIGURE 4

Observe from the graph in Figure 4 that the function values $f(x)$ approach 0 as $|x|$ increases without bound. For this reason the x axis is a *horizontal asymptote* of the graph. To define a horizontal asymptote we use the notation $f(x) \to b^+$ to mean that $f(x)$ approaches b through values greater than b, and $f(x) \to b^-$ to mean that $f(x)$ approaches b through values less than b. The notation $x \to +\infty$ indicates that x is *increasing without bound*, and $x \to -\infty$ means that x is *decreasing without bound*.

Definition

> **Horizontal Asymptote of the Graph of a Rational Function**
> The line $y = b$ is said to be a **horizontal asymptote** of the graph of the rational function f if at least one of the following statements is true:
> (i) $f(x) \rightarrow b^{+}$ as $x \rightarrow +\infty$
> (ii) $f(x) \rightarrow b^{+}$ as $x \rightarrow -\infty$
> (iii) $f(x) \rightarrow b^{-}$ as $x \rightarrow +\infty$
> (iv) $f(x) \rightarrow b^{-}$ as $x \rightarrow -\infty$

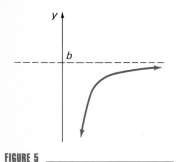

FIGURE 5

Illustration 2

In Figure 5 is a sketch of the graph of a function for which statement (iii) of the above definition is true, and in Figure 6 statement (ii) is true. Both statements (i) and (iv) are true for the graph of the function shown in Figure 7. The graph in Figure 7 also has the line $x = a$ as a vertical asymptote because $f(x) \rightarrow +\infty$ as $x \rightarrow a^{+}$ and $f(x) \rightarrow -\infty$ as $x \rightarrow a^{-}$.   ■

Theorem 2

> The graph of a rational function of the form
>
> $$\frac{a_n x^n + a_{n-1} x^{n-1} + \cdots + a_1 x + a_0}{b_m x^m + b_{m-1} x^{m-1} + \cdots + b_1 x + b_0}$$
>
> has
> (i) the x axis as a horizontal asymptote if $n < m$;
> (ii) the line $y = \dfrac{a_n}{b_m}$ as a horizontal asymptote if $n = m$;
> (iii) no horizontal asymptote if $n > m$.

FIGURE 6

The proof of Theorem 2 is omitted, but the following two illustrations should make parts (i) and (ii) plausible. Later, in Example 4, is a function for which part (iii) applies.

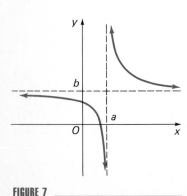

FIGURE 7

Illustration 3

Let the function g be defined by

$$g(x) = \frac{4x}{x^2 - 25}$$

Since the degree of the numerator, which is 1, is less than the degree of the denominator, which is 2, g is a rational function for which part (i) of Theorem 2 should apply. To see this, let us divide the numerator and denominator by x^2. We then have

$$g(x) = \frac{\dfrac{4}{x}}{1 - \dfrac{25}{x^2}}$$

As $x \to +\infty$ $\dfrac{4}{x} \to 0^+$ and $\dfrac{25}{x^2} \to 0^+$

Therefore, as $x \to +\infty$, $g(x) \to 0^+$.

As $x \to -\infty$ $\dfrac{4}{x} \to 0^-$ and $\dfrac{25}{x^2} \to 0^+$

Therefore, as $x \to -\infty$, $g(x) \to 0^-$.

Thus, from the definition, the line $y = 0$ is a horizontal asymptote of the graph of g. Function g is discussed in Example 3 later in this section, where a sketch of its graph appears in Figure 9.   ■

Illustration 4

Let the function h be defined by

$$h(x) = \frac{3x^2}{2x^2 - 32}$$

The degree of the numerator equals the degree of the denominator. Hence h is a rational function for which part (ii) of Theorem 2 should apply. If we divide the numerator and denominator by x^2, we get

$$h(x) = \frac{3}{2 - \dfrac{32}{x^2}}$$

As $x \to +\infty$ or $x \to -\infty$ $\dfrac{32}{x^2} \to 0^+$

Therefore, as $x \to +\infty$ or $x \to -\infty$, $h(x) \to \dfrac{3}{2}^+$. From the definition it follows that the line $y = \dfrac{3}{2}$ is a horizontal asymptote. Function h is discussed in Example 4 later in this section, where a sketch of its graph is in Figure 10.   ■

Example   2   Draw a sketch of the graph of the function f defined by

$$f(x) = \frac{2x - 3}{x + 1}$$

Solution   The domain of f is the set of all real numbers except -1. Because $f(0) = -3$, the y intercept is -3. The x intercept is $\frac{3}{2}$, which is obtained by setting the numerator equal to 0. Because f is neither even nor odd, there is no symmetry with respect to the y axis or origin.

By setting the denominator equal to 0 we obtain the line $x = -1$ as a vertical asymptote. Because the degrees of the numerator and denominator are equal, it follows from Theorem 2(ii) that a horizontal asymptote is the line $y = 2$. Some points on the graph, determined by computing $f(x)$ for selected values of x, are given by Table 4. We plot these points and, with the asymptotes as guides, obtain the sketch of the graph of f in Figure 8.

TABLE 4

x	0	2	4	6	8	10	−2	−4	−6	−8	−10	−12
$f(x) = \dfrac{2x - 3}{x + 1}$	−3	$\dfrac{1}{3}$	1	$\dfrac{9}{7}$	$\dfrac{13}{9}$	$\dfrac{17}{11}$	7	$\dfrac{11}{3}$	3	$\dfrac{19}{7}$	$\dfrac{23}{9}$	$\dfrac{27}{11}$

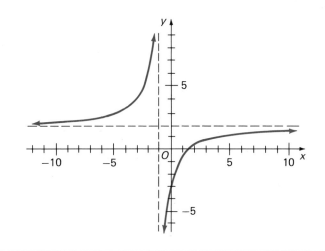

FIGURE 8 ◼

Example 3   Draw a sketch of the graph of the function g defined by

$$g(x) = \frac{4x}{x^2 - 25}$$

Solution   Factoring the denominator, we have

$$g(x) = \frac{4x}{(x - 5)(x + 5)}$$

The domain of g is the set of all real numbers except 5 and −5. Because $g(0) = 0$, the graph has an intercept at the origin. Because $g(-x) = -g(x)$, g is an odd function and the graph is symmetric with respect to the origin.

Setting the denominator equal to 0, we obtain the vertical asymptotes $x = 5$ and $x = -5$. From Theorem 2(i), the x axis is a horizontal asymptote (we also showed this fact in Illustration 3). Table 5 gives a few points on the graph. We plot these points and use the asymptotes as guides to draw the portion of the graph in the first and fourth quadrants. Using symmetry, we complete the graph in the second and third quadrants. See Figure 9.

TABLE 5

x	0	1	2	3	4	6	8	10
$g(x) = \dfrac{4x}{x^2 - 25}$	0	$-\dfrac{1}{6}$	$-\dfrac{8}{21}$	$-\dfrac{3}{4}$	$-\dfrac{16}{9}$	$\dfrac{24}{11}$	$\dfrac{32}{39}$	$\dfrac{8}{15}$

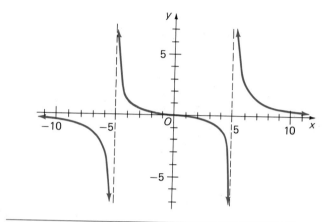

FIGURE 9

Example 4 Draw a sketch of the graph of the function h defined by

$$h(x) = \frac{3x^2}{2x^2 - 32}$$

Solution We factor the denominator and obtain

$$h(x) = \frac{3x^2}{2(x - 4)(x + 4)}$$

The domain of h is the set of all real numbers except 4 and -4. Because $h(0) = 0$, the graph has an intercept at the origin. Because $h(-x) = h(x)$, h is an even function and the graph is symmetric with respect to the y axis.

The vertical asymptotes are obtained by equating the denominator to 0. They are $x = 4$ and $x = -4$. As demonstrated in Illustration 4, the line $y = \frac{3}{2}$ is a horizontal asymptote, which also follows from Theorem 2(ii). A few points on the graph are obtained from Table 6. We plot these points and use the asymptotes as guides to draw the portion of the graph in the first and fourth quadrants. From properties of symmetry we complete the graph in the second and third quadrants. See Figure 10.

TABLE 6

x	1	2	3	5	6	8	10
$h(x) = \dfrac{3x^2}{2x^2 - 32}$	$-\dfrac{1}{10}$	$-\dfrac{1}{2}$	$-\dfrac{27}{14}$	$\dfrac{25}{6}$	$\dfrac{27}{10}$	2	$\dfrac{25}{14}$

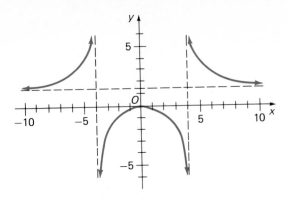

Illustration 5

Let the function f be defined by

$$f(x) = \frac{x^2 - 16}{x - 3}$$

Dividing the numerator by the denominator, we have the following computation.

$$
\begin{array}{r}
x + 3 \\
x - 3 \overline{)x^2 + 0x - 16} \\
\underline{x^2 - 3x } \\
3x - 16 \\
\underline{3x - 9} \\
-7
\end{array}
$$

Therefore

$$f(x) = x + 3 - \frac{7}{x - 3}$$

Because, as $x \to +\infty$ or $x \to -\infty$, $\dfrac{7}{x - 3} \to 0$, it follows from the above equation that

$$\text{as} \quad x \to +\infty \quad \text{or} \quad x \to -\infty, \quad f(x) \to x + 3$$

For this reason, the line $y = x + 3$ is an *oblique asymptote* of the graph of f. Refer to Figure 11, obtained in Example 5, where the graph of f is discussed further. ■

    If a line is an asymptote of a graph but is neither horizontal nor vertical, it is called an **oblique asymptote.** The graph of any rational function of the form $P(x)/Q(x)$, where the degree of $P(x)$ is one more than the degree of $Q(x)$, will have an oblique asymptote. To find it in such a case, we proceed as in Illustration 5: Divide the polynomial in the numerator by the polynomial in the denominator and obtain the sum of a linear function and a rational function. As $|x|$ increases without bound, the values of the original function approach the values of the linear function. The oblique asymptote is the graph of that linear function.

Example   5 Draw a sketch of the graph of the function f defined by

$$f(x) = \frac{x^2 - 16}{x - 3}$$

Solution   The domain of f is the set of all real numbers except 3. Because $f(0) = \frac{16}{3}$, the y intercept of the graph is $\frac{16}{3}$. The x intercepts of the graph are obtained by setting the numerator equal to 0. Doing this, we obtain $x = \pm 4$. Because f is neither even nor odd, it is not symmetric with respect to the y axis or the origin.

By setting the denominator equal to 0, we obtain the line $x = 3$ as a vertical asymptote. Because the degree of the numerator is greater than the degree of the denominator, it follows from Theorem 2(iii) that there are no horizontal asymptotes. There is an oblique asymptote, since the degree of the numerator is one more than the degree of the denominator. As we showed in Illustration 5, this is the line $y = x + 3$.

We find a few points on the graph. The results of the computation of $f(x)$ for selected values of x appear in Table 7. We plot these points and, using the asymptotes as guides, we obtain the sketch of the graph of f shown in Figure 11.

TABLE 7

x	-6	-4	-2	-1	0	1	2	4	5	6
$f(x) = \dfrac{x^2 - 16}{x - 3}$	$-\dfrac{20}{9}$	0	$\dfrac{12}{5}$	$\dfrac{15}{4}$	$\dfrac{16}{3}$	$\dfrac{15}{2}$	12	0	$\dfrac{9}{2}$	$\dfrac{20}{3}$

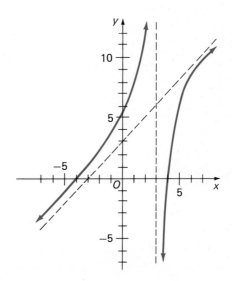

FIGURE 11

For a complete discussion of graphs of rational functions, techniques of calculus are needed. At this stage, however, the following steps are recommended for obtaining sketches of such graphs.

1. Find any intercepts.
2. Test for symmetry with respect to the y axis and the origin.
3. Find any vertical asymptotes by applying Theorem 1.
4. Find any horizontal asymptotes by applying Theorem 2.
5. If the degree of the numerator is 1 more than the degree of the denominator, find an oblique asymptote by the method of Illustration 5.
6. Plot a few points on the graph. Select as many points as are necessary to complete the sketch from the information obtained in Steps 1–5.

EXERCISES 4.6

In the following exercises, a rational function is defined. (a) Determine the domain of the function. (b) Find any intercepts of the graph of the function. (c) Test for symmetry of the graph with respect to the y axis and the origin. (d) Find the vertical and horizontal asymptotes of the graph if there are any. Find an oblique asymptote if there is one. (e) Draw a sketch of the graph.

1. $f(x) = \dfrac{1}{x}$

2. $f(x) = \dfrac{1}{x - 3}$

3. $g(x) = \dfrac{4}{x - 2}$

4. $g(x) = \dfrac{2}{x + 4}$

5. $f(x) = \dfrac{1 + x}{3 - x}$

6. $f(x) = \dfrac{x - 3}{x + 1}$

7. $h(x) = \dfrac{2x - 4}{x + 4}$

8. $g(x) = \dfrac{3x + 6}{x - 2}$

9. $f(x) = \dfrac{4}{x^2}$

10. $h(x) = -\dfrac{2}{x^2}$

11. $g(x) = -\dfrac{2}{x^3}$

12. $f(x) = \dfrac{3}{x^3}$

13. $f(x) = \dfrac{-1}{(x + 2)^2}$

14. $f(x) = \dfrac{4}{(x - 3)^2}$

15. $f(x) = \dfrac{5x}{x^2 - 4}$

16. $g(x) = \dfrac{7x}{x^2 - 9}$

17. $h(x) = \dfrac{9x}{16 - x^2}$

18. $f(x) = \dfrac{2x}{1 - x^2}$

19. $g(x) = \dfrac{2x^2}{x^2 - 9}$

20. $h(x) = \dfrac{3x^2}{x^2 - 4}$

21. $f(x) = \dfrac{x^2 + 1}{x^2 - 1}$

22. $f(x) = \dfrac{x^2 + 12}{x^2 - 16}$

23. $f(x) = \dfrac{x + 1}{x^2 + x - 6}$

24. $f(x) = \dfrac{x - 5}{x^2 - 8x + 12}$

25. $g(x) = \dfrac{x^2 - 9}{x - 2}$

26. $g(x) = \dfrac{x^2 - 25}{x - 4}$

27. $h(x) = \dfrac{x^2 + 4}{x}$

28. $h(x) = \dfrac{2x^2 + 2}{x}$

29. $f(x) = \dfrac{4}{x^2 + 4}$

30. $f(x) = \dfrac{6}{2x^2 + 2}$

31. $g(x) = \dfrac{2x^4}{x^4 + 1}$

32. $f(x) = \dfrac{3x^2}{x^2 + 4}$

4.7   Inverse Functions

You are already familiar with *inverse operations*. Addition and subtraction are inverse operations; so are multiplication and division, or raising to powers and extracting roots. One of a pair of inverse operations essentially "undoes" the other. For instance, if 4 is added to x, the sum is $x + 4$; if 4 is then subtracted from this sum, the difference is x. In the following illustration we use pairs of functions associated with inverse operations.

Illustration 1

We compute composite function values for some specific functions f and g.

a) Let $f(x) = x + 4$ and $g(x) = x - 4$. Then

$$f(g(x)) = f(x - 4) \qquad g(f(x)) = g(x + 4)$$
$$= (x - 4) + 4 \qquad\qquad = (x + 4) - 4$$
$$= x \qquad\qquad\qquad = x$$

b) Let $f(x) = 2x$ and $g(x) = \dfrac{x}{2}$. Then

$$f(g(x)) = f\left(\frac{x}{2}\right) \qquad g(f(x)) = g(2x)$$
$$= 2\left(\frac{x}{2}\right) \qquad\qquad = \frac{2x}{2}$$
$$= x \qquad\qquad\qquad = x$$

c) Let $f(x) = x^3$ and $g(x) = \sqrt[3]{x}$. Then

$$f(g(x)) = f(\sqrt[3]{x}) \qquad g(f(x)) = g(x^3)$$
$$= (\sqrt[3]{x})^3 \qquad\qquad = \sqrt[3]{x^3}$$
$$= x \qquad\qquad\qquad = x$$ ■

Each pair of functions f and g in Illustration 1 satisfies the following two equations:

and
$$f(g(x)) = x \qquad \text{for } x \text{ in the domain of } g$$
$$g(f(x)) = x \qquad \text{for } x \text{ in the domain of } f$$

Observe that for the functions f and g in these two equations the composite functions $f(g(x))$ and $g(f(x))$ are equal, a relationship that is not generally true for arbitrary functions f and g. You will learn subsequently (in Illustration 5) that each pair of functions in Illustration 1 is a set of *inverse functions,* and that is the reason the two equations are satisfied.

We lead up to the formal definition of the *inverse of a function* by considering some particular functions. A sketch of the graph of the function defined by

$$f(x) = x^2$$

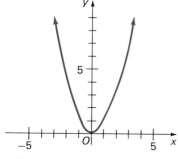

FIGURE 1

is shown in Figure 1. The domain of f is the set of real numbers, and the range of f is the interval $[0, +\infty)$. To each value of x in the domain there corresponds one and only one number in the range. For instance, because $f(2) = 4$, the number in the range that corresponds to the number 2 in the domain is 4. However, because $f(-2) = 4$, the number corresponding to the number -2 in the domain is also the number 4 in the range. So 4 is the function value of two distinct numbers in the domain. Furthermore, every number except 0 in the range of this function is the function value of two distinct numbers in the domain. In particular, $\frac{25}{4}$ is the function value of both $\frac{5}{2}$ and $-\frac{5}{2}$, 1 is the function value of both 1 and -1, and 9 is the function value of both 3 and -3.

A different situation occurs with the function g defined by

$$g(x) = x^3 \qquad -2 \leq x \leq 2$$

The domain of g is the closed interval $[-2, 2]$, and the range is $[-8, 8]$. A sketch of the graph of g is shown in Figure 2. This function is one for which a number in its range is the function value of one and only one number in the domain. Such a function is called *one-to-one*.

Definition

> **One-to-One Function**
> A function f is said to be one-to-one if and only if whenever a and b are any two distinct numbers in the domain of f, then $f(a) \neq f(b)$.

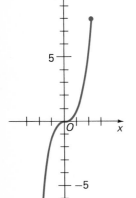

FIGURE 2

Illustration   2

As discussed above, for the function f defined by $f(x) = x^2$, every number except 0 in its range is the function value of two distinct numbers in the domain. For instance, 2 and -2 are two distinct numbers in the domain such that $f(2) = 4$ and $f(-2) = 4$. Therefore, by the above definition, this function is not one-to-one. ∎

You learned in Section 4.1 that a vertical line can intersect the graph of a function in at most one point. For a one-to-one function, it is also true that a horizontal line can intersect the graph in at most one point. Notice that this is the situation for the one-to-one function defined by $g(x) = x^3$, where $-2 \leq x \leq 2$, whose graph appears in Figure 2. Furthermore, observe that for the function defined by $f(x) = x^2$, which is not one-to-one, any horizontal line above the x axis intersects the graph in two points (see Figure 3). Thus we have the following geometric test for determining whether a function is one-to-one.

Horizontal Line Test

> A function is one-to-one if and only if every horizontal line intersects the graph of the function in no more than one point.

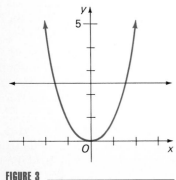

FIGURE 3

For our purposes, the horizontal line test is the most convenient method for determining whether a function is one-to-one.

Example   1   For each of the following functions use the horizontal line test to determine whether it is one-to-one.

a) $f(x) = 4x - 3$ b) $f(x) = (x + 1)^4$ c) $f(x) = x$ d) $f(x) = \dfrac{4}{x - 2}$

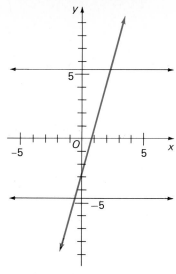

Solution   a) This function is linear and its graph is the line appearing in Figure 4. Because any horizontal line intersects the graph in exactly one point, the function is one-to-one.

b) We obtained the graph of this function in Example 2(a) of Section 4.5. It is reproduced here in Figure 5. Note that any horizontal line above the x axis intersects the graph in two points. Therefore the function is not one-to-one.

c) The graph of the absolute value function appears in Figure 6. Observe that any horizontal line above the x axis intersects this graph in two points. Thus the absolute value function is not one-to-one.

d) Figure 7 shows the graph of the given rational function. The line $x = 2$ is a vertical asymptote and the x axis is a horizontal asymptote. Any horizontal line, except the x axis, intersects the graph in exactly one point. Therefore the function is one-to-one.

FIGURE 4 _____

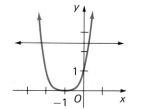

FIGURE 5 _____

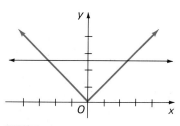

FIGURE 6 _____

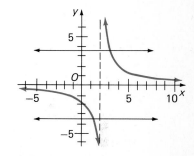

FIGURE 7 _____ ∎

Illustration 3

We have indicated that the horizontal line test shows that the function g defined by

$$g(x) = x^3 \qquad -2 \leq x \leq 2$$

is one-to-one. In the equation defining $g(x)$, if we replace $g(x)$ by y, we have

$$y = x^3 \qquad -2 \leq x \leq 2$$

If we solve this equation for x, we obtain

$$x = \sqrt[3]{y} \qquad -8 \leq y \leq 8$$

which defines a function G where

$$G(y) = \sqrt[3]{y} \qquad -8 \leq y \leq 8$$ ∎

    The function G of Illustration 3 is called the *inverse function* of g. In the following formal definition of an inverse function we use the notation f^{-1} to denote the inverse function of f. This notation is read "f inverse" and it should not be confused with the use of -1 as an exponent.

Definition

> **Inverse Function**
>
> If f is a one-to-one function that is the set of ordered pairs (x, y), then there is a function f^{-1}, called the **inverse function of f**, where f^{-1} is the set of ordered pairs (y, x) defined by
>
> $$x = f^{-1}(y) \qquad \text{if and only if} \qquad y = f(x)$$
>
> The domain of f^{-1} is the range of f and the range of f^{-1} is the domain of f.

A crucial part of the above definition is that f be a one-to-one function. This requirement is necessary so that for each value of y, $f^{-1}(y)$ is unique.

We eliminate y from the equations of the definition by writing the equation

$$f^{-1}(y) = x$$

and replacing y by $f(x)$. We obtain

$$f^{-1}(f(x)) = x \tag{1}$$

where x is in the domain of f.

We eliminate x from the same pair of equations by writing the equation

$$f(x) = y$$

and replacing x by $f^{-1}(y)$. We get

$$f(f^{-1}(y)) = y$$

where y is in the domain of f^{-1}. Because the symbol used for the independent variable is arbitrary, we can replace y by x to obtain

$$f(f^{-1}(x)) = x \tag{2}$$

where x is in the domain of f^{-1}.

From Equations (1) and (2) we see that if the inverse function of f is f^{-1}, then the inverse function of f^{-1} is f.

Illustration 4

In Illustration 3 the function G defined by

$$G(y) = \sqrt[3]{y} \qquad -8 \leq y \leq 8$$

is the inverse function of g defined by

$$g(x) = x^3 \qquad -2 \leq x \leq 2$$

Therefore g^{-1} can be written in place of G and we have

$$g^{-1}(y) = \sqrt[3]{y} \quad -8 \le y \le 8$$

or, equivalently, if we replace y by x,

$$g^{-1}(x) = \sqrt[3]{x} \quad -8 \le x \le 8$$

Observe that the domain of g is $[-2, 2]$, which is the range of g^{-1}; also the range of g is $[-8, 8]$, which is the domain of g^{-1}. ∎

If a function f has an inverse, then f^{-1} can be found by the method used in the following illustration.

Illustration 5

Each of the functions f in Illustration 1 is one-to-one. Therefore, $f^{-1}(x)$ exists. For each function we compute $f^{-1}(x)$ from the definition of $f(x)$ by substituting y for $f(x)$ and solving the resulting equation for x. This procedure gives us the equation $x = f^{-1}(y)$. We then have the definition of $f^{-1}(y)$, from which we obtain $f^{-1}(x)$.

a) $f(x) = x + 4$ b) $f(x) = 2x$ c) $f(x) = x^3$

   $y = x + 4$ $y = 2x$ $y = x^3$

   $x = y - 4$ $x = \dfrac{y}{2}$ $x = \sqrt[3]{y}$

   $f^{-1}(y) = y - 4$ $f^{-1}(y) = \sqrt[3]{y}$

   $f^{-1}(x) = x - 4$ $f^{-1}(y) = \dfrac{y}{2}$ $f^{-1}(x) = \sqrt[3]{x}$

   $f^{-1}(x) = \dfrac{x}{2}$

Observe that the function f^{-1} in each part is the function g in the corresponding part of Illustration 1. ∎

Example 2

Find $f^{-1}(x)$ for the function f of Example 1(a). Verify Equations (1) and (2) for f and f^{-1}, and draw a sketch of the graph of f^{-1}.

Solution In Example 1(a) we showed by the horizontal line test that the function f defined by

$$f(x) = 4x - 3$$

is one-to-one. Therefore f^{-1} exists. To find $f^{-1}(x)$, we write the equation

$$y = 4x - 3$$

and solve for x. We obtain

$$x = \frac{y + 3}{4}$$

Therefore

$$f^{-1}(y) = \frac{y + 3}{4} \quad \Leftrightarrow \quad f^{-1}(x) = \frac{x + 3}{4}$$

We verify Equations (1) and (2).

$$f^{-1}(f(x)) = f^{-1}(4x - 3) \qquad f(f^{-1}(x)) = f\left(\frac{x + 3}{4}\right)$$

$$= \frac{(4x - 3) + 3}{4} \qquad\qquad = 4\left(\frac{x + 3}{4}\right) - 3$$

$$= \frac{4x}{4} \qquad\qquad\qquad = (x + 3) - 3$$

$$= x \qquad\qquad\qquad\qquad = x$$

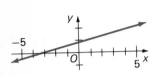

FIGURE 8

Figure 8 shows a sketch of the graph of f^{-1}.   ∎

Figure 4 shows a sketch of the graph of the function f of Example 2, and in Figure 8 there is a sketch of the graph of f^{-1}. The graphs of f and f^{-1} are shown on the same set of axes in Figure 9. It appears that if $Q(u, v)$ is on the graph of f, then the point $R(v, u)$ is on the graph of f^{-1}. We now demonstrate that this situation actually occurs.

In general, if Q is the point (u, v) and R is the point (v, u), the line segment QR is perpendicular to the line $y = x$ and is bisected by it. We state that the point Q is a *reflection of the point R* with respect to the line $y = x$, and the point R is a *reflection of the point Q* with respect to the line $y = x$. If x and y are interchanged in the equation $y = f(x)$, we obtain the equation $x = f(y)$, and the graph of the equation $x = f(y)$ is said to be a *reflection of the graph* of the equation $y = f(x)$ with respect to the line $y = x$. Because the equation $x = f(y)$ is equivalent to the equation $y = f^{-1}(x)$, we conclude that the graph of $y = f^{-1}(x)$ is a reflection of the graph of $y = f(x)$ with respect to the line $y = x$. Therefore, if a function has an inverse function, the graphs of the functions are reflections of each other with respect to the line $y = x$.

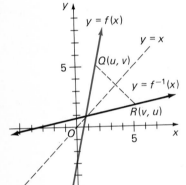

FIGURE 9

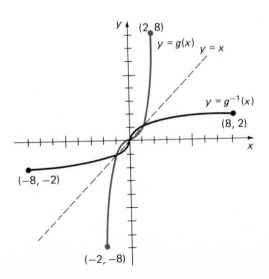

FIGURE 10

Illustration 6

Functions g and g^{-1} of Illustration 4 are defined by

$$g(x) = x^3 \qquad -2 \le x \le 2$$

and

$$g^{-1}(x) = \sqrt[3]{x} \qquad -8 \le x \le 8$$

Sketches of the graphs of g and g^{-1} appear on the same set of axes in Figure 10. Observe that the graphs are reflections of each other with respect to line $y = x$. ∎

Example 3 Let f be the function defined by

$$f(x) = x^2 \qquad x \ge 0$$

a) Show by the horizontal line test that f is one-to-one.
b) Find $f^{-1}(x)$.
c) Draw sketches of the graphs of f and f^{-1} on the same set of axes.

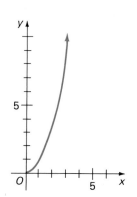

FIGURE 11 _____

Solution a) A sketch of the graph of f appears in Figure 11. Observe that the x axis and any horizontal line above it intersect the graph in exactly one point. Any horizontal line below the x axis does not intersect the graph. Therefore, by the horizontal line test, f is one-to-one.

b) Because f is one-to-one it has an inverse function f^{-1}. To find $f^{-1}(x)$, we replace $f(x)$ by y in the given equation and we have

$$y = x^2 \qquad x \ge 0$$

We solve this equation for x. Because $x \ge 0$, we obtain

$$x = \sqrt{y}$$

Thus

$$f^{-1}(y) = \sqrt{y}$$

To use x as the independent variable, we replace y by x and get

$$f^{-1}(x) = \sqrt{x}$$

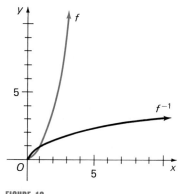

FIGURE 12 _____

c) Sketches of the graphs of f and f^{-1} on the same set of axes appear in Figure 12. ∎

EXERCISES 4.7

In Exercises 1 through 14, draw a sketch of the graph of the function and use the horizontal line test to determine whether it is one-to-one.

1. $f(x) = 2x + 3$
2. $g(x) = 8 - 4x$
3. $f(x) = x^2 - 6$
4. $f(x) = 4 - x^2$

5. $g(x) = 4 - x^3$
6. $h(x) = x^3 + 1$
7. $f(x) = (x - 3)^4$
8. $g(x) = x^4 - 3$
9. $h(x) = \sqrt{x + 3}$
10. $f(x) = \sqrt{1 - x^2}$
11. $f(x) = \dfrac{x + 5}{x - 4}$
12. $g(x) = \dfrac{3}{(x - 1)^2}$

13. $g(x) = |x - 2|$ 14. $f(x) = 5$

In Exercises 15 through 20, (a) draw a sketch of the graph of f and use the horizontal line test to prove that f is one-to-one; (b) find $f^{-1}(x)$; (c) verify Equations (1) and (2) of this section for f and f^{-1}.

15. $f(x) = 3 - 4x$ 16. $f(x) = 3x - 2$

17. $f(x) = x^3 + 2$ 18. $f(x) = (x + 2)^3$

19. $f(x) = \dfrac{1}{x + 1}$ 20. $f(x) = \dfrac{x + 5}{x - 1}$

In Exercises 21 through 28, a one-to-one function f is given. (a) State the range of f. (b) Determine f^{-1} and state the domain of f^{-1}.

21. $f(x) = x^2 - 5, x \geq 0$ 22. $f(x) = 2 - x^2, x \leq 0$

23. $f(x) = \sqrt{x^2 - 9}, x \geq 3$ 24. $f(x) = -\sqrt{x^2 - 9}, x \geq 3$

25. $f(x) = \sqrt{x^2 - 9}, x \leq -3$

26. $f(x) = -\sqrt{x^2 - 9}, x \leq -3$

27. $f(x) = \frac{1}{8}x^3, -1 \leq x \leq 1$

28. $f(x) = (2x + 1)^3, -\frac{1}{2} \leq x \leq \frac{1}{2}$

In Exercises 29 through 36, (a) use the horizontal line test to show that f is one-to-one; (b) find $f^{-1}(x)$; (c) draw sketches of the graphs of f and f^{-1} on the same set of axes.

29. $f(x) = 2x + 5$ 30. $f(x) = 6 - \frac{1}{2}x$

31. $f(x) = (x + 1)^3$ 32. $f(x) = (1 - x)^3$

33. $f(x) = (x - 2)^2, x \geq 2$ 34. $f(x) = (x + 3)^2, x \leq -3$

35. $f(x) = 4 - x^2, x \leq 0$ 36. $f(x) = x^2 - 9, x \geq 0$

37. If x degrees is the Celsius temperature, then the number of degrees in the Fahrenheit temperature can be expressed as a function of x. If f is this function, then $f(x)$ degrees is the Fahrenheit temperature and

$$f(x) = 32 + \tfrac{9}{5}x$$

Determine the inverse function f^{-1} that expresses the number of degrees in the Celsius temperature as a function of the number of degrees in the Fahrenheit temperature.

38. If $f(t)$ dollars is the amount in t years of an investment of \$1000 at 12 percent simple interest, then

$$f(t) = 1000(1 + 0.12t)$$

Determine the inverse function f^{-1} that expresses the number of years that \$1000 has been invested at 12 percent simple interest as a function of the amount of the investment.

39. Given that $f(x) = \sqrt{16 - x^2}, 0 \leq x \leq 4$, show that f is its own inverse function.

40. The function f defined by

$$f(x) = \frac{x + 5}{x + k}$$

where k is a constant, is one-to-one. Find the value of k so that f will be its own inverse function.

REVIEW EXERCISES FOR CHAPTER 4

In Exercises 1 through 20, draw a sketch of the graph of the function and determine its domain and range.

1. $f = \{(x, y) | y = 4 - 2x\}$ 2. $g = \{(x, y) | y = 3x + 2\}$

3. $g = \{(x, y) | y = x^2 - 4\}$ 4. $f = \{(x, y) | y = 9 - x^2\}$

5. $h = \{(x, y) | y = \sqrt{x^2 - 16}\}$ 6. $H = \{(x, y) | y = \sqrt{1 - x^2}\}$

7. $F = \{(x, y) | y = \sqrt{16 - x^2}\}$ 8. $G = \{(x, y) | y = \sqrt{x^2 - 1}\}$

9. $f(x) = |5 - x|$ 10. $g(x) = |x + 4|$

11. $g(x) = \dfrac{x^2 - 16}{x + 4}$ 12. $f(x) = \dfrac{x^2 + x - 6}{x - 2}$

13. $G(x) = \begin{cases} x - 4 & \text{if } x \neq -4 \\ 3 & \text{if } x = -4 \end{cases}$

14. $F(x) = \begin{cases} x + 3 & \text{if } x \neq 2 \\ 1 & \text{if } x = 2 \end{cases}$

15. $F(x) = \begin{cases} 3 - x & \text{if } x < 0 \\ 3 + 2x & \text{if } 0 \leq x \end{cases}$

16. $G(x) = \begin{cases} 3x + 2 & \text{if } x \leq 0 \\ 4 - 2x & \text{if } 0 < x \end{cases}$

17. $h(x) = \begin{cases} x^2 - 1 & \text{if } x \leq 0 \\ x - 1 & \text{if } 0 < x \end{cases}$

18. $H(x) = \begin{cases} x^2 & \text{if } x < -1 \\ (x + 2)^2 & \text{if } -1 \leq x \end{cases}$

19. $f(x) = \begin{cases} x + 4 & \text{if } x < -4 \\ \sqrt{16 - x^2} & \text{if } -4 \leq x \leq 4 \\ 4 - x & \text{if } 4 < x \end{cases}$

20. $g(x) = \begin{cases} x^2 - 4 & \text{if } x \leq 2 \\ 4 - x & \text{if } 2 < x \leq 4 \\ x - 8 & \text{if } 4 < x \end{cases}$

21. Given $f(x) = 3x^2 - x + 5$, find: (a) $f(-3)$; (b) $f(-x^2)$;
(c) $-[f(x)]^2$; (d) $\dfrac{f(x + h) - f(x)}{h}$, $h \neq 0$.

22. Given $g(x) = \sqrt{x + 3}$, find: (a) $g(-2)$; (b) $g(x^2)$;
(c) $[g(x)]^2$; (d) $\dfrac{g(x + h) - g(x)}{h}$, $h \neq 0$.

23. Given $g(x) = \sqrt{1 - x}$, find $\dfrac{g(x + h) - g(x)}{h}$, $h \neq 0$.

24. Determine whether the function is even, odd, or neither:
(a) $f(x) = 2x^3 - 3x$; (b) $g(x) = 5x^4 + 2x^2 - 1$;
(c) $h(x) = 3x^5 - 2x^3 + x^2 - x$; (d) $F(x) = \dfrac{x^2 + 1}{x^3 - x}$

In Exercises 25 through 30, for the functions f and g, define the following functions and determine the domain of the resulting function: (a) $f + g$; (b) $f - g$; (c) $f \cdot g$; (d) f/g; (e) g/f; (f) $f \circ g$; (g) $g \circ f$.

25. $f(x) = x^2 - 4$ and $g(x) = 4x - 3$
26. $f(x) = \sqrt{x}$ and $g(x) = x^2 + 1$
27. $f(x) = \sqrt{x + 2}$ and $g(x) = x^2 - 4$
28. $f(x) = x^2 - 9$ and $g(x) = \sqrt{x + 5}$
29. $f(x) = \dfrac{1}{x - 3}$ and $g(x) = \dfrac{x}{x + 1}$
30. $f(x) = \sqrt{x}$ and $g(x) = \dfrac{1}{x^2}$

In Exercises 31 through 34, draw a sketch of the graph of the function and determine from the graph which of the following statements characterizes the zeros of the function: (a) two real zeros; (b) one real zero of multiplicity two; or (c) two imaginary zeros.

31. $f(x) = x^2 - 6x$
32. $f(x) = 2x^2 + 8x + 11$
33. $g(x) = -x^2 + 8x - 16$
34. $g(x) = -\frac{1}{2}x^2 + 2x + 2$

In Exercises 35 through 38, find either a maximum or minimum value of the function.

35. $f(x) = 4 - 4x - 2x^2$
36. $g(x) = 3x^2 + 6x + 7$
37. $g(x) = \frac{1}{4}(x^2 + 2x - 11)$
38. $f(x) = \frac{1}{6}(15 + 6x - x^2)$

In Exercises 39 through 48, draw a sketch of the graph of the polynomial function.

39. $f(x) = (x - 2)^2$
40. $f(x) = -2(x + 3)^2$
41. $P(x) = (x + 1)^3$
42. $P(x) = x^3 + 4$
43. $f(x) = x^4 - 3$
44. $g(x) = (x - 1)^4$
45. $h(x) = x^3 - 6x^2 + 9x + 6$
46. $f(x) = x^4 - 8x^2 + 9$
47. $g(x) = x^4 - 14x^2 - 24x$
48. $h(x) = x^3 + 3x^2 - 4$

In Exercises 49 through 60, a rational function is defined. (a) Determine the domain of the function. (b) Find any intercepts of the graph of the function. (c) Test for symmetry of the graph with respect to the y axis and the origin. (d) Find the vertical and horizontal asymptotes of the graph if there are any. Find an oblique asymptote if there is one. (e) Draw a sketch of the graph.

49. $f(x) = \dfrac{2}{x - 5}$
50. $f(x) = \dfrac{x + 2}{x - 3}$
51. $g(x) = \dfrac{8}{x^2}$
52. $g(x) = -\dfrac{4}{x^3}$
53. $f(x) = \dfrac{16}{(x + 2)^3}$
54. $f(x) = \dfrac{6}{(x + 1)^2}$
55. $f(x) = \dfrac{3x}{x^2 - 1}$
56. $g(x) = \dfrac{x^2 + 4}{x^2 - 4}$
57. $g(x) = \dfrac{4x^2}{9 - x^2}$
58. $f(x) = \dfrac{5x}{x^2 - 16}$
59. $h(x) = \dfrac{x^2 - 4}{x - 1}$
60. $h(x) = \dfrac{3x^2 - 12}{x}$

In Exercises 61 through 66, (a) use the horizontal line test to prove that f is one-to-one; (b) find $f^{-1}(x)$; (c) verify that $f^{-1}(f(x)) = x$ and $f(f^{-1}(x)) = x$.

61. $f(x) = 5x - 2$
62. $f(x) = 4 - 3x$
63. $f(x) = 8 - x^3$
64. $f(x) = x^3 + 1$
65. $f(x) = \dfrac{x - 1}{x + 2}$
66. $f(x) = \dfrac{-4}{x + 1}$

In Exercises 67 through 72, (a) use the horizontal line test to prove that f is one-to-one; (b) find $f^{-1}(x)$; (c) draw sketches of the graphs of f and f^{-1} on the same set of axes.

67. $f(x) = 5 - 4x$
68. $f(x) = 3x - 2$
69. $f(x) = (8 - x)^3$
70. $f(x) = (x + 8)^3$
71. $f(x) = x^2 - 4$, $x \geq 0$
72. $f(x) = 1 - x^2$, $x \leq 0$

73. The distance a body falls from rest is directly proportional to the square of the time it has been falling, and a body falls 64 ft in 2 sec.
(a) Express the number of feet in the distance a body falls from rest as a function of the number of seconds it has been falling.
(b) How far will a body fall from rest in $\frac{5}{2}$ sec?

74. Boyle's law states that at a constant temperature the volume of a gas is inversely proportional to the pressure of the gas, and a gas occupies 100 m^3 at a pressure of 24 kg/cm^2.
(a) Express the number of cubic meters occupied by a gas as a function of the number of kilograms per square centimeter in its pressure.
(b) What is the volume of a gas when its pressure is 16 kg/cm^2?

75. If a pond can support a maximum of 10,000 fish, the rate of growth of the fish population is jointly proportional to the number of fish present and the difference between 10,000 and the number present.
    (a) Given that the rate of growth is 90 fish per week when 1000 fish are present, express the rate of population growth as a function of the number present.
    (b) Find the rate of population growth when 2000 fish are present.
76. Find two positive numbers whose sum is 12 and such that their product is a maximum.
77. Find two positive numbers whose sum is 12 and such that the sum of their squares is a minimum.
78. Show that among all the rectangles having a perimeter of 36 in., the square of side 9 in. has the greatest area.
79. A school-sponsored trip that can accommodate up to 250 students will cost each student $15 if not more than 150 students make the trip; however, the cost per student will be reduced 5 cents for each student in excess of 150 until the cost reaches $10 per student.
    (a) Given that x students make the trip, express the number of dollars in the gross income as a function of x.
    (b) What is the domain of the resulting function?
    (c) How many students should make the trip for the school to receive the largest gross income?
80. A wholesaler offers to deliver to a dealer 300 chairs at $90 per chair and to reduce the price per chair on the entire order by 25 cents for each additional chair over 300.
    (a) Given that x chairs are ordered, express the number of dollars in the dealer's cost as a function of x.
    (b) What is the domain of the resulting function?
    (c) Find the dollar total involved in the largest possible transaction between the wholesaler and the dealer under these circumstances.
81. In a town of population 11,000 the growth rate of an epidemic is jointly proportional to the number of people infected and the number of people not infected.
    (a) Given that the epidemic is growing at the rate of $f(x)$ people per day when x people are infected, write an equation defining $f(x)$.

    (b) What is the domain of the function f in part (a)?
    (c) Determine the number of people infected when the epidemic is growing at a maximum rate.
82. A carpenter can sell all the end tables that are made at a price of $64 per table. If x tables are built and sold each week, then the number of dollars in the total cost of the week's production is $x^2 + 15x + 225$.
    (a) Express the number of dollars in the carpenter's weekly profit as a function of x.
    (b) How many tables should be constructed each week in order for the carpenter to have the greatest weekly total profit?
    (c) What is the greatest weekly total profit?
83. A Norman window consists of a rectangle surmounted by a semicircle. Suppose a particular Norman window is to have a perimeter of 200 in. Furthermore, assume that the amount of light transmitted by the window is directly proportional to the area of the window.
    (a) Let r inches be the radius of the semicircle; express the amount of light transmitted by the window as a function of r.
    (b) What is the domain of the resulting function?

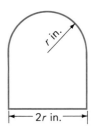

84. Do Exercise 83 with this change: The region bounded by the semicircle transmits only half as much light per square inch of area as the region bounded by the rectangle.
85. Refer to Exercise 83. A particular Norman window is to have a perimeter of 200 in. What must be the radius of the semicircle so that the window will admit the most light?
86. Refer to Exercise 84. Do Exercise 85 if the region bounded by the semicircle transmits only half as much light per square inch of area as the region bounded by the rectangle.

CHAPTER 5

Exponential and Logarithmic Functions

Introduction

Until now we have considered only algebraic functions. Functions that are not algebraic are called transcendental. Examples of transcendental functions are the exponential and logarithmic functions, which we define in this chapter. Applications of these functions arise in all of the sciences: physical, life, and social.

In Section 5.1 we discuss irrational exponents and introduce the number e by considering interest on an investment at a rate compounded continuously. We follow this presentation in Section 5.2 by a treatment of exponential functions. We illustrate exponential growth by an increase in the number of bacteria present in a culture, and exponential decay by a decrease in the value of a piece of equipment. We use the learning curve to demonstrate bounded growth. In Section 5.3, a logarithmic function is defined as the inverse of an exponential function. Properties of logarithmic functions are given in Section 5.4. We solve exponential and logarithmic equations in Section 5.5.

5.1 Exponents and the Number e

In this chapter you need to be familiar with the laws of rational exponents. Thus you may wish to review Sections 1.2 and 1.6. In those sections we defined the power of a real number when the exponent is a rational number. In particular, 2^x has been defined for any rational value of x. For instance,

$$2^5 = 2 \cdot 2 \cdot 2 \cdot 2 \cdot 2 \qquad 2^0 = 1 \qquad 2^{-3} = \frac{1}{2^3} \qquad 2^{2/3} = \sqrt[3]{2^2}$$
$$= 32 \qquad\qquad\qquad\qquad\qquad = \frac{1}{8} \qquad\qquad = \sqrt[3]{4}$$

It is not quite so simple to define 2^x when x is irrational. For example, what is meant by $2^{\sqrt{3}}$? The definition of an irrational exponent requires a knowledge of more advanced mathematics than is covered in this book. However, we can give an intuitive argument that irrational powers of positive numbers can exist by showing how $2^{\sqrt{3}}$ can be interpreted. To do this, we make use of the following theorem, which is stated without proof.

Theorem 1

> If r and s are rational numbers, then
>
> (i) if $b > 1$, $r < s$ implies $b^r < b^s$;
> (ii) if $0 < b < 1$, $r < s$ implies $b^r > b^s$.

We demonstrate this theorem in the following illustration.

Illustration 1

In part (i) of Theorem 1, $b > 1$. Let $b = 4$; then, because $2 < 3$, $4^2 < 4^3$.

     In part (ii), $0 < b < 1$. Let $b = \frac{1}{3}$; then, because $2 < 3$, $(\frac{1}{3})^2 > (\frac{1}{3})^3$. ■

A decimal approximation for $\sqrt{3}$ can be obtained accurate to any number of decimal places desired. To four decimal places, $\sqrt{3} \approx 1.7321$. Because $1 < 1.7 < 2$, then from Theorem 1(i),

$$2^1 < 2^{1.7} < 2^2 \qquad \Leftrightarrow \qquad 2 < 2^{1.7} < 4$$

Because $1.7 < 1.73 < 1.8$, then

$$2^{1.7} < 2^{1.73} < 2^{1.8} \qquad \Leftrightarrow \qquad 3.2 < 2^{1.73} < 3.5$$

Because $1.73 < 1.732 < 1.74$, then

$$2^{1.73} < 2^{1.732} < 2^{1.74} \qquad \Leftrightarrow \qquad 3.32 < 2^{1.732} < 3.34$$

Because $1.732 < 1.7321 < 1.733$, then

$$2^{1.732} < 2^{1.7321} < 2^{1.733} \qquad \Leftrightarrow \qquad 3.322 < 2^{1.7321} < 3.324$$

and so on. In each inequality there is a power of 2 for which the exponent is a decimal approximation of the value of $\sqrt{3}$. In each successive inequality the exponent contains one more decimal place than the exponent in the previous inequality. By following this procedure indefinitely, the difference between the left member of the inequality and the right member can be made as small as we please. Hence our intuition leads us to expect that there is a value of $2^{\sqrt{3}}$ that satisfies each successive inequality as the procedure is continued indefinitely. A similar discussion could be given for any irrational power of any positive number. Furthermore, Theorem 1 is valid if r and s are any real numbers.

The laws of exponents that are valid for rational exponents also hold if the exponents are any real numbers. These laws are summarized in the following theorem.

Theorem 2

If a and b are any positive numbers, and x and y are any real numbers, then

(i) $a^x a^y = a^{x+y}$

(ii) $\dfrac{a^x}{a^y} = a^{x-y}$

(iii) $(a^x)^y = a^{xy}$

(iv) $(ab)^x = a^x b^x$

(v) $\left(\dfrac{a}{b}\right)^x = \dfrac{a^x}{b^x}$

The proofs of properties (i) through (v) for real number exponents are beyond the scope of this book.

Example   **1**   Simplify each of the following by applying laws of exponents.

a) $2^{\sqrt{3}} \cdot 2^{\sqrt{12}}$ b) $(7^{\sqrt{5}})^{\sqrt{20}}$

Solution

a) $2^{\sqrt{3}} \cdot 2^{\sqrt{12}} = 2^{\sqrt{3}} \cdot 2^{2\sqrt{3}}$ b) $(7^{\sqrt{5}})^{\sqrt{20}} = 7^{\sqrt{5} \cdot \sqrt{20}}$

$\qquad\qquad\quad = 2^{\sqrt{3} + 2\sqrt{3}} \qquad\qquad\qquad\qquad = 7^{\sqrt{100}}$

$\qquad\qquad\quad = 2^{3\sqrt{3}} \qquad\qquad\qquad\qquad\quad = 7^{10}$ ∎

We can apply exponents to help compute the interest on an investment. This application will lead us to the definition of e, a number that arises not only in business and economics but in the physical and life sciences as well.

If money is loaned at the interest rate of 0.12 (that is, 12 percent) per year, then the borrower's debt at the end of a year is $1.12 for each $1 borrowed. In general, if the interest rate is i (that is, $100i$ percent) per year, then for each dollar borrowed the repayment at the end of a year is $(1 + i)$ dollars. If P dollars is borrowed, then the debt is $P(1 + i)$ dollars at the end of a year.

We shall consider several types of interest. **Simple interest** is interest due only on the original amount that is borrowed. In this case no interest is paid on any accrued interest. For example, suppose that 10 percent simple interest on $100 is due annually. Then the lender would receive $10 at the end of each year.

Suppose now that P dollars is invested at a simple interest rate of $100i$ percent. Then the interest earned at the end of the year is Pi dollars. If no withdrawals are made for n years, the total interest earned is Pni dollars, and if A dollars is the total amount on deposit at the end of n years,

$$A = P + Pni$$
$$= P(1 + ni)$$

Simple interest is sometimes used for short-term investments or loans of a period of possibly 30, 60, or 90 days. In such cases, in order to simplify calculations, a year is considered as having 360 days, and each month is assumed to contain 30 days; then 30 days is equivalent to $\frac{1}{12}$ of a year.

Example   **2**   A loan of $500 is made for a period of 90 days at a simple interest rate of 16 percent annually. Determine the amount to be repaid at the end of 90 days.

Solution   We are given $P = 500$, $i = 0.16$, and $n = \frac{90}{360}$; that is, $n = \frac{1}{4}$. Therefore, if A dollars is the amount to be repaid,

$$
\begin{aligned}
A &= P(1 + ni) \\
&= 500[1 + \tfrac{1}{4}(0.16)] \\
&= 520
\end{aligned}
$$

Thus the amount to be repaid is \$520.   ■

Rates of interest are customarily stated as annual rates, but often the interest is computed more than once a year. When the interest for each period is added to the principal and itself earns interest, we have **compound interest.** Whenever the word "interest" is used without an adjective, it is customarily assumed to be compound interest. If the interest is compounded m times per year, then the annual rate must be divided by m to determine the interest for each period. For example, if \$100 is deposited in a savings account that pays 8 percent compounded quarterly, then the number of dollars in the account at the end of the first 3-month period will be

$$
100 \left(1 + \frac{0.08}{4} \right) = 100 \,(1.02)
$$

For the second quarter we consider the principal to be $100(1.02)$. Therefore the number of dollars in the account at the end of the second 3-month period will be

$$
[100(1.02)](1.02) = 100(1.02)^2
$$

At the end of the third 3-month period the number of dollars in the account will be

$$
[100(1.02)^2](1.02) = 100(1.02)^3
$$

and so on. At the end of the nth 3-month period the number of dollars in the account will be

$$
100(1.02)^n
$$

More generally, we have the following theorem.

Theorem 3

If P dollars is invested at an annual interest rate of $100i$ percent compounded m times per year, and if A_n dollars is the amount of the investment at the end of n interest periods, then

$$
A_n = P \left(1 + \frac{i}{m} \right)^n
$$

The proof of this theorem involves mathematical induction, which is discussed in Section 11.2.

If t is the number of years for which P dollars is invested at an interest rate of $100i$ percent compounded m times per year, then the number of interest periods n is mt. Letting A dollars be the total amount at t years, we can write the formula of Theorem 3 as

$$A = P\left(1 + \frac{i}{m}\right)^{mt} \tag{1}$$

Example 3 Suppose that $400 is deposited into a savings account that pays 8 percent interest per year compounded semiannually. If no withdrawals and no additional deposits are made, what is the amount on deposit at the end of 3 years?

Solution The interest is compounded twice a year; so $m = 2$. Because the time is 3 years, $t = 3$. Furthermore, $P = 400$ and $i = 0.08$. Therefore, if A dollars is the amount on deposit at the end of 3 years, we have from Formula (1)

$$A = P\left(1 + \frac{i}{m}\right)^{mt}$$

$$= 400\left(1 + \frac{0.08}{2}\right)^{2(3)}$$

$$= 400\,(1.04)^6$$

$$= 506.13$$

The amount on deposit at the end of 3 years is therefore $506.13. ∎

Formula (1) gives the number of dollars in the total amount after t years if P dollars is invested at a rate of $100i$ percent compounded m times per year. Now let us imagine the interest is continuously compounding. That is, suppose in Formula (1) that the number of interest periods per year increases without bound. Thus we are concerned with the behavior of A as $m \to +\infty$. Because $mt = \frac{m}{i}\,(it)$, we can write Formula (1) as

$$A = P\left[\left(1 + \frac{i}{m}\right)^{m/i}\right]^{it}$$

In this equation let $x = m/i$; then

$$A = P\left[\left(1 + \frac{1}{x}\right)^{x}\right]^{it} \tag{2}$$

Because "$m \to +\infty$" is equivalent to "$x \to +\infty$," let us examine

$$\left(1 + \frac{1}{x}\right)^x \quad \text{as} \quad x \to +\infty$$

You can use a calculator to compute values of $\left(1 + \frac{1}{x}\right)^x$ as x takes on larger and larger numbers. Refer to Table 1 for some of these values.

TABLE 1

x	10	100	1,000	10,000	100,000
$\left(1 + \dfrac{1}{x}\right)^x$	2.5937	2.7048	2.7169	2.7181	2.7183

This table leads us to suspect that $\left(1 + \frac{1}{x}\right)^x$ probably approaches a finite number as x increases without bound. This is indeed the case, and the finite number is denoted by the letter e. The letter e was chosen because of the Swiss mathematician and physicist Leonhard Euler (1707–1783). The number e is an irrational number. Its value can be expressed to any required degree of accuracy by using infinite series, which are studied in calculus. To seven decimal places, the value of e is 2.7182818. We write

$$e \approx 2.7182818.$$

In the back of the book Table V gives values of powers of e. Such values can also be obtained from a calculator with an $\boxed{e^x}$ key. If your calculator has no $\boxed{e^x}$ key, powers of e can be obtained by pressing the $\boxed{\text{INV}}$ key followed by the $\boxed{\text{ln}}$ key. The reason that these two successive operations give a value for e^x will be apparent when you study Section 5.3.

Returning now to the discussion of interest compounding continuously, we have from the above argument and Formula (2)

$$A = Pe^{it} \tag{3}$$

where A dollars is the amount after t years if P dollars is invested at a rate of $100i$ percent compounded continuously. This value of A is an upper bound for the amount given by Formula (1) when interest is compounded frequently and can be used as an approximation in such a situation. This fact is demonstrated in the following illustration, where we compare the amount at the end of 1 year when interest is compounded continuously with the corresponding amounts obtained when interest is compounded monthly, semimonthly, and daily.

Illustration 2

Suppose that $5000 is borrowed at an interest rate of 12 percent compounded monthly, and the loan is to be repaid in one payment at the end of the year. If A dollars is the amount to be repaid, then from Formula (1), with $P = 5000$, $i = 0.12$, $m = 12$, and $t = 1$, we have

$$A = 5000(1.01)^{12}$$
$$= 5634.13$$

If the interest rate of 12 percent is compounded semimonthly instead of monthly, then from Formula (1), with $m = 24$, we have

$$A = 5000\left(1 + \frac{0.12}{24}\right)^{24}$$
$$= 5635.80$$

If the interest rate of 12 percent is compounded daily, then from Formula (1) with $m = 365$, we have

$$A = 5000\left(1 + \frac{0.12}{365}\right)^{365}$$
$$= 5637.37$$

Now suppose that the interest is compounded continuously at 12 percent. Because $P = 5000$, $i = 0.12$, and $t = 1$, we have from Formula (3)

$$A = 5000e^{0.12}$$

From Table V, $e^{0.12} \approx 1.1275$. Thus

$$A = 5000(1.1275)$$
$$= 5637.5$$

If a calculator is used for the power of e, we obtain $e^{0.12} \approx 1.1274968$, and with this value, $A = 5637.48$.

Because $5637.5 is the amount when interest is compounded continuously at 12 percent, it is an upper bound for the amount regardless of how often interest is compounded. ∎

If, in Formula (3), $P = 1$, $i = 1$, and $t = 1$, we get

$$A = e$$

which gives a justification for the economist's interpretation of the number e as the yield on an investment of $1 for a year at an interest rate of 100 percent compounded continuously.

Example 4   A bank advertises that interest on savings accounts is computed at 6 percent per year compounded daily. Suppose that $100 is deposited into a savings account at this bank. Find (a) an approximate amount at the end of 1 year by taking the interest rate as 6 percent compounded continuously and (b) the exact amount at the end of 1 year by considering an annual interest rate of 6 percent compounded 365 times per year.

Solution   a) From Formula (3), with $P = 100$, $i = 0.06$ and $t = 1$, we have, if A dollars is the amount,

$$A = 100e^{0.06}$$
$$= 106.18$$

Thus $106.18 is an approximate amount on deposit at the end of 1 year.

b) From Formula (1), with $P = 100$, $i = 0.06$, $m = 365$, and $t = 1$, we have, if A_{365} dollars is the amount,

$$A_{365} = 100\left(1 + \frac{0.06}{365}\right)^{365}$$
$$= 100(1.0001644)^{365}$$
$$= 106.18$$

Therefore the exact amount on deposit at the end of 1 year is $106.18. ■

EXERCISES 5.1

In Exercises 1 through 8, simplify the expression by applying laws of exponents. In Exercises 3 and 4, x is a real number.

1. (a) $3^{\sqrt{2}} \cdot 3^{\sqrt{50}}$; (b) $(e^{\sqrt{2}})^{\sqrt{50}}$

2. (a) $2^{\sqrt{12}} \cdot 2^{\sqrt{27}}$; (b) $(e^{\sqrt{12}})^{\sqrt{27}}$

3. (a) $(5^{\sqrt{15}})^{\sqrt{6}}$; (b) $5^{\sqrt[3]{x}} \cdot 5^{\sqrt[3]{x^2}}$

4. (a) $(10^{\sqrt{10}})^{\sqrt{5}}$; (b) $(10^{\sqrt{3x}})^{\sqrt{15x}}$

5. (a) $\dfrac{4^{\sqrt{32}}}{2^{\sqrt{18}}}$; (b) $\dfrac{250^{\sqrt{5}}}{10^{\sqrt{20}}}$ 6. (a) $\dfrac{3^{\sqrt{45}}}{9^{\sqrt{20}}}$; (b) $\dfrac{14^{\sqrt{98}}}{28^{\sqrt{72}}}$

7. $\dfrac{e^2 \cdot e^{\sqrt{54}}}{e^{\sqrt{24}}}$ 8. $\left(\dfrac{e^{\sqrt{14}}}{e^3 \cdot e^{\sqrt{7}}}\right)^2$

9. A loan of $2000 is made at a simple interest rate of 12 percent annually. Determine the amount to be repaid if the period of the loan is
(a) 90 days; (b) 6 months; (c) 1 year.

10. Solve Exercise 9 if the loan is for $1500 and the rate is 10 percent annually.

11. A man borrowed $10,000 at an annual interest rate of 9 percent with the understanding that interest was to be paid monthly. However, the borrower did not make the monthly interest payments and so the principal with interest at 9 percent compounded monthly was due at the end of the year. What was the amount due?

12. On his twenty-fifth birthday a man inherits $5000. If he invests this amount at 8 percent compounded annually, how much will he receive when he retires at the age of 65?

13. Determine the amount at the end of 4 years of an investment of $1000 given that the annual interest rate is 8 percent and
(a) simple interest is earned;
(b) interest is compounded annually;
(c) interest is compounded semiannually;
(d) interest is compounded quarterly;
(e) interest is compounded continuously.

14. Find the amount of an investment of $500 at the end of 2 years, given that the annual interest rate is 6 percent and
    (a) simple interest is earned;
    (b) interest is compounded annually;
    (c) interest is compounded semiannually;
    (d) interest is compounded monthly;
    (e) interest is compounded continuously.

15. An investment of $5000 earns interest at the rate of 16 percent per year and the interest is paid once at the end of the year. Find the interest earned during the first year, given that
    (a) interest is compounded quarterly;
    (b) interest is compounded continuously.

16. A loan of $2000 is repaid in one payment at the end of a year with interest at an annual rate of 12 percent. Determine the total amount repaid, given that
    (a) interest is compounded quarterly;
    (b) interest is compounded continuously.

17. A deposit of $1000 is made at a savings bank that advertises that interest on accounts is computed at an annual rate of 9 percent compounded daily. Find
    (a) an approximate amount at the end of 1 year by taking the interest rate as 9 percent compounded continuously;
    (b) the exact amount at the end of 1 year by considering an annual interest rate of 9 percent compounded 365 times per year.

18. Solve Exercise 17 if the bank advertises that interest is computed at an annual rate of 7 percent compounded daily. Take the rate as
    (a) 7 percent compounded continuously;
    (b) an annual rate of 7 percent compounded 365 times per year.

19. How much should be deposited in a savings account now if it is desired to have $5000 in the account at the end of 5 years and if the annual interest rate is 8 percent compounded quarterly?

20. At the end of 4 years a savings account had a balance of $3000. One deposit was made at the be-

ginning of the 4-year period and no withdrawals were made. How much was the original deposit if the interest rate was 6 percent per year compounded monthly?

21. Solve Exercise 19, given that the annual interest rate is 8 percent compounded continuously.

22. Solve Exercise 20, given that the interest rate was 6 percent per year compounded continuously.

Exercises 23 through 25 are based on the following: In calculus it is shown that

$$(1 + z)^{1/z} \to e \qquad \text{as} \qquad z \to 0$$

Observe that $(1 + z)^{1/z}$ can be obtained from $\left(1 + \dfrac{1}{x}\right)^x$ by letting $x = \dfrac{1}{z}$.

23. Use a calculator to compute the values of $(1 + z)^{1/z}$ when $z = 0.001$ and $z = -0.001$. Then obtain an approximation of the number e to three decimal places by using these values to find the average value of $(1.001)^{1000}$ and $(0.999)^{-1000}$.

24. Use a calculator to compute the values of $(1 + z)^{1/z}$ when $z = 0.0001$ and $z = -0.0001$. Then obtain an approximation of the number e to four decimal places by using these values to find the average value of $(1.0001)^{10,000}$ and $(0.9999)^{-10,000}$.

25. Draw a sketch of the graph of the function defined by

$$f(x) = (1 + x)^{1/x}$$

on the interval $[-0.5, 0.5]$ by assuming that f is continuous (there are no breaks in the graph) on the interval and plotting the points for which x has the values $-0.5, -0.1, -0.01, 0.01, 0.1,$ and 0.5. Use a calculator for the function values. Use different scales on the two axes. On the x axis take tick marks at every 0.1 unit and on the y axis take tick marks at every unit. The y coordinate of the point at which the graph intersects the y axis is the number e.

5.2   Exponential Functions

Exponential functions occur in many diverse fields such as biology, sociology, business, economics, chemistry, physics, and engineering. Examples 2, 3, and 4 and Exercises 23 through 32 give some of these applications.

If $b > 0$, then for each real number x there corresponds a unique real number b^x. Thus we can make the following definition.

Definition

> **Exponential Function with Base b**
> If $b > 0$ and $b \neq 1$, then the **exponential function with base b** is the function f defined by
> $$f(x) = b^x$$
> The domain of f is the set of real numbers, and the range is the set of positive numbers.

Note: If $b = 1$, then b^x becomes 1^x, and since $1^x = 1$ for any x, we have a constant function. For this reason, we impose the condition that $b \neq 1$ in the preceding definition.

In the following two illustrations we consider the graphs of the exponential functions with bases 2 and $\frac{1}{2}$.

Illustration   1

The exponential function with base 2 is defined by

$$F(x) = 2^x$$

Table 1 gives some rational values of x with the corresponding function values.

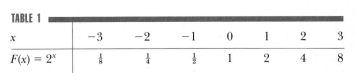

TABLE 1							
x	-3	-2	-1	0	1	2	3
$F(x) = 2^x$	$\frac{1}{8}$	$\frac{1}{4}$	$\frac{1}{2}$	1	2	4	8

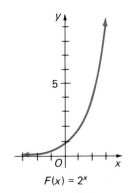

$F(x) = 2^x$

FIGURE 1

A sketch of the graph is shown in Figure 1; it is drawn by plotting the points whose coordinates are given by Table 1 and connecting these points with a curve. Observe from the graph that the function values increase as x increases. This fact follows from Theorem 1(i) of Section 5.1, with r and s real numbers.

Note that

$$2^x \to 0^+ \quad \text{as} \quad x \to -\infty$$

that is, 2^x approaches zero through values greater than zero as x decreases without bound. Therefore, the x axis is a horizontal asymptote of the graph of F. Furthermore,

$$2^x \to +\infty \quad \text{as} \quad x \to +\infty$$

That is, 2^x increases without bound as x increases without bound.

Illustration 2

The exponential function with base $\frac{1}{2}$ is defined by

$$G(x) = \left(\frac{1}{2}\right)^x$$

In Table 2 there are some rational values of x with the corresponding function values.

TABLE 2

x	-3	-2	-1	0	1	2	3
$G(x) = \left(\dfrac{1}{2}\right)^x$	8	4	2	1	$\dfrac{1}{2}$	$\dfrac{1}{4}$	$\dfrac{1}{8}$

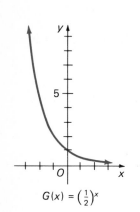

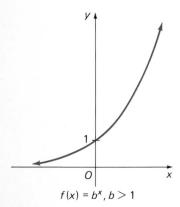

$G(x) = \left(\frac{1}{2}\right)^x$

FIGURE 2

By plotting the points whose coordinates are given in the table and connecting these points with a curve, we obtain the sketch of the graph appearing in Figure 2. The function values decrease as x increases, as indicated by the graph. This fact follows from Theorem 1(ii) of Section 5.1 with r and s real numbers. Because

$$\left(\frac{1}{2}\right)^x \to 0^+ \qquad \text{as} \qquad x \to +\infty$$

the x axis is a horizontal asymptote of the graph of G. Furthermore,

$$\left(\frac{1}{2}\right)^x \to +\infty \qquad \text{as} \qquad x \to -\infty$$

that is, $\left(\frac{1}{2}\right)^x$ increases without bound as x decreases without bound. ∎

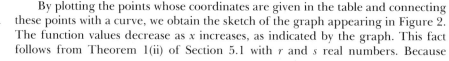

$f(x) = b^x, b > 1$

FIGURE 3

Figure 3 shows a sketch of the graph of the exponential function with base b, when $b > 1$. The function values increase as x increases. In Figure 4 there is a sketch of the graph of the exponential function with base b when $0 < b < 1$. For this function, the function values decrease as x increases.

From the definition of the exponential function with base b it follows that b can be any positive number except 1. When b is e, we have the *natural exponential function.*

Definition

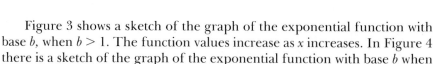

Natural Exponential Function
The **natural exponential function** is the function f defined by

$$f(x) = e^x$$

The domain of the natural exponential function is the set of real numbers, and its range is the set of positive numbers.

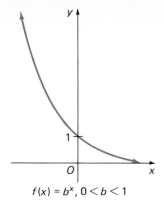

$f(x) = b^x, 0 < b < 1$

FIGURE 4

Example   1   Draw a sketch of the graph of the natural exponential function.

Solution   Table 3 gives some values of x and the corresponding function values of the natural exponential function. The approximation of the powers of e can be obtained from a calculator or from Table V in the back of the book.

TABLE 3

x	0	0.5	1	1.5	2	2.5	−0.5	−1	−2
e^x	1	1.6	2.7	4.5	7.4	12.2	0.6	0.4	0.1

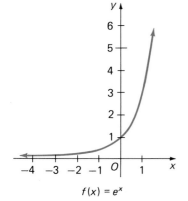

$f(x) = e^x$

FIGURE 5

The points whose coordinates are given by Table 3 are plotted, and these points are connected with a curve to yield the sketch of the graph of the natural exponential function shown in Figure 5.   ∎

Exponential growth and *exponential decay* give mathematical models involving powers of e. A function defined by an equation of the form

$$f(t) = Be^{kt} \qquad t \geq 0 \qquad (1)$$

where B and k are positive constants, is said to describe **exponential growth.** If you study calculus you will learn that such a function results when the rate of growth of a quantity is proportional to its size. To draw a sketch of the graph of $f(t) = Be^{kt}$, note that $f(0) = B$ and $f(t)$ is always positive. Furthermore,

$$Be^{kt} \to +\infty \qquad \text{as} \qquad t \to +\infty$$

Thus Be^{kt} increases without bound as t increases without bound. A sketch of the graph of Equation (1) appears in Figure 6.

The following example involves exponential growth in biology.

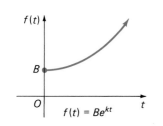

$f(t) = Be^{kt}$

FIGURE 6

Example   2   In a particular bacterial culture, if $f(t)$ bacteria are present at t minutes, then

$$f(t) = Be^{0.04t}$$

where B is a constant. If there are 1500 bacteria present initially, how many bacteria will be present after 1 hour?

Solution   Because there are 1500 bacteria present initially, $f(0) = 1500$.

From the equation defining $f(t)$,

$$f(0) = Be^{0.04(0)}$$
$$1500 = Be^0$$
$$1500 = B$$

Thus with $B = 1500$, we have

$$f(t) = 1500e^{0.04t}$$

The number of bacteria present after 1 hour is $f(60)$, and from the above equation

$$f(60) = 1500e^{0.04(60)}$$
$$= 1500e^{2.4}$$
$$= 1500(11.023)$$
$$= 16{,}535$$

Therefore there are 16,535 bacteria present in the culture after 1 hour. ∎

A function defined by an equation of the form

$$F(t) = Be^{-kt} \qquad t \geq 0 \tag{2}$$

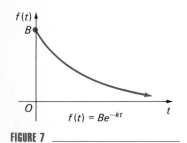

$f(t) = Be^{-kt}$

FIGURE 7

where B and k are positive constants, is said to describe **exponential decay.** Exponential decay occurs when the rate of decrease of a quantity is proportional to its size, as is shown in calculus. For instance, it is known from experiments that the rate of decay of radium is proportional to the amount of radium present at a given instant. A sketch of the graph of Equation (2) appears in Figure 7. Observe that

$$Be^{-kt} \to 0^+ \qquad \text{as} \qquad t \to +\infty$$

The next example involves exponential decay in which the value of some equipment is decreasing exponentially.

Example   **3**   If $V(t)$ dollars is the value of a certain piece of equipment t years after its purchase, then

$$V(t) = Be^{-0.20t}$$

where B is a constant. If the equipment was purchased for $8000, what will be its value in 2 years?

Solution   Because the equipment was purchased for $8000, $V(0) = 8000$. From the equation defining $V(t)$,

$$V(0) = Be^{-0.20(0)}$$
$$8000 = Be^0$$
$$8000 = B$$

Therefore with $B = 8000$, we have

$$V(t) = 8000e^{-0.20t}$$

The number of dollars in the value of the equipment after 2 years is $V(2)$, and from the above equation

$$\begin{aligned} V(2) &= 8000e^{-0.20(2)} \\ &= 8000e^{-0.40} \\ &= 8000(0.670320) \\ &= 5362.56 \end{aligned}$$

Thus the value of the equipment in 2 years will be \$5362.56.   ■

Another mathematical model involving powers of e is given by the function defined by

$$f(t) = A(1 - e^{-kt}) \qquad t \geq 0$$

where A and k are positive constants. This function describes **bounded growth.** Because

$$A(1 - e^{-kt}) = A - Ae^{-kt}$$

and $Ae^{-kt} \to 0^+$ as $t \to +\infty$, it follows that

$$A(1 - e^{-kt}) \to A^- \qquad \text{as} \qquad t \to +\infty$$

Therefore the line A units above the t axis is a horizontal asymptote of the graph of f. Also note that

$$\begin{aligned} f(0) &= A(1 - e^0) \\ &= 0 \end{aligned}$$

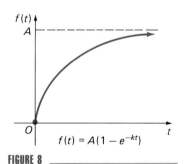

$$f(t) = A(1 - e^{-kt})$$

FIGURE 8

From this information we obtain the sketch of the graph of f shown in Figure 8. This graph is sometimes called a **learning curve.** The name appears appropriate when $f(t)$ represents someone's competence in performing a job. As his or her experience increases, competence increases rapidly at first and then slows down, because additional experience has little effect on the skill with which the task is performed.

Example   4   A typical worker at a certain factory can produce $f(t)$ units per day after t days on the job, where

$$f(t) = 50(1 - e^{-0.34t})$$

(a) How many units per day can the worker produce after 7 days on the job? (b) How many units per day can the worker eventually be expected to produce?

Solution a) We wish to find $f(7)$. From the equation defining $f(t)$,

$$f(7) = 50(1 - e^{-0.34(7)})$$
$$= 50(1 - e^{-2.38})$$
$$= 50(1 - 0.093)$$
$$= 45$$

Thus the worker can produce 45 units per day after 7 days on the job.

b) Because $e^{-0.34t} \to 0^+$ as $t \to +\infty$, it follows that

$$50(1 - e^{-0.34t}) \to 50^- \qquad \text{as} \qquad t \to +\infty$$

Thus we can conclude that the worker can eventually be expected to produce 50 units per day. ∎

Bounded growth is also described by a function defined by

$$f(t) = A - Be^{-kt} \qquad t \geq 0, \quad A > B$$

where A, B, and k are positive constants. For this function $f(0) = A - B$. A sketch of the graph appears in Figure 9.

In calculus it can be shown that bounded growth occurs when a quantity grows at a rate proportional to the difference between a fixed number A and the size of the quantity. In this case A serves as an upper bound. Exercises 30, 31, and 32 involve applications of bounded growth.

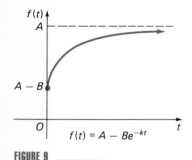

FIGURE 9

EXERCISES 5.2

In Exercises 1 through 20, draw a sketch of the graph of the exponential function. In Exercises 9 through 20, use a calculator or Table V for powers of e.

1. $f(x) = 3^x$
2. $g(x) = 4^x$
3. $F(x) = 3^{-x}$
4. $G(x) = 4^{-x}$
5. $g(x) = (\frac{1}{5})^x$
6. $f(x) = 10^x$
7. $f(x) = 2^{x+1}$
8. $g(x) = 3^{x-1}$
9. $F(x) = e^{-x}$
10. $G(x) = e^{2x}$
11. $f(x) = -e^x$
12. $g(x) = 2e^x$
13. $F(x) = e^{x-1}$
14. $G(x) = e^{x+2}$
15. $f(x) = e^x - 1$
16. $g(x) = e^x + 2$
17. $f(x) = 10e^{0.2x}$
18. $g(x) = 10e^{0.1x}$
19. $F(x) = 10e^{-0.1x}$
20. $G(x) = 10e^{-0.2x}$
21. Draw sketches of the graphs of $y = 3^x$ and $x = 3^y$ on the same coordinate axes.
22. Draw sketches of the graphs of $y = e^x$ and $x = e^y$ on the same coordinate axes.
23. The value of a particular machine t years after its purchase is $V(t)$ dollars, where $V(t) = ke^{-0.30t}$ and k

is a constant. If the machine was purchased 8 years ago for $10,000, what is its value now?
24. If $f(t)$ grams of a radioactive substance are present after t seconds, then $f(t) = ke^{-0.3t}$ where k is a constant. If 100 grams of the substance are present initially, how much is present after 5 sec?
25. If $P(h)$ pounds per square foot is the atmospheric pressure at a height h feet above sea level, then $P(h) = ke^{-0.00003h}$, where k is a constant. Given that the atmospheric pressure at sea level is 2116 lb/ft^2, find the atmospheric pressure outside of an airplane that is 10,000 ft high.
26. Suppose $f(t)$ is the number of bacteria present in a certain culture at t minutes and $f(t) = ke^{0.035t}$, where k is a constant. If 5000 bacteria are present after 10 min have elapsed, how many bacteria were present initially?
27. The population of a particular town is increasing at a rate proportional to its size. If this rate is 6

percent and if the population after t years is $P(t)$, then $P(t) = ke^{0.06t}$, where k is a constant. If the current population is 10,000, what is the expected population (a) after 10 years and (b) after 20 years?

28. In 1980 it was estimated that for the succeeding 20 years the population of a particular town was expected to be $f(t)$ people t years from 1980, where $f(t) = C \cdot 10^{kt}$, and C and k are constants. If the actual population in 1980 was 1000 and in 1985 it was 4000, what is the expected population in 1995?

29. A historically important abstract painting was purchased in 1928 for $200 and its value has doubled every 10 years since its purchase.
    (a) Let $f(t)$ dollars be the value t years after its purchase; define $f(t)$.
    (b) What was the value of the painting in 1988?

30. After t hours of practice typing, it was determined that a certain person could type $f(t)$ words per minute, where $f(t) = 90(1 - e^{-0.03t})$.
    (a) Draw a sketch of the graph of f and observe the behavior of $f(t)$ as t increases without bound.
    (b) How many words per minute can the person type after 30 hours of practice?
    (c) How many words per minute can the person eventually be expected to type?

31. The efficiency of a typical worker at a certain factory is given by the function defined by the equation $f(t) = 100 - 60e^{-0.2t}$, where the worker can complete $f(t)$ units of work per day after being on the job for t months.
    (a) Draw a sketch of the graph of f and observe the behavior of $f(t)$ as t increases without bound.
    (b) How many units per day can be completed by a beginning worker?
    (c) How many units per day can be completed by a worker having a year's experience?
    (d) How many units per day can the typical worker eventually be expected to complete?

32. The resale value of a certain piece of equipment is $f(t)$ dollars t years after its purchase, where $f(t) = 1200 + 8000e^{-0.25t}$.
    (a) Draw a sketch of the graph of f and observe the behavior of $f(t)$ as t increases without bound.
    (b) What is the value of the equipment when it is purchased?
    (c) What is the value of the equipment 10 years after its purchase?
    (d) What is the anticipated scrap value of the equipment after a long period of time?

5.3   Logarithmic Functions

In Example 2 of Section 5.2 we had the equation

$$f(t) = 1500e^{0.04t}$$

where $f(t)$ is the number of bacteria present in a certain culture at t minutes. Suppose that we wish to find in how many minutes there will be 30,000 bacteria present. If T is the number of minutes to be determined, then we have the equation

$$30,000 = 1500e^{0.04T}$$

In this equation the unknown T appears in an exponent. At the present we cannot solve such an equation, but the concept of a logarithm will give us the means to do so. We will develop this concept and then return to the problem in Example 3.

Observe from Figures 3 and 4 in Section 5.2 that a horizontal line intersects the graph of the exponential function with base b in not more than one point. Therefore this function is one-to-one. Thus it has an inverse function, which is given in the following definition.

Definition

> **Logarithmic Function with Base _b_**
> The **logarithmic function with base _b_** is the inverse of the exponential function with base _b_.

We use the notation $\log_b$ to denote the logarithmic function with base b. The function values of the function $\log_b$ are denoted by $\log_b(x)$ or more simply $\log_b x$ (read "logarithm with base b of x"). Therefore, because $\log_b$ and the exponential function with base b are inverse functions,

$$y = \log_b x \quad \text{if and only if} \quad x = b^y \tag{1}$$

The domain of the exponential function with base b is the set of real numbers, and its range is the set of positive numbers. Therefore the domain of $\log_b$ is the set of positive numbers and the range is the set of real numbers.

The two equations appearing in statement (1) are equivalent. We make use of this fact in the following two illustrations.

Illustration 1

$$3^2 = 9 \quad \Leftrightarrow \quad \log_3 9 = 2 \qquad\qquad 2^3 = 8 \quad \Leftrightarrow \quad \log_2 8 = 3$$

$$\left(\frac{1}{16}\right)^{1/2} = \frac{1}{4} \quad \Leftrightarrow \quad \log_{1/16}\frac{1}{4} = \frac{1}{2} \qquad\qquad 5^{-2} = \frac{1}{25} \quad \Leftrightarrow \quad \log_5 \frac{1}{25} = -2 \quad \blacksquare$$

Illustration 2

$$\log_{10} 10{,}000 = 4 \quad \Leftrightarrow \quad 10^4 = 10{,}000 \qquad\qquad \log_8 2 = \frac{1}{3} \quad \Leftrightarrow \quad 8^{1/3} = 2$$

$$\log_6 1 = 0 \quad \Leftrightarrow \quad 6^0 = 1 \qquad\qquad \log_9 \frac{1}{3} = -\frac{1}{2} \quad \Leftrightarrow \quad 9^{-1/2} = \frac{1}{3} \quad \blacksquare$$

Example 1   Find the value of each of the following logarithms.

a) $\log_7 49$ b) $\log_5 \sqrt{5}$ c) $\log_6 \frac{1}{6}$ d) $\log_3 81$ e) $\log_{10} 0.001$

Solution   In each part we let y represent the given logarithm and obtain an equivalent equation in exponential form. We then solve for y by making use of the fact that if $b > 0$ and $b \neq 1$, then

$$b^y = b^n \qquad \text{implies} \qquad y = n$$

a) Let $\log_7 49 = y$. This equation is equivalent to $7^y = 49$. Because $49 = 7^2$, we have

$$7^y = 7^2$$

Therefore $y = 2$; that is, $\log_7 49 = 2$.

b) Let $\log_5 \sqrt{5} = y$. Therefore $5^y = \sqrt{5}$ or, equivalently,

$$5^y = 5^{1/2}$$

Hence $y = \frac{1}{2}$; that is, $\log_5 \sqrt{5} = \frac{1}{2}$.

c) Let $\log_6 \frac{1}{6} = y$. Thus $6^y = \frac{1}{6}$ or, equivalently,

$$6^y = 6^{-1}$$

Therefore $y = -1$; that is, $\log_6 \frac{1}{6} = -1$.

d) Let $\log_3 81 = y$. Thus $3^y = 81$ or, equivalently,

$$3^y = 3^4$$

Hence $y = 4$; that is, $\log_3 81 = 4$.

e) Let $\log_{10} 0.001 = y$. Then $10^y = 0.001$. Because $10^{-3} = 0.001$, we have

$$10^y = 10^{-3}$$

Therefore $y = -3$; that is, $\log_{10} 0.001 = -3$. ∎

Example 2 Solve the given equation for either x or b.

a) $\log_6 x = 2$ b) $\log_{27} x = \frac{2}{3}$ c) $\log_b 4 = \frac{1}{3}$ d) $\log_b 81 = -2$

Solution

a) $\log_6 x = 2 \iff 6^2 = x$

$$x = 36$$

b) $\log_{27} x = \frac{2}{3} \iff 27^{2/3} = x$

$$x = (\sqrt[3]{27})^2$$
$$x = 3^2$$
$$x = 9$$

c) $\log_b 4 = \frac{1}{3} \iff b^{1/3} = 4$

$$(b^{1/3})^3 = 4^3$$
$$b = 64$$

d) $\log_b 81 = -2 \iff b^{-2} = 81$

$$(b^{-2})^{-1/2} = 81^{-1/2}$$
$$b = \frac{1}{81^{1/2}}$$
$$b = \frac{1}{9} \quad ∎$$

From Statement (1), $b^y = x$ and $y = \log_b x$ are equivalent equations. If in the first of these equations we replace y by $\log_b x$, we obtain

$$b^{\log_b x} = x \tag{2}$$

where $b > 0$, $b \neq 1$, and $x > 0$.

From Equation (2) we observe that *a logarithm is an exponent;* that is, $\log_b x$ is the exponent to which b must be raised to yield x.

Illustration 3

From Equation (2) it follows that

$$3^{\log_3 7} = 7 \qquad \text{and} \qquad 10^{\log_{10} 5} = 5 \qquad \blacksquare$$

If we write the equations in Statement (1) as $\log_b x = y$ and $x = b^y$ and replace x in the first of these equations by b^y, we obtain

$$\log_b b^y = y \tag{3}$$

where $b > 0$, $b \neq 1$, and y is any real number.

Illustration 4

From Equation (3) it follows that

$$\log_2 2^{-5} = -5 \qquad \text{and} \qquad \log_{10} 10^3 = 3 \qquad \blacksquare$$

Equations (2) and (3) were obtained from Statement (1) as a result of the fact that the logarithmic and exponential functions are inverses of each other.

Recall from Section 4.7 that the graphs of a function and its inverse are reflections of each other with respect to the line $y = x$. We use this fact to obtain the graph of a logarithmic function from the graph of the corresponding exponential function.

Illustration 5

The graph of $y = 2^x$ appears as the black curve in Figure 1. The graph of $y = \log_2 x$ is shown in color in the figure. The two graphs are reflections of each other with respect to the line $y = x$. $\blacksquare$

In Figure 1 we have a special case of the graph of the logarithmic function with base b where $b > 1$. In Figure 2 we have the general case.

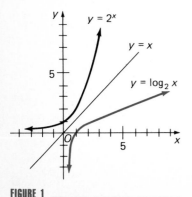

FIGURE 1

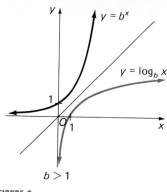

FIGURE 2

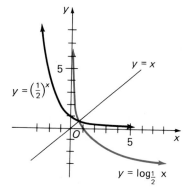

FIGURE 3

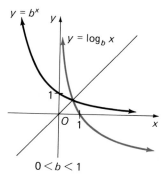

FIGURE 4

This figure shows in color a sketch of the graph of

$$f(x) = \log_b x \qquad b > 1$$

This graph is symmetric, with respect to the line $y = x$, to the graph of the exponential function with base b ($b > 1$), a sketch of which also appears in Figure 2.

Illustration   6

Figure 3 shows the graph of $y = \left(\dfrac{1}{2}\right)^x$ in black. The graph of $y = \log_{1/2} x$, in color, is obtained from this graph by a reflection of $y = \left(\dfrac{1}{2}\right)^x$ with respect to the line $y = x$.
■

The graph in Figure 3 is a special case of the graph of the logarithmic function with base b where $0 < b < 1$. For the general case refer to Figure 4, showing a sketch of the graph of

$$f(x) = \log_b x \qquad 0 < b < 1$$

in color. Also in Figure 4 is a sketch of the graph of the exponential function with base b ($0 < b < 1$), and we observe that the two graphs are symmetric with respect to the line $y = x$.

From the sketches of the graphs of $\log_b$ in Figures 2 and 4, we note the following properties of the logarithmic function with base b.

1. If $b > 1$, $\log_b x$ increases as x increases. If $0 < b < 1$, $\log_b x$ decreases as x increases.
2. If $b > 1$, $\log_b x$ is positive if $x > 1$, and $\log_b x$ is negative if $0 < x < 1$. If $0 < b < 1$, $\log_b x$ is negative if $x > 1$, and $\log_b x$ is positive if $0 < x < 1$. Furthermore, $\log_b x$ is not defined if x is nonpositive.
3. The only zero of the function $\log_b$ is 1; that is, $\log_b x = 0$ if and only if $x = 1$.
4. $\log_b x = 1$ if and only if $x = b$.
5. If $b > 1$, $\log_b x \to -\infty$ as $x \to 0^+$; and if $0 < b < 1$, $\log_b x \to +\infty$ as $x \to 0^+$.

Property 3 gives the equation

$$\log_b 1 = 0$$

and property 4 gives

$$\log_b b = 1$$

The logarithmic function with base e is called the *natural logarithmic function*.

Definition

> **Natural Logarithmic Function**
> The **natural logarithmic function** is the inverse of the natural exponential function.

The natural logarithmic function can be denoted by $\log_e$, but a more customary notation is ln. The function values of ln are denoted by ln x (read "natural logarithm of x"). Because ln and the natural exponential function are inverse functions,

$$y = \ln x \quad \text{if and only if} \quad x = e^y \tag{4}$$

Natural logarithmic function values to four decimal places are given in Table IV in the back of the book. A calculator with an $\boxed{\textbf{ln}}$ key can also be used to obtain these values. Recall from Section 5.1 that if your calculator has no $\boxed{e^x}$ key, you can obtain values of e^x by pressing the $\boxed{\textbf{INV}}$ key followed by the $\boxed{\textbf{ln}}$ key. This procedure is based on the fact that the natural exponential function and the natural logarithmic function are inverses of each other.

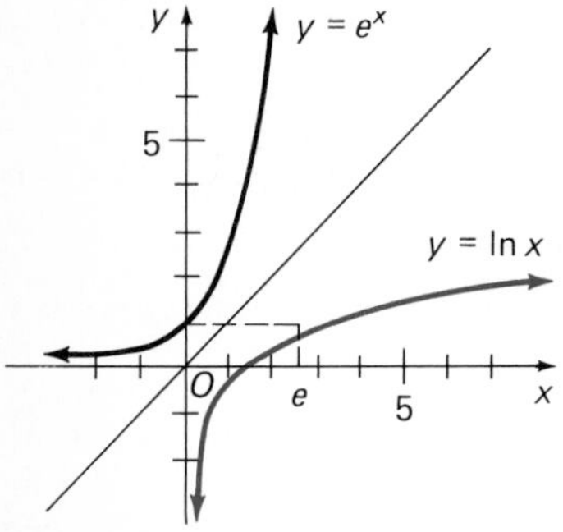

A sketch of the graph of the natural logarithmic function appears in Figure 5. In the figure the graph of ln is shown in color and a sketch of the graph of the natural exponential function is also shown. The two graphs are symmetric with respect to the line $y = x$. In Figure 5 observe that e is the number whose natural logarithm is 1; that is,

$$\ln e = 1$$

This follows from statement (4) because ln e is equivalent to $e = e^1$. If in (2) and (3), $b = 3$, we have, respectively,

$$e^{\ln x} = x \quad \text{and} \quad \ln e^x = x$$

Example   3   In Example 2 of Section 5.2 we obtained the equation

$$f(t) = 1500e^{0.04t}$$

FIGURE 5

where $f(t)$ is the number of bacteria present in a certain culture at t minutes when there are 1500 bacteria present initially. Determine how many minutes elapse until there are 30,000 bacteria present in the culture.

Solution Let T represent the number of minutes that elapse until there are 30,000 bacteria present. Then in the given equation we substitute T for t and we have

$$f(T) = 1500e^{0.04T}$$

Because $f(T) = 30,000$,

$$30,000 = 1500e^{0.04T}$$
$$20 = e^{0.04T}$$

Because the equation $x = e^y$ is equivalent to the equation $y = \ln x$, it follows that $20 = e^{0.04T}$ is equivalent to

$$0.04T = \ln 20$$

From Table IV or a calculator, $\ln 20 = 2.9957$; thus

$$0.04T = 2.9957$$
$$T = \frac{2.9957}{0.04}$$
$$T = 74.9$$

Therefore 1 hr, 14 min, and 54 sec elapse until there are 30,000 bacteria present. ■

Example **4** If \$1000 is deposited into a savings account that pays an annual interest rate of 6 percent compounded continuously, and no withdrawals or additional deposits are made, how long will it take until there is \$1500 on deposit?

Solution Let T years be how long it will take \$1000 to accumulate to \$1500 at an annual interest rate of 6 percent compounded continuously. From formula (3) of Section 5.1, $A = Pe^{it}$. Thus

$$1500 = 1000e^{0.06T}$$
$$e^{0.06T} = 1.5$$
$$0.06T = \ln 1.5$$
$$T = \frac{\ln 1.5}{0.06}$$
$$T = \frac{0.4055}{0.06}$$
$$T = 6.758$$

Because 6.758 years is equivalent to 6 years, 9 months, and 3 days, it will take that long until there is \$1500 on deposit. ■

Example 5   In Example 4 of Section 5.2 we had the equation

$$f(t) = 50(1 - e^{-0.34t})$$

where a typical worker at a certain factory can produce $f(t)$ units per day after t days on the job. After how many days on the job will the worker produce 40 units per day?

Solution   Let T represent the number of days on the job necessary for the worker to produce 40 units per day. Substituting T for t in the given equation, we have

$$f(T) = 50(1 - e^{-0.34T})$$

Because $f(T) = 40$,

$$40 = 50(1 - e^{-0.34T})$$
$$0.80 = 1 - e^{-0.34T}$$
$$e^{-0.34T} = 0.2$$
$$-0.34T = \ln 0.2$$
$$-0.34T = -1.6094$$
$$T = \frac{-1.6094}{-0.34}$$
$$T = 4.7$$

Therefore after 5 days on the job the worker will produce 40 units per day. ∎

EXERCISES 5.3

In Exercises 1 through 4, express the relationship in the equation by using logarithmic notation.

1. (a) $3^4 = 81$; (b) $5^3 = 125$; (c) $10^{-3} = 0.001$
2. (a) $2^5 = 32$; (b) $7^2 = 49$; (c) $5^{-2} = \frac{1}{25}$
3. (a) $8^{2/3} = 4$; (b) $625^{-3/4} = \frac{1}{125}$; (c) $2^0 = 1$
4. (a) $81^{3/4} = 27$; (b) $10^0 = 1$; (c) $64^{-2/3} = \frac{1}{16}$

In Exercises 5 through 8, express the relationship in the equation by using exponential notation.

5. (a) $\log_8 64 = 2$; (b) $\log_3 81 = 4$; (c) $\log_2 1 = 0$
6. (a) $\log_{10} 10{,}000 = 4$; (b) $\log_5 125 = 3$;
    (c) $\log_{10} 1 = 0$
7. (a) $\log_8 2 = \frac{1}{3}$; (b) $\log_{1/3} 9 = -2$; (c) $\log_9 \frac{1}{3} = -\frac{1}{2}$
8. (a) $\log_{32} 2 = \frac{1}{5}$; (b) $\log_{1/2} 64 = -6$; (c) $\log_{16} \frac{1}{8} = -\frac{3}{4}$

In Exercises 9 through 12, find the value of the logarithm.

9. (a) $\log_{10} 100$; (b) $\log_{27} 9$; (c) $\log_2 \frac{1}{8}$
10. (a) $\log_4 64$; (b) $\log_6 6$; (c) $\log_3 \frac{1}{81}$
11. (a) $\log_8 \frac{1}{2}$; (b) $\log_{27} \frac{1}{81}$; (c) $\ln \sqrt{e}$
12. (a) $\log_7 7$; (b) $\log_{1/4} \frac{1}{32}$; (c) $\ln e^2$

In Exercises 13 through 16, solve the equation for x.

13. (a) $\log_7 x = 3$; (b) $\log_{1/3} x = -4$
14. (a) $\log_4 x = 3$; (b) $\log_{1/4} x = -3$
15. (a) $\log_2 x = \frac{3}{2}$; (b) $\log_{1/4} x = \frac{1}{2}$
16. (a) $\log_3 x = \frac{2}{3}$; (b) $\log_{1/9} x = \frac{5}{2}$

In Exercises 17 through 20, solve the equation for b.

17. (a) $\log_b 144 = 2$; (b) $\log_b 6 = \frac{1}{3}$
18. (a) $\log_b 1000 = 3$; (b) $\log_b 3 = \frac{1}{4}$
19. (a) $\log_b 0.01 = -2$; (b) $\log_b \frac{1}{4} = -\frac{2}{3}$
20. (a) $\log_b 27 = -3$; (b) $\log_b 0.001 = -\frac{3}{2}$

In Exercises 21 through 24, simplify the expression.

21. (a) $\log_6 (\log_5 5)$; (b) $\log_2 (\log_9 81)$

22. (a) $\log_5 (\log_2 32)$; (b) $\log_2 (\log_3 81)$
23. (a) $\log_2 (\log_2 256)$; (b) $\log_b (\log_b b)$, $b > 0$
24. (a) $\log_3 (\log_3 3)$; (b) $\log_b(\log_a a^b)$, $a > 0$ and $b > 0$

In Exercises 25 through 32, draw a sketch of the graph of the function.

25. $f(x) = \log_{10} x$ 26. $f(x) = \log_3 x$
27. $g(x) = \log_{1/3} x$ 28. $g(x) = \log_{1/10} x$
29. $F(x) = \log_3 x^2$ 30. $G(x) = \log_2 |x|$
31. $f(x) = \ln(x - 1)$ 32. $g(x) = \ln(x + 1)$
33. $g(x) = \ln x + 2$ 34. $f(x) = \ln x - 3$

35. The value of a machine purchased 8 years ago for $10,000 is $V(t)$ dollars t years after its purchase, where $V(t) = 10,000e^{-0.3t}$. If the machine will be replaced when its value is $500, when will a new machine be purchased?

36. If $f(t)$ grams of a radioactive substance are present after t seconds and 100 grams are present initially, then $f(t) = 100e^{-0.3t}$. After how many seconds will there be only 10 grams of the substance present?

37. The current population of a particular town is 10,000 and it is increasing at a rate proportional to its size. If this rate is 6 percent and if the population after t years is $P(t)$, then $P(t) = 10,000e^{0.06t}$ When is the population expected to be 45,000?

38. In a certain culture there are 2000 bacteria present initially, and after t minutes $f(t)$ bacteria are present, where $f(t) = 2000e^{0.035t}$. When will there be 10,000 bacteria in the culture?

39. How long will it take for $500 to accumulate to $900 if money is invested at 9 percent compounded continuously?

40. Solve Exercise 39, given that money is invested at 12 percent compounded continuously.

41. How long will it take for an investment to double itself if interest is paid at the rate of 8 percent compounded continuously?

42. How long will it take for an investment to triple itself if interest is paid at the rate of 12 percent compounded continuously?

43. The efficiency of a typical worker at a certain factory is given by $f(t) = 100 - 60e^{-0.2t}$, where the worker can complete $f(t)$ units of work per day after being on the job for t months. After how many months' experience is the typical worker expected to complete 70 units per day?

44. The resale value of a certain piece of equipment is $f(t)$ dollars t years after its purchase, where $f(t) = 1200 + 8000e^{-0.25t}$. How long after its purchase will the resale value of the equipment be $2000?

5.4   Properties of Logarithmic Functions

Before the arrival of electronic calculators, logarithms were used to perform tedious calculations involving products, quotients, powers, and roots. For this purpose logarithms to the base 10 were applied, and the procedure involved the use of tables of these logarithms. This application of logarithms is now obsolete. Today we are concerned primarily with the properties of logarithmic functions and their use in solving exponential and logarithmic equations discussed in Section 5.5. The computations in the illustrations, examples, and exercises in this section are presented only to demonstrate these properties and are not intended to advocate such an application of them.

The three theorems of this section concern properties of logarithms that follow from corresponding laws of exponents. After the statement of each theorem an illustration is given to show the law of exponents involved. In the proofs use is made of the fact that

$$x = b^y \quad \Leftrightarrow \quad y = \log_b x$$

We refer to the equation $x = b^y$ as the *exponential form* of the equation $y = \log_b x$, and we refer to the equation $y = \log_b x$ as the *logarithmic form* of the equation $x = b^y$.

Theorem 1

> If $b > 0$, $b \neq 1$, and u and v are positive numbers, then
> $$\log_b uv = \log_b u + \log_b v$$

Illustration 1

Suppose in the statement of Theorem 1 that b is 2, u is 4, and v is 8. Then

$$
\begin{aligned}
\log_b uv &= \log_2 4 \cdot 8 & \qquad \log_b u + \log_b v &= \log_2 4 + \log_2 8 \\
&= \log_2 2^2 \cdot 2^3 & &= \log_2 2^2 + \log_2 2^3 \\
&= \log_2 2^{2+3} & &= 2 + 3 \\
&= \log_2 2^5 & &= 5 \\
&= 5 \quad \text{(because } \log_b b^y = y)
\end{aligned}
$$

Therefore, when b is 2, u is 4, and v is 8, Theorem 1 is valid. ∎

Proof of Theorem 1   Let

$$r = \log_b u \qquad \text{and} \qquad s = \log_b v$$

The exponential forms of these equations are, respectively,

$$u = b^r \qquad \text{and} \qquad v = b^s$$

Therefore

$$uv = b^r \cdot b^s$$
$$uv = b^{r+s}$$

The logarithmic form of this equation is

$$\log_b uv = r + s$$

Substituting $\log_b u$ for r and $\log_b v$ for s, we have

$$\log_b uv = \log_b u + \log_b v$$ ∎

Illustration 2

If we are given $\log_{10} 2 = 0.3010$ and $\log_{10} 3 = 0.4771$, we can apply Theorem 1 to find $\log_{10} 6$.

$$
\begin{aligned}
\log_{10} 6 &= \log_{10} (2 \cdot 3) \\
&= \log_{10} 2 + \log_{10} 3 \\
&= 0.3010 + 0.4771 \\
&= 0.7781
\end{aligned}
$$ ∎

Because $\log_{10} 2$, $\log_{10} 3$, and $\log_{10} 6$ are irrational numbers, the values given for them in Illustration 2 are only decimal approximations. However, in computations such as Illustration 2 we shall use the equals symbol.

Theorem 2

If $b > 0$, $b \neq 1$, and u and v are positive numbers, then

$$\log_b \frac{u}{v} = \log_b u - \log_b v$$

Illustration 3

Suppose in the statement of Theorem 2 that b is 2, u is 128, and v is 16. Then

$$\log_b \frac{u}{v} = \log_2 \frac{128}{16} \qquad \log_b u - \log_b v = \log_2 128 - \log_2 16$$
$$= \log_2 \frac{2^7}{2^4} \qquad\qquad\qquad = \log_2 2^7 - \log_2 2^4$$
$$= \log_2 2^{7-4} \qquad\qquad\qquad = 7 - 4$$
$$= \log_2 2^3 \qquad\qquad\qquad\quad = 3$$
$$= 3$$

Hence when b is 2, u is 128, and v is 16, Theorem 2 holds.   ■

Proof of Theorem 2   As in the proof of Theorem 1, let

$$r = \log_b u \qquad \text{and} \qquad s = \log_b v$$

The exponential forms of these equations are, respectively,

$$u = b^r \qquad \text{and} \qquad v = b^s$$

Hence

$$\frac{u}{v} = \frac{b^r}{b^s}$$

$$\frac{u}{v} = b^{r-s}$$

The logarithmic form of this equation is

$$\log_b \frac{u}{v} = r - s$$

Since $\log_b u = r$ and $\log_b v = s$, we have

$$\log_b \frac{u}{v} = \log_b u - \log_b v$$

■

Illustration 4

From Theorem 2 it follows that

$$\log_{10} \frac{3}{2} = \log_{10} 3 - \log_{10} 2$$

Substituting the values of $\log_{10} 3$ and $\log_{10} 2$ given in Illustration 2, we obtain

$$\log_{10} \frac{3}{2} = 0.4771 - 0.3010$$

$$= 0.1761$$

Theorem 3

> If $b > 0$, $b \neq 1$, n is any real number, and u is a positive number, then
>
> $$\log_b u^n = n \log_b u$$

Illustration 5

Suppose in the statement of Theorem 3 that b is 2, n is 3, and u is 4. Then

$$
\begin{aligned}
\log_b u^n &= \log_2 4^3 & n \log_b u &= 3 \log_2 4 \\
&= \log_2 (2^2)^3 & &= 3 \log_2 2^2 \\
&= \log_2 2^{2 \cdot 3} & &= 3 \cdot 2 \\
&= \log_2 2^6 & &= 6 \\
&= 6
\end{aligned}
$$

Thus, when b is 2, n is 3, and u is 4, Theorem 3 is valid. ■

Proof of Theorem 3   Let

$$r = \log_b u \qquad \Leftrightarrow \qquad u = b^r$$

Then

$$u^n = (b^r)^n$$
$$u^n = b^{nr}$$

The logarithmic form of this equation is

$$\log_b u^n = nr$$

Because $\log_b u = r$, we have

$$\log_b u^n = n \log_b u \qquad\qquad ■$$

Illustration   6

Because $\log_{10} 2 = 0.3010$, it follows from Theorem 3 that

$$
\begin{aligned}
\log_{10} 32 &= \log_{10} 2^5 & \log_{10} \sqrt[3]{2} &= \log_{10} 2^{1/3} \\
&= 5 \log_{10} 2 & &= \tfrac{1}{3} \log_{10} 2 \\
&= 5(0.3010) & &= \tfrac{1}{3}(0.3010) \\
&= 1.5050 & &= 0.1003
\end{aligned}
$$

■

Example   1   Express each of the following in terms of logarithms of x, y, and z, each of which represents a positive number.

a) $\log_b x^2 y^3 z^4$ b) $\log_b \dfrac{x}{yz^2}$ c) $\log_b \sqrt[5]{\dfrac{xy^2}{z^3}}$

Solution   a) By Theorem 1,

$$
\log_b x^2 y^3 z^4 = \log_b x^2 + \log_b y^3 + \log_b z^4
$$

Applying Theorem 3 to each of the logarithms on the right side, we obtain

$$
\log_b x^2 y^3 z^4 = 2 \log_b x + 3 \log_b y + 4 \log_b z
$$

b) From Theorem 2

$$
\log_b \frac{x}{yz^2} = \log_b x - \log_b yz^2
$$

Applying Theorems 1 and 3 to the second logarithm on the right side, we have

$$
\log_b \frac{x}{yz^2} = \log_b x - (\log_b y + \log_b z^2)
$$

$$
= \log_b x - \log_b y - 2 \log_b z
$$

c) From Theorem 3

$$
\log_b \sqrt[5]{\frac{xy^2}{z^3}} = \frac{1}{5} \log_b \frac{xy^2}{z^3}
$$

Applying Theorem 2 to the right side, we obtain

$$
\log_b \sqrt[5]{\frac{xy^2}{z^3}} = \frac{1}{5} (\log_b xy^2 - \log_b z^3)
$$

$$
= \frac{1}{5} (\log_b x + \log_b y^2 - \log_b z^3)
$$

$$
= \frac{1}{5} (\log_b x + 2 \log_b y - 3 \log_b z)
$$

$$
= \frac{1}{5} \log_b x + \frac{2}{5} \log_b y - \frac{3}{5} \log_b z
$$

■

Example 2 Write each of the following expressions as a single logarithm with a coefficient of 1.

a) $\log_b x + 2 \log_b y - 3 \log_b z$

b) $\frac{1}{3}(\log_b 4 - \log_b 3 + 2 \log_b x - \log_b y)$

Solution

a) $\log_b x + 2 \log_b y - 3 \log_b z = (\log_b x + \log_b y^2) - \log_b z^3$

$$= \log_b xy^2 - \log_b z^3$$

$$= \log_b \frac{xy^2}{z^3}$$

b) $\dfrac{1}{3}(\log_b 4 - \log_b 3 + 2 \log_b x - \log_b y)$

$$= \frac{1}{3}[(\log_b 4 + \log_b x^2) - (\log_b 3 + \log_b y)]$$

$$= \frac{1}{3}[\log_b 4x^2 - \log_b 3y]$$

$$= \frac{1}{3} \log_b \frac{4x^2}{3y}$$

$$= \log_b \sqrt[3]{\frac{4x^2}{3y}} \qquad \blacksquare$$

Example 3 Given

$$\log_{10} 2 = 0.3010 \qquad \log_{10} 3 = 0.4771 \qquad \log_{10} 7 = 0.8451$$

use the properties of logarithms from Theorems 1, 2, and 3 to find the value of each of the following logarithms.

a) $\log_{10} 5$ b) $\log_{10} 28$

c) $\log_{10} 2100$ d) $\log_{10} \sqrt[3]{4.2}$

Solution In addition to the given logarithms, we can easily determine the logarithm with base 10 of any integer power of 10; for instance, $\log_{10} 10 = 1$, $\log_{10} 10^2 = 2$, $\log_{10} 10^3 = 3$, $\log_{10} 10^{-1} = -1$, and so on.

a) $\log_{10} 5 = \log_{10} \frac{10}{2}$

$$= \log_{10} 10 - \log_{10} 2$$

$$= 1 - 0.3010$$

$$= 0.6990$$

b) $\log_{10} 28 = \log_{10} 2^2 \cdot 7$

$\qquad = \log_{10} 2^2 + \log_{10} 7$

$\qquad = 2 \log_{10} 2 + \log_{10} 7$

$\qquad = 2(0.3010) + 0.8451$

$\qquad = 0.6020 + 0.8451$

$\qquad = 1.4471$

c) $\log_{10} 2100 = \log_{10} 3 \cdot 7 \cdot 10^2$

$\qquad = \log_{10} 3 + \log_{10} 7 + \log_{10} 10^2$

$\qquad = 0.4771 + 0.8451 + 2$

$\qquad = 3.3222$

d) $\log_{10} \sqrt[3]{4.2} = \log_{10} \left(\dfrac{2 \cdot 3 \cdot 7}{10} \right)^{1/3}$

$\qquad = \dfrac{1}{3} (\log_{10} 2 + \log_{10} 3 + \log_{10} 7 - \log_{10} 10)$

$\qquad = \dfrac{1}{3} (0.3010 + 0.4771 + 0.8451 - 1)$

$\qquad = \dfrac{1}{3} (0.6232)$

$\qquad = 0.2077$

■

Example   4   Use the values of $\log_{10} 2$ and $\log_{10} 7$ given in Example 3 to find the value of each of the following.

a) $\log_{10} \dfrac{7}{2}$

b) $\dfrac{\log_{10} 7}{\log_{10} 2}$

Solution

a) $\log_{10} \dfrac{7}{2} = \log_{10} 7 - \log_{10} 2$ b) $\dfrac{\log_{10} 7}{\log_{10} 2} = \dfrac{0.8451}{0.3010}$

$\qquad = 0.8451 - 0.3010$ $\qquad\qquad\qquad = 2.808$

$\qquad = 0.5441$

■

Compare the computations in parts (a) and (b) of Example 4. In part (a) there is the logarithm of a quotient, which upon applying Theorem 2 is the difference of two logarithms. In part (b) we have the quotient of the logarithms of two numbers. The computation is performed by dividing 0.8451 by 0.3010.

EXERCISES 5.4

In Exercises 1 through 8, express the logarithm in terms of logarithms of x, y, and z, where the variables represent positive numbers.

1. (a) $\log_b(5xy)$; (b) $\log_b\left(\dfrac{y}{z}\right)$; (c) $\log_b\left(\dfrac{x}{yz}\right)$

2. (a) $\log_b(3xyz)$; (b) $\log_b\left(\dfrac{xy}{z}\right)$; (c) $\log_b(x^4y^2)$

3. (a) $\log_b(xy^5)$; (b) $\log_b\sqrt{xy}$

4. (a) $\log_b(z^{1/3})$; (b) $\log_b\sqrt[3]{yz^2}$

5. (a) $\log_b(x^{1/3}z^3)$; (b) $\log_b\left(\dfrac{xy^{1/2}}{z^4}\right)$

6. (a) $\log_b(x^2y^3z)$; (b) $\log_b\left(\dfrac{y^2}{x^5z^{1/4}}\right)$

7. (a) $\log_b\sqrt[3]{\dfrac{x^2}{yz^2}}$; (b) $\log_b(\sqrt[3]{x^2}\sqrt{yz})$

8. (a) $\log_b\sqrt[5]{\dfrac{x^3y^4}{z^2}}$; (b) $\log_b(\sqrt[4]{xy^3}\sqrt{z})$

In Exercises 9 through 12, write the expression as a single logarithm with a coefficient of 1.

9. (a) $4\log_{10}x + \tfrac{1}{2}\log_{10}y$;
   (b) $\tfrac{3}{4}\log_b x - 6\log_b y - \tfrac{4}{5}\log_b z$

10. (a) $5\log_{10}x + \tfrac{1}{2}\log_{10}y - \tfrac{1}{3}\log_{10}z$;
    (b) $\tfrac{2}{3}\log_b x - 4\log_b y + \log_b z$

11. (a) $\log_{10}g + 2\log_{10}t - \log_{10}2$;
    (b) $\ln\pi + \ln h + 2\ln r - \ln 3$

12. (a) $\log_{10}4 + \log_{10}\pi + 3\log_{10}r - \log_{10}3$;
    (b) $\ln 2 + \ln\pi + \tfrac{1}{2}\ln t - \tfrac{1}{2}\ln g$

In Exercises 13 through 20, determine the value of the logarithm by using the following: $\log_{10}2 = 0.3010$, $\log_{10}3 = 0.4771$, and $\log_{10}7 = 0.8451$.

13. (a) $\log_{10}14$; (b) $\log_{10}15$

14. (a) $\log_{10}18$; (b) $\log_{10}42$

15. (a) $\log_{10}63$; (b) $\log_{10}140$

16. (a) $\log_{10}120$; (b) $\log_{10}0.21$

17. (a) $\log_{10}\sqrt[3]{10.5}$; (b) $\log_{10}\left(\dfrac{\sqrt[5]{49}}{36^2}\right)$

18. (a) $\log_{10}\sqrt[3]{126}$; (b) $\log_{10}\left(\dfrac{14}{\sqrt[3]{84}}\right)$

19. (a) $\log_{10}\dfrac{2}{3}$; (b) $\dfrac{\log_{10}2}{\log_{10}3}$

20. (a) $\log_{10}\dfrac{7}{5}$; (b) $\dfrac{\log_{10}7}{\log_{10}5}$

In Exercises 21 through 24, determine the value of the logarithm by using the following: $\ln 2 = 0.6931$, $\ln 3 = 1.0986$, and $\ln 5 = 1.6094$.

21. (a) $\ln 300$; (b) $\ln 7.5$

22. (a) $\ln 90$; (b) $\ln 1.2$

23. (a) $\ln\left(\dfrac{\sqrt{10}}{\sqrt[3]{3}}\right)$; (b) $\ln\dfrac{5}{3}e^2$

24. (a) $\ln\left(\dfrac{\sqrt[3]{5}}{\sqrt{6}}\right)$; (b) $\ln\dfrac{2}{5}e^3$

In Exercises 25 through 27, f is an exponential function; that is, $f(x) = b^x$. Use laws of exponents to prove the equality.

25. $f(x + y) = f(x) \cdot f(y)$ 26. $f(x - y) = \dfrac{f(x)}{f(y)}$

27. $f(nx) = [f(x)]^n$

In Exercises 28 through 30, g is a logarithmic function; that is, $g(x) = \log_b x$. Use properties of logarithms to prove the equality.

28. $g(xy) = g(x) + g(y)$ 29. $g\left(\dfrac{x}{y}\right) = g(x) - g(y)$

30. $g(x^n) = ng(x)$

5.5   Exponential and Logarithmic Equations

In Section 5.4 we learned that $\log_b u^n = n\log_b u$. Because of this fact, logarithms are applied to solve an **exponential equation,** one in which a variable occurs in an exponent. Since we express numbers in decimal notation, logarithms with the base 10, called **common logarithms,** are often used for this purpose. When writing common logarithms, it is customary to omit the subscript 10. Thus $\log x$ is understood to represent the number $\log_{10}x$, and the function "log" denotes the logarithmic function with base 10. Hence

$$\log x = y \quad \Leftrightarrow \quad 10^y = x$$

Illustration 1

$$\begin{array}{lll}
\log 10 = 1 & \text{because} & 10^1 = 10 \\
\log 100 = 2 & \text{because} & 10^2 = 100 \\
\log 1000 = 3 & \text{because} & 10^3 = 1000 \\
\log 10{,}000 = 4 & \text{because} & 10^4 = 10{,}000
\end{array}$$

and so on. Furthermore,

$$\begin{array}{lll}
\log 1 = 0 & \text{because} & 10^0 = 1 \\
\log 0.1 = -1 & \text{because} & 10^{-1} = 0.1 \\
\log 0.01 = -2 & \text{because} & 10^{-2} = 0.01 \\
\log 0.001 = -3 & \text{because} & 10^{-3} = 0.001 \\
\log 0.0001 = -4 & \text{because} & 10^{-4} = 0.0001
\end{array}$$

and so on. ∎

Common logarithms can be found on a scientific calculator by using the key labeled log .

To solve an exponential equation, consider the equivalent equation obtained by equating the common (or natural) logarithms of the two sides and then solving the resulting equation, as demonstrated in the following illustration.

Illustration 2

The exponential equation

$$3^x = 16$$

is equivalent to the equation

$$x = \log_3 16$$

To obtain a real number value for x we equate the common logarithms of the two sides of the given equation and get

$$\log 3^x = \log 16$$
$$x \log 3 = \log 16$$
$$x = \frac{\log 16}{\log 3}$$
$$x = \frac{1.2041}{0.47712}$$
$$x = 2.5237$$

Therefore the solution set is {2.5237}. Note that 2.5237 is an approximation, to five significant digits, of the value of x. The exact value of x is given by either $\log_3 16$ or $\dfrac{\log 16}{\log 3}$.

The given equation can also be solved by equating the natural logarithms of the two sides. If this is done, the solution has the following form:

$$\ln 3^x = \ln 16$$

$$x \ln 3 = \ln 16$$

$$x = \frac{\ln 16}{\ln 3}$$

$$x = \frac{2.7726}{1.0986}$$

$$x = 2.5237 \qquad \blacksquare$$

Example 1 Find the solution set of the equation

$$5^{3x-1} = 0.08$$

Solution Equating the common logarithms of both sides of the given equation, we have

$$\log 5^{3x-1} = \log 0.08$$

$$(3x - 1) \log 5 = \log 0.08$$

$$3x \log 5 - \log 5 = \log 0.08$$

$$3x \log 5 = \log 0.08 + \log 5$$

$$x = \frac{\log 0.08 + \log 5}{3 \log 5}$$

$$x = \frac{-1.0969 + 0.6990}{3(0.6990)}$$

$$x = \frac{-0.3979}{2.0970}$$

$$x = -0.1897$$

Thus the solution set is {−0.1897}. $\qquad \blacksquare$

Example 2 Find the solution set of the equation

$$7^x = 3^{x+1}$$

Solution We equate the common logarithms of both sides of the given equation and get

$$\log 7^x = \log 3^{x+1}$$
$$x \log 7 = (x + 1) \log 3$$
$$x \log 7 = x \log 3 + \log 3$$
$$x \log 7 - x \log 3 = \log 3$$
$$x(\log 7 - \log 3) = \log 3$$
$$x = \frac{\log 3}{\log 7 - \log 3}$$
$$x = \frac{0.4771}{0.8451 - 0.4771}$$
$$x = \frac{0.4771}{0.3680}$$
$$x = 1.296$$

Thus the solution set is {1.296}.   ■

The logarithm of a number to any base can be found by solving an exponential equation. The next example shows the method.

Example   3   Find the value of $\log_4 19$.

Solution   Let

$$y = \log_4 19$$

Writing this equation in exponential form, we have

$$4^y = 19$$

Therefore

$$\log 4^y = \log 19$$
$$y \log 4 = \log 19$$
$$y = \frac{\log 19}{\log 4}$$
$$y = \frac{1.2788}{0.6021}$$
$$y = 2.124$$

Thus, to four significant digits, $\log_4 19 = 2.124$.   ■

The following two examples and Exercises 29 through 42 give applications of exponential equations.

Example 4 Suppose on January 1, 1988, the population of a certain city was 800,000. From then until the year 2000 the population is expected to increase at the rate of 3.5 percent per year. Therefore t years after January 1, 1988, the population is expected to be $800{,}000(1.035)^t$, where $0 \le t \le 12$. When would you predict the population will be one million?

Solution We wish to determine the value of t such that

$$800{,}000(1.035)^t = 1{,}000{,}000$$

$$(1.035)^t = \frac{1{,}000{,}000}{800{,}000}$$

$$(1.035)^t = 1.25$$

Because the variable t appears in the exponent, we equate the logarithms of the two sides and obtain

$$\log(1.035)^t = \log 1.25$$

$$t \log 1.035 = \log 1.25$$

$$t = \frac{\log 1.25}{\log 1.035}$$

$$t = \frac{0.0969}{0.0149}$$

$$t = 6.5$$

Six and one-half years from January 1, 1988 is July 1, 1994, which is when the population will be one million. ∎

Example 5 If $1000 is deposited in a savings account that pays an annual interest rate of 6 percent compounded quarterly and no withdrawals or additional deposits are made, how long will it take until there is $1500 on deposit?

Solution From Theorem 3 of Section 5.1,

$$A_n = P\left(1 + \frac{i}{m}\right)^n$$

where A_n dollars is the amount at the end of n interest periods of an investment of P dollars at an annual interest rate of $100i$ percent, compounded m times per year. In this problem $A_n = 1500$, $P = 1000$, $i = 0.06$, and $m = 4$.

We wish to find n. From the formula,

$$1500 = 1000\left(1 + \frac{0.06}{4}\right)^n$$

$$1.5 = (1.015)^n$$

$$\log 1.5 = \log(1.015)^n$$

$$\log 1.5 = n \log 1.015$$

$$n = \frac{\log 1.5}{\log 1.015}$$

$$n = \frac{0.1761}{0.006466}$$

$$n = 27.23$$

Because n is the number of interest periods, and interest is compounded quarterly, it will take 28 quarters until there is $1500 on deposit. ∎

Compare the above example with Example 4 in Section 5.3, which involves the same data except that interest is compounded continuously instead of quarterly.

An equation involving logarithms is called a **logarithmic equation.** Because the domain of a logarithmic function is restricted so that logarithms of positive numbers are obtained, you must *check* any possible solution in the given equation. The next three examples involve the solution of a logarithmic equation.

Example   6   Find the solution set of the equation

$$\log_{10}(x + 3) = 2$$

Solution   The exponential form of the given equation is

$$x + 3 = 10^2$$
$$x = 100 - 3$$
$$x = 97$$

For this value of x, the given equation becomes $\log_{10} 100 = 2$, which is true. Therefore the solution set is $\{97\}$. ∎

Example   7   Find the solution set of the equation

$$\log_2(x + 4) - \log_2(x - 3) = 3$$

Solution   Because the difference of two logarithms is the logarithm of a quotient, we have

$$\log_2 \frac{x+4}{x-3} = 3$$

Writing this equation in the equivalent exponential form, we get

$$\frac{x+4}{x-3} = 2^3$$
$$x + 4 = 8(x-3)$$
$$x + 4 = 8x - 24$$
$$-7x = -28$$
$$x = 4$$

Replacing x by 4 on the left side of the given equation, we obtain $\log_2 8 - \log_2 1$. Because $\log_2 8 = 3$ and $\log_2 1 = 0$, the solution checks. Thus the solution set is $\{4\}$. ∎

Example   8   Find the solution set of the equation

$$\log_3 x + \log_3(2x - 3) = 3$$

Solution   The left side of the given equation is the sum of two logarithms; writing this as the logarithm of a product, we have

$$\log_3 x(2x - 3) = 3$$

The exponential form of this equation is

$$2x^2 - 3x = 3^3$$
$$2x^2 - 3x - 27 = 0$$
$$(x + 3)(2x - 9) = 0$$
$$x + 3 = 0 \qquad 2x - 9 = 0$$
$$x = -3 \qquad\qquad x = \tfrac{9}{2}$$

When $x = -3$, neither $\log_3 x$ nor $\log_3(2x - 3)$ is defined; hence we reject the root -3. When $x = \tfrac{9}{2}$, the left side of the given equation is

$$\log_3 \tfrac{9}{2} + \log_3 6 = \log_3 \left(\tfrac{9}{2} \cdot 6\right)$$
$$= \log_3 27$$
$$= 3$$

Therefore the solution set is $\{\tfrac{9}{2}\}$. ∎

Example   9   Find the solution set of the equation

$$3^x - 3^{-x} = 4$$

Solution   The given equation may be written as

$$3^x - \frac{1}{3^x} = 4$$

$$3^{2x} - 1 = 4(3^x)$$

$$(3^x)^2 - 4(3^x) - 1 = 0$$

By letting $u = 3^x$, we obtain the quadratic equation

$$u^2 - 4u - 1 = 0$$

We solve this equation by the quadratic formula.

$$u = \frac{4 \pm \sqrt{16 + 4}}{2}$$

$$= \frac{4 \pm 2\sqrt{5}}{2}$$

$$= 2 \pm \sqrt{5}$$

Replacing u by 3^x, we get the two equations

$$3^x = 2 + \sqrt{5} \qquad \text{and} \qquad 3^x = 2 - \sqrt{5}$$

The second equation has no solution because $3^x > 0$, and $2 - \sqrt{5} < 0$. Therefore

$$3^x = 2 + \sqrt{5}$$

Equating the common logarithms of the two members of this equation, we have

$$\log 3^x = \log(2 + \sqrt{5})$$

$$x \log 3 = \log(2 + \sqrt{5})$$

$$x = \frac{\log(2 + \sqrt{5})}{\log 3}$$

$$x = \frac{\log(2 + 2.236)}{\log 3}$$

$$x = \frac{\log 4.236}{\log 3}$$

$$x = \frac{0.6270}{0.4771}$$

$$x = 1.314$$

Therefore the solution set is $\{1.314\}$.   ■

EXERCISES 5.5

In Exercises 1 through 12, find the solution set of the equation. Express the results to four significant digits.

1. $4^x = 7$
2. $3^x = 25$
3. $5^{2x} = 4$
4. $100^x = 65$
5. $3^{2+x} = 5^x$
6. $10^{3x-2} = 37$
7. $3^{x+1} = 4^{x-1}$
8. $3^{2x+1} = 5^{3x-1}$
9. $(1.02)^x = 1.892$
10. $(1.04)^x = 0.932$
11. $e^{3x} = 21$
12. $10^{3x} = 57$

In Exercises 13 through 20, find the value of the logarithm to four significant digits.

13. $\log_3 12$
14. $\log_5 200$
15. $\log_2 18$
16. $\log_6 54$
17. $\log_4 155$
18. $\log_8 28$
19. $\log_{100} 75$
20. $\log_{20} 100$

In Exercises 21 through 28, find the solution set of the equation.

21. $\log_5(4x - 3) = 2$
22. $\log_2(2 - 3x) = -3$
23. $\log_{10} x + 3 \log_{10} 2 = 3$
24. $\log_{10} x + \log_{10}(x + 15) = 2$
25. $\log_3(x + 6) - \log_3(x - 2) = 2$
26. $\log_2(11 - x) = \log_2(x + 1) + 3$
27. $\log_2(x + 1) + \log_2(3x - 5) = \log_2(5x - 3) + 2$
28. $\log_3(2x - 1) - \log_3(5x + 2) = \log_3(x - 2) - 2$
29. For the city of Example 4, when would you predict the population will be 900,000?
30. Would you expect the population of the city of Example 4 to reach 1,200,000 before the year 2000? Prove your answer.
31. How long will it take $1000 to triple itself if it is earning interest at an annual rate of 6 percent compounded semiannually?
32. How long will it take an investment to double itself if the annual interest rate is 8 percent compounded quarterly?
33. Refer to Exercise 25 of Exercises 5.2. At what height is the atmospheric pressure 500 lb/ft^2?
34. For the town in Exercise 28 of Exercises 5.2, when is the population expected to be 21,000?
35. When is the value of the abstract painting in Exercise 29 of Exercises 5.2 expected to be $18,000?

36. What will be the age of the man in Exercise 12 of Exercises 5.1 when there is $50,000 in the investment account?
37. It is determined statistically that the population of a certain city t years from now will be $f(t)$, where $f(t) = 40,000e^{kt}$ and k is a constant. If the population is expected to be 60,000 in 40 years, when is it expected to be 80,000?
38. After how many hours of practice typing can the person in Exercise 30 of Exercises 5.2 type 60 words per minute?
39. After how many months on the job can the worker in Exercise 31 of Exercises 5.2 complete 80 units per day?
40. There was a time when a United States government bond sold at $74 to be redeemed 12 years later at a maturity value of $100. Determine the annual rate of interest, compounded monthly, that was earned.
41. In a certain speculative investment, a piece of real estate was purchased 3 years ago for $20,000 and sold today for $100,000. What is the annual rate of interest, compounded monthly, that has been earned?
42. A simple electric circuit containing no condensers, a resistance of R ohms, and an inductance of L henrys has the electromotive force cut off when the current is I_0 amperes. The current dies down so that at t sec the current is i amperes, and $i = I_0 e^{-(R/L)t}$. Use natural logarithms to solve this equation for t in terms of i and the constants R, L, and I_0.

In Exercises 43 through 48, find the solution set of the equation. Express the results to four significant digits.

43. $10^x - 10^{-x} = 2$
44. $10^x + 10^{-x} = 4$
45. $4^x - 4^{-x} = 3$
46. $5^x - 5^{-x} = 8$
47. $\frac{1}{2}(e^x + e^{-x}) = 4$
48. $\frac{1}{2}(e^x - e^{-x}) = 3$

In Exercises 49 and 50, solve for x in terms of y.

49. $y = \dfrac{10^x - 10^{-x}}{10^x + 10^{-x}}$

50. $y = \dfrac{e^x + e^{-x}}{e^x - e^{-x}}$

REVIEW EXERCISES FOR CHAPTER 5

In Exercises 1 through 26, draw a sketch of the graph of the function.

1. $f(x) = 6^x$
2. $g(x) = 5^x$
3. $F(x) = 6^{-x}$
4. $G(x) = 5^{-x}$
5. $g(x) = 2^{x/2}$
6. $f(x) = 3^{x/3}$
7. $G(x) = 2^{-x/2}$
8. $F(x) = 3^{-x/3}$
9. $f(x) = 3e^x$
10. $g(x) = 2e^{-x}$
11. $g(x) = -e^{-x}$
12. $f(x) = -2e^{3x}$
13. $G(x) = 4e^{-2x}$
14. $F(x) = 5e^{-0.1x}$
15. $f(x) = e^{x-2}$
16. $g(x) = e^{x+1}$
17. $F(x) = e^x - 3$
18. $G(x) = e^x + 1$
19. $f(x) = 2 \ln x$
20. $g(x) = 4 \log_2 x$
21. $g(x) = -\frac{1}{2} \log_{10} x$
22. $f(x) = 3 \ln(-x)$
23. $f(x) = 3 - \ln x$
24. $g(x) = \ln(x + 2)$
25. $g(x) = 3 \ln(x - 2)$
26. $f(x) = 2 \ln x + 1$

In Exercises 27 through 30, solve the equation for x, y, or b.

27. (a) $\log_5 x = 4$; (b) $\log_8 16 = y$
28. (a) $\log_9 x = -\frac{3}{2}$; (b) $\log_{27} 81 = y$
29. (a) $\log_b 4 = -\frac{1}{3}$; (b) $\ln x = -2$
30. (a) $\log_b 256 = \frac{4}{3}$; (b) $\ln x = \frac{3}{4}$

In Exercises 31 through 34, express the logarithm in terms of logarithms of x, y, and z, where the variables represent positive numbers.

31. $\log_b (x^3 y^2 \sqrt{z})$
32. $\log_b (\sqrt[3]{xyz^5})$
33. $\log_b \left(\dfrac{x\sqrt[3]{y}}{z^4} \right)$
34. $\log_b \sqrt[5]{\dfrac{x^4 y^2}{z^3}}$

In Exercises 35 through 38, write the expression as a single logarithm with a coefficient of 1.

35. $\frac{2}{3} \log_{10} y - 4 \log_{10} x - \frac{1}{3} \log_{10} z$
36. $\ln k + 4 \ln L - \ln b - 3 \ln d$
37. $\ln 4 + \ln \pi + 2 \ln r + \ln h - \ln 3$
38. $\log_b 3 + \frac{1}{2} \log_b x + 3 \log_b z - \frac{1}{3} \log_b y - \log_b 2$

In Exercises 39 through 46, find the solution set of the equation. Express the results to three significant digits.

39. $5^x = 26$
40. $3^{x-2} = 8$
41. $(21.6)^x = 104$
42. $e^{-4x} = 0.231$
43. $e^{x/3} = 14.8$
44. $8^x - 8^{-x} = 8$
45. $2^x + 2^{-x} = 6$
46. $\dfrac{e^x - e^{-x}}{e^x + e^{-x}} = \dfrac{1}{2}$

In Exercises 47 through 50, find the value of the logarithm to four significant digits.

47. $\log_8 7$
48. $\log_7 100$
49. $\log_2 38$
50. $\log_5 e$

In Exercises 51 through 54, find the solution set of the equation.

51. $\log_4(2x + 3) - 2 \log_4 x = 2$
52. $\log_3(2x - 3) + \log_3(x + 3) = 4$
53. $\log_{10} x + \log_{10}(x - 200) - \log_{10} 4 = 5 - \log_{10} 5$
54. $\log_2(x + 2) - 3 = \log_2 3 - \log_2 x$
55. Given that $\ln i = \ln I - \dfrac{Rt}{L}$, show that $i = Ie^{-RT/L}$.
56. A loan of $1000 at an annual interest rate of 16 percent is repaid in one payment at the end of a year. Find the total amount repaid, given that
   (a) simple interest is earned;
   (b) interest is compounded quarterly;
   (c) interest is compounded monthly.
57. An investment of $8000 earns interest at the rate of 12 percent per year, and the interest is paid once at the end of the year. Find the interest for the first year, given that
   (a) simple interest is earned;
   (b) interest is compounded semiannually;
   (c) interest is compounded quarterly.
58. Do Exercise 56, given that interest is compounded continuously.
59. Do Exercise 57, given that interest is compounded continuously.
60. An amount of $500 is deposited in a savings account and earns interest for 7 years at a rate of 6 percent. There are no withdrawals or additional deposits. How much is in the account after 7 years if
   (a) interest is compounded quarterly;
   (b) interest is compounded continuously?
61. A house purchased 10 years ago was sold for $100,000. If the annual interest rate was determined to be 20 percent compounded quarterly, what was the purchase price of the house to the nearest thousand dollars?
62. Interest on a savings account is computed at 8 percent per year compounded continuously. If one wishes to have $1000 in the account at the

end of a year by making a single deposit now, what should the amount of the deposit be?

63. How long will it take for an investment to double itself if interest is earned at the rate of 12 percent per year compounded (a) quarterly and (b) continuously?

64. How long will it take for a deposit of $500 in a savings account to accumulate to $600 if interest is computed at 8 percent per year compounded (a) quarterly and (b) continuously?

65. If A milligrams of radium are present after t years, then $A = ke^{-0.0004t}$, where k is a constant. Furthermore, 60 mg of radium are present now.
   (a) How much radium will be present 100 years from now?
   (b) How long will it take until there are only 50 mg of radium present?

66. In t minutes there will be $f(t)$ bacteria present in a certain culture, where $f(t) = ke^{0.03t}$ and k is a constant. If 60,000 bacteria are present initially,
   (a) how many will be present in 15 min, and
   (b) after how many minutes will there be 200,000 present?

The functions S and C in Exercises 67 through 72 are important in calculus. They are called *hyperbolic functions* and are defined by

$$S(x) = \tfrac{1}{2}(e^x - e^{-x}) \qquad \text{and} \qquad C(x) = \tfrac{1}{2}(e^x + e^{-x})$$

In Exercises 67 and 68, determine the indicated function values.

67. (a) $S(0)$; (b) $C(1)$;
   (c) $S(-1)$; (d) $S(3.5)$;
   (e) $C(-2)$

68. (a) $C(0)$; (b) $S(1)$;
   (c) $C(3)$; (d) $S(-2.5)$;
   (e) $C(-2.5)$

69. (a) Prove that S is an odd function. (b) Draw a sketch of the graph of S.

70. (a) Prove that C is an even function. (b) Draw a sketch of the graph of C.

71. Find $S^{-1}(x)$. (*Hint:* Let $y = S(x)$, write the equation as quadratic in e^x, and solve for e^x by the quadratic formula.)

72. Find $C^{-1}(x)$. Use a hint similar to that for Exercise 71.

CHAPTER 6

Trigonometric Functions

Introduction

We now continue the discussion of transcendental (nonalgebraic) functions that we began in Chapter 5 with the exponential and logarithmic functions. Here we introduce the trigonometric functions.

In Greek, the word *trigónon* means triangle and *metron* means measure. Thus *trigonometry* means measurement of triangles. It was used in ancient times in surveying, navigation, and astronomy to find relationships between the lengths of sides of triangles and measurement of angles. It is still used for these purposes, and in such instances, the trigonometric functions have angle measurements as their domains. But beyond that use, in modern times there are applications of trigonometry that involve periodically repetitive phenomena such as wave motion, alternating electric current, vibrating strings, oscillating pendulums, business cycles, and biological rhythms. These applications of trigonometry utilize a functional relationship between sets of real numbers. If you study calculus, you will learn that trigonometric functions of real numbers are important in that subject.

Angles and their measurement are presented in Section 6.1, and we define the six trigonometric functions of angles in Section 6.2. In Section 6.3 we determine values of these functions for any angle. The sine and cosine functions of real numbers are introduced in Section 6.4, and sketches of their graphs are obtained in Section 6.5. These graphs are called sine waves. Other sine waves are also discussed in Section 6.5. The other four trigonometric functions of real numbers appear along with their graphs in Section 6.6. We present applications pertaining to the solution of

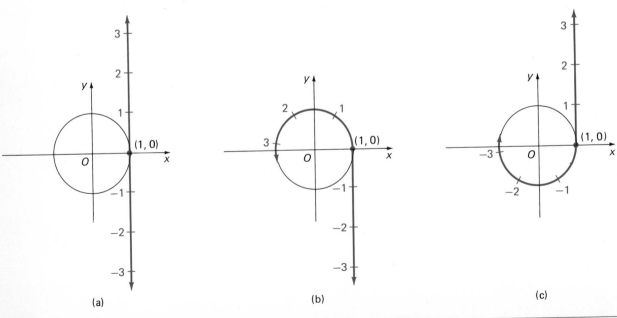

(a)   (b)   (c)

FIGURE 1

288

right triangles in Section 6.7. Simple harmonic motion provides an application of sine waves, and this topic appears in Section 6.8. In Supplementary Section 6.9 we are concerned with more graphs involving sine and cosine functions.

6.1 Angles and Their Measurement

Before discussing *angles* we introduce the concept of the *length of arc* on the unit circle; later we shall use this idea to define the measurement of an angle.

In Section 3.2 we learned that the graph of the equation

$$x^2 + y^2 = 1$$

is the unit circle having its center at the origin and radius 1. Let us denote the unit circle by U. We shall show that there is a one-to-one correspondence between the lengths of all arcs of U, starting at the initial point $(1, 0)$, and the elements of the set R of real numbers. We begin by imagining the real number line "wrapped around" U, so that the number 0 (zero) on the real number line coincides with the point $(1, 0)$ on U. See Figure 1(a)–(c).

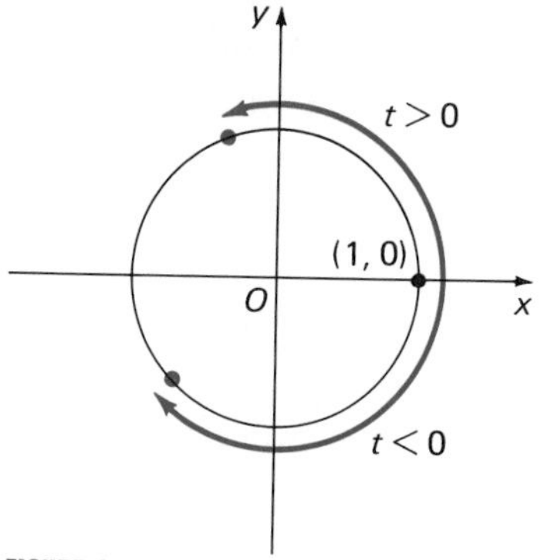

Figure 1(a) shows U and the real number line tangent to U at the point $(1, 0)$. In Figure 1(b), the positive side of the real number line is "wrapped around" U in a counterclockwise sense, and in Figure 1(c) the negative side is "wrapped around" U in a clockwise sense. If we consider an arc with initial point at $(1, 0)$ to have its terminal point in a counterclockwise direction from $(1, 0)$, a positive real number represents the length of arc; if we consider the arc to have its terminal point in a clockwise direction from $(1, 0)$, a negative real number represents the length of arc. See Figure 2.

Because the circumference of a circle is given by $2\pi r$, where r is the measure of the radius, the circumference of U is 2π. Thus the distance one-half of the way around U is π, the distance one-fourth of the way around is $\frac{1}{2}\pi$, the distance one-eighth of the way around is $\frac{1}{4}\pi$, and so on. Figure 3 shows some terminal points of arcs on U where the length of arc is measured in the counterclockwise direction from the point $(1, 0)$. The corresponding arc length in terms of π is indicated in the figure at the terminal point; it is a positive number. Figure 4 shows the same terminal points of arcs of U as in Figure 3, but in this case the length of arc is measured in the clockwise direction from the point $(1, 0)$; therefore the corresponding arc length is a negative number.

A length of arc of U is often given in terms of π. However, when decimals are used, we can approximate π by 3.14 and write $\pi \approx 3.14$. Thus

$$\frac{1}{2}\pi \approx 1.57 \qquad \frac{3}{2}\pi \approx 4.71 \qquad 2\pi \approx 6.28$$
$$\frac{1}{4}\pi \approx 0.79 \qquad -\frac{1}{3}\pi \approx -1.05 \qquad -\frac{3}{4}\pi \approx -2.36$$

and so on.

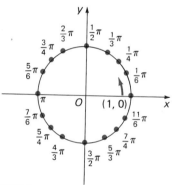

FIGURE 2 _____

FIGURE 3 _____

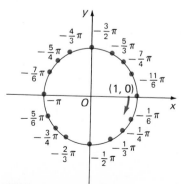

FIGURE 4 _____

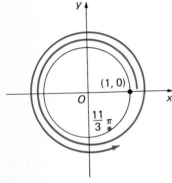

FIGURE 5

Example   **1**   Show by a figure the location on U of the terminal point of the arc having initial point at $(1, 0)$ and having the given arc length; also state the quadrant in which the terminal point lies.

a) $\frac{1}{12}\pi$ b) 2 c) $-\frac{5}{8}\pi$ d) -3

Solution   Refer to Figure 5.

a) Because $0 < \frac{1}{12}\pi < \frac{1}{2}\pi$, the terminal point is in the first quadrant.

b) Because $1.57 < 2 < 3.14$, the terminal point is in the second quadrant.

c) Because $-\pi < -\frac{5}{8}\pi < -\frac{1}{2}\pi$, the terminal point is in the third quadrant.

d) Because $-3.14 < -3 < -1.57$, the terminal point is in the third quadrant.   ■

We have imagined the real number line as "wrapped around" U. Thus, if the length of an arc from $(1, 0)$ is more than 2π or less than -2π, the wrapped part of the number line will traverse more than the circumference of U.

Illustration   1

a) In Figure 6 there is an arc of length $\frac{11}{3}\pi$. Because

$$\tfrac{11}{3}\pi = 2\pi + \tfrac{5}{3}\pi \qquad \text{and} \qquad \tfrac{3}{2}\pi < \tfrac{5}{3}\pi < 2\pi$$

the terminal point of this arc is in the fourth quadrant.

b) Figure 7 shows an arc of length -10.34. Because

$$-10.34 = -6.28 + (-4.06) \qquad \text{and} \qquad -4.71 < -4.06 < -3.14$$

the terminal point of this arc is in the second quadrant.   ■

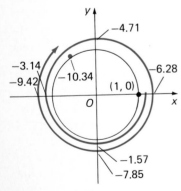

FIGURE 6

FIGURE 7

With t representing the length of an arc of U with initial point at $(1, 0)$, we have demonstrated that t can be any real number.

In geometry an **angle** is thought of as the union of two rays called the **sides,** having a common endpoint called the **vertex.** An angle having its vertex at the origin and having one side, called the **initial side,** lying on the positive side of the x axis is said to be in **standard position.** Any angle is congruent to some angle in standard position. Figure 8 shows an angle AOB in standard position with OA as the initial side. The other side, OB, is called the **terminal side.** The angle AOB can be formed by rotating the side OA to the side OB, and under such a rotation the point A moves to the point B along the circumference of a circle having its center at O and radius $|OA|$. The angle is **positive** if OA is rotated in a counterclockwise direction to OB and **negative** if it is rotated in a clockwise direction.

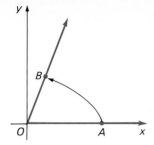

FIGURE 8

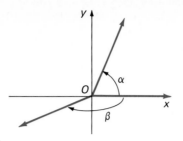

FIGURE 9

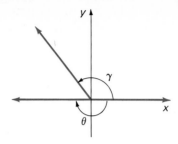

FIGURE 10

Greek letters are often used to represent angles, and the direction of rotation is indicated by an arc with an arrow at its endpoint. Figure 9 shows a positive angle α and a negative angle β. Figure 10 shows a positive angle γ and a negative angle θ.

We consider an angle to be in the quadrant containing its terminal side. However, if the terminal side lies on an axis, the angle is said to be **quadrantal.** In Figure 9, α is in the first quadrant and β is in the third quadrant. In Figure 10, γ is in the second quadrant and θ is a quadrantal angle.

Consider an angle θ in standard position, and let its terminal side intersect the unit circle U at P, the terminal point of the arc of length t measured along U from the point $(1, 0)$. See Figure 11. Because t is a real number, there is a one-to-one correspondence between the set R of real numbers and all angles θ in standard position. Figure 12 shows the angles when t is $\frac{1}{3}\pi$, $\frac{1}{2}\pi$, $-\frac{3}{4}\pi$, and $-\frac{3}{2}\pi$. The number t corresponding to the angle θ is a measure of the size of the angle. Furthermore, when t is positive, the angle has a counterclockwise rotation, and when t is negative, the angle's rotation

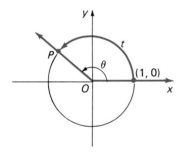

FIGURE 11

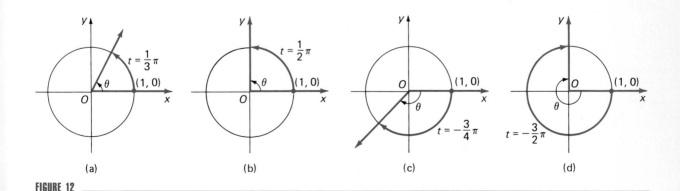

FIGURE 12

is clockwise. The measurement of the angle for which $t = 1$ is called a *radian*. See Figure 13.

Definition

> **Radian**
>
> If an angle has its vertex at the center of the unit circle U and intercepts on U an arc of length 1, the angle has a measurement of 1 **radian.**

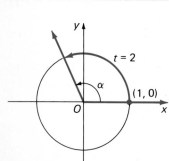

FIGURE 13 _____

Illustration 1

The radian measures of the angles in Figure 12 are $\frac{1}{3}\pi$, $\frac{1}{2}\pi$, $-\frac{3}{4}\pi$, and $-\frac{3}{2}\pi$. ∎

If the measurement of an angle θ is t radians, we write

$$m^R(\theta) = t$$

This equality is sometimes read as "the radian measure of angle θ is t."

Illustration 2

a) If an angle is denoted by α, and we have

$$m^R(\alpha) = 2$$

the equation means that the radian measure of α is 2. Figure 14 shows this angle α that intercepts on U an arc of length 2 and the rotation for α is counterclockwise.

b) If β denotes an angle and

$$m^R(\beta) = -3.25$$

then the radian measure of β is -3.25, β intercepts in U an arc of length -3.25, and the rotation for β is clockwise. See Figure 15. ∎

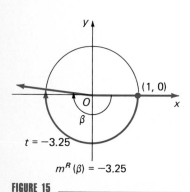

$$m^R(\alpha) = 2$$

FIGURE 14 _____

$$m^R(\beta) = -3.25$$

FIGURE 15 _____

Because the circumference of the unit circle is 2π, one complete revolution of the terminal side from the initial side in the counterclockwise direction generates an angle of radian measure 2π. See Figure 16, where we have designated the angle by writing its radian measure. More than one complete revolution in the counterclockwise direction generates an angle of radian measure greater than 2π. For instance, in Figure 17 is an angle having radian measure $\frac{9}{4}\pi$. Observe that the angle of radian measure $\frac{1}{4}\pi$ has the same terminal side. Another angle having this same terminal side is the one for which the radian measure is $-\frac{7}{4}\pi$. In fact, there are an unlimited number of angles having this terminal side. These angles are called *coterminal*. In general, coterminal angles are those in standard position that have the same terminal side.

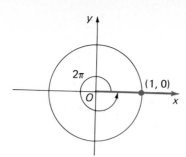

FIGURE 16 _____

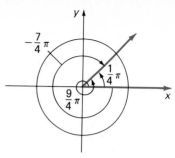

FIGURE 17 _____

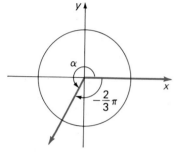

FIGURE 18 _____

Example   2 Find the radian measure of the smallest positive angle that is coterminal with the angle having the given radian measure and draw a sketch showing both angles.

a) $-\frac{2}{3}\pi$ b) $\frac{11}{4}\pi$ c) 7.15   d) -0.54

Solution a) An angle of radian measure $-\frac{2}{3}\pi$ appears in Figure 18 where the angle is designated by its radian measure. In the figure the required angle is designated by α, and $m^R(\alpha) = 2\pi - \frac{2}{3}\pi$. Therefore $m^R(\alpha) = \frac{4}{3}\pi$.

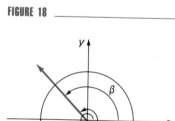

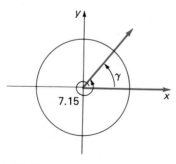

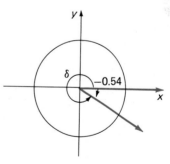

FIGURE 19 _____   FIGURE 20 _____   FIGURE 21 _____

b) Figure 19 shows an angle of radian measure $\frac{11}{4}\pi$. If β is the required angle, $m^R(\beta) = \frac{11}{4}\pi - 2\pi$. Hence $m^R(\beta) = \frac{3}{4}\pi$.

c) Let γ be the required angle. See Figure 20. To the nearest hundredth, 2π is 6.28. Thus $m^R(\gamma) = 7.15 - 6.28$, and so $m^R(\gamma) = 0.87$.

d) See Figure 21. Let δ be the required angle. Then $m^R(\delta) = 6.28 - 0.54$. Therefore $m^R(\delta) = 5.74$. ∎

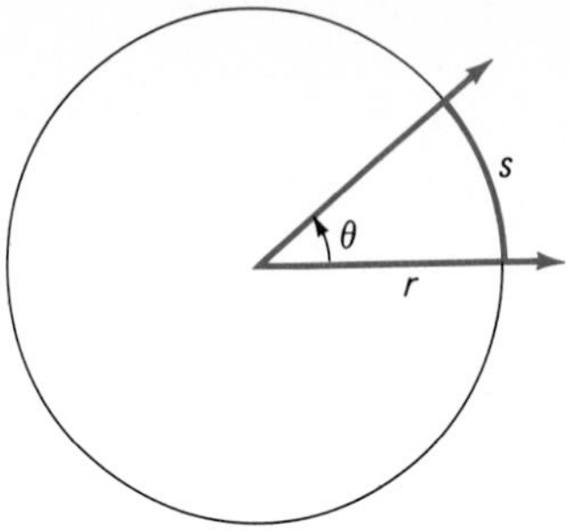

FIGURE 22 _____

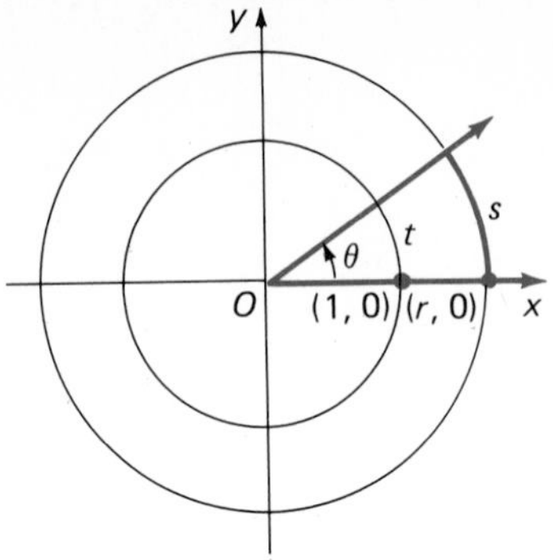

FIGURE 23 _____

A **central angle** of a circle is one whose vertex is at the center of the circle. Figure 22 shows a circle of radius r and a central angle θ. The angle intercepts on the circle an arc of length s. We now proceed to obtain a formula relating r, s, and t, the radian measure of θ.

We construct a rectangular cartesian coordinate system such that θ is in standard position. Assume $r > 1$, so that the unit circle U is within the given circle. See Figure 23. The length of the arc intercepted by θ on U is t. From a theorem in geometry, the ratio of the arc lengths t and s is equal to the ratio of the radii of the two circles. Thus

$$\frac{t}{s} = \frac{1}{r}$$

$$rt = s$$

We have proved the following theorem.

Theorem 1

> If r is the radius of a circle and t is the radian measure of a central angle that intercepts on the circle an arc of length s, then
>
> $$s = rt$$

Illustration 3

If in the formula of Theorem 1, $r = 3$ and $t = 2$, then $s = 6$. This result states that on a circle of radius 3 units, a central angle of 2 radians intercepts on the circle an arc of length 6 units. Refer to Figure 24.   ■

Observe that if $r = 1$ in the formula of Theorem 1, then $s = t$. Therefore on the unit circle U the length of the intercepted arc is the radian measure of the central angle. Also observe that if $t = 1$, the formula be-

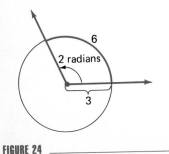

FIGURE 24 _____

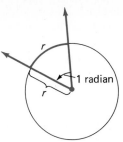
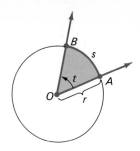

FIGURE 25 _____   **FIGURE 26** _____

comes $s = r$. This equation states that an arc of a circle equal in length to the radius subtends a central angle of 1 radian. See Figure 25.

A **sector** of a circle is the region bounded by an arc of the circle and the sides of a central angle. For the circle in Figure 26, r units is the radius, t is the radian measure of the central angle, s units is the length of the intercepted arc, and the shaded region is the sector AOB. From geometry, the ratio of the area of the sector to the area of the circle (given by πr^2) is equal to the ratio of the length of the intercepted arc to the circumference of the circle (given by $2\pi r$). Thus, if K square units is the area of the sector, then

$$\frac{K}{\pi r^2} = \frac{s}{2\pi r}$$

$$K = \tfrac{1}{2}rs$$

Replacing s by rt (from Theorem 1), we obtain

$$K = \tfrac{1}{2}r^2t$$

Example 3 A circle of radius 6 in. has a sector whose central angle has radian measure $\tfrac{1}{3}\pi$. Find the arc length and the area of the sector.

Solution Refer to Figure 27. We are given $r = 6$ and $t = \tfrac{1}{3}\pi$. With these values

$$s = rt$$
$$= 6(\tfrac{1}{3}\pi)$$
$$= 2\pi$$

Thus the arc length is 2π in.

If K square inches is the area of the sector, then

$$K = \tfrac{1}{2}r^2t$$
$$= \tfrac{1}{2}(6)^2(\tfrac{1}{3}\pi)$$
$$= 6\pi$$

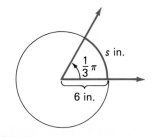

FIGURE 27 _____   Hence the area of the sector is 6π in^2. ∎

The *degree* is another unit of angle measurement. If a circle has a central angle subtended by an arc whose length is $\frac{1}{360}$ of the circumference of the circle, the angle is said to have **degree measure** 1. A measurement of 1 degree is written 1°. See Figure 28. Observe that the definition of degree measure is independent of the radius of the circle. Figure 29 shows some angles in standard position and their measurements in degrees.

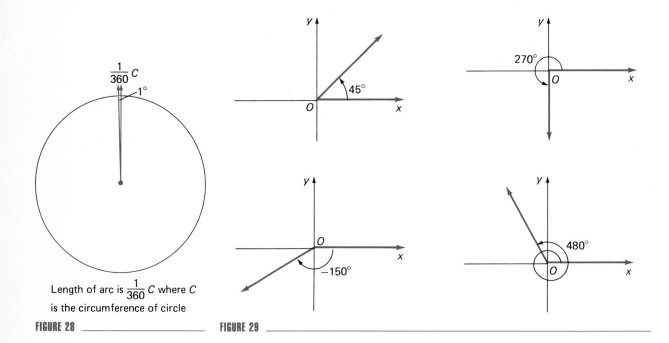

$\frac{1}{360}$ C

Length of arc is $\frac{1}{360}$ C where C is the circumference of circle

FIGURE 28

FIGURE 29

An angle formed by one complete revolution, so that the terminal side coincides with the initial side, has degree measure of 360 and radian measure of 2π. It follows that

$$180° \sim \pi \text{ radians}$$

where the symbol $\sim$ (read "corresponds to") indicates that the given measurements are for congruent angles. We thus have the following correspondence between angle measurements in degrees and those in radians:

$$1° \sim \frac{\pi}{180} \text{ radian} \quad \text{and} \quad 1 \text{ radian} \sim \frac{180°}{\pi}$$

If 3.1416 is taken as an approximation for π, we obtain

$$1° \sim 0.017453 \text{ radians} \quad \text{and} \quad 1 \text{ radian} \sim 57.296°$$

From this correspondence between degrees and radians, the measurement of an angle can be converted from one system of units to the other.

Illustration 4

a) $162° \sim 162 \cdot \dfrac{\pi}{180}$ radians   b) $\dfrac{5\pi}{12}$ radians $\sim \dfrac{5\pi}{12} \cdot \dfrac{180°}{\pi}$

$\qquad 162° \sim \dfrac{9\pi}{10}$ radians $\qquad\qquad\qquad \dfrac{5\pi}{12}$ radians $\sim 75°$ ∎

TABLE 1 ▬▬▬

Degree Measure	Radian Measure
30	$\dfrac{1}{6}\pi$
45	$\dfrac{1}{4}\pi$
60	$\dfrac{1}{3}\pi$
90	$\dfrac{1}{2}\pi$
120	$\dfrac{2}{3}\pi$
135	$\dfrac{3}{4}\pi$
150	$\dfrac{5}{6}\pi$
180	π
270	$\dfrac{3}{2}\pi$
360	2π

Table 1 gives the corresponding degree and radian measures for certain angles.

We use the notation $m°(\theta)$ to indicate the degree measure of an angle θ. It follows from the above discussion that $m°(\theta)$ and $m^R(\theta)$ are related by the equation

$$m°(\theta) = \frac{180}{\pi} m^R(\theta)$$

$\Leftrightarrow$

$$m^R(\theta) = \frac{\pi}{180} m°(\theta)$$

Example 4   Find the degree measure to the nearest hundredth of a degree for the angle having the given radian measure (let $\pi \approx 3.1416$).

a) $m^R(\alpha) = \frac{5}{7}\pi$ b) $m^R(\beta) = 0.3826$

Solution

a) $m°(\alpha) = \dfrac{180}{\pi} \cdot \dfrac{5\pi}{7}$ b) $m°(\beta) = \dfrac{180}{\pi}(0.3826)$

$\qquad = \dfrac{900}{7} \qquad\qquad\qquad\quad \approx 21.92$

$\qquad \approx 128.57$

Therefore angle α has a measurement of 128.57° and angle β has a measurement of 21.92°.   ∎

As in Example 4, we have been using decimals for measurements of less than 1 degree. Another way of dealing with such measurements is to use minutes and seconds. One minute is $\frac{1}{60}$ of a degree; that is, 60 minutes is equivalent to 1 degree. Also, 1 second is $\frac{1}{60}$ of a minute; thus 60 seconds is equivalent to 1 minute. Obviously, 1 second is $\frac{1}{3600}$ of a degree, and 3600 seconds is equivalent to 1 degree. The symbol for minutes is ', and the symbol for seconds is ". Thus $\theta = 26°14'46''$ means that θ is an angle having a measurement of 26 degrees, 14 minutes, and 46 seconds.

Because of the wide use of calculators, we will express angle measurements of less than 1 degree by decimals. If you are given an angle measurement in minutes and seconds, you can convert it easily to a form using decimals. For example, to the nearest hundredth of a degree,

$$26°14'46'' = (26 + \tfrac{14}{60} + \tfrac{46}{3600})°$$
$$= (26 + 0.233 + 0.013)°$$
$$= 26.25°$$

Example 5   Find the distance on the surface of the earth from a point having latitude 38.40° N to the closest point on the equator. Assume that the earth is a sphere of radius 3960 mi.

Solution   Refer to Figure 30, where C is at the center of the earth, P is the given point, and E is the point on the equator closest to P. We wish to find the length of the arc from E to P on the circle with center at C and radius 3960 mi. Let s miles be this length. From Theorem 1 $s = rt$, where $r = 3960$ and t is the radian measure of the angle at C. We first compute t by converting 38.40° to radians:

$$38.40° \sim 38.40 \left(\frac{\pi}{180} \right)$$
$$\approx 0.6702$$

Thus

$$s = rt$$
$$= 3960(0.6702)$$
$$= 2654$$

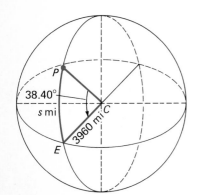

FIGURE 30 _____   Therefore the distance is 2654 mi.   ■

EXERCISES 6.1

In Exercises 1 through 8, show by a figure the location on U of the terminal point of the arc having initial point at $(1, 0)$ and having the given arc length; also state the quadrant in which the terminal point lies.

1. (a) $\tfrac{1}{7}\pi$; (b) $\tfrac{3}{5}\pi$
2. (a) $\tfrac{1}{5}\pi$; (b) $\tfrac{9}{8}\pi$
3. (a) 1.23; (b) 5
4. (a) 3; (b) -0.25
5. (a) $\tfrac{17}{6}\pi$; (b) $-\tfrac{8}{7}\pi$
6. (a) $\tfrac{29}{8}\pi$; (b) $-\tfrac{8}{5}\pi$
7. (a) -2; (b) 12.2
8. (a) 10.6; (b) -4

In Exercises 9 through 14, show by a diagram the angle that has the given radian measure.

9. (a) $\tfrac{1}{6}\pi$; (b) $\tfrac{2}{3}\pi$; (c) π; (d) $\tfrac{5}{4}\pi$; (e) $\tfrac{11}{6}\pi$
10. (a) $\tfrac{1}{4}\pi$; (b) $\tfrac{5}{6}\pi$; (c) $\tfrac{4}{3}\pi$; (d) $\tfrac{3}{2}\pi$; (e) $\tfrac{7}{4}\pi$
11. (a) $-\tfrac{1}{4}\pi$; (b) $-\tfrac{1}{2}\pi$; (c) $-\tfrac{5}{6}\pi$; (d) $-\tfrac{5}{4}\pi$; (e) $-\tfrac{5}{3}\pi$
12. (a) $-\tfrac{1}{6}\pi$; (b) $-\tfrac{3}{4}\pi$; (c) $-\pi$; (d) $-\tfrac{4}{3}\pi$; (e) $-\tfrac{11}{6}\pi$
13. (a) 0.78; (b) 3; (c) -2; (d) 5.20; (e) -4.35
14. (a) 2.35; (b) -1; (c) 4; (d) -1.80; (e) -6.18

In Exercises 15 through 18, find the radian measure of the smallest positive angle that is coterminal with the angle having the given radian measure and draw a sketch showing both angles.

15. (a) $-\frac{3}{4}\pi$; (b) $\frac{19}{6}\pi$; (c) $\frac{7}{2}\pi$; (d) $-\frac{7}{3}\pi$
16. (a) $-\frac{7}{6}\pi$; (b) $\frac{8}{3}\pi$; (c) $\frac{11}{4}\pi$; (d) $-\frac{7}{2}\pi$
17. (a) 7.28; (b) 9; (c) -4.25; (d) -11
18. (a) 8.28; (b) -2.63; (c) -14; (d) 10

In Exercises 19 through 22, find the equivalent radian measurement for the angle having the given degree measurement.

19. (a) 60°; (b) 135°; (c) 210°; (d) $-150°$
20. (a) 45°; (b) 120°; (c) 240°; (d) $-225°$
21. (a) 20°; (b) 450°; (c) $-75°$; (d) 100°
22. (a) 15°; (b) 540°; (c) $-48°$; (d) 2°

In Exercises 23 through 26, find the equivalent degree measurement for the angle having the given radian measurement.

23. (a) $\frac{1}{4}\pi$; (b) $\frac{2}{3}\pi$; (c) $\frac{11}{6}\pi$; (d) $-\frac{1}{2}\pi$
24. (a) $\frac{1}{6}\pi$; (b) $\frac{4}{3}\pi$; (c) $\frac{3}{4}\pi$; (d) -5π
25. (a) $\frac{1}{2}$; (b) -2; (c) 4.78; (d) 0.23
26. (a) $\frac{1}{3}$; (b) 0.2; (c) -2.75; (d) 5.66

In Exercises 27 and 28, convert the angle measurement to a form using decimals to the nearest hundredth of a degree and then find the equivalent radian measurement.

27. (a) 35°22′12″; (b) 102°31′27″
28. (a) 68°53′48″; (b) 251°8′14″

In Exercises 29 through 32, find the degree measurement of the smallest positive angle that is coterminal with the angle having the given degree measurement and draw a sketch showing both angles.

29. (a) $-45°$; (b) 510°; (c) $-540°$; (d) $-120°$
30. (a) $-30°$; (b) 585°; (c) $-240°$; (d) $-630°$
31. (a) 382.56°; (b) $-118.24°$; (c) $-253.85°$; (d) $-302.36°$
32. (a) 496.58°; (b) $-28.16°$; (c) $-342.15°$; (d) $-197.74°$

In Exercises 33 through 38, find (a) the arc length and (b) the area of the sector of the circle having the radius and central angle α.

33. Radius is 9 in.; $m^R(\alpha) = \frac{2}{3}\pi$
34. Radius is 8 cm; $m^R(\alpha) = \frac{1}{4}\pi$
35. Radius is 6 cm; $m°(\alpha) = 135$

36. Radius is 12 in.; $m°(\alpha) = 120$
37. Radius is 4.72 in.; $m°(\alpha) = 22.14$
38. Radius is 8.53 cm; $m°(\alpha) = 80.35$

In Exercises 39 through 41, assume the radius of the earth is 3960 mi.

39. A point P on the surface of the earth is 1500 mi north of the point on the equator closest to it. Find the latitude of the point P in degree measurement.

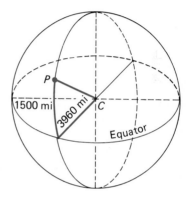

40. One nautical mile can be defined as the distance on the surface of the earth from a point having latitude 1′ N to the closest point on the equator. Show that 1 nautical mile is approximately 1.15 ordinary miles.

41. Two points A and B on the surface of the earth are on the same circle, which is a meridian having center at C, where C is the center of the earth. If A has latitude 10° N and B has latitude 4.6° S, what is the distance between A and B?

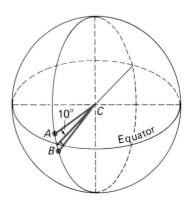

42. The end of a pendulum of length 40 cm travels an arc length of 5 cm as it swings through an angle α. Find (a) $m^R(\alpha)$ and (b) $m°(\alpha)$.

43. An automobile tire has diameter of 36 in. How many revolutions will the wheel make as the automobile travels 1 mi (5280 ft)?

44. If the minute hand of a clock has a length of 6 in., how far does its tip travel in 18 min?

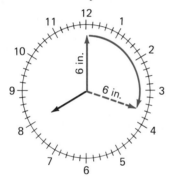

45. If the hour hand of a clock has a length of 4 in., how far does its tip travel in 1 hr and 20 min?

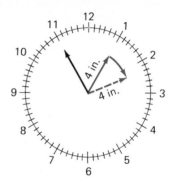

46. A pulley having a diameter of 36 cm is turned by a belt that moves at the rate of 5 m/sec. How many revolutions does the pulley make per second?

47. One *mil* is the measurement of the central angle of a circle that intercepts on the circle an arc equal in length to $\frac{1}{6400}$ of the circumference of the circle. Determine the number of mils in angle α if (a) $m°(\alpha) = 34.4$ and (b) $m^R(\alpha) = 2.3$.

48. If the measurement of angle α is 34 mils (see Exercise 47), find (a) $m°(\alpha)$ and (b) $m^R(\alpha)$.

6.2   Trigonometric Functions of Angles

So far the functions we have studied have been algebraic (such as polynomial and rational functions), exponential, and logarithmic. The domains of these functions are sets of real numbers. In this section we introduce the six *trigonometric functions*, whose domains are sets of angles. To lead up to the definition of these functions, let θ be an angle in standard position. Associated with any point P on the terminal side of θ are three real numbers: x, the abscissa of P; y, the ordinate of P; and r, the distance $|\overline{OP}|$. In Figure 1, θ is an angle in the second quadrant and the point $P(x, y)$ is any point other than the origin on the terminal side. From the distance formula, $r = \sqrt{x^2 + y^2}$. For a particular angle θ there are six ratios that can be formed from these three numbers. These ratios are called *sine* (abbreviated *sin*), *cosine* (abbreviated *cos*), *tangent* (abbreviated *tan*), *cotangent* (abbreviated *cot*), *secant* (abbreviated *sec*), and *cosecant* (abbreviated *csc*).

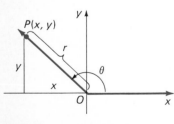

FIGURE 1

Definition

The Six Trigonometric Ratios

If θ is an angle in standard position, $P(x, y)$ is any point other than the origin on the terminal side of θ, and $r = \sqrt{x^2 + y^2}$, then

$$\sin \theta = \frac{y}{r} \qquad\qquad \cos \theta = \frac{x}{r} \qquad\qquad \tan \theta = \frac{y}{x} \quad \text{if} \quad x \neq 0$$

$$\csc \theta = \frac{r}{y} \quad \text{if} \quad y \neq 0 \qquad \sec \theta = \frac{r}{x} \quad \text{if} \quad x \neq 0 \qquad \cot \theta = \frac{x}{y} \quad \text{if} \quad y \neq 0$$

When $x = 0$, $\tan \theta$ and $\sec \theta$ are not defined, and when $y = 0$, $\csc \theta$ and $\cot \theta$ are not defined.

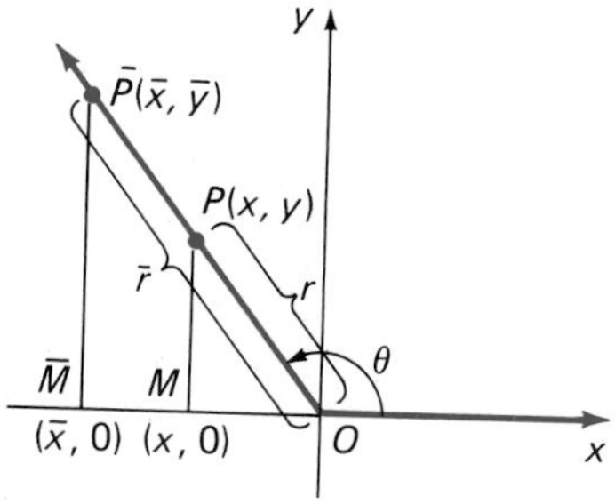

FIGURE 2

Suppose in the above definition we select the point $\bar{P}(\bar{x}, \bar{y})$ on the terminal side of θ instead of $P(x, y)$. See Figure 2. Because triangles OMP and $O\bar{M}\bar{P}$ are similar,

$$\frac{y}{r} = \frac{\bar{y}}{\bar{r}} \qquad \frac{x}{r} = \frac{\bar{x}}{\bar{r}} \qquad \frac{y}{x} = \frac{\bar{y}}{\bar{x}} \qquad \frac{r}{y} = \frac{\bar{r}}{\bar{y}} \qquad \frac{r}{x} = \frac{\bar{r}}{\bar{x}} \qquad \frac{x}{y} = \frac{\bar{x}}{\bar{y}}$$

Therefore the value of each ratio is determined only by the angle θ and not by the particular point $P(x, y)$ on the terminal side; that is, the value of the ratio is a *function* of the angle θ. Hence we call these ratios **trigonometric functions** of θ. Because the trigonometric functions are not algebraic, they are transcendental.

Example 1 Find the values of the six trigonometric functions of the angle θ in standard position, if $P(4, 3)$ is on the terminal side of θ.

Solution Figure 3 shows an angle θ with the point $P(4, 3)$ on its terminal side. Because $x = 4$ and $y = 3$, $r = \sqrt{4^2 + 3^2}$; that is, $r = 5$. Therefore, from the definition

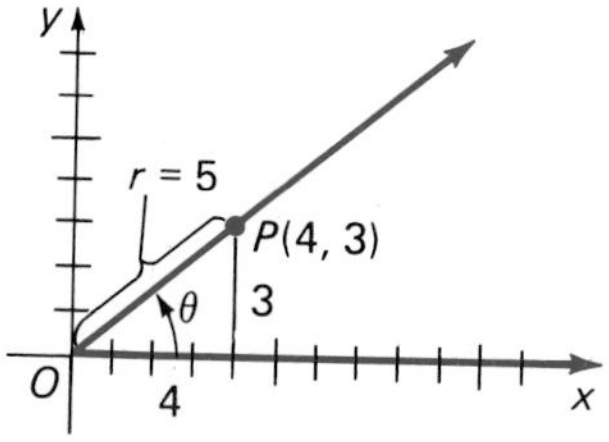

$$\sin \theta = \frac{y}{r} \qquad\qquad \cos \theta = \frac{x}{r} \qquad\qquad \tan \theta = \frac{y}{x}$$

$$= \frac{3}{5} \qquad\qquad\qquad = \frac{4}{5} \qquad\qquad\qquad = \frac{3}{4}$$

$$\csc \theta = \frac{r}{y} \qquad\qquad \sec \theta = \frac{r}{x} \qquad\qquad \cot \theta = \frac{x}{y}$$

$$= \frac{5}{3} \qquad\qquad\qquad = \frac{5}{4} \qquad\qquad\qquad = \frac{4}{3}$$

FIGURE 3

Example 2 If θ is an angle in standard position and the point $P(5, -12)$ is on the terminal side of θ, what are the values of the six trigonometric functions of θ?

Solution In Figure 4 there is an angle θ with the point $P(5, -12)$ on its terminal side. Because $r = \sqrt{x^2 + y^2}$, we have $r = \sqrt{5^2 + (-12)^2}$, which is $\sqrt{169}$; thus $r = 13$. Therefore, from the definition, with $x = 5$, $y = -12$, and $r = 13$, we obtain

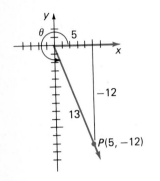

$$\sin \theta = \frac{y}{r} \qquad \cos \theta = \frac{x}{r} \qquad \tan \theta = \frac{y}{x}$$

$$= -\frac{12}{13} \qquad\qquad = \frac{5}{13} \qquad\qquad = -\frac{12}{5}$$

$$\csc \theta = \frac{r}{y} \qquad \sec \theta = \frac{r}{x} \qquad \cot \theta = \frac{x}{y}$$

$$= -\frac{13}{12} \qquad\qquad = \frac{13}{5} \qquad\qquad = -\frac{5}{12}$$ ∎

FIGURE 4

Observe from the definition that when the functions are defined, $\csc \theta$, $\sec \theta$, and $\cot \theta$ are the reciprocals of $\sin \theta$, $\cos \theta$, and $\tan \theta$, respectively; that is

$$\csc \theta = \frac{1}{\sin \theta} \qquad \sec \theta = \frac{1}{\cos \theta} \qquad \cot \theta = \frac{1}{\tan \theta}$$

These equations are true for all values of θ for which each side is defined, and they are called the **reciprocal identities.**

We now obtain two more identities that follow from the definition of the trigonometric functions. Because

$$\sin \theta = \frac{y}{r} \qquad \text{and} \qquad \cos \theta = \frac{x}{r}$$

then

$$\text{if} \quad x \neq 0 \quad \frac{\sin \theta}{\cos \theta} = \frac{\dfrac{y}{r}}{\dfrac{x}{r}} \qquad \text{and} \qquad \text{if} \quad y \neq 0 \quad \frac{\cos \theta}{\sin \theta} = \frac{\dfrac{x}{r}}{\dfrac{y}{r}}$$

$$= \frac{y}{x} \qquad\qquad\qquad\qquad = \frac{x}{y}$$

$$= \tan \theta \qquad\qquad\qquad\qquad = \cot \theta$$

We have proved the following **quotient identities:**

$$\tan \theta = \frac{\sin \theta}{\cos \theta} \quad \text{if} \quad \cos \theta \neq 0 \quad \text{and} \quad \cot \theta = \frac{\cos \theta}{\sin \theta} \quad \text{if} \quad \sin \theta \neq 0$$

Illustration 1

For the angle θ in Example 2

$$\sin \theta = -\tfrac{12}{13} \quad \text{and} \quad \cos \theta = \tfrac{5}{13}$$

From the reciprocal identities

$$\csc \theta = \frac{1}{\sin \theta} \quad \text{and} \quad \sec \theta = \frac{1}{\cos \theta}$$

$$= -\frac{13}{12} \qquad\qquad\qquad = \frac{13}{5}$$

From the quotient identities

$$\tan \theta = \frac{\sin \theta}{\cos \theta} \quad \text{and} \quad \cot \theta = \frac{\cos \theta}{\sin \theta}$$

$$= \frac{-\frac{12}{13}}{\frac{5}{13}} \qquad\qquad\qquad = \frac{\frac{5}{13}}{-\frac{12}{13}}$$

$$= -\frac{12}{5} \qquad\qquad\qquad = -\frac{5}{12}$$

Observe that these results agree with those in Example 2.   ■

Each of the trigonometric functions of an angle θ is given by two of the variables x, y, and r associated with θ. Because r is always positive, the sign $(+ \text{ or } -)$ of a trigonometric function is determined by the signs of x and y, and therefore by the quadrant containing θ. Figure 5(a)–(d) shows angles θ in each of the four quadrants. Above the coordinates x and y of the point P

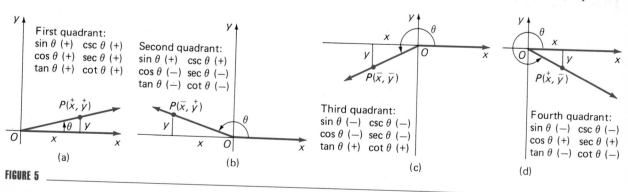

FIGURE 5

are their signs, which depend on the quadrant in which P lies. Also indicated in the figure is the sign of the function in the particular quadrant. Observe that when the sign of a function is determined for a specific quadrant, its reciprocal function will have the same sign in that quadrant. Table 1 summarizes the results.

TABLE 1

Quadrant Containing θ	$\sin \theta$	$\cos \theta$	$\tan \theta$	$\cot \theta$	$\sec \theta$	$\csc \theta$
First	+	+	+	+	+	+
Second	+	−	−	−	−	+
Third	−	−	+	+	−	−
Fourth	−	+	−	−	+	−

Example 3 Determine the quadrant containing angle θ.

a) $\sin \theta < 0$ and $\tan \theta > 0$ b) $\cos \theta > 0$ and $\csc \theta < 0$

Solution a) From Table 1, $\sin \theta < 0$ for the third and fourth quadrants, and $\tan \theta > 0$ for the first and third quadrants. For both conditions to hold, θ must be in the third quadrant.

b) From the table, $\cos \theta > 0$ for the first and fourth quadrants, and $\csc \theta < 0$ for the third and fourth quadrants. Thus θ is in the fourth quadrant. ■

If the value of one of the trigonometric functions of an angle is given, as well as the quadrant containing the angle, we can determine the values of the other five functions. The following illustration and example demonstrate the procedure.

Illustration 2

Suppose $\tan \theta = -\dfrac{8}{15}$ and θ is in the second quadrant. From the definition, if $P(x, y)$ is any point on the terminal side of θ, $\tan \theta = \dfrac{y}{x}$. Therefore there is a point on the terminal side of θ whose ordinate is 8 and whose abscissa is -15.

Refer to Figure 6, showing an angle θ having the point $P(-15, 8)$ on the terminal side. Because $x = -15$, $y = 8$, and

$$r = \sqrt{x^2 + y^2}$$
$$= \sqrt{(-15)^2 + 8^2}$$
$$= \sqrt{289}$$
$$= 17$$

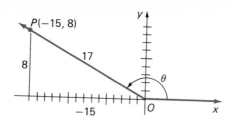

FIGURE 6

we have from the definition

$$\sin \theta = \frac{8}{17} \qquad \cos \theta = \frac{-15}{17}$$

$$= -\frac{15}{17}$$

From the reciprocal identities

$$\csc \theta = \frac{1}{\sin \theta} \qquad \sec \theta = \frac{1}{\cos \theta} \qquad \cot \theta = \frac{1}{\tan \theta}$$

$$= \frac{17}{8} \qquad \qquad = -\frac{17}{15} \qquad \qquad = -\frac{15}{8}$$ ∎

Example   4 If $\cos \theta = -\frac{2}{3}$ and $\cot \theta > 0$, find the other five trigonometric functions of θ.

Solution   Because $\cos \theta < 0$ if θ is in the second or third quadrants and $\cot \theta > 0$ if θ is in the first or third quadrants, we conclude that θ must be in the third quadrant. Since $\cos \theta = -\frac{2}{3}$ and $\cos \theta = x/r$, there is a point on the terminal side of θ for which $x = -2$ and $r = 3$. Furthermore, because $x^2 + y^2 = r^2$ and $y < 0$ (since θ is in the third quadrant),

$$(-2)^2 + y^2 = 3^2$$
$$y^2 = 9 - 4$$
$$y^2 = 5$$
$$y = -\sqrt{5}$$

Figure 7 shows an angle θ in the third quadrant and the point $P(-2, -\sqrt{5})$ on the terminal side. From the definition

$$\sin \theta = \frac{-\sqrt{5}}{3} \qquad \qquad \tan \theta = \frac{-\sqrt{5}}{-2}$$

$$= -\frac{\sqrt{5}}{3} \qquad \qquad = \frac{\sqrt{5}}{2}$$

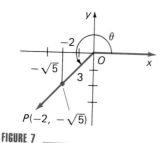

FIGURE 7

From the reciprocal identities

$$\csc \theta = -\frac{3}{\sqrt{5}} \qquad \sec \theta = -\frac{3}{2} \qquad \cot \theta = \frac{2}{\sqrt{5}} \qquad \blacksquare$$

If θ is an angle in standard position, $P(x, y)$ is any point other than the origin on the terminal side of θ, and $r = |\overline{OP}|$, then

$$y^2 + x^2 = r^2$$

Because $r \neq 0$, we can divide on both sides of this equation by r^2 and obtain

$$\left(\frac{y}{r}\right)^2 + \left(\frac{x}{r}\right)^2 = 1$$

Thus, because $\sin \theta = \dfrac{y}{r}$ and $\cos \theta = \dfrac{x}{r}$,

$$(\sin \theta)^2 + (\cos \theta)^2 = 1$$

Instead of $(\sin \theta)^2$ and $(\cos \theta)^2$, it is customary to write $\sin^2\theta$ and $\cos^2\theta$. Because the equation is true for all θ, we have the identity

$$\sin^2 \theta + \cos^2 \theta = 1$$

It is called the **fundamental Pythagorean identity** because the distance formula (obtained from the Pythagorean theorem) was used to derive the equation $x^2 + y^2 = r^2$. The identity shows the relationship between the sine and cosine values and can be used to compute one of them when the other is known. Then the other four functions can be obtained from the reciprocal and quotient identities.

Example 5 Use a calculator to find to four significant digits the other five functions of θ if θ is in the second quadrant and

$$\sin \theta = 0.2716$$

Solution From the fundamental Pythagorean identity

$$\sin^2 \theta + \cos^2 \theta = 1$$

Because $\sin \theta = 0.2716$, we have

$$(0.2716)^2 + \cos^2 \theta = 1$$
$$\cos^2 \theta = 1 - (0.2716)^2$$

Taking the square root on both sides of this equation and using the minus sign, because $\cos \theta$ is negative when θ is in the second quadrant, we have

$$\cos \theta = -\sqrt{1 - (0.2716)^2}$$

Doing the computation with a calculator, we obtain

$$\cos \theta \approx -0.9624$$

From the quotient identities

$$\tan \theta = \frac{\sin \theta}{\cos \theta} \qquad \cot \theta = \frac{\cos \theta}{\sin \theta}$$

$$\approx \frac{0.2716}{-0.9624} \qquad \approx \frac{-0.9624}{0.2716}$$

$$\approx -0.2822 \qquad \approx -3.543$$

From the reciprocal identities

$$\csc \theta = \frac{1}{\sin \theta} \qquad \sec \theta = \frac{1}{\cos \theta}$$

$$\approx \frac{1}{0.2716} \qquad \approx \frac{1}{-0.9624}$$

$$\approx 3.682 \qquad \approx -1.039 \qquad \blacksquare$$

EXERCISES   6.2

In Exercises 1 through 18, θ is an angle in standard position, and the point P is on the terminal side of θ. Use the definition to find the six trigonometric functions of θ.

1. $P(3, 4)$
2. $P(8, 15)$
3. $P(-5, 12)$
4. $P(6, -8)$
5. $P(-8, -15)$
6. $P(-12, -5)$
7. $P(4, -3)$
8. $P(1, 1)$
9. $P(6, -3)$
10. $P(-6, 2)$
11. $P(0, -4)$
12. $P(0, 2)$
13. $P(3, 0)$
14. $P(1, -\sqrt{3})$
15. $P(-2, 2)$
16. $P(-4, 0)$
17. $P(2\sqrt{3}, -2)$
18. $P(-3\sqrt{3}, -3)$

In Exercises 19 through 24, determine the quadrant containing the angle θ.

19. (a) $\cos \theta > 0$ and $\tan \theta < 0$;
    (b) $\sin \theta < 0$ and $\cot \theta > 0$
20. (a) $\cos \theta < 0$ and $\tan \theta > 0$;
    (b) $\sin \theta > 0$ and $\cot \theta < 0$

21. (a) $\tan \theta < 0$ and $\csc \theta > 0$;
    (b) $\cot \theta > 0$ and $\sec \theta < 0$
22. (a) $\tan \theta > 0$ and $\csc \theta < 0$;
    (b) $\cot \theta < 0$ and $\sec \theta > 0$
23. (a) $\cos \theta < 0$ and $\cot \theta < 0$;
    (b) $\sin \theta < 0$ and $\tan \theta < 0$
24. (a) $\sin \theta < 0$ and $\cos \theta < 0$;
    (b) $\tan \theta < 0$ and $\sec \theta < 0$

In Exercises 25 through 36, find the exact values of the other five trigonometric functions of θ.

25. $\sin \theta = \frac{5}{13}$ and $\cos \theta > 0$
26. $\cos \theta = \frac{3}{5}$ and $\sin \theta > 0$
27. $\cos \theta = -\frac{4}{5}$ and $\sin \theta > 0$
28. $\sin \theta = -\frac{12}{13}$ and $\cos \theta > 0$
29. $\tan \theta = \frac{15}{8}$ and $\sec \theta < 0$
30. $\cot \theta = \frac{4}{3}$ and $\csc \theta < 0$
31. $\cot \theta = -\frac{5}{12}$ and $\csc \theta < 0$
32. $\tan \theta = -\frac{8}{15}$ and $\sec \theta < 0$

33. $\csc \theta = -2$ and $\cot \theta > 0$
34. $\sec \theta = \sqrt{2}$ and $\cot \theta < 0$
35. $\tan \theta = -1$ and $\sin \theta > 0$
36. $\cos \theta = -\frac{1}{2}$ and $\tan \theta > 0$

In Exercises 37 through 44, use a calculator to find, to four significant digits, approximate values of the other five trigonometric functions of θ.

37. $\cos \theta = 0.7816$ and θ is in the first quadrant.
38. $\sin \theta = 0.1234$ and θ is in the first quadrant.
39. $\sin \theta = -0.4178$ and θ is in the third quadrant.
40. $\cos \theta = -0.8245$ and θ is in the second quadrant.
41. $\cos \theta = 0.6453$ and θ is in the fourth quadrant.
42. $\sin \theta = -0.3709$ and θ is in the third quadrant.
43. $\sin \theta = 0.2990$ and θ is in the second quadrant.
44. $\cos \theta = 0.7685$ and θ is in the fourth quadrant.

6.3   Trigonometric Function Values

When considering a trigonometric function of an angle θ, often the measurement of the angle is used in place of θ. For instance, if the measurement of an angle θ is $60°$ or, equivalently, $\frac{1}{3}\pi$ radians, then in place of $\sin \theta$ we could write $\sin 60°$ or $\sin \frac{1}{3}\pi$. Notice that when the measurement of the angle is in degrees, the degree symbol is written. However, when there is no symbol attached, the measurement of the angle is in radians. For instance, $\cos 2°$ means the cosine of an angle having degree measure 2, while $\cos 2$ means the cosine of an angle having radian measure 2.

The equation

$$\theta = 60° \qquad \Leftrightarrow \qquad \theta = \tfrac{1}{3}\pi$$

means that θ is an angle having measurement $60°$ or, equivalently, $\frac{1}{3}\pi$ radians. The inequality

$$0° < \theta < 45° \qquad \Leftrightarrow \qquad 0 < \theta < \tfrac{1}{4}\pi$$

means that θ is an angle whose degree measurement is between $0°$ and $45°$ or, equivalently, whose radian measurement is between 0 radians and $\frac{1}{4}\pi$ radians.

In applications of trigonometry we are often given an angle and wish to determine the value of one or more trigonometric functions of the angle. We can *approximate* the value by using a calculator or a table, as we discuss later in this section. But first we show how to determine *exact* values of the trigonometric functions of certain angles by applying theorems from plane geometry.

There is a theorem stating that in a right triangle having acute angles of measurements $30°$ and $60°$, the length of the side opposite the $30°$ angle is one-half the length of the hypotenuse. Figure 1 shows a $30°$ angle in standard position. Because any point on the terminal side may be selected to determine the trigonometric functions, let us choose the point having an ordinate of 1; thus we select the point $P(x, 1)$. In the right triangle OMP of Figure 1, the side opposite the $30°$ angle has length 1 unit. Therefore the

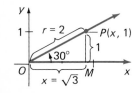

FIGURE 1

length of the hypotenuse is 2 units; thus $r = 2$. Because $x^2 + y^2 = r^2$,

$$x^2 + 1^2 = 2^2$$
$$x^2 + 1 = 4$$
$$x^2 = 3$$
$$x = \sqrt{3}$$

We reject the negative square root of 3 because $P(x, 1)$ is in the first quadrant. From the definition,

$$\sin 30° = \frac{1}{2} \qquad \cos 30° = \frac{\sqrt{3}}{2} \qquad \tan 30° = \frac{1}{\sqrt{3}}$$

$$\csc 30° = 2 \qquad \sec 30° = \frac{2}{\sqrt{3}} \qquad \cot 30° = \sqrt{3}$$

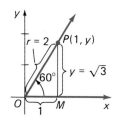

FIGURE 2

A 60° angle in standard position appears in Figure 2. For this angle we choose the point $P(1, y)$, having an abscissa of 1, on the terminal side of the angle. In the right triangle OMP of Figure 2, the angle at P is 30°; therefore $r = 2$. From the Pythagorean theorem, we obtain $y = \sqrt{3}$. Thus

$$\sin 60° = \frac{\sqrt{3}}{2} \qquad \cos 60° = \frac{1}{2} \qquad \tan 60° = \sqrt{3}$$

$$\csc 60° = \frac{2}{\sqrt{3}} \qquad \sec 60° = 2 \qquad \cot 60° = \frac{1}{\sqrt{3}}$$

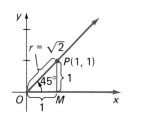

FIGURE 3

If an acute angle in a right triangle is 45°, then the other acute angle is also 45°; so the triangle is isosceles. Thus for an angle of 45° in standard position, as shown in Figure 3, we select the point $P(1, 1)$ on the terminal side. Applying the Pythagorean theorem to the right triangle OMP in Figure 3, we have

$$r^2 = 1^2 + 1^2$$
$$r^2 = 2$$
$$r = \sqrt{2}$$

We reject the negative square root of 2 because $r > 0$. Hence

$$\sin 45° = \frac{1}{\sqrt{2}} \qquad \cos 45° = \frac{1}{\sqrt{2}} \qquad \tan 45° = 1$$

$$\csc 45° = \sqrt{2} \qquad \sec 45° = \sqrt{2} \qquad \cot 45° = 1$$

Table 1 summarizes the values of the trigonometric functions of 30°, 45°, and 60°. You should be able to determine these exact values when needed. It is not necessary to memorize them if you recall the methods of obtaining them from constructing an angle in standard position and choosing a suitable point on the terminal side.

TABLE 1

θ (degrees)	θ (radians)	$\sin \theta$	$\cos \theta$	$\tan \theta$	$\cot \theta$	$\sec \theta$	$\csc \theta$
30°	$\frac{1}{6}\pi$	$\frac{1}{2}$	$\frac{\sqrt{3}}{2}$	$\frac{1}{\sqrt{3}}$	$\sqrt{3}$	$\frac{2}{\sqrt{3}}$	2
45°	$\frac{1}{4}\pi$	$\frac{1}{\sqrt{2}}$	$\frac{1}{\sqrt{2}}$	1	1	$\sqrt{2}$	$\sqrt{2}$
60°	$\frac{1}{3}\pi$	$\frac{\sqrt{3}}{2}$	$\frac{1}{2}$	$\sqrt{3}$	$\frac{1}{\sqrt{3}}$	2	$\frac{2}{\sqrt{3}}$

By methods similar to those used for 30°, 45°, and 60°, we can find the trigonometric functions of multiples of these angles provided they are not quadrantal angles. The following illustration and examples demonstrate the procedure.

Illustration 1

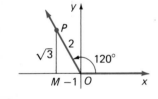

FIGURE 4 _____

To find the values of the trigonometric functions of 120°, we first construct the angle in standard position as shown in Figure 4. From a point P on the terminal side we draw a perpendicular line intersecting the x axis at the point M. Observe that in right triangle OMP, the acute angle at O in the triangle is 60° and the acute angle at P in the triangle is 30°. Therefore, if we choose P as the point for which $r = 2$, then $x = -1$ and $y = \sqrt{3}$. Thus

$$\sin 120° = \frac{\sqrt{3}}{2} \qquad \cos 120° = -\frac{1}{2} \qquad \tan 120° = -\sqrt{3}$$

$$\csc 120° = \frac{2}{\sqrt{3}} \qquad \sec 120° = -2 \qquad \cot 120° = -\frac{1}{\sqrt{3}}$$

Note that because 120° is a second-quadrant angle, the sine and cosecant are positive, and the other four functions are negative. ■

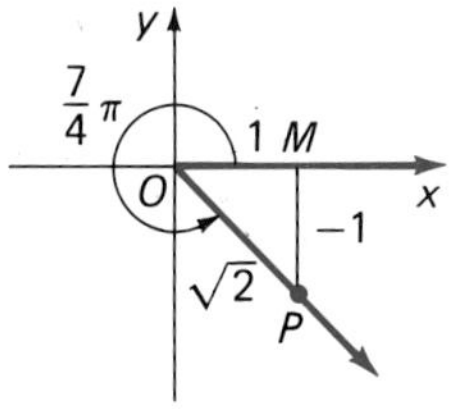

Example 1   Find the trigonometric functions of $\frac{7}{4}\pi$.

FIGURE 5 _____

Solution   Figure 5 shows an angle of measurement $\frac{7}{4}\pi$ or, equivalently, 315° in standard position. We choose point P on the terminal side and form right triangle OMP by drawing a perpendicular line from P to the x axis. Each of the acute angles in this right triangle is 45°. Therefore the lengths of the sides opposite these angles are equal. Thus select P so that $x = 1$ and

$y = -1$. Then $r = \sqrt{2}$. Therefore

$$\sin\frac{7}{4}\pi = -\frac{1}{\sqrt{2}} \qquad \cos\frac{7}{4}\pi = \frac{1}{\sqrt{2}} \qquad \tan\frac{7}{4}\pi = -1$$

$$\csc\frac{7}{4}\pi = -\sqrt{2} \qquad \sec\frac{7}{4}\pi = \sqrt{2} \qquad \cot\frac{7}{4}\pi = -1 \qquad\blacksquare$$

Observe in Example 1 that we have a fourth-quadrant angle; thus the cosine and secant are positive while the other four functions are negative. In the next example, the angle is in the third quadrant; so the tangent and cotangent are the only functions having positive values.

Example   2 Find the trigonometric functions of $-150°$.

Solution   In Figure 6 there is an angle of $-150°$ in standard position, with point P on its terminal side. We form right triangle OMP by drawing a perpendicular line from P to the x axis. The acute angle at O in this triangle is $30°$. Therefore, if we select P so that $r = 2$, then $y = -1$, and $x = -\sqrt{3}$. Hence

$$\sin(-150°) = -\frac{1}{2} \qquad \cos(-150°) = -\frac{\sqrt{3}}{2} \qquad \tan(-150°) = \frac{1}{\sqrt{3}}$$

$$\csc(-150°) = -2 \qquad \sec(-150°) = -\frac{2}{\sqrt{3}} \qquad \cot(-150°) = \sqrt{3} \qquad\blacksquare$$

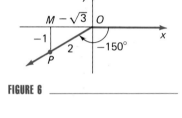

FIGURE 6

So far we have not considered the possibility that the terminal side of an angle could lie on either the x axis or the y axis, in which case it is a quadrantal angle. The formulas defining the trigonometric functions are also valid for these angles. We now use them to compute the trigonometric functions of $0°$, $90°$, $180°$, and $270°$. Figure 7 shows each of these angles in standard position. For each angle we choose as the point P on the terminal side the one for which $r = 1$. For $0°$ in Figure 7(a), $x = 1$, $y = 0$, and $r = 1$.

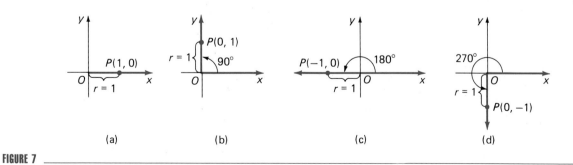

(a)   (b)   (c)   (d)

FIGURE 7

Applying the definition, we have

$$\sin 0° = \frac{y}{r} \qquad \cos 0° = \frac{x}{r} \qquad \tan 0° = \frac{y}{x} \qquad \sec 0° = \frac{r}{x}$$

$$= \frac{0}{1} \qquad\qquad = \frac{1}{1} \qquad\qquad = \frac{0}{1} \qquad\qquad = \frac{1}{1}$$

$$= 0 \qquad\qquad = 1 \qquad\qquad = 0 \qquad\qquad = 1$$

Neither cot 0° nor csc 0° is defined because when using the formulas for cotangent and cosecant, we obtain 0 in the denominator.

For 90° in Figure 7(b), $x = 0$, $y = 1$, and $r = 1$. Thus

$$\sin 90° = \frac{1}{1} \qquad \cos 90° = \frac{0}{1} \qquad \cot 90° = \frac{0}{1} \qquad \csc 90° = \frac{1}{1}$$

$$= 1 \qquad\qquad = 0 \qquad\qquad = 0 \qquad\qquad = 1$$

Because $x = 0$, neither tan 90° nor sec 90° is defined.

Figure 7(c) shows angle 180° with $x = -1$, $y = 0$, and $r = 1$. Therefore

$$\sin 180° = \frac{0}{1} \qquad \cos 180° = \frac{-1}{1} \qquad \tan 180° = \frac{0}{-1} \qquad \sec 180° = \frac{1}{-1}$$

$$= 0 \qquad\qquad = -1 \qquad\qquad = 0 \qquad\qquad = -1$$

Neither cot 180° nor csc 180° is defined because $y = 0$.

An angle of 270° appears in Figure 7(d) with $x = 0$, $y = -1$, and $r = 1$. Hence

$$\sin 270° = \frac{-1}{1} \qquad \cos 270° = \frac{0}{1} \qquad \cot 270° = \frac{0}{-1} \qquad \csc 270° = \frac{1}{-1}$$

$$= -1 \qquad\qquad = 0 \qquad\qquad = 0 \qquad\qquad = -1$$

Because $x = 0$, neither tan 270° nor sec 270° is defined.

Table 2 summarizes the values of the trigonometric functions of 0°, 90°, 180°, and 270°.

TABLE 2

θ (degrees)	θ (radians)	$\sin\theta$	$\cos\theta$	$\tan\theta$	$\cot\theta$	$\sec\theta$	$\csc\theta$
0°	0	0	1	0	Undefined	1	Undefined
90°	$\frac{1}{2}\pi$	1	0	Undefined	0	Undefined	1
180°	π	0	−1	0	Undefined	−1	Undefined
270°	$\frac{3}{2}\pi$	−1	0	Undefined	0	Undefined	−1

Other quadrantal angles are coterminal with one of the angles in Table 2, and their function values are obtained in a similar manner.

Geometric methods of obtaining function values of most angles are not applicable. There are other ways of finding them, but they are beyond the scope of this text. For our purposes, approximate values of the trigonometric functions can be found by a calculator, or they are available from tables like Tables VI and VII in the back of the book.

A scientific calculator has $\boxed{\text{sin}}$, $\boxed{\text{cos}}$, and $\boxed{\text{tan}}$ keys to find values of the trigonometric functions of angles. There is also a key that sets the calculator in the *degree* or *radian mode*. The number of significant digits shown in the display will vary according to the calculator used. In this book, when a trigonometric function value is obtained from a calculator, we shall round off the numeral to four significant digits to be consistent with the value obtained from a table.

Illustration 2

a) To evaluate cos 1.384 with a calculator, first set it in the radian mode. Then enter 1.384 in the display and press the $\boxed{\text{cos}}$ key. By rounding off the result in the display to four significant digits, you will read

$$\cos 1.384 \approx 0.1857$$

b) If a calculator is used to determine sin 41.3°, set it in the degree mode, enter 41.3 in the display, press the $\boxed{\text{sin}}$ key, and read

$$\sin 41.3° \approx 0.6600$$

The calculator does not have keys labeled cot, sec, and csc. To obtain values of cot θ, sec θ, and csc θ, we use the reciprocal identities

$$\cot \theta = \frac{1}{\tan \theta} \qquad \sec \theta = \frac{1}{\cos \theta} \qquad \csc \theta = \frac{1}{\sin \theta}$$

Illustration 3

a) To find cot 49.8° with a calculator, we apply the equality

$$\cot 49.8° = \frac{1}{\tan 49.8°}$$

With the calculator in the degree mode, enter 49.8 in the display and press the $\boxed{\text{tan}}$ key to obtain tan 49.8° ≈ 1.183340. Then press the $\boxed{\text{1/x}}$ key and read from the display, to four significant digits,

$$\cot 49.8° \approx 0.8451$$

Here are some words of caution. Be sure that you press the $\boxed{\text{1/x}}$ key *after* the $\boxed{\text{tan}}$ key. If you press the $\boxed{\text{1/x}}$ key *before* the $\boxed{\text{tan}}$ key, you will be computing tan (1/49.8)°, which is not equal to 1/(tan 49.8°).

b)   To use a calculator to find $\sec \frac{5}{12}\pi$, set it in the radian mode. First compute $\frac{5}{12}\pi \approx 1.309$. Then

$$\sec \tfrac{5}{12}\pi \approx \sec 1.309$$

$$= \frac{1}{\cos 1.309}$$

$$\approx 3.864 \qquad \blacksquare$$

The trigonometric function values of any angle can be determined by knowing the function values of acute angles (an angle having positive radian measure less than $\frac{1}{2}\pi$). The procedure involves the concept of a *reference angle*.

Definition

> **Reference Angle**
> The reference angle associated with a nonquadrantal angle θ in standard position is the acute angle $\bar{\theta}$ formed by the terminal side of θ and the x axis.

In Figure 8 there are four angles in standard position and each is in a different quadrant. The reference angle $\bar{\theta}$ associated with the angle θ is indicated in the figure.

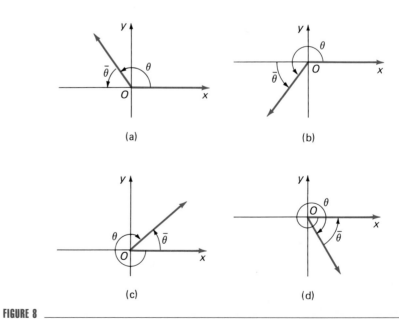

FIGURE 8

Illustration 4

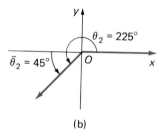

$$\bar{\theta}_1 = \frac{1}{3}\pi \qquad \theta_1 = \frac{2}{3}\pi$$

(a)

Figure 9 shows four angles in standard position with $\theta_1 = \frac{2}{3}\pi$, $\theta_2 = 225°$, $\theta_3 = 5.25$, and $\theta_4 = 432°$, and their associated reference angles $\bar{\theta}_1$, $\bar{\theta}_2$, $\bar{\theta}_3$, and $\bar{\theta}_4$, respectively. The computation of the measurements of these reference angles is as follows.

a) $\bar{\theta}_1 = \pi - \frac{2}{3}\pi$
   $= \frac{1}{3}\pi$

b) $\bar{\theta}_2 = 225° - 180°$
   $= 45°$

c) $\bar{\theta}_3 \approx 2(3.14) - 5.25$
   ≈ 1.03

d) $\bar{\theta}_4 = 432° - 360°$
   $= 72°$

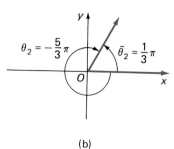

(b)

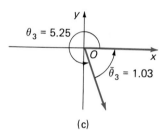

(c)

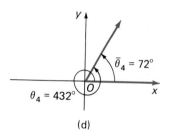

(d)

FIGURE 9

The next illustration shows how to obtain the reference angle associated with a negative angle.

Illustration 5

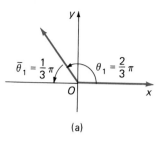

(a)

Figure 10 shows the four negative angles $\theta_1 = -150°$, $\theta_2 = -\frac{5}{3}\pi$, $\theta_3 = -202.4°$, and $\theta_4 = -7.36$. The computations of the respective reference angles are as follows.

a) $\bar{\theta}_1 = 180° - 150°$
   $= 30°$

b) $\bar{\theta}_2 = 2\pi - \frac{5}{3}\pi$
   $= \frac{1}{3}\pi$

c) $\bar{\theta}_3 = 202.4° - 180°$
   $= 22.4°$

d) $\bar{\theta}_4 \approx 7.36 - 2(3.14)$
   ≈ 1.08

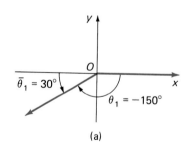

(b)

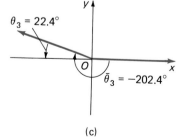

(c)

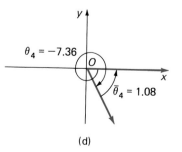

(d)

FIGURE 10

We now show how reference angles are used to find values of the trigonometric functions. Consider an angle θ in standard position and choose on the terminal side of θ the point $P(x, y)$, where $r = |\overline{OP}|$. The associated reference angle is $\overline{\theta}$. Construct the angle $\overline{\theta}$ in standard position; this is a first-quadrant angle. Select on the terminal side of $\overline{\theta}$ the point $\overline{P}(\overline{x}, \overline{y})$ so that $\overline{r} = r$. Figure 11 shows angle θ in the second, third, and fourth quadrants. Because of the symmetry of points P and $\overline{P}$ with respect to either the origin or one of the coordinate axes, we have for each position of $P(x, y)$

$$|x| = \overline{x} \qquad \text{and} \qquad |y| = \overline{y}$$

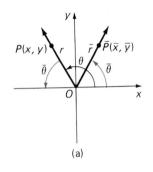

(a)

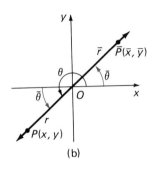

(b)

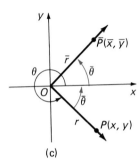

(c)

FIGURE 11

Therefore

$$|\sin \theta| = \frac{|y|}{r} \qquad\qquad |\cos \theta| = \frac{|x|}{r} \qquad\qquad |\tan \theta| = \frac{|y|}{|x|}$$

$$= \frac{\overline{y}}{\overline{r}} \qquad\qquad\qquad = \frac{\overline{x}}{\overline{r}} \qquad\qquad\qquad = \frac{\overline{y}}{\overline{x}}$$

$$= \sin \overline{\theta} \qquad\qquad\quad = \cos \overline{\theta} \qquad\qquad\quad = \tan \overline{\theta}$$

In a similar manner

$$|\csc \theta| = \csc \overline{\theta} \qquad |\sec \theta| = \sec \overline{\theta} \qquad |\cot \theta| = \cot \overline{\theta}$$

From these relationships, we conclude that a trigonometric function value of any angle θ can be found by determining the corresponding function value of the reference angle $\overline{\theta}$ and prefixing the appropriate algebraic sign (+ or −). The procedure can be summarized as follows:

1. Determine the reference angle $\overline{\theta}$ associated with θ.

2. Find the value of the corresponding trigonometric function of $\overline{\theta}$. This can be an exact value if $\overline{\theta}$ is 30°, 60°, or 45°, or it can be an approximate value obtained from a calculator or a table.

3. Affix the proper algebraic sign for the particular function by noting which quadrant contains θ.

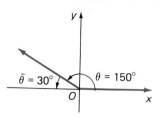

FIGURE 12

Illustration 6

To find tan 150° by the above procedure, we first determine the reference angle associated with 150°. See Figure 12 and observe that if $\theta = 150°$, then

$$\bar{\theta} = 180° - 150°$$
$$= 30°$$

Because the angle 150° is in the second quadrant, tan 150° is negative. Therefore

$$\tan 150° = -\tan 30°$$
$$= -\frac{1}{\sqrt{3}}$$

■

It is not necessary to use reference angles when trigonometric function values are obtained from a calculator. However, since tables give function values of acute angles only, reference angles must be employed when using tables for nonacute angles.

The following examples and Exercises 39 through 54 give practice in using reference angles. Such practice is encouraged because of the importance of reference angles in later work in mathematics.

Example 3 Approximate each of the following values by first expressing it in terms of a function of the associated reference angle

a) cos 116°24'   b) tan(−105.3°)   c) csc(−16°48')

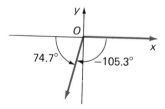

FIGURE 13

Solution a) See Figure 13, showing an angle of 116°24'. The reference angle is 180° − 116°24' = 63°36'. We have

$$\cos 116°24' = -\cos 63°36'$$
$$= -\cos 63.6°$$

From a calculator or Table VII, cos 63.6° ≈ 0.4446. Therefore

$$\cos 116°24' \approx -0.4446$$

b) In Figure 14 there is an angle of −105.3°, which has a reference angle of 180° − 105.3° = 74.7°. Then

$$\tan(-105.3°) = \tan 74.7°$$

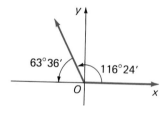

FIGURE 14

From a calculator or Table VII, tan 74.7° ≈ 3.6554. Hence

$$\tan(-105.3°) \approx 3.6554$$

c) An angle of −16°48' appears in Figure 15. The reference angle is 16°48' or, equivalently, 16.8°. Then

$$\csc(-16.8°) = -\csc 16.8°$$

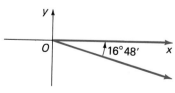

FIGURE 15

From a calculator, with the equality csc 16.8° = 1/(sin 16.8°), or from

Table VII, csc 16.8° ≈ 3.4598. Therefore

$$\csc(-16.8°) \approx -3.4598 \qquad \blacksquare$$

In the next example, because 3.14 (three significant digits) is used for an approximation of π, the results are expressed with only three significant digits.

Example 4 Approximate the value of each of the following by using Table VI in the back of the book.

a) cos 4.46   b) sin(−3.56)

Solution a) Figure 16 shows an angle $\theta = 4.46$. Then

$$\bar{\theta} \approx 4.46 - 3.14$$
$$= 1.32$$

Because the angle of 4.46 radians is in the third quadrant, cos 4.46 is negative. Thus

$$\cos 4.46 = -\cos \bar{\theta}$$
$$\approx -\cos 1.32$$

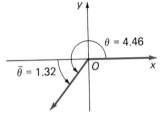

From Table VI, cos 1.32 ≈ 0.248. Therefore

$$\cos 4.46 \approx -0.248$$

FIGURE 16

b) The angle $\theta = -3.56$ appears in Figure 17. Thus

$$\bar{\theta} \approx 3.56 - 3.14$$
$$= 0.42$$

The angle of −3.56 radians is in the second quadrant; therefore sin(−3.56) is positive. Hence

$$\sin(-3.56) = \sin \bar{\theta}$$
$$\approx \sin 0.42$$

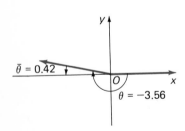

From Table F, sin 0.42 ≈ 0.408. Therefore

$$\sin(-3.56) \approx 0.408 \qquad \blacksquare$$

FIGURE 17

If we obtain cos 4.46 directly from a calculator, we read

$$\cos 4.46 \approx -0.250$$

This result is different from the one in part (a) of Example 4. The discrepancy in the two answers occurs because when finding the reference angle, we used 3.14 as an approximation of π, while the computation by a calculator utilizes a better approximation of π.

EXERCISES   6.3

In Exercises 1 through 4, find the trigonometric function value by first drawing a sketch showing the angle in standard position. Do not use Table 1.

1. (a) $\sin 60°$; (b) $\cos 30°$; (c) $\tan 45°$
2. (a) $\sin 30°$; (b) $\cos 45°$; (c) $\tan 60°$
3. (a) $\sin \frac{1}{4}\pi$; (b) $\tan \frac{1}{3}\pi$; (c) $\cot \frac{1}{6}\pi$
4. (a) $\cos \frac{1}{3}\pi$; (b) $\tan \frac{1}{6}\pi$; (c) $\cot \frac{1}{4}\pi$

In Exercises 5 through 20, draw a sketch showing the angle in standard position and use the method of Examples 1 and 2 to find the six trigonometric functions of the angle.

5. $135°$	6. $150°$
7. $210°$	8. $225°$
9. $\frac{5}{3}\pi$	10. $\frac{4}{3}\pi$
11. $-\frac{1}{6}\pi$	12. $\frac{11}{6}\pi$
13. $-120°$	14. $-225°$
15. $-\frac{5}{4}\pi$	16. $-\frac{1}{4}\pi$
17. $-210°$	18. $-60°$
19. $480°$	20. $-\frac{17}{6}\pi$

In Exercises 21 through 26, draw a sketch showing the quadrantal angle in standard position. Find the values of the four trigonometric functions that are defined. Which two trigonometric functions are not defined?

21. $360°$	22. $-180°$
23. $-90°$	24. $-\frac{3}{2}\pi$
25. 3π	26. $\frac{7}{2}\pi$

In Exercises 27 through 34, use a calculator to approximate to four significant digits the trigonometric function value.

27. (a) $\sin 32°24'$; (b) $\cos 57°36'$; (c) $\tan 73°42'$; (d) $\cot 16°18'$; (e) $\sec 44°30'$; (f) $\csc 45°30'$
28. (a) $\sin 81°6'$; (b) $\cos 8°54'$; (c) $\tan 25°48'$; (d) $\cot 64°12'$; (e) $\sec 52°18'$; (f) $\csc 37°42'$
29. (a) $\sin 75.4°$; (b) $\cos 22.3°$; (c) $\tan 34.8°$; (d) $\cot 88.1°$; (e) $\sec 52.6°$; (f) $\csc 2.7°$
30. (a) $\sin 1.5°$; (b) $\cos 48.2°$; (c) $\tan 76.9°$; (d) $\cot 14.6°$; (e) $\sec 33.7°$; (f) $\csc 80.4°$
31. (a) $\sin 0.34$; (b) $\cos 0.34$; (c) $\tan 1.26$; (d) $\cot 1.26$; (e) $\sec 1.05$; (f) $\csc 0.21$
32. (a) $\sin 1.42$; (b) $\cos 1.42$; (c) $\tan 0.57$; (d) $\cot 0.57$; (e) $\sec 0.18$; (f) $\csc 1.33$
33. (a) $\sin \frac{5}{11}\pi$; (b) $\cos \frac{1}{8}\pi$; (c) $\tan \frac{2}{9}\pi$; (d) $\cot \frac{3}{7}\pi$; (e) $\sec \frac{1}{10}\pi$; (f) $\csc \frac{7}{15}\pi$

34. (a) $\sin \frac{2}{5}\pi$; (b) $\cos \frac{2}{7}\pi$; (c) $\tan \frac{3}{10}\pi$; (d) $\cot \frac{5}{12}\pi$; (e) $\sec \frac{3}{8}\pi$; (f) $\csc \frac{2}{11}\pi$

In Exercises 35 through 38, find the reference angle for the given angle.

35. (a) $\theta = 300°$; (b) $\theta = 415°$; (c) $\theta = \frac{5}{6}\pi$; (d) $\theta = \frac{7}{4}\pi$; (e) $\theta = 3.74$; (f) $\theta = 6.59$
36. (a) $\theta = 135°$; (b) $\theta = 390°$; (c) $\theta = \frac{4}{3}\pi$; (d) $\theta = \frac{11}{6}\pi$; (e) $\theta = 4.93$; (f) $\theta = 7.65$
37. (a) $\theta = 117°24'$; (b) $\theta = -52.7°$; (c) $\theta = \frac{11}{9}\pi$; (d) $\theta = -\frac{13}{8}\pi$; (e) $\theta = -0.23$; (f) $\theta = 8.00$
38. (a) $\theta = -8°54'$; (b) $\theta = 129.2°$; (c) $\theta = \frac{12}{7}\pi$; (d) $\theta = -\frac{7}{10}\pi$; (e) $\theta = -12.00$; (f) $\theta = 10.35$

In Exercises 39 through 42, express the trigonometric function value in terms of a function of the associated reference angle; then determine the exact value.

39. (a) $\sin 135°$; (b) $\cos 210°$; (c) $\tan(-240°)$; (d) $\cot 330°$; (e) $\csc(-120°)$
40. (a) $\sin 315°$; (b) $\cos 120°$; (c) $\tan 210°$; (d) $\cot(-225°)$; (e) $\sec 240°$
41. (a) $\sin(-\frac{2}{3}\pi)$; (b) $\cos \frac{7}{4}\pi$; (c) $\tan \frac{5}{4}\pi$; (d) $\sec(-\frac{7}{6}\pi)$; (e) $\csc(-\frac{5}{8}\pi)$
42. (a) $\cos \frac{11}{6}\pi$; (b) $\tan \frac{2}{3}\pi$; (c) $\cot(-\frac{5}{8}\pi)$; (d) $\sec(-\frac{3}{4}\pi)$; (e) $\csc(-\frac{7}{4}\pi)$

In Exercises 43 through 54, express the trigonometric function value in terms of a trigonometric function of the associated reference angle. Then approximate the value by using either a calculator or a table in the back of the book.

43. (a) $\sin 124°18'$; (b) $\cos 243°36'$; (c) $\tan(-15°6')$
44. (a) $\cos 151°12'$; (b) $\sin(-98°24')$; (c) $\tan 196°42'$
45. (a) $\cos(-172.4°)$; (b) $\sin(-263.8°)$; (c) $\tan 200.2°$
46. (a) $\sin(-193.7°)$; (b) $\cos(-10.9°)$; (c) $\tan 108.3°$
47. (a) $\cot(-169°30')$; (b) $\sec 292.6°$; (c) $\csc 175.5°$
48. (a) $\cot 348.1°$; (b) $\sec(-304.9°)$; (c) $\csc(-126.5°)$
49. (a) $\sin 2.53$; (b) $\cos 5.46$; (c) $\tan(-3.00)$
50. (a) $\sin 4.28$; (b) $\cos(-0.73)$; (c) $\tan 2.00$
51. (a) $\sin(-6.10)$; (b) $\cos 11.63$; (c) $\tan(-4.17)$
52. (a) $\sin 4.92$; (b) $\cos(-4.92)$; (c) $\tan 9.40$
53. (a) $\sin 36.00$; (b) $\cos 21.00$; (c) $\tan(-27.00)$
54. (a) $\sin(-18.00)$; (b) $\cos 30.00$; (c) $\tan(-11.00)$

In Exercises 55 through 62, find an approximation of the trigonometric function value by two methods: (i) use a calculator to find the value directly; (ii) first compute the reference angle and express the function value in terms of a function of the reference angle; then use a calculator.

55. (a) sin 132.4°; (b) csc 132.4°; (c) cos 3.62;
    (d) sec 3.62
56. (a) cos 235.2°; (b) sec 235.2°; (c) sin 1.84;
    (d) csc 1.84
57. (a) cos(−156.1°); (b) sec(−156.1°); (c) sin(−4.32);
    (d) csc(−4.32)

58. (a) sin(−346.7°); (b) csc(−346.7°); (c) cos(−2.36);
    (d) sec(−2.36)
59. (a) tan 420.6°; (b) cot 420.6°; (c) tan 6.00;
    (d) cot 6.00
60. (a) tan 285.3°; (b) cot 285.3°; (c) tan 2.50;
    (d) cot 2.50
61. (a) cos $\frac{9}{10}\pi$; (b) sec $\frac{9}{10}\pi$; (c) tan($-\frac{13}{5}\pi$);
    (d) cot($-\frac{13}{5}\pi$)
62. (a) sin $\frac{12}{7}\pi$; (b) csc $\frac{12}{7}\pi$; (c) tan($-\frac{32}{15}\pi$);
    (d) cot($-\frac{32}{15}\pi$)

6.4 The Sine and Cosine of Real Numbers

To discuss the sine and cosine of real numbers we use the unit circle U, shown in Figure 1 and introduced in Section 6.1. Recall from that section that the radian measure of an angle θ in standard position is equal to the length of arc t measured along U from the point $(1, 0)$ to the point P where the terminal side of θ intersects U. See Figure 2. The *sine* and *cosine* of a real number are then defined to be the corresponding trigonometric functions of an angle of that number of radians.

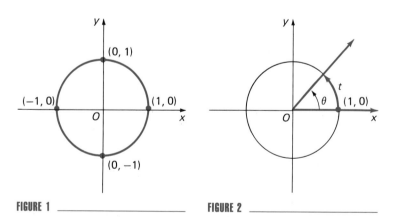

FIGURE 1 ─────────────   FIGURE 2 ──────────────

Definition

Sine and Cosine of a Real Number
If t is a real number and θ is an angle having radian measure t, then the sine and cosine of t are defined as follows:

$$\sin t = \sin \theta \qquad \cos t = \cos \theta$$

In Section 6.3 when we considered trigonometric functions of angles measured in radians, no symbol was attached to the measurement. For instance, $\cos \frac{1}{4}\pi$ indicated the cosine of the angle of radian measure $\frac{1}{4}\pi$. Now $\cos \frac{1}{4}\pi$ also indicates the cosine of the real number $\frac{1}{4}\pi$, which has the same numerical value. Similarly, $\sin 2$ represents either the sine of the angle of radian measure 2 or the sine of the real number 2, both of which have the same numerical value.

Refer to the unit circle U appearing in Figure 3, where the point (x, y) is associated with a real number t; that is, t is the arc length from the point $(1, 0)$ to (x, y) and t radians is the measurement of angle θ. In Figure 3(a), (x, y) is in the first quadrant and $t > 0$; in Figure 3(b), (x, y) is in the second quadrant and $t < 0$; in Figure 3(c), (x, y) is in the fourth quadrant and $t > 0$; in Figure 3(d), (x, y) is in the third quadrant and $t < 0$. Observe that

$$\sin t = \frac{y}{1} \qquad \cos t = \frac{x}{1}$$
$$= y \qquad\qquad = x$$

Therefore the x and y coordinates of the point on U associated with the real number t are $\cos t$ and $\sin t$, respectively. This fact is valid for any position

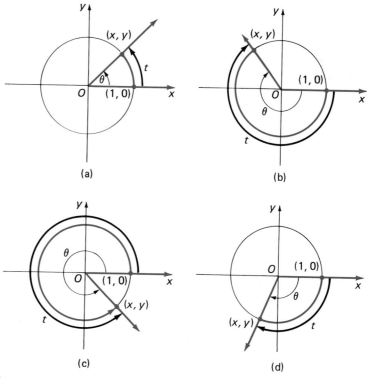

(a)

(b)

(c)

(d)

FIGURE 3

of the point (x, y) and any real number t associated with it. We state these results as a theorem.

Theorem 1

If t is any real number, and t is the length of arc on U with initial point $(1, 0)$ and terminal point (x, y), then

$$\sin t = y \quad \text{and} \quad \cos t = x$$

From Theorem 1, the domain of the sine and cosine functions is the set of all real numbers. To determine the ranges of these functions, note that because (x, y) is a point on U

$$|y| \leq 1 \quad \text{and} \quad |x| \leq 1$$

Therefore the range of each function is the closed interval $[-1, 1]$; that is,

$$-1 \leq \sin t \leq 1 \quad \text{and} \quad -1 \leq \cos t \leq 1$$

Because $\sin t$ and $\cos t$ are coordinates of a point on a circle, the sine and cosine of a real number are sometimes called **circular functions.**

Figure 4 shows the unit circle U on which there are tick marks for every 0.1 unit. The circumference of U is $2\pi \approx 6.28$. For any real number t, we can approximate $\sin t$ and $\cos t$ by obtaining approximate values of the rectangular cartesian coordinates of the point whose arc length from $(1, 0)$ is t. We do this for three values of t in the following illustration.

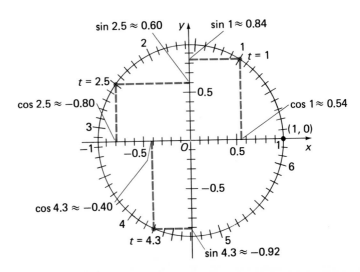

FIGURE 4

Illustration 1

Figure 4 shows the points on U whose arc lengths from $(1, 0)$ are $t = 1$, $t = 2.5$, and $t = 4.3$. The sine and cosine of these values of t are approximated by the point's y and x coordinates, respectively. From the figure, we obtain

$$\sin 1 \approx 0.84 \qquad \cos 1 \approx 0.54$$
$$\sin 2.5 \approx 0.60 \qquad \cos 2.5 \approx -0.80$$
$$\sin 4.3 \approx -0.92 \qquad \cos 4.3 \approx -0.40$$

∎

Exact values of the sine and cosine for certain real numbers can be obtained from the coordinates of points on the unit circle. In particular, refer to Figure 5, which shows arcs for which t is 0, $\frac{1}{2}\pi$, π, and $\frac{3}{2}\pi$. These values of t are called **quadrantal numbers.**

When $t = 0$, the arc length is 0; so the initial and terminal points of the arc are both at $(1, 0)$. Thus

$$\sin 0 = 0 \qquad \text{and} \qquad \cos 0 = 1$$

When $t = \frac{1}{2}\pi$, the terminal point of the arc is at $(0, 1)$. Therefore

$$\sin \tfrac{1}{2}\pi = 1 \qquad \text{and} \qquad \cos \tfrac{1}{2}\pi = 0$$

When $t = \pi$, the terminal point of the arc is at $(-1, 0)$. Hence

$$\sin \pi = 0 \qquad \text{and} \qquad \cos \pi = -1$$

When $t = \frac{3}{2}\pi$, the terminal point of the arc is at $(0, -1)$. Thus

$$\sin \tfrac{3}{2}\pi = -1 \qquad \text{and} \qquad \cos \tfrac{3}{2}\pi = 0$$

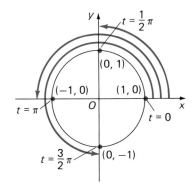

FIGURE 5

We summarize these results in Table 1. They agree with the corresponding trigonometric functions found in Section 6.3 for quadrantal angles having radian measures 0, $\frac{1}{2}\pi$, π, and $\frac{3}{2}\pi$, respectively. Also in Section 6.3 we determined the trigonometric functions of angles having radian measures $\frac{1}{6}\pi$, $\frac{1}{4}\pi$, and $\frac{1}{3}\pi$. The sine and cosine of the real numbers $\frac{1}{6}\pi$, $\frac{1}{4}\pi$, and $\frac{1}{3}\pi$ can also be obtained from coordinates of points on U. We show how this is done for the real number $\frac{1}{4}\pi$.

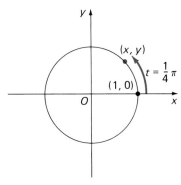

FIGURE 6

TABLE 1

t	(x, y)	$\sin t$	$\cos t$
0	$(1, 0)$	0	1
$\frac{1}{2}\pi$	$(0, 1)$	1	0
π	$(-1, 0)$	0	-1
$\frac{3}{2}\pi$	$(0, -1)$	-1	0

The distance one-eighth of the way around U is $\frac{1}{4}\pi$. Refer to Figure 6, which shows the terminal point (x, y) for an arc on U having length $\frac{1}{4}\pi$. At this point $x = y$. Substituting y for x in the equation of U, which is

$x^2 + y^2 = 1$, we have

$$y^2 + y^2 = 1$$

$$2y^2 = 1$$

$$y^2 = \frac{1}{2}$$

$$y = \frac{1}{\sqrt{2}}$$

We reject the negative square root of $\frac{1}{2}$ because the point is in the first quadrant. Since $x = y$,

$$x = \frac{1}{\sqrt{2}}$$

Therefore

$$\sin \frac{1}{4}\pi = \frac{1}{\sqrt{2}} \qquad \text{and} \qquad \cos \frac{1}{4}\pi = \frac{1}{\sqrt{2}}$$

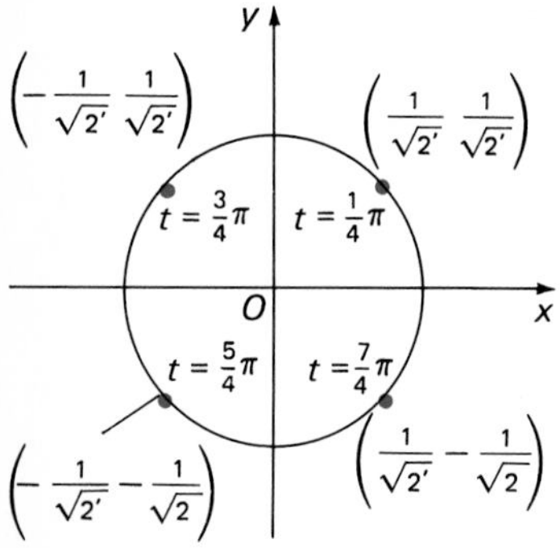

From these results and because U is symmetric with respect to both the coordinate axes and the origin, we obtain the coordinates of the points for which t is $\frac{3}{4}\pi$, $\frac{5}{4}\pi$, and $\frac{7}{4}\pi$. See Figure 7. Thus

$$\sin \frac{3}{4}\pi = \frac{1}{\sqrt{2}} \qquad \text{and} \qquad \cos \frac{3}{4}\pi = -\frac{1}{\sqrt{2}}$$

$$\sin \frac{5}{4}\pi = -\frac{1}{\sqrt{2}} \qquad \text{and} \qquad \cos \frac{5}{4}\pi = -\frac{1}{\sqrt{2}}$$

$$\sin \frac{7}{4}\pi = -\frac{1}{\sqrt{2}} \qquad \text{and} \qquad \cos \frac{7}{4}\pi = \frac{1}{\sqrt{2}}$$

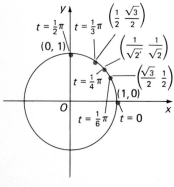

FIGURE 7

In Exercises 9 and 10 you are asked to obtain the exact values of the sine and cosine of the real numbers $\frac{1}{6}\pi$ and $\frac{1}{3}\pi$ from coordinates of points on U.

In Figure 8 and Table 2, we have the sine and cosine of the real numbers 0, $\frac{1}{6}\pi$, $\frac{1}{4}\pi$, $\frac{1}{3}\pi$, and $\frac{1}{2}\pi$. Of course, they agree with the results in Sec-

FIGURE 8

TABLE 2

t	(x, y)	$\sin t$	$\cos t$
0	$(1, 0)$	0	1
$\dfrac{1}{6}\pi$	$\left(\dfrac{\sqrt{3}}{2}, \dfrac{1}{2}\right)$	$\dfrac{1}{2}$	$\dfrac{\sqrt{3}}{2}$
$\dfrac{1}{4}\pi$	$\left(\dfrac{1}{\sqrt{2}}, \dfrac{1}{\sqrt{2}}\right)$	$\dfrac{1}{\sqrt{2}}$	$\dfrac{1}{\sqrt{2}}$
$\dfrac{1}{3}\pi$	$\left(\dfrac{1}{2}, \dfrac{\sqrt{3}}{2}\right)$	$\dfrac{\sqrt{3}}{2}$	$\dfrac{1}{2}$
$\dfrac{1}{2}\pi$	$(0, 1)$	1	0

tion 6.3 for the sine and cosine of the angle having the corresponding radian measure. The sine and cosine of integer multiples of the values of t in Table 2 can be obtained from U by symmetry.

Example   1   Determine the sine and cosine of each of the following real numbers from the coordinates of a point on the unit circle U.

   a) $\frac{4}{3}\pi$ b) $-\frac{1}{6}\pi$

Solution   a) The point on U for $t = \frac{4}{3}\pi$ is in the third quadrant, and it is symmetric with respect to the origin to the first quadrant point for which $t = \frac{1}{3}\pi$. Therefore

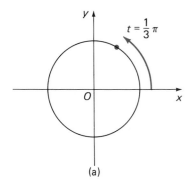
(a)

$$\sin\frac{4}{3}\pi = -\frac{\sqrt{3}}{2} \quad \text{and} \quad \cos\frac{4}{3}\pi = -\frac{1}{2}$$

b) The point on U for $t = -\frac{1}{6}\pi$ is in the fourth quadrant, and it is symmetric with respect to the x axis to the point in the first quadrant for which $t = \frac{1}{6}\pi$. Hence

$$\sin\left(-\frac{1}{6}\pi\right) = -\frac{1}{2} \quad \text{and} \quad \cos\left(-\frac{1}{6}\pi\right) = \frac{\sqrt{3}}{2} \quad \blacksquare$$

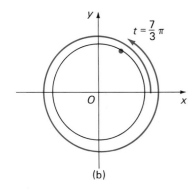
(b)

Approximate values of the sines and cosines of real numbers may be obtained from a calculator set in the radian mode. These values may also be found from a table such as Table VI in the back of the book, which is, of course, the same table used to find approximate trigonometric functions of angles measured in radians.

Because the circumference of the unit circle is 2π, two arcs having initial point at $(1, 0)$ and differing in length by an integer multiple of 2π have the same terminal point on U. For example, refer to Figure 9. Figure 9(a) shows an arc on U of length $\frac{1}{3}\pi$. In Figure 9(b) there is an arc of length $\frac{7}{3}\pi = \frac{1}{3}\pi + 2\pi$, and this arc has the same terminal point. In Figure 9(c) there is an arc of length $-\frac{5}{3}\pi = \frac{1}{3}\pi + (-1)(2\pi)$, and this arc also has the same terminal point. As a matter of fact, any arc for which $t = \frac{1}{3}\pi + k(2\pi)$, where k is any integer, will have the same terminal point. Because the coordinates of the terminal point of the arc determine the sine and cosine of the arc length, it follows that

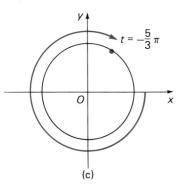
(c)

$$\sin\left(\frac{1}{3}\pi + k \cdot 2\pi\right) = \sin\frac{1}{3}\pi \quad \text{and} \quad \cos\left(\frac{1}{3}\pi + k \cdot 2\pi\right) = \cos\frac{1}{3}\pi(1)$$

FIGURE 9

where k is any integer.

Illustration 2

$$\sin\frac{1}{3}\pi = \frac{\sqrt{3}}{2} \qquad\qquad \cos\frac{1}{3}\pi = \frac{1}{2}$$

$$\sin\frac{7}{3}\pi = \sin\left(\frac{1}{3}\pi + 2\pi\right) \qquad\qquad \cos\frac{7}{3}\pi = \cos\left(\frac{1}{3}\pi + 2\pi\right)$$

$$= \frac{\sqrt{3}}{2} \qquad\qquad = \frac{1}{2}$$

$$\sin\frac{13}{3}\pi = \sin\left(\frac{1}{3}\pi + 4\pi\right) \qquad\qquad \cos\frac{13}{3}\pi = \cos\left(\frac{1}{3}\pi + 4\pi\right)$$

$$= \frac{\sqrt{3}}{2} \qquad\qquad = \frac{1}{2}$$

$$\sin\left(-\frac{5}{3}\pi\right) = \sin\left(\frac{1}{3}\pi + (-1)2\pi\right) \qquad \cos\left(-\frac{5}{3}\pi\right) = \cos\left(\frac{1}{3}\pi + (-1)2\pi\right)$$

$$= \frac{\sqrt{3}}{2} \qquad\qquad = \frac{1}{2}$$

and so on. ∎

Since an arc of length $t + k \cdot 2\pi$, where k is any integer, has the same terminal point as an arc of length t, it follows that Equations (1) are valid if $\frac{1}{3}\pi$ is replaced by any real number t. We state this result as a theorem.

Theorem 2

If t is a real number and k is any integer, then

$$\sin(t + k \cdot 2\pi) = \sin t \qquad \text{and} \qquad \cos(t + k \cdot 2\pi) = \cos t$$

The property of sine and cosine given in Theorem 2 is called *periodicity*.

Definition

Periodic Function

A function f is said to be **periodic** if there exists a positive real number p such that whenever x is in the domain of f, then $x + p$ is also in the domain of f, and

$$f(x + p) = f(x)$$

The smallest such positive real number p is called the **period** of f.

By comparing this definition with Theorem 2, it follows that the sine and cosine are periodic. The period is 2π. It is because of their periodicity that the sine and cosine functions have important applications in connection with periodically repetitive phenomena such as wave motion, vibrations, and business cycles. We discuss some of these applications in Section 6.8.

Example 2   Use the periodicity of the sine and cosine functions as well as the values of $\sin t$ and $\cos t$ when $0 \le t < 2\pi$ to find each of the following values.

     a) $\sin \frac{17}{4}\pi$ b) $\cos(-\frac{7}{6}\pi)$ c) $\sin \frac{15}{2}\pi$

Solution

     a) $\displaystyle \sin \frac{17}{4}\pi = \sin\left(\frac{1}{4}\pi + 2 \cdot 2\pi\right)$

$$= \sin \frac{1}{4}\pi$$

$$= \frac{1}{\sqrt{2}}$$

     b) $\displaystyle \cos\left(-\frac{7}{6}\pi\right) = \cos\left(\frac{5}{6}\pi + (-1)2\pi\right)$

$$= \cos \frac{5}{6}\pi$$

$$= -\frac{\sqrt{3}}{2}$$

     c) $\displaystyle \sin \frac{15}{2}\pi = \sin\left(\frac{3}{2}\pi + 3 \cdot 2\pi\right)$

$$= \sin \frac{3}{2}\pi$$

$$= -1$$ ∎

EXERCISES 6.4

In Exercises 1 through 8, use a figure similar to Figure 4 to show on the unit circle U the point whose arc length from $(1, 0)$ is t. Then approximate the value of $\sin t$ and $\cos t$ to two decimal places.

1. $t = 2$
2. $t = 3$
3. $t = 5.2$
4. $t = 4.6$
5. $t = -3$
6. $t = -2$
7. $t = -6.1$
8. $t = -0.8$

9. Show that the point of intersection of the unit circle U and the line $y = \dfrac{1}{\sqrt{3}}x$ is the terminal point of the arc whose length from $(1, 0)$ is $\frac{1}{6}\pi$. Then by

substituting $\dfrac{1}{\sqrt{3}}x$ for y in the equation of U, find the coordinates of this point and hence $\sin \frac{1}{6}\pi$ and $\cos \frac{1}{6}\pi$.

10. Show that the point of intersection of the unit circle U and the line $y = \sqrt{3}x$ is the terminal point of the arc whose length from $(1, 0)$ is $\frac{1}{3}\pi$. Then by substituting $\sqrt{3}x$ for y in the equation of U, find the coordinates of this point and hence $\sin \frac{1}{3}\pi$ and $\cos \frac{1}{3}\pi$.

In Exercises 11 through 24, determine the sine and cosine of the real number from the coordinates of a point on the unit circle U.

11. (a) $-\frac{1}{2}\pi$; (b) $\frac{5}{4}\pi$ 12. (a) $-\frac{3}{2}\pi$; (b) $\frac{7}{4}\pi$
13. (a) $\frac{7}{6}\pi$; (b) $-\frac{1}{3}\pi$ 14. (a) $\frac{4}{3}\pi$; (b) $-\frac{1}{6}\pi$
15. (a) $\frac{2}{3}\pi$; (b) $-\frac{7}{6}\pi$ 16. (a) $\frac{11}{6}\pi$; (b) $-\frac{5}{3}\pi$
17. (a) 4π; (b) $\frac{5}{2}\pi$ 18. (a) 3π; (b) $\frac{7}{2}\pi$
19. (a) 5π; (b) $\frac{9}{4}\pi$ 20. (a) 6π; (b) $-\frac{9}{4}\pi$
21. (a) $\frac{11}{3}\pi$; (b) $-\frac{11}{4}\pi$ 22. (a) $\frac{19}{4}\pi$; (b) $-\frac{13}{6}\pi$
23. (a) $\frac{23}{6}\pi$; (b) $-\frac{14}{3}\pi$ 24. (a) $\frac{20}{3}\pi$; (b) $-\frac{27}{2}\pi$

In Exercises 25 through 30, use the periodicity of the sine and cosine functions as well as the values of $\sin t$ and $\cos t$ when $0 \le t < 2\pi$ to find the function value.

25. (a) $\sin \frac{13}{6}\pi$; (b) $\cos \frac{8}{3}\pi$; (c) $\sin(-\frac{17}{4}\pi)$
26. (a) $\sin \frac{14}{3}\pi$; (b) $\cos \frac{21}{4}\pi$; (c) $\cos(-\frac{11}{6}\pi)$
27. (a) $\sin \frac{13}{2}\pi$; (b) $\cos 7\pi$; (c) $\cos(-\frac{31}{6}\pi)$
28. (a) $\sin \frac{11}{2}\pi$; (b) $\cos 8\pi$; (c) $\sin(-\frac{17}{3}\pi)$

29. (a) $\sin \frac{23}{4}\pi$; (b) $\cos(-4\pi)$; (c) $\sin(-\frac{9}{2}\pi)$
30. (a) $\cos(-\frac{15}{4}\pi)$; (b) $\cos(-5\pi)$; (c) $\sin \frac{31}{2}\pi$

In Exercises 31 through 38, $f(t) = \sin t$, $g(t) = \cos t$, $\alpha(t) = \frac{1}{6}t$; $\beta(t) = \frac{1}{4}t$, and $\gamma(t) = \frac{1}{3}t$. Compute the composite function value.

31. (a) $f(\beta(\pi))$; (b) $g(\alpha(2\pi))$
32. (a) $f(\gamma(\pi))$; (b) $g(\beta(2\pi))$
33. (a) $f(\alpha(3\pi))$; (b) $g(\gamma(2\pi))$
34. (a) $f(\alpha(\pi))$; (b) $g(\beta(3\pi))$
35. (a) $f(\gamma(3\pi))$; (b) $g(\beta(-\pi))$
36. (a) $f(\alpha(-\pi))$; (b) $g(\gamma(4\pi))$
37. (a) $f(\beta(5\pi))$; (b) $g(\alpha(-4\pi))$
38. (a) $f(\gamma(5\pi))$; (b) $g(\beta(-3\pi))$

39. The quotient $\dfrac{\sin t}{t}$ arises in calculus. When $t = 0$, this quotient is not defined. However, in calculus it is necessary to know the behavior of this quotient as t gets closer and closer to zero. Use a calculator to find this quotient when (a) $t = 0.1$; (b) $t = 0.08$; (c) $t = 0.06$; (d) $t = 0.04$; (e) $t = 0.02$; (f) $t = 0.01$; (g) $t = 0.001$. (h) What values does the quotient appear to be approaching as t gets closer and closer to zero?

40. The quotient $\dfrac{1 - \cos t}{t}$ arises in calculus and it is not defined when $t = 0$. Apply the instructions of Exercise 39 to this quotient.

6.5   Graphs of the Sine and Cosine and Other Sine Waves

We now obtain the graphs of the sine and cosine functions. You will observe that the periodicity of these functions plays an important part.

In our previous discussions of graphs, the coordinate axes were denoted by the symbols x and y. However, in our treatment of the trigonometric functions of real numbers, x and y are used for the coordinates of the terminal point of the arc of length t on the unit circle U. Therefore in this chapter, where we discuss graphs of the trigonometric functions, we shall denote the horizontal axis by t and the vertical axis by $f(t)$. Often on the t axis it is convenient to use rational multiples of π for the tick marks. With $\pi \approx 3.14$, we show in Figure 1 where these tick marks occur in relation to those for integers.

Consider first the graph of the sine function. Let

$$f(t) = \sin t$$

Because the sine function is periodic with period 2π, it is sufficient to deter-

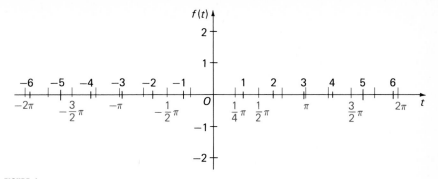

FIGURE 1 _____

mine the portion of the graph for $0 \le t \le 2\pi$. This portion will then be repeated in intervals of length 2π on the t axis. Because $-1 \le \sin t \le 1$, the greatest value $\sin t$ assumes is 1 and the least value it assumes is -1. Table 1 summarizes the behavior of $\sin t$ in each of the four quadrants as t increases from 0 to 2π.

TABLE 1

As t increases from	$\sin t$ goes from
0 to $\frac{1}{2}\pi$	0 to 1
$\frac{1}{2}\pi$ to π	1 to 0
π to $\frac{3}{2}\pi$	0 to -1
$\frac{3}{2}\pi$ to 2π	-1 to 0

We next plot some specific points on the graph in the interval $[0, 2\pi]$. Table 2 contains the values of $\sin t$ for every $\frac{1}{4}\pi$ units, and Figure 2 shows the points whose coordinates are the number pairs $(t, \sin t)$ given in the table, where we use $\dfrac{1}{\sqrt{2}} \approx 0.7$. It is proved in calculus that the sine function is

TABLE 2

t	$\sin t$	t	$\sin t$
0	0	$\dfrac{5}{4}\pi$	$-\dfrac{1}{\sqrt{2}}$
$\dfrac{1}{4}\pi$	$\dfrac{1}{\sqrt{2}}$	$\dfrac{3}{2}\pi$	-1
$\dfrac{1}{2}\pi$	1	$\dfrac{7}{4}\pi$	$-\dfrac{1}{\sqrt{2}}$
$\dfrac{3}{4}\pi$	$\dfrac{1}{\sqrt{2}}$	2π	0
π	0		

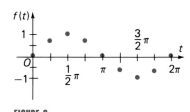

FIGURE 2 _____

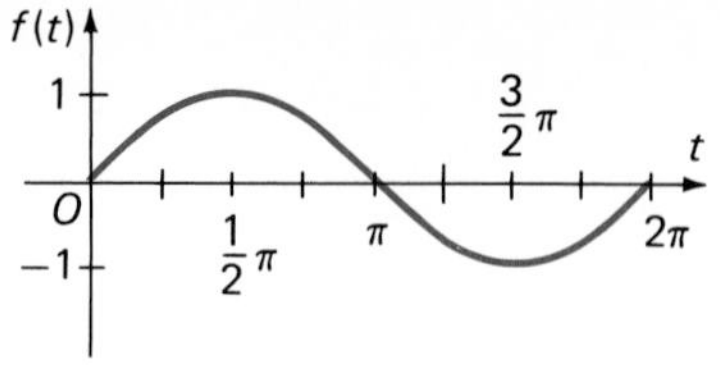

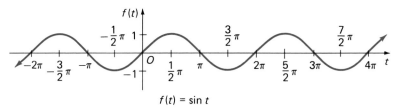

FIGURE 3 _____

continuous, which indicates that there are no "breaks" in its graph. There-fore, we may connect the points in Figure 2 with a smooth curve and obtain the sketch shown in Figure 3. To ensure the accuracy of the sketch, the points for which t is $\frac{1}{6}\pi, \frac{1}{3}\pi, \frac{2}{3}\pi, \frac{5}{6}\pi$, and so on can also be plotted. Function values from a calculator or a table can be used to obtain still other points.

Now that we have the portion of the graph over the interval $[0, 2\pi]$, this portion is repeated for every interval on the t axis of length 2π: $[2\pi, 4\pi]$, $[4\pi, 6\pi]$, $[-2\pi, 0]$, $[-4\pi, -2\pi]$, and so on. Figure 4 shows a sketch of the complete graph; it continues indefinitely to the left and right for t any real number. This graph is called the **sine curve.** It is also referred to as a **sine wave.** The portion of the graph over one period is called a **cycle.**

![Figure 4: graph of f(t) = sin t showing the sine wave from -2π to 4π with marked points at -2π, -3/2π, -π, -1/2π, O, 1/2π, π, 3/2π, 2π, 5/2π, 3π, 7/2π, 4π]

$$f(t) = \sin t$$

FIGURE 4 _____

The graph of the cosine function is obtained in a manner similar to that used for the graph of the sine. Let

$$f(t) = \cos t$$

Because the cosine is periodic with period 2π, we consider first the behav-ior of $\cos t$ for $t \in [0, 2\pi]$ as summarized in Table 3.

TABLE 3 ▬▬▬▬▬▬▬▬▬▬▬▬

As t increases from	$\cos t$ goes from
0 to $\frac{1}{2}\pi$	1 to 0
$\frac{1}{2}\pi$ to π	0 to -1
π to $\frac{3}{2}\pi$	-1 to 0
$\frac{3}{2}\pi$ to 2π	0 to 1

TABLE 4

t	$\cos t$
0	1
$\dfrac{1}{4}\pi$	$\dfrac{1}{\sqrt{2}}$
$\dfrac{1}{2}\pi$	0
$\dfrac{3}{4}\pi$	$-\dfrac{1}{\sqrt{2}}$
π	-1
$\dfrac{5}{4}\pi$	$-\dfrac{1}{\sqrt{2}}$
$\dfrac{3}{2}\pi$	0
$\dfrac{7}{4}\pi$	$\dfrac{1}{\sqrt{2}}$
2π	1

In Table 4 there are values of $\cos t$ for every $\frac{1}{4}\pi$ units in the interval $[0, 2\pi]$, and Figure 5 shows the points having as coordinates the number pairs $(t, \cos t)$. Because the cosine function is continuous, we connect the points with a smooth curve to get the sketch appearing in Figure 6. By repeating this portion of the graph for every interval on the t axis of length 2π, we obtain the sketch shown in Figure 7. As with the sine, additional points can be plotted by using some of the known exact values of the cosine or by taking values from a calculator or table.

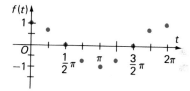

FIGURE 5

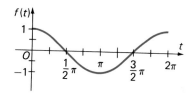

FIGURE 6

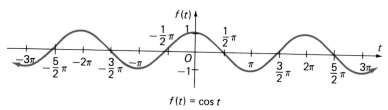

$f(t) = \cos t$

FIGURE 7

Observe that the graph of the sine function is symmetric with respect to the origin, and the graph of the cosine function is symmetric with respect to the $f(t)$ axis. These symmetry properties follow from the identities

$$\sin(-t) = -\sin t \tag{1}$$

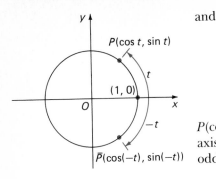

FIGURE 8

and

$$cos(-t) = \cos t \qquad (2)$$

To justify these identities refer to Figure 8. Observe that because points $P(\cos t, \sin t)$ and $\bar{P}(\cos(-t), \sin(-t))$ are symmetric with respect to the x axis, (1) and (2) are obtained. From (1) we can conclude that the sine is an odd function, and from (2) it follows that the cosine is an even function.

Notice that the graphs of the sine and cosine have the same shape. Actually, the graph of the cosine can be obtained by "shifting" the graph of the sine a distance of $\frac{1}{2}\pi$ units to the left. Thus the graph of the cosine is also referred to as a sine wave, and again the portion of the graph for one period is a cycle.

Other sine waves are obtained from functions defined by equations of the form

$$f(t) = a \sin b(t - c) \qquad (3)$$

and

$$f(t) = a \cos b(t - c) \qquad (4)$$

where a, b, and c are real numbers, $a \neq 0$ and $b \neq 0$. To learn how the values of the constants a, b, and c affect the appearance of the sine wave defined by one of these equations, we consider special cases.

The function defined by

$$f(t) = a \sin t \qquad (5)$$

is the special case of (3) where $c = 0$ and $b = 1$. The ordinate of a point on the graph of (5) is a times the corresponding ordinate on the graph of $f(t) = \sin t$. Because

$$-1 \leq \sin t \leq 1$$

the maximum value of $f(t)$ is $|a|$, and the minimum value is $-|a|$. The number $|a|$ is called the **amplitude** of the sine wave.

Illustration 1

Suppose we wish to draw a sketch of the graph of

$$f(t) = 3 \sin t$$

The amplitude of the graph is 3, and each ordinate is 3 times the corresponding ordinate of the graph of $f(t) = \sin t$. In Figure 9 the graph of $f(t) = \sin t$ is shown by the black curve. By multiplying the ordinates on this curve by 3, we obtain the ordinates on the required curve, shown in color in the figure.

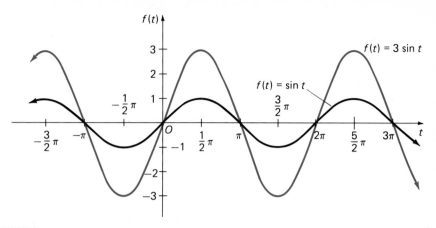

FIGURE 9   ■

The discussion in the paragraph preceding Illustration 1 also applies to graphs of functions defined by equations of the form

$$f(t) = a \cos t \tag{6}$$

If $a < 0$ in either (5) or (6), then the ordinates of points on the graph are the negatives of the corresponding ordinates on the graphs of $f(t) = |a| \sin t$ or $f(t) = |a| \cos t$, respectively.

Example   1   Draw sketches of the graphs of the functions defined by the following equations.

   a) $f(t) = -2 \cos t$ b) $f(t) = \frac{1}{2} \cos t$

Solution   a) The equation $f(t) = -2 \cos t$ is of the form of (6), where $a = -2$. The amplitude of the graph is $|-2| = 2$. Each ordinate of the graph is -2 times the corresponding ordinate of the graph of $f(t) = \cos t$. The required curve appears in color in Figure 10, where the graph of $f(t) = \cos t$ is indicated by a black curve.

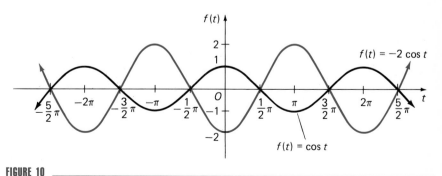

FIGURE 10

b) The graph of the function defined by $f(t) = \frac{1}{2} \cos t$ has amplitude $\frac{1}{2}$. Each ordinate of the graph is one-half times the corresponding ordinate of the graph of $f(t) = \cos t$. Figure 11 shows the required curve in color and the black curve denotes the graph of $f(t) = \cos t$.

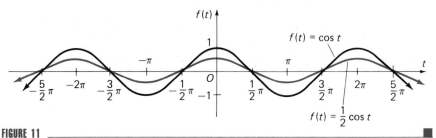

FIGURE 11

Consider now an equation of the form

$$f(t) = \sin bt \tag{7}$$

This function is the special case of $f(t) = a \sin b(t - c)$ where $c = 0$ and $a = 1$. Because the sine function has period 2π, we obtain one cycle of the graph of f for bt in the interval $[0, 2\pi]$; that is,

$$0 \le bt \le 2\pi$$

If $b > 0$, this inequality is equivalent to

$$0 \le t \le \frac{2\pi}{b}$$

and if $b < 0$, it is equivalent to

$$\frac{2\pi}{b} \le t \le 0$$

Therefore for either a positive or negative value of b we obtain one cycle of the graph of (7) on an interval of length $\dfrac{2\pi}{|b|}$. Hence the function defined by (7) is periodic with period $\dfrac{2\pi}{|b|}$. The same argument applies to the function defined by $f(t) = \cos bt$. This result is stated as a theorem.

Theorem 1

The period P of a periodic function defined by either

$$f(t) = \sin bt \quad \text{or} \quad f(t) = \cos bt$$

where $b \ne 0$, is given by

$$P = \frac{2\pi}{|b|}$$

Illustration 2

To obtain the graph of

$$f(t) = \cos 2t$$

we first apply Theorem 1 to find the period P. Because $b = 2$, $P = \pi$. The amplitude is 1. A sketch of the graph appears in Figure 12.

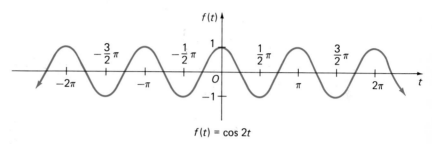

$f(t) = \cos 2t$

FIGURE 12

Illustration 3

For the graph of

$$f(t) = \cos \frac{1}{2}t$$

the period is P, where from Theorem 1

$$P = \frac{2\pi}{\frac{1}{2}}$$

$$= 4\pi$$

The amplitude is 1. One cycle of this sine wave on the interval $[0, 4\pi]$ is shown in Figure 13.

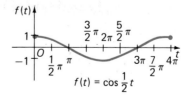

$f(t) = \cos \frac{1}{2}t$

FIGURE 13

Example   2   Draw a sketch of the graph of the function defined by

$$f(t) = \sin 3t$$

Solution   The amplitude is 1. Because $b = 3$, from Theorem 1 the period is

$\frac{2}{3}\pi$. In Figure 14, the units on the t axis are marked off on every interval of length $\frac{1}{3}\pi$. The figure shows a sketch of the graph.

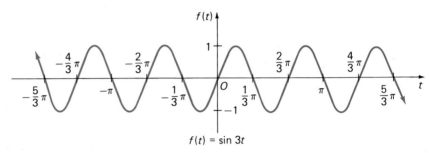

$$f(t) = \sin 3t$$

FIGURE 14

Example 3 Draw one cycle of the sine wave defined by

$$f(t) = \sin \frac{2}{3}t$$

Solution The amplitude is 1. If the period is P, then from Theorem 1

$$P = \frac{2\pi}{\frac{2}{3}}$$

$$= 3\pi$$

Therefore one cycle of the sine wave is over the interval $[0, 3\pi]$. This cycle appears in Figure 15.

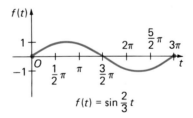

$$f(t) = \sin \frac{2}{3}t$$

FIGURE 15

The period of the graph of $f(t) = \sin t$ is 2π. In Illustration 2 for the graph of $f(t) = \cos 2t$, the period is π, and in Example 2 the period of the graph of $f(t) = \sin 3t$ is $\frac{2}{3}\pi$. For the graphs of $f(t) = \sin bt$ and $f(t) = \cos bt$, as b increases when $b > 0$, the period decreases and the cycles of the sine wave get closer together. Notice in Example 3 that the period of the graph of $f(t) = \sin \frac{2}{3}t$ is 3π, and in Illustration 3 the period of the graph of $f(t) = \cos \frac{1}{2}t$ is 4π. As b decreases when $b > 0$, the period for the graphs of $f(t) = \sin bt$ and $f(t) = \cos bt$ gets larger. In particular, the graph of

$f(t) = \sin \frac{1}{8}t$ has a period of 16π. Thus in the interval $[0, 16\pi]$ there is only one cycle of this sine wave.

What about the graphs of $f(t) = \sin bt$ and $f(t) = \cos bt$ when $b < 0$? To answer this question, we make use of identities (1) and (2). For example, the graph of $f(t) = \sin(-3t)$ is the same as the graph of $f(t) = -\sin 3t$, and the graph of $f(t) = \cos(-3t)$ is the same as the graph of $f(t) = \cos 3t$.

If in the equation $f(t) = a \sin b(t - c)$, $a = 1$ and $b = 1$, we have an equation of the form

$$f(t) = \sin(t - c) \tag{8}$$

The graph of this equation can be obtained from the graph of $f(t) = \sin t$ by a translation of axes to the new origin $(c, 0)$. Therefore, if $c > 0$, the graph of $f(t) = \sin t$ is *shifted* c units to the right to obtain the graph of (8). If $c < 0$, the graph of $f(t) = \sin t$ is shifted $|c|$ units to the left to obtain the graph of (8). The absolute value of c is called the **phase shift** of the graph of (8). The phase shift of a graph is also the phase shift of the corresponding function.

Illustration 4

Consider the graph of

$$f(t) = \sin(t - \tfrac{1}{4}\pi)$$

Here $c = \frac{1}{4}\pi$; thus the phase shift of this graph is $\frac{1}{4}\pi$. We obtain the graph by shifting the graph of $f(t) = \sin t$ a distance of $\frac{1}{4}\pi$ units to the right because $c > 0$. Figure 16 shows a sketch of this graph in color. The graph of $f(t) = \sin t$ on the interval $[0, 2\pi]$ is indicated by a black curve in the figure.

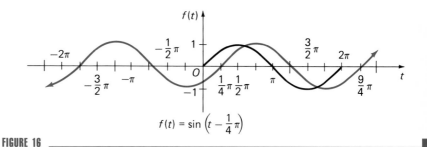

$$f(t) = \sin\left(t - \frac{1}{4}\pi\right)$$

FIGURE 16

Illustration 5

For the graph of

$$f(t) = \sin(t + \tfrac{1}{2}\pi)$$

$c = -\frac{1}{2}\pi$. Thus the phase shift is $\frac{1}{2}\pi$, and the graph is obtained by shifting the graph of $f(t) = \sin t$ a distance of $\frac{1}{2}\pi$ units to the left since $c < 0$. Figure 17 shows a sketch of the graph in color as well as the graph of $f(t) = \sin t$ on the interval $[0, 2\pi]$, indicated by a black curve.

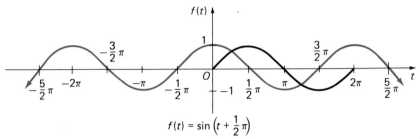

$$f(t) = \sin\left(t + \frac{1}{2}\pi\right)$$

FIGURE 17

We now apply the properties of the graphs of (5), (7), and (8) to obtain the graphs of the functions defined by

$$f(t) = a \sin b(t - c) \qquad \text{and} \qquad f(t) = a \cos b(t - c)$$

Example 4 Draw a sketch of the graph of

$$f(t) = \tfrac{1}{2} \sin(t + \pi)$$

Solution This equation is of the form of $f(t) = a \sin b(t - c)$, where $a = \frac{1}{2}$, $b = 1$, and $c = -\pi$. Because $a = \frac{1}{2}$, the amplitude of the graph is $\frac{1}{2}$. Because $b = 1$, the period of the graph is 2π. Because $c = -\pi$, the phase shift is π and the required graph is obtained by shifting the graph of $f(t) = \frac{1}{2} \sin t$ a distance of π units to the left. In Figure 18 the required graph appears in color, and the graph of $f(t) = \frac{1}{2} \sin t$ on the interval $[0, 2\pi]$ is shown as a black curve.

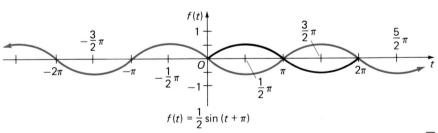

$$f(t) = \frac{1}{2} \sin(t + \pi)$$

FIGURE 18

Example 5 Draw a sketch of the graph of

$$f(t) = 4 \cos(2t - \tfrac{1}{2}\pi)$$

Solution   We write the given equation in the form $f(t) = a \cos b(t - c)$ by factoring 2 from the expression $2t - \frac{1}{2}\pi$. We obtain

$$f(t) = 4 \cos 2(t - \tfrac{1}{4}\pi)$$

Because $a = 4$, the amplitude of the graph is 4. The period of the graph is π because $b = 2$. Since $c = \frac{1}{4}\pi$, the phase shift is $\frac{1}{4}\pi$ and the required graph is obtained by shifting the graph of $f(t) = 4 \cos 2t$ a distance of $\frac{1}{4}\pi$ units to the right. Figure 19 shows the required graph in color and the graph of $f(t) = 4 \cos 2t$ on the interval $[0, \pi]$ is represented by a black curve.

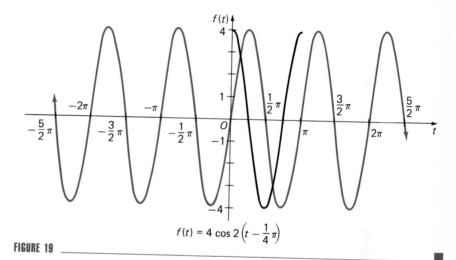

$$f(t) = 4 \cos 2\left(t - \frac{1}{4}\pi\right)$$

FIGURE 19

The function in the next example differs from those previously discussed because the period is a rational number instead of a multiple of π. In such a case, when drawing a sketch of the graph, it is more convenient to use rational numbers for the tick marks on the t axis.

Example   6   Draw a sketch of the graph of the function defined by

$$f(t) = 2 \sin \tfrac{1}{2}\pi t$$

Solution   We compare the given equation with

$$f(t) = a \sin bt$$

Because $a = 2$, the amplitude is 2. Because $b = \frac{1}{2}\pi$, if P is the period, then

$$P = \frac{2\pi}{\frac{1}{2}\pi}$$

$$= 4$$

A sketch of the graph appears in Figure 20.

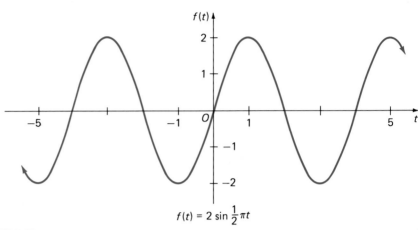

$$f(t) = 2 \sin \frac{1}{2}\pi t$$

FIGURE 20 _____ ■

EXERCISES 6.5

In Exercises 1 through 4, find the period of the function.

1. (a) $\sin 3t$; (b) $3 \cos 6t$; (c) $-\cos \frac{1}{4}t$; (d) $2 \sin \frac{1}{5}t$
2. (a) $\cos 5t$; (b) $-4 \sin 4t$; (c) $\sin \frac{5}{6}t$; (d) $\frac{1}{2} \cos \frac{1}{3}t$
3. (a) $\cos 4\pi t$; (b) $\frac{1}{4} \sin \frac{1}{3}\pi t$
4. (a) $\sin \frac{2}{3}\pi t$; (b) $-2 \cos 3\pi t$

In Exercises 5 through 12, draw a sketch of the graph of the function.

5. (a) $f(t) = 2 \sin t$; (b) $g(t) = \sin 2t$
6. (a) $f(t) = 3 \cos t$; (b) $g(t) = \cos 3t$
7. (a) $f(t) = \frac{1}{2} \cos t$; (b) $g(t) = \cos \frac{1}{2}t$
8. (a) $f(t) = \frac{1}{3} \sin t$; (b) $g(t) = \sin \frac{1}{3}t$
9. (a) $f(t) = -\sin t$; (b) $g(t) = \sin(-t)$
10. (a) $f(t) = -\cos t$; (b) $g(t) = \cos(-t)$
11. (a) $f(t) = 3 \cos \frac{1}{4}t$; (b) $g(t) = \frac{1}{4} \sin 3t$
12. (a) $f(t) = 4 \cos \frac{1}{2}t$; (b) $g(t) = \frac{1}{2} \sin 4t$

In Exercises 13 through 16, draw one cycle of the sine wave.

13. $f(t) = \cos \frac{2}{3}t$
14. $f(t) = \sin \frac{2}{5}t$
15. $f(t) = 5 \sin \frac{4}{5}t$
16. $f(t) = 3 \cos \frac{4}{3}t$

In Exercises 17 through 38, draw a sketch of the graph of the function.

17. $f(t) = 6 \cos \pi t$
18. $f(t) = 4 \sin \pi t$
19. $f(t) = 2 \cos \frac{1}{2}\pi t$
20. $f(t) = 3 \sin \frac{1}{3}\pi t$
21. $f(t) = \frac{1}{2} \sin 2\pi t$
22. $f(t) = \frac{1}{2} \cos \frac{1}{2}\pi t$
23. $f(t) = \cos(t + \frac{1}{2}\pi)$
24. $f(t) = \cos(t + \frac{1}{4}\pi)$
25. $f(t) = 2 \sin(t - \frac{1}{4}\pi)$
26. $f(t) = 3 \sin(t - \frac{1}{2}\pi)$
27. $f(t) = \cos(t - \frac{1}{3}\pi)$
28. $f(t) = \sin(t + \frac{1}{3}\pi)$
29. $f(t) = -3 \cos(t + \frac{1}{6}\pi)$
30. $f(t) = 6 \cos(t - \frac{1}{6}\pi)$
31. $f(t) = 2 \sin(\frac{3}{2}\pi - t)$
32. $f(t) = \frac{1}{2} \cos(\pi - t)$
33. $f(t) = 4 \sin(2t - \pi)$
34. $f(t) = 2 \cos(4t + \pi)$
35. $f(t) = 5 \cos(3t + \frac{1}{2}\pi)$
36. $f(t) = -2 \sin(3t - \frac{1}{2}\pi)$
37. $f(t) = 2 \sin(\frac{1}{3}\pi t + \frac{1}{2}\pi)$
38. $f(t) = 3 \cos(\frac{1}{2}\pi t + \frac{1}{3}\pi)$

6.6 The Tangent, Cotangent, Secant, and Cosecant of Real Numbers

In Section 6.4 we defined the sine and cosine of a real number t in terms of the sine and cosine of an angle having radian measure t. We then showed that these functions of a real number are coordinates of a point on the unit circle. We stated that for this reason the sine and cosine of a real number are sometimes called circular functions. We now define the other four trigonometric (circular) functions of a real number as quotients involving the sine and cosine.

Definition

> **Tangent, Cotangent, Secant, and Cosecant of a Real Number**
> If t is a real number, then
>
> $$\tan t = \frac{\sin t}{\cos t} \quad \text{if} \quad \cos t \neq 0$$
>
> $$\cot t = \frac{\cos t}{\sin t} \quad \text{if} \quad \sin t \neq 0$$
>
> $$\sec t = \frac{1}{\cos t} \quad \text{if} \quad \cos t \neq 0$$
>
> $$\csc t = \frac{1}{\sin t} \quad \text{if} \quad \sin t \neq 0$$

The equations in the above definition are, of course, the quotient and reciprocal identities introduced in Section 6.2. We shall use these equations to determine the domains of the four functions.

For the domain of the tangent and secant we must exclude the values of t for which $\cos t = 0$. These numbers are $\pm\frac{1}{2}\pi$, $\pm\frac{3}{2}\pi$, $\pm\frac{5}{2}\pi$, and so on. Thus the domain of the tangent and secant is the set of all real numbers except $\frac{1}{2}\pi + k\pi = (2k + 1)\frac{1}{2}\pi$, where k is any integer. Note that $(2k + 1)\frac{1}{2}\pi$ is an odd multiple of $\frac{1}{2}\pi$.

The values of t for which $\sin t = 0$ are excluded from the domain of the cotangent and cosecant. These numbers are 0, $\pm\pi$, $\pm2\pi$, $\pm3\pi$, and so on. Therefore the domain of the cotangent and cosecant is the set of all real numbers except $k\pi$, where k is any integer.

We state these results formally as the following theorem.

Theorem 1

> The domain of the tangent and secant functions of real numbers is
>
> $$\{t \mid t \neq \tfrac{1}{2}\pi + k\pi, \text{ where } k \text{ is any integer}\}$$
>
> The domain of the cotangent and cosecant functions of real numbers is
>
> $$\{t \mid t \neq k\pi, \text{ where } k \text{ is any integer}\}$$

Exact values of the tangent, cotangent, secant, and cosecant for the numbers $0, \frac{1}{6}\pi, \frac{1}{4}\pi, \frac{1}{3}\pi, \frac{1}{2}\pi, \frac{2}{3}\pi, \frac{3}{4}\pi$, and so on, are, of course, the same as for the angles having the corresponding radian measure. Approximate values can be obtained from a calculator set in the radian mode or from a table such as Table VI in the back of the book.

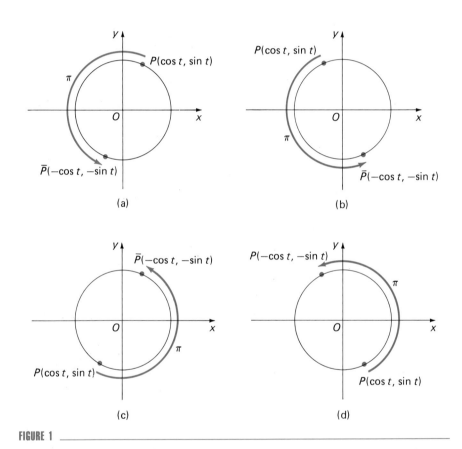

FIGURE 1

To show that the tangent function is periodic, we consider the points $P(\cos t, \sin t)$ and $\bar{P}(-\cos t, -\sin t)$ on the unit circle U. These are endpoints of a diameter. See Figure 1, showing the point P in each of the four quadrants. Observe that in each case P is the terminal point of an arc of length t and $\bar{P}$ is the terminal point of an arc of length $t + \pi$. From the definition of tangent

$$\tan t = \frac{\sin t}{\cos t} \quad \text{if} \quad \cos t \neq 0$$

Because $\bar{P}(-\cos t, -\sin t)$ is the terminal point of an arc of length $t + \pi$,

$$\tan(t + \pi) = \frac{-\sin t}{-\cos t}$$

$$= \frac{\sin t}{\cos t} \quad \text{if} \quad \cos t \neq 0$$

Therefore

$$\tan(t + \pi) = \tan t \quad \text{if} \quad \cos t \neq 0$$

In a similar way we can show that

$$\cot(t + \pi) = \cot t \quad \text{if} \quad \sin t \neq 0$$

From these formulas it follows that the tangent and cotangent are periodic functions with period π. The following theorem can be proved by repeated applications of the formulas.

Theorem 2

> If t is a real number and k is any integer, then
>
> $$\tan(t + k\pi) = \tan t \quad \text{if} \quad t \neq \tfrac{1}{2}\pi + k\pi$$
> $$\cot(t + k\pi) = \cot t \quad \text{if} \quad t \neq k\pi$$

Illustration 1

a) $\tan \tfrac{5}{4}\pi = \tan(\tfrac{1}{4}\pi + \pi)$
$\quad\quad = \tan \tfrac{1}{4}\pi$
$\quad\quad = 1$

b) $\cot(-\tfrac{19}{6}\pi) = \cot(\tfrac{5}{6}\pi + (-4)\pi)$
$\quad\quad\quad = \cot \tfrac{5}{6}\pi$
$\quad\quad\quad = -\sqrt{3}$

■

The procedure for obtaining the graph of the tangent function is similar to that used for the graphs of the sine and cosine. Because the period of the tangent is π, we first determine the portion of the graph on the interval $[0, \pi]$. To get some specific points on the graph, we use some special

TABLE 1

t	$\tan t$
0	0
$\dfrac{1}{6}\pi$	$\dfrac{1}{\sqrt{3}}$
$\dfrac{1}{4}\pi$	1
$\dfrac{1}{3}\pi$	$\sqrt{3}$
$\dfrac{1}{2}\pi$	Undefined
$\dfrac{2}{3}\pi$	$-\sqrt{3}$
$\dfrac{3}{4}\pi$	-1
$\dfrac{5}{6}\pi$	$-\dfrac{1}{\sqrt{3}}$
π	0

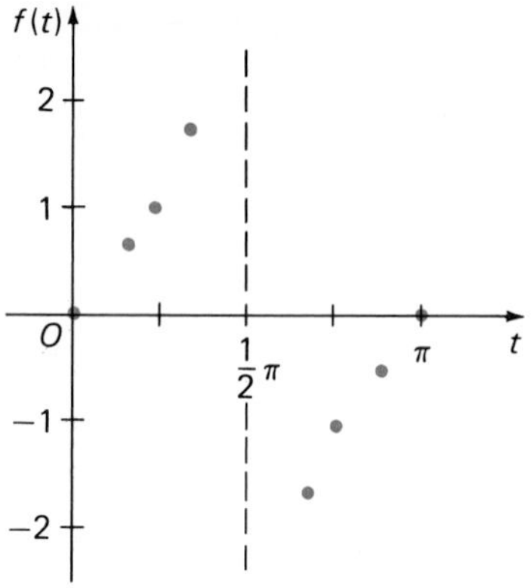

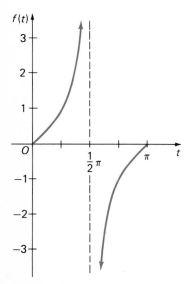

FIGURE 2

values of $\tan t$ listed in Table 1. Figure 2 shows the points having as coordinates the number pairs $(t, \tan t)$ given in the table, where $\sqrt{3} \approx 1.7$ and $\dfrac{1}{\sqrt{3}} \approx 0.6$.

Because $\tan \frac{1}{2}\pi$ is not defined, there is no point on the graph for $t = \frac{1}{2}\pi$. But what happens as t approaches $\frac{1}{2}\pi$? To answer this question, we use the identity

$$\tan t = \frac{\sin t}{\cos t}$$

As t approaches $\frac{1}{2}\pi$, $\sin t$ approaches 1 and $\cos t$ approaches 0. More specifically, as t approaches $\frac{1}{2}\pi$ through values less than $\frac{1}{2}\pi$ (that is, $t \to \frac{1}{2}\pi^{-}$), $\cos t$ approaches 0 through positive values; thus $\tan t$ increases without bound. So we write

$$\tan t \to +\infty \qquad \text{as} \qquad t \to \tfrac{1}{2}\pi^{-}$$

As t approaches $\frac{1}{2}\pi$ through values greater than $\frac{1}{2}\pi$ (that is, $t \to \frac{1}{2}\pi^{+}$), $\cos t$ approaches 0 through negative values; hence $\tan t$ decreases without bound. We write

$$\tan t \to -\infty \qquad \text{as} \qquad t \to \tfrac{1}{2}\pi^{+}$$

Therefore the line $t = \frac{1}{2}\pi$ is a vertical asymptote of the graph. With this information we connect the points in Figure 2 with a smooth curve and obtain the sketch shown in Figure 3. As an aid in drawing the sketch, additional points can be plotted from values of $\tan t$ obtained from a calculator or Table VI.

The portion of the graph over the interval $[0, \pi]$ is repeated for every interval on the t axis of length π: $[\pi, 2\pi]$, $[2\pi, 3\pi]$, $[-\pi, 0]$, $[-2\pi, -\pi]$, and

FIGURE 3

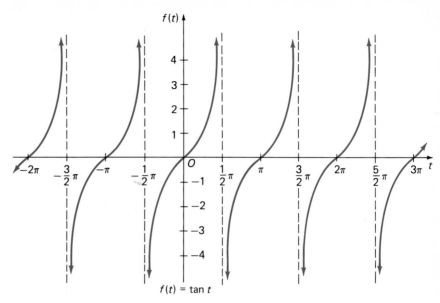

$f(t) = \tan t$

FIGURE 4

so on. Figure 4 shows a sketch of the complete graph. The vertical asymptotes are lines having the equations

$$t = \tfrac{1}{2}\pi + k\pi$$
$$\Leftrightarrow \qquad t = (2k + 1)\tfrac{1}{2}\pi$$

where k is any integer. The values of t at which the graph intersects the t axis are $t = k\pi$, where k is any integer.

Observe that the graph of the tangent function is symmetric with respect to the origin because the tangent is an odd function; that is,

$$\tan(-t) = -\tan t$$

This identity can be obtained from identities (1) and (2) in Section 6.5 as follows:

$$\tan(-t) = \frac{\sin(-t)}{\cos(-t)}$$
$$= \frac{-\sin t}{\cos t}$$
$$= -\tan t$$

The graph of the cotangent function can be obtained by plotting points as we did for the tangent. However, an easier method is one that makes use of the identity

$$\cot t = \frac{1}{\tan t}$$

if $t \neq \frac{1}{2}k\pi$, where k is any integer. At a value of t in the domains of both functions, an ordinate on the graph of the cotangent can be found by taking the reciprocal of the corresponding ordinate on the graph of the tangent.

If $t = \frac{1}{2}k\pi$, where k is an even integer, then $\tan t = 0$, and $\cot t$ is undefined. Therefore there is no point on the graph of the cotangent for these values of t. When $t = \frac{1}{2}k\pi$, where k is an odd integer, $\cot t = 0$. Thus the graph of the cotangent intersects the t axis for these values of t. We use the graph of the tangent as an aid by first sketching it with a black curve as indicated in Figure 5. Then by taking reciprocals of the ordinates we obtain points on the graph of the cotangent, a sketch of which is shown in color in the figure. Observe that the vertical asymptotes of the graph of the cotangent are the lines having the equations $t = k\pi$, where k is any integer.

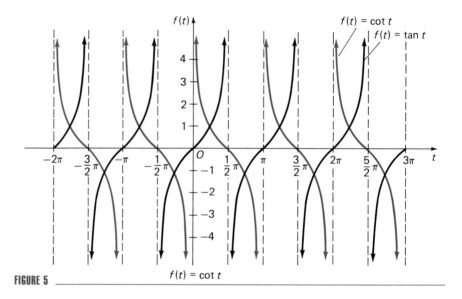

FIGURE 5

From their graphs it is apparent that the range of the tangent and cotangent is the set R of real numbers.

Theorem 1 of Section 6.5 stated that a function defined by either $f(t) = \sin bt$ or $f(t) = \cos bt$, where $b \neq 0$, has period $\dfrac{2\pi}{|b|}$. The proof was based on the fact that the sine and cosine have period 2π. A similar proof can be given for the following theorem about the period of a function defined by either $f(t) = \tan bt$ or $f(t) = \cot bt$. The period of these functions is $\dfrac{\pi}{|b|}$ because the period of tangent and cotangent is π.

Theorem 3

> The period P of a periodic function defined by either
>
> $$f(t) = \tan bt \qquad \text{or} \qquad f(t) = \cot bt$$
>
> where $b \neq 0$, is given by
>
> $$P = \frac{\pi}{|b|}$$

Example 1 Draw a sketch of the graph of

$$f(t) = \tan 3t$$

where t is any number in the interval $[0, \pi]$ at which the function is defined. Write equations of the asymptotes of the graph on the interval.

Solution From Theorem 3, with $b = 3$, the period of f is $\frac{1}{3}\pi$. To find the intersections of the graph with the t axis, we solve the equation $\tan 3t = 0$ and obtain $3t = k\pi$, where k is any integer. Therefore on $[0, \pi]$ the intersections with the t axis are at 0, $\frac{1}{3}\pi$, $\frac{2}{3}\pi$, and π.

To find equations of the asymptotes, we observe that $\tan 3t$ is not defined when

$$3t = \tfrac{1}{2}\pi + k\pi$$

where k is any integer. Because t is in $[0, \pi]$, $3t$ is in $[0, 3\pi]$. Thus the equations of the asymptotes on $[0, \pi]$ are given by

$$\Leftrightarrow \qquad \begin{array}{ccc} 3t = \tfrac{1}{2}\pi & 3t = \tfrac{3}{2}\pi & 3t = \tfrac{5}{2}\pi \\ t = \tfrac{1}{6}\pi & t = \tfrac{1}{2}\pi & t = \tfrac{5}{6}\pi \end{array}$$

With this information and by plotting a few points we obtain the graph shown in Figure 6. For this graph the interval $[0, \pi]$ on the t axis is marked off every $\frac{1}{6}\pi$ units. ∎

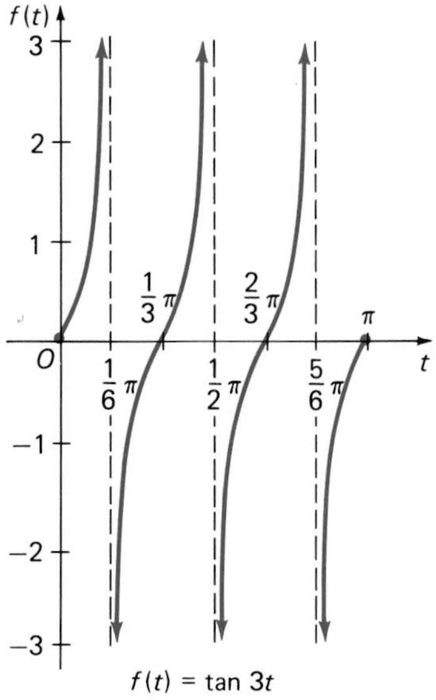

The next example involves the graph of a function of the form

$$f(t) = a \cot bt$$

An ordinate on the graph is a times the corresponding ordinate on the graph of $f(t) = \cot bt$. Observe that because $f(t)$ has no maximum or minimum value, there is no amplitude of the graph.

Example 2 Draw a sketch of the graph of

$$f(t) = 2 \cot \tfrac{1}{3}t$$

over an interval whose length is the period.

$f(t) = \tan 3t$

FIGURE 6

$f(t) = 2 \cot \frac{1}{3}t$

FIGURE 7

Solution   From Theorem 3, with $b = \frac{1}{3}$, the period of f is $\dfrac{\pi}{\frac{1}{3}} = 3\pi$. Therefore, we obtain a sketch of the graph over the interval $[0, 3\pi]$.

The graph intersects the t axis when $\cot \frac{1}{3}t = 0$, that is, when

$$\tfrac{1}{3}t = \tfrac{1}{2}\pi + k\pi$$

where k is any integer. Because $\frac{1}{3}t$ is in $[0, \pi]$, the only point of intersection with the t axis is given by

$$\tfrac{1}{3}t = \tfrac{1}{2}\pi$$
$$t = \tfrac{3}{2}\pi$$

For the asymptotes, we determine where $\cot \frac{1}{3}t$ is not defined. This happens when

$$\tfrac{1}{3}t = k\pi$$

where k is any integer. On the interval $[0, 3\pi]$, the asymptotes are given by

$$\tfrac{1}{3}t = 0 \qquad \tfrac{1}{3}t = \pi$$
$$\Leftrightarrow \qquad t = 0 \qquad t = 3\pi$$

We use the above information and plot a few points to obtain the required sketch, which appears in Figure 7.   ∎

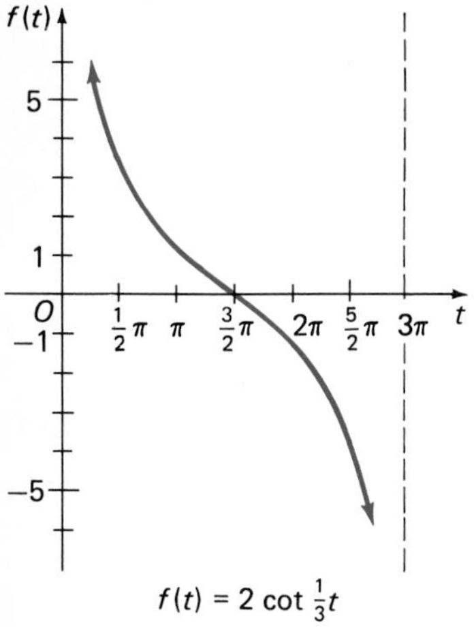

We now show that the secant is periodic. Because

$$\sec t = \frac{1}{\cos t} \qquad \text{if} \qquad \cos t \neq 0$$

then

$$\sec(t + 2\pi) = \frac{1}{\cos(t + 2\pi)} \qquad \text{if} \qquad \cos(t + 2\pi) \neq 0$$

But $\cos(t + 2\pi) = \cos t$. Thus

$$\sec(t + 2\pi) = \frac{1}{\cos t} \qquad \text{if} \qquad \cos t \neq 0$$

Because $\dfrac{1}{\cos t} = \sec t$, we have

$$\sec(t + 2\pi) = \sec t \qquad \text{if} \qquad \cos t \neq 0$$

The same procedure can be used to show that

$$\csc(t + 2\pi) = \csc t \qquad \text{if} \qquad \sin t \neq 0$$

We conclude from these formulas that the secant and cosecant are periodic functions with period 2π. We can prove the following theorem by repeated applications of the formulas.

Theorem 4

If t is a real number and k is any integer, then

$$\sec(t + k \cdot 2\pi) = \sec t \quad \text{if} \quad t \neq \tfrac{1}{2}\pi + k\pi$$
$$\csc(t + k \cdot 2\pi) = \csc t \quad \text{if} \quad t \neq k\pi$$

Illustration 2

a) $\sec \tfrac{13}{4}\pi = \sec(\tfrac{5}{4}\pi + 2\pi)$ b) $\csc(-\tfrac{19}{6}\pi) = \csc(\tfrac{5}{6}\pi + (-2)(2\pi))$

$\qquad\quad = \sec \tfrac{5}{4}\pi$ $\qquad\qquad\qquad\qquad\quad = \csc \tfrac{5}{6}\pi$

$\qquad\quad = -\sqrt{2}$ $\qquad\qquad\qquad\qquad\qquad = 2$ ∎

The graphs of the secant and cosecant functions are obtained in a manner similar to that used for the cotangent. For the secant, we apply the identity

$$\sec t = \frac{1}{\cos t}$$

if $t \neq \tfrac{1}{2}\pi + k\pi$, where k is any integer. On the graph of $f(t) = \sec t$, each ordinate is the reciprocal of the corresponding ordinate of the graph of $f(t) = \cos t$, except for values of t for which $\cos t = 0$. The graph does not intersect the t axis. As a matter of fact, because $|\sec t| \geq 1$, there are no ordinates of the graph between -1 and 1.

A sketch of the graph of $f(t) = \sec t$ appears in color in Figure 8, where the graph of $f(t) = \cos t$ is indicated by a black curve. The vertical asymp-

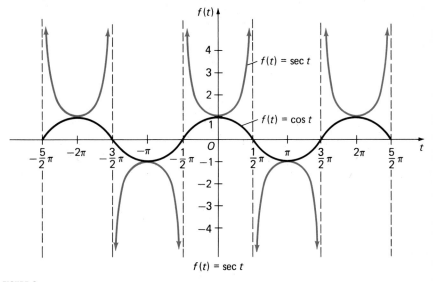

FIGURE 8

totes of the graph of the secant are the lines having the equations

$$t = \tfrac{1}{2}\pi + k\pi$$

where k is any integer.

For the graph of the cosecant, we use the identity

$$\csc t = \frac{1}{\sin t}$$

if $t \neq k\pi$, where k is any integer. Each ordinate of the graph of the cosecant is the reciprocal of the corresponding ordinate of the graph of the sine except for values of t for which $\sin t = 0$. As with the secant, there are no ordinates of the graph between -1 and 1.

Figure 9 shows a sketch of the graph of the cosecant function in color. The graph of the sine is represented by a black curve. The vertical asymptotes of the graph of the cosecant are the lines having the equations

$$t = k\pi$$

where k is any integer. We observe from their graphs that the range of the secant and cosecant is $(-\infty, -1] \cup [1, +\infty)$.

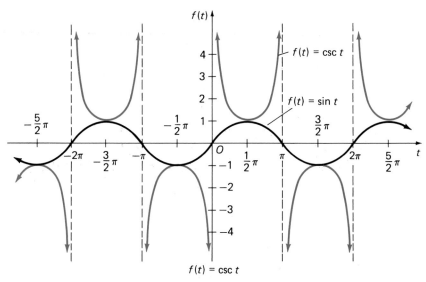

FIGURE 9

Because the secant and cosecant are reciprocals of the cosine and sine, respectively, we have the following theorem, corresponding to Theorem 1 in Section 6.5.

Theorem 5

The period P of a periodic function defined by either

$$f(t) = \sec bt \quad \text{or} \quad f(t) = \csc bt$$

where $b \neq 0$, is given by

$$P = \frac{2\pi}{|b|}$$

Example   3   Draw a sketch of the graph of

$$f(t) = -2 \csc \tfrac{1}{2}t$$

over an interval whose length is the period.

Solution   From Theorem 5, with $b = \tfrac{1}{2}$, the period of f is $\dfrac{2\pi}{\tfrac{1}{2}} = 4\pi$. Thus we obtain a sketch of the graph over the interval $[0, 4\pi]$.

To find the vertical asymptotes, we determine where $\csc \tfrac{1}{2}t$ is not defined. We therefore let $\tfrac{1}{2}t = k\pi$, where k is any integer. On the interval $[0, 4\pi]$, the asymptotes are

$$\Leftrightarrow \quad \begin{array}{ccc} \tfrac{1}{2}t = 0 & \tfrac{1}{2}t = \pi & \tfrac{1}{2}t = 2\pi \\ t = 0 & t = 2\pi & t = 4\pi \end{array}$$

Because $|\csc \tfrac{1}{2}t| \geq 1$, $f(t) \leq -2$ or $f(t) \geq 2$. Furthermore, $f(t) = -2$ when $\csc \tfrac{1}{2}t = 1$, that is, when

$$\tfrac{1}{2}t = \tfrac{1}{2}\pi \quad \Leftrightarrow \quad t = \pi$$

And $f(t) = 2$ when $\csc \tfrac{1}{2}t = -1$, that is, when

$$\tfrac{1}{2}t = \tfrac{3}{2}\pi \quad \Leftrightarrow \quad t = 3\pi$$

With the above information and by plotting a few points, we obtain the required sketch shown in Figure 10.   ∎

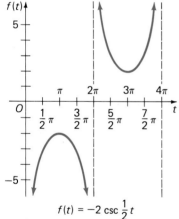

$$f(t) = -2 \csc \tfrac{1}{2}t$$

FIGURE 10

Example   4   Draw a sketch of the graph of

$$f(t) = \sec(t - \tfrac{1}{2}\pi)$$

Solution   The graph of the given equation can be obtained from the graph of $f(t) = \sec t$ by a translation of axes to the new origin $(\tfrac{1}{2}\pi, 0)$. Thus the graph of the secant function is shifted a distance of $\tfrac{1}{2}\pi$ units to the right to get the required sketch, appearing in Figure 11. The figure also shows the graph of $f(t) = \sec t$ on the interval $[0, 2\pi]$, represented by a black curve.

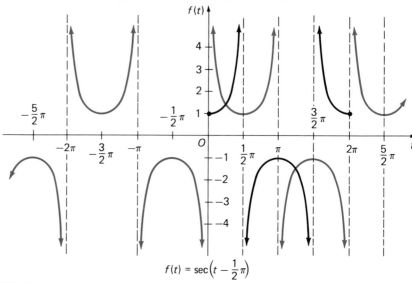

$$f(t) = \sec\left(t - \frac{1}{2}\pi\right)$$

FIGURE 11

EXERCISES   6.6

In Exercises 1 and 2, use the periodicity of the tangent and cotangent functions as well as the values of tan t and cot t when $0 \leq t < \pi$ to find the function value.

1. (a) $\tan \frac{7}{4}\pi$; (b) $\cot(-\frac{1}{4}\pi)$; (c) $\tan(-\frac{7}{3}\pi)$; (d) $\cot \frac{11}{3}\pi$; (e) $\tan 5\pi$; (f) $\cot(-\frac{9}{2}\pi)$

2. (a) $\tan \frac{4}{3}\pi$; (b) $\cot(-\frac{2}{3}\pi)$; (c) $\tan(-\frac{11}{4}\pi)$; (d) $\cot \frac{13}{4}\pi$; (e) $\tan \frac{11}{2}\pi$; (f) $\cot(-6\pi)$

In Exercises 3 and 4, find the period of the function.

3. (a) $\tan 4t$; (b) $3 \cot \frac{1}{2}t$; (c) $2 \tan \frac{1}{3}\pi t$; (d) $-4 \cot 2\pi t$

4. (a) $\tan \frac{3}{4}t$; (b) $-2 \cot 2t$; (c) $\frac{1}{2} \tan 3\pi t$; (d) $5 \cot \frac{1}{2}\pi t$

In Exercises 5 and 6, use the periodicity of the secant and cosecant functions as well as the values of sec t and csc t when $0 \leq t < 2\pi$ to find the function value.

5. (a) $\sec \frac{9}{4}\pi$; (b) $\csc \frac{10}{3}\pi$; (c) $\sec(-\frac{1}{6}\pi)$; (d) $\csc(-\frac{5}{4}\pi)$; (e) $\sec 7\pi$; (f) $\csc(-\frac{9}{2}\pi)$

6. (a) $\sec \frac{17}{6}\pi$; (b) $\csc \frac{15}{4}\pi$; (c) $\sec(-\frac{2}{3}\pi)$; (d) $\csc(-\frac{11}{6}\pi)$; (e) $\sec(-10\pi)$; (f) $\csc \frac{9}{2}\pi$

In Exercises 7 and 8, find the period of the function.

7. (a) $\sec 5t$; (b) $-2 \csc \frac{3}{4}t$; (c) $3 \sec 2\pi t$; (d) $\csc \frac{2}{3}\pi t$
8. (a) $\sec 4t$; (b) $\csc \frac{1}{4}t$; (c) $-4 \sec \frac{4}{3}\pi t$; (d) $\frac{1}{3} \csc \frac{5}{2}\pi t$

In Exercises 9 through 24, (a) draw a sketch of the graph of the function defined by the equation, where t is any number in the indicated interval at which the function is defined; (b) write equations of the asymptotes of the graph on the interval.

9. $f(t) = 2 \tan t$; $[0, 2\pi]$ 10. $f(t) = 3 \cot t$; $(0, 2\pi)$
11. $f(t) = \cot 2t$; $(0, \pi)$ 12. $f(t) = \tan 3t$; $[0, \pi]$
13. $f(t) = \tan \frac{1}{2}t$; $[-2\pi, 2\pi]$
14. $f(t) = \frac{1}{2} \cot 4t$; $(-\frac{1}{2}\pi, \frac{1}{2}\pi)$
15. $f(t) = -2 \cot 3t$; $(0, \pi)$
16. $f(t) = -3 \tan \frac{1}{3}t$; $[0, 3\pi]$
17. $f(t) = 3 \sec t$; $[0, 2\pi]$ 18. $f(t) = 2 \csc t$; $(0, 2\pi)$
19. $f(t) = \csc 3t$; $(0, 2\pi)$ 20. $f(t) = \sec 2t$; $[-\pi, \pi]$
21. $f(t) = \frac{1}{2} \cot \pi t$; $(0, 2)$ 22. $f(t) = \tan \frac{1}{2}\pi t$; $[0, 4]$
23. $f(t) = 2 \sec \frac{1}{2}\pi t$; $[0, 2]$
24. $f(t) = 3 \csc \frac{1}{3}\pi t$; $(-3, 3)$

In Exercises 25 through 32, draw a sketch of the graph of the function over an interval whose length is the period.

25. $f(t) = 3 \tan 2t$ 26. $f(t) = 2 \cot 3t$
27. $f(t) = 2 \cot \frac{1}{2}t$ 28. $f(t) = -\tan \frac{1}{4}t$
29. $f(t) = -\sec 4t$ 30. $f(t) = 3 \csc \frac{1}{3}t$
31. $f(t) = \frac{1}{2} \tan \frac{1}{4}\pi t$ 32. $f(t) = \frac{1}{2} \sec 3\pi t$

In Exercises 33 through 42, draw a sketch of the graph of the function.

33. $f(t) = \tan(t + \frac{1}{4}\pi)$ 34. $f(t) = \tan(t - \frac{1}{4}\pi)$

35. $f(t) = \cot(t - \frac{1}{6}\pi)$ 36. $f(t) = \cot(t + \frac{1}{3}\pi)$

37. $f(t) = 2\sec(t - \pi)$

38. $f(t) = \frac{1}{2}\csc(t - \frac{1}{2}\pi)$

39. $f(t) = \frac{1}{3}\csc(t + \frac{1}{2}\pi)$

40. $f(t) = 3\sec(t + \frac{1}{6}\pi)$

41. $f(t) = \frac{1}{4}\tan(\frac{1}{4}\pi - t)$

42. $f(t) = 2\cot(\frac{1}{2}\pi - t)$

6.7 Solutions of Right Triangles

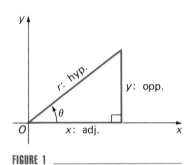

FIGURE 1

The use of trigonometry to find relationships between lengths of sides and measurements of angles in triangles has applications in fields such as engineering, physics, astronomy, and navigation. In this section we deal with right triangles, and triangles not containing a right angle are discussed in Sections 8.1 and 8.2.

Because an acute angle is a positive angle of degree measure less than 90, it can be an angle in a right triangle, and the trigonometric ratios of an acute angle can be expressed as ratios of the measures of the sides of a right triangle. Figure 1 shows a right triangle having an acute angle θ and a rectangular cartesian coordinate system placed so that θ is in standard position.

The following theorem is the result of applying the definition of the six trigonometric functions where

x is the measure of the side adjacent to θ (abbreviated **adj**)

y is the measure of the side opposite θ (abbreviated **opp**)

r is the measure of the hypotenuse (abbreviated **hyp**)

Theorem 1

If θ is an acute angle in a right triangle, then

$$\sin\theta = \frac{\text{opp}}{\text{hyp}} \qquad \csc\theta = \frac{\text{hyp}}{\text{opp}}$$

$$\cos\theta = \frac{\text{adj}}{\text{hyp}} \qquad \sec\theta = \frac{\text{hyp}}{\text{adj}}$$

$$\tan\theta = \frac{\text{opp}}{\text{adj}} \qquad \cot\theta = \frac{\text{adj}}{\text{opp}}$$

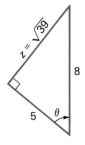

FIGURE 2

Illustration   1

Figure 2 shows a right triangle with an acute angle θ. The length of the hypotenuse is 8 units, and the length of the side adjacent to θ is 5 units. We can find the values of the six trigonometric functions of θ by the formulas of Theorem 1; but first we must compute the length of the side opposite θ. If z is this length, then from the Pythagorean theorem we have

$$z^2 + 5^2 = 8^2$$
$$z^2 + 25 = 64$$
$$z^2 = 39$$
$$z = \sqrt{39}$$

Therefore

$$\sin \theta = \frac{\text{opp}}{\text{hyp}} \qquad\qquad \csc \theta = \frac{\text{hyp}}{\text{opp}}$$
$$= \frac{\sqrt{39}}{8} \qquad\qquad\qquad = \frac{8}{\sqrt{39}}$$

$$\cos \theta = \frac{\text{adj}}{\text{hyp}} \qquad\qquad \sec \theta = \frac{\text{hyp}}{\text{adj}}$$
$$= \frac{5}{8} \qquad\qquad\qquad = \frac{8}{5}$$

$$\tan \theta = \frac{\text{opp}}{\text{adj}} \qquad\qquad \cot \theta = \frac{\text{adj}}{\text{opp}}$$
$$= \frac{\sqrt{39}}{5} \qquad\qquad\qquad = \frac{5}{\sqrt{39}} \qquad\qquad ■$$

The vertices of a right triangle are usually denoted by A, B, and C, where C is used for the vertex at which the 90° angle appears. The measures of the sides opposite A, B, and C are designated by a, b, and c, respectively; thus c represents the measure of the hypotenuse, and a and b represent the measures of the legs. The acute angles at vertices A and B are denoted by α and β, respectively. See Figure 3.

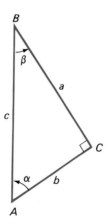

FIGURE 3

The following theorem is the result of applying the ratios of Theorem 1.

Theorem 2

If α and β are the two acute angles of a right triangle, a and b are, respectively, the measures of the two sides opposite these angles, and c is the measure of the hypotenuse; then

$$\sin \alpha = \frac{a}{c} \qquad \sin \beta = \frac{b}{c}$$

$$\cos \alpha = \frac{b}{c} \qquad \cos \beta = \frac{a}{c}$$

$$\tan \alpha = \frac{a}{b} \qquad \tan \beta = \frac{b}{a}$$

$$\cot \alpha = \frac{b}{a} \qquad \cot \beta = \frac{a}{b}$$

$$\sec \alpha = \frac{c}{b} \qquad \sec \beta = \frac{c}{a}$$

$$\csc \alpha = \frac{c}{a} \qquad \csc \beta = \frac{c}{b}$$

From the formulas of Theorem 2 observe that

$$\sin \alpha = \cos \beta \qquad \text{and} \qquad \cos \alpha = \sin \beta \qquad (1)$$

$$\tan \alpha = \cot \beta \qquad \text{and} \qquad \cot \alpha = \tan \beta \qquad (2)$$

$$\sec \alpha = \csc \beta \qquad \text{and} \qquad \csc \alpha = \sec \beta \qquad (3)$$

Because $\alpha + \beta = 90°$,

$$\beta = 90° - \alpha \qquad (4)$$

Substituting from (4) into (1), (2), and (3), we have

$$\sin \alpha = \cos(90° - \alpha) \qquad \text{and} \qquad \cos \alpha = \sin(90° - \alpha) \qquad (5)$$

$$\tan \alpha = \cot(90° - \alpha) \qquad \text{and} \qquad \cot \alpha = \tan(90° - \alpha) \qquad (6)$$

$$\sec \alpha = \csc(90° - \alpha) \qquad \text{and} \qquad \csc \alpha = \sec(90° - \alpha) \qquad (7)$$

Because of Equations (5), we say that the sine and cosine are **cofunctions** of each other. Furthermore, because of (6), the tangent and cotangent are cofunctions of each other, and because of (7), the secant and cosecant are cofunctions of each other. When the sum of two acute angles is 90°, we say that the two angles are **complementary** and that each angle is the **complement** of the other. Therefore Equations (5), (6), and (7) state that *any trigonometric function of an acute angle is the cofunction of its comple-ment.* This fact is used in the construction of tables of trigonometric func-tions such as Tables VI and VII in the back of the book.

Illustration 2

a) $\cos(90° - 23.4°) = \cos 66.6°$
$$= 0.3971$$
$$= \sin 23.4°$$

b) $\tan(90° - 39.8°) = \tan 50.2°$
$$= 1.2002$$
$$= \cot 39.8°$$

c) $\csc(90° - 81.3°) = \csc 8.7°$
$$= 6.6111$$
$$= \sec 81.3°$$

■

Example 1   Express each of the following trigonometric function values as the function value of a positive angle less than 45°.

a) $\sin 72.1°$ b) $\cos 45.5°$ c) $\cot 89.7°$

Solution

a) $\sin 72.1° = \cos(90° - 72.1°)$
$$= \cos 17.9°$$

b) $\cos 45.5° = \sin(90° - 45.5°)$
$$= \sin 44.5°$$

c) $\cot 89.7° = \tan(90° - 89.7°)$
$$= \tan 0.3°$$

■

The formulas of Theorem 1 can be used to *solve a right triangle,* which means to find the measures of its sides and acute angles. In finding such solutions, the accuracy of the given data determines the accuracy of the results of the computations. Table 1 gives the relationship between the accuracies of the measures of the sides and the measurements of the acute angles in degrees. Although we are dealing with approximate numbers, we shall use the equals symbol with the understanding that the equality is valid only for the number of significant digits warranted by Table 1.

Table 1

Number of Significant Digits in Measures of Sides	Measurements of Acute Angles
4	To nearest 0.01° or 1′
3	To nearest 0.1° or 10′
2	To nearest degree

When solving a triangle, it is recommended that you first draw a sketch approximately to scale to gain an understanding of the problem. Doing this enables you to notice an error if the computed value is unreasonably large or much too small.

Example   2   Solve the right triangle for which $\alpha = 24.2°$ and $c = 16.3$.

Solution   Figure 4 shows the triangle. We wish to determine β, a, and b. Because $\alpha + \beta = 90°$,

$$\beta = 90° - \alpha$$
$$= 90° - 24.2°$$
$$= 65.8°$$

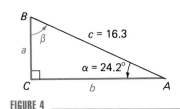

FIGURE 4

To find a we need a formula containing a and the given values of c and α. The formulas for sin α and csc α involve these quantities. If we use the sine, we have

$$\sin 24.2° = \frac{a}{16.3}$$
$$a = 16.3 \sin 24.2°$$

From Table VII or a calculator, sin $24.2° \approx 0.4099$. Therefore

$$a = 16.3(0.4099)$$
$$= 6.68$$

To find b we wish to use a formula involving α, c, and b. From the formula for cosine, we have

$$\cos 24.2° = \frac{b}{16.3}$$
$$b = 16.3 \cos 24.2°$$
$$b = 16.3(0.9121)$$
$$b = 14.9$$

We could have obtained the value for b by using the computed value of a and cot α. By this method we have

$$\cot 24.2° = \frac{b}{6.68}$$
$$b = 6.68 \cot 24.2°$$
$$b = 6.68(2.225)$$
$$b = 14.9$$

Observe that by computing b by the two methods we have a check on the work.   ■

In the next example, where we are given the measures of two sides of a right triangle, it is necessary to determine an acute angle from one of the function values. Before doing this example, we demonstrate how a calculator can be used in such a situation.

Illustration   3

Suppose θ is an acute angle and

$$\cos \theta = 0.3254$$

The problem of finding θ from $\cos \theta$ is the *inverse* of that in determining $\cos \theta$ when θ is given. On some calculators this inverse process is performed by using a $\boxed{\cos^{-1}}$ key. In this case, to determine θ, first enter 0.3254 in the display and then, with the calculator in the degree mode, press the $\boxed{\cos^{-1}}$ key to obtain

$$\theta = 71.0°$$

If there is no $\boxed{\cos^{-1}}$ key, then with the calculator in the degree mode and 0.3254 in the display, press the $\boxed{\text{INV}}$ key followed by the $\boxed{\cos}$ key to get the same result. Similar procedures are used if another function of θ is given.   ∎

The $\boxed{\sin^{-1}}$, $\boxed{\cos^{-1}}$, and $\boxed{\tan^{-1}}$ keys on the calculator refer to *inverse trigonometric functions*, which are discussed in Section 7.5.

Example   3   Solve the right triangle for which $a = 32.46$ and $b = 25.78$.

Solution   The triangle appears in Figure 5. The unknowns are α, β, and c. We first determine one of the angles. To solve for α, we use the tangent function and have

$$\tan \alpha = \frac{32.46}{25.78}$$

$$\tan \alpha = 1.259$$

To find α, we can use a calculator or Table VII with interpolation. We obtain

$$\alpha = 51.54°$$

To find β, we also use the tangent function.

$$\tan \beta = \frac{25.78}{32.46}$$

$$\tan \beta = 0.7942$$

$$\beta = 38.46°$$

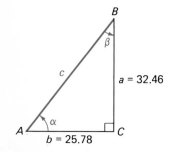

FIGURE 5

Because $\alpha + \beta = 90°$, we could have computed β by subtracting the value of α from 90°. However, by finding β from the given information we have a check on our work by showing that

$$\alpha + \beta = 51.54° + 38.46°$$
$$= 90°$$

To solve for c, we can use any of the functions of α and β. Using $\sin \alpha$, we have

$$\sin 51.54° = \frac{32.46}{c}$$

$$c = \frac{32.46}{\sin 51.54°}$$

$$c = \frac{32.46}{0.7830}$$

$$c = 41.45 \qquad \blacksquare$$

The area of a triangle can be found from the formula

$$K = \tfrac{1}{2}bh$$

where K square units is the area, b units is the length of the base, and h units is the length of the altitude. In a right triangle the sides (legs) opposite the acute angles are a base and altitude.

Illustration   4

a)   If K_1 square units is the area of the triangle of Example 2, then

$$K_1 = \tfrac{1}{2}ba$$
$$= \tfrac{1}{2}(14.9)(6.68)$$
$$= 49.8$$

b)   If K_2 square units is the area of the triangle of Example 3, then

$$K_2 = \tfrac{1}{2}ba$$
$$= \tfrac{1}{2}(25.78)(32.46)$$
$$= 418.4 \qquad \blacksquare$$

We now discuss some applications involving the solution of right triangles. In the following example, the length of a line segment is given as 50.00 m, which indicates that the measurement is accurate to the nearest one-hundredth of a meter. That is, there are four significant digits.

Example   4   To determine the distance across a lake, a surveyor selected two points P and Q, one on each shore and directly opposite each other. On

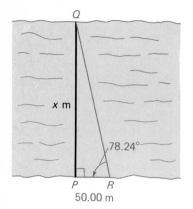

FIGURE 6

the shore containing P, another point R was chosen, 50.00 m from P, so that the line segment PR was perpendicular to the line segment PQ. The angle having the sides PR and RQ was measured to be 78.24°. What is the distance across the lake?

Solution   Refer to Figure 6, showing the right triangle having its right angle at P. The distance across the lake is x meters. Then

$$\tan 78.24° = \frac{x}{50.00}$$

$$x = 50.00 \tan 78.24°$$

$$x = (50.00)(4.8035)$$

$$x = 240.2$$

Therefore the distance across the lake is 240.2 m.   ■

A line segment from an observation point O to a point P being observed is called the **line of sight** of P. The angle, having its vertex at O, made by a horizontal ray and the line of sight is called the **angle of elevation** of P or the **angle of depression** of P, according as P is above or below the point O. See Figure 7.

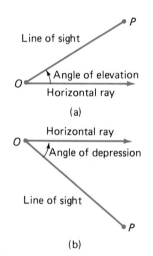

FIGURE 7

Example   5   At a point P the angle of elevation of the top of a hill is 36.3°. At a point Q on the same horizontal line as P and the foot of the hill and 60.0 m from P, the angle of elevation is 24.5°. Find the height of the hill.

Solution   See Figure 8, where point R is at the top of the hill. The height of the hill is x meters. Let y meters be the distance from P to the foot of the hill. There are two right triangles. From the right triangle having a vertex at P, we have

$$\cot 36.3° = \frac{y}{x}$$

$$y = x \cot 36.3°$$

From the right triangle having a vertex at Q, we obtain

$$\cot 24.5° = \frac{60.0 + y}{x}$$

$$60.0 + y = x \cot 24.5°$$

We replace y in this equation by $x \cot 36.3°$ and obtain

$$60.0 + x \cot 36.3° = x \cot 24.5°$$

$$60.0 = x(\cot 24.5° - \cot 36.3°)$$

$$x = \frac{60.0}{\cot 24.5° - \cot 36.3°}$$

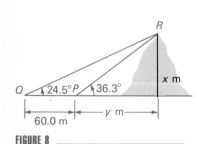

FIGURE 8

$$x = \frac{60.0}{2.194 - 1.361}$$

$$x = \frac{60.0}{0.833}$$

$$x = 72.0$$

Thus the height of the hill is 72.0 m.   ∎

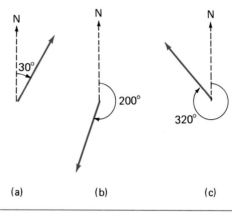

(a)   (b)   (c)

FIGURE 9

In navigation, the **course** of a ship or airplane is the angle measured in degrees clockwise from the north to the direction in which the carrier is traveling. The angle is considered positive even though it is in the clockwise sense. Figure 9 shows courses of 30°, 200°, and 320°. The **bearing** of a particular location P from an observer at O is the angle measured in degrees clockwise from the north to the line segment OP. Refer to Figure 10(a), which shows two points A and B. The bearing from A to B is the angle shown in Figure 10(b), and the bearing from B to A is the angle shown in Figure 10(c). Be sure to distinguish the two situations. In this case θ is the bearing from A to B, and $180° + \theta$ is the bearing from B to A.

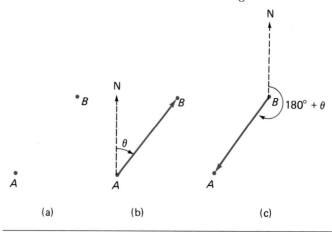

(a)   (b)   (c)

FIGURE 10

Example   **6**   A navigator on a ship sailing on a course of 338° at 12 knots (nautical miles per hour) observes a lighthouse, due north of the ship. Fifteen minutes later the lighthouse is due east of the ship. How far is the lighthouse from the ship at that time?

Solution   Refer to Figure 11, where x nautical miles is the distance to be determined, C is the position of the lighthouse, A is the ship's position at the time of the initial observation, and B is the ship's position 15 min later. Because the ship is traveling at 12 knots, it covers a distance of 3 nautical miles in 15 min. Because the course is 338°, the angle at A in the triangle is $360° - 338°$ or 22°. From the right triangle, we have

$$\sin 22° = \frac{x}{3}$$
$$x = 3 \sin 22°$$
$$x = 3(0.3746)$$
$$x = 1.1$$

Therefore the lighthouse is 1.1 nautical miles from the ship at the indicated time.   ■

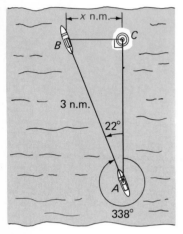

FIGURE 11

EXERCISES   6.7

In Exercises 1 through 4, a right triangle is shown with an acute angle θ and the lengths of two sides. Find the values of the six trigonometric functions of θ.

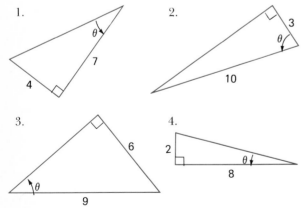

In Exercises 5 through 8, express the trigonometric function value as a function of a positive angle not greater than 45°.

5. (a) sin 60°; (b) cos 84.3°; (c) tan 49.8°;
    (d) cot 52.1°; (e) csc 67.5°

6. (a) sin 79.6°; (b) cos 46°; (c) tan 64.3°;
    (d) cot 88.8°; (e) sec 59.7°
7. (a) sin 47°18′; (b) cos 71°42′; (c) tan 46°;
    (d) cot 55°6′; (e) sec 64°30′
8. (a) sin 62°12′; (b) cos 80°54′; (c) tan 75°24′;
    (d) cot 60°; (e) csc 49°36′

In Exercises 9 through 12, solve the right triangle where the right angle is at vertex C. For the trigonometric function values, do not use tables or a calculator. Express the results to the number of significant digits justified by the given information.

9. $\alpha = 60°$, $c = 2.7$ 11. $\beta = 45°$, $a = 56$
10. $\alpha = 45°$, $a = 34$ 12. $\beta = 30°$, $c = 8.2$

In Exercises 13 through 34, solve the right triangle where the right angle is at vertex C. Express the results to the number of significant digits justified by the given information.

13. $\alpha = 24°$, $a = 16$ 14. $\alpha = 65°$, $b = 6.3$
15. $\beta = 71°$, $c = 44$ 16. $\beta = 10°$, $b = 22$
17. $b = 26$, $c = 38$ 18. $a = 3.4$, $b = 5.7$
19. $\beta = 37.4°$, $a = 4.18$ 20. $\alpha = 52.3°$, $c = 48.5$

21. $\alpha = 16.9°$, $a = 136$
22. $\beta = 29.7°$, $a = 0.534$
23. $a = 63.6$, $b = 58.1$
24. $a = 154$, $c = 393$
25. $\alpha = 58.43°$, $c = 625.3$
26. $\beta = 18.63°$, $c = 52.10$
27. $\beta = 40.92°$, $b = 36.72$
28. $\alpha = 70.25°$, $a = 6584$
29. $a = 312.7$, $c = 809.0$
30. $b = 4.218$, $c = 6.759$
31. $\beta = 65°18'$, $c = 39.2$
32. $\alpha = 35°48'$, $c = 25.0$
33. $\alpha = 29°36'$, $b = 287$
34. $\beta = 81°12'$, $a = 43.6$

In Exercises 35 through 38, find the area of the triangle of the indicated exercise.

35. Exercise 15
36. Exercise 22
37. Exercise 29
38. Exercise 32

39. From the top of a cliff 126 m high, the angle of depression of a boat is 20.7°. How far is the boat from the foot of the cliff?

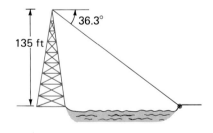

40. A tower 135 ft high stands on the water's edge of a lake. From the top of the tower, the angle of depression of an object on the water's edge on the other side of the lake is 36.3°. What is the distance across the lake?

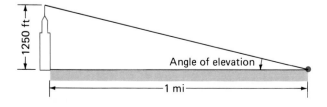

41. The Empire State Building is 1250 ft tall. What is the angle of elevation of the top from a point on

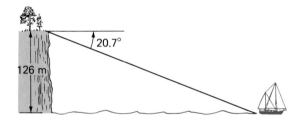

the ground 1 mi (5280 ft) from the base of the building?

42. If the angle of elevation of the sun is 42°, what is the length of the shadow on the level ground of a man who is 6.1 ft tall?

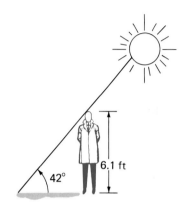

43. Two ships leave a port at the same time. The first ship sails on a course of 35° at 15 knots while the second ship sails on a course of 125° at 20 knots. Find after 2 hr (a) the distance between the ships, (b) the bearing from the first ship to the second, and (c) the bearing from the second ship to the first.

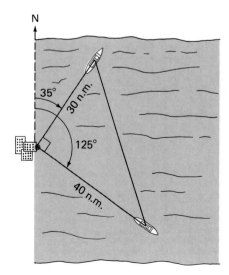

44. A ship leaves a port and sails for 4 hours on a course of 78° at 18 knots. Then the ship changes its course to 168° and sails for 6 hours at 16 knots. After the 10 hours (a) what is the distance of the ship from the port and (b) what is the bearing from the port to the ship?

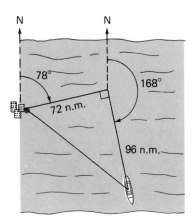

45. Points *A* and *B* are on the same horizontal line with the foot of a hill and the angles of depression of these points from the top of the hill are 30.2° and 22.5°, respectively. If the distance between *A* and *B* is 75.0 m, what is the height of the hill?

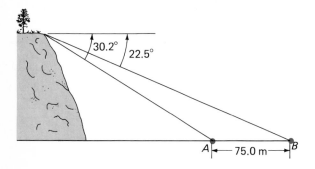

46. From the top of a building 60 ft high, the angle of elevation of the top of a vertical pole is 14°. At the bottom of the building the angle of elevation of the top of the pole is 28°. Find (a) the height

of the pole and (b) the distance of the pole from the building.

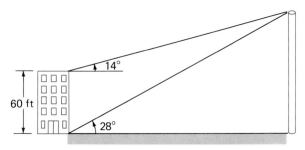

47. From the top of a mountain 532 m higher than a nearby river, the angle of depression of a point *P* on the closer bank of the river is 52.6°, and the angle of depression of a point *Q* directly opposite *P* on the other side is 34.5°. Points *P*, *Q*, and the foot of the mountain are on the same horizontal line. Find the distance across the river from *P* to *Q*.

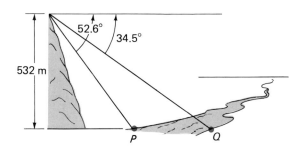

48. Point *T* is at the top of a mountain. From a point *P* on the ground, the angle of elevation of *T* is 16.3°. From a point *Q* on the same horizontal line with *P* and the foot of the mountain, the angle of elevation of *T* is 28.7°. What is the height of the mountain if the distance between *P* and *Q* is 125 m?

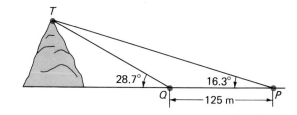

6.8 Some Applications of the Sine and Cosine to Periodic Phenomena

The functions discussed in Section 6.5 and defined by

$$f(t) = a \sin b(t - c) \tag{1}$$

and

$$f(t) = a \cos b(t - c) \tag{2}$$

are mathematical models describing **simple harmonic motion,** either vibrating or oscillating.

An example of simple harmonic motion occurs when a weight is suspended from a spring and is vibrating vertically. Let $f(t)$ centimeters be the directed distance of the weight from its central, or rest, position after t seconds of time. See Figure 1, where a positive value of $f(t)$ indicates that the weight is above its central position. If on a rectangular cartesian coordinate system the function values $f(t)$ are plotted for specific values of t, then if friction is neglected, the resulting graph will have an equation of the form of (1) or (2). The constants a, b, and c are determined by the weight and the spring as well as by how the weight is set into motion. For instance, the further the weight is pulled down before it is released, the greater will be a, the amplitude of the motion. Furthermore, the stiffer the spring, the more rapidly the weight will vibrate and thus the smaller will be P, the period of the motion; recall that the constants b and P are related by the equation $P = 2\pi/|b|$. The **frequency** of a simple harmonic motion is the number of vibrations, or oscillations, per unit of time. Thus, if n is the frequency of the motion, $n = 1/P$.

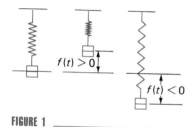

$f(t) > 0$

$f(t) < 0$

FIGURE 1

Illustration 1

A weight is vibrating vertically according to the equation

$$f(t) = 8 \cos \tfrac{1}{3}\pi t \tag{3}$$

where $f(t)$ centimeters is the directed distance of the weight from its central position (the origin) at t seconds and the positive direction is upward. Equation (3) is the special case of (2) where $a = 8$, $b = \tfrac{1}{3}\pi$, and $c = 0$. Therefore the motion is simple harmonic. Because the amplitude is 8, the maximum displacement is 8 cm. Because $b = \tfrac{1}{3}\pi$, the period P is given by

$$P = \frac{2\pi}{|b|}$$

$$= \frac{2\pi}{\tfrac{1}{3}\pi}$$

$$= 6$$

Therefore it takes 6 sec for one complete vibration of the weight. The frequency n is given by

$$n = \frac{1}{P}$$

$$= \frac{1}{6}$$

Thus there is $\frac{1}{6}$ of a vibration per second. From (3) we obtain values of $f(t)$ for the particular values of t shown in Table 1. From these values we can discuss the motion of the weight.

TABLE 1

t	0	$\frac{1}{2}$	1	$\frac{3}{2}$	2	$\frac{5}{2}$	3	$\frac{7}{2}$	4	$\frac{9}{2}$	5	$\frac{11}{2}$	6
$f(t)$	8	$4\sqrt{3} \approx 6.9$	4	0	-4	$-4\sqrt{3} \approx -6.9$	-8	-6.9	-4	0	4	6.9	8

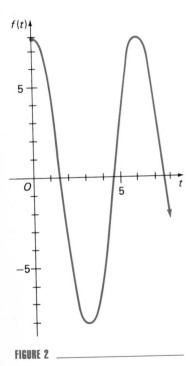

Initially the weight is 8 cm above the origin, the central position. In the first $\frac{1}{2}$ sec the weight moves downward 1.1 cm to a point 6.9 cm above the origin. Then in the next $\frac{1}{2}$ sec the weight moves downward 2.9 cm to a position 4 cm above the origin. In the third $\frac{1}{2}$ sec the weight moves downward a distance of 4 cm to its central position. Thus the speed of the weight is increasing in the first $\frac{3}{2}$ sec. In the next $\frac{3}{2}$ sec the motion of the weight continues downward while its speed is decreasing until after a total of 3 sec the weight is 8 cm below its central position. Then the weight reverses its direction and its speed increases until it attains its central position, following which the speed decreases until it is back to its starting position after a total of 6 sec. The weight then reverses its direction and the motion down and up is repeated indefinitely. A sketch of the graph of (3) appears in Figure 2. ■

FIGURE 2

So far we have neglected friction, which would cause the weight eventually to come to rest. We discuss *damped harmonic motion*, for which friction is taken into account, in the next section.

Example   **1**   A weight is suspended from a spring and is vibrating vertically. Suppose the weight passes through its central position as it is rising when $t = 2.5$, then attains a maximum upward displacement of 4 cm and passes through its central position as it is descending when $t = 3.5$. The motion is simple harmonic and is described by an equation of the form

$$f(t) = a \sin b(t - c)$$

where $f(t)$ centimeters is the directed distance of the weight from its central position after t seconds and the positive direction is upward. Find this equation.

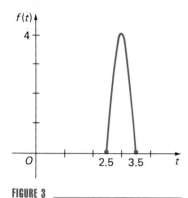

FIGURE 3

Solution   The motion from $t = \frac{5}{2}$ to $t = \frac{7}{2}$ is indicated in Figure 3. Because the maximum upward displacement is 4 cm, the amplitude of the motion is 4; thus $a = 4$. One-half cycle of the motion is completed between $t = \frac{5}{2}$ and $t = \frac{7}{2}$. Therefore, if the period is P,

$$\frac{P}{2} = \frac{7}{2} - \frac{5}{2}$$

$$P = 2$$

Because $P = 2\pi/|b|$, we have, if $b > 0$,

$$\frac{2\pi}{b} = 2$$

$$b = \pi$$

The graph of the required function can be thought of as a sine wave shifted $\frac{5}{2}$ units to the right from the origin. Thus $c = \frac{5}{2}$. Therefore from the equation $f(t) = a \sin b(t - c)$ with $a = 4$, $b = \pi$, and $c = \frac{5}{2}$ we obtain

$$f(t) = 4 \sin \pi(t - \tfrac{5}{2})$$ ∎

Example   2   Draw a sketch of the graph of the function in Example 1 and discuss the motion.

Solution   The equation is

$$f(t) = 4 \sin \pi(t - \tfrac{5}{2})$$

The period is 2. Thus it takes 2 sec for one complete vibration of the weight. The frequency n is given by $n = 1/P$; thus $n = \frac{1}{2}$. Therefore there is $\frac{1}{2}$ of a vibration per second. Table 2 gives values of $f(t)$ for every $\frac{1}{4}$ unit when t is in the interval $[0, 2]$. The portion of the graph on this interval is drawn from these values and the graph repeats this behavior on every interval of 2 units. See Figure 4.

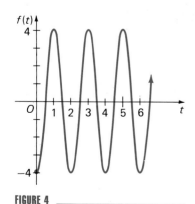

FIGURE 4

TABLE 2

t	0	$\frac{1}{4}$	$\frac{1}{2}$	$\frac{3}{4}$	1	$\frac{5}{4}$	$\frac{3}{2}$	$\frac{7}{4}$	2
$f(t)$	-4	$-2\sqrt{2} \approx -2.8$	0	$2\sqrt{2} \approx 2.8$	4	2.8	0	-2.8	-4

Initially the weight is 4 cm below the central position. In the first $\frac{1}{4}$ sec the weight moves upward a distance of 1.2 cm to a point 2.8 cm below the central position. In the second $\frac{1}{4}$ sec the weight moves upward a distance of 2.8 cm to its central position. The speed is increasing in the first $\frac{1}{2}$ sec. In the next $\frac{1}{4}$ sec the motion of the weight continues upward while its speed is decreasing until after a total of 1 sec the weight is 4 cm above its central position. The weight then reverses its direction and its speed increases until it attains its central position, after which the speed decreases until it has returned to its starting position after a total of 2 sec. This motion is repeated indefinitely every 2 sec. ∎

The following two examples show applications of the periodicity of the sine and cosine in business and electricity.

Example   3   A company that sells men's overcoats starts its fiscal year on July 1. For the three fiscal years beginning on July 1, 1987, the profit from sales was given approximately by

$$P(t) = 20{,}000(1 - \cos \tfrac{1}{6}\pi t) \qquad 0 \le t \le 36$$

where $P(t)$ dollars per month was the profit t months since July 1, 1987. (a) Find the period of P. Find the profit per month on (b) October 1, 1987; (c) November 1, 1987; (d) December 1, 1987; (e) January 1, 1988; (f) April 1, 1988; and (g) July 1, 1988.

Solution   a) The period of P is the same as the period of the function $\cos \tfrac{1}{6}\pi t$, which is $\dfrac{2\pi}{\frac{1}{6}\pi} = 12$.

For parts (b)–(g), the profits per month on October 1, 1987; November 1, 1987; December 1, 1987; January 1, 1988; April 1, 1988; and July 1, 1988, were, respectively, $P(3)$, $P(4)$, $P(5)$, $P(6)$, $P(9)$, and $P(12)$ dollars.

b) $P(3) = 20{,}000(1 - \cos \tfrac{1}{2}\pi)$
$ = 20{,}000(1 - 0)$
$ = 20{,}000$

On October 1, 1987, the profit per month was $20,000.

c) $P(4) = 20{,}000(1 - \cos \tfrac{2}{3}\pi)$
$ = 20{,}000(1 + \tfrac{1}{2})$
$ = 30{,}000$

On November 1, 1987, the profit per month was $30,000.

d) $P(5) = 20{,}000(1 - \cos \tfrac{5}{6}\pi)$
$ = 20{,}000\left(1 + \dfrac{\sqrt{3}}{2}\right)$
$ = 37{,}321$

On December 1, 1987, the profit per month was $37,321.

e) $P(6) = 20{,}000(1 - \cos \pi)$
$ = 20{,}000(1 + 1)$
$ = 40{,}000$

On January 1, 1988, the profit per month was $40,000.

f) $P(9) = 20,000(1 - \cos \frac{3}{2}\pi)$
$= 20,000(1 - 0)$
$= 20,000$

On April 1, 1988, the profit per month was $20,000.

g) $P(12) = 20,000(1 - \cos 2\pi)$
$= 20,000(1 - 1)$
$= 0$

On July 1, 1988, there was no profit per month.   ■

Example   4   The electromotive force for an electric circuit with a simplified generator is $E(t)$ volts at t seconds, where

$$E(t) = 50 \sin 120 \pi t$$

(a) Find the period of E. Find the electromotive force at (b) 0.02 sec and (c) 0.2 sec. (d) Draw one cycle of the sine wave defined by the given equation.

Solution   a) If the period is P, then

$$P = \frac{2\pi}{120\pi}$$

$$= \frac{1}{60}$$

b) The electromotive force at 0.02 sec is $E(0.02)$ volts and

$$\begin{aligned}
E(0.02) &= 50 \sin 120\pi(0.02) \\
&= 50 \sin 2.4\pi \\
&= 50 \sin(0.4\pi + 2\pi) \\
&= 50 \sin 0.4\pi \\
&= 50(0.9511) \\
&= 47.55
\end{aligned}$$

The electromotive force at 0.02 sec is 47.55 volts.

c) The electromotive force at 0.2 sec is $E(0.2)$ volts and

$$\begin{aligned}
E(0.2) &= 50 \sin 120\pi(0.2) \\
&= 50 \sin 24\pi \\
&= 50 \sin[0 + 12(2\pi)] \\
&= 50 \sin 0 \\
&= 0
\end{aligned}$$

The electromotive force at 0.2 sec is 0.

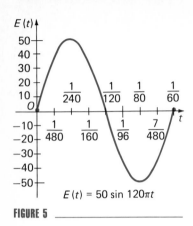

TABLE 3

t	0	$\frac{1}{480}$	$\frac{1}{240}$	$\frac{1}{160}$	$\frac{1}{120}$	$\frac{1}{96}$	$\frac{1}{80}$	$\frac{7}{480}$	$\frac{1}{60}$
$E(t)$	0	35.4	50	35.4	0	-35.4	-50	-35.4	0

d) From part (a) the period of E is $\frac{1}{60}$. The amplitude of E is 50. We use different scales on the two axes; on the t axis we take tick marks at every $\frac{1}{480}$ unit, and on the $E(t)$ axis we take tick marks at every 10 units. Table 3 gives values of $E(t)$ for every $\frac{1}{480}$ unit in the interval $[0, \frac{1}{60}]$. The cycle appears in Figure 5. ■

FIGURE 5

EXERCISES 6.8

In Exercises 1 through 6, the equation describes the simple harmonic motion of a weight suspended from a spring and vibrating vertically, where $f(t)$ centimeters is the directed distance of the weight from its central position (the origin) at t seconds and the positive direction is upward. For each motion, determine (a) the amplitude; (b) the period; (c) the frequency; and (d) the positions of the weight at t_1, t_2, t_3, and t_4 seconds.

1. $f(t) = 3 \cos \frac{1}{6}\pi t$; $t_1 = 0$, $t_2 = 2$, $t_3 = 4$, $t_4 = 6$
2. $f(t) = 2 \sin \frac{1}{2}\pi t$; $t_1 = 1$, $t_2 = 2$, $t_3 = 3$, $t_4 = 4$
3. $f(t) = 5 \sin 2t$; $t_1 = 0$, $t_2 = \frac{1}{4}\pi$, $t_3 = \frac{1}{2}\pi$, $t_4 = \frac{3}{4}\pi$
4. $f(t) = 6 \cos 3t$; $t_1 = 0$, $t_2 = \frac{1}{6}\pi$, $t_3 = \frac{1}{3}\pi$, $t_4 = \frac{1}{2}\pi$
5. $f(t) = 8 \cos \pi(2t - \frac{1}{3})$; $t_1 = 0$, $t_2 = \frac{1}{6}$, $t_3 = \frac{1}{3}$, $t_4 = \frac{1}{2}$
6. $f(t) = 3 \cos \pi(3t + \frac{1}{2})$; $t_1 = 0$, $t_2 = \frac{1}{6}$, $t_3 = \frac{1}{3}$, $t_4 = \frac{1}{2}$

In Exercises 7 through 10, draw a sketch of the graph of the function in the indicated exercise and discuss the motion.

7. Exercise 1
8. Exercise 2
9. Exercise 5
10. Exercise 6
11. A weight suspended from a spring is lifted up to a point 2 cm above its central position and then released. It takes $\frac{1}{2}$ sec for the weight to complete one vibration. (a) Write an equation defining $f(t)$, where $f(t)$ centimeters is the directed distance of the weight from its central position t seconds after the start of the motion and the positive direction is upward. Determine the position of the weight (b) $\frac{1}{12}$ sec after the start of the motion and (c) $\frac{1}{4}$ sec after the start of the motion.

12. A weight suspended from a spring is set into vibratory motion by pulling it down 4 cm from its central position and then releasing it. It takes 1.5 sec for the weight to complete one vibration. (a) Write an equation defining $f(t)$, where $f(t)$ centimeters is the directed distance of the weight from its central position t seconds after the start of the motion and the positive direction is upward. Determine the position of the weight (b) 1 sec after the start of the motion and (c) 2 sec after the start of the motion.

13. A weight is suspended from a spring and is vibrating vertically. The weight passes through its central position as it is rising, when $t = 2$, then attains a maximum displacement of 9 cm and passes through its central position as it is descending, when $t = 3.2$. The motion is simple harmonic and is described by an equation of the form $f(t) = a \cos b(t - c)$ where $f(t)$ centimeters is the directed distance of the weight from its central position after t seconds and the positive direction is upward. Find this equation.

14. A weight is suspended from a spring and is vibrating vertically. Suppose the weight passes through its central position at 3 sec and 7 sec. Between these times the weight attains twice a maximum displacement of 10 cm above its central position and attains once a maximum displacement of 10 cm below its central position. The motion is simple harmonic and is described by an equation of the form $f(t) = a \sin b(t - c)$, where $f(t)$ centi-

meters is the directed distance of the weight from its central position at t seconds and the positive direction is upward. Find this equation.

15. In a particular city on each of the dates January 15, 16, and 17, the temperature varied from $-5°$ Celsius at 2 A.M. to 5° Celsius at 2 P.M. With the assumption that the graph is a sine wave, (a) write an equation defining $T(t)$ as a function of t where $T(t)$ degrees Celsius was the temperature t hours since 2 A.M., January 15 and $0 \leq t \leq 48$. What was the temperature at (b) 6 A.M., January 15; (c) 12 noon, January 15; (d) 4 P.M., January 15; and (e) 10 P.M., January 15? (f) Draw a sketch of the graph of T.

16. In a certain city, at any particular time of day from October 1 through October 4, the Fahrenheit temperature was $T(t)$ degrees at t hours since midnight, September 30, where

$$T(t) = 60 - 15 \sin \tfrac{1}{12}\pi(8 - t) \qquad 0 \leq t \leq 96$$

(a) Determine the period of T. Find the temperature at (b) 8 A.M., October 1; (c) 12 noon, October 1; (d) 2 P.M., October 1; (e) 6 P.M., October 1; and (f) midnight, October 1. (g) Draw a sketch of the graph of T.

17. Suppose the motion of a particle along a straight line is simple harmonic and is described by an equation of the form $S(t) = a \sin b(t - c)$ where $S(t)$ centimeters is the displacement of the particle from a fixed point (the origin) at t seconds. Then,

it is proved in calculus, if $V(t)$ centimeters per second is the velocity of the particle at t seconds $V(t) = ab \cos b(t - c)$ and if $A(t)$ centimeters per second per second is the acceleration of the particle at t seconds $A(t) = -ab^2 \sin b(t - c)$. If an equation describing the motion is $S(t) = 2 \sin \tfrac{1}{3}\pi(t - 1)$, what are the particle's position, velocity, and acceleration at (a) 0 sec; (b) 1 sec; (c) 2 sec; (d) 3 sec; and (e) 4 sec?

18. Do Exercise 17 if $S(t) = 4 \sin \tfrac{1}{6}\pi(t + 2)$.

19. In an electric circuit the electromotive force is $E(t)$ volts at t seconds, where $E(t) = 2 \cos 50\pi t$. (a) Determine the period of E. Find the electromotive force at (b) 0.02 sec; (c) 0.03 sec; (d) 0.04 sec; and (e) 0.06 sec.

20. Do Exercise 19 if $E(t) = 40 \sin 120\pi t$.

21. An alternating current of electricity is described by $I(t) = 10 \sin 2800t$, where $I(t)$ amperes is the current at t seconds. (a) Determine the period of I. Find the current at (b) 0.001 sec; (c) 0.003 sec; (d) 0.005 sec; and (e) 0.01 sec.

22. Do exercise 21 if $I(t) = 2 \sin 3000t$.

23. A wave produced by a simple sound has the equation $F(t) = 0.02 \sin 1500\pi t$, where $F(t)$ dynes per square centimeter is the difference between the atmospheric pressure and the air pressure at the eardrum at t seconds. (a) Determine the period of F. Find the difference between the atmospheric pressure and the air pressure at the eardrum at (b) $\tfrac{1}{9}$ sec; (c) $\tfrac{1}{8}$ sec; (d) $\tfrac{1}{7}$ sec; and (e) $\tfrac{1}{6}$ sec.

24. Do Exercise 23 if $F(t) = 0.003 \sin 1800\pi t$.

6.9   Other Graphs Involving Sine and Cosine Functions (Supplementary)

The sum of sine and cosine functions appears often in various applications of mathematics, particularly in the study of sound, heat, and electricity.

We first discuss a procedure for sketching the graph of the sum of two functions. Let h be the sum of the functions f and g, having the same domain, so that

$$h(x) = f(x) + g(x)$$

for every x in the common domain. The graph of h can be obtained from the graphs of f and g by the *graphical addition of ordinates*.

Figure 1 shows sketches of the graphs of f and g on the same set of coordinate axes. For every $\bar{x}$ in the domain of h,

$$h(\bar{x}) = f(\bar{x}) + g(\bar{x})$$

Thus at $x = \bar{x}$ the ordinate of the point on the graph of h is the sum of the ordinates of the corresponding points on the graphs of f and g. Observe that the point $(\bar{x}, h(\bar{x}))$ on the graph of h can be obtained by measuring the directed distance $g(\bar{x})$ with either a ruler or compass and adding this directed distance to the directed distance $f(\bar{x})$. In Figure 1, because $g(\bar{x}_1)$ is positive the point $(\bar{x}_1, h(\bar{x}_1))$ is above the point $(\bar{x}_1, f(\bar{x}_1))$, and because $g(\bar{x}_2)$ is negative the point $(\bar{x}_2, h(\bar{x}_2))$ is below the point $(\bar{x}_2, f(\bar{x}_2))$. Observe that when $g(x) = 0$ the graph of h intersects the graph of f, and when $f(x) = 0$ the graph of h intersects the graph of g. Furthermore, when $f(x)$ and $g(x)$ have the same absolute value and are opposite in sign, $h(x) = 0$, and the graph of h intersects the x axis.

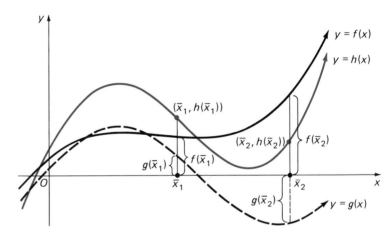

FIGURE 1

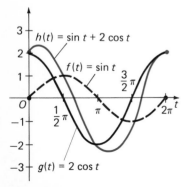

FIGURE 2

Illustration   1

Consider the function h defined by

$$h(t) = \sin t + 2 \cos t$$

We obtain a sketch of the graph of h for $0 \le t \le 2\pi$ by first drawing sketches of the graphs of $f(t) = \sin t$ and $g(t) = 2 \cos t$ on the interval $[0, 2\pi]$. Points on the graph of h are then obtained by the graphical addition of ordinates. Figure 2 shows these sketches with the sketch of h on $[0, 2\pi]$ appearing in color.

The points on h can also be obtained by computing function values of h. In particular,

$$h(0) = f(0) + g(0) \qquad h(\tfrac{1}{4}\pi) = f(\tfrac{1}{4}\pi) + g(\tfrac{1}{4}\pi) \qquad h(\tfrac{3}{4}\pi) = f(\tfrac{3}{4}\pi) + g(\tfrac{3}{4}\pi)$$
$$= 0 + 2 \qquad\qquad\qquad = \frac{1}{\sqrt{2}} + \sqrt{2} \qquad\qquad = \frac{1}{\sqrt{2}} - \sqrt{2}$$
$$= 2 \qquad\qquad\qquad\quad \approx 0.7 + 1.4 \qquad\qquad \approx 0.7 - 1.4$$
$$h(\pi) = f(\pi) + g(\pi) \qquad\quad = 2.1 \qquad\qquad\qquad = -0.7$$
$$= 0 - 2$$
$$= -2$$

$$h(2\pi) = f(2\pi) + g(2\pi) \qquad h(\tfrac{5}{4}\pi) = f(\tfrac{5}{4}\pi) + g(\tfrac{5}{4}\pi) \qquad h(\tfrac{7}{4}\pi) = f(\tfrac{7}{4}\pi) + g(\tfrac{7}{4}\pi)$$
$$= 0 + 2 \qquad\qquad\qquad = -\frac{1}{\sqrt{2}} - \sqrt{2} \qquad\qquad = -\frac{1}{\sqrt{2}} + \sqrt{2}$$
$$= 2 \qquad\qquad\qquad\quad \approx -0.7 - 1.4 \qquad\qquad \approx -0.7 + 1.4$$
$$= -2.1 \qquad\qquad\qquad = 0.7$$

Example   1   Draw a sketch of the graph of $f(t) = \cos 2t + 2 \cos t$.

Solution   Let

$$F(t) = \cos 2t \qquad \text{and} \qquad G(t) = 2 \cos t$$

The period of F is π and the period of G is 2π. Therefore the period of f is 2π. Recall that $\cos(-t) = \cos t$. Therefore

$$f(-t) = \cos(-2t) + 2 \cos(-t)$$
$$= \cos 2t + 2 \cos t$$
$$= f(t)$$

Thus f is an even function, and its graph is symmetric with respect to the $f(t)$ axis. Therefore we begin by sketching the graph of f on $[0, \pi]$. Figure 3 shows the graphs of F and G on $[0, \pi]$. The graph of f on $[0, \pi]$, shown in color, is obtained by graphical addition of ordinates.

Figure 4 shows the complete sketch. The portion of the graph on $[-\pi, 0]$ is obtained from the portion on $[0, \pi]$ by properties of symmetry. The remainder of the sketch follows because f has period 2π. Observe that the scales on the axes of Figure 4 are smaller than those on the axes of Figure 3 so that more of the graph can be shown.

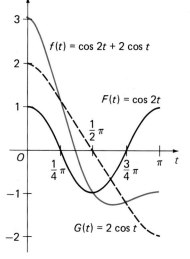

FIGURE 3

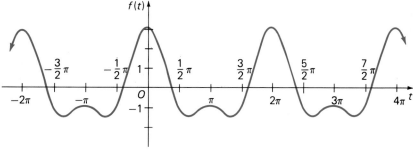

FIGURE 4

The function in the next example involves the sum of the identity function and the cosine function.

Example 2 Draw a sketch of the graph of $f(t) = t + \cos t$.

Solution Let

$$F(t) = t \qquad \text{and} \qquad G(t) = \cos t$$

We draw sketches of the graphs of F and G on the same set of coordinate axes. See Figure 5, where the graphs of F and G are indicated by black curves. The graph of f is obtained by graphical addition of ordinates.

First plot the points at values of t for which $\cos t = 0$. At these points the graph of f intersects the graph of F. Then plot the point for $t = 0$, which is where the graph of f intersects the graph of G. We obtain other points on the graph of f by choosing some arbitrary values of t. The graph of f is shown in color in Figure 5.

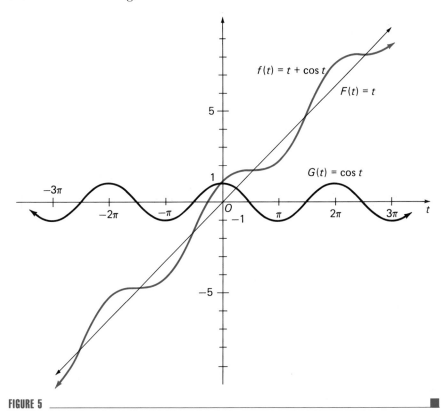

FIGURE 5

We saw that simple harmonic motion continues indefinitely, repeating a cycle every interval of length a period. For instance, in Example 1 of Section 6.8 the weight suspended from a spring moves vertically upward and downward and one complete oscillation occurs every interval of 2 sec. In practice, however, friction would cause the amplitude of the motion to decrease until the weight finally came to rest. This is the case of *damped harmonic motion,* which can be described by the product of a sine function and a nonconstant function called a *damping factor.* The damping factor is what causes the decrease in amplitude. An important damping factor is an exponential function whose values approach zero as the independent variable increases without bound. The following example illustrates the effect of this factor.

Example   **3**   A damped harmonic motion is described by

$$f(t) = e^{-t} \sin 2t$$

Draw a sketch of the graph of f for $t \geq 0$.

Solution   The given function f is the product of two functions. Because $|\sin 2t| \leq 1$ and $e^{-t} > 0$ for all t, it follows that

$$|f(t)| \leq e^{-t}$$

for all t. Thus

$$-e^{-t} \leq f(t) \leq e^{-t}$$

for all t. Let

$$F(t) = -e^{-t} \quad \text{and} \quad G(t) = e^{-t}$$

Then

$$F(t) \leq f(t) \leq G(t)$$

and the graph of f lies between the graphs of F and G. When $\sin 2t = 1$, the graph of f intersects the graph of G; these are the points for which $t = \frac{1}{4}\pi + k\pi$, $k \in J$. When $\sin 2t = -1$, the graph of f intersects the graph of F; these are the points for which $t = \frac{3}{4}\pi + k\pi$, $k \in J$. Furthermore, when $\sin 2t = 0$, the graph of f intersects the t axis; these are the points for which $t = \frac{1}{2}k\pi$, $k \in J$. In Figure 6 the graphs of F and G are indicated by black curves, and the graph of f appears in color.

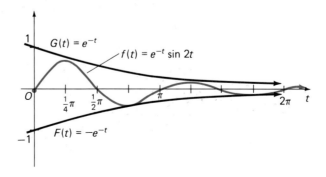

FIGURE 6

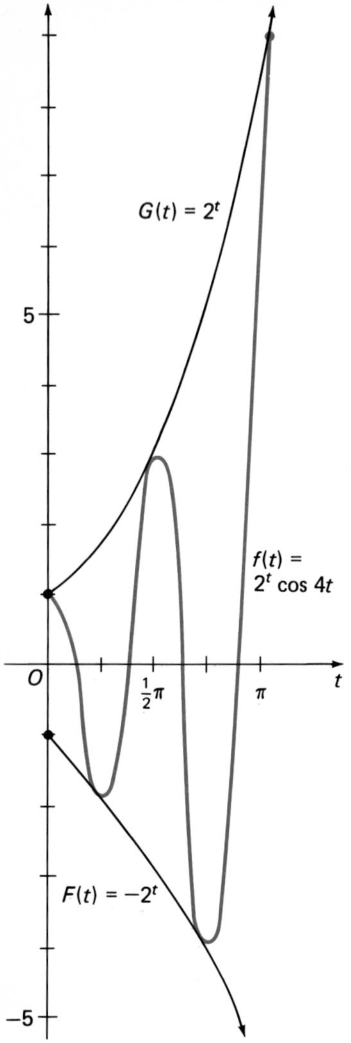

Damped harmonic motion is important in the design of buildings, bridges, and vehicles. For instance, in order to damp the oscillations when an automobile encounters a bump in the road, shock absorbers are used.

For damped harmonic motion, the amplitude decreases to zero as the time increases. If the amplitude increases without bound as time increases, then **resonance** occurs. The following example gives a mathematical model describing resonance.

Example   4   Resonance is described by

$$f(t) = 2^t \cos 4t$$

Draw a sketch of the graph of f on $[0, \pi]$.

Solution   The given function f is the product of two functions. Because $|\cos 4t| \leq 1$ and $2^t > 0$ for all t, then

$$|f(t)| \leq 2^t$$
$$-2^t \leq f(t) \leq 2^t$$

for all t. Let

$$F(t) = -2^t \quad \text{and} \quad G(t) = 2^t$$

Then

$$F(t) \leq f(t) \leq G(t)$$

and the graph of f lies between the graphs of F and G. When $\cos 4t = 1$, the graph of f intersects the graph of G. On $[0, \pi]$ these are the points where t is 0, $\frac{1}{2}\pi$, and π. When $\cos 4t = -1$, the graph of f intersects the graph of F and this occurs on $[0, \pi]$ when t is $\frac{1}{4}\pi$ and $\frac{3}{4}\pi$. Because $\cos 4t = 0$ when t is $\frac{1}{8}\pi$, $\frac{3}{8}\pi$, $\frac{5}{8}\pi$, and $\frac{7}{8}\pi$, the graph of f intersects the t axis for these values of t. In Figure 7 the graphs of F and G are indicated by black curves and the graph of f is shown in color.   ∎

Example   5   Draw a sketch of the graph of the function defined by

$$f(t) = \frac{1}{t} \sin t$$

for $t \in [-\pi, 0) \cup (0, \pi]$.

Solution   Observe that $f(0)$ does not exist. Thus the graph does not intersect the vertical axis. Furthermore, because $\sin(-t) = -\sin t$, we have

$$f(-t) = \frac{1}{-t} \sin(-t)$$

$$= -\frac{1}{t}(-\sin t)$$

$$= \frac{1}{t} \sin t$$

$$= f(t)$$

$G(t) = 2^t$

$f(t) = 2^t \cos 4t$

$F(t) = -2^t$

FIGURE 7

Therefore f is an even function and its graph is symmetric with respect to the vertical axis. We shall first obtain the portion of the graph for $t \in (0, \pi]$.

If $t > 0$, then because $|\sin t| \leq 1$,

$$-\frac{1}{t} \leq \frac{1}{t} \sin t \leq \frac{1}{t}$$

Let

$$F(t) = -\frac{1}{t} \quad \text{and} \quad G(t) = \frac{1}{t}$$

Then if $t > 0$,

$$F(t) \leq f(t) \leq G(t)$$

and for positive values of t the graph of f lies between the graphs of F and G. Because $\sin \frac{1}{2}\pi = 1$, the graph of f intersects the graph of G at $t = \frac{1}{2}\pi$. Because $\sin \pi = 0$, the graph of f intersects the t axis at $t = \pi$. Also, when $0 < t < \pi$, $\sin t > 0$. Therefore on the interval $(0, \pi)$ the graph lies above the t axis.

The graph of f on $(0, \pi]$ is drawn as shown in Figure 8. The portion of the graph on $[-\pi, 0)$ follows from properties of symmetry. Observe that there is an open dot on the vertical axis, which indicates that $f(0)$ is not defined. ∎

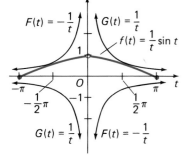

FIGURE 8

For the function of Example 5, notice in Figure 8 that as t gets closer and closer to zero through either positive or negative values, then $f(t)$ appears to be approaching 1. In Exercise 39 of Exercises 6.4 you were asked to compute values of the quotient $\dfrac{\sin t}{t}$ for small values of t. These results also suggested that $f(t)$ approaches 1 as t approaches zero, a fact with important consequences in calculus.

EXERCISES 6.9

In Exercises 1 through 32, draw a sketch of the graph of the function.

1. $f(t) = \cos t + 2 \sin t$
2. $f(t) = \sin t + 3 \cos t$
3. $f(t) = 3 \sin t + 2 \cos t$
4. $f(t) = 4 \cos t + 2 \sin t$
5. $f(t) = 3 + \cos t$
6. $f(t) = 4 - \cos t$
7. $f(t) = 2 - 3 \sin t$
8. $f(t) = 3 + 2 \cos t$
9. $f(t) = 3 \cos t - 2 \sin t$
10. $f(t) = 2 \sin t - 3 \cos t$
11. $f(t) = \sin 2t + 2 \sin t$
12. $f(t) = \cos 2t - 2 \cos t$
13. $f(t) = 2 \cos t - \sin \frac{1}{2}t$
14. $f(t) = \cos \frac{1}{2}t + \frac{1}{2} \sin t$
15. $f(t) = \sin \pi t + 3 \cos \pi t$
16. $f(t) = 2 \cos \pi t + 3 \sin \pi t$
17. $f(t) = \cos 2\pi t - 2 \cos \pi t$
18. $f(t) = \frac{1}{2} \sin 2\pi t - 2 \sin \frac{1}{2}\pi t$
19. $f(t) = t + \sin t$
20. $f(t) = \cos t - t$

21. $f(t) = 2t - \cos t$
22. $f(t) = \frac{1}{2}t - \sin t$
23. $f(t) = t^2 - \cos 2t$
24. $f(t) = 2 \sin t - \frac{1}{2}t^2$
25. $f(t) = \sin \frac{1}{2}\pi t + t$
26. $f(t) = \cos 2\pi t + t$
27. $f(t) = 1 + \tan t$
28. $f(t) = \cot t - 1$
29. $f(t) = \csc t - 1$
30. $f(t) = 3 \sec t - 2$
31. $f(t) = \tan t + \cot t$
32. $f(t) = \tan t - \cot t$

In Exercises 33 through 38, a damped harmonic motion is described by the equation. Draw a sketch of the graph for $t \geq 0$.

33. $f(t) = e^{-t} \sin 4t$
34. $f(t) = e^{-t} \cos 2t$
35. $f(t) = 2^{-t} \cos 2t$
36. $f(t) = e^{-2t} \sin t$
37. $f(t) = e^{-t/2} \cos 4t$
38. $f(t) = 3^{-t} \sin 4t$

In Exercises 39 through 44, resonance is described by the equation. Draw a sketch of the graph for t in the indicated interval.

39. $f(t) = 2^t \sin 4t$; $t \in [0, \pi]$
40. $f(t) = e^t \cos 8t$; $t \in [0, \frac{1}{2}\pi]$
41. $f(t) = t \cos t$; $t \in [0, 2\pi]$
42. $f(t) = t \sin t$; $t \in [0, 2\pi]$

43. $f(t) = t^2 \sin t$; $t \in [0, 2\pi]$
44. $f(t) = t^2 \cos t$; $t \in [0, 2\pi]$
45. Let $f(t) = \cos^2 t$ and $g(t) = \frac{1}{2} + \frac{1}{2} \cos 2t$. On separate coordinate axes, draw sketches of the graphs of f and g for $t \in [0, 2\pi]$. How do the graphs compare?
46. Let $f(t) = \sin^2 t$ and $g(t) = \frac{1}{2} - \frac{1}{2} \cos 2t$. On separate coordinate axes, draw sketches of the graphs of f and g for $t \in [0, 2\pi]$. How do the graphs compare?

REVIEW EXERCISES FOR CHAPTER 6

In Exercises 1 through 4, show by a diagram an angle in standard position that has the given degree measurement. Also find the equivalent radian measurement of the angle.

1. (a) 30°; (b) 225°
2. (a) 15°; (b) 330°
3. (a) −120°; (b) 100°
4. (a) −135°; (b) 250°

In Exercises 5 through 8, show by a diagram an angle in standard position that has the given radian measurement. Also find the equivalent degree measurement of the angle.

5. (a) $\frac{1}{3}\pi$; (b) $\frac{7}{4}\pi$
6. (a) $\frac{5}{6}\pi$; (b) $\frac{5}{4}\pi$
7. (a) $-\frac{1}{3}$; (b) 2.36
8. (a) $\frac{5}{6}$; (b) −1.48

In Exercises 9 and 10, find (a) the arc length and (b) the area of the sector of the circle having the radius and central angle α.

9. Radius is 12 cm; $m°(\alpha) = 150$
10. Radius is 6.48 in.; $m°(\alpha) = 27.25$

In Exercises 11 through 14, θ is an angle in standard position and the point P is on the terminal side of θ; find the six trigonometric functions of θ.

11. $P(-8, 15)$
12. $P(3, -4)$
13. $P(-5, -12)$
14. $P(-1, 0)$

In Exercises 15 through 18, find the exact values of the other five trigonometric functions of θ by the method of Section 6.2.

15. $\sin \theta = \frac{3}{5}$ and $\cos \theta < 0$
16. $\tan \theta = -\frac{5}{12}$ and $\cos \theta > 0$
17. $\cot \theta = -\frac{8}{15}$ and $\sin \theta < 0$
18. $\cos \theta = -\frac{15}{17}$ and $\sin \theta < 0$

In Exercises 19 through 26, draw a sketch showing the angle in standard position, and find the exact values of the six trigonometric functions of the angle.

19. 45°
20. 60°
21. $\frac{2}{3}\pi$
22. $\frac{3}{4}\pi$
23. −150°
24. −135°
25. $-\frac{9}{4}\pi$
26. $-\frac{1}{6}\pi$

In Exercises 27 through 30, draw a sketch showing the quadrantal angle in standard position. Find the exact values of the four trigonometric functions that are defined. Which two trigonometric functions are not defined?

27. 270°
28. 90°
29. $-\pi$
30. 2π

In Exercises 31 and 32, find the reference angle for the given angle.

31. (a) $\theta = 315°$; (b) $\theta = 218°$; (c) $\theta = \frac{13}{6}\pi$; (d) $\theta = -4.27$; (e) $\theta = -\frac{5}{9}\pi$; (f) $\theta = 11.23$
32. (a) $\theta = 150°$; (b) $\theta = 291°$; (c) $\theta = -\frac{3}{4}\pi$; (d) $\theta = 7.65$; (e) $\theta = \frac{11}{10}\pi$; (f) $\theta = -216.3°$

In Exercises 33 and 34, express the trigonometric function value in terms of a function of the associated reference angle; then determine the exact value.

33. (a) $\sin 150°$; (b) $\cos 225°$; (c) $\tan(-240°)$; (d) $\cot \frac{7}{4}\pi$; (e) $\sec(-\frac{2}{3}\pi)$; (f) $\csc \frac{11}{6}\pi$
34. (a) $\sin 240°$; (b) $\cos(-45°)$; (c) $\tan 120°$; (d) $\cot(-\frac{3}{4}\pi)$; (e) $\sec(-\frac{11}{6}\pi)$; (f) $\csc \frac{2}{3}\pi$

In Exercises 35 through 38, express the trigonometric function value in terms of a trigonometric function of the associated reference angle. Then approximate the value by using either Table VI or Table VII in the back of the book.

35. (a) $\sin 136°24'$; (b) $\cos(-104.8°)$; (c) $\tan 327°6'$

36. (a) $\sin 263°12'$; (b) $\cos 348.9°$; (c) $\tan(-191.7°)$
37. (a) $\sin 3.57$; (b) $\cos 2.42$; (c) $\tan(-14.86)$
38. (a) $\sin 5.85$; (b) $\cos(-11.43)$; (c) $\tan 4.26$

In Exercises 39 through 44, (a) determine the sine and cosine of the real number from the coordinates of a point on the unit circle U; (b) from the values in (a) find the tangent, cotangent, secant, and cosecant of the real number.

39. (a) $\frac{1}{4}\pi$; (b) $\frac{1}{6}\pi$; (c) 0 40. (a) $\frac{1}{3}\pi$; (b) $\frac{1}{2}\pi$; (c) π
41. (a) $\frac{3}{4}\pi$; (b) $-\frac{5}{6}\pi$; (c) 3π
42. (a) $-\frac{1}{4}\pi$; (b) $\frac{2}{3}\pi$; (c) -6π
43. (a) $-\frac{4}{3}\pi$; (b) $\frac{17}{4}\pi$; (c) $-\frac{9}{2}\pi$
44. (a) $\frac{19}{6}\pi$; (b) $-\frac{15}{4}\pi$; (c) $\frac{15}{2}\pi$

In Exercises 45 and 46, use the periodicity of the sine, cosine, secant, and cosecant functions as well as values of $\sin t$, $\cos t$, $\sec t$, and $\csc t$ when $0 \le t < 2\pi$ to find the function value.

45. (a) $\sin \frac{21}{4}\pi$; (b) $\cos \frac{14}{3}\pi$; (c) $\sec(-\frac{11}{6}\pi)$;
    (d) $\sin(-5\pi)$; (e) $\cos \frac{11}{2}\pi$; (f) $\csc \frac{13}{2}\pi$
46. (a) $\sin \frac{8}{3}\pi$; (b) $\cos \frac{13}{6}\pi$; (c) $\csc(-\frac{17}{4}\pi)$; (d) $\sin 7\pi$;
    (e) $\cos(-\frac{9}{2}\pi)$; (f) $\sec(-4\pi)$

In Exercises 47 and 48, use the periodicity of the tangent and cotangent functions as well as the values of $\tan t$ and $\cot t$ when $0 \le t < \pi$ to find the function value.

47. (a) $\tan \frac{5}{4}\pi$; (b) $\cot \frac{5}{3}\pi$; (c) $\tan(-\frac{1}{2}\pi)$; (d) $\tan \frac{14}{3}\pi$;
    (e) $\cot(-\frac{11}{6}\pi)$; (f) $\cot(-6\pi)$
48. (a) $\tan \frac{13}{6}\pi$; (b) $\cot(-\frac{1}{4}\pi)$; (c) $\cot(-\pi)$;
    (d) $\tan(-\frac{4}{3}\pi)$; (e) $\cot \frac{17}{4}\pi$; (f) $\tan(-\frac{7}{2}\pi)$

In Exercises 49 through 52, find the period of the function.

49. (a) $\sin 5t$; (b) $\cos \frac{1}{5}t$; (c) $3 \tan \frac{5}{4}t$; (d) $5 \sec \frac{3}{4}t$
50. (a) $\cos 4t$; (b) $\sin \frac{1}{4}t$; (c) $7 \cot \frac{2}{3}t$; (d) $2 \csc \frac{7}{3}t$
51. (a) $2 \cos 6\pi t$; (b) $\sin \frac{3}{2}\pi t$; (c) $5 \tan 2\pi t$; (d) $\frac{1}{2} \cot \frac{1}{3}\pi t$
52. (a) $4 \sin 5\pi t$; (b) $\frac{2}{3} \cos \frac{3}{2}\pi t$; (c) $\frac{1}{4} \tan \pi t$; (d) $3 \cot \frac{3}{4}\pi t$

In Exercises 53 through 76, draw a sketch of the graph of the function.

53. $f(t) = \cos 2t$
54. $f(t) = \sin 3t$
55. $g(t) = 2 \cos t$
56. $g(t) = 3 \sin t$
57. $f(t) = \sin 2\pi t$
58. $f(t) = \cos \frac{1}{2}\pi t$
59. $g(t) = 2 \sin \pi t$
60. $g(t) = \frac{1}{2} \cos \pi t$
61. $f(t) = \sin(t - \frac{1}{2}\pi)$
62. $f(t) = \cos(t - \pi)$
63. $g(t) = \cos(t + \frac{1}{3}\pi)$
64. $g(t) = \sin(t + \frac{1}{6}\pi)$
65. $f(t) = \tan 2t$
66. $f(t) = \cot \frac{1}{2}t$
67. $g(t) = \frac{1}{2} \cot t$
68. $g(t) = 2 \tan t$

69. $f(t) = \tan 3\pi t$
70. $f(t) = \cot 2\pi t$
71. $f(t) = 3 \csc t$
72. $f(t) = 2 \sec t$
73. $g(t) = 2 \sec \frac{1}{3}\pi t$
74. $g(t) = 3 \csc \frac{1}{2}\pi t$
75. $f(t) = \cot(t + \frac{1}{3}\pi)$
76. $f(t) = \frac{1}{2} \tan(t - \frac{1}{4}\pi)$

In Exercises 77 and 78, express the trigonometric function value as a function of a positive angle not greater than 45°.

77. (a) $\sin 75°$; (b) $\cos 62.3°$; (c) $\tan 47.2°$;
    (d) $\cot 87.6°$; (e) $\sec 55.1°$; (f) $\csc 78.9°$
78. (a) $\sin 80.5°$; (b) $\cos 57.4°$; (c) $\tan 67°$;
    (d) $\cot 46.8°$; (e) $\sec 72.2°$; (f) $\csc 89.4°$

In Exercises 79 and 80, solve the right triangle, where the right angle is at vertex C. For the trigonometric function values, do not use tables or a calculator. Express the results to the number of significant digits justified by the given information.

79. $\alpha = 30°$, $b = 16$ 80. $\beta = 45°$, $c = 7.4$

In Exercises 81 through 84, solve the right triangle, where the right angle is at vertex C. Express the results to the number of significant digits justified by the given information.

81. $a = 4.8$, $b = 3.2$ 82. $\alpha = 23.2°$, $a = 12.2$
83. $\beta = 51.34°$, $c = 832.7$ 84. $\beta = 23.59°$, $a = 46.10$

In Exercises 85 and 86, find the area of the triangle of the indicated exercise.

85. Exercise 81   86. Exercise 84
87. An automobile tire has diameter of 30 in. How many revolutions per minute will the wheel make when the automobile maintains a speed of 30 mi/hr?
88. A tower is 150 ft high and from its top the angle of depression of an object on the ground is 36.4°. (a) Determine the distance from the base of the tower to the object. (b) How far is the object from the top of the tower?

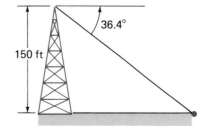

89. From the roof of a building 70 ft high, the angle of elevation of the top of a pole is 11.2°. From the bottom of the building the angle of elevation of the top of the pole is 33.4°. Find (a) the height of the pole and (b) the distance from the building to the pole.

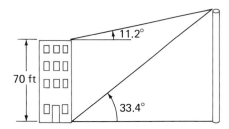

90. At a point P south of a building, the angle of elevation of the top of the building is 58°. At a point Q, 250 ft west of P, the angle of elevation is 27°. Find the height of the building.

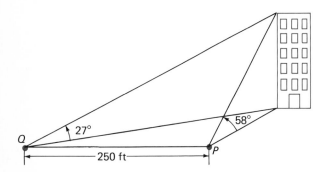

91. A weight suspended from a spring is vibrating vertically. The directed distance of the weight from its central position (the origin) at t seconds is $f(t)$ centimeters, with the positive direction upward. If $f(t) = 2 \sin 3t$, (a) determine the period of f, and find the position of the weight at (b) 0 sec; (c) 1 sec; (d) 2 sec; (e) 5 sec.

92. A weight suspended from a spring is vibrating vertically. The directed distance of the weight from its central position (the origin) at t seconds is $f(t)$ centimeters, with the positive direction upward, where $f(t) = 3 \cos 4t$. (a) Determine the period of f. Find the position of the weight at (b) 0 sec; (c) 2 sec; (d) 4 sec; (e) 8 sec.

93. A weight suspended from a spring is set into vibratory motion by pulling it down 6 cm from its central position and then releasing it. It takes 1.8 sec for the weight to complete one vibration. (a) Write an equation defining $f(t)$, where $f(t)$ centimeters is the directed distance of the weight from its central position t seconds after the start of the motion and the positive direction is upward. Determine the position of the weight when it has been in motion (b) $\frac{1}{2}$ sec; (c) 1 sec; (d) 2 sec; (e) 4 sec.

94. A weight is suspended from a spring and is vibrating vertically. Suppose the weight passes through its central position at 4 sec and 8 sec. Between these times the weight attains once a maximum displacement of 12 cm above its central position and attains twice a maximum displacement of 12 cm below its central position. The motion is simple harmonic and is described by an equation of the form $f(t) = a \sin b(t - c)$, where $f(t)$ centimeters is the directed distance of the weight from its central position at t seconds and the positive direction is upward. Find this equation.

95. In an electric circuit the electromotive force is $E(t)$ volts at t seconds, where $E(t) = 4 \cos 100\pi t$. (a) Determine the period of E. Find the electromotive force at (b) 0.01 sec; (c) 0.02 sec; (d) 0.05 sec; (e) 0.08 sec.

96. Do Exercise 95 if $E(t) = 6 \sin 50\pi t$.

Exercises 97 through 104 pertain to Supplementary Section 6.9. In these exercises draw a sketch of the graph of the function defined by the equation.

97. $f(t) = 3 \sin t + \cos t$ 98. $f(t) = 4 \cos t - \sin t$

99. $f(t) = \sin \pi t + t$ 100. $f(t) = \frac{3}{2}t - \cos t$

101. $f(t) = e^{-t} \sin 2t; \ t \geq 0$ 102. $f(t) = 3^{-2t} \cos t; \ t \geq 0$

103. $f(t) = 2^t \sin 8t; \ t \in [0, \frac{1}{2}\pi]$

104. $f(t) = e^t \cos 4t; \ t \in [0, \pi]$

C H A P T E R

7

Analytic Trigonometry

Introduction

Analytic trigonometry includes the study of trigonometric identities, inverse trigonometric functions, and trigonometric equations.

The eight fundamental identities are discussed in Section 7.1, and in Section 7.2 we apply them to prove other identities. These proofs will help you acquire the manipulative skills that may be necessary to write a complicated trigonometric expression in simpler form. In Sections 7.3 and 7.4, we derive other important identities and use them to simplify trigonometric expressions. The identities in Supplementary Section 7.7 are not utilized as much as those in the other sections, but we present them here to give you a complete set of formulas for later work. All of the identities are listed on the endpapers.

In Section 7.5, we define the inverses of the sine, cosine, and tangent functions and obtain their graphs. We discuss methods of solving trigonometric equations in Section 7.6. Throughout the chapter, the treatment is augmented by a number of applications in the examples and exercises.

7.1 The Eight Fundamental Identities

Recall from Section 2.1 that an identity is an equation for which the solution set is the same as the domain of the variable. For trigonometric identities we will use symbols such as s, t, u, and v to represent real numbers and α, β, γ, and θ to represent angles. When we use symbols such as x, y, and z it is understood that they can be either real numbers or angles.

From Chapter 6 we have the following reciprocal and quotient identities:

$$\csc x = \frac{1}{\sin x} \qquad \sec x = \frac{1}{\cos x} \qquad \cot x = \frac{1}{\tan x}$$

$$\tan x = \frac{\sin x}{\cos x} \qquad \cot x = \frac{\cos x}{\sin x}$$

We also have the fundamental Pythagorean identity:

$$\sin^2 x + \cos^2 x = 1$$

If $\cos x \neq 0$, we can divide on each side by $\cos^2 x$ and obtain

$$\frac{\sin^2 x}{\cos^2 x} + \frac{\cos^2 x}{\sin^2 x} = \frac{1}{\cos^2 x}$$

$$\left(\frac{\sin x}{\cos x}\right)^2 + 1 = \left(\frac{1}{\cos x}\right)^2$$

$$\tan^2 x + 1 = \sec^2 x$$

We can divide on each side of the fundamental Pythagorean identity by

$\sin^2 x$ provided that $\sin x \neq 0$. Doing this, we get

$$\frac{\sin^2 x}{\sin^2 x} + \frac{\cos^2 x}{\sin^2 x} = \frac{1}{\sin^2 x}$$

$$1 + \left(\frac{\cos x}{\sin x}\right)^2 = \left(\frac{1}{\sin x}\right)^2$$

$$1 + \cot^2 x = \csc^2 x$$

Therefore we have the following two identities that are valid when the functions are defined:

$$\tan^2 x + 1 = \sec^2 x$$
$$1 + \cot^2 x = \csc^2 x$$

These identities are also called Pythagorean identities. The relationships given by the Pythagorean, reciprocal, and quotient identities constitute the **eight fundamental trigonometric identities.** Because of their importance, we list them together in Table 1. Identities I through III and VI through VIII are each written in three equivalent forms. They should be recognized in any of their forms. In the table, the symbol x represents either a real number or an angle, and k is an integer.

TABLE 1.   The Eight Fundamental Trigonometric Identities

I	$\sin x \csc x = 1 \leftrightarrow \csc x = \dfrac{1}{\sin x} \leftrightarrow \sin x = \dfrac{1}{\csc x}, \; x \neq k\pi$
II	$\cos x \sec x = 1 \leftrightarrow \sec x = \dfrac{1}{\cos x} \leftrightarrow \cos x = \dfrac{1}{\sec x}, \; x \neq \frac{1}{2}\pi + k\pi$
III	$\tan x \cot x = 1 \leftrightarrow \cot x = \dfrac{1}{\tan x} \leftrightarrow \tan x = \dfrac{1}{\cot x}, \; x \neq \frac{1}{2}k\pi$
IV	$\tan x = \dfrac{\sin x}{\cos x}, \; x \neq \frac{1}{2}\pi + k\pi$
V	$\cot x = \dfrac{\cos x}{\sin x}, \; x \neq k\pi$
VI	$\sin^2 x + \cos^2 x = 1 \leftrightarrow \sin^2 x = 1 - \cos^2 x \leftrightarrow \cos^2 x = 1 - \sin^2 x$
VII	$1 + \tan^2 x = \sec^2 x \leftrightarrow \tan^2 x = \sec^2 x - 1 \leftrightarrow \sec^2 x - \tan^2 x = 1$
VIII	$1 + \cot^2 x = \csc^2 x \leftrightarrow \cot^2 x = \csc^2 x - 1 \leftrightarrow \csc^2 x - \cot^2 x = 1$

In Chapter 6, when one trigonometric function was given, we obtained the values of the other five in a geometric manner. The same results can be obtained by using the fundamental identities as shown in Example 1.

Example   1   Given that $\sin x = -\frac{3}{5}$, use the fundamental identities to find the exact values of the other five trigonometric functions of x if $\cos x > 0$.

Solution Because $\sin x < 0$ and $\cos x > 0$, x is in the fourth quadrant. From fundamental identity VI and because $\cos x > 0$,

$$\cos x = \sqrt{1 - \sin^2 x}$$
$$= \sqrt{1 - \left(-\frac{3}{5}\right)^2}$$
$$= \sqrt{\frac{16}{25}}$$
$$= \frac{4}{5}$$

From fundamental identity IV,

$$\tan x = \frac{\sin x}{\cos x}$$
$$= \frac{-\frac{3}{5}}{\frac{4}{5}}$$
$$= -\frac{3}{4}$$

From fundamental identities I through III,

$$\cot x = \frac{1}{\tan x} \qquad \sec x = \frac{1}{\cos x} \qquad \csc x = \frac{1}{\sin x}$$
$$= -\frac{4}{3} \qquad\qquad = \frac{5}{4} \qquad\qquad = -\frac{5}{3}$$ ∎

Example 2 In each of the following, use the fundamental identities and a calculator to find to four significant digits the indicated function value.

a) $\tan \theta$ if $\sec \theta = 1.734$ and $0 < \theta < \frac{1}{2}\pi$

b) $\csc t$ if $\tan t = 5.582$ and $\pi < t < \frac{3}{2}\pi$

Solution a) From fundamental identity VII and because $\tan \theta > 0$ in the first quadrant,

$$\tan \theta = \sqrt{\sec^2 \theta - 1}$$
$$= \sqrt{(1.734)^2 - 1}$$
$$\approx 1.417$$

b) From fundamental identity III,

$$\cot t = \frac{1}{\tan t}$$
$$= \frac{1}{5.582}$$
$$\approx 0.1791$$

From fundamental identity VIII and because $\csc t < 0$ in the third quadrant,

$$\csc t = -\sqrt{\cot^2 t + 1}$$
$$\approx -\sqrt{(0.1791)^2 + 1}$$
$$\approx -1.016 \qquad \blacksquare$$

Observe that in the eight fundamental identities, the sine and cosine appear more often than any of the other four functions. Identities I, II, IV, and V enable us to express the other four functions in terms of either the sine or cosine or both.

Example   3   Write the following expression in terms of $\sin x$ and simplify: $\tan^2 x \csc x$.

Solution

$$\tan^2 x \csc x = \left(\frac{\sin x}{\cos x}\right)^2 \cdot \frac{1}{\sin x}$$
$$= \frac{\sin^2 x}{\cos^2 x \sin x}$$
$$= \frac{\sin x}{\cos^2 x}$$
$$= \frac{\sin x}{1 - \sin^2 x} \qquad \blacksquare$$

Example   4   Write the following expression in terms of $\cos \theta$ and simplify:

$$\frac{\sec^2 \theta - 1}{\sin^2 \theta}$$

Solution

$$\frac{\sec^2 \theta - 1}{\sin^2 \theta} = \frac{\tan^2 \theta}{\sin^2 \theta}$$
$$= \frac{\left(\dfrac{\sin \theta}{\cos \theta}\right)^2}{\sin^2 \theta}$$
$$= \frac{\sin^2 \theta}{\cos^2 \theta} \cdot \frac{1}{\sin^2 \theta}$$
$$= \frac{1}{\cos^2 \theta} \qquad \blacksquare$$

EXERCISES 7.1

In Exercises 1 through 16, use the fundamental identities to find the exact values of the other five trigonometric functions.

1. $\sin \theta = \frac{5}{13}$ and $\cos \theta > 0$
2. $\sin \theta = \frac{4}{5}$ and $\cos \theta > 0$
3. $\cos t = \frac{4}{5}$ and $\sin t > 0$
4. $\cos t = \frac{5}{13}$ and $\sin t > 0$
5. $\cos x = \frac{8}{17}$ and $\tan x < 0$
6. $\sin x = -\frac{12}{13}$ and $\cot x > 0$
7. $\sin u = -\frac{5}{13}$ and $\cos u > 0$
8. $\cos v = -\frac{3}{5}$ and $\sin v > 0$
9. $\csc \beta = -\frac{6}{5}$ and $\cot \beta > 0$
10. $\sec \alpha = -\frac{4}{3}$ and $\tan \alpha < 0$
11. $\cot y = -\frac{3}{4}$ and $\csc y > 0$
12. $\tan z = \frac{8}{15}$ and $\sec z < 0$
13. $\tan \theta = -\frac{12}{5}$ and $\sec \theta < 0$
14. $\cot \gamma = -\frac{12}{5}$ and $\csc \gamma < 0$
15. $\csc x = -\dfrac{\sqrt{34}}{3}$ and $\cos x < 0$
16. $\sec t = -\frac{7}{4}$ and $\sin t < 0$

In Exercises 17 through 28, use the fundamental identities and a calculator to find to four significant digits the indicated function value.

17. Find $\tan x$ if $\sec x = 1.642$ and $0 < x < \frac{1}{2}\pi$.
18. Find $\cot x$ if $\csc x = 2.276$ and $0 < x < \frac{1}{2}\pi$.
19. Find $\csc t$ if $\cot t = 0.4955$ and $\pi < t < \frac{3}{2}\pi$.
20. Find $\sec t$ if $\tan t = -0.5128$ and $\frac{1}{2}\pi < t < \pi$.
21. Find $\sin \theta$ if $\sec \theta = -2.305$ and $\frac{1}{2}\pi < \theta < \pi$.
22. Find $\cos \theta$ if $\csc \theta = -3.694$ and $\pi < \theta < \frac{3}{2}\pi$.
23. Find $\csc y$ if $\tan y = -0.4190$ and $\frac{3}{2}\pi < y < 2\pi$.
24. Find $\sec z$ if $\cot z = 6.471$ and $\pi < z < \frac{3}{2}\pi$.

25. Find $\tan v$ if $\sin v = 0.7139$ and $\frac{1}{2}\pi < v < \pi$.
26. Find $\cot \beta$ if $\cos \beta = 0.2113$ and $\frac{3}{2}\pi < \beta < 2\pi$.
27. Find $\cos \alpha$ if $\cot \alpha = -1.805$ and $\frac{3}{2}\pi < \alpha < 2\pi$.
28. Find $\sin u$ if $\tan u = -0.9740$ and $\frac{1}{2}\pi < u < \pi$.

In Exercises 29 through 42, write the given expression in terms of either sine or cosine, as indicated, and simplify.

29. $\tan t \csc t$ in terms of $\cos t$
30. $\cot \theta \sec \theta$ in terms of $\sin \theta$
31. $\cot^2 x \sec x$ in terms of $\cos x$
32. $\cot^2 y \csc y$ in terms of $\sin y$
33. $\dfrac{\sin^3 \theta}{\tan^2 \theta}$ in terms of $\sin \theta$
34. $\dfrac{\cos^3 t}{\cot^2 t}$ in terms of $\cos t$
35. $\dfrac{1 - \sec^2 z}{\sec^2 z}$ in terms of $\sin z$
36. $\dfrac{1 - \csc^2 \alpha}{\csc^2 \alpha}$ in terms of $\cos \alpha$
37. $\dfrac{\csc^2 \gamma - 1}{\cos \gamma \cot \gamma}$ in terms of $\sin \gamma$
38. $\dfrac{\sec^2 x - 1}{\sin x \tan x}$ in terms of $\cos x$
39. $\dfrac{\cot^2 y + 1}{\sin y}$ in terms of $\cos y$
40. $\dfrac{\tan^2 z + 1}{\cos z}$ in terms of $\sin z$
41. $\dfrac{\tan^2 \beta - \csc^2 \beta}{\sec^2 \beta}$ in terms of $\sin \beta$
42. $\dfrac{\sec^2 \theta - \cot^2 \theta}{\csc^2 \theta}$ in terms of $\cos \theta$

7.2 Proving Trigonometric Identities

It is sometimes necessary to convert a trigonometric expression from one form to an equivalent one. To perfect that skill, we now apply the fundamental identities and algebraic manipulations to verify other identities. There is no general method that can be applied to verify, or prove, an identity. As we present the examples, we shall give suggestions that will be helpful in determining the best approach to use for a proof. A familiarity with the eight fundamental identities in their various forms is crucial.

If one side of an identity is in a more complicated form than the other, it may be desirable to start with it and transform it to the simpler form on the other side. Bear this simpler form in mind as you proceed. It represents your objective.

Example   1   Prove the identity

$$\frac{1 + \sin x}{\cos x} = \sec x + \tan x$$

Solution   Because the left side is a fraction, we consider it as more complicated than the right side. We therefore begin with the left side, which we write as the sum of two fractions. We then apply two fundamental identities to obtain the right side.

$$\frac{1 + \sin x}{\cos x} = \frac{1}{\cos x} + \frac{\sin x}{\cos x}$$
$$= \sec x + \tan x \qquad \blacksquare$$

It is often expedient to convert an expression to one containing only the sine and cosine.

Example   2   Prove the identity

$$\frac{\csc x + \sec x}{1 + \tan x} = \csc x$$

Solution   Because the left side is the more complicated one, we start with it. We use fundamental identities to express it in terms of the sine and cosine, and then apply algebraic processes to get the right side.

$$\frac{\csc x + \sec x}{1 + \tan x} = \frac{\dfrac{1}{\sin x} + \dfrac{1}{\cos x}}{1 + \dfrac{\sin x}{\cos x}}$$

$$= \frac{\sin x \cos x \left(\dfrac{1}{\sin x} + \dfrac{1}{\cos x} \right)}{\sin x \cos x \left(1 + \dfrac{\sin x}{\cos x} \right)}$$

$$= \frac{\cos x + \sin x}{\sin x \cos x + \sin^2 x}$$

$$= \frac{\cos x + \sin x}{\sin x (\cos x + \sin x)}$$

$$= \frac{1}{\sin x}$$

$$= \csc x \qquad \blacksquare$$

Example **3** Prove the identity

$$(1 + \sec \theta)(1 - \cos \theta) = \tan \theta \sin \theta$$

Solution The left side is the more complicated one. We start with it and begin by performing the indicated multiplication.

$$
\begin{aligned}
(1 + \sec \theta)(1 - \cos \theta) &= 1 + \sec \theta - \cos \theta - \sec \theta \cos \theta \\
&= 1 + \frac{1}{\cos \theta} - \cos \theta - 1 \\
&= \frac{1 - \cos^2 \theta}{\cos \theta} \\
&= \frac{\sin^2 \theta}{\cos \theta} \\
&= \frac{\sin \theta}{\cos \theta}(\sin \theta) \\
&= \tan \theta \sin \theta
\end{aligned}
$$ ∎

The verification of the identity in the next example involves a procedure different from those used in the preceding examples. We transform each side separately into the same equivalent form.

Example **4** Prove the identity

$$\frac{1 + \cot y}{\csc y} = \frac{1 + \tan y}{\sec y}$$

Solution We start with the left side and express $\cot y$ and $\csc y$ in terms of $\sin y$ and $\cos y$.

$$
\begin{aligned}
\frac{1 + \cot y}{\csc y} &= \frac{1 + \dfrac{\cos y}{\sin y}}{\dfrac{1}{\sin y}} \\
&= \frac{\sin y \left(1 + \dfrac{\cos y}{\sin y}\right)}{\sin y \left(\dfrac{1}{\sin y}\right)} \\
&= \sin y + \cos y
\end{aligned}
$$

We now express the right side in terms of $\sin y$ and $\cos y$.

$$\frac{1 + \tan y}{\sec y} = \frac{1 + \dfrac{\sin y}{\cos y}}{\dfrac{1}{\cos y}}$$

$$= \frac{\cos y\left(1 + \dfrac{\sin y}{\cos y}\right)}{\cos y\left(\dfrac{1}{\cos y}\right)}$$

$$= \cos y + \sin y$$

Because each side is equal to $\sin y + \cos y$, we can conclude that

$$\frac{1 + \cot y}{\csc y} = \frac{1 + \tan y}{\sec y}$$

■

As shown in the following example, sometimes it may be advantageous to convert an expression to one involving only a single function.

Example   5   Prove the identity

$$(1 - \tan \beta)^3 = \frac{\cot \beta - 1}{\cot \beta}(\sec^2 \beta - 2 \tan \beta)$$

Solution   The right side is more complicated. We start with it, and because the left side involves only the tangent, we express the right side in terms of $\tan \beta$.

$$\frac{\cot \beta - 1}{\cot \beta}(\sec^2 \beta - 2 \tan \beta) = \frac{\dfrac{1}{\tan \beta} - 1}{\dfrac{1}{\tan \beta}}(1 + \tan^2 \beta - 2 \tan \beta)$$

$$= \frac{\tan \beta\left(\dfrac{1}{\tan \beta} - 1\right)}{\tan \beta\left(\dfrac{1}{\tan \beta}\right)}(1 - 2 \tan \beta + \tan^2 \beta)$$

$$= \frac{1 - \tan \beta}{1}(1 - \tan \beta)^2$$

$$= (1 - \tan \beta)^3$$

■

Example 6 Prove the identity

$$\frac{\sin x}{1 + \cos x} = \frac{1 - \cos x}{\sin x}$$

Solution Each side of the identity involves a fraction containing only the sine and cosine. On the left the binomial $1 + \cos x$ appears in the denominator, and on the right the binomial $1 - \cos x$ appears in the numerator. We can start with the left side and obtain a factor of $1 - \cos x$ in the numerator by multiplying the numerator and denominator by $1 - \cos x$. This operation is equivalent to multiplying the fraction by 1. Therefore we have

$$\frac{\sin x}{1 + \cos x} = \frac{\sin x(1 - \cos x)}{(1 + \cos x)(1 - \cos x)}$$

$$= \frac{\sin x(1 - \cos x)}{1 - \cos^2 x}$$

$$= \frac{\sin x(1 - \cos x)}{\sin^2 x}$$

$$= \frac{1 - \cos x}{\sin x} \qquad \blacksquare$$

Following is a summary of the suggestions we have given for proving an identity. Review these suggestions before doing the exercises.

1. Start with the more complicated side and transform it to the simpler form on the other side. See Examples 1, 2, 3, and 5.

2. Instead of applying suggestion 1, it may be more convenient to transform each side separately into the same equivalent form. See Example 4.

3. Often it is desirable to convert an expression to one containing only the sine and cosine. See Examples 2, 3, and 4.

4. Instead of applying suggestion 3, it may be advantageous to convert an expression to one involving only a single function, provided no radicals are introduced. See Example 5.

5. Consider the possibilities of applying algebraic processes such as: multiplying; factoring; combining fractions into a single fraction; writing a single fraction, having more than one term in the numerator, into a sum of fractions. See Examples 1, 3, and 5.

6. To obtain a particular factor in the numerator or denominator of a fraction, you may multiply the numerator and denominator by this desired factor. See Example 6.

To prove that an equation is not an identity, it is only necessary to find one element in the domain of the variable for which the equality is not true. By doing this, you have found a *counterexample*.

Illustration 1

Consider the equation

$$4 \sin t \cos t = \sec t$$

If $t = 0$, the left side is

$$4 \sin 0 \cos 0 = 4(0)(1)$$
$$= 0$$

and the right side is

$$\sec 0 = 1$$

Therefore the equation is not an identity. ■

Illustration 2

The equation

$$\sqrt{1 - \cos^2 x} = \sin x$$

is not an identity, because if x is any number for which $\sin x < 0$, the right side is negative and the left side is nonnegative. In particular, if $x = \frac{7}{6}\pi$, the left side is

$$\sqrt{1 - \cos^2 \frac{7}{6}\pi} = \sqrt{1 - \left(-\frac{\sqrt{3}}{2}\right)^2}$$

$$= \sqrt{1 - \frac{3}{4}}$$

$$= \sqrt{\frac{1}{4}}$$

$$= \frac{1}{2}$$

and the right side is

$$\sin \frac{7}{6}\pi = -\frac{1}{2}$$

■

EXERCISES 7.2

In Exercises 1 through 6, write the expression in terms of either $\sin x$ or $\cos x$, or both, in a simplified form.

1. (a) $\cos x \tan x$; (b) $\dfrac{\cos x}{\sec x}$

2. (a) $\sin x \cot x$; (b) $\dfrac{\sin x}{\csc x}$

3. (a) $\sec x \cot x$; (b) $\dfrac{\cot^2 x}{\csc^2 x}$

4. (a) $\csc x \tan x$; (b) $\dfrac{\tan^2 x}{\sec^2 x}$

5. (a) $\tan x + \cot x$; (b) $\sin x + \cos^2 x \csc x$

6. (a) $\sec^2 x + \csc^2 x$; (b) $\cos x + \sin^2 x \sec x$

In Exercises 7 through 12, prove that the first expression is equivalent to the second expression for all values of the variable for which both expressions are defined.

7. (a) $\cos \theta \csc \theta$, $\cot \theta$; (b) $\sin \theta \sec \theta \cot \theta$, 1

8. (a) $\sin \theta \sec \theta$, $\tan \theta$; (b) $\cos \theta \csc \theta \tan \theta$, 1

9. (a) $\dfrac{\sec \alpha}{\csc \alpha}$, $\tan \alpha$; (b) $\dfrac{\tan \alpha}{\sin \alpha}$, $\sec \alpha$

10. (a) $\dfrac{\csc \alpha}{\sec \alpha}$, $\cot \alpha$; (b) $\dfrac{\cot \alpha}{\cos \alpha}$, $\csc \alpha$

11. (a) $\dfrac{1 + \cot^2 \beta}{\sec^2 \beta}$, $\cot^2 \beta$;

  (b) $(1 + \sin \beta)(1 - \sin \beta)$, $\cos^2 \beta$

12. (a) $\dfrac{1 + \tan^2 \beta}{\csc^2 \beta}$, $\tan^2 \beta$;

  (b) $(1 + \cos \beta)(1 - \cos \beta)$, $\sin^2 \beta$

In Exercises 13 through 42, prove the identity.

13. $\sec^2 x + \csc^2 x = \sec^2 x \cdot \csc^2 x$

14. $(\tan x + \cot x)^2 = \sec^2 x \cdot \csc^2 x$

15. $\tan^2 x - \sin^2 x = \tan^2 x \cdot \sin^2 x$

16. $\cot^2 x - \cos^2 x = \cos^2 x \cdot \cot^2 x$

17. $\dfrac{\sec \theta + 1}{\sec \theta - 1} = \dfrac{1 + \cos \theta}{1 - \cos \theta}$

18. $\dfrac{\csc \theta + 1}{\csc \theta - 1} = \dfrac{1 + \sin \theta}{1 - \sin \theta}$

19. $\dfrac{1 - \cot^2 \theta}{1 + \cot^2 \theta} = \sin^2 \theta - \cos^2 \theta$

20. $\dfrac{1}{\tan \theta + \cot \theta} = \sin \theta \cos \theta$

21. $\dfrac{\csc^2 \alpha - 1}{\sec^2 \alpha - 1} = \cot^4 \alpha$

22. $\dfrac{1}{\csc \alpha - 1} - \dfrac{1}{\csc \alpha + 1} = 2 \tan^2 \alpha$

23. $\dfrac{1}{1 + \sin \alpha} + \dfrac{1}{1 - \sin \alpha} = 2 \sec^2 \alpha$

24. $\dfrac{1}{1 + \cos \alpha} + \dfrac{1}{1 - \cos \alpha} = 2 \csc^2 \alpha$

25. $\dfrac{\sin \beta}{1 + \cos \beta} + \dfrac{1 + \cos \beta}{\sin \beta} = 2 \csc \beta$

26. $\dfrac{\cos \beta}{1 + \sin \beta} + \dfrac{\cos \beta}{1 - \sin \beta} = 2 \sec \beta$

27. $\dfrac{\cos \beta}{1 + \sin \beta} = \dfrac{1 - \sin \beta}{\cos \beta}$

28. $\dfrac{\sec \beta + 1}{\tan \beta} = \dfrac{\tan \beta}{\sec \beta - 1}$

29. $\dfrac{\cos t}{\sec t - \tan t} = 1 + \sin t$

30. $\sin t \tan t = \sec t - \cos t$

31. $\dfrac{\tan y}{\tan^2 y - 1} = \dfrac{1}{\tan y - \cot y}$

32. $\cot y \csc y = \dfrac{1}{\sec y - \cos y}$

33. $(\tan x + \cot x)^2 = \sec^2 x + \csc^2 x$

34. $\sec x + \tan x = \dfrac{1}{\sec x - \tan x}$

35. $\csc^4 \theta - \cot^4 \theta = \csc^2 \theta + \cot^2 \theta$

36. $\cos^4 \theta - \sin^4 \theta = \cos^2 \theta - \sin^2 \theta$

37. $\tan^4 \alpha + \tan^2 \alpha = \sec^4 \alpha - \sec^2 \alpha$

38. $\csc^4 \alpha - \csc^2 \alpha = \cot^4 \alpha + \cot^2 \alpha$

39. $\sin^3 t + \cos^3 t + \sin t \cos^2 t + \sin^2 t \cos t = \sin t + \cos t$

40. $\dfrac{\sin^3 t + \cos^3 t}{\sin t + \cos t} = 1 - \sin t \cos t$

41. $\dfrac{\tan^3 x + \sin x \sec x - \sin x \cos x}{\sec x - \cos x} = \tan x \sec x + \sin x$

42. $\dfrac{2 \tan x}{1 - \tan^2 x} + \dfrac{1}{2 \cos^2 x - 1} = \dfrac{\cos x + \sin x}{\cos x - \sin x}$

In Exercises 43 through 46, show that the equation is not an identity by finding a counterexample.

43. $\sin t \cos t + 1 = \sin t + \cos t$

44. $\sin t \sec t + 1 = 2 \sin t + \sec t$

45. $\sec y = 1 + \tan^2 y$ 46. $\tan^2 x + \tan x = 2 \tan^3 x$

In Exercises 47 through 54, determine whether or not the equation is an identity.

47. $\dfrac{1}{1 + \cos \theta} - \dfrac{1}{1 - \cos \theta} = \dfrac{2}{\sec \theta - \cos \theta}$

48. $\dfrac{1}{1 - \sin \theta} - \dfrac{1}{1 + \sin \theta} = \dfrac{2}{\tan \theta \cos \theta}$

49. $\dfrac{1}{1 - \cos \theta} - \dfrac{1}{1 + \cos \theta} = 2 \cot \theta \csc \theta$

50. $\tan \theta \sec \theta - \sin \theta = \tan \theta$

51. $1 - \tan^3 x = (2 \tan x - \sec^2 x)(\tan x - 1)$

52. $\csc^4 x(1 - \cos^4 x) = 1 + 2 \cot^2 x$

53. $\dfrac{1 + \cos x}{1 - \cos x} + \dfrac{1 + \sin x}{1 - \sin x} = \dfrac{2(\cos x - \csc x)}{\cot x - \cos x - \csc x + 1}$

54. $\dfrac{1 + \cos x}{1 - \cos x} + \dfrac{1 - \sin x}{1 + \sin x} = \dfrac{2(\sin x + \sec x)}{\tan x - \sin x + \sec x - 1}$

7.3 Sum and Difference Identities

We now obtain identities that express trigonometric function values of the sum or difference of two real numbers (or angles) in terms of trigonometric function values of each real number (or angle). We begin by deriving a formula for $\cos(u - v)$, where u and v are real numbers. In our discussion, we take $\frac{1}{2}\pi < u < \pi$ and $0 < v < \frac{1}{2}\pi$. However, a similar argument can be used for u and v in any quadrants. Figure 1 shows the unit circle U and the point $A(1, 0)$. Point P_1 is the terminal point of the arc having initial point at A and length u, and P_2 is the terminal point of the arc having initial point at A and length v. Therefore, from Chapter 6, P_1 is the point $(\cos u, \sin u)$ and P_2 is the point $(\cos v, \sin v)$. The length of the arc from P_2 to P_1 is $u - v$. Let P_3 on the unit circle U be the terminal point of the arc having initial point at A and length $u - v$. Thus point P_3 is $(\cos(u - v), \sin(u - v))$. Because the length of the arc from P_2 to P_1 is the same as the length of the arc from A to P_3, it follows from geometry that $|\overline{P_2P_1}| = |\overline{AP_3}|$; hence

$$|\overline{P_2P_1}|^2 = |\overline{AP_3}|^2 \tag{1}$$

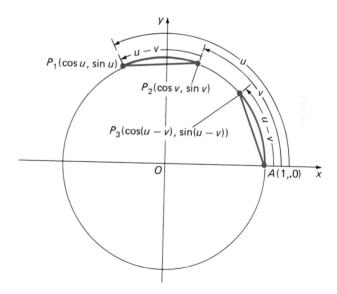

FIGURE 1

From the distance formula,

$$
\begin{aligned}
|\overline{P_2P_1}|^2 &= (\cos u - \cos v)^2 + (\sin u - \sin v)^2 \\
&= \cos^2 u - 2\cos u \cos v + \cos^2 v + \sin^2 u - 2\sin u \sin v + \sin^2 v \\
&= (\cos^2 u + \sin^2 u) + (\cos^2 v + \sin^2 v) - 2(\cos u \cos v + \sin u \sin v) \\
&= 1 + 1 - 2(\cos u \cos v + \sin u \sin v) \\
&= 2 - 2(\cos u \cos v + \sin u \sin v) \tag{2}
\end{aligned}
$$

Also from the distance formula,

$$
\begin{aligned}
|\overline{AP_3}|^2 &= [\cos(u - v) - 1]^2 + [\sin(u - v) - 0]^2 \\
&= \cos^2(u - v) - 2\cos(u - v) + 1 + \sin^2(u - v) \\
&= [\cos^2(u - v) + \sin^2(u - v)] + 1 - 2\cos(u - v) \\
&= 1 + 1 - 2\cos(u - v) \\
&= 2 - 2\cos(u - v) \tag{3}
\end{aligned}
$$

Substituting from (2) and (3) into (1), we have

$$
\begin{aligned}
2 - 2(\cos u \cos v + \sin u \sin v) &= 2 - 2\cos(u - v) \\
\cos(u - v) &= \cos u \cos v + \sin u \sin v
\end{aligned}
$$

This formula is an identity because it is true for all real numbers. It is also true for all angles. It is called the cosine difference identity, and we write it now in terms of x and y.

Cosine Difference Identity

$$\cos(x - y) = \cos x \cos y + \sin x \sin y$$

If in this identity $-y$ is substituted for y, we have

$$\cos[x - (-y)] = \cos x \cos(-y) + \sin x \sin(-y)$$

From Chapter 6, $\sin(-y) = -\sin y$ and $\cos(-y) = \cos y$. Therefore we have the following identity.

Cosine Sum Identity

$$\cos(x + y) = \cos x \cos y - \sin x \sin y$$

In the following example we apply the difference and sum identities to compute certain exact function values. This particular application of the identities is not important in itself. Its purpose is to give practice in using the formulas so that you gain familiarity with them.

Example   **1**   Find the exact value of (a) $\cos 15°$ and (b) $\cos \frac{5}{12}\pi$.

Solution   a) Because $15° = 60° - 45°$ and the exact function values of $60°$ and $45°$ are known, we use the cosine difference identity and get

$$
\begin{aligned}
\cos 15° &= \cos(60° - 45°) \\
&= \cos 60° \cos 45° + \sin 60° \sin 45° \\
&= \frac{1}{2} \cdot \frac{1}{\sqrt{2}} + \frac{\sqrt{3}}{2} \cdot \frac{1}{\sqrt{2}} \\
&= \frac{1 + \sqrt{3}}{2\sqrt{2}}
\end{aligned}
$$

b) Because $\frac{5}{12}\pi = \frac{1}{6}\pi + \frac{1}{4}\pi$ and the exact function values of $\frac{1}{6}\pi$ and $\frac{1}{4}\pi$ are known, we use the cosine sum identity and get

$$\cos \tfrac{5}{12}\pi = \cos(\tfrac{1}{6}\pi + \tfrac{1}{4}\pi)$$
$$= \cos \tfrac{1}{6}\pi \cos \tfrac{1}{4}\pi - \sin \tfrac{1}{6}\pi \sin \tfrac{1}{4}\pi$$
$$= \frac{\sqrt{3}}{2} \cdot \frac{1}{\sqrt{2}} - \frac{1}{2} \cdot \frac{1}{\sqrt{2}}$$
$$= \frac{\sqrt{3} - 1}{2\sqrt{2}} \qquad \blacksquare$$

The cosine sum and difference identities can be used to obtain other identities. If in the cosine difference identity we let $x = \tfrac{1}{2}\pi$, we get

$$\cos(\tfrac{1}{2}\pi - y) = \cos \tfrac{1}{2}\pi \cos y + \sin \tfrac{1}{2}\pi \sin y$$
$$= (0) \cos y + (1) \sin y$$
$$= \sin y$$

Now in this formula let $y = \tfrac{1}{2}\pi - x$. We have then

$$\cos[\tfrac{1}{2}\pi - (\tfrac{1}{2}\pi - x)] = \sin(\tfrac{1}{2}\pi - x)$$
$$\cos x = \sin(\tfrac{1}{2}\pi - x)$$

From a fundamental identity,

$$\tan(\tfrac{1}{2}\pi - x) = \frac{\sin(\tfrac{1}{2}\pi - x)}{\cos(\tfrac{1}{2}\pi - x)}$$

Replacing $\sin(\tfrac{1}{2}\pi - x)$ by $\cos x$ and $\cos(\tfrac{1}{2}\pi - x)$ by $\sin x$, we have

$$\tan(\tfrac{1}{2}\pi - x) = \frac{\cos x}{\sin x}$$
$$= \cot x$$

We have obtained the following three identities, called the cofunction identities.

Cofunction Identities

$$\cos(\tfrac{1}{2}\pi - x) = \sin x$$
$$\sin(\tfrac{1}{2}\pi - x) = \cos x$$
$$\tan(\tfrac{1}{2}\pi - x) = \cot x$$

Observe that these formulas for angles appeared in Chapter 6. However, there the angle was restricted to being acute.

We now derive the sine sum and difference identities. From the first cofunction identity, with x replaced by $x + y$, we have

$$\sin(x + y) = \cos[\tfrac{1}{2}\pi - (x + y)]$$
$$= \cos[(\tfrac{1}{2}\pi - x) - y]$$

We now apply the cosine difference identity and get

$$\sin(x + y) = \cos(\tfrac{1}{2}\pi - x)\cos y + \sin(\tfrac{1}{2}\pi - x)\sin y$$

Using cofunction identities on the right side, we obtain the following sine sum identity.

Sine Sum Identity

$$\sin(x + y) = \sin x \cos y + \cos x \sin y$$

If in this identity we replace y by $-y$, we get

$$\sin[x + (-y)] = \sin x \cos(-y) + \cos x \sin(-y)$$

Replacing $\cos(-y)$ by $\cos y$ and $\sin(-y)$ by $-\sin y$, we have the following sine difference identity.

Sine Difference Identity

$$\sin(x - y) = \sin x \cos y - \cos x \sin y$$

Example 2 Given that $\sin \alpha = \frac{24}{25}$, where α is in the first quadrant, and $\sin \beta = \frac{4}{5}$, where β is in the second quadrant, find

a) $\sin(\alpha + \beta)$

b) $\cos(\alpha + \beta)$

c) the quadrant containing $\alpha + \beta$.

Solution In order to use the sine sum identity and the cosine sum identity, we need to determine $\cos \alpha$ and $\cos \beta$. To find $\cos \alpha$, we use the identity $\sin^2 \alpha + \cos^2 \alpha = 1$ with $\sin \alpha = \frac{24}{25}$ and $\cos \alpha > 0$ because α is in the first quadrant:

$$\left(\tfrac{24}{25}\right)^2 + \cos^2 \alpha = 1$$
$$\cos^2 \alpha = 1 - \tfrac{576}{625}$$
$$\cos^2 \alpha = \tfrac{49}{625}$$
$$\cos \alpha = \tfrac{7}{25}$$

To find $\cos \beta$, we use the identity $\sin^2 \beta + \cos^2 \beta = 1$ with $\sin \beta = \frac{4}{5}$ and $\cos \beta < 0$ because β is in the second quadrant:

$$\left(\tfrac{4}{5}\right)^2 + \cos^2 \beta = 1$$
$$\cos^2 \beta = 1 - \tfrac{16}{25}$$
$$\cos^2 \beta = \tfrac{9}{25}$$
$$\cos \beta = -\tfrac{3}{5}$$

a) From the sine sum identity,

$$\sin(\alpha + \beta) = \sin \alpha \cos \beta + \cos \alpha \sin \beta$$
$$= \left(\tfrac{24}{25}\right)\left(-\tfrac{3}{5}\right) + \left(\tfrac{7}{25}\right)\left(\tfrac{4}{5}\right)$$
$$= -\tfrac{72}{125} + \tfrac{28}{125}$$
$$= -\tfrac{44}{125}$$

b) From the cosine sum identity,

$$\cos(\alpha + \beta) = \cos \alpha \cos \beta - \sin \alpha \sin \beta$$
$$= \left(\tfrac{7}{25}\right)\left(-\tfrac{3}{5}\right) - \left(\tfrac{24}{25}\right)\left(\tfrac{4}{5}\right)$$
$$= -\tfrac{21}{125} - \tfrac{96}{125}$$
$$= -\tfrac{117}{125}$$

c) Because $\sin(\alpha + \beta) < 0$ and $\cos(\alpha + \beta) < 0$, we conclude that $\alpha + \beta$ is in the third quadrant. ∎

Observe in Example 2 that $\sin(\alpha + \beta)$ and $\cos(\alpha + \beta)$ were obtained without finding the actual values of α and β.

To express $\tan(x + y)$ in terms of $\tan x$ and $\tan y$, we begin with the fundamental identity that states that the tangent is the quotient of the sine and cosine. We then use the sine and cosine sum identities. We have

$$\tan(x + y) = \frac{\sin(x + y)}{\cos(x + y)}$$

$$= \frac{\sin x \cos y + \cos x \sin y}{\cos x \cos y - \sin x \sin y}$$

So that the identity involves $\tan x$ and $\tan y$, we divide the numerator and denominator by $\cos x \cos y$, with the assumption that $\cos x \cos y \neq 0$. Thus

$$\tan(x + y) = \frac{\dfrac{\sin x}{\cos x} \cdot \dfrac{\cos y}{\cos y} + \dfrac{\cos x}{\cos x} \cdot \dfrac{\sin y}{\cos y}}{\dfrac{\cos x}{\cos x} \cdot \dfrac{\cos y}{\cos y} - \dfrac{\sin x}{\cos x} \cdot \dfrac{\sin y}{\cos y}}$$

$$= \frac{\tan x \cdot 1 + 1 \cdot \tan y}{1 \cdot 1 - \tan x \cdot \tan y}$$

We have therefore obtained the following tangent sum identity.

Tangent Sum Identity

$$\tan(x + y) = \frac{\tan x + \tan y}{1 - \tan x \tan y}$$

If in this identity y is replaced by $-y$, we get

$$\tan[x + (-y)] = \frac{\tan x + \tan(-y)}{1 - \tan x \tan(-y)}$$

In Chapter 6, we proved that $\tan(-y) = -\tan y$. Making this substitution in the above equation, we have

$$\tan(x - y) = \frac{\tan x + (-\tan y)}{1 - \tan x(-\tan y)}$$

from which the following tangent difference identity is obtained.

Tangent Difference Identity

$$\tan(x - y) = \frac{\tan x - \tan y}{1 + \tan x \tan y}$$

In the derivation of the tangent sum identity, the restriction that $\cos x \cos y \neq 0$ indicates that the identity does not hold if either x or y is $\frac{1}{2}\pi + k\pi$, where k is any integer. Observe that if you attempt to apply either the tangent sum identity or the tangent difference identity when x or y has a value of $\frac{1}{2}\pi + k\pi$, the right side involves $\tan(\frac{1}{2}\pi + k\pi)$, which is not defined.

Example 3 Verify the identity

$$\tan(x - \tfrac{1}{4}\pi) = \frac{\tan x - 1}{\tan x + 1}$$

Solution From the tangent difference identity

$$\tan(x - \tfrac{1}{4}\pi) = \frac{\tan x - \tan\frac{1}{4}\pi}{1 + \tan x \tan\frac{1}{4}\pi}$$

Because $\tan\frac{1}{4}\pi = 1$, we get

$$\tan(x - \tfrac{1}{4}\pi) = \frac{\tan x - 1}{1 + \tan x(1)}$$

$$= \frac{\tan x - 1}{\tan x + 1} \qquad \blacksquare$$

Identities that express a trigonometric function of $\frac{1}{2}k\pi \pm x$, where k is any integer, in terms of a function of x are called **reduction formulas.** In particular, the cofunction identities are reduction formulas. Some others are

$$\sin(\tfrac{1}{2}\pi + x) = \cos x \qquad \cos(\tfrac{1}{2}\pi + x) = -\sin x \qquad \tan(\tfrac{1}{2}\pi + x) = -\cot x$$
$$\sin(\pi - x) = \sin x \qquad \cos(\pi - x) = -\cos x \qquad \tan(\pi - x) = -\tan x$$
$$\sin(\pi + x) = -\sin x \qquad \cos(\pi + x) = -\cos x \qquad \tan(\pi + x) = \tan x$$
$$\sin(\tfrac{3}{2}\pi - x) = -\cos x \qquad \cos(\tfrac{3}{2}\pi - x) = -\sin x \qquad \tan(\tfrac{3}{2}\pi - x) = \cot x$$
$$\sin(\tfrac{3}{2}\pi + x) = -\cos x \qquad \cos(\tfrac{3}{2}\pi + x) = \sin x \qquad \tan(\tfrac{3}{2}\pi + x) = -\cot x$$
$$\sin(2\pi - x) = -\sin x \qquad \cos(2\pi - x) = \cos x \qquad \tan(2\pi - x) = -\tan x$$

and so on. The following illustrations give the proofs of some of these formulas. The other formulas are proved in a similar way. You are asked to supply some of these proofs in Exercises 21 through 24.

Illustration 1

From the sine difference and sine sum identities

$$\begin{aligned}\sin(\pi - x) &= \sin \pi \cos x - \cos \pi \sin x\\ &= (0) \cos x - (-1) \sin x\\ &= \sin x\end{aligned}$$

$$\begin{aligned}\sin(\tfrac{3}{2}\pi + x) &= \sin \tfrac{3}{2}\pi \cos x + \cos \tfrac{3}{2}\pi \sin x\\ &= (-1) \cos x + (0) \sin x\\ &= -\cos x\end{aligned}$$

■

Illustration 2

From the cosine difference and cosine sum identities,

$$\begin{aligned}\cos(\pi - x) &= \cos \pi \cos x + \sin \pi \sin x\\ &= (-1) \cos x + (0) \sin x\\ &= -\cos x\end{aligned}$$

$$\begin{aligned}\cos(\tfrac{3}{2}\pi + x) &= \cos \tfrac{3}{2}\pi \cos x - \sin \tfrac{3}{2}\pi \sin x\\ &= (0) \cos x - (-1) \sin x\\ &= \sin x\end{aligned}$$

■

Illustration 3

From the tangent difference identity,

$$\begin{aligned}\tan(\pi - x) &= \frac{\tan \pi - \tan x}{1 + \tan \pi \tan x}\\ &= \frac{0 - \tan x}{1 + (0) \tan x}\\ &= -\tan x\end{aligned}$$

From a fundamental identity and the results of Illustrations 1 and 2,

$$\begin{aligned}\tan(\tfrac{3}{2}\pi + x) &= \frac{\sin(\frac{3}{2}\pi + x)}{\cos(\frac{3}{2}\pi + x)}\\ &= \frac{-\cos x}{\sin x}\\ &= -\cot x\end{aligned}$$

We cannot use the tangent sum identity to find $\tan(\frac{3}{2}\pi + x)$. An attempt to use it would result in an expression containing $\tan \frac{3}{2}\pi$, which is not defined. ■

In physics, when discussing electricity, heat, and dynamics, it is sometimes necessary to write an expression of the form

$$A \sin bt + B \cos bt$$

in the form

$$a \sin(bt + c)$$

so that the amplitude, period, frequency, phase shift, and graph of the corresponding function can be obtained more easily. We shall now obtain a theorem that is used in such a case. Suppose the function f is defined by the equation

$$f(t) = A \sin bt + B \cos bt \tag{4}$$

where A, B, and b are constants, and $A \neq 0$. We can write $f(t)$ in the form

$$f(t) = \sqrt{A^2 + B^2} \left(\frac{A}{\sqrt{A^2 + B^2}} \sin bt + \frac{B}{\sqrt{A^2 + B^2}} \cos bt \right) \tag{5}$$

Let c be a real number such that

$$\tan c = \frac{B}{A} \tag{6}$$

and

$$\cos c = \frac{A}{\sqrt{A^2 + B^2}} \quad \text{and} \quad \sin c = \frac{B}{\sqrt{A^2 + B^2}} \tag{7}$$

You are asked to prove that Equation (6) follows from (7) in Exercise 42. Substituting from (7) into (5), we have

$$f(t) = \sqrt{A^2 + B^2} \, (\cos c \sin bt + \sin c \cos bt) \tag{8}$$

Observe that the expression in parentheses is $\sin(bt + c)$. If we let

$$a = \sqrt{A^2 + B^2}$$

then (8) can be written in the form

$$f(t) = a \sin(bt + c) \tag{9}$$

By equating the right sides of (4) and (9), we have the following theorem.

Theorem 1

> If t is any real number, A, B, and b are constants, and $A \neq 0$,
> $$A \sin bt + B \cos bt = a \sin(bt + c)$$
> where $a = \sqrt{A^2 + B^2}$ and c satisfies (6) and (7).

Example   4 Given

$$f(t) = 2 \sin t + 2\sqrt{3} \cos t$$

a) Define $f(t)$ by an equation of the form $f(t) = a \sin(t + c)$.

b) Determine the amplitude, period, and phase shift of f and draw a sketch of the graph.

Solution   a) From Theorem 1,

$$2 \sin t + 2\sqrt{3} \cos t = a \sin(t + c)$$

where $a = \sqrt{2^2 + (2\sqrt{3})^2}$; that is, $a = 4$; furthermore, $\tan c = \dfrac{2\sqrt{3}}{2}$; that is, $\tan c = \sqrt{3}$. Thus we take $c = \frac{1}{3}\pi$. Hence

$$f(t) = 4 \sin(t + \tfrac{1}{3}\pi)$$

b) The amplitude of f is 4. Because the coefficient of t is 1, the period is $2\pi/1 = 2\pi$. The phase shift is $\frac{1}{3}\pi$, and the graph of $f(t) = 4 \sin t$ is shifted $\frac{1}{3}\pi$ units to the left to obtain the required graph. Figure 2 shows a sketch of this graph in color and the graph of the function defined by $f(t) = 4 \sin t$ is indicated by a black curve.

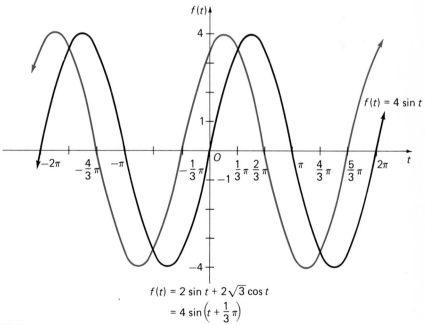

$$f(t) = 2 \sin t + 2\sqrt{3} \cos t$$
$$= 4 \sin\left(t + \frac{1}{3}\pi\right)$$

FIGURE 2

It can be shown that the sum of two sine functions having the same period is a sine function having that common period. This fact has important applications in physics in the fields of sound and electricity. The next example gives a particular case.

Example 5 At a particular point in space two atmospheric waves produce pressures of $F(t)$ dyn/cm^2 and $G(t)$ dyn/cm^2 at t seconds, where

$$F(t) = 0.04 \sin 400\pi t \quad \text{and} \quad G(t) = 0.04 \sin(400\pi t + \tfrac{1}{3}\pi).$$

Show that the sum of these two pressures is a sine function having the same period as F and G.

Solution If f is the sum of F and G, then

$$f(t) = F(t) + G(t)$$
$$= 0.04 \sin 400\pi t + 0.04 \sin(400\pi t + \tfrac{1}{3}\pi)$$
$$= 0.04 \sin 400\pi t + 0.04[\sin 400\pi t \cos \tfrac{1}{3}\pi + \cos 400\pi t \sin \tfrac{1}{3}\pi]$$
$$= 0.04 \sin 400\pi t + 0.04\left[(\sin 400\pi t)\left(\frac{1}{2}\right) + (\cos 400\pi t)\left(\frac{\sqrt{3}}{2}\right)\right]$$
$$= 0.04 \sin 400\pi t + 0.02 \sin 400\pi t + 0.02(\sqrt{3}) \cos 400\pi t$$
$$= 0.06 \sin 400\pi t + 0.02(\sqrt{3}) \cos 400\pi t$$

From Theorem 1,

$$f(t) = a \sin(400\pi t + c)$$

where

$$a = \sqrt{(0.06)^2 + (0.02)^2(\sqrt{3})^2} \quad \text{and} \quad \tan c = \frac{0.02(\sqrt{3})}{0.06}$$
$$= \sqrt{(0.02)^2(3)^2 + (0.02)^2(3)}$$
$$= 0.02\sqrt{12} \qquad\qquad \tan c = \frac{\sqrt{3}}{3}$$
$$= 0.04\sqrt{3} \qquad\qquad\qquad c = \tfrac{1}{6}\pi$$

Therefore

$$f(t) = 0.04\sqrt{3} \sin(400\pi t + \tfrac{1}{6}\pi)$$

Observe that the period of each of the functions F, G, and f is

$$\frac{2\pi}{400\pi} = \frac{1}{200}$$

■

EXERCISES 7.3

In Exercises 1 through 6, find the exact value.

1. (a) $\sin 15°$; (b) $\cos 165°$
2. (a) $\sin 75°$; (b) $\cos 105°$

3. (a) $\sin(-\tfrac{7}{12}\pi)$; $\cos \tfrac{23}{12}\pi$
4. (a) $\sin \tfrac{11}{12}\pi$; (b) $\cos(-\tfrac{11}{12}\pi)$
5. (a) $\tan 105°$; (b) $\tan \tfrac{13}{12}\pi$

6. (a) $\tan 345°$; (b) $\tan \frac{5}{12}\pi$

In Exercises 7 through 12, find the exact value.

7. (a) $\sin 32° \cos 58° + \cos 32° \sin 58°$;
   (b) $\cos 116° \cos 64° - \sin 116° \sin 64°$
8. (a) $\sin 110° \cos 20° - \cos 110° \sin 20°$;
   (b) $\sin 110° \sin 70° - \cos 110° \cos 70°$
9. (a) $\sin 85° \sin 25° + \cos 85° \cos 25°$;
   (b) $\sin 70° \cos 205° - \cos 70° \sin 205°$
10. (a) $\cos 50° \cos 275° + \sin 50° \sin 275°$;
    (b) $\cos 50° \sin 100° + \sin 50° \cos 100°$
11. (a) $\dfrac{\tan 20° + \tan 25°}{1 - \tan 20° \tan 25°}$;
    (b) $\dfrac{\tan 200° - \tan 80°}{1 + \tan 200° \tan 80°}$
12. (a) $\dfrac{\tan 110° + \tan 100°}{1 - \tan 110° \tan 100°}$;
    (b) $\dfrac{\tan 65° - \tan 110°}{1 + \tan 65° \tan 110°}$

In Exercises 13 through 16, find (a) $\sin(\alpha + \beta)$, (b) $\cos(\alpha + \beta)$, (c) $\sin(\alpha - \beta)$, (d) $\cos(\alpha - \beta)$, (e) the quadrant containing $\alpha + \beta$, and (f) the quadrant containing $\alpha - \beta$.

13. $\sin \alpha = \frac{12}{13}$, α in the first quadrant; $\sin \beta = \frac{7}{25}$, β in the first quadrant.
14. $\cos \alpha = \frac{24}{25}$, α in the first quadrant; $\cos \beta = \frac{4}{5}$, β in the first quadrant.
15. $\sin \alpha = -\frac{3}{5}$, α in the fourth quadrant; $\cos \beta = -\frac{12}{13}$, β in the third quadrant.
16. $\cos \alpha = -\frac{5}{13}$, α in the second quadrant; $\sin \beta = -\frac{7}{25}$, β in the fourth quadrant.

In Exercises 17 through 20, find (a) $\tan(x + y)$, (b) $\tan(x - y)$, (c) the quadrant containing $x + y$, and (d) the quadrant containing $x - y$.

17. $\tan x = \frac{7}{3}$, $0 < x < \frac{1}{2}\pi$; $\tan y = \frac{3}{4}$, $0 < y < \frac{1}{2}\pi$
18. $\tan x = \frac{2}{3}$, $0 < x < \frac{1}{2}\pi$; $\tan y = \frac{5}{6}$, $0 < y < \frac{1}{2}\pi$
19. $\tan x = \frac{3}{10}$, $\pi < x < \frac{3}{2}\pi$; $\tan y = -\frac{5}{4}$, $\frac{1}{2}\pi < y < \pi$
20. $\tan x = -\frac{7}{2}$, $\frac{3}{2}\pi < x < 2\pi$; $\tan y = \frac{4}{5}$, $\pi < y < \frac{3}{2}\pi$

In Exercises 21 through 24, use the sum and difference identities to prove the reduction formula.

21. (a) $\sin(\frac{1}{2}\pi + x) = \cos x$; (b) $\sin(\pi + x) = -\sin x$;
    (c) $\sin(\frac{3}{2}\pi - x) = -\cos x$
22. (a) $\cos(\frac{1}{2}\pi + x) = -\sin x$; (b) $\cos(\pi + x) = -\cos x$;
    (c) $\cos(\frac{3}{2}\pi - x) = -\sin x$
23. (a) $\sin(2\pi - x) = -\sin x$; (b) $\tan(\pi + x) = \tan x$;
    (c) $\tan(\frac{1}{2}\pi + x) = -\cot x$

24. (a) $\cos(2\pi - x) = \cos x$; (b) $\tan(2\pi - x) = -\tan x$;
    (c) $\tan(\frac{3}{2}\pi - x) = \cot x$

In Exercises 25 through 30, prove the identity.

25. $\cot(\alpha + \beta) = \dfrac{\cot \alpha \cot \beta - 1}{\cot \beta + \cot \alpha}$

26. $\cot(\alpha - \beta) = \dfrac{\cot \alpha \cot \beta + 1}{\cot \beta - \cot \alpha}$

27. $\sec(\alpha + \beta) = \dfrac{\sec \alpha \sec \beta}{1 - \tan \alpha \tan \beta}$

28. $\csc(\alpha + \beta) = \dfrac{\csc \alpha \csc \beta}{\cot \beta + \cot \alpha}$

29. $\tan(\frac{1}{4}\pi + x) = \dfrac{1 + \tan x}{1 - \tan x}$

30. $\tan(x - \frac{3}{4}\pi) = \dfrac{1 + \tan x}{1 - \tan x}$

In Exercises 31 through 36, simplify the expression by writing it as $\pm\sin kx$, $\pm\cos kx$, or $\pm\tan kx$, where k is a positive integer.

31. (a) $\cos 8x \cos x + \sin 8x \sin x$;
    (b) $\sin 5x \cos 2x + \cos 5x \sin 2x$
32. (a) $\cos 3x \cos 2x - \sin 3x \sin 2x$;
    (b) $\sin 5x \cos 4x - \cos 5x \sin 4x$
33. (a) $\sin 4x \sin 6x - \cos 4x \cos 6x$;
    (b) $\sin 3x \cos 4x - \cos 3x \sin 4x$
34. (a) $\sin x \cos 3x + \cos x \sin 3x$;
    (b) $\sin x \sin 5x + \cos x \cos 5x$
35. (a) $\dfrac{\tan 8x + \tan 2x}{1 - \tan 8x \tan 2x}$;
    (b) $\dfrac{\tan \frac{1}{3}x - \tan \frac{7}{3}x}{1 + \tan \frac{1}{3}x \tan \frac{7}{3}x}$
36. (a) $\dfrac{\tan \frac{1}{4}x + \tan \frac{3}{4}x}{1 - \tan \frac{1}{4}x \tan \frac{3}{4}x}$;
    (b) $\dfrac{\tan 4x - \tan 5x}{1 + \tan 4x \tan 5x}$

In Exercises 37 through 40, prove the identity.

37. $\dfrac{\sin(u + v) + \sin(u - v)}{\cos(u + v) + \cos(u - v)} = \tan u$

38. $\dfrac{\sin(u - v)}{\sin u \sin v} = \cot v - \cot u$

39. $\cos(u + v) \cos(u - v) = \cos^2 u - \sin^2 v$

40. $\cos u \sin(u + v) - \sin u \cos(u + v) = \sin v$

41. Prove that if α, β, and γ are angles in a triangle, then
    (a) $\sin \alpha \cos \beta + \cos \alpha \sin \beta = \sin \gamma$ and
    (b) $\cos \alpha \cos \beta - \sin \alpha \sin \beta = -\cos \gamma$

42. Prove that Equation (6) follows from (7).

In Exercises 43 through 46, (a) express $f(t)$ in the form $a \sin(t + c)$, (b) determine the amplitude, period, and phase shift of f, and (c) draw a sketch of the graph.

43. $f(t) = \sqrt{3} \sin t + \cos t$ 44. $f(t) = 2 \sin t + 2 \cos t$
45. $f(t) = 3 \sin t - 3 \cos t$
46. $f(t) = 3 \sin t - 3\sqrt{3} \cos t$
47. A weight is suspended from a spring and vibrating vertically according to the equation

$$f(t) = 3 \sin 6t + 4 \cos 6t$$

where $f(t)$ centimeters is the directed distance of the weight from its central position t seconds after the start of the motion.
  (a) Define $f(t)$ by an equation of the form
$$f(t) = a \sin(bt + c)$$
  (b) Determine the amplitude, period, and frequency of f.
48. Do Exercise 47 if $f(t) = -6 \sin 4t - 8 \cos 4t$.
49. Two atmospheric waves at a given point in space produce pressures of $F(t)$ dynes/cm^2 and $G(t)$ dynes/cm^2 at t seconds, where

$$F(t) = 0.06 \sin 2\pi nt$$

and
$$G(t) = 0.04 \sin(2\pi nt - \tfrac{3}{4}\pi)$$

Define the sum of F and G by an equation of the form
$$f(t) = a \sin(2\pi nt + c)$$

50. In an electric circuit the electromotive force is $E(t)$ volts, where

$$E(t) = 50 + 6 \sin 60\pi t + 3.2 \cos 60\pi t$$
$$- 0.6 \sin 120\pi t + 0.8 \cos 120\pi t$$

Define $E(t)$ by an equation of the form

$$E(t) = a_0 + a_1 \sin(60\pi t + c_1) + a_2 \sin(120\pi t + c_2)$$

51. A particle is moving along a straight line according to the equation of motion

$$s = \sin(4t + \tfrac{1}{3}\pi) + \sin(4t + \tfrac{1}{6}\pi)$$

where s centimeters is the directed distance of the particle from the origin at t seconds.
  (a) Show that the motion is simple harmonic by defining s by an equation of the form

$$s = a \sin(bt + c)$$

  (b) Find the amplitude and frequency of the motion.
52. Do Exercise 51 if $s = 6 \sin(t + \tfrac{1}{3}\pi) + 4 \sin(t - \tfrac{1}{6}\pi)$.
53. A flagpole 15 ft high is situated on top of a building 10 ft high. At a point on the ground x feet from the base of the building, the flagpole and the building subtend equal angles. Determine x.

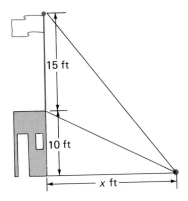

15 ft

10 ft

x ft

7.4 Double-Measure and Half-Measure Identities

The double-measure identities are special cases of the sine sum, cosine sum, and tangent sum identities obtained in Section 7.3. First of all, in the sine sum identity,

$$\sin(x + y) = \sin x \cos y + \cos x \sin y$$

let $y = x$, and we have

$$\sin(x + x) = \sin x \cos x + \cos x \sin x$$

from which we have the following sine double-measure identity.

Sine Double-Measure Identity

$$\sin 2x = 2 \sin x \cos x$$

We proceed in a similar fashion with the cosine sum identity.

$$\cos(x + y) = \cos x \cos y - \sin x \sin y$$
$$\cos(x + x) = \cos x \cos x - \sin x \sin x$$

We have then the cosine double-measure identity.

Cosine Double-Measure Identity

$$\cos 2x = \cos^2 x - \sin^2 x$$

If x is an angle, the double-measure identities are referred to as *double-angle identities*.

There are two other forms for the cosine double-measure identity. If we substitute $1 - \sin^2 x$ for $\cos^2 x$ in the formula for $\cos 2x$, we get

$$\cos 2x = (1 - \sin^2 x) - \sin^2 x$$
$$= 1 - 2 \sin^2 x$$

Furthermore, if we substitute $1 - \cos^2 x$ for $\sin^2 x$ in the cosine double-measure identity, we have

$$\cos 2x = \cos^2 x - (1 - \cos^2 x)$$
$$= 2 \cos^2 x - 1$$

We state these two results formally.

Alternative Forms of Cosine Double-Measure Identity

$$\cos 2x = 1 - 2 \sin^2 x$$
$$\cos 2x = 2 \cos^2 x - 1$$

Example   1   Given that $\sin t = \frac{3}{5}$ and $\frac{1}{2}\pi < t < \pi$, find $\sin 2t$ and $\cos 2t$.

Solution   We first find $\cos t$ from the fundamental identity

$$\sin^2 t + \cos^2 t = 1$$

Because $\sin t = \frac{3}{5}$, we have

$$(\tfrac{3}{5})^2 + \cos^2 t = 1$$
$$\cos^2 t = 1 - \tfrac{9}{25}$$
$$\cos^2 t = \tfrac{16}{25}$$

Because $\frac{1}{2}\pi < t < \pi$, $\cos t < 0$. Therefore, when we solve for $\cos t$ by taking the square root, we reject the positive value. Thus

$$\cos t = -\tfrac{4}{5}$$

Therefore

$$\sin 2t = 2 \sin t \cos t \qquad \cos 2t = \cos^2 t - \sin^2 t$$
$$= 2(\tfrac{3}{5})(-\tfrac{4}{5}) \qquad\qquad = (-\tfrac{4}{5})^2 - (\tfrac{3}{5})^2$$
$$= -\tfrac{24}{25} \qquad\qquad\qquad = \tfrac{7}{25} \qquad\qquad ■$$

The next example shows how another multiple-measure identity follows from the double-measure identities and the sine sum identity.

Example   2 Obtain an identity for $\sin 3x$ in terms of $\sin x$.

Solution

$$\sin 3x = \sin(2x + x)$$
$$= \sin 2x \cos x + \cos 2x \sin x$$
$$= (2 \sin x \cos x) \cos x + (1 - 2 \sin^2 x) \sin x$$
$$= 2 \sin x \cos^2 x + \sin x - 2 \sin^3 x$$
$$= 2 \sin x(1 - \sin^2 x) + \sin x - 2 \sin^3 x$$
$$= 2 \sin x - 2 \sin^3 x + \sin x - 2 \sin^3 x$$
$$= 3 \sin x - 4 \sin^3 x \qquad\qquad ■$$

To derive the formula expressing $\tan 2x$ in terms of $\tan x$, we start with the tangent sum identity.

$$\tan(x + y) = \frac{\tan x + \tan y}{1 - \tan x \tan y}$$

We let $y = x$ and get

$$\tan(x + x) = \frac{\tan x + \tan x}{1 - \tan x \tan x}$$

from which we have the following tangent double-measure identity.

Tangent Double-Measure Identity

$$\tan 2x = \frac{2 \tan x}{1 - \tan^2 x}$$

This identity does not hold if $x = \frac{1}{4}\pi + \frac{1}{2}k\pi$, where k is an integer, because for these values of x the denominator is zero. The identity also does not hold if $x = \frac{1}{2}\pi + k\pi$, where k is an integer, because for these values $\tan x$ does not exist.

Example 3   Obtain an identity for $\tan 4\theta$ in terms of $\tan \theta$.

Solution   From the tangent double-measure identity with $x = 2\theta$, we have

$$\tan 2(2\theta) = \frac{2 \tan 2\theta}{1 - \tan^2 2\theta}$$

On the right side of this identity we apply the tangent double-measure identity with $x = \theta$, and we have

$$\tan 4\theta = \frac{2\left(\dfrac{2 \tan \theta}{1 - \tan^2 \theta}\right)}{1 - \left(\dfrac{2 \tan \theta}{1 - \tan^2 \theta}\right)^2}$$

$$= \frac{\dfrac{4 \tan \theta}{1 - \tan^2 \theta}}{\dfrac{(1 - \tan^2 \theta)^2 - 4 \tan^2 \theta}{(1 - \tan^2 \theta)^2}}$$

$$= \frac{(1 - \tan^2 \theta)^2\left(\dfrac{4 \tan \theta}{1 - \tan^2 \theta}\right)}{(1 - \tan^2 \theta)^2\left[\dfrac{(1 - 2 \tan^2 \theta + \tan^4 \theta) - 4 \tan^2 \theta}{(1 - \tan^2 \theta)^2}\right]}$$

$$= \frac{4 \tan \theta(1 - \tan^2 \theta)}{1 - 6 \tan^2 \theta + \tan^4 \theta} \qquad \blacksquare$$

Useful identities expressing $\sin^2 x$, $\cos^2 x$, and $\tan^2 x$ in terms of $\cos 2x$ are obtained from the alternative forms of the cosine double-measure identity.

$$\cos 2x = 1 - 2 \sin^2 x \qquad \cos 2x = 2 \cos^2 x - 1$$
$$2 \sin^2 x = 1 - \cos 2x \qquad 2 \cos^2 x = 1 + \cos 2x$$
$$\sin^2 x = \frac{1 - \cos 2x}{2} \qquad \cos^2 x = \frac{1 + \cos 2x}{2}$$

$$\tan^2 x = \frac{\sin^2 x}{\cos^2 x}$$

$$= \frac{\dfrac{1 - \cos 2x}{2}}{\dfrac{1 + \cos 2x}{2}}$$

$$= \frac{1 - \cos 2x}{1 + \cos 2x}$$

We have then the following three identities.

Identities for $\sin^2 x$, $\cos^2 x$, and $\tan^2 x$ in Terms of $\cos 2x$

$$\sin^2 x = \frac{1 - \cos 2x}{2}$$

$$\cos^2 x = \frac{1 + \cos 2x}{2}$$

$$\tan^2 x = \frac{1 - \cos 2x}{1 + \cos 2x}$$

In the first two of these identities, x can be any real number or angle. In the third identity, because $1 + \cos 2x$ cannot be zero, $x \neq \frac{1}{2}\pi + k\pi$, where k is any integer. If in each of the identities we let $x = \frac{1}{2}y$, then $2x = y$, and we obtain the following half-measure identities.

Half-Measure Identities

$$\sin^2 \tfrac{1}{2}y = \frac{1 - \cos y}{2}$$

$$\cos^2 \tfrac{1}{2}y = \frac{1 + \cos y}{2}$$

$$\tan^2 \tfrac{1}{2}y = \frac{1 - \cos y}{1 + \cos y}$$

When applying these identities to find $\sin \frac{1}{2}y$, $\cos \frac{1}{2}y$, and $\tan \frac{1}{2}y$, the sign ($+$ or $-$) is determined by the value of $\frac{1}{2}y$.

Illustration 1

To find the exact value of $\cos \frac{1}{8}\pi$, we use the identity for $\cos^2 \frac{1}{2}y$ with $y = \frac{1}{4}\pi$:

$$\cos^2 \frac{1}{8}\pi = \frac{1 + \cos \dfrac{1}{4}\pi}{2}$$

$$\cos^2 \frac{1}{8}\pi = \frac{1 + \dfrac{\sqrt{2}}{2}}{2}$$

$$\cos^2 \frac{1}{8}\pi = \frac{2 + \sqrt{2}}{4}$$

Because $0 < \frac{1}{8}\pi < \frac{1}{2}\pi$, $\cos \frac{1}{8}\pi > 0$. Therefore

$$\cos \frac{1}{8}\pi = \frac{\sqrt{2 + \sqrt{2}}}{2}$$

■

There are two other identities for $\tan \frac{1}{2}y$. One of them is derived by starting with the identity

$$\tan \tfrac{1}{2}y = \frac{\sin \tfrac{1}{2}y}{\cos \tfrac{1}{2}y}$$

and multiplying the numerator and denominator by $2 \sin \frac{1}{2}y$. We then have

$$\tan \tfrac{1}{2}y = \frac{2 \sin^2 \tfrac{1}{2}y}{2 \sin \tfrac{1}{2}y \cos \tfrac{1}{2}y} \qquad (1)$$

From the identity for $\sin^2 \frac{1}{2}y$, we obtain

$$2 \sin^2 \tfrac{1}{2}y = 1 - \cos y \qquad (2)$$

and from the sine double-measure identity with $x = \frac{1}{2}y$, we have

$$2 \sin \tfrac{1}{2}y \cos \tfrac{1}{2}y = \sin y \qquad (3)$$

Substituting from (2) and (3) in (1), we get

$$\tan \tfrac{1}{2}y = \frac{1 - \cos y}{\sin y}$$

Another identity involving $\tan \frac{1}{2}y$ is obtained from this one by multiplying the numerator and denominator by $1 + \cos y$.

$$
\begin{aligned}
\tan \tfrac{1}{2}y &= \frac{(1 - \cos y)(1 + \cos y)}{\sin y(1 + \cos y)} \\
&= \frac{1 - \cos^2 y}{\sin y(1 + \cos y)} \\
&= \frac{\sin^2 y}{\sin y(1 + \cos y)} \\
&= \frac{\sin y}{1 + \cos y}
\end{aligned}
$$

We state these two identities formally.

Tangent Half-Measure Identities

$$\tan \tfrac{1}{2}y = \frac{1 - \cos y}{\sin y}$$

$$\tan \tfrac{1}{2}y = \frac{\sin y}{1 + \cos y}$$

In the first of these identities y can be any real number or angle except $k\pi$, where k is any integer. In the second y can be any real number or angle except $(2k + 1)\pi$, where k is any integer. Note that $(2k + 1)\pi$ is an odd multiple of π.

When y represents an angle in the half-measure identities, they are also referred to as *half-angle identities.*

Example 4 Use the half-measure identities to find the exact value of (a) $\cos 105°$ and (b) $\tan 22.5°$.

Solution a) From the identity

$$\cos^2 \tfrac{1}{2}y = \frac{1 + \cos y}{2}$$

with $y = 210°$, we have

$$\cos^2 105° = \frac{1 + \cos 210°}{2}$$

Because $105°$ is a second-quadrant angle, $\cos 105°$ is negative. Therefore, with $\cos 210° = -\tfrac{1}{2}\sqrt{3}$, we have

$$\cos 105° = -\sqrt{\frac{1 - \dfrac{\sqrt{3}}{2}}{2}}$$

$$= -\frac{\sqrt{2 - \sqrt{3}}}{2}$$

b) From the identity

$$\tan \tfrac{1}{2}y = \frac{1 - \cos y}{\sin y}$$

with $y = 45°$, we have

$$\tan 22.5° = \frac{1 - \cos 45°}{\sin 45°}$$

$$= \frac{1 - \dfrac{1}{\sqrt{2}}}{\dfrac{1}{\sqrt{2}}}$$

$$= \sqrt{2} - 1 \qquad ■$$

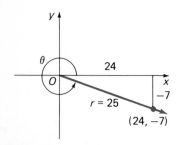

FIGURE 1

Example 5 Given that $\tan \theta = -\tfrac{7}{24}$ and $270° < \theta < 360°$, find $\tan \tfrac{1}{2}\theta$.

Solution Figure 1 shows angle θ with the point $(24, -7)$ chosen on its terminal side. From the Pythagorean theorem, $r = 25$. Then

$$\sin \theta = -\tfrac{7}{25} \quad \text{and} \quad \cos \theta = \tfrac{24}{25}$$

Therefore

$$\tan \tfrac{1}{2}\theta = \frac{1 - \cos \theta}{\sin \theta}$$

$$= \frac{1 - \frac{24}{25}}{-\frac{7}{25}}$$

$$= -\tfrac{1}{7}$$ ∎

EXERCISES 7.4

In Exercises 1 through 8, find (a) $\sin 2t$, (b) $\cos 2t$, and (c) $\tan 2t$.

1. $\sin t = \frac{4}{5}$ and $0 < t < \frac{1}{2}\pi$
2. $\cos t = \frac{7}{25}$ and $0 < t < \frac{1}{2}\pi$
3. $\cos t = -\frac{5}{13}$ and $\frac{1}{2}\pi < t < \pi$
4. $\sin t = \frac{3}{5}$ and $\frac{1}{2}\pi < t < \pi$
5. $\tan t = \frac{8}{15}$ and $\sin t < 0$
6. $\tan t = -\frac{12}{5}$ and $\sin t < 0$
7. $\sin t = -\frac{7}{25}$ and $\cos t > 0$
8. $\cos t = -\frac{15}{17}$ and $\tan t > 0$

In Exercises 9 through 16, use the half-measure identities to find the exact value.

9. $\sin \frac{1}{8}\pi$
10. $\cos \frac{1}{12}\pi$
11. $\cos 165°$
12. $\sin 112.5°$
13. $\tan \frac{5}{12}\pi$
14. $\tan \frac{3}{8}\pi$
15. $\tan 112.5°$
16. $\tan 105°$

In Exercises 17 through 22, use a half-measure identity to find the indicated function value.

17. $\cos t = \frac{1}{3}$ and $0 < t < \frac{1}{2}\pi$; find $\sin \frac{1}{2}t$.
18. $\cos t = \frac{1}{2}$ and $0 < t < \frac{1}{2}\pi$; find $\cos \frac{1}{2}t$.
19. $\sin t = \frac{24}{25}$ and $\frac{1}{2}\pi < t < \pi$; find $\cos \frac{1}{2}t$.
20. $\sin t = \frac{4}{5}$ and $\cos t < 0$; find $\tan \frac{1}{2}t$.
21. $\tan t = -\frac{5}{12}$ and $\sin t > 0$; find $\tan \frac{1}{2}t$.
22. $\tan t = \frac{24}{7}$ and $\sin t < 0$; find $\cos \frac{1}{2}t$.

In Exercises 23 through 32, simplify the expression by writing it as $a \sin kx$, $a \cos kx$, or $a \tan kx$, where a is an integer and k is a positive integer.

23. (a) $\cos^2 x - \sin^2 x$; (b) $\cos^2 2x - \sin^2 2x$;
    (c) $2 \sin \frac{3}{2}x \cos \frac{3}{2}x$
24. (a) $1 - 2 \sin^2 x$; (b) $1 - 2 \sin^2 3x$; (c) $1 - 2 \cos^2 x$
25. (a) $2 \sin x \cos x$; (b) $4 \sin 4x \cos 4x$;
    (c) $2 \sin \dfrac{x}{2} \cos \dfrac{x}{2}$

26. (a) $6 \sin 2x \cos 2x$; (b) $2 \sin 6x \cos 6x$;
    (c) $2 \sin \frac{3}{2}x \cos \frac{3}{2}x$
27. (a) $\dfrac{2 \tan x}{1 - \tan^2 x}$; (b) $\dfrac{4 \tan 2x}{\tan^2 2x - 1}$; (c) $\dfrac{6 \tan \frac{3}{2}x}{1 - \tan^2 \frac{3}{2}x}$
28. (a) $\dfrac{2 \tan \frac{1}{2}x}{1 - \tan^2 \frac{1}{2}x}$; (b) $\dfrac{4 \tan 4x}{1 - \tan^2 4x}$; (c) $\dfrac{8 \tan 3x}{\tan^2 3x - 1}$
29. (a) $\dfrac{\sin 2t}{1 + \cos 2t}$; (b) $\dfrac{\sin 4t}{\cos 4t + 1}$; (c) $\dfrac{\cos 4t - 1}{\sin 4t}$
30. (a) $\dfrac{1 - \cos 2t}{\sin 2t}$; (b) $\dfrac{1 - \cos 6t}{\sin 6t}$; (c) $\dfrac{\cos 8t - 1}{\sin 8t}$
31. (a) $\dfrac{\cos 2t \sin 4t}{1 + \cos 4t}$; (b) $\pm\sqrt{\dfrac{1 - \cos 8t}{2}}$;
    (c) $\pm\sqrt{2 - 2 \cos 8t}$
32. (a) $\dfrac{\cos 3t(1 - \cos 6t)}{\sin 6t}$; (b) $\pm\sqrt{\dfrac{1 + \cos 4t}{2}}$;
    (c) $\pm\sqrt{2 + 2 \cos 4t}$

In Exercises 33 through 40, obtain an identity for the first function value in terms of the second function value.

33. $\cos 3x$; $\cos x$
34. $\cos 4x$; $\cos x$
35. $\sin 4x$; $\sin x$
36. $\sin 5x$; $\sin x$
37. $\tan 3x$; $\tan x$
38. $\cot 2x$; $\cot x$
39. $\sec 2x$; $\sec x$
40. $\csc 2x$; $\csc x$

In Exercises 41 through 48, prove the identity.

41. $\dfrac{2 \tan x}{1 + \tan^2 x} = \sin 2x$
42. $\dfrac{\tan x + \cot x}{\cot x - \tan x} = \sec 2x$
43. $\dfrac{2}{1 + \cos 2t} = \sec^2 t$
44. $\dfrac{1 + \cos 2t}{1 - \cos 2t} = \cot^2 t$
45. $\cos^4 \theta - \sin^4 \theta = \cos 2\theta$
46. $\dfrac{1 - \tan^2 \theta}{1 + \tan^2 \theta} = \cos 2\theta$
47. $\dfrac{1 + \sin 2\beta + \cos 2\beta}{1 + \sin 2\beta - \cos 2\beta} = \cot \beta$

48. $\dfrac{1 - \cos 8x}{8} = \sin^2 2x \cos^2 2x$

49. If α and β are the acute angles in a right triangle, verify that $\sin 2\alpha = \sin 2\beta$.

50. If K square units is the area of a right triangle, c units is the length of the hypotenuse, and α is an acute angle, verify that $K = \frac{1}{4}c^2 \sin 2\alpha$.

51. A particle is moving along a straight line according to the equation of motion $s = 12 \cos^2 4t - 6$, where s centimeters is the directed distance of the particle from the origin at t seconds. (a) Show that the motion is simple harmonic by defining s by an equation of the form $s = a \sin(bt + c)$. (b) Find the amplitude and frequency of the motion.

52. Do Exercise 51 if $s = 5 - 10 \sin^2 2t$.

53. A pendulum of length 10 cm has swung so that θ is the radian measure of the angle formed by the pendulum and a vertical line. Show that the number of centimeters in the vertical height of the end of the pendulum above its lowest position is $20 \sin^2 \frac{1}{2}\theta$.

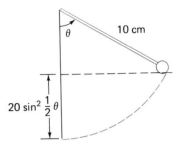

7.5   Inverse Trigonometric Functions

Before our treatment of inverse trigonometric functions, you may wish to review Section 4.7, where we introduced inverse functions. We showed there that it was necessary for a function to be one-to-one in order for it to have an inverse. We used the horizontal line test to determine whether a function is one-to-one. This test states that if every horizontal line intersects the graph of a function in no more than one point, then the function is one-to-one.

Figure 1 shows a sketch of the graph of the sine function. This function is not one-to-one, because every number in its range is the function value of more than one number in its domain. Therefore the sine function does not have an inverse. However, observe in Figure 1 that on the interval $[-\frac{1}{2}\pi, \frac{1}{2}\pi]$, every horizontal line intersects this portion of the graph in no more than one point. Thus from the horizontal line test, the function F for which

$$F(x) = \sin x \qquad \text{and} \qquad -\tfrac{1}{2}\pi \le x \le \tfrac{1}{2}\pi \qquad (1)$$

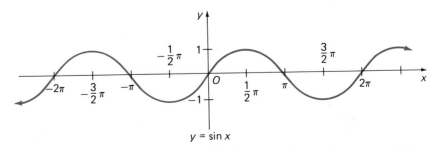

$y = \sin x$

FIGURE 1

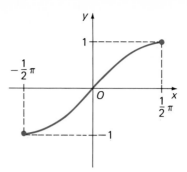

FIGURE 2

is one-to-one and therefore has an inverse function. The domain of F is $[-\frac{1}{2}\pi, \frac{1}{2}\pi]$, and its range is $[-1, 1]$. A sketch of the graph of F appears in Figure 2. The inverse of the function defined by (1) is called the *inverse sine function* and is denoted by the symbol $\sin^{-1}$. Following is the formal definition.

Definition

> **Inverse Sine Function**
> The **inverse sine function**, denoted by $\sin^{-1}$, is defined as follows:
>
> $$y = \sin^{-1} x \qquad \text{if and only if} \qquad x = \sin y \quad \text{and} \quad -\tfrac{1}{2}\pi \le y \le \tfrac{1}{2}\pi$$

The domain of $\sin^{-1}$ is the closed interval $[-1, 1]$, and the range is the closed interval $[-\frac{1}{2}\pi, \frac{1}{2}\pi]$.

To draw a sketch of the graph of the inverse sine function, let

$$y = \sin^{-1} x$$

Table 1 gives some values of x and y satisfying this equation. A sketch of the graph appears in Figure 3.

TABLE 1

x	y
-1	$-\dfrac{1}{2}\pi$
$-\dfrac{\sqrt{3}}{2}$	$-\dfrac{1}{3}\pi$
$-\dfrac{1}{2}$	$-\dfrac{1}{6}\pi$
0	0
$\dfrac{1}{2}$	$\dfrac{1}{6}\pi$
$\dfrac{\sqrt{3}}{2}$	$\dfrac{1}{3}\pi$
1	$\dfrac{1}{2}\pi$

FIGURE 3

The use of the symbol -1 to represent the inverse sine function makes it necessary to denote the reciprocal of $\sin x$ by $(\sin x)^{-1}$ to avoid confusion. A similar convention is applied when using any negative exponent with a trigonometric function. For instance,

$$\frac{1}{\sin x} = (\sin x)^{-1} \qquad \frac{1}{\sin^2 x} = (\sin x)^{-2} \qquad \frac{1}{\cos^3 x} = (\cos x)^{-3}$$

and so on.

The terminology **arc sine** is sometimes used in place of inverse sine, and the notation arc $\sin x$ can be used instead of $\sin^{-1} x$. This notation probably comes from the fact that if $t = $ arc sin u, then $\sin t = u$, and t units is the length of the arc on the unit circle for which the sine is u.

Illustration 1

a) $\sin^{-1} \dfrac{1}{\sqrt{2}} = \dfrac{1}{4} \pi$ b)   arc sin $\dfrac{1}{\sqrt{2}} = \dfrac{1}{4} \pi$

c) $\sin^{-1} \left(-\dfrac{1}{\sqrt{2}} \right) = -\dfrac{1}{4} \pi$ d)   arc sin $\left(-\dfrac{1}{\sqrt{2}} \right) = -\dfrac{1}{4} \pi$ ■

Illustration 2

a)   To find $\sin^{-1}(0.8724)$ from Table VI, look for 0.8724 in the sine column of the table and observe that $\sin 1.06 \approx 0.8724$. Therefore

$$\sin^{-1}(0.8724) \approx 1.06$$

b)   If a calculator is used to determine $\sin^{-1}(0.8724)$, first set the calculator in the radian mode. Obtain 0.8724 in the display and then press the $\boxed{\sin^{-1}}$ key (or the $\boxed{\text{INV}}$ key followed by the $\boxed{\sin}$ key) and read 1.06.   ■

From the definition of the inverse sine function

$$\sin(\sin^{-1} x) = x \qquad \text{for } x \text{ in } [-1, 1]$$
$$\sin^{-1}(\sin y) = y \qquad \text{for } y \text{ in } [-\tfrac{1}{2}\pi, \tfrac{1}{2}\pi]$$

Observe that $\sin^{-1}(\sin y) \neq y$ if y is not in the interval $[-\tfrac{1}{2}\pi, \tfrac{1}{2}\pi]$. For example,

$$\sin^{-1} \left(\sin \frac{3}{4} \pi \right) = \sin^{-1} \frac{1}{\sqrt{2}} \qquad \text{and} \qquad \sin^{-1} \left(\sin \frac{7}{4} \pi \right) = \sin^{-1} \left(-\frac{1}{\sqrt{2}} \right)$$

$$= \frac{1}{4} \pi \qquad\qquad\qquad\qquad = -\frac{1}{4} \pi$$

Example 1   Find: (a) $\cos[\sin^{-1}(-\tfrac{1}{2})]$; (b) $\sin^{-1}(\cos \tfrac{2}{3}\pi)$.

Solution   Because the range of the inverse sine function is $[-\tfrac{1}{2}\pi, \tfrac{1}{2}\pi]$, $\sin^{-1}(-\tfrac{1}{2}) = -\tfrac{1}{6}\pi$.

a) $\cos\left[\sin^{-1}\left(-\dfrac{1}{2}\right)\right] = \cos\left(-\dfrac{1}{6}\pi\right)$ b) $\sin^{-1}\left(\cos\dfrac{2}{3}\pi\right) = \sin^{-1}\left(-\dfrac{1}{2}\right)$

$$= \dfrac{\sqrt{3}}{2} \qquad\qquad\qquad = -\dfrac{1}{6}\pi \qquad ■$$

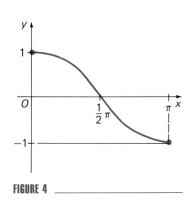

FIGURE 4

The cosine does not have an inverse function because it also is not one-to-one. To define the inverse cosine function, we restrict the cosine to the interval $[0, \pi]$. Consider the function G defined by

$$G(x) = \cos x \qquad \text{and} \qquad 0 \le x \le \pi$$

The range of G is the closed interval $[-1, 1]$, and the domain is the closed interval $[0, \pi]$. A sketch of the graph of G is shown in Figure 4. From the horizontal line test, we observe that G is one-to-one. Therefore it has an inverse function, called the *inverse cosine function* and denoted by $\cos^{-1}$.

Definition

> **Inverse Cosine Function**
> The **inverse cosine function,** denoted by $\cos^{-1}$, is defined as follows:
> $$y = \cos^{-1} x \qquad \text{if and only if} \qquad x = \cos y \quad \text{and} \quad 0 \le y \le \pi$$

The domain of $\cos^{-1}$ is the closed interval $[-1, 1]$, and the range is the closed interval $[0, \pi]$.

Table 2 gives some values of x and y satisfying the equation

$$y = \cos^{-1} x$$

From these values the sketch of the graph of the inverse cosine function shown in Figure 5 is obtained.

TABLE 2

x	y
-1	π
$-\dfrac{\sqrt{3}}{2}$	$\dfrac{5}{6}\pi$
$-\dfrac{1}{2}$	$\dfrac{2}{3}\pi$
0	$\dfrac{1}{2}\pi$
$\dfrac{1}{2}$	$\dfrac{1}{3}\pi$
$\dfrac{\sqrt{3}}{2}$	$\dfrac{1}{6}\pi$
1	0

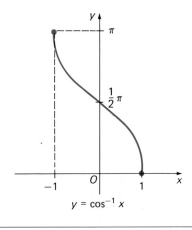

$$y = \cos^{-1} x$$

FIGURE 5

The inverse cosine function is also called the **arc cosine** function, and the notation arc cos x can be used in place of $\cos^{-1} x$.

Illustration 3

a) $\cos^{-1} \dfrac{1}{\sqrt{2}} = \dfrac{1}{4}\pi$

b) arc cos $\dfrac{1}{\sqrt{2}} = \dfrac{1}{4}\pi$

c) $\cos^{-1} \left(-\dfrac{1}{\sqrt{2}}\right) = \dfrac{3}{4}\pi$

d) arc cos $\left(-\dfrac{1}{\sqrt{2}}\right) = \dfrac{3}{4}\pi$ ■

Illustration 4

a) Suppose we wish to find $\cos^{-1}(-0.6137)$ by using a table. If we let $t = \cos^{-1}(-0.6137)$, then $\frac{1}{2}\pi < t < \pi$. We first find a value of $\overline{t}$ such that $\overline{t} = \cos^{-1}(0.6137)$. So we look for 0.6137 in the cosine column of Table VI and find cos $0.91 \approx 0.6137$. Thus $\overline{t} \approx 0.91$. Therefore $t \approx 3.14 - 0.91$; that is, $t \approx 2.23$. Hence

$$\cos^{-1}(-0.6137) \approx 2.23$$

b) If a calculator is used to determine $\cos^{-1}(-0.6137)$, set it in the radian mode, obtain -0.6137 in the display, press the $\boxed{\cos^{-1}}$ key (or the $\boxed{\text{INV}}$ key followed by the $\boxed{\cos}$ key), and obtain 2.23. ■

From the definition of the inverse cosine function,

$$\cos(\cos^{-1} x) = x \qquad \text{for } x \text{ in } [-1, 1]$$
$$\cos^{-1}(\cos y) = y \qquad \text{for } y \text{ in } [0, \pi]$$

Notice there is again a restriction on y in order to have the equality $\cos^{-1}(\cos y) = y$. For example, because $\frac{3}{4}\pi$ is in $[0, \pi]$,

$$\cos^{-1}(\cos \tfrac{3}{4}\pi) = \tfrac{3}{4}\pi$$

However,

$$\cos^{-1}\left(\cos \frac{5}{4}\pi\right) = \cos^{-1}\left(-\frac{1}{\sqrt{2}}\right) \qquad \text{and} \qquad \cos^{-1}\left(\frac{7}{4}\pi\right) = \cos^{-1}\left(\frac{1}{\sqrt{2}}\right)$$

$$= \frac{3}{4}\pi \qquad\qquad\qquad\qquad = \frac{1}{4}\pi$$

Example 2 Find the exact value of $\sin[2 \cos^{-1}(-\frac{3}{5})]$.

Solution Because we wish to obtain trigonometric functions of the number $\cos^{-1}(-\frac{3}{5})$, we shall let t represent this number:

$$t = \cos^{-1}(-\tfrac{3}{5})$$

Because the range of the inverse cosine function is $[0, \pi]$ and cos t is negative, t is in the second quadrant. Thus

$$\cos t = -\tfrac{3}{5} \quad \text{and} \quad \tfrac{1}{2}\pi < t < \pi$$

We wish to find the exact value of $\sin 2t$. From the sine double-measure identity, $\sin 2t = 2 \sin t \cos t$. Thus we need to compute $\sin t$. From the identity $\sin^2 t + \cos^2 t = 1$, and because $\sin t > 0$ since t is in $(\tfrac{1}{2}\pi, \pi)$, $\sin t = \sqrt{1 - \cos^2 t}$. Thus

$$\sin t = \sqrt{1 - (-\tfrac{3}{5})^2}$$
$$= \tfrac{4}{5}$$

Therefore

$$\sin 2t = 2 \sin t \cos t$$
$$= 2(\tfrac{4}{5})(-\tfrac{3}{5})$$
$$= -\tfrac{24}{25}$$

from which we conclude that

$$\sin[(2 \cos^{-1}(-\tfrac{3}{5})] = -\tfrac{24}{25}$$ ∎

To obtain the inverse tangent function, we first restrict the tangent function to the open interval $(-\tfrac{1}{2}\pi, \tfrac{1}{2}\pi)$. We let H be the function defined by

$$H(x) = \tan x \quad \text{and} \quad -\tfrac{1}{2}\pi < x < \tfrac{1}{2}\pi$$

The domain of H is the open interval $(-\tfrac{1}{2}\pi, \tfrac{1}{2}\pi)$, and the range is the set R of real numbers. A sketch of its graph appears in Figure 6. From the horizontal line test, H is one-to-one. Therefore it has an inverse function, which is called the *inverse tangent function* and is denoted by $\tan^{-1}$.

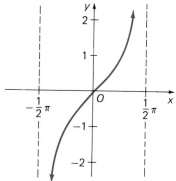

FIGURE 6

Definition

Inverse Tangent Function
The **inverse tangent function,** denoted by $\tan^{-1}$, is defined as follows:

$$y = \tan^{-1} x \quad \text{if and only if} \quad x = \tan y \quad \text{and} \quad -\tfrac{1}{2}\pi < y < \tfrac{1}{2}\pi.$$

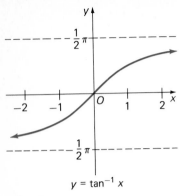

$$y = \tan^{-1} x$$

FIGURE 7

The domain of $\tan^{-1}$ is the set R of real numbers, and the range is the open interval $(-\frac{1}{2}\pi, \frac{1}{2}\pi)$. A sketch of its graph is shown in Figure 7.

The inverse tangent function is sometimes referred to as the arc tangent function, and then arc tan x is used instead of $\tan^{-1} x$.

Illustration 5

a) $\tan^{-1} \sqrt{3} = \frac{1}{3}\pi$ b) $\text{arc tan}\left(-\frac{1}{\sqrt{3}}\right) = -\frac{1}{6}\pi$ c) $\tan^{-1} 0 = 0$ ■

From the definition of the inverse tangent function,

$$\tan(\tan^{-1} x) = x \qquad \text{for } x \text{ in } R$$
$$\tan^{-1}(\tan y) = y \qquad \text{for } y \text{ in } (-\tfrac{1}{2}\pi, \tfrac{1}{2}\pi)$$

Illustration 6

$$\tan^{-1}(\tan \tfrac{1}{4}\pi) = \tfrac{1}{4}\pi \qquad \text{and} \qquad \tan^{-1}(\tan(-\tfrac{1}{4}\pi)) = -\tfrac{1}{4}\pi$$

However,

$$\tan^{-1}(\tan \tfrac{3}{4}\pi) = \tan^{-1}(-1) \qquad \text{and} \qquad \tan^{-1}(\tan \tfrac{5}{4}\pi) = \tan^{-1} 1$$
$$= -\tfrac{1}{4}\pi \qquad\qquad\qquad = \tfrac{1}{4}\pi$$ ■

Example 3 Find the exact value of

$$\sec[\text{arc tan}(-3)]$$

Solution We shall do this problem by letting arc tan(-3) be an angle. Let

$$\theta = \text{arc tan}(-3)$$

Because the range of the arc tangent function is $(-\frac{1}{2}\pi, \frac{1}{2}\pi)$, and because $\tan \theta$ is negative, $-\frac{1}{2}\pi < \theta < 0$. Thus

$$\tan \theta = -3 \qquad \text{and} \qquad -\tfrac{1}{2}\pi < \theta < 0$$

Figure 8 shows an angle θ that satisfies these requirements. Observe that the point P selected on the terminal side of θ is $(1, -3)$. From the Pythagorean theorem, r is $\sqrt{1^2 + (-3)^2} = \sqrt{10}$. Therefore $\sec \theta = \sqrt{10}$. Hence

$$\sec[\text{arc tan}(-3)] = \sqrt{10}$$ ■

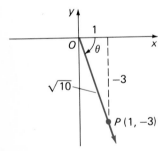

FIGURE 8

Example 4 A picture 7 ft high is placed on a wall with its base 9 ft above the level of the eye of an observer. Suppose the observer is x feet from the wall and θ is the radian measure of the angle subtended at the observer's eye by the picture. (a) Define θ as a function of x. Find θ when (b) $x = 10$; (c) $x = 12$; and (d) $x = 15$.

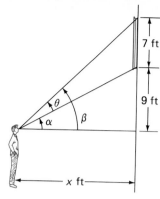

FIGURE 9

Solution a) In Figure 9, α is the radian measure of the angle subtended at the observer's eye by the portion of the wall above eye level and below the picture. Furthermore,

$$\alpha + \theta = \beta$$

So

$$\theta = \beta - \alpha$$

From the tangent difference identity,

$$\tan \theta = \frac{\tan \beta - \tan \alpha}{1 + \tan \beta \tan \alpha}$$

Observe from Figure 9 that

$$\tan \alpha = \frac{9}{x} \qquad \text{and} \qquad \tan \beta = \frac{16}{x}$$

Substituting these values into the expression for $\tan \theta$, we obtain

$$\tan \theta = \frac{\dfrac{16}{x} - \dfrac{9}{x}}{1 + \dfrac{16}{x} \cdot \dfrac{9}{x}}$$

$$= \frac{16x - 9x}{x^2 + 144}$$

$$= \frac{7x}{x^2 + 144}$$

Therefore

$$\theta = \tan^{-1} \frac{7x}{x^2 + 144}$$

b) When $x = 10$,

$$\theta = \tan^{-1} \frac{70}{100 + 144}$$

$$= \tan^{-1} \frac{70}{244}$$

$$= \tan^{-1} 0.2869$$

$$= 0.2794$$

c) When $x = 12$,

$$\theta = \tan^{-1} \frac{84}{144 + 144}$$

$$= \tan^{-1} \frac{84}{288}$$

$$= \tan^{-1} 0.2917$$

$$= 0.2838$$

d) When $x = 15$,

$$\theta = \tan^{-1} \frac{105}{225 + 144}$$

$$= \tan^{-1} \frac{105}{369}$$

$$= \tan^{-1} 0.2846$$

$$= 0.2772 \qquad \blacksquare$$

In Example 4, when x is large (that is, when the observer is far away from the wall), θ is small. As the observer gets closer to the wall, θ increases until it reaches a maximum value. Then as the observer gets even closer to the wall, θ gets smaller. In calculus we can find the value of x that will make θ a maximum. So we can determine how far from the wall the observer should stand in order for the angle subtended at the observer's eye by the picture to be the greatest. When θ is a maximum, the observer has the "best view" of the picture. It turns out that this "best view" occurs when he or she is 12 ft from the wall.

The remaining inverse trigonometric functions, $\cot^{-1}$, $\sec^{-1}$, and $\csc^{-1}$, are not used as much as the other three and are not discussed here. However, they are defined in a similar manner and you are asked to give these definitions as well as draw sketches of their graphs in Exercises 53 through 55.

EXERCISES   7.5

In Exercises 1 through 6, find the exact value.

1. (a) $\sin^{-1}\frac{1}{2}$; (b) $\sin^{-1}(-\frac{1}{2})$; (c) $\cos^{-1}\frac{1}{2}$; (d) $\cos^{-1}(-\frac{1}{2})$

2. (a) $\sin^{-1}\dfrac{\sqrt{3}}{2}$; (b) $\sin^{-1}\left(-\dfrac{\sqrt{3}}{2}\right)$; (c) $\cos^{-1}\dfrac{\sqrt{3}}{2}$; (d) $\cos^{-1}\left(-\dfrac{\sqrt{3}}{2}\right)$

3. (a) $\tan^{-1}\dfrac{1}{\sqrt{3}}$; (b) $\tan^{-1}(-\sqrt{3})$

4. (a) $\tan^{-1}\sqrt{3}$; (b) $\tan^{-1}\left(-\dfrac{1}{\sqrt{3}}\right)$

5. (a) $\sin^{-1}1$; (b) $\cos^{-1}(-1)$; (c) $\tan^{-1}1$; (d) $\sin^{-1}0$
6. (a) $\cos^{-1}1$; (b) $\sin^{-1}(-1)$; (c) $\tan^{-1}(-1)$; (d) $\cos^{-1}0$

In Exercises 7 through 10, use a calculator to approximate the function value.

7. (a) $\sin^{-1}0.4882$; (b) $\sin^{-1}(-0.4882)$; (c) $\cos^{-1}0.4882$; (d) $\cos^{-1}(-0.4882)$
8. (a) $\sin^{-1}0.2764$; (b) $\sin^{-1}(-0.2764)$; (c) $\cos^{-1}0.2764$; (d) $\cos^{-1}(-0.2764)$
9. (a) $\tan^{-1}0.4346$; (b) $\tan^{-1}(-0.4346)$
10. (a) $\tan^{-1}2.733$; (b) $\tan^{-1}(-2.733)$

In Exercises 11 through 14, use Table VI to approximate the function values of the indicated exercise.

11. Exercise 7   12. Exercise 8
13. Exercise 9   14. Exercise 10
15. Given that $x = \arcsin\frac{1}{3}$, find the exact value of each of the following: (a) $\cos x$; (b) $\tan x$; (c) $\cot x$; (d) $\sec x$; (e) $\csc x$.
16. Given that $x = \arccos\frac{2}{3}$, find the exact value of each of the following: (a) $\sin x$; (b) $\tan x$; (c) $\cot x$; (d) $\sec x$; (e) $\csc x$.
17. Do Exercise 15 if $x = \arcsin(-\frac{1}{3})$.

18. Do Exercise 16 if $x = \arccos(-\frac{2}{3})$.
19. Given that $y = \tan^{-1}(-2)$, find the exact value of each of the following: (a) $\sin y$; (b) $\cos y$; (c) $\cot y$; (d) $\sec y$; (e) $\csc y$.
20. Given that $t = \tan^{-1}\frac{1}{2}$, find the exact value of each of the following: (a) $\sin t$; (b) $\cos t$; (c) $\cot t$; (d) $\sec t$; (d) $\csc t$.

In Exercises 21 through 42, find the exact value.

21. (a) $\sin^{-1}(\sin\frac{1}{6}\pi)$; (b) $\sin^{-1}[\sin(-\frac{1}{6}\pi)]$; (c) $\sin^{-1}(\sin\frac{5}{6}\pi)$; (d) $\sin^{-1}(\sin\frac{11}{6}\pi)$
22. (a) $\sin^{-1}(\sin\frac{1}{3}\pi)$; (b) $\sin^{-1}[\sin(-\frac{1}{3}\pi)]$; (c) $\sin^{-1}(\sin\frac{2}{3}\pi)$; (d) $\sin^{-1}(\sin\frac{5}{3}\pi)$
23. (a) $\cos^{-1}(\cos\frac{1}{3}\pi)$; (b) $\cos^{-1}[\cos(-\frac{1}{3}\pi)]$; (c) $\cos^{-1}(\cos\frac{2}{3}\pi)$; (d) $\cos^{-1}(\cos\frac{4}{3}\pi)$
24. (a) $\cos^{-1}(\cos\frac{1}{4}\pi)$; (b) $\cos^{-1}[\cos(-\frac{1}{4}\pi)]$; (c) $\cos^{-1}(\cos\frac{3}{4}\pi)$; (d) $\cos^{-1}(\cos\frac{5}{4}\pi)$
25. (a) $\tan^{-1}(\tan\frac{1}{6}\pi)$; (b) $\tan^{-1}[\tan(-\frac{1}{3}\pi)]$; (c) $\tan^{-1}(\tan\frac{7}{6}\pi)$; (d) $\tan^{-1}[\tan(-\frac{4}{3}\pi)]$
26. (a) $\tan^{-1}(\tan\frac{1}{3}\pi)$; (b) $\tan^{-1}[\tan(-\frac{1}{6}\pi)]$; (c) $\tan^{-1}(\tan\frac{4}{3}\pi)$; (d) $\tan^{-1}[\tan(-\frac{7}{6}\pi)]$
27. (a) $\sin^{-1}(\tan\frac{1}{4}\pi)$; (b) $\tan^{-1}[\sin(-\frac{1}{2}\pi)]$; (c) $\tan(\sin^{-1}\frac{1}{2}\sqrt{3})$; (d) $\sin(\tan^{-1}\frac{1}{2}\sqrt{3})$
28. (a) $\cos^{-1}(\tan\frac{1}{4}\pi)$; (b) $\tan^{-1}(\cos\pi)$; (c) $\cos(\tan^{-1}\sqrt{3})$; (d) $\tan(\cos^{-1}\sqrt{3})$
29. (a) $\arcsin(\cos\frac{1}{3}\pi)$; (b) $\arccos(\sin\frac{1}{3}\pi)$; (c) $\cos[\arcsin(-\frac{1}{2})]$; (d) $\sin[\arccos(-\frac{1}{2})]$
30. (a) $\arcsin(\cos\frac{1}{6}\pi)$; (b) $\arccos(\sin\frac{1}{6}\pi)$; (c) $\cos\left[\arcsin\left(-\dfrac{\sqrt{3}}{2}\right)\right]$; (d) $\sin\left[\arccos\left(-\dfrac{\sqrt{3}}{2}\right)\right]$
31. (a) $\arcsin(\cos\frac{1}{2}\pi)$; (b) $\arccos(\sin\frac{1}{2}\pi)$; (c) $\tan(\arccos 1)$; (d) $\cos(\arctan 1)$
32. (a) $\arcsin(\cos\pi)$; (b) $\arccos(\sin\pi)$; (c) $\tan[\arccos(-1)]$; (d) $\cos[\arctan(-1)]$
33. $\cos[2\sin^{-1}(-\frac{5}{13})]$ 34. $\tan[2\cos^{-1}(-\frac{4}{5})]$
35. $\sin(\arcsin\frac{2}{3} + \arccos\frac{1}{3})$

36. $\cos[\arcsin(-\tfrac{1}{2}) + \arcsin\tfrac{1}{4}]$
37. $\cos[\sin^{-1}\tfrac{2}{3} + 2\sin^{-1}(-\tfrac{1}{3})]$
38. $\tan[\tan^{-1}(-\tfrac{2}{5}) - \cos^{-1}(-\tfrac{1}{2}\sqrt{2})]$
39. $\tan(\arctan\tfrac{3}{4} - \arcsin\tfrac{1}{2})$
40. $\tan[\arccos\tfrac{3}{5} + \arcsin(-\tfrac{12}{13})]$
41. $\cos(\sin^{-1}\tfrac{1}{3} - \tan^{-1}\tfrac{1}{2})$
42. $\sin[\cos^{-1}(-\tfrac{2}{3}) + 2\sin^{-1}(-\tfrac{1}{3})]$

In Exercises 43 through 50, draw a sketch of the graph of the equation.

43. $y = \tfrac{1}{2}\sin^{-1}x$ 44. $y = \sin^{-1}\tfrac{1}{2}x$
45. $y = \arctan 2x$ 46. $y = 2\arctan x$
47. $y = \cos^{-1}3x$ 48. $y = \tfrac{1}{2}\cos^{-1}2x$
49. $y = 2\tan^{-1}\tfrac{1}{2}x$ 50. $y = \sin^{-1}\tfrac{1}{3}x$

51. Prove: $\cos^{-1}\dfrac{3}{\sqrt{10}} + \cos^{-1}\dfrac{2}{\sqrt{5}} = \dfrac{1}{4}\pi.$
52. Prove: $2\tan^{-1}\tfrac{1}{3} - \tan^{-1}(-\tfrac{1}{7}) = \tfrac{1}{4}\pi.$
53. (a) Define $\cot^{-1}$ by first restricting the domain of the cotangent function to the interval $(0, \pi)$. (b) With your definition in part (a) draw a sketch of the graph of $y = \cot^{-1}x$.
54. (a) Define $\sec^{-1}$ by first restricting the domain of the secant function to $[0, \tfrac{1}{2}\pi) \cup [\pi, \tfrac{3}{2}\pi)$; this domain is chosen so that in calculus certain computations are simplified. (b) With your definition in part (a) draw a sketch of the graph of $y = \sec^{-1}x$.
55. (a) Define $\csc^{-1}$ by first restricting the domain of the cosecant function to $(-\pi, -\tfrac{1}{2}\pi] \cup (0, \tfrac{1}{2}\pi]$; this domain is chosen so that in calculus certain relationships are valid. (b) With your definition in part (a) draw a sketch of the graph of $y = \csc^{-1}x$.

56. A picture w feet high is placed on a wall with its base z feet above the level of the eye of an observer. If the observer is x feet from the wall and θ is the radian measure of the angle subtended at the observer's eye by the picture, show that

$$\theta = \tan^{-1}\left(\frac{wx}{x^2 + wz + z^2}\right)$$

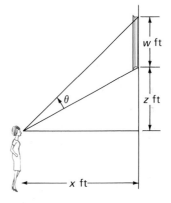

57. (a) Use your definition of $\cot^{-1}$ in Exercise 53 to show that in Example 4 another equation defining θ in terms of x is

$$\theta = \cot^{-1}\frac{x}{16} - \cot^{-1}\frac{x}{9}$$

Use the equation in part (a) to find θ when (b) $x = 10$; (c) $x = 12$; and (d) $x = 15$.

7.6   Trigonometric Equations

In the first four sections of this chapter we were concerned with trigonometric identities, which are equations that are satisfied by all real numbers, or angles, for which each member of the equation is defined. We now discuss conditional trigonometric equations, which are satisfied by only particular values of the variable.

    The methods used to find solutions of trigonometric equations are similar to those used to solve algebraic equations. However, here we first solve for a particular trigonometric function value. As with algebraic equations, we obtain a succession of equivalent equations until we have one for which the trigonometric function value is apparent. Trigonometric identities are helpful in securing the equivalent equations.

Illustration 1

To solve the equation

$$2 \sin x - 1 = 0$$

for $0 \le x \le \frac{1}{2}\pi$, we first solve for $\sin x$

$$2 \sin x = 1$$
$$\sin x = \tfrac{1}{2}$$
$$x = \sin^{-1} \tfrac{1}{2}$$
$$x = \tfrac{1}{6}\pi$$ ∎

Example 1 Find the solution of the equation if $0 \le x \le \frac{1}{2}\pi$.

a) $\tan^2 x - 3 = 0$ b) $2 \cos^2 x - 1 = 0$

Solution

a) $\tan^2 x - 3 = 0$
   $\tan^2 x = 3$

Because $0 \le x \le \frac{1}{2}\pi$, $\tan x > 0$; thus

$$\tan x = \sqrt{3}$$
$$x = \tan^{-1} \sqrt{3}$$
$$x = \tfrac{1}{3}\pi$$

b) $2 \cos^2 x - 1 = 0$
   $\cos^2 x = \tfrac{1}{2}$

Because $0 \le x \le \frac{1}{2}\pi$, $\cos x > 0$; thus

$$\cos x = \frac{1}{\sqrt{2}}$$
$$x = \cos^{-1} \frac{1}{\sqrt{2}}$$
$$x = \frac{1}{4}\pi$$ ∎

Illustration 2

To solve the equation

$$\cot^2 x - 1 = 0$$

for $0 \le x \le \pi$, we first solve for $\cot x$.

$$\cot^2 x = 1$$
$$\cot x = \pm 1$$

The value of x in $[0, \pi]$ for which $\cot x = 1$ is $\frac{1}{4}\pi$. The value of x in $[0, \pi]$ for which $\cot x = -1$ is $\frac{3}{4}\pi$. Therefore the solution set of the given equation is $\{\frac{1}{4}\pi, \frac{3}{4}\pi\}$. ∎

Illustration 3

Suppose we wish to find the solutions of the equation

$$2 \sin^2 t - \cos t - 1 = 0 \qquad 0 \le t < 2\pi$$

We first replace $\sin^2 t$ by $1 - \cos^2 t$, and then solve the resulting quadratic equation by factoring the left side.

$$2 \sin^2 t - \cos t - 1 = 0$$
$$2(1 - \cos^2 t) - \cos t - 1 = 0$$
$$2 - 2 \cos^2 t - \cos t - 1 = 0$$
$$2 \cos^2 t + \cos t - 1 = 0$$
$$(2 \cos t - 1)(\cos t + 1) = 0$$
$$2 \cos t - 1 = 0 \qquad \cos t + 1 = 0$$
$$\cos t = \tfrac{1}{2} \qquad \cos t = -1$$

The values of t in $[0, 2\pi)$ for which $\cos t = \tfrac{1}{2}$ are $\tfrac{1}{3}\pi$ and $\tfrac{5}{3}\pi$. The value of t in $[0, 2\pi)$ for which $\cos t = -1$ is π. Thus the solution set of the given equation is $\{\tfrac{1}{3}\pi, \pi, \tfrac{5}{3}\pi\}$. ■

Illustration 4

To solve the equation

$$8 \sin^2 \theta + 6 \sin \theta - 9 = 0$$

for $0° \leq \theta < 360°$, we factor the left side and set each factor equal to zero.

$$8 \sin^2 \theta + 6 \sin \theta - 9 = 0$$
$$(2 \sin \theta + 3)(4 \sin \theta - 3) = 0$$
$$2 \sin \theta + 3 = 0 \qquad 4 \sin \theta - 3 = 0$$
$$\sin \theta = -\tfrac{3}{2} \qquad \sin \theta = \tfrac{3}{4}$$
$$\sin \theta = 0.75$$

The solution set of $\sin \theta = -\tfrac{3}{2}$ is $\varnothing$ because $|\sin \theta| \leq 1$. For the equation $\sin \theta = 0.75$, we have both a first-quadrant and a second-quadrant angle. The first-quadrant angle is $\sin^{-1} 0.75 \approx 48.6°$, obtained from either Table VII or a calculator. The second-quadrant angle is $180° - 48.6° = 131.4°$. Therefore the solution set is $\{48.6°, 131.4°\}$. ■

Example 2 Solve the equation

$$\sec^2 x - \tan x = 1$$

if $0 \leq x < 2\pi$.

Solution We use the identity $\sec^2 x = 1 + \tan^2 x$ and solve the resulting quadratic equation by factoring.

$$\sec^2 x - \tan x = 1$$
$$(1 + \tan^2 x) - \tan x = 1$$
$$\tan^2 x - \tan x = 0$$
$$\tan x(\tan x - 1) = 0$$
$$\tan x - 1 = 0$$
$$\tan x = 0 \qquad\qquad \tan x = 1$$
$$x = 0 \qquad x = \pi \qquad x = \tfrac{1}{4}\pi \qquad x = \tfrac{5}{4}\pi$$

The solution set is $\{0, \tfrac{1}{4}\pi, \pi, \tfrac{5}{4}\pi\}$. ■

Illustration 5

Suppose we wish to determine all real-number solutions of the equation of Illustration 3. That is, we have

$$2 \sin^2 t - \cos t - 1 = 0 \qquad t \in R$$

We proceed as in Illustration 3, and for $t \in [0, 2\pi)$ the solutions are $\frac{1}{3}\pi$, $\frac{5}{3}\pi$, and π. Because the period of the cosine is 2π, all real-number solutions are obtained by adding to each of these numbers $k \cdot 2\pi$, where k is any integer. Thus the solution set for $t \in R$ is

$$\{t | t = \tfrac{1}{3}\pi + k \cdot 2\pi\} \cup \{t | t = \tfrac{5}{3}\pi + k \cdot 2\pi\} \cup \{t | t = \pi + k \cdot 2\pi\} \qquad k \in J \qquad \blacksquare$$

Illustration 6

To obtain all angles θ that are solutions of the equation of Illustration 4, we first find those θ for which

$$8 \sin^2 \theta + 6 \sin \theta - 9 = 0$$

and $0° \leq \theta < 360°$. In Illustration 4 they were found to be $48.6°$ and $131.4°$. Thus if θ can be any angle, the solution set is

$$\{\theta | \theta \approx 48.6° + k \cdot 360°\} \cup \{\theta | \theta \approx 131.4° + k \cdot 360°\} \qquad k \in J \qquad \blacksquare$$

Illustration 7

The equation of Example 2 is

$$\sec^2 x - \tan x = 1$$

and if $x \in [0, 2\pi)$, the solution set is $\{0, \frac{1}{4}\pi, \pi, \frac{5}{4}\pi\}$. If there is no restriction on x, then because the period of the tangent is π, all the solutions of this equation are in the set

$$\{x | x = k\pi\} \cup \{x | x = \tfrac{1}{4}\pi + k\pi\} \qquad k \in J \qquad \blacksquare$$

Example 3 Find all solutions of the equation

$$4 \sin 2\theta - 3 \cos \theta = 0$$

if θ is any angle. Express the solutions in degree measurement.

Solution We begin by applying the sine double-measure identity.

$$4 \sin 2\theta - 3 \cos \theta = 0$$
$$4(2 \sin \theta \cos \theta) - 3 \cos \theta = 0$$
$$\cos \theta (8 \sin \theta - 3) = 0$$
$$\cos \theta = 0 \qquad 8 \sin \theta - 3 = 0$$
$$\sin \theta = \tfrac{3}{8}$$
$$\sin \theta = 0.375$$

The solution set of $\cos \theta = 0$ is $\{\theta | \theta = 90° + k \cdot 180°\}$, $k \in J$. To obtain a first-quadrant angle for which $\sin \theta = 0.375$, we can use Table VII or a calculator and get $\theta \approx 22.0°$. A second-quadrant angle is $180° - 22.0°$ or $158.0°$. Therefore the solution set of $\sin \theta = 0.375$ is

$$\{\theta | \theta \approx 22.0° + k \cdot 360°\} \cup \{\theta | \theta \approx 158.0° + k \cdot 360°\} \qquad k \in J$$

The solution set of the given equation is then

$$\{\theta | \theta = 90° + k \cdot 180°\} \cup \{\theta | \theta \approx 22.0° + k \cdot 360°\} \cup \{\theta | \theta \approx 158.0° + k \cdot 360°\}$$

where $k \in J$.   ■

Example   4   An electric generator produces a 30-cycle alternating current described by the equation

$$I(t) = 40 \sin 60\pi \left(t - \tfrac{7}{72}\right)$$

where $I(t)$ amperes is the current at t seconds. Find the smallest positive value of t for which the current is 20 amperes.

Solution   If $I(t)$ is replaced by 20 in the given equation, we have

$$40 \sin 60\pi(t - \tfrac{7}{72}) = 20$$

$$\sin 60\pi(t - \tfrac{7}{72}) = \tfrac{1}{2}$$

$$60\pi(t - \tfrac{7}{72}) = \tfrac{1}{6}\pi + k \cdot 2\pi \qquad \text{or} \qquad 60\pi(t - \tfrac{7}{72}) = \tfrac{5}{6}\pi + k \cdot 2\pi \qquad k \in J$$

$$t - \tfrac{7}{72} = \tfrac{1}{360} + k \cdot \tfrac{1}{30} \qquad\qquad\qquad t - \tfrac{7}{72} = \tfrac{1}{72} + k \cdot \tfrac{1}{30}$$

$$t = \tfrac{1}{10} + k \cdot \tfrac{1}{30} \qquad\qquad\qquad\qquad t = \tfrac{1}{9} + k \cdot \tfrac{1}{30}$$

$$t = \tfrac{3}{30} + k \cdot \tfrac{1}{30} \qquad\qquad\qquad\qquad t = \tfrac{10}{90} + k \cdot \tfrac{3}{90}$$

The smallest positive value of t obtained from $t = \tfrac{3}{30} + k \cdot \tfrac{1}{30}$ occurs when $k = -2$; this value of t is $\tfrac{1}{30}$. The smallest positive value of t obtained from $t = \tfrac{10}{90} + k \cdot \tfrac{3}{90}$ occurs when $k = -3$; this value of t is $\tfrac{1}{90}$. Thus $\tfrac{1}{90}$ is the smallest positive value of t for which the current is 20 amperes.   ■

Example   5   Find the solutions of the equation

$$3 \tan t - \cot t - 5 = 0$$

if $0 \le t < 2\pi$.

Solution   By making the substitution $\cot t = \dfrac{1}{\tan t}$, we obtain an equivalent equation containing only one function:

$$3 \tan t - \cot t - 5 = 0$$

$$3 \tan t - \frac{1}{\tan t} - 5 = 0$$

$$3 \tan^2 t - 1 - 5 \tan t = 0$$

$$3 \tan^2 t - 5 \tan t - 1 = 0$$

This is a quadratic equation, which we solve by applying the quadratic formula.

$$\tan t = \frac{-b \pm \sqrt{b^2 - 4ac}}{2a}$$

$$= \frac{-(-5) \pm \sqrt{(-5)^2 - 4(3)(-1)}}{2(3)}$$

$$= \frac{5 \pm \sqrt{37}}{6}$$

$$\approx \frac{5 \pm 6.083}{6}$$

$$\tan t \approx \frac{5 + 6.083}{6} \qquad \tan t \approx \frac{5 - 6.083}{6}$$

$$\approx 1.847 \qquad\qquad \approx -0.180$$

For $\tan t \approx 1.847$, from Table VI or a calculator, $\tan^{-1} 1.847 \approx 1.07$. Thus one solution is $t \approx 1.07$. Because t is in $[0, 2\pi)$, another solution is $t \approx 3.14 + 1.07$; that is, $t \approx 4.21$. For $\tan t \approx -0.180$, we first determine from Table VI or a calculator a $\bar{t}$ for which $\bar{t} \approx \tan^{-1} 0.180$; it is $\bar{t} \approx 0.18$. There are two values of t in $[0, 2\pi)$ for which $\tan t \approx -0.180$; they are $t \approx 3.14 - 0.18$, or $t \approx 2.96$, and $t \approx 6.28 - 0.18$, or $t \approx 6.10$. Thus the solution set of the given equation is $\{1.07, 2.96, 4.21, 6.10\}$. ∎

The equation in the next example involves a trigonometric function of a multiple measure.

Example 6 Find all values of x in $[0, 2\pi)$ for which

$$\tan 3x = 1$$

Solution If $0 \le x < 2\pi$, then $0 \le 3x < 6\pi$. Therefore, to find all the values of x in $[0, 2\pi)$ that are solutions of the equation, we must first determine all the values of $3x$ in $[0, 6\pi)$ for which

$$\tan 3x = 1$$

They are

$$3x = \tfrac{1}{4}\pi \qquad 3x = \tfrac{5}{4}\pi \qquad 3x = \tfrac{9}{4}\pi \qquad 3x = \tfrac{13}{4}\pi \qquad 3x = \tfrac{17}{4}\pi \qquad 3x = \tfrac{21}{4}\pi$$

Dividing by 3 on both sides of each of these equations, we obtain all the solutions of the given equation in $[0, 2\pi)$. They are

$$x = \tfrac{1}{12}\pi \qquad x = \tfrac{5}{12}\pi \qquad x = \tfrac{3}{4}\pi \qquad x = \tfrac{13}{12}\pi \qquad x = \tfrac{17}{12}\pi \qquad x = \tfrac{7}{4}\pi$$ ∎

Illustration 8

To determine all solutions of the equation in Example 6, we first obtain the value of $3x$ in $[0, \pi)$ that satisfies the equation. It is

$$3x = \frac{1}{4}\pi$$

Because the period of the tangent function is π, all solutions of the equation are given by

$$3x = \frac{1}{4}\pi + k \cdot \pi \qquad k \in J$$

$$\Leftrightarrow$$

$$x = \frac{1}{12}\pi + k \cdot \frac{1}{3}\pi \qquad k \in J \qquad \blacksquare$$

Example 7 Find the solutions of the equation

$$2 \sin 2\theta \cos 3\theta + \cos 3\theta = 0$$

if $0° \leq \theta < 360°$.

Solution

$$2 \sin 2\theta \cos 3\theta + \cos 3\theta = 0$$
$$\cos 3\theta (2 \sin 2\theta + 1) = 0$$
$$\cos 3\theta = 0 \qquad 2 \sin 2\theta + 1 = 0$$
$$\sin 2\theta = -\tfrac{1}{2}$$

We must first find the solutions of the equation $\cos 3\theta = 0$. Because $0° \leq \theta < 360°$, then $0° \leq 3\theta < 1080°$. The solutions are

$$3\theta = 90° \qquad 3\theta = 270° \qquad 3\theta = 450° \qquad 3\theta = 630° \qquad 3\theta = 810° \qquad 3\theta = 990°$$

Therefore

$$\theta = 30° \qquad \theta = 90° \qquad \theta = 150° \qquad \theta = 210° \qquad \theta = 270° \qquad \theta = 330°$$

We now find the solutions of the equation $\sin 2\theta = -\tfrac{1}{2}$. Since $0° \leq \theta < 360°$, then $0° \leq 2\theta < 720°$. The solutions are

$$2\theta = 210° \qquad 2\theta = 330° \qquad 2\theta = 570° \qquad 2\theta = 690°$$

Hence

$$\theta = 105° \qquad \theta = 165° \qquad \theta = 285° \qquad \theta = 345°$$

The solution set of the given equation then is

$$\{30°, \ 90°, \ 105°, \ 150°, \ 165°, \ 210°, \ 270°, \ 285°, \ 330°, \ 345°\} \qquad \blacksquare$$

EXERCISES 7.6

In Exercises 1 through 10, find the solution of the equation if $0 \leq x \leq \frac{1}{2}\pi$.

1. (a) $\sin x - 1 = 0$; (b) $2 \cos x - 1 = 0$
2. (a) $\cos x - 1 = 0$; (b) $\tan x - 1 = 0$
3. (a) $2 \sin^2 x - 1 = 0$; (b) $3 \cot^2 x - 1 = 0$
4. (a) $\cot^2 x - 3 = 0$; (b) $4 \cos^2 x - 3 = 0$
5. (a) $\sec^2 x - 1 = 0$; (b) $\csc^2 x - 2 = 0$
6. (a) $\csc^2 x - 1 = 0$; (b) $\sec^2 x - 2 = 0$
7. (a) $\sin x \cos x = 0$; (b) $\cos x \cot x = 0$
8. (a) $\tan x \sec x = 0$; (b) $\sin x \tan x = 0$
9. (a) $4 \sin^3 x - 3 \sin x = 0$; (b) $\tan^3 x - \tan x = 0$
10. (a) $4 \cos^3 x - \cos x = 0$; (b) $3 \csc^3 x - 4 \csc x = 0$

In Exercises 11 through 18, find the solutions of the equation if $0 \leq t < 2\pi$.

11. (a) $4 \sin^2 t - 1 = 0$; (b) $\tan^2 t - 1 = 0$
12. (a) $\sec^2 t - 4 = 0$; (b) $3 \tan^2 t - 1 = 0$
13. (a) $2 \cos^2 t + 3 \cos t + 1 = 0$;
    (b) $2 \sin^2 t - 5 \sin t - 3 = 0$
14. (a) $2 \sin^2 t + \sin t - 1 = 0$;
    (b) $2 \cos^2 t + 3 \cos t - 2 = 0$
15. (a) $\sec^2 t - 2 \tan t = 0$; (b) $\tan^2 t - \sec t = 1$
16. (a) $\tan t + \cot t + 2 = 0$; (b) $\cot^2 t + \csc t = 1$
17. (a) $\sin t + \cos t = 0$; (b) $2 \sin t + \sec t = 0$
18. (a) $\tan t - \cot t = 0$; (b) $\tan t + \sec t = 0$

In Exercises 19 through 22, find all real-number solutions of the equations of the indicated exercise.

19. Exercise 11   20. Exercise 14
21. Exercise 15   22. Exercise 16

In Exercises 23 through 28, find all angles θ that are solutions of the equation if $0° \leq \theta < 360°$.

23. $9 \cos^2 \theta + 6 \cos \theta - 8 = 0$
24. $5 \sin^2 \theta - 11 \sin \theta + 2 = 0$
25. $2 \sin 2\theta - 3 \sin \theta = 0$
26. $2 \cos 2\theta + 3 \cos \theta + 1 = 0$
27. $\tan \theta - 3 \cot \theta = 2$
28. $2 \tan^2 \theta - \sec \theta = 1$

In Exercises 29 through 34, find all angles θ that are solutions of the equation of the indicated exercise. Express the solutions in degree measurement.

29. Exercise 23   30. Exercise 24
31. Exercise 25   32. Exercise 26
33. Exercise 27   34. Exercise 28

In Exercises 35 through 38, find the solutions of the equation if $0 \leq t < 2\pi$.

35. $\sin^2 t + 5 \cos t + 2 = 0$
36. $3 \sec^2 t + \tan t - 5 = 0$
37. $3 \tan^2 t - \tan t - 3 = 0$
38. $10 \cos^2 t - 4 \cos t - 5 = 0$

In Exercises 39 through 46, find all values of x in $[0, 2\pi)$ that are solutions of the equation.

39. (a) $\tan 3x = -1$; (b) $\sin 3x = \frac{1}{2}$
40. (a) $\cot 3x = \sqrt{3}$; (b) $\cos 3x = -\frac{1}{2}$
41. (a) $\cot 4x = -\sqrt{3}$; (b) $\sec 5x = 2$
42. (a) $\cot 4x = -1$; (b) $\csc 5x = \sqrt{2}$
43. (a) $\sin^2 2x = 1$; (b) $\cos \frac{1}{2}x = -\frac{1}{2}$
44. (a) $\cos^2 2x = \frac{1}{4}$; (b) $\sin \frac{1}{3}x = 1$
45. (a) $\tan^2 \frac{1}{2}x = 3$; (b) $\csc \frac{1}{3}x = 2$
46. (a) $\sec^2 \frac{1}{2}x = 2$; (b) $\cot \frac{1}{2}x = -1$

In Exercises 47 through 50, find all real-number solutions of the equations of the indicated exercise.

47. Exercise 39   48. Exercise 40
49. Exercise 43   50. Exercise 44

In Exercises 51 through 54, find all angles θ that are solutions of the equation if $0° \leq \theta < 360°$.

51. $2 \cos 3\theta \sin 2\theta - \sin 2\theta = 0$
52. $2 \sin 3\theta \cos 3\theta = -1$
53. $\tan 2\theta + 5 = 3 \sec^2 2\theta$
54. $\cot^2 4\theta - 1 = \csc 4\theta$

In Exercises 55 through 60, determine whether the equality is an equation or an identity. If it is an equation, solve it for all values of x in $[0, 2\pi)$. If it is an identity, prove it.

55. $\cot x + \tan x = \csc x \sec x$
56. $1 - \tan^2 x = \tan x \sec x$
57. $\sin 3x - \sin x = \cos 2x$
58. $\cos 4x + \cos 2x = 2 \cos x \cos 3x$
59. $\sin^2 4x + \cos^2 2x = 1$
60. $\tan 2x - \tan x = \tan x \sec 2x$
61. In an electric circuit the electromotive force is $E(t)$ volts, where $E(t) = 2 \cos 50\pi t$. Find the smallest positive value of t for which the electromotive force is (a) 2 volts; (b) 1 volt; (c) -2 volts; and (d) -1 volt.
62. Do Exercise 61 if $E(t) = 4 \sin 120\pi t$.

63. A weight is suspended from a spring and vibrating vertically according to the equation

$$f(t) = 10 \sin \tfrac{3}{4}\pi(t - 3)$$

where $f(t)$ centimeters is the directed distance of the weight from its central position at t seconds, and the positive direction is upward. Determine the smallest positive value of t for which the displacement of the weight above its central position is (a) 5 cm and (b) 6 cm.

64. Do Exercise 63 if $f(t) = 10 \cos \tfrac{5}{6}\pi(t - \tfrac{2}{5})$.

65. A weight is suspended from a spring and vibrating vertically according to the equation

$$y = 2 \sin 4\pi(t + \tfrac{1}{8})$$

where y centimeters is the directed distance of the weight from its central position t seconds after the start of the motion and the positive direction is upward. (a) Solve the equation for t. (b) Use the equation in part (a) to determine the smallest three positive values of t for which the weight is 1 cm above its central position.

66. A 60-cycle alternating current is described by the equation $x = 20 \sin 120\pi(t - \tfrac{11}{720})$, where x amperes is the current at t seconds. (a) Solve the equation for t. (b) Use the equation in part (a) to determine the smallest three positive values of t for which the current is 10 amperes.

7.7   Identities for the Product, Sum, and Difference of Sine and Cosine (Supplementary)

In certain computations it is necessary to write an expression involving the product of sine and cosine functions as a sum or difference. The tools for doing this are provided by the product sine and cosine identities, which follow from the sine and cosine sum and difference identities.

The sine sum and difference identities are

$$\sin(x + y) = \sin x \cos y + \cos x \sin y$$
$$\sin(x - y) = \sin x \cos y - \cos x \sin y$$

If we add corresponding terms of these two equations, we obtain

$$\sin(x + y) + \sin(x - y) = 2 \sin x \cos y$$

and if we subtract terms of the second equation from corresponding terms of the first, we get

$$\sin(x + y) - \sin(x - y) = 2 \cos x \sin y$$

These results give the following identities.

Product Sine and Cosine Identities

$$\sin x \cos y = \tfrac{1}{2}[\sin(x + y) + \sin(x - y)]$$
$$\cos x \sin y = \tfrac{1}{2}[\sin(x + y) - \sin(x - y)]$$

These two identities express the product of a sine and cosine function as the sum or difference of two sine functions. They are valid for all real numbers and angles x and y.

Illustration 1

a) From the identity $\sin x \cos y = \frac{1}{2}[\sin(x + y) + \sin(x - y)]$,

$$\sin 5t \cos 3t = \frac{1}{2}[\sin(5t + 3t) + \sin(5t - 3t)]$$
$$= \frac{1}{2}(\sin 8t + \sin 2t)$$

b) From the identity $\cos x \sin y = \frac{1}{2}[\sin(x + y) - \sin(x - y)]$,

$$\cos 3t \sin 5t = \frac{1}{2}[\sin(3t + 5t) - \sin(3t - 5t)]$$
$$= \frac{1}{2}[\sin 8t - \sin(-2t)]$$
$$= \frac{1}{2}[\sin 8t - (-\sin 2t)]$$
$$= \frac{1}{2}(\sin 8t + \sin 2t)$$

$\blacksquare$

The cosine sum and difference identities are

$$\cos(x + y) = \cos x \cos y - \sin x \sin y$$
$$\cos(x - y) = \cos x \cos y + \sin x \sin y$$

Adding corresponding terms of these two equations, we have

$$\cos(x + y) + \cos(x - y) = 2 \cos x \cos y$$

and subtracting terms of the first equation from corresponding terms of the second, we get

$$\cos(x - y) - \cos(x + y) = 2 \sin x \sin y$$

From these two results we have the following identities.

Product Sine and Cosine Identities

$$\cos x \cos y = \frac{1}{2}[\cos(x + y) + \cos(x - y)]$$
$$\sin x \sin y = \frac{1}{2}[\cos(x - y) - \cos(x + y)]$$

The first identity expresses the product of two cosine functions as the sum of two cosine functions; the second identity expresses the product of two sine functions as the difference of two cosine functions. They are both valid for all real numbers and all angles x and y.

Illustration 2

a) From the identity $\cos x \cos y = \frac{1}{2}[\cos(x + y) + \cos(x - y)]$,

$$\cos 4\theta \cos 2\theta = \frac{1}{2}[\cos(4\theta + 2\theta) + \cos(4\theta - 2\theta)]$$
$$= \frac{1}{2}(\cos 6\theta + \cos 2\theta)$$

b) From the identity $\sin x \sin y = \frac{1}{2}[\cos(x - y) - \cos(x + y)]$,

$$\sin 4\theta \sin 2\theta = \frac{1}{2}[\cos(4\theta - 2\theta) - \cos(4\theta + 2\theta)]$$
$$= \frac{1}{2}(\cos 2\theta - \cos 6\theta)$$

$\blacksquare$

Example   1   Find the exact value of

$$\sin \tfrac{25}{24}\pi \, \cos \tfrac{5}{24}\pi$$

Solution   From the identity $\sin x \cos y = \tfrac{1}{2}[\sin(x + y) + \sin(x - y)]$,

$$\sin \tfrac{25}{24}\pi \, \cos \tfrac{5}{24}\pi = \tfrac{1}{2}[\sin(\tfrac{25}{24}\pi + \tfrac{5}{24}\pi) + \sin(\tfrac{25}{24}\pi - \tfrac{5}{24}\pi)]$$

$$= \tfrac{1}{2}(\sin \tfrac{30}{24}\pi + \sin \tfrac{20}{24}\pi)$$

$$= \tfrac{1}{2}(\sin \tfrac{5}{4}\pi + \sin \tfrac{5}{6}\pi)$$

$$= \frac{1}{2}\left(-\frac{\sqrt{2}}{2} + \frac{1}{2}\right)$$

$$= \frac{1 - \sqrt{2}}{4}$$

∎

Example   2   Verify the identity

$$(\sin 2t)(1 + 2 \cos t) = \sin t + \sin 2t + \sin 3t$$

Solution   We start with the left side. After removing parentheses we apply the product sine and cosine identity for $\sin x \cos y$.

$$(\sin 2t)(1 + 2 \cos t) = \sin 2t + 2 \sin 2t \cos t$$

$$= \sin 2t + 2 \cdot \tfrac{1}{2}[\sin(2t + t) + \sin(2t - t)]$$

$$= \sin 2t + (\sin 3t + \sin t)$$

$$= \sin t + \sin 2t + \sin 3t$$

∎

The product sine and cosine identities can be used to write a sum or difference of sine and cosine functions as a product. We make the substitutions

$$x + y = w \qquad \text{and} \qquad x - y = z \tag{1}$$

Then

$$(x + y) + (x - y) = w + z$$

$$x = \frac{w + z}{2} \tag{2}$$

Also

$$(x + y) - (x - y) = w - z$$

$$y = \frac{w - z}{2} \tag{3}$$

Substituting from (1), (2), and (3) in the four product sine and cosine identities, we obtain the following identities.

Sum and Difference Sine and Cosine Identities

$$\sin w + \sin z = 2 \sin\left(\frac{w+z}{2}\right) \cos\left(\frac{w-z}{2}\right)$$

$$\sin w - \sin z = 2 \cos\left(\frac{w+z}{2}\right) \sin\left(\frac{w-z}{2}\right)$$

$$\cos w + \cos z = 2 \cos\left(\frac{w+z}{2}\right) \cos\left(\frac{w-z}{2}\right)$$

$$\cos w - \cos z = -2 \sin\left(\frac{w+z}{2}\right) \sin\left(\frac{w-z}{2}\right)$$

The first and third of these identities are called the *sum sine* and *sum cosine identities*, respectively. The second and fourth are called the *difference sine* and *difference cosine identities*, respectively. They are valid for all real numbers and angles w and z.

Illustration 3

To write $\sin 8x + \sin 4x$ as a product, we apply the sum sine identity and obtain

$$\sin 8x + \sin 4x = 2 \sin\left(\frac{8x+4x}{2}\right) \cos\left(\frac{8x-4x}{2}\right)$$

$$= 2 \sin 6x \cos 2x \qquad \blacksquare$$

Example 3   Verify the identity

$$\frac{\cos 2y + \cos 4y}{\sin 4y - \sin 2y} = \cot y$$

Solution   We begin with the left side and apply the sum cosine identity to the numerator and the difference sine identity to the denominator.

$$\frac{\cos 2y + \cos 4y}{\sin 4y - \sin 2y} = \frac{2 \cos\left(\dfrac{2y+4y}{2}\right) \cos\left(\dfrac{2y-4y}{2}\right)}{2 \cos\left(\dfrac{4y+2y}{2}\right) \sin\left(\dfrac{4y-2y}{2}\right)}$$

$$= \frac{2 \cos 3y \cos(-y)}{2 \cos 3y \sin y}$$

$$= \frac{\cos(-y)}{\sin y}$$

$$= \frac{\cos y}{\sin y}$$

$$= \cot y \qquad \blacksquare$$

In Example 5 of Section 7.3 we showed that the sum of two sine functions having the same period was a sine function having that common period. The following example involves the addition of two sine functions having different periods but the same amplitude.

Example 4 Given

$$f(t) = \sin 44\pi t + \sin 36\pi t$$

a) Express $f(t)$ as the product of sine and cosine functions.

b) Draw a sketch of the graph of f over one period of the cosine function found in part (a).

Solution a) Applying the sum sine identity on the right side of the given equation, we have

$$f(t) = 2 \sin\left(\frac{44\pi t + 36\pi t}{2}\right) \cos\left(\frac{44\pi t - 36\pi t}{2}\right)$$

$$= 2 \sin 40\pi t \cos 4\pi t$$

b) The period of $\cos 4\pi t$ is $\dfrac{2\pi}{4\pi}$, or $\frac{1}{2}$. Therefore we wish to obtain a sketch of the graph of f on $[0, \frac{1}{2}]$. Because $-1 \le \sin 40\pi t \le 1$, the graph of f lies between the graphs of

$$g(t) = 2 \cos 4\pi t \qquad \text{and} \qquad h(t) = -2 \cos 4\pi t$$

which are shown as dashed curves in Figure 1. The graph of f intersects

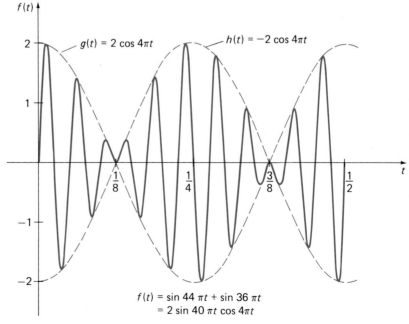

FIGURE 1

the graph of g or h when $\sin 40\pi t = \pm 1$; that is, when $40\pi t = \frac{1}{2}\pi + k \cdot \pi$, $k \in J$; or, equivalently, when $t = \frac{1}{80} + k \cdot \frac{1}{40}$, $k \in J$. With this information, a sketch of the graph of f on $[0, \frac{1}{2}]$ is drawn and appears as the solid curve in Figure 1.   ∎

The graph of the function in Example 4 demonstrates the *principle of superposition*. This kind of behavior occurs whenever we have the sum of two sine or cosine functions of the same amplitude and for which the difference of the frequencies is small compared with the sum of the frequencies. For instance, in Example 4, if we let

$$F(t) = \sin 44\pi t \qquad\qquad G(t) = \sin 36\pi t$$
$$= \sin 2\pi(22)t \qquad\qquad = \sin 2\pi(18)t$$

the frequencies of F and G are $n_1 = 22$ and $n_2 = 18$, respectively; $n_1 - n_2 = 4$ and $n_1 + n_2 = 40$. Thus the graph of $F + G$ has the appearance shown in Figure 1, where the amplitude is changing according to a specific pattern.

The principle of superposition of sound waves produce what physicists call *beats*. The concept of the superposition of two or more waves has applications in the fields of television, radio, and telephone communications as well as electricity involving current with a high frequency.

EXERCISES 7.7

In Exercises 1 through 4, write the indicated product as a sum or difference of function values.

1. (a) $\sin 4x \cos x$; (b) $\sin 4x \sin x$; (c) $\cos 4x \cos x$
2. (a) $\sin 7x \cos 3x$; (b) $\sin 7x \sin 3x$;
   (c) $\cos 7x \cos 3x$
3. (a) $\sin 2x \cos 6x$; (b) $\sin 2x \sin 6x$;
   (c) $\cos 2x \cos 6x$
4. (a) $\sin 4x \cos 5x$; (b) $\sin 4x \sin 5x$;
   (c) $\cos 4x \cos 5x$

In Exercises 5 through 8, find the exact value.

5. (a) $\sin \frac{7}{24}\pi \cos \frac{1}{24}\pi$; (b) $\cos \frac{1}{8}\pi \cos \frac{7}{8}\pi$
6. (a) $\sin \frac{3}{8}\pi \cos \frac{1}{8}\pi$; (b) $\sin \frac{1}{24}\pi \sin \frac{5}{24}\pi$
7. (a) $\cos \frac{9}{8}\pi \sin \frac{3}{8}\pi$; (b) $\sin \frac{5}{24}\pi \sin \frac{25}{24}\pi$
8. (a) $\cos \frac{31}{24}\pi \sin \frac{13}{24}\pi$; (b) $\cos \frac{3}{8}\pi \cos \frac{11}{8}\pi$

In Exercises 9 through 12, write the indicated sum or difference as a product of function values.

9. (a) $\sin 6t + \sin 2t$; (b) $\sin 6t - \sin 2t$;
   (c) $\cos 6t + \cos 2t$; (d) $\cos 6t - \cos 2t$
10. (a) $\sin 5t + \sin 3t$; (b) $\sin 5t - \sin 3t$;
    (c) $\cos 5t + \cos 3t$; (d) $\cos 5t - \cos 3t$

11. (a) $\sin 3\theta + \sin 7\theta$; (b) $\sin 3\theta - \sin 7\theta$;
    (c) $\cos 3\theta + \cos 7\theta$; (d) $\cos 3\theta - \cos 7\theta$
12. (a) $\sin 4\theta + \sin 6\theta$; (b) $\sin 4\theta - \sin 6\theta$;
    (c) $\cos 4\theta + \cos 6\theta$; (d) $\cos 4\theta - \cos 6\theta$

In Exercises 13 through 16, find the exact value.

13. (a) $\sin 75° + \sin 15°$; (b) $\cos 75° - \cos 15°$
14. (a) $\cos 105° + \cos 15°$; (b) $\sin 105° - \sin 15°$
15. (a) $\sin 165° + \cos 195°$; (b) $\sin 165° - \cos 195°$
16. (a) $\sin 195° + \cos 345°$; (b) $\sin 195° - \cos 345°$

In Exercises 17 through 20, verify the equality.

17. $\dfrac{\sin 37° + \sin 23°}{\cos 37° + \cos 23°} = \dfrac{1}{\sqrt{3}}$

18. $\dfrac{\cos 62° - \cos 28°}{\sin 62° - \sin 28°} = -1$

19. $\dfrac{\sin 144° - \sin 126°}{\cos 144° - \cos 126°} = 1$

20. $\dfrac{\sin 140° - \sin 20°}{\cos 140° + \cos 20°} = \sqrt{3}$

In Exercises 21 through 34, prove the identity.

21. $\dfrac{\sin 3x + \sin 7x}{\cos 3x + \cos 7x} = \tan 5x$

22. $\dfrac{\cos 6x - \cos 2x}{\sin 6x - \sin 2x} = -\tan 4x$

23. $\dfrac{\sin 2\alpha - \sin 2\beta}{\sin 2\alpha + \sin 2\beta} = \dfrac{\tan(\alpha - \beta)}{\tan(\alpha + \beta)}$

24. $\dfrac{\sin 2\alpha + \sin 2\beta}{\cos 2\alpha + \cos 2\beta} = \tan(\alpha + \beta)$

25. $\dfrac{\cos t - \cos 5t}{\cos t \sin t} = 4 \sin 3t$

26. $\dfrac{\sin t - \sin 3t}{\sin^2 t - \cos^2 t} = 2 \sin t$

27. $\dfrac{\sin \theta - \sin 7\theta}{2 \sin^2 2\theta - 1} = 2 \sin 3\theta$

28. $\dfrac{\sin 6\theta + \sin 4\theta}{\sin \theta \sin 5\theta} = 2 \cot \theta$

29. $\dfrac{\cos 2y - \cos 3y}{\sin 3y + \sin 2y} = \dfrac{\sin y}{1 + \cos y}$

30. $\dfrac{\sin 5y - \sin 4y}{\cos 5y + \cos 4y} = \dfrac{1 - \cos y}{\sin y}$

31. $\dfrac{\sin x + \sin 2x + \sin 3x}{\cos x + \cos 2x + \cos 3x} = \tan 2x$

32. $\dfrac{\sin x + \sin 3x + \sin 5x + \sin 7x}{\cos x + \cos 3x + \cos 5x + \cos 7x} = \tan 4x$

33. $\sin^2 t \cos^4 t = \dfrac{2 + \cos 2t - 2 \cos 4t - \cos 6t}{32}$

34. $\sin^3 t \cos^3 t = \dfrac{3 \sin 2t - \sin 6t}{32}$

In Exercises 35 through 38, use the method of Example 4 to draw a sketch of the graph of the function on the indicated interval.

35. $f(t) = \sin 21\pi t + \sin 19\pi t$; $[0, 2]$
36. $f(t) = \cos 9\pi t - \cos 11\pi t$; $[0, 2]$
37. $f(t) = \cos 80\pi t + \cos 100\pi t$; $[0, \frac{1}{5}]$
38. $f(t) = \sin 95t + \sin 105t$; $[0, \frac{2}{5}\pi]$

REVIEW EXERCISES FOR CHAPTER 7

In Exercises 1 through 4, use the fundamental identities to find the exact values of the other five trigonometric functions.

1. $\sin \theta = \frac{8}{17}$ and $\cos \theta < 0$
2. $\cos t = \frac{12}{13}$ and $\sin t < 0$
3. $\tan x = -\dfrac{\sqrt{21}}{2}$ and $\sec x > 0$
4. $\csc \alpha = -\dfrac{\sqrt{10}}{3}$ and $\cot \alpha > 0$

In Exercises 5 through 10, use the fundamental identities and a calculator to find to four significant digits the indicated function value.

5. If $\sin x = 0.6156$ and $0 < x < \frac{1}{2}\pi$, find (a) $\cos x$ and (b) $\tan x$.
6. If $\cos \theta = 0.3652$ and $0 < \theta < \frac{1}{2}\pi$, find (a) $\sin \theta$ and (b) $\cot \theta$.
7. If $\sec t = -3.893$ and $\pi < t < \frac{3}{2}\pi$, find (a) $\tan t$ and (b) $\cos t$.
8. If $\csc y = 2.709$ and $\frac{1}{2}\pi < y < \pi$, find (a) $\cot y$ and (b) $\sin y$.

9. If $\cot \beta = -0.7124$ and $\frac{3}{2}\pi < \beta < 2\pi$, find (a) $\csc \beta$ and (b) $\sec \beta$.
10. If $\tan t = -8.915$ and $-\frac{1}{2}\pi < t < 0$, find (a) $\sec t$ and (b) $\csc t$.

In Exercises 11 through 14, write the expression in terms of either sine or cosine, as indicated, and simplify.

11. $\dfrac{\tan^2 t + 1}{\cos t}$ in terms of $\cos t$

12. $\dfrac{1 - \csc^2 x}{\sin x \cot x}$ in terms of $\cos x$

13. $\dfrac{\sec^2 \theta - 1}{\cos \theta \tan \theta}$ in terms of $\sin \theta$

14. $\dfrac{\sin u}{1 + \cot^2 u}$ in terms of $\sin u$

In Exercises 15 through 38, prove the identity.

15. $\sin x \sec x = \tan x$

16. $\dfrac{\sin \theta}{\tan \theta} = \cos \theta$

17. $\cos^2 \theta(\tan^2 \theta + 1) = 1$

18. $\sec^2 t \cot^2 t = 1 + \cot^2 t$

19. $\sin x(\csc x - \sin x) = \cos^2 x$

20. $\csc x(\csc x - \sin x) = \cot^2 x$

21. $\dfrac{1 - \tan^2 x}{1 + \tan^2 x} = 1 - 2 \sin^2 x$

22. $\dfrac{1 - \cot^2 \theta}{1 + \cot^2 \theta} = 2 \cos^2 \theta - 1$

23. $\dfrac{\sin \alpha}{\csc \alpha - \cot \alpha} = 1 + \cos \alpha$

24. $\dfrac{\cos y}{\sec y - \cos y} = \cot^2 y$

25. $\dfrac{\sec t - 1}{\cos t} = \dfrac{\tan^2 t}{1 + \cos t}$

26. $\dfrac{\tan x}{\sec x - \cos x} = \dfrac{\sec x}{\tan x}$

27. $\sec^2 \theta - \csc^2 \theta = \tan^2 \theta - \cot^2 \theta$

28. $\sec^4 \beta - \tan^4 \beta = 1 + 2 \tan^2 \beta$

29. $\dfrac{\sin x}{1 + \cos x} = \csc x - \cot x$

30. $\dfrac{\sin x}{1 - \cos x} = \csc x + \cot x$

31. $\dfrac{\cos y}{1 - \tan y} + \dfrac{\sin y}{1 - \cot y} = \sin y + \cos y$

32. $\dfrac{\sec^2 \theta + \csc^2 \theta}{\sec^2 \theta - \csc^2 \theta} = \dfrac{\tan^2 \theta + 1}{\tan^2 \theta - 1}$

33. $\dfrac{\sin(\alpha + \beta)}{\sin(\alpha - \beta)} = \dfrac{\tan \alpha + \tan \beta}{\tan \alpha - \tan \beta}$

34. $\dfrac{\cos(\alpha + \beta)}{\cos(\alpha - \beta)} = \dfrac{\cot \alpha - \tan \beta}{\cot \alpha + \tan \beta}$

35. $\tan(\theta - \tfrac{1}{4}\pi) = \dfrac{\tan \theta - 1}{\tan \theta + 1}$

36. $\sin(\tfrac{1}{4}\pi + x) \sin(\tfrac{1}{4}\pi - x) = \tfrac{1}{2} \cos 2x$

37. $\tan \tfrac{1}{2}(x - y) = \dfrac{\sin x - \sin y}{\cos x + \cos y}$

38. $\tan \tfrac{1}{2}(x + y) = \dfrac{\sin x + \sin y}{\cos x + \cos y}$

39. Compute the exact value of cos 75° in two ways: (a) use the cosine sum identity; (b) use a half-angle identity.

40. Compute the exact value of sin 105° in two ways: (a) use the sine sum identity; (b) use a half-angle identity.

41. Compute the exact value of $\tan \tfrac{7}{12}\pi$ in two ways: (a) use the tangent sum identity; (b) use a half-measure identity.

42. If $\tan x = \tfrac{4}{5}$, $0 < x < \tfrac{1}{2}\pi$, and $\tan y = -\tfrac{2}{3}$, where $\tfrac{1}{2}\pi < y < \pi$, find: (a) $\tan(x + y)$; (b) $\tan(x - y)$; (c) the quadrant containing $x + y$; (d) the quadrant containing $x - y$.

43. If $\sin \alpha = \tfrac{5}{13}$ with α in the second quadrant and $\cos \beta = -\tfrac{7}{25}$ with β in the third quadrant, find: (a) $\sin(\alpha + \beta)$; (b) $\cos(\alpha + \beta)$; (c) $\sin(\alpha - \beta)$; (d) $\cos(\alpha - \beta)$; (e) the quadrant containing $\alpha + \beta$; (f) the quadrant containing $\alpha - \beta$.

44. If $\tan t = \tfrac{24}{7}$ and $\cos t > 0$, find $\sin \tfrac{1}{2}t$.

45. If $\cos t = -\tfrac{3}{5}$ and $\sin t < 0$, find $\tan \tfrac{1}{2}t$.

46. If $f(t) = \sin t + \sqrt{3} \cos t$, (a) express $f(t)$ in the form $a \sin(t + c)$, (b) determine the amplitude, period, and phase shift of f, and (c) draw a sketch of the graph.

In Exercises 47 through 54, find the solutions of the equation if $0 \le t < 2\pi$.

47. (a) $2 \cos t - 1 = 0$; (b) $\tan^2 t - 1 = 0$

48. (a) $\cot t + 1 = 0$; (b) $2 \sin^2 t - 1 = 0$

49. $2 \sin^2 t + 3 \sin t - 2 = 0$

50. $\tan^2 t - 3 \sec t + 3 = 0$

51. $\tan t - 2 = 3 \cot t$

52. $6 \cos^2 t - \sin t - 4 = 0$

53. (a) $4 \cos^2 3t - 3 = 0$; (b) $\cot^2 2t = 4$

54. (a) $4 \sin^2 2t - 1 = 0$; (b) $\tan^2 3t = 2$

In Exercises 55 and 56, find all real-number solutions of the equations of the indicated exercise.

55. Exercise 47   56. Exercise 48

In Exercises 57 through 60, find all angles θ that are solutions of the equation if $0° \le \theta < 360°$.

57. (a) $\tan \theta - \cot \theta = 0$; (b) $2 \sin^2 \tfrac{1}{2}\theta = 1$

58. (a) $2 \cos \theta + \sec \theta - 3 = 0$; (b) $\cot^2 \tfrac{1}{2}\theta = 3$

59. $\tan^2 \theta + 3 \sec \theta + 3 = 0$

60. $2 \sin 2\theta + 1 = \csc 2\theta$

In Exercises 61 and 62, find all angles θ that are solutions of the equations of the indicated exercise. Express the solutions in degree measurement.

61. Exercise 57   62. Exercise 58

In Exercises 63 through 66, determine whether the equality is an equation or an identity. If it is an equation, solve it for all values of x in $[0, 2\pi)$. If it is an identity, prove it.

63. $\sin 2x - \cos 2x \tan x = \tan x$

64. $\cos 2x = 2 \sin x \cos x$

65. $\cos^2 2x + 3 \sin 2x = 3$
66. $\cot x - \tan x = 2 \cot 2x$

In Exercises 67 and 68, use a calculator to approximate the function value.

67. (a) $\sin^{-1}(0.6032)$; (b) $\sin^{-1}(-0.6032)$;
    (c) $\cos^{-1}(0.6032)$; (d) $\cos^{-1}(-0.6032)$;
    (e) $\tan^{-1}(0.6032)$; (f) $\tan^{-1}(-0.6032)$
68. (a) $\sin^{-1}(0.4833)$; (b) $\sin^{-1}(-0.4833)$;
    (c) $\cos^{-1}(0.4833)$; (d) $\cos^{-1}(-0.4833)$;
    (e) $\tan^{-1}(0.4833)$; (f) $\tan^{-1}(-0.4833)$

In Exercises 69 and 70 use Table VI to approximate the function values of the indicated exercise.

69. Exercise 67   70. Exercise 68
71. Given $x = \arc \sin \frac{3}{5}$ and $y = \arc \cos(-\frac{4}{5})$, find the exact value of each of the following: (a) $\cos x$; (b) $\sin y$; (c) $\tan x$; (d) $\tan y$.
72. Given $x = \cos^{-1} \frac{7}{25}$ and $y = \tan^{-1}(-\frac{7}{24})$, find the exact value of each of the following: (a) $\sin x$; (b) $\sin y$; (c) $\cos y$; (d) $\tan x$.

In Exercises 73 and 74, find the exact value of the quantity.

73. (a) $\sin[2 \cos^{-1}(-\frac{12}{13})]$; (b) $\tan[\cos^{-1} \frac{3}{5} + \sin^{-1}(-\frac{7}{25})]$
74. (a) $\tan[2 \sin^{-1}(-\frac{24}{25})]$; (b) $\cos[\tan^{-1} \frac{4}{3} - \sin^{-1}(-\frac{5}{13})]$
75. At a particular point in space two atmospheric waves produce pressures of $F(t)$ dynes/cm^2 and $G(t)$ dynes/cm^2 at t seconds, where

$$F(t) = 0.02 \sin(200\pi t + \tfrac{2}{3}\pi)$$
and
$$G(t) = 0.04 \sin(200\pi t - \tfrac{1}{3}\pi)$$

Define the sum of F and G by an equation of the form $f(t) = a \sin(200\pi t + c)$.
76. A weight is suspended from a spring and vibrating vertically according to the equation

$$f(t) = -4 \sin 10t - 3 \cos 10t$$

where $f(t)$ centimeters is the directed distance of the weight from its central position t seconds after the start of the motion and the positive direction is upward. (a) Define $f(t)$ by an equation of the form

$$f(t) = a \sin(bt + c)$$

(b) Determine the amplitude, period, and frequency of f.

77. A particle is moving along a straight line according to the equation of motion

$$s = \sin(6t - \tfrac{1}{3}\pi) + \sin(6t + \tfrac{1}{6}\pi)$$

where s centimeters is the directed distance of the particle from the origin at t seconds. (a) Show that the motion is simple harmonic by defining s by an equation of the form $s = a \sin(bt + c)$. (b) Determine the amplitude and frequency of the motion.
78. Do Exercise 77 if $s = 8 \cos^2 6t - 4$.
79. A weight is suspended from a spring and vibrating vertically according to the equation

$$y = 4 \sin 2\pi(t + \tfrac{1}{6})$$

where y centimeters is the directed distance of the weight from its central position t seconds after the start of the motion and the positive direction is upward. (a) Solve the equation for t. Use the equation in part (a) to determine the smallest positive value of t for which the displacement of the weight above its central position is (b) 2 cm and (c) 3 cm.
80. In an electric circuit the electromotive force is $E(t)$ volts, where $E(t) = 20 \cos 120\pi t$. Find the smallest positive value of t for which the electromotive force is (a) 10 volts; (b) 5 volts; (c) -10 volts; (d) -5 volts.
81. A picture 5 ft high is placed on a wall with its base 4 ft above the level of the eye of an observer. Let θ be the radian measure of the angle subtended at the observer's eye by the picture when the observer is x feet from the wall. (a) Define θ as a function of x. Find θ when (b) $x = 5$, (c) $x = 6$, and (d) $x = 7$.

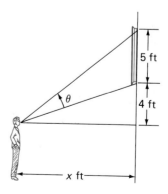

Exercises 82 through 88 pertain to Supplementary Section 7.7.

82. Find the exact value:
  (a) $\sin \frac{9}{8}\pi \cos \frac{3}{8}\pi$;
  (b) $\sin 195° - \cos 165°$

In Exercises 83 through 87, prove the identity.

83. $2 \cos 2\theta \cos \theta - \cos 3\theta = \cos \theta$

84. $2 \cos 2t \sin t - \sin 3t = -\sin t$

85. $\dfrac{\sin x + \sin 3x + \sin 5x}{\cos x + \cos 3x + \cos 5x} = \tan 3x$

86. $\dfrac{\sin \theta + \sin 2\theta + \sin 3\theta}{\cos \theta + \cos 2\theta + \cos 3\theta} = \tan 2\theta$

87. $\sin 2x + \sin 2y - \sin 2(x + y)$
  $= 4 \sin x \sin y \sin (x + y)$

88. Given $f(t) = \sin 41\pi t + \sin 39\pi t$.
  (a) Express $f(t)$ as the product of sine and cosine functions.
  (b) Draw a sketch of the graph of f over one period of the cosine function found in part (a).

CHAPTER 8

Applications of Trigonometry

Introduction

We apply the law of sines in Section 8.1 to solve oblique triangles for which we are given either the measures of two angles and a side or the measures of two sides and an angle opposite one of them. In Section 8.2, we apply the law of cosines to solve oblique triangles in other circumstances. We present a brief treatment of vectors and some of their applications in Section 8.3. Section 8.4 is devoted to polar coordinates. We introduce them mainly to represent a complex number in polar form as shown in Section 8.5. Then, in that section, we use the polar form to obtain the product and quotient of complex numbers. We state De Moivre's theorem in Section 8.6 and employ it to determine powers and roots of complex numbers expressed in polar form. In Supplementary Section 8.7, we discuss graphs of some polar equations.

8.1 The Law of Sines

An oblique triangle is one that does not contain a right angle. This section and the next are devoted to solving such triangles, which means finding the measures of the sides and angles.

As with a right triangle, the vertices of oblique triangles are labeled A, B, and C, and the measures of the sides opposite them are designated by a, b, and c, respectively. The angles at the vertices A, B, and C are denoted by α, β, and γ, respectively. Figure 1 shows an oblique triangle having all acute angles. An oblique triangle having at vertex A an obtuse angle (an angle whose degree measure is between 90 and 180) appears in Figure 2.

To solve an oblique triangle, we must know the measure of one side and any two other measures. In this section we consider two cases. In the first we are given the measures of two angles and a side. In the second we are given the measures of two sides and an angle opposite one of them.

For each of the triangles in Figures 1 and 2 choose a rectangular cartesian coordinate system so that the origin is at the vertex A and the positive x axis is along the side AB. The triangles and the coordinate axes are shown in Figure 3(a) and (b). In each triangle the angle α is in standard position. Also for each triangle, a line segment is drawn through C parallel to the y axis and intersecting the x axis at D. Let $h = |\overline{DC}|$. In either case

$$\sin \alpha = \frac{h}{b}$$

$$h = b \sin \alpha \qquad (1)$$

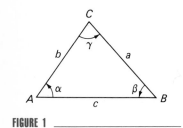

FIGURE 1 _____

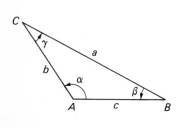

FIGURE 2 _____

440

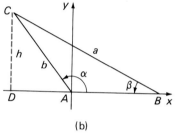

(a)

(b)

FIGURE 3

Also from right triangle BDC, in either case

$$\sin \beta = \frac{h}{a}$$

$$h = a \sin \beta \qquad (2)$$

Substituting from (2) into (1), we have

$$a \sin \beta = b \sin \alpha$$

$$\frac{a}{\sin \alpha} = \frac{b}{\sin \beta} \qquad (3)$$

If the coordinate axes are chosen so that the origin is at A and the positive x axis is along the side AC, then by a similar argument, we obtain

$$\frac{a}{\sin \alpha} = \frac{c}{\sin \gamma} \qquad (4)$$

Observe that (4) holds if the triangle is a right triangle. That is, if γ is $90°$, then because $\sin 90° = 1$, (4) becomes

$$\frac{a}{\sin \alpha} = \frac{c}{1}$$

$$\sin \alpha = \frac{a}{c}$$

From (3) and (4) we have the following theorem, known as the *law of sines*.

The Law of Sines

> If α, β, and γ are the angles of any triangle, and a, b, and c are, respectively, the measures of the sides opposite these angles, then
>
> $$\frac{a}{\sin \alpha} = \frac{b}{\sin \beta} = \frac{c}{\sin \gamma}$$

In the following example the law of sines is applied to solve a triangle when two angles and a side are known.

Example 1 Solve the triangle for which $\alpha = 51.2°$, $\beta = 48.6°$, and $a = 23.5$.

Solution Figure 4 shows the triangle. Because $\alpha + \beta + \gamma = 180°$,

$$\gamma = 180° - \alpha - \beta$$
$$= 180° - 51.2° - 48.6°$$
$$= 80.2°$$

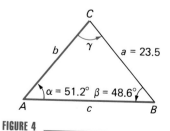

FIGURE 4

From the law of sines,

$$\frac{23.5}{\sin 51.2°} = \frac{b}{\sin 48.6°}$$

$$b = \frac{23.5(\sin 48.6°)}{\sin 51.2°}$$

$$b = \frac{23.5(0.7501)}{0.7793}$$

$$b = 22.6$$

Also from the law of sines,

$$\frac{23.5}{\sin 51.2°} = \frac{c}{\sin 80.2°}$$

$$c = \frac{23.5(\sin 80.2°)}{\sin 51.2°}$$

$$c = \frac{23.5(0.9854)}{0.7793}$$

$$c = 29.7$$

∎

Example   2   On a hill, inclined at an angle of 14.2° with the horizontal, stands a vertical tower. At a point P, 62.5 m down the hill from the foot of the tower, the angle of elevation of the top of the tower is 43.6°. How tall is the tower?

Solution   See Figure 5, where x is the number of meters in the height of the tower, the top of the tower is denoted by T, and F represents the point at the foot of the tower. We have an oblique triangle with vertices at P, T. and F. The angle at P in the triangle is found by computing 43.6° − 14.2°, which is 29.4°. The angle at T in the triangle is found by computing 90° − 43.6°, which is 46.4°. Thus we know the measures of two angles and a side of the triangle. From the law of sines,

$$\frac{x}{\sin 29.4°} = \frac{62.5}{\sin 46.4°}$$

$$x = \frac{62.5(\sin 29.4°)}{\sin 46.4°}$$

$$x = \frac{62.5(0.4909)}{0.7242}$$

$$x = 42.4$$

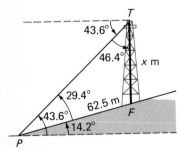

FIGURE 5

Therefore the tower is 42.4 m tall.

∎

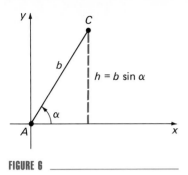

FIGURE 6

The law of sines can also be used when the measures of two sides and the angle opposite one of them are given. However, in such a case we do not always have a unique triangle. Suppose, for instance, we are given a, b, and α, where α is an acute angle. To construct a triangle having the given measurements, let the angle be in standard position on a rectangular coordinate system so that the vertex A is at the origin. See Figure 6. Because b is given, a line segment AC of length b units is marked off on the terminal side of α. Thus the position of vertex C is determined. The side BC of length a units is to be opposite vertex A, and the vertex B should be on the x axis. To locate the possible position of B, first draw the perpendicular line segment from C to the x axis. If this line segment has length h units, then

$$\sin \alpha = \frac{h}{b}$$

$$h = b \sin \alpha$$

The position of the vertex B will depend on the relationship between a and $b \sin \alpha$. There are four possibilities: $a < b \sin \alpha$; $a = b \sin \alpha$; $b \sin \alpha < a < b$; and $a \geq b$. We consider each possibility separately and follow each discussion with an illustration involving a particular set of values of a, b, and α.

Possibility 1: $a < b \sin \alpha$ See Figure 7. A side BC of length a units does not intersect the x axis. Therefore there is no triangle possible.

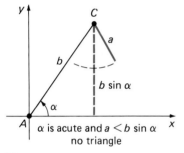

α is acute and $a < b \sin \alpha$
no triangle

FIGURE 7

Illustration   1

Let $a = 2.3$, $b = 4.5$, and $\alpha = 42°$. From the law of sines,

$$\frac{a}{\sin \alpha} = \frac{b}{\sin \beta}$$

$$\frac{2.3}{\sin 42°} = \frac{4.5}{\sin \beta}$$

$$\sin \beta = \frac{4.5(\sin 42°)}{2.3}$$

$$\sin \beta = \frac{4.5(0.6691)}{2.3}$$

$$\sin \beta = 1.309$$

Because $|\sin \beta|$ cannot be greater than 1, this equation has no solution. Thus there is no triangle satisfying the given information.

Observe that this set of values satisfies Possibility 1 because

$$b \sin \alpha = 4.5(\sin 42°)$$
$$= 3.01$$

Furthermore, $a = 2.3$ and $2.3 < 3.01$.   ■

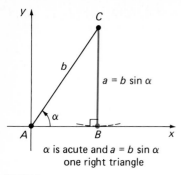

α is acute and $a = b \sin \alpha$
one right triangle

FIGURE 8

Possibility 2: $a = b \sin \alpha$　See Figure 8. The perpendicular distance from C to the x axis is a units, and so at vertex B there is a right angle. Hence there is one right triangle.

Illustration 2

Suppose $a = 2.0$, $b = 4.0$, and $\alpha = 30°$. Then from the law of sines,

$$\frac{a}{\sin \alpha} = \frac{b}{\sin \beta}$$

$$\frac{2.0}{\sin 30°} = \frac{4.0}{\sin \beta}$$

$$\sin \beta = \frac{4.0(\sin 30°)}{2.0}$$

$$\sin \beta = \frac{4.0(\frac{1}{2})}{2.0}$$

$$\sin \beta = 1$$

Therefore $\beta = 90°$ and the triangle is a right triangle.

Possibility 2 holds because for this set of values $b \sin \alpha = 4.0(\sin 30°)$; that is, $b \sin \alpha = 2$ and $a = 2$. ∎

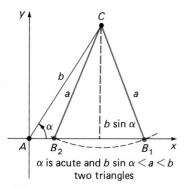

α is acute and $b \sin \alpha < a < b$
two triangles

FIGURE 9

Possibility 3: $b \sin \alpha < a < b$　See Figure 9. There are two possible positions of vertex B on the x axis shown in the figure as B_1 and B_2. Thus there are two triangles possible.

Illustration 3

Let $a = 25.2$, $b = 30.5$, and $\alpha = 54.2°$. From the law of sines,

$$\frac{a}{\sin \alpha} = \frac{b}{\sin \beta}$$

$$\frac{25.2}{\sin 54.2°} = \frac{30.5}{\sin \beta}$$

$$\sin \beta = \frac{30.5(\sin 54.2°)}{25.2}$$

$$\sin \beta = \frac{30.5(0.8111)}{25.2}$$

$$\sin \beta = 0.9817$$

There are two angles β having degree measure between 0 and 180 for which $\sin \beta = 0.9817$. Let the acute angle be β_1. From a calculator or Table VII,

$$\beta_1 = 79.0°$$

If the obtuse angle is β_2, then because β_1 is the reference angle of β_2, we have

$$\beta_2 = 180° - 79.0°$$
$$= 101.0°$$

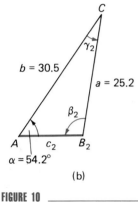

(a)

(b)

FIGURE 10

Therefore there are two triangles. Figure 10(a) shows the triangle for which $a = 25.2$, $b = 30.5$, $\alpha = 54.2°$, and $\beta_1 = 79.0°$. Figure 10(b) shows the triangle for which $a = 25.2$, $b = 30.5$, $\alpha = 54.2°$, and $\beta_2 = 101.0°$. We have two triangles to solve.

For the triangle of Figure 10(a), we compute γ_1 and c_1.

$$\begin{aligned}\gamma_1 &= 180° - \alpha - \beta_1 \\ &= 180° - 54.2° - 79.0° \\ &= 46.8°\end{aligned}$$

From the law of sines,

$$\frac{c_1}{\sin \gamma_1} = \frac{b}{\sin \beta_1}$$

$$\frac{c_1}{\sin 46.8°} = \frac{30.5}{\sin 79.0°}$$

$$c_1 = \frac{30.5(\sin 46.8°)}{\sin 79.0°}$$

$$c_1 = \frac{30.5(0.7290)}{0.9816}$$

$$c_1 = 22.7$$

We now compute γ_2 and c_2 for the triangle of Figure 10(b).

$$\begin{aligned}\gamma_2 &= 180° - \alpha - \beta_2 \\ &= 180° - 54.2° - 101.0° \\ &= 24.8°\end{aligned}$$

From the law of sines,

$$\frac{c_2}{\sin \gamma_2} = \frac{b}{\sin \beta_2}$$

$$\frac{c_2}{\sin 24.8°} = \frac{30.5}{\sin 101.0°}$$

$$c_2 = \frac{30.5(\sin 24.8°)}{\sin 101.0°}$$

$$c_2 = \frac{30.5(0.4195)}{0.9816}$$

$$c_2 = 13.0$$

The given set of values satisfies Possibility 3 because

$$\begin{aligned}b \sin \alpha &= 30.5 \sin 54.2° \\ &= 24.7\end{aligned}$$

Furthermore, $a = 25.2$, $b = 30.5$, and $24.7 < 25.2 < 30.5$.   ∎

Possibility 4: $a \geq b$ See Figure 11. There is only one potential position of B on the x axis. Therefore there is one triangle possible. If $a = b$, the triangle is isosceles.

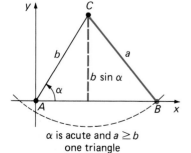

α is acute and $a \geq b$
one triangle

FIGURE 11

Illustration 4

Suppose $a = 5.21$, $b = 3.06$, and $\alpha = 47.6°$. From the law of sines,

$$\frac{a}{\sin \alpha} = \frac{b}{\sin \beta}$$

$$\frac{5.21}{\sin 47.6°} = \frac{3.06}{\sin \beta}$$

$$\sin \beta = \frac{3.06(\sin 47.6°)}{5.21}$$

$$\sin \beta = \frac{3.06(0.7385)}{5.21}$$

$$\sin \beta = 0.4337$$

There are two angles, having degree measure between 0 and 180, whose sine has a value 0.4337. But because $a > b$, it follows that $\alpha > \beta$; therefore $47.6° > \beta$. Thus we have only one value for β. From a calculator or Table VII,

$$\beta = 25.7°$$

We compute γ and c for this triangle.

$$\gamma = 180° - \alpha - \beta$$
$$= 180° - 47.6° - 25.7°$$
$$= 106.7°$$

From the law of sines,

$$\frac{c}{\sin \gamma} = \frac{a}{\sin \alpha}$$

$$\frac{c}{\sin 106.7°} = \frac{5.21}{\sin 47.6°}$$

$$c = \frac{5.21(\sin 106.7°)}{\sin 47.6°}$$

$$c = \frac{5.21(0.9578)}{0.7385}$$

$$c = 6.76$$

It is apparent that the given set of values satisfies Possibility 4 because $a = 5.21$, $b = 3.06$, and $5.21 > 3.06$. ■

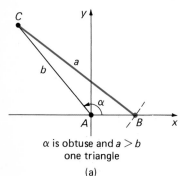

α is obtuse and $a > b$
one triangle

(a)

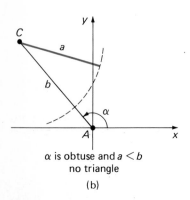

α is obtuse and $a < b$
no triangle

(b)

FIGURE 12 _____

If an angle of a triangle is obtuse, the measure of the side opposite it must be greater than the measures of the other sides. Therefore, if a, b, and α are given and α is obtuse, then one triangle is possible if and only if $a > b$. Figure 12(a) shows one triangle where α is obtuse and $a > b$. Figure 12(b) indicates there is no triangle when α is obtuse and $a < b$.

Because of the various situations that can occur, the case when two sides and the angle opposite one of them are given is called the *ambiguous case*. The same results are obtained if the known sides and the angle opposite one of them are represented by other symbols, such as c, b, and γ; b, a, and β; and so on.

Example 3 In parts (a) through (d) we are given the measures of two sides of a triangle and the angle opposite one of them; thus we have the ambiguous case. Determine the number of triangles that are possible.

a) $c = 2.0$, $b = 6.0$, $\gamma = 30°$ b) $b = 32.4$, $a = 20.6$, $\beta = 52.1°$

c) $a = 10.3$, $c = 16.5$, $\alpha = 23.8°$ d) $b = 32$, $c = 25$, $\beta = 114°$

Solution a) The given angle is γ and $c < b$. The measure of the side opposite the given angle is less than the other given measure of a side. Therefore we first compute $b \sin \gamma$.

$$b \sin \gamma = 6.0 \sin 30°$$
$$= 6.0(\tfrac{1}{2})$$
$$= 3.0$$

Because $c = 2.0$,

$$c < b \sin \gamma$$

Thus we have Possibility 1. Therefore there is no triangle.

b) The given angle is β and $b > a$. Here the measure of the side opposite the given angle is greater than the other given measure of a side. Therefore, from Possibility 4, there is one triangle.

c) The given angle is α and $a < c$. The measure of the side opposite the given angle is less than the other given measure of a side. So we compute $c \sin \alpha$.

$$c \sin \alpha = 16.5 \sin 23.8°$$
$$= 16.5(0.4035)$$
$$= 6.66$$

Because $a = 10.3$, $c = 16.5$, and $6.66 < 10.3 < 16.5$, then

$$c \sin \alpha < a < c$$

Therefore, from Possibility 3, there are two triangles.

d) The given angle is β, which is obtuse. Because $b = 32$ and $c = 25$, the measure of the side opposite the given angle is greater than the other given measure of a side. Hence there is one triangle. ∎

Example 4 A ladder 35.4 ft long is leaning against an embankment inclined 62.5° to the horizontal. If the bottom of the ladder is 10.2 ft from the embankment, what is the distance from the top of the ladder down the embankment to the ground?

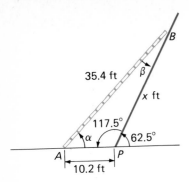

FIGURE 13

Solution See Figure 13, where x feet is the distance to be determined, B is the point at the top of the ladder, A is the point at the bottom of the ladder, and P is the point at the bottom of the embankment. The angle at P in the triangle is found by computing $180° - 62.5°$, which is $117.5°$. We have an oblique triangle for which the measures of two sides and the angle opposite one of them are known. Thus we have the ambiguous case. Because the known angle is obtuse and the measure of the side opposite this angle is greater than the other given measure of a side, there is one triangle.

Before we can use the law of sines to determine x, we must know the measurement of the angle at A in the triangle. Let α be this angle and let β be the angle at B in the triangle. We first find β from the law of sines:

$$\frac{35.4}{\sin 117.5°} = \frac{10.2}{\sin \beta}$$

$$\sin \beta = \frac{10.2(\sin 117.5°)}{35.4}$$

$$\sin \beta = \frac{10.2(0.8870)}{35.4}$$

$$\sin \beta = 0.2556$$

$$\beta = 14.8°$$

Because $\alpha + \beta + 117.5° = 180°$,

$$\alpha = 180° - \beta - 117.5°$$
$$= 180° - 14.8° - 117.5°$$
$$= 47.7°$$

From the law of sines,

$$\frac{x}{\sin 47.7°} = \frac{35.4}{\sin 117.5°}$$

$$x = \frac{35.4(\sin 47.7°)}{\sin 117.5°}$$

$$x = \frac{35.4(0.7396)}{0.8870}$$

$$x = 29.5$$

Therefore the distance from the top of the ladder down the embankment to the ground is 29.5 ft. ∎

Figure 14 shows an oblique triangle with the customary symbols denoting the vertices, angles, and measures of the sides. If the base is considered to be side AB, and h units is the length of the altitude, then if K square units is the area of the triangle,

$$K = \tfrac{1}{2}ch$$

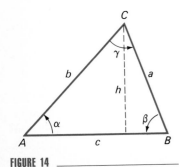

FIGURE 14

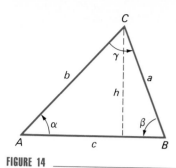

FIGURE 14

Because $h = b \sin \alpha$, we have

$$K = \tfrac{1}{2}c(b \sin \alpha)$$
$$K = \tfrac{1}{2}bc \sin \alpha \tag{5}$$

Another equation giving h is $h = a \sin \beta$. Substituting this value of h in $K = \tfrac{1}{2}ch$, we get

$$K = \tfrac{1}{2}c(a \sin \beta)$$
$$K = \tfrac{1}{2}ac \sin \beta \tag{6}$$

If the base is considered to be side AC, then by a similar argument, we obtain

$$K = \tfrac{1}{2}ab \sin \gamma \tag{7}$$

From formulas (5), (6), and (7), we have the following theorem.

Area of a Triangle

> The measure of the area of a triangle is one-half the product of the measures of two sides and the sine of the angle included between the two sides.

Illustration 5

We compute the area of the triangle in Example 1 by each of the three formulas (5), (6), and (7). If K square units is the area of the triangle,

$$\begin{aligned} K &= \tfrac{1}{2}bc \sin \alpha \\ &= \tfrac{1}{2}(22.6)(29.7)(\sin 51.2°) \\ &= \tfrac{1}{2}(22.6)(29.7)(0.7793) \\ &= 262 \end{aligned} \qquad \begin{aligned} K &= \tfrac{1}{2}ac \sin \beta \\ &= \tfrac{1}{2}(23.5)(29.7)(\sin 48.6°) \\ &= \tfrac{1}{2}(23.5)(29.7)(0.7501) \\ &= 262 \end{aligned}$$

$$\begin{aligned} K &= \tfrac{1}{2}ab \sin \gamma \\ &= \tfrac{1}{2}(23.5)(22.6)(\sin 80.2°) \\ &= \tfrac{1}{2}(23.5)(22.6)(0.9854) \\ &= 262 \end{aligned}$$

■

EXERCISES 8.1

In Exercises 1 through 4, two angles of a triangle and the measure of the side opposite one of them are given. Find the measure of the side opposite the other given angle, but do not use a calculator or tables for the trigonometric function values. Express the results to the number of significant digits justified by the given information.

1. $a = 4.6$, $\alpha = 45°$, $\beta = 60°$

2. $b = 23$, $\beta = 45°$, $\gamma = 30°$
3. $c = 88$, $\alpha = 30°$, $\gamma = 120°$
4. $a = 9.5$, $\alpha = 135°$, $\beta = 30°$

In Exercises 5 through 12, solve the triangle. Express the results to the number of significant digits justified by the given information.

5. $\alpha = 34°$, $\beta = 71°$, $a = 24$

6. $\alpha = 62°$, $\gamma = 55°$, $a = 8.3$
7. $\beta = 48.6°$, $\gamma = 61.4°$, $c = 53.2$
8. $\alpha = 26.5°$, $\beta = 32.7°$, $b = 187$
9. $\alpha = 73.2°$, $\gamma = 23.8°$, $b = 2.30$
10. $\beta = 84.6°$, $\gamma = 51.9°$, $a = 46.4$
11. $\alpha = 52°42'$, $\beta = 75°36'$, $b = 408$
12. $\beta = 101°6'$, $\gamma = 23°24'$, $c = 0.149$

In Exercises 13 through 24 because the measures of two sides of a triangle and the angle opposite one of them are given, we have the ambiguous case. Determine the number of triangles that satisfy the given set of conditions and solve each triangle. Express the results to the number of significant digits justified by the given information.

13. $a = 6.4$, $b = 4.7$, $\alpha = 42°$
14. $b = 27$, $a = 46$, $\beta = 38°$
15. $b = 17$, $c = 34$, $\beta = 30°$
16. $c = 18.3$, $b = 12.5$, $\gamma = 58.3°$
17. $c = 42.5$, $a = 68.0$, $\gamma = 35.2°$
18. $a = 245$, $b = 302$, $\alpha = 136.4°$
19. $b = 846$, $a = 431$, $\beta = 116.4°$
20. $a = 40.2$, $b = 52.4$, $\alpha = 41.5°$
21. $a = 54.0$, $c = 83.7$, $\alpha = 43.6°$
22. $c = 9.04$, $a = 3.52$, $\gamma = 128.1°$
23. $b = 3.562$, $c = 4.210$, $\beta = 50.23°$
24. $b = 5649$, $a = 6382$, $\beta = 59.43°$

In Exercises 25 through 32, find the area of the triangle of the indicated exercise.

25. Exercise 5   26. Exercise 6
27. Exercise 9   28. Exercise 10
29. Exercise 13   30. Exercise 16
31. Exercise 19   32. Exercise 22

33. A building is located at the end of a street that is inclined at an angle of 8.4° with the horizontal. At a point P, 210 m down the street from the building, the angle subtended by the building is 15.6°. How tall is the building?

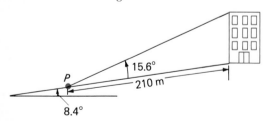

34. A flagpole is situated at the top of a building 115 ft tall. From a point in the same horizontal plane as the base of the building the angles of elevation of the top and bottom of the flagpole are 63.2° and 58.6°, respectively. How tall is the flagpole?

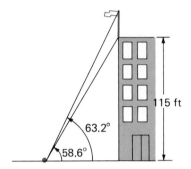

35. To determine the distance across a straight river a surveyor chooses two points P and Q on the bank, where the distance between P and Q is 200 m. At each of these points a point R on the opposite bank is sighted. The angle having sides PQ and PR is measured to be 63.1° and the angle having sides PQ and QR is measured to be 80.4°. What is the distance across the river?

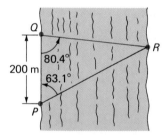

36. A triangular parcel of land with vertices at R, S, and T was to be enclosed by a fence, but it was discovered that the surveyor's mark at S was missing. From a deed to the property, it was learned that the distance from T to R is 324 m, the dis-

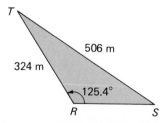

tance from T to S is 506 m, and the angle at R in the triangle is 125.4°. Determine the location of S by finding the distance from R to S.

37. A ramp is inclined at an angle of 41.3° with the ground. One end of a board, 20.6 ft in length, is located on the ground at a point P, 12.2 ft from the base Q of the ramp, and the other end rests on the ramp at point R. Find the distance from point Q up the ramp to point R.

38. At a particular instant, when an airplane was directly above a straight-line road connecting two small towns, the angles of depression of these towns were 10.2° and 8.7°. (a) Find the straight-line distances from the airplane to each of the towns at this instant given that the towns are 8.45 km apart. (b) Determine the height of the airplane at this instant.

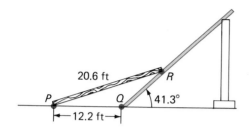

8.2   The Law of Cosines

A unique triangle is determined if the measures of two sides and the angle included between them are known. We should therefore be able to solve a triangle when we are given a, b, and γ; or a, c, and β; or b, c, and α. The solution cannot be obtained from using the law of sines exclusively. Nevertheless, there is a theorem, called the *law of cosines*, that can be applied, and we now discuss it.

Suppose a, b, and γ are known. Figure 1 shows a triangle with a rectangular coordinate system chosen so that γ is in standard position. In the figure γ is an obtuse angle, but the discussion is also valid if γ is acute. The vertex B is at $(a, 0)$. To determine the coordinates of A, let the point be (x, y). Then

$$\cos \gamma = \frac{x}{b} \quad \text{and} \quad \sin \gamma = \frac{y}{b}$$

Therefore

$$x = b \cos \gamma \quad \text{and} \quad y = b \sin \gamma$$

From the distance formula applied to side BA,

$$c^2 = (x - a)^2 + (y - 0)^2$$

Substituting $b \cos \gamma$ for x and $b \sin \gamma$ for y, we get

$$c^2 = (b \cos \gamma - a)^2 + (b \sin \gamma - 0)^2$$
$$= b^2 \cos^2 \gamma - 2ab \cos \gamma + a^2 + b^2 \sin^2 \gamma$$
$$= b^2(\cos^2 \gamma + \sin^2 \gamma) - 2ab \cos \gamma + a^2$$

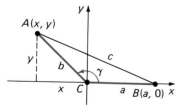

FIGURE 1

Because $\cos^2 \gamma + \sin^2 \gamma = 1$, we have

$$c^2 = a^2 + b^2 - 2ab \cos \gamma$$

This equation gives one form of the law of cosines. Two other forms are obtained in a similar manner by having either β or α in standard position on a rectangular coordinate system. We now state the law formally.

The Law of Cosines

If α, β, and γ are the angles of any triangle and a, b, and c are, respectively, the measures of the sides opposite these angles, then

$$c^2 = a^2 + b^2 - 2ab \cos \gamma$$
$$b^2 = a^2 + c^2 - 2ac \cos \beta$$
$$a^2 = b^2 + c^2 - 2bc \cos \alpha$$

Observe that if $\gamma = 90°$, we have a right triangle, and from the law of cosines,

$$\begin{aligned} c^2 &= a^2 + b^2 - 2ab \cos 90° \\ &= a^2 + b^2 - 2ab(0) \\ &= a^2 + b^2 \end{aligned}$$

which is the Pythagorean theorem. Rather than memorizing the separate forms of the law of cosines, think of it as a generalized version of the Pythagorean theorem that states:

The square of the measure of any side of a triangle is equal to the sum of the squares of the measures of the other two sides minus twice the product of the measures of the other two sides and the cosine of the angle between them.

Example **1** Solve the triangle for which $a = 24.0$, $c = 32.0$, and $\beta = 64.0°$.

Solution The triangle appears in Figure 2. From the law of cosines,

$$\begin{aligned} b^2 &= a^2 + c^2 - 2ac \cos \beta \\ b^2 &= (24.0)^2 + (32.0)^2 - 2(24.0)(32.0) \cos 64.0° \\ b^2 &= 576 + 1024 - 1536(0.4384) \\ b^2 &= 926.6 \\ b &= 30.4 \end{aligned}$$

Because we now have values for b and β, we can use the law of sines.

$$\frac{a}{\sin \alpha} = \frac{b}{\sin \beta}$$

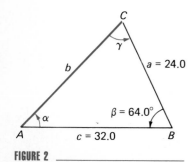

FIGURE 2

$$\frac{24.0}{\sin \alpha} = \frac{30.4}{\sin 64.0°}$$

$$\sin \alpha = \frac{24.0(\sin 64.0°)}{30.4}$$

$$\sin \alpha = \frac{24.0(0.8988)}{30.4}$$

$$\sin \alpha = 0.7096$$

There are two angles in a triangle for which the sine is 0.7096: 45.2° and 134.8°. However, because $a < b$, then $\alpha < \beta$. Therefore, we reject 134.8° and

$$\alpha = 45.2°$$

Because $\alpha + \beta + \gamma = 180°$,

$$\gamma = 180° - 45.2° - 64.0°$$
$$= 70.8°$$

The law of sines can be applied to find γ. If this method is used, there is a check on the work by verifying that $\alpha + \beta + \gamma = 180°$. ∎

Example   2   Two ships leave the same port at the same time. One ship sails on a course of 125° at 18 knots while the other sails on a course of 230° at 24 knots. Find after 3 hours (a) the distance between the ships and (b) the bearing from the first ship to the second.

Solution   a)   Refer to Figure 3. The port is at point P. After 3 hours the first ship is at point A and the second ship is at point B. Because the first ship is traveling at 18 knots, the distance from P to A is 54 nautical miles. The second ship is traveling at 24 knots, and so the distance from P to B is 72 nautical miles. The angle at P in the triangle is $230° - 125° = 105°$. Let the distance between the two ships after 3 hours be x nautical miles. From the law of cosines,

$$x^2 = (72)^2 + (54)^2 - 2(72)(54) \cos 105°$$

The reference angle of 105° is 75° and cos 105° is negative. Hence

$$\cos 105° = -\cos 75°$$
$$= -0.2588$$

Therefore

$$x^2 = 5184 + 2916 - 7776(-0.2588)$$
$$x^2 = 10{,}112$$
$$x = 100$$

Thus after 3 hours the distance between the two ships is 100 nautical miles.

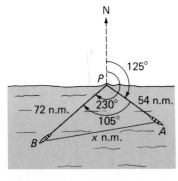

FIGURE 3

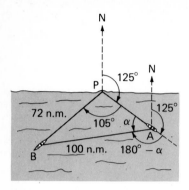

FIGURE 4

b) The bearing from the first ship to the second ship after 3 hours is the bearing from A to B. See Figure 4. To find the bearing, we must first determine α, the angle at A in the triangle. From the law of sines,

$$\frac{72}{\sin \alpha} = \frac{100}{\sin 105°}$$

$$\sin \alpha = \frac{72(\sin 105°)}{100}$$

$$\sin \alpha = \frac{72(0.9659)}{100}$$

$$\sin \alpha = 0.6954$$

$$\alpha = 44°$$

From Figure 4 we observe that the bearing from A to B is $125° + (180° - \alpha)$, which is $125° + (180° - 44°)$ or $261°$. ∎

Because the law of cosines involves the measures of the three sides and one angle of any triangle, it can be used to find an angle of a triangle when the measures of the three sides are known. For instance, one form of the law of cosines is

$$c^2 = a^2 + b^2 - 2ab \cos \gamma$$

If we solve this equation for $\cos \gamma$, we obtain

$$2ab \cos \gamma = a^2 + b^2 - c^2$$

$$\cos \gamma = \frac{a^2 + b^2 - c^2}{2ab}$$

Similarly, the other two forms of the law of cosines can be used to solve for $\cos \beta$ and $\cos \alpha$, and we have

$$\cos \beta = \frac{a^2 + c^2 - b^2}{2ac}$$

$$\cos \alpha = \frac{b^2 + c^2 - a^2}{2bc}$$

If you wish to determine the three angles of a triangle when the measures of the three sides are given, first you should use the law of cosines to find the largest angle, which is the one opposite the longest side. This is so that you can determine whether the angle is obtuse (when its cosine is negative) or acute (when its cosine is positive). In either case the other two angles will be acute and can be found from the law of sines. You cannot assume that the other two angles will be acute if the first angle found is not the largest.

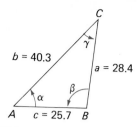

FIGURE 5

Example   3   Solve the triangle for which $a = 28.4$, $b = 40.3$, and $c = 25.7$.

Solution   Figure 5 shows the triangle. We wish to find α, β, and γ. Because β is opposite the longest side, we find it first. From the law of cosines,

$$\cos \beta = \frac{a^2 + c^2 - b^2}{2ac}$$

$$= \frac{(28.4)^2 + (25.7)^2 - (40.3)^2}{2(28.4)(25.7)}$$

$$= \frac{807 + 660 - 1624}{1460}$$

$$= \frac{-157}{1460}$$

$$= -0.1075$$

Because $\cos \beta < 0$, β is an obtuse angle. You can find it directly from a calculator or use Table VII to find the reference angle $\bar{\beta}$ for which $\cos \bar{\beta} = 0.1075$. From Table VII, $\bar{\beta} = 83.8°$. Therefore $\beta = 180° - 83.8°$; that is, $\beta = 96.2°$.

We use the law of sines to compute α and γ.

$$\frac{\sin \alpha}{28.4} = \frac{\sin 96.2°}{40.3} \qquad \frac{\sin \gamma}{25.7} = \frac{\sin 96.2°}{40.3}$$

$$\sin \alpha = \frac{28.4(\sin 96.2°)}{40.3} \qquad \sin \gamma = \frac{25.7(\sin 96.2°)}{40.3}$$

$$\sin \alpha = \frac{28.4(0.9942)}{40.3} \qquad \sin \gamma = \frac{25.7(0.9942)}{40.3}$$

$$\sin \alpha = 0.7006 \qquad \sin \gamma = 0.6340$$

$$\alpha = 44.5° \qquad \gamma = 39.3°$$

Of course, we could have computed just one of these angles by the law of sines and then found the other angle from the fact that the sum of the degree measures is 180. However, by computing them separately, we have a check:

$$\alpha + \beta + \gamma = 44.5° + 96.2° + 39.3°$$
$$= 180°$$

∎

Example   4   The lengths of two sides of a parallelogram are 7.4 cm and 9.2 cm, and one of the diagonals has a length of 6.2 cm. Find the area of the parallelogram.

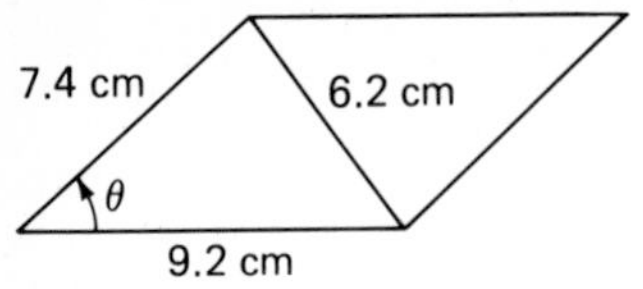

7.4 cm   6.2 cm

θ

9.2 cm

FIGURE 6

Solution The parallelogram appears in Figure 6. Because the diagonal divides the parallelogram into two congruent triangles, we first find the area of one of these triangles. Recall that the measure of the area of a triangle is one-half the product of the measures of two sides and the sine of the angle included between the two sides. To use this fact we find the angle opposite the diagonal of length 6.2 cm. If θ is this angle, then from the law of cosines,

$$\cos \theta = \frac{(9.2)^2 + (7.4)^2 - (6.2)^2}{2(9.2)(7.4)}$$

$$\cos \theta = \frac{84.6 + 54.8 - 38.4}{136}$$

$$\cos \theta = \frac{101}{136}$$

$$\cos \theta = 0.743$$

$$\theta = 42°$$

We now can find the area of the triangle formed by two sides and the given diagonal. If K square centimeters is the area of this triangle,

$$K = \tfrac{1}{2}(7.4)(9.2) \sin 42°$$
$$= 34(0.6691)$$
$$= 23$$

Because $2K = 46$, the area of the parallelogram is 46 cm^2. ∎

EXERCISES 8.2

In Exercises 1 through 4, measures of two sides of a triangle and the angle included between them are given. Find the measure of the third side, but do not use a calculator or tables for the trigonometric function values. Express the results to the number of significant digits justified by the given information.

1. $a = 4.5$, $b = 6.3$, $\gamma = 60°$
2. $b = 26$, $c = 37$, $\alpha = 45°$
3. $a = 15$, $c = 22$, $\beta = 135°$
4. $a = 1.4$, $b = 2.1$, $\gamma = 120°$

In Exercises 5 through 12, solve the triangle. Express the results to the number of significant digits justified by the given information.

5. $b = 3.4$, $c = 2.8$, $\alpha = 82°$
6. $a = 43$, $c = 32$, $\beta = 59°$
7. $a = 11.2$, $b = 15.3$, $\gamma = 116.4°$
8. $a = 40.2$, $b = 45.3$, $\gamma = 72.2°$
9. $a = 2045$, $c = 3126$, $\beta = 10.52°$
10. $b = 182.4$, $c = 245.1$, $\alpha = 126.81°$
11. $a = 5.26$, $b = 3.74$, $\gamma = 135°12'$
12. $a = 325$, $c = 108$, $\beta = 18°36'$

In Exercises 13 through 18, find the area of the triangle of the indicated exercise.

13. Exercise 1
14. Exercise 4
15. Exercise 7
16. Exercise 8
17. Exercise 9
18. Exercise 10

In Exercises 19 through 26, solve the triangle. Express the results to the number of significant digits justified by the given information.

19. $a = 5.2$, $b = 7.1$, $c = 3.5$
20. $a = 8.4$, $b = 2.7$, $c = 7.3$
21. $a = 20.7$, $b = 10.2$, $c = 24.3$
22. $a = 1.24$, $b = 1.56$, $c = 1.38$
23. $a = 408$, $b = 256$, $c = 283$
24. $a = 11.3$, $b = 25.0$, $c = 27.6$

25. $a = 66.92$, $b = 53.46$, $c = 15.78$
26. $a = 718.5$, $b = 634.2$, $c = 528.4$
27. A triangle has sides of lengths 34 cm, 23 cm, and 42 cm. (a) Find the measurement of the smallest angle. (b) Determine the area of the triangle.
28. A triangle has sides of lengths 2.8 in., 3.2 in., and 4.1 in. What is (a) the measurement of the largest angle and (b) the area of the triangle?
29. A parallelogram has sides of lengths 10.3 cm and 23.2 cm, and one of the angles is 54.2°. What is (a) the length of the longer diagonal and (b) the area of the parallelogram?
30. The sides of a parallelogram have lengths of 15.6 cm and 33.0 cm. If one of the angles is 42.6°, find (a) the length of the shorter diagonal and (b) the area of the parallelogram.
31. At 9 A.M. a boat leaves a pier on a course of 63.2° at 8 knots. At 10 A.M. another boat leaves the same pier on a course of 108.4° at 10 knots. At 12 noon (a) what is the distance between the boats and (b) what is the bearing from the first boat to the second?

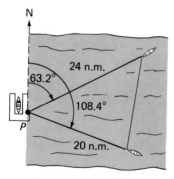

32. A point P is 1.4 km from one end of a lake and 2.2 km from the other end. If at P the lake subtends an angle of 54°, what is the length of the lake?

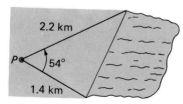

33. Two points P and Q are on opposite sides of a building. To determine the distance between these points, a third point R is selected where the distance from P to R is 50.2 m and the distance from Q to R is 61.4 m. The angle formed by the line segments PR and QR is measured as 62.5°. Determine the distance from P to Q.

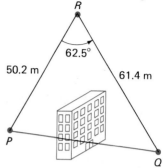

34. Two straight roads intersect at a point P and make an angle of 42.6° there. At a point R on one road is a building that is 368 m from P and at a point S on the other road is a building that is 426 m from P. Determine the direct distance from R to S.

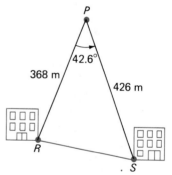

35. A tower 23.5 m tall makes an angle of 110.2° with

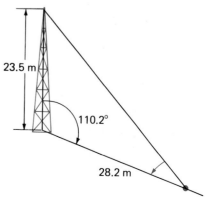

the inclined road on which it is located. Determine the angle subtended by the tower at a point down the road 28.2 m from its foot.

36. A triangular field has sides of lengths 212 m, 255 m, and 168 m. Determine the area of the field.

37. A ladder 24 ft long is leaning against a sloping embankment. The foot of the ladder is 11 ft from the base of the embankment, and the distance from the top of the ladder down the embankment to the ground is 16 ft. What is the angle at which the embankment is inclined to the horizontal?

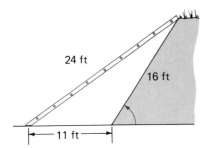

38. Two straight-line flight patterns intersect each other at an angle of 50.6°. At a particular time an airplane on one flight pattern is 53.4 mi from the intersection and a plane on the other pattern is 63.9 mi from the intersection. What is the distance between the planes at this time? There are two solutions.

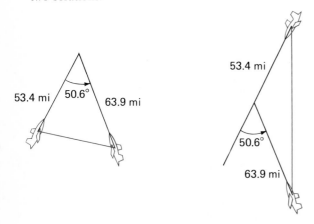

39. An airplane leaves an airport on a course of 310°. After flying 150 mi, it must return to the airport. Because of a navigational error, the plane flies 150 mi on a course of 115°. After flying the 300 mi, (a) how far is the plane from the airport

and (b) what is the bearing from the plane to the airport?

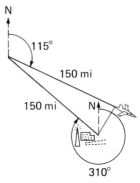

40. On a particular day the distance from the earth at E to the sun at S was $(9.2)10^7$ mi and the distance from Mars at M to the sun at S was $(1.4)10^8$ mi. If the angle between line segments ES and MS was 59°, what was the distance between the earth and Mars on that day?

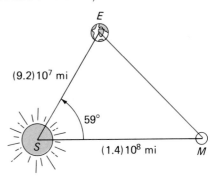

41. From the law of cosines, $\cos \alpha = \dfrac{b^2 + c^2 - a^2}{2bc}$. Use this equation to prove that

$$1 + \cos \alpha = \frac{(b + c + a)(b + c - a)}{2bc}$$

and

$$1 - \cos \alpha = \frac{(a - b + c)(a + b - c)}{2bc}$$

42. If K square units is the area of a triangle having angle α included between the sides of lengths b units and c units, then $K = \frac{1}{2}bc \sin \alpha$. Use this equation to prove that

$$K = \sqrt{\tfrac{1}{2}bc(1 + \cos \alpha) \cdot \tfrac{1}{2}bc(1 - \cos \alpha)}$$

43. Use the results of Exercises 41 and 42 to prove
Heron's Formula:

$$K = \sqrt{s(s-a)(s-b)(s-c)}$$

where $s = \frac{1}{2}(a + b + c)$. Heron's Formula is used to compute the area of a triangle when only the lengths of the three sides are known.

In Exercises 44 through 46, use Heron's Formula from Exercise 43 to find the area of the triangle for the given values of a, b, and c.

44. $a = 18.7$, $b = 12.6$, $c = 17.9$
45. $a = 325$, $b = 236$, $c = 411$
46. $a = 1.847$, $b = 2.112$, $c = 1.903$

In Exercises 47 and 48, use Heron's Formula from Exercise 43 to find the area of the triangle having vertices at the given points.

47. $(-2, 1)$, $(2, -3)$, and $(5, 4)$

48. $(0, -3)$, $(2, 4)$, and $(5, 2)$
49. If one acre is equivalent to 4840 yd^2, use Heron's Formula in Exercise 43 to find the number of acres in the area of the triangular field having sides of lengths 453 yd, 592 yd, and 700 yd.
50. If r units is the radius of the inscribed circle of a triangle having sides of lengths a units, b units, and c units, use Heron's Formula in Exercise 43 to show that

$$r = \sqrt{\frac{(s-a)(s-b)(s-c)}{s}}$$

where $s = \frac{1}{2}(a + b + c)$.

In Exercises 51 and 52, use the result of Exercise 50 to find the radius of the inscribed circle of the triangle in the indicated exercise.

51. Exercise 45
52. Exercise 46

8.3 Vectors

The applications of mathematics are often concerned with quantities that possess both magnitude and direction. An example of such a quantity is *velocity*. For instance an airplane's velocity has magnitude (the speed of the airplane) and direction, which determines the course of the airplane. Other examples of vector quantities are *force, displacement,* and *acceleration.* Such quantities may be represented geometrically by a *direct line segment.* Physicists and engineers refer to a directed line segment as a *vector* and the quantities that have both magnitude and direction are called **vector quantities.** In contrast, a quantity that has magnitude but not direction is called a **scalar quantity.** Examples of scalar quantities are length, area, volume, and speed. The study of vectors is called **vector analysis.**

The approach to vector analysis can be on either a geometric or an analytic basis. If the geometric approach is taken, we first define a directed line segment as a line segment from a point P to a point Q and denote this directed line segment by $\overrightarrow{PQ}$. The point P is called the **initial point,** and the point Q is called the **terminal point.** Two directed line segments $\overrightarrow{PQ}$ and $\overrightarrow{RS}$ are said to be **equal** if they have the same *length* and *direction,* and we write $\overrightarrow{PQ} = \overrightarrow{RS}$ (see Figure 1). The directed line segment $\overrightarrow{PQ}$ is called the **vector** from P to Q. A vector is denoted by a single letter, set in boldface type, such as **A.** In some books a letter in lightface type, with an arrow above it, is used to indicate a vector, for example $\overrightarrow{A}$.

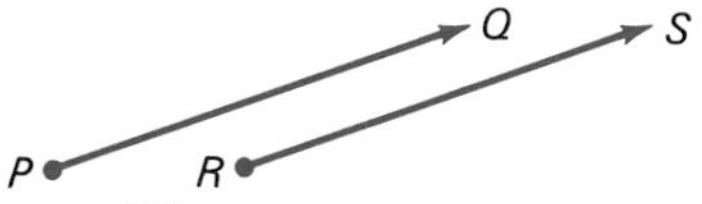

Continuing with the geometric approach to vector analysis, note that if the directed line segment $\overrightarrow{PQ}$ is the vector **A,** and $\overrightarrow{PQ} = \overrightarrow{RS}$, the directed line segment $\overrightarrow{RS}$ is also the vector **A.** Then a vector is considered to remain

$\overrightarrow{PQ} = \overrightarrow{RS}$

FIGURE 1

unchanged if it is moved parallel to itself. With this interpretation of a vector, we can assume for convenience that every vector has its initial point at some fixed reference point. By taking this point as the origin of a rectangular cartesian coordinate system, a vector can be defined analytically in terms of real numbers. Such a definition permits the study of vector analysis from a purely mathematical viewpoint.

We use the analytic approach; however, the geometric interpretation is given for illustrative purposes. A vector in the plane is denoted by an ordered pair of real numbers. The notation $\langle x, y \rangle$ is used instead of (x, y) to avoid confusing a vector with a point. Following is the formal definition.

Definition

Vector
A **vector in the plane** is an ordered pair of real numbers $\langle x, y \rangle$. The numbers x and y are called the **components** of the vector.

There is a one-to-one correspondence between the vectors $\langle x, y \rangle$ in the plane and the points (x, y) in the plane. Let the vector **A** be the ordered pair of real numbers $\langle a_1, a_2 \rangle$. If A is the point (a_1, a_2), then the vector **A** may be represented geometrically by the directed line segment $\overrightarrow{OA}$. Such a directed line segment is called a **representation** of vector **A**. Any directed line segment that is equal to $\overrightarrow{OA}$ is also a representation of vector **A**. The particular representation of a vector that has its initial point at the origin is called the **position representation** of the vector.

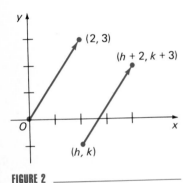

FIGURE 2

Illustration 1

The vector $\langle 2, 3 \rangle$ has as its position representation the directed line segment from the origin to the point $(2, 3)$. The representation of the vector $\langle 2, 3 \rangle$ whose initial point is (h, k) has as its terminal point $(h + 2, k + 3)$; refer to Figure 2. ∎

The vector $\langle 0, 0 \rangle$ is called the **zero vector,** and it is denoted by **0**; that is,

$$\mathbf{0} = \langle 0, 0 \rangle$$

Any point is a representation of the zero vector.

The **magnitude** of a vector **A** is the length of any of its representations and is denoted by $|\mathbf{A}|$. The **direction** of a nonzero vector is the direction of any of its representations.

Theorem 1

If **A** is the vector $\langle a_1, a_2 \rangle$, then
$$|\mathbf{A}| = \sqrt{a_1{}^2 + a_2{}^2}$$

Proof Because $|\mathbf{A}|$ is the length of any of the representations of $\mathbf{A}$, it will be the length of the position representation, which is the distance from the origin to the point (a_1, a_2). From the distance formula

$$|\mathbf{A}| = \sqrt{(a_1 - 0)^2 + (a_2 - 0)^2}$$
$$= \sqrt{a_1{}^2 + a_2{}^2} \qquad\blacksquare$$

Observe that $|\mathbf{A}|$ is a nonnegative number and is not a vector. From Theorem 1, it follows that

$$|\mathbf{0}| = 0$$

that is, the magnitude of the zero vector is 0.

Illustration 2

If $\mathbf{A} = \langle -3, 5\rangle$, then

$$|\mathbf{A}| = \sqrt{(-3)^2 + 5^2}$$
$$= \sqrt{34} \qquad\blacksquare$$

The **direction angle** of any nonzero vector is the angle θ measured from the positive side of the x axis counterclockwise to the position representation of the vector. If θ is measured in radians, $0 \le \theta < 2\pi$. If $\mathbf{A} = \langle a_1, a_2\rangle$, then

$$\tan\theta = \frac{a_2}{a_1} \qquad\text{if}\qquad a_1 \ne 0$$

If $a_1 = 0$ and $a_2 > 0$, then $\theta = \frac{1}{2}\pi$; if $a_1 = 0$ and $a_2 < 0$, then $\theta = \frac{3}{2}\pi$. Figures 3 through 5 show the direction angle θ for specific vectors whose position representations are drawn.

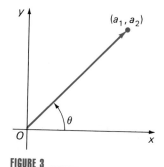

FIGURE 3 _____

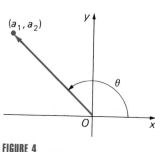

FIGURE 4 _____

FIGURE 5 _____

Example   1 Find the radian measure of the direction angle of each of the following vectors.

  a) $\langle -1, 1\rangle$ b) $\langle 0, -5\rangle$ c) $\langle 1, -2\rangle$

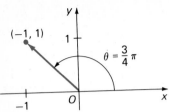

FIGURE 6

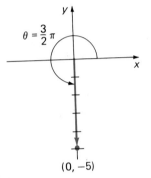

FIGURE 7

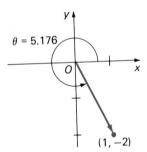

FIGURE 8

Solution   a) Figure 6 shows the position representation of $\langle -1, 1 \rangle$.

$$\tan \theta = \frac{1}{-1}$$

$$= -1$$

Because $\frac{1}{2}\pi < \theta < \pi$, $\theta = \frac{3}{4}\pi$.

b) The position representation of $\langle 0, -5 \rangle$ appears in Figure 7. Because $a_1 = 0$, $\tan \theta$ does not exist. Therefore, because $a_2 < 0$, $\theta = \frac{3}{2}\pi$.

c) The position representation of $\langle 1, -2 \rangle$ is shown in Figure 8.

$$\tan \theta = \frac{-2}{1}$$

$$= -2$$

Because $\frac{3}{2}\pi < \theta < 2\pi$, $\theta = 5.176$.   ■

Observe that if $\mathbf{A} = \langle a_1, a_2 \rangle$ and θ is the direction angle of $\mathbf{A}$, then

$$a_1 = |\mathbf{A}| \cos \theta \quad \text{and} \quad a_2 = |\mathbf{A}| \sin \theta \tag{1}$$

If the vector $\mathbf{A} = \langle a_1, a_2 \rangle$, then the representation of $\mathbf{A}$ whose initial point is (x, y) has as its endpoint $(x + a_1, y + a_2)$. Figure 9 illustrates five representations of the vector $\mathbf{A} = \langle a_1, a_2 \rangle$. In each case $\mathbf{A}$ translates the point (x_i, y_i) into the point $(x_i + a_1, y_i + a_2)$.

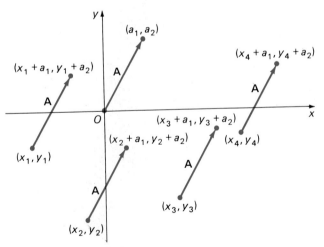

FIGURE 9

The following definition gives the method for adding two vectors.

Definition

> **Sum of Two Vectors**
> The sum of two vectors $\mathbf{A} = \langle a_1, a_2 \rangle$ and $\mathbf{B} = \langle b_1, b_2 \rangle$ is the vector $\mathbf{A} + \mathbf{B}$ defined by
>
> $$\mathbf{A} + \mathbf{B} = \langle a_1 + b_1, a_2 + b_2 \rangle$$

Illustration 3

If $\mathbf{A} = \langle 3, -1 \rangle$ and $\mathbf{B} = \langle -4, 5 \rangle$, then

$$\mathbf{A} + \mathbf{B} = \langle 3 + (-4), -1 + 5 \rangle$$
$$= \langle -1, 4 \rangle \qquad \blacksquare$$

The geometric interpretation of the sum of two vectors is shown in Figure 10. Let $\mathbf{A} = \langle a_1, a_2 \rangle$ and $\mathbf{B} = \langle b_1, b_2 \rangle$, and let P be the point (x, y). Then $\mathbf{A}$ translates the point P into the point $(x + a_1, y + a_2) = Q$. The vector $\mathbf{B}$ translates the point Q into the point $((x + a_1) + b_1, (y + a_2) + b_2)$ or, equivalently, $(x + (a_1 + b_1), y + (a_2 + b_2)) = R$. Furthermore,

$$\mathbf{A} + \mathbf{B} = \langle a_1 + b_1, a_2 + b_2 \rangle$$

Therefore $\mathbf{A} + \mathbf{B}$ translates P into $(x + (a_1 + b_1), y + (a_2 + b_2)) = R$. Thus in Figure 10 $\overrightarrow{PQ}$ is a representation of the vector $\mathbf{A}$, $\overrightarrow{QR}$ is a representation of the vector $\mathbf{B}$, and $\overrightarrow{PR}$ is a representation of the vector $\mathbf{A} + \mathbf{B}$. The representations of the vectors $\mathbf{A}$ and $\mathbf{B}$ are adjacent sides of a parallelogram, and the representation of the vector $\mathbf{A} + \mathbf{B}$ is a diagonal of the parallelogram. This diagonal is called the **resultant** of the vectors $\mathbf{A}$ and $\mathbf{B}$. The rule for the addition of vectors is sometimes referred to as the **parallelogram law.**

Force is a vector quantity where the magnitude is expressed in force units and the direction angle is determined by the direction of the force. It is shown in physics that two forces applied to an object at a particular point can be replaced by an equivalent force that is their resultant.

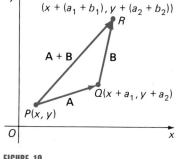

FIGURE 10

Example 2 Two forces of magnitudes 200 lb and 250 lb make an angle of 60° with each other and are applied to an object at the same point. Find (a) the magnitude of the resultant force and (b) the angle it makes with the force of 200 lb.

Solution Refer to Figure 11, where the axes are chosen so that the position representation of the force of 200 lb is along the positive side of the x axis. The vector $\mathbf{A}$ represents this force and $\mathbf{A} = \langle 200, 0 \rangle$. The vector $\mathbf{B}$ represents the force of 250 lb. From formulas (1), if $\mathbf{B} = \langle b_1, b_2 \rangle$, then

$$b_1 = 250 \cos 60° \qquad b_2 = 250 \sin 60°$$
$$= 125 \qquad\qquad = 216.5$$

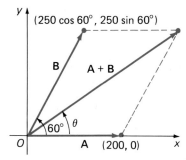

FIGURE 11

Thus $\mathbf{B} = \langle 125, 216.5 \rangle$. The resultant force is $\mathbf{A} + \mathbf{B}$, and

$$\mathbf{A} + \mathbf{B} = \langle 200, 0 \rangle + \langle 125, 216.5 \rangle$$
$$= \langle 325, 216.5 \rangle$$

a) $|\mathbf{A} + \mathbf{B}| = \sqrt{(325)^2 + (216.5)^2}$
$\qquad\qquad = 390.5$

b) If θ is the angle the vector $\mathbf{A} + \mathbf{B}$ makes with $\mathbf{A}$, then

$$\tan \theta = \frac{216.5}{325}$$

$$\tan \theta = 0.6662$$

$$\theta = 33.67°$$

           ■

The following illustration gives an alternative solution for Example 2.

Illustration 4

From Figure 11, we have the triangle shown in Figure 12. Applying the law of cosines to this triangle, we obtain

$$|\mathbf{A} + \mathbf{B}|^2 = (200)^2 + (250)^2 - 2(200)(250) \cos 120°$$
$$|\mathbf{A} + \mathbf{B}|^2 = 40,000 + 62,500 - 100,000(-\tfrac{1}{2})$$
$$|\mathbf{A} + \mathbf{B}|^2 = 152,500$$
$$|\mathbf{A} + \mathbf{B}| = \sqrt{152,500}$$
$$|\mathbf{A} + \mathbf{B}| = 390.5$$

We can compute θ by applying the law of sines to the triangle of Figure 12:

$$\frac{\sin \theta}{250} = \frac{\sin 120°}{390.5}$$

$$\sin \theta = \frac{250 \sin 120°}{390.5}$$

$$\sin \theta = 0.5544$$

$$\theta = 33.67°$$

           ■

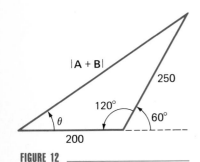

FIGURE 12

We mentioned previously that velocity is a vector quantity while speed is a scalar quantity. If $\mathbf{V}$ is a velocity vector, then $|\mathbf{V}|$ is a speed. In the next example, pertaining to marine navigation, we refer to a boat's velocity relative to the water and the velocity of the current. The resultant of these two velocities is the velocity of the boat relative to the land.

Example   3   A boat leaves the south bank of a river with a compass heading of north and traveling 8 mi/hr relative to the water. If the velocity of the current is 3 mi/hr toward the east, what is the speed of the boat relative to the land and what is its course?

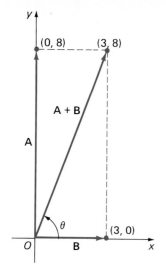

FIGURE 13 _____

Solution   See Figure 13, showing the position representations of vectors **A,** **B,** and **A** + **B**. The vector **A** represents the velocity of the boat relative to the water. Because **A** has a magnitude of 8 and a direction angle of 90°, **A** = $\langle 0, 8 \rangle$. The vector **B** represents the velocity of the current relative to the land, which has a magnitude of 3 and a direction angle of 0°. Thus **B** = $\langle 3, 0 \rangle$. The resultant of **A** and **B** is **A** + **B**, which is the velocity of the boat relative to the land:

$$\mathbf{A} + \mathbf{B} = \langle 0, 8 \rangle + \langle 3, 0 \rangle$$
$$= \langle 3, 8 \rangle$$
$$|\mathbf{A} + \mathbf{B}| = \sqrt{3^2 + 8^2}$$
$$= \sqrt{73}$$
$$\approx 8.54$$

If θ is the direction angle of **A** + **B,** then

$$\tan \theta = \tfrac{8}{3}$$
$$\tan \theta = 2.667$$
$$\theta = 69.4°$$

Thus the boat is traveling at a speed of 8.54 mi/hr relative to the land in the direction of 69.4° with respect to the south bank or, equivalently, a course of 20.6°. ■

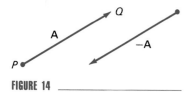

FIGURE 14 _____

If **A** is the vector $\langle a_1, a_2 \rangle$, then the **negative** of **A,** denoted by $-\mathbf{A}$, is the vector $\langle -a_1, -a_2 \rangle$. If the directed line segment $\overrightarrow{PQ}$ is a representation of the vector **A,** then the directed line segment $\overrightarrow{QP}$ is a representation of $-\mathbf{A}$. Any directed line segment that is parallel to $\overrightarrow{PQ}$, has the same length as $\overrightarrow{PQ}$, and has a direction opposite to that of $\overrightarrow{PQ}$ is also a representation of $-\mathbf{A}$. See Figure 14.

We now define subtraction of two vectors.

Definition

> **Difference of Two Vectors**
>
> The difference of the two vectors **A** and **B,** denoted by **A** − **B,** is the vector obtained by adding **A** to the negative of **B;** that is,
>
> $$\mathbf{A} - \mathbf{B} = \mathbf{A} + (-\mathbf{B})$$

From the definition, if   **A** = $\langle a_1, a_2 \rangle$ and   **B** = $\langle b_1, b_2 \rangle$,   then $-\mathbf{B} = \langle -b_1, -b_2 \rangle$, and

$$\mathbf{A} - \mathbf{B} = \langle a_1 - b_1, a_2 - b_2 \rangle$$

Illustration 5

If $\mathbf{A} = \langle 4, -2 \rangle$ and $\mathbf{B} = \langle 6, -3 \rangle$, then

$$\begin{aligned}
\mathbf{A} - \mathbf{B} &= \langle 4, -2 \rangle - \langle 6, -3 \rangle \\
&= \langle 4, -2 \rangle + \langle -6, 3 \rangle \\
&= \langle -2, 1 \rangle
\end{aligned}$$
■

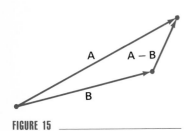

FIGURE 15

To interpret the difference of two vectors geometrically, let the representations of the vectors $\mathbf{A}$ and $\mathbf{B}$ have the same initial point. Then the directed line segment from the endpoint of the representation of $\mathbf{B}$ to the endpoint of the representation of $\mathbf{A}$ is a representation of the vector $\mathbf{A} - \mathbf{B}$. This obeys the parallelogram law $\mathbf{B} + (\mathbf{A} - \mathbf{B}) = \mathbf{A}$. See Figure 15.

The following example, involving the difference of two vectors, is concerned with air navigation. The *air speed* of a plane refers to its speed relative to the air, and the *ground speed* is its speed relative to the ground. When there is a wind, the velocity of the plane relative to the ground is the resultant of the vector representing the wind's velocity and the vector representing the velocity of the plane relative to the air.

Example 4 An airplane can fly at an air speed of 300 mi/hr. If there is a wind blowing toward the east at 50 mi/hr, what should be the plane's compass heading in order for its course to be 30°? What will be the plane's ground speed if it flies this course?

Solution Refer to Figure 16, showing position representations of the vectors $\mathbf{A}$ and $\mathbf{B}$ as well as a representation of $\mathbf{A} - \mathbf{B}$. The vector $\mathbf{A}$ represents the velocity of the plane relative to the ground on a course of 30°. The direction angle of $\mathbf{A}$ is 60°, which is 90° − 30°. The vector $\mathbf{B}$ represents the velocity of the wind. Because $\mathbf{B}$ has a magnitude of 50 and a direction angle of 0°, $\mathbf{B} = \langle 50, 0 \rangle$. The vector $\mathbf{A} - \mathbf{B}$ represents the velocity of the plane relative to the air; thus $|\mathbf{A} - \mathbf{B}| = 300$. Let θ be the direction angle of $\mathbf{A} - \mathbf{B}$. From Figure 16 we obtain the triangle shown in Figure 17. Applying the law of sines to this triangle, we get

$$\frac{\sin \phi}{50} = \frac{\sin 60°}{300}$$

$$\sin \phi = \frac{50 \sin 60°}{300}$$

$$\sin \phi = 0.1443$$

$$\phi = 8.3°$$

FIGURE 16

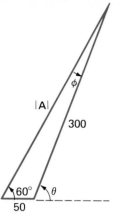

FIGURE 17

Therefore

$$\theta = 60° + 8.3°$$
$$= 68.3°$$

Again applying the law of sines to the triangle in Figure 17, we have

$$\frac{|\mathbf{A}|}{\sin(180° - \theta)} = \frac{300}{\sin 60°}$$

$$|\mathbf{A}| = \frac{300 \sin 111.7°}{\sin 60°}$$

$$|\mathbf{A}| = 322$$

The plane's compass heading should be $90° - \theta$, which is $21.7°$, and if the plane flies this course, its ground speed will be 322 mi/hr. ∎

EXERCISES   8.3

In Exercises 1 and 2, draw the position representation of the vector **A** and also the particular representation through the point P; find the magnitude of **A**.

1. (a) $\mathbf{A} = \langle 3, 4 \rangle$, $P = (2, 1)$;
   (b) $\mathbf{A} = \langle 0, -2 \rangle$, $P = (-3, 4)$
2. (a) $\mathbf{A} = \langle -2, 5 \rangle$, $P = (3, -4)$:
   (b) $\mathbf{A} = \langle 4, 0 \rangle$, $P = (2, 6)$

In Exercises 3 and 4, find the vector **A** having $\overrightarrow{PQ}$ as a representation. Draw $\overrightarrow{PQ}$ and the position representation of **A**.

3. (a) $P = (3, 7)$, $Q = (5, 4)$;
   (b) $P = (-3, 5)$, $Q = (-5, -2)$
4. (a) $P = (5, 4)$, $Q = (3, 7)$;
   (b) $P = (-5, -3)$, $Q = (0, 3)$

In Exercises 5 and 6, find the point S so that $\overrightarrow{PQ}$ and $\overrightarrow{RS}$ are each representations of the same vector.

5. (a) $P = (2, 5)$, $Q = (1, 6)$, $R = (-3, 2)$;
   (b) $P = (0, 3)$, $Q = (5, -2)$, $R = (7, 0)$
6. (a) $P = (-1, 4)$, $Q = (2, -3)$, $R = (-5, -2)$;
   (b) $P = (-2, 0)$, $Q = (-3, -4)$, $R = (4, 2)$

In Exercises 7 and 8, find the sum of the pairs of vectors and illustrate geometrically.

7. (a) $\langle 2, 4 \rangle$, $\langle -3, 5 \rangle$; (b) $\langle -3, 0 \rangle$, $\langle 4, -5 \rangle$
8. (a) $\langle 0, 3 \rangle$, $\langle -2, 3 \rangle$; (b) $\langle 2, 5 \rangle$, $\langle 2, 5 \rangle$

In Exercises 9 and 10, subtract the second vector from the first and illustrate geometrically.

9. (a) $\langle 4, 5 \rangle$, $\langle -3, 2 \rangle$; (b) $\langle -3, -4 \rangle$, $\langle 6, 0 \rangle$
10. (a) $\langle 0, 5 \rangle$, $\langle 2, 8 \rangle$; (b) $\langle 3, 7 \rangle$, $\langle 3, 7 \rangle$

In Exercises 11 and 12, let $A = \langle 2, 4 \rangle$, $B = \langle 4, -3 \rangle$, and $C = \langle -3, 2 \rangle$.

11. (a) Find $\mathbf{A} + \mathbf{B}$; (b) find $|\mathbf{C}|$.
12. (a) Find $\mathbf{A} - \mathbf{B}$; (b) find $|\mathbf{C} - \mathbf{B}|$.
13. Given $\mathbf{A} = \langle 3, 2 \rangle$; $\mathbf{C} = \langle 8, 8 \rangle$; $\mathbf{A} + \mathbf{B} = \mathbf{C}$, find $|\mathbf{B}|$.
14. Given $\mathbf{A} = \langle 2, -5 \rangle$, $\mathbf{B} = \langle 3, 1 \rangle$, and $\mathbf{C} = \langle -4, 2 \rangle$.
    (a) Find $\mathbf{A} + (\mathbf{B} + \mathbf{C})$ and illustrate geometrically.
    (b) Find $(\mathbf{A} + \mathbf{B}) + \mathbf{C}$ and illustrate geometrically.

In Exercises 15 through 18, find the components of the vector having the given magnitude and direction angle.

15. $18; \frac{1}{5}\pi$ 16. $24; 41.2°$
17. $35; 250°$ 18. $110; \frac{1}{3}\pi$
19. Two forces of magnitudes 60 lb and 80 lb make an angle of 30° with each other and are applied to an object at the same point. Find (a) the magnitude of the resultant force and (b) to the nearest degree, the angle it makes with the force of 60 lb. Use the method of Example 2.
20. Two forces of magnitudes 340 lb and 475 lb make an angle of 34.6° with each other and are applied to an object at the same point. Find (a) the magnitude of the resultant force and (b) to the nearest tenth of a degree the angle it makes with the force of 475 lb. Use the method of Example 2.
21. Do Exercise 19 by the method of Illustration 4.
22. Do Exercise 20 by the method of Illustration 4.

23. A force of magnitude 112 lb and one of 84 lb are applied to an object at the same point, and the resultant force has a magnitude of 162 lb. Find to the nearest tenth of a degree the angle made by the resultant force with the force of 112 lb.

24. A force of 22 lb and a force of 34 lb are applied to an object at the same point and make an angle of θ with each other. If the resultant force has a magnitude of 46 lb, find θ to the nearest degree.

25. A swimmer who can swim at a speed of 1.5 mi/hr relative to the water leaves the south bank of a river and is headed north directly across the river. If the river's current is toward the east at 0.8 mi/hr, (a) in what direction is the swimmer going? (b) What is the swimmer's speed relative to the land? (c) If the distance across the river is 1 mi, how far down the river does the swimmer reach the north bank?

26. Suppose the swimmer in Exercise 25 wishes to reach the point directly north across the river. (a) In what direction should the swimmer head? (b) What will be the swimmer's speed relative to the land if this direction is taken?

27. In an airplane that has an air speed of 250 mi/hr, a pilot wishes to fly due north. There is a wind blowing at 60 mi/hr toward the east. (a) What should be the plane's compass heading? (b) What will be the plane's ground speed if it flies this course?

28. A plane has an air speed of 350 mi/hr. In order for the actual course of the plane to be due north, the compass heading is 340°. The wind is blowing from the west. (a) What is the magnitude of the wind's velocity? (b) What is the plane's ground speed?

29. A boat can travel 15 knots relative to the water. On a river whose current is 3 knots toward the west the boat has a compass heading of south. What is the speed of the boat relative to the land and what is its course?

30. Let $\overrightarrow{PQ}$ be a representation of vector **A**, $\overrightarrow{QR}$ be a representation of vector **B**, and $\overrightarrow{RS}$ be a representation of vector **C**. Prove that if $\overrightarrow{PQ}$, $\overrightarrow{QR}$, and $\overrightarrow{RS}$ are sides of a triangle, then $\mathbf{A} + \mathbf{B} + \mathbf{C} = \mathbf{0}$.

8.4 Polar Coordinates

Until now we have located a point in a plane by its rectangular cartesian coordinates. There are other coordinate systems that give the position of a point in a plane. The **polar coordinate system** is one of them, and it is important because certain curves have simpler equations in that system. In polar coordinates all three conics (the parabola, ellipse, and hyperbola) have one equation. This equation is applied in the derivation of Kepler's laws in physics and in the study of the motion of planets in astronomy.

Cartesian coordinates are numbers, the abscissa and ordinate, and these numbers are directed distances from two fixed lines. Polar coordinates consist of a directed distance and the measure of an angle which is taken relative to a fixed point and a fixed ray (or half line). The fixed point is called the **pole** (or origin), designated by the letter O. The fixed ray is called the **polar axis** (or polar line), which we label OA. The ray OA is usually drawn horizontally and to the right, and it extends indefinitely. See Figure 1.

Let P be any point in the plane distinct from O. Let θ be the radian measure of the directed angle AOP, positive when measured counterclockwise and negative when measured clockwise, having as its initial side the ray OA and as its terminal side the ray OP. Then if r is the undirected distance from O to P (i.e., $r = |\overline{OP}|$), one set of polar coordinates of P is given by r and θ, and we write these coordinates as (r, θ).

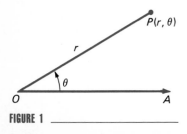

FIGURE 1

Example 1 Plot each of the following points having the given set of polar coordinates:

a) $(2, \frac{1}{4}\pi)$ b) $(5, \frac{1}{2}\pi)$ c) $(1, \frac{2}{3}\pi)$ d) $(3, \frac{7}{6}\pi)$

e) $(4, -\frac{1}{3}\pi)$ f) $(\frac{5}{2}, -\pi)$ g) $(2, -\frac{5}{4}\pi)$

Solution a) The point $(2, \frac{1}{4}\pi)$ is determined by first drawing the angle with radian measure $\frac{1}{4}\pi$ having its vertex at the pole and its initial side along the polar axis. The point on the terminal side that is 2 units from the pole is $(2, \frac{1}{4}\pi)$. See Figure 2(a). In a similar manner we obtain the points appearing in Figure 2(b)–(g).

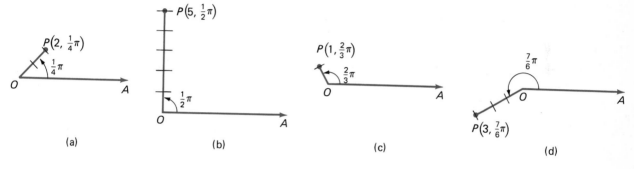

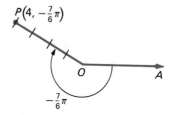

FIGURE 2

Illustration 1

Figure 3 shows the point $(4, \frac{5}{6}\pi)$. Another set of polar coordinates for this point is $(4, -\frac{7}{6}\pi)$; see Figure 4. Furthermore, the polar coordinates $(4, \frac{17}{6}\pi)$ also yield the same point, as shown in Figure 5.

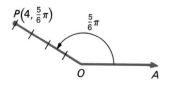

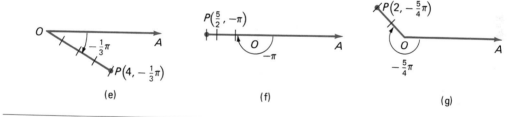

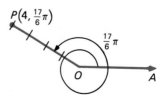

FIGURE 3   FIGURE 4   FIGURE 5

Actually, the coordinates $(4, \frac{5}{6}\pi + 2k\pi)$, where k is any integer, give the same point as $(4, \frac{5}{6}\pi)$. So a given point has an unlimited number of sets of polar coordinates. This is unlike the rectangular cartesian coordinate system, in which there is a one-to-one correspondence between the coordinates and the position of points in the plane. There is no such one-to-one correspondence between the polar coordinates and the position of points in the plane. A further example is obtained by considering sets of polar coordinates for the pole. If $r = 0$ and θ is any real number, we have the pole, which is therefore designated by $(0, \theta)$.

There are polar coordinates for which r is negative. In this case, instead of being on the terminal side of the angle, the point is on the extension of the terminal side, which is the ray from the pole extending in the direction opposite to the terminal side. So if P is on the extension of the terminal side of the angle of radian measure θ, a set of polar coordinates of P is (r, θ), where $r = -|\overline{OP}|$.

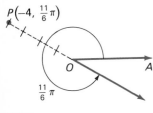

FIGURE 6

Illustration 2

The point $(-4, -\frac{1}{6}\pi)$, shown in Figure 6, is the same point as $(4, \frac{5}{6}\pi)$, $(4, -\frac{7}{6}\pi)$, and $(4, \frac{17}{6}\pi)$ in Illustration 1. Still another set of polar coordinates for this point is $(-4, \frac{11}{6}\pi)$; see Figure 7. ∎

FIGURE 7

The angle is usually measured in radians; thus a set of polar coordinates of a point is an ordered pair of real numbers. For each ordered pair of real numbers there is a unique point having this set of polar coordinates. However, we have seen that a particular point can be given by an unlimited number of ordered pairs of real numbers. If the point P is not the pole, and r and θ are restricted so that $r > 0$ and $0 \le \theta < 2\pi$, then there is a unique set of polar coordinates for P.

Example 2

a) Plot the point having polar coordinates $(3, -\frac{2}{3}\pi)$. Find another set of polar coordinates of this point for which (b) $r < 0$ and $0 < \theta < 2\pi$; (c) $r > 0$ and $0 < \theta < 2\pi$; (d) $r < 0$ and $-2\pi < \theta < 0$.

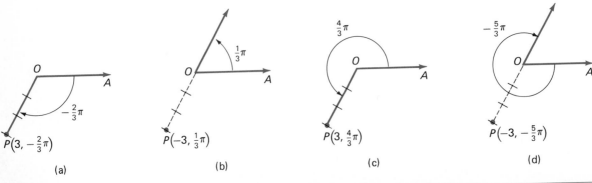

FIGURE 8

Solution   a) The point is plotted by drawing the angle of radian measure $-\frac{2}{3}\pi$ in a clockwise direction from the polar axis. Because $r > 0$, P is on the terminal side of the angle, three units from the pole; see Figure 8(a).

The answers to (b), (c), and (d) are, respectively, $(-3, \frac{1}{3}\pi)$, $(3, \frac{4}{3}\pi)$, and $(-3, -\frac{5}{3}\pi)$. They are illustrated in Figure 8(b), (c), and (d). ∎

Often we wish to refer to both the rectangular cartesian coordinates and the polar coordinates of a point. To do this we take the origin of the first system and the pole of the second system coincident, the polar axis as the positive side of the x axis, and the ray for which $\theta = \frac{1}{2}\pi$ as the positive side of the y axis.

Suppose that P is a point whose representation in the rectangular cartesian coordinate system is (x, y) and (r, θ) is a polar coordinate representation of P. We distinguish two cases: $r > 0$ and $r < 0$. In the first case, if $r > 0$, then the point P is on the terminal side of the angle of radian measure θ, and $r = |\overline{OP}|$. Such a case is shown in Figure 9. Then

$$\cos \theta = \frac{x}{|\overline{OP}|} \qquad \sin \theta = \frac{y}{|\overline{OP}|}$$

$$= \frac{x}{r} \qquad\qquad = \frac{y}{r}$$

Thus

$$x = r \cos \theta \qquad \text{and} \qquad y = r \sin \theta \tag{1}$$

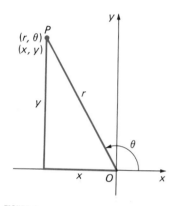

FIGURE 9

In the second case, if $r < 0$, the point P is on the extension of the terminal side and $r = -|\overline{OP}|$. See Figure 10. Then if Q is the point $(-x, -y)$,

$$\cos \theta = \frac{-x}{|\overline{OQ}|} \qquad \sin \theta = \frac{-y}{|\overline{OQ}|}$$

$$= \frac{-x}{|\overline{OP}|} \qquad = \frac{-y}{|\overline{OP}|}$$

$$= \frac{-x}{-r} \qquad = \frac{-y}{-r}$$

$$= \frac{x}{r} \qquad\qquad = \frac{y}{r}$$

Hence

$$x = r \cos \theta \qquad \text{and} \qquad y = r \sin \theta$$

These equations are the same as Equations (1); thus they hold in all cases.

From Equations (1) we can obtain the rectangular cartesian coordinates of a point when its polar coordinates are known. Also, from the equations we can find a polar equation of a curve if a rectangular cartesian equation is given.

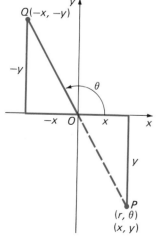

FIGURE 10

To obtain equations that give a set of polar coordinates of a point when its rectangular cartesian coordinates are known, we square on both sides of each equation in (1) and get

$$x^2 = r^2 \cos^2 \theta \quad \text{and} \quad y^2 = r^2 \sin^2 \theta$$

Equating the sum of the left members of the above to the sum of the right members we have

$$x^2 + y^2 = r^2 \cos^2 \theta + r^2 \sin^2 \theta$$
$$x^2 + y^2 = r^2(\sin^2 \theta + \cos^2 \theta)$$
$$x^2 + y^2 = r^2$$

$$\Leftrightarrow \qquad \boxed{r = \pm\sqrt{x^2 + y^2}} \qquad (2)$$

From the equations in (1) and dividing, we have

$$\frac{r \sin \theta}{r \cos \theta} = \frac{y}{x}$$

$$\Leftrightarrow \qquad \boxed{\tan \theta = \frac{y}{x}} \qquad (3)$$

Illustration 3

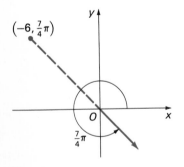

FIGURE 11

The point whose polar coordinates are $(-6, \frac{7}{4}\pi)$ is plotted in Figure 11. We find its rectangular cartesian coordinates. From (1),

$$
\begin{aligned}
x &= r \cos \theta & y &= r \sin \theta \\
&= -6 \cos \tfrac{7}{4}\pi & &= -6 \sin \tfrac{7}{4}\pi \\
&= -6 \cdot \frac{\sqrt{2}}{2} & &= -6\left(-\frac{\sqrt{2}}{2}\right) \\
&= -3\sqrt{2} & &= 3\sqrt{2}
\end{aligned}
$$

So the point is $(-3\sqrt{2}, 3\sqrt{2})$. ∎

The graph of an equation in polar coordinates r and θ consists of all those points and only those points having at least one pair of coordinates that satisfy the equation. If an equation of a graph is given in polar coordinates, it is called a *polar equation* to distinguish it from a *cartesian equation*, which is the term used when an equation is given in rectangular cartesian coordinates. In Supplementary Section 8.7 we discuss methods of obtaining the graph of a polar equation.

Example   3   Given that a polar equation of a graph is

$$r^2 = 4 \sin 2\theta$$

find a cartesian equation.

Solution   Because $\sin 2\theta = 2 \sin \theta \cos \theta$ we have $\sin 2\theta = 2(y/r)(x/r)$. With this substitution and $r^2 = x^2 + y^2$, we obtain from the given polar equation

$$x^2 + y^2 = 4(2)\frac{y}{r} \cdot \frac{x}{r}$$

$$x^2 + y^2 = \frac{8xy}{r^2}$$

$$x^2 + y^2 = \frac{8xy}{x^2 + y^2}$$

$$(x^2 + y^2)^2 = 8xy$$ ∎

Example   4   Find (r, θ) if $r > 0$ and $0 \le \theta < 2\pi$, for the point whose rectangular cartesian coordinate representation is $(-\sqrt{3}, -1)$.

Solution   The point $(-\sqrt{3}, -1)$ is plotted in Figure 12. From (2), because $r > 0$,

$$r = \sqrt{3 + 1}$$
$$= 2$$

From (3), $\tan \theta = -1/(-\sqrt{3})$, and since $\pi < \theta < \frac{3}{2}\pi$,

$$\theta = \tfrac{7}{6}\pi$$

So the point is $(2, \tfrac{7}{6}\pi)$. ∎

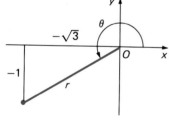

FIGURE 12

Example   5   Find a polar equation of the graph whose cartesian equation is

$$x^2 + y^2 - 4x = 0$$

Solution   Substituting $x = r \cos \theta$ and $y = r \sin \theta$ in

$$x^2 + y^2 - 4x = 0$$

we have

$$r^2 \cos^2 \theta + r^2 \sin^2 \theta - 4r \cos \theta = 0$$
$$r^2 - 4r \cos \theta = 0$$
$$r(r - 4 \cos \theta) = 0$$

Therefore

$$r = 0 \qquad \text{or} \qquad r - 4 \cos \theta = 0$$

The graph of $r = 0$ is the pole. However, the pole is a point on the graph of $r - 4 \cos \theta = 0$ because $r = 0$ when $\theta = \frac{1}{2}\pi$. Therefore a polar equation of the graph is

$$r = 4 \cos \theta$$

The graph of $x^2 + y^2 - 4x = 0$ is a circle. The equation may be written in the form

$$(x - 2)^2 + y^2 = 4$$

which is an equation of the circle with center at $(2, 0)$ and radius 2. ∎

EXERCISES   8.4

In Exercises 1 through 4, plot the point having the given set of polar coordinates.

1. (a) $(3, \frac{1}{6}\pi)$; (b) $(2, \frac{2}{3}\pi)$; (c) $(1, \pi)$; (d) $(4, \frac{5}{4}\pi)$;
   (e) $(5, \frac{11}{6}\pi)$
2. (a) $(4, \frac{1}{3}\pi)$; (b) $(3, \frac{3}{4}\pi)$; (c) $(1, \frac{7}{6}\pi)$; (d) $(2, \frac{3}{2}\pi)$;
   (e) $(5, \frac{5}{3}\pi)$
3. (a) $(1, -\frac{1}{4}\pi)$; (b) $(3, -\frac{5}{8}\pi)$; (c) $(-1, \frac{1}{4}\pi)$;
   (d) $(-3, \frac{5}{8}\pi)$; (e) $(-2, -\frac{1}{2}\pi)$
4. (a) $(5, -\frac{2}{3}\pi)$; (b) $(2, -\frac{7}{6}\pi)$; (c) $(-5, \frac{2}{3}\pi)$;
   (d) $(-2, \frac{7}{6}\pi)$; (e) $(-4, -\frac{5}{4}\pi)$

In Exercises 5 through 10, plot the point having the given set of polar coordinates; then find another set of polar coordinates for the same point for which (a) $r < 0$ and $0 \leq \theta < 2\pi$; (b) $r > 0$ and $-2\pi < \theta \leq 0$; (c) $r < 0$ and $-2\pi < \theta \leq 0$.

5. $(4, \frac{1}{4}\pi)$ 6. $(3, \frac{5}{6}\pi)$
7. $(2, \frac{1}{2}\pi)$ 8. $(3, \frac{3}{2}\pi)$
9. $(\sqrt{2}, \frac{7}{4}\pi)$ 10. $(2, \frac{4}{3}\pi)$
11. Plot the point having the polar coordinates
    $(2, -\frac{1}{4}\pi)$. Find another set of polar coordinates for
    this point for which
    (a) $r < 0$ and $0 \leq \theta < 2\pi$;
    (b) $r < 0$ and $-2\pi < \theta \leq 0$;
    (c) $r > 0$ and $2\pi \leq \theta < 4\pi$.
12. Plot the point having the polar coordinates
    $(-3, -\frac{2}{3}\pi)$. Find another set of polar coordinates
    for this point for which
    (a) $r > 0$ and $0 \leq \theta < 2\pi$;
    (b) $r > 0$ and $-2\pi < \theta \leq 0$;
    (c) $r < 0$ and $2\pi \leq \theta < 4\pi$.

In Exercises 13 through 20, plot the point having the given set of polar coordinates; then give two other sets of polar coordinates of the same point, one with the same value of r and one with an r having opposite sign.

13. $(3, -\frac{2}{3}\pi)$ 14. $(\sqrt{2}, -\frac{1}{4}\pi)$
15. $(-4, \frac{5}{6}\pi)$ 16. $(-2, \frac{4}{3}\pi)$
17. $(-2, -\frac{5}{4}\pi)$ 18. $(-3, -\pi)$
19. $(2, 6)$ 20. $(5, \frac{1}{6}\pi)$

In Exercises 21 and 22, find the rectangular cartesian coordinates of the points whose polar coordinates are given.

21. (a) $(3, \pi)$; (b) $(\sqrt{2}, -\frac{3}{4}\pi)$; (c) $(-4, \frac{2}{3}\pi)$;
    (d) $(-1, -\frac{7}{6}\pi)$
22. (a) $(-2, -\frac{1}{2}\pi)$; (b) $(-1, \frac{1}{4}\pi)$; (c) $(2, -\frac{7}{6}\pi)$; (d) $(2, \frac{7}{4}\pi)$

In Exercises 23 and 24, find a set of polar coordinates of the points whose rectangular cartesian coordinates are given. Take $r > 0$ and $0 \leq \theta < 2\pi$.

23. (a) $(1, -1)$; (b) $(-\sqrt{3}, 1)$; (c) $(2, 2)$; (d) $(-5, 0)$
24. (a) $(3, -3)$; (b) $(-1, \sqrt{3})$; (c) $(0, -2)$;
    (d) $(-2, -2\sqrt{3})$

In Exercises 25 through 34, find a polar equation of the graph having the given cartesian equation.

25. $x^2 + y^2 = a^2$ 26. $x + y = 1$
27. $y^2 = 4(x + 1)$ 28. $x^3 = 4y^2$
29. $x^2 = 6y - y^2$ 30. $x^2 - y^2 = 16$
31. $(x^2 + y^2)^2 = 4(x^2 - y^2)$ 32. $2xy = a^2$
33. $x^3 + y^3 - 3axy = 0$ 34. $y = \dfrac{2x}{x^2 + 1}$

In Exercises 35 through 44, find a cartesian equation of the graph having the given polar equation.

35. $r^2 = 2 \sin 2\theta$ 36. $r^2 \cos 2\theta = 10$
37. $r^2 = \cos \theta$ 38. $r^2 = 4 \cos 2\theta$
39. $r^2 = \theta$ 40. $r = 2 \sin 3\theta$
41. $r \cos \theta = -1$ 42. $r^6 = r^2 \cos^2 \theta$
43. $r = \dfrac{6}{2 - 3 \sin \theta}$ 44. $r = \dfrac{4}{3 - 2 \cos \theta}$

8.5 Polar Form of Complex Numbers

Complex numbers have arisen naturally in problem solving in previous chapters of this book. They were introduced in Section 1.8 and occurred as roots of equations in Chapter 2. If you continue your study of mathematics to more advanced courses, you will learn that complex numbers have significant importance in both the theoretical and applied aspects of the subject and that *imaginary* numbers can be put to *real* use. In this section we give a geometric representation of complex numbers and show how these numbers can be expressed in terms of polar coordinates. Then, in Section 8.6, we apply the *polar form* to compute powers and roots of complex numbers.

The set of complex numbers can be represented by points in a rectangular coordinate system. In such a representation, the horizontal axis is called the **real axis,** and the vertical axis is called the **imaginary axis.** The points of the plane, called the **complex plane,** are then placed in one-to-one correspondence with the complex numbers. The geometric representation of the complex number $a + bi$ is the point $P(a, b)$ in the complex plane, and the point is called the **graph** of the number. Refer to Figure 1.

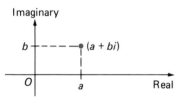

FIGURE 1

Example   1 Show the geometric representation of each of the following complex numbers as a point in a rectangular coordinate system: $3 + 5i$; $-3 + 5i$; $-3 - 5i$; $3 - 5i$; i; $-2i$; 4; and -6.

Solution   The points are shown in Figure 2.   ∎

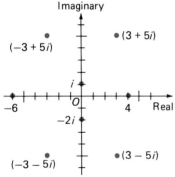

FIGURE 2

Observe that any real number is represented by a point on the real axis and any pure imaginary number is represented by a point on the imaginary axis. The geometric representations of a complex number $a + bi$ and its conjugate $a - bi$ are points that are symmetric with respect to the real axis. See Figure 3.

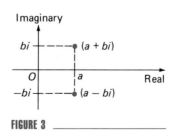

FIGURE 3

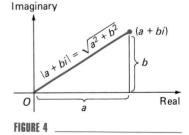

FIGURE 4

If $a + bi$ is an arbitrary complex number, then the distance in the complex plane from the origin to the graph of $a + bi$ is $\sqrt{a^2 + b^2}$. Refer to Figure 4. This number is called the *absolute value* or *modulus* of $a + bi$.

Definition

Absolute Value of a Complex Number
The **absolute value** of the complex number $a + bi$, denoted by $|a + bi|$, is given by

$$|a + bi| = \sqrt{a^2 + b^2}$$

Example   2   Write each of the following numbers without absolute value bars.

a) $|6 - i|$ b) $|-2 + 2i|$ c) $|-4 - 3i|$ d) $|5i|$

Solution

a) $|6 - i| = \sqrt{6^2 + (-1)^2}$
$ = \sqrt{37}$

b) $|-3 + 2i| = \sqrt{(-3)^2 + 2^2}$
$ = \sqrt{13}$

c) $|-4 - 3i| = \sqrt{(-4)^2 + (-3)^2}$
$ = \sqrt{25}$
$ = 5$

d) $|5i| = \sqrt{0^2 + 5^2}$
$ = \sqrt{25}$
$ = 5$ ∎

We do not refer to one complex number with nonzero imaginary part being greater than, or less than, another complex number. However, we can use the symbols $<$ and $>$ with the absolute values of complex numbers because they are real numbers. If $|z_1| < |z_2|$, then the point in the complex plane representing z_1 is closer to the origin than the point representing z_2. Furthermore, if $|z_1| = |z_2|$, the points representing z_1 and z_2 are at the same distance from the origin of the complex plane; that is, they lie on a circle with center at the origin.

If the point (a, b) representing $a + bi$ is expressed in polar coordinates, with $r \geq 0$, as indicated in Figure 5, then

$$a = r \cos \theta \quad \text{and} \quad b = r \sin \theta$$

Therefore

$$a + bi = r \cos \theta + (r \sin \theta)i$$

$$a + bi = r(\cos \theta + i \sin \theta)$$

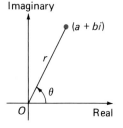

FIGURE 5

The right side of this equation is called the **polar form** (or **trigonometric form**) of the complex number $a + bi$. It is sometimes abbreviated as r cis θ, where *cis* comes from cosine, i, and sine. In contrast, $a + bi$ is called the **cartesian form** (or **algebraic form**).

Because we required r to be nonnegative,

$$r = |a + bi|$$

The angle θ is called an **argument** (or **amplitude**) of $a + bi$. The argument may be measured in either degrees or radians. Because

$$\sin(\theta + 2k\pi) = \sin\theta \qquad \text{and} \qquad \cos(\theta + 2k\pi) = \cos\theta$$

where $k \in J$, then if θ is an argument of $a + bi$, so is an angle $\theta + 2k\pi$, $k \in J$. The argument θ, for which $0 \leq \theta < 2\pi$, is called the **principal argument**. When the principal argument is used, the complex number is said to be in **standard polar form.**

Illustration 1

Let us express the complex number $\sqrt{3} + i$ in standard polar form. Figure 6 shows the geometric representation of this number. We first determine r and θ from the formulas

$$r = |a + bi| \qquad \cos\theta = \frac{a}{r} \qquad \sin\theta = \frac{b}{r}$$

$$= \sqrt{a^2 + b^2}$$

Because $a = \sqrt{3}$ and $b = 1$,

$$r = |\sqrt{3} + i| \qquad \cos\theta = \frac{\sqrt{3}}{2} \qquad \sin\theta = \frac{1}{2}$$

$$= \sqrt{(\sqrt{3})^2 + 1^2}$$

$$= 2$$

From the values of $\sin\theta$ and $\cos\theta$, we obtain $\theta = \frac{1}{6}\pi$. Therefore

$$|\sqrt{3} + i| = r(\cos\theta + i\sin\theta)$$
$$= 2(\cos\tfrac{1}{6}\pi + i\sin\tfrac{1}{6}\pi)$$

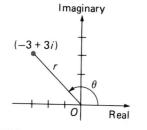

FIGURE 6

Illustration 2

To express the complex number $-3 + 3i$ in standard polar form, we first determine r and θ with $a = -3$ and $b = 3$. See Figure 7. We have

$$r = \sqrt{a^2 + b^2} \qquad \cos\theta = \frac{a}{r} \qquad \sin\theta = \frac{b}{r}$$

$$= \sqrt{(-3)^2 + 3^2} \qquad = \frac{-3}{3\sqrt{2}} \qquad = \frac{3}{3\sqrt{2}}$$

$$= 3\sqrt{2} \qquad = -\frac{1}{\sqrt{2}} \qquad = \frac{1}{\sqrt{2}}$$

From the values of $\cos\theta$ and $\sin\theta$, it follows that $\theta = \frac{3}{4}\pi$. Thus

$$-3 + 3i = r(\cos\theta + i\sin\theta)$$
$$= 3\sqrt{2}(\cos\tfrac{3}{4}\pi + i\sin\tfrac{3}{4}\pi)$$

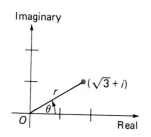

FIGURE 7

If a complex number is real, then in cartesian form it is $a + 0i$. For this number, $r = |a|$. If $a > 0$, $\cos \theta = 1$, $\sin \theta = 0$, and so $\theta = 0$; if $a < 0$, $\cos \theta = -1$, $\sin \theta = 0$, and so $\theta = \pi$. Therefore, in polar form

$$a = |a|(\cos 0 + i \sin 0) \quad \text{if } a > 0 \qquad a = |a|(\cos \pi + i \sin \pi) \quad \text{if } a < 0$$

For the real number 0, the polar form is $0(\cos \theta + i \sin \theta)$ where θ is any angle.

For the pure imaginary number $0 + bi$, $r = |b|$. If $b > 0$, $\cos \theta = 0$, $\sin \theta = 1$, and so $\theta = \frac{1}{2}\pi$; if $b < 0$, $\cos \theta = 0$, $\sin \theta = -1$, and so $\theta = \frac{3}{2}\pi$. Hence in polar form

$$bi = |b|(\cos \tfrac{1}{2}\pi + i \sin \tfrac{1}{2}\pi) \quad \text{if } b > 0 \qquad bi = |b|(\cos \tfrac{3}{2}\pi + i \sin \tfrac{3}{2}\pi) \quad \text{if } b < 0$$

Example 3 Express the following numbers in standard polar form.

a) $4 - 4\sqrt{3}i$ b) $2 + 5i$

Solution We use the formulas

$$r = |a + bi| \qquad \cos \theta = \frac{a}{r} \qquad \sin \theta = \frac{b}{r}$$
$$= \sqrt{a^2 + b^2}$$

a) If $a + bi = 4 - 4\sqrt{3}i$, then $a = 4$ and $b = -4\sqrt{3}$. The point representing the complex number is shown in Figure 8. From the formulas we get

$$r = \sqrt{16 + 48} \qquad \cos \theta = \frac{4}{8} \qquad \sin \theta = \frac{-4\sqrt{3}}{8}$$
$$= 8 \qquad\qquad\quad = \frac{1}{2} \qquad\qquad = -\frac{\sqrt{3}}{2}$$

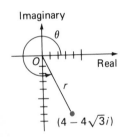

Imaginary

θ

Real

r

$(4 - 4\sqrt{3}i)$

FIGURE 8

From the values of $\cos \theta$ and $\sin \theta$ we determine that $\theta = \frac{5}{3}\pi$. Therefore

$$4 - 4\sqrt{3}i = r(\cos \theta + i \sin \theta)$$
$$= 8(\cos \tfrac{5}{3}\pi + i \sin \tfrac{5}{3}\pi)$$

b) If $a + bi = 2 + 5i$, then $a = 2$ and $b = 5$. See Figure 9. We compute r, $\cos \theta$, and $\sin \theta$ from the formulas and obtain

$$r = \sqrt{4 + 25} \qquad \cos \theta = \frac{2}{\sqrt{29}} \qquad \sin \theta = \frac{5}{\sqrt{29}}$$
$$= \sqrt{29}$$

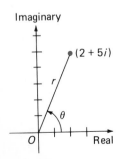

Imaginary

$(2 + 5i)$

r

θ

O Real

FIGURE 9

Because $\cos \theta > 0$ and $\sin \theta > 0$, θ is in the first quadrant. Furthermore, because $\tan \theta = \dfrac{b}{a}$, $\tan \theta = \frac{5}{2}$. Thus $\theta = \arctan \frac{5}{2}$. Therefore

$$2 + 5i = r(\cos \theta + i \sin \theta)$$
$$= \sqrt{29}[\cos(\arctan \tfrac{5}{2}) + i \sin(\arctan \tfrac{5}{2})]$$

Because arc tan $\frac{5}{2} \approx 1.19$ (or 68.2°), we can write

$$2 + 5i \approx \sqrt{29}(\cos 1.19 + i \sin 1.19)$$

or

$$2 + 5i \approx \sqrt{29}(\cos 68.2° + i \sin 68.2°)$$ ∎

The next illustration shows the conversion of the polar form of a complex number to the equivalent cartesian form.

Illustration 3

The graph of the complex number $6(\cos \frac{7}{6}\pi + i \sin \frac{7}{6}\pi)$ appears in Figure 10.

$$6\left(\cos \frac{7}{6}\pi + i \sin \frac{7}{6}\pi\right) = 6\left[-\frac{\sqrt{3}}{2} + i\left(-\frac{1}{2}\right)\right]$$

$$= -3\sqrt{3} - 3i$$ ∎

FIGURE 10

To find formulas for the product and quotient of two complex numbers when the numbers are expressed in polar form, we apply trigonometric identities. The following theorem summarizes the results.

Theorem 1

If

$$z_1 = r_1(\cos \theta_1 + i \sin \theta_1) \quad \text{and} \quad z_2 = r_2(\cos \theta_2 + i \sin \theta_2),$$

then

(i) $z_1 \cdot z_2 = r_1 r_2[\cos(\theta_1 + \theta_2) + i \sin(\theta_1 + \theta_2)]$

(ii) $\dfrac{z_1}{z_2} = \dfrac{r_1}{r_2}[\cos(\theta_1 - \theta_2) + i \sin(\theta_1 - \theta_2)]$

Proof of (i)

$$z_1 \cdot z_2 = r_1(\cos \theta_1 + i \sin \theta_1) \cdot r_2(\cos \theta_2 + i \sin \theta_2)$$
$$= r_1 r_2[\cos \theta_1 \cdot \cos \theta_2 + \cos \theta_1(i \sin \theta_2) + (i \sin \theta_1) \cos \theta_2 + (i \sin \theta_1)(i \sin \theta_2)]$$
$$= r_1 r_2[\cos \theta_1 \cos \theta_2 + i \cos \theta_1 \sin \theta_2 + i \sin \theta_1 \cos \theta_2 + i^2 \sin \theta_1 \sin \theta_2]$$
$$= r_1 r_2[(\cos \theta_1 \cos \theta_2 - \sin \theta_1 \sin \theta_2) + i(\cos \theta_1 \sin \theta_2 + \sin \theta_1 \cos \theta_2)]$$
$$= r_1 r_2[\cos(\theta_1 + \theta_2) + i \sin(\theta_1 + \theta_2)]$$ ∎

The proof of part (ii) is similar.

Illustration 4

Let $z_1 = -3\sqrt{3} - 3i$ and $z_2 = 4 - 4\sqrt{3}i$. From Illustration 3,

$$z_1 = 6(\cos \frac{7}{6}\pi + i \sin \frac{7}{6}\pi)$$

and from Example 3(a),

$$z_2 = 8(\cos \tfrac{5}{3}\pi + i \sin \tfrac{5}{3}\pi)$$

Applying Theorem 1(i), we obtain

$$z_1 \cdot z_2 = 6 \cdot 8[\cos(\tfrac{7}{6}\pi + \tfrac{5}{3}\pi) + i \sin(\tfrac{7}{6}\pi + \tfrac{5}{3}\pi)]$$
$$= 48(\cos \tfrac{17}{6}\pi + i \sin \tfrac{17}{6}\pi)$$
$$= 48\left(-\frac{\sqrt{3}}{2} + \frac{1}{2}i\right)$$
$$= -24\sqrt{3} + 24i$$

We can check this result by using the procedure of Section 1.8. We have

$$z_1 \cdot z_2 = (-3\sqrt{3} - 3i)(4 - 4\sqrt{3}i)$$
$$= -12\sqrt{3} + 36i - 12i + 12\sqrt{3}i^2$$
$$= -24\sqrt{3} + 24i$$

From Theorem 1(ii) we get

$$\frac{z_1}{z_2} = \frac{6}{8}\left[\cos\left(\frac{7}{6}\pi - \frac{5}{3}\pi\right) + i \sin\left(\frac{7}{6}\pi - \frac{5}{3}\pi\right)\right]$$
$$= \tfrac{3}{4}[\cos(-\tfrac{1}{2}\pi) + i \sin(-\tfrac{1}{2}\pi)]$$
$$= \tfrac{3}{4}[0 + i(-1)]$$
$$= -\tfrac{3}{4}i$$

As a check, we have

$$\frac{z_1}{z_2} = \frac{-3\sqrt{3} - 3i}{4 - 4\sqrt{3}i}$$
$$= \frac{(-3\sqrt{3} - 3i)(4 + 4\sqrt{3}i)}{(4 - 4\sqrt{3}i)(4 + 4\sqrt{3}i)}$$
$$= \frac{-12\sqrt{3} - 36i - 12i - 12\sqrt{3}i^2}{16 - 48i^2}$$
$$= \frac{-48i}{64}$$
$$= -\frac{3}{4}i$$

■

Example 4 Express

$$\tfrac{2}{3}(\cos \tfrac{2}{9}\pi + i \sin \tfrac{2}{9}\pi) \cdot 6(\cos \tfrac{19}{36}\pi + i \sin \tfrac{19}{36}\pi)$$

in the form $a + bi$

Solution   From Theorem 1(i)

$$\tfrac{2}{3}(\cos \tfrac{2}{9}\pi + i \sin \tfrac{2}{9}\pi) \cdot 6(\cos \tfrac{19}{36}\pi + i \sin \tfrac{19}{36}\pi)$$

$$= \tfrac{2}{3} \cdot 6[\cos(\tfrac{2}{9}\pi + \tfrac{19}{36}\pi) + i \sin(\tfrac{2}{9}\pi + \tfrac{19}{36}\pi)]$$

$$= 4(\cos \tfrac{3}{4}\pi + i \sin \tfrac{3}{4}\pi)$$

$$= 4\left[-\frac{\sqrt{2}}{2} + i\left(\frac{\sqrt{2}}{2}\right) \right]$$

$$= -2\sqrt{2} + 2\sqrt{2}i$$

■

Example   5   Express the following quotient in the form $a + bi$:

$$\frac{4(\cos 345° + i \sin 345°)}{5(\cos 105° + i \sin 105°)}$$

Solution   From Theorem 1(ii)

$$\frac{4(\cos 345° + i \sin 345°)}{5(\cos 105° + i \sin 105°)} = \frac{4}{5}\left[\cos(345° - 105°) + i \sin(345° - 105°) \right]$$

$$= \frac{4}{5}\left[\cos 240° + i \sin 240° \right]$$

$$= \frac{4}{5}\left[-\frac{1}{2} + i\left(-\frac{\sqrt{3}}{2}\right) \right]$$

$$= -\frac{2}{5} - \frac{2\sqrt{3}}{5}i$$

■

EXERCISES   8.5

In Exercises 1 through 8, show the geometric representation of the complex number as a point in the complex plane.

1. (a) $4 + 5i$; (b) $7 - 8i$ 2. (a) $7 + i$; (b) $-4 + 9i$
3. (a) $-1 + 6i$; (b) $-3 - i$
4. (a) $2 - 6i$; (b) $-5 - 3i$
5. (a) 2; (b) $-6i$ 6. (a) -4; (b) $3i$
7. (a) $2 - 6i$; (b) $2 + 6i$
8. (a) $-4 + 3i$; (b) $-4 - 3i$

In Exercises 9 through 16, write the expression without absolute value bars.

9. (a) $|5 + 2i|$; (b) $|-1 + 2i|$
10. (a) $|8 - 3i|$; (b) $|-2 - 5i|$
11. (a) $|3|$; (b) $|3i|$ 12. (a) $|5|$; (b) $|5i|$
13. (a) $|-2|$; (b) $|-2i|$ 14. (a) $|-7|$; (b) $|-7i|$
15. (a) $|-6 + 8i|$; (b) $|-6 - 8i|$
16. (a) $|6 + i|$; (b) $|6 - i|$

In Exercises 17 through 22, show the geometric representation of the complex number in the complex plane, and write it in cartesian form.

17. (a) $3(\cos \tfrac{1}{3}\pi + i \sin \tfrac{1}{3}\pi)$; (b) $3(\cos \tfrac{4}{3}\pi + i \sin \tfrac{4}{3}\pi)$
18. (a) $4(\cos \tfrac{1}{4}\pi + i \sin \tfrac{1}{4}\pi)$; (b) $4(\cos \tfrac{3}{4}\pi + i \sin \tfrac{3}{4}\pi)$
19. (a) $6(\cos 150° + i \sin 150°)$;
    (b) $6(\cos 330° + i \sin 330°)$
20. (a) $5(\cos 210° + i \sin 210°)$;
    (b) $5(\cos 300° + i \sin 300°)$
21. (a) $2(\cos \tfrac{1}{2}\pi + i \sin \tfrac{1}{2}\pi)$; (b) $\tfrac{2}{3}(\cos 180° + i \sin 180°)$
22. (a) $3(\cos 0° + i \sin 0°)$; (b) $\tfrac{1}{2}(\cos \tfrac{3}{2}\pi + i \sin \tfrac{3}{2}\pi)$

In Exercises 23 through 26, express the complex number in standard polar form.

23. (a) $4 - 4i$; (b) 6; (c) i
24. (a) $-5 + 5i$; (b) 0; (c) $-4i$
25. (a) $-3\sqrt{3} + 3i$; (b) -7; (c) $-7i$
26. (a) $2 - 2i$; (b) 5; (c) $5i$

In Exercises 27 through 30, express the complex number in standard polar form where the argument is written with (a) inverse function notation, (b) radian measurement to two decimal places, and (c) degree measurement to one-tenth of a degree.

27. $3 + 4i$

28. $5 + i$

29. $1 - 2i$

30. $-3 + 2i$

In Exercises 31 through 38, express the product in the form $a + bi$.

31. $2(\cos 20° + i \sin 20°) \cdot 5(\cos 70° + i \sin 70°)$
32. $(\cos 70° + i \sin 70°) \cdot 4(\cos 110° + i \sin 110°)$
33. $3(\cos \frac{1}{18}\pi + i \sin \frac{1}{18}\pi) \cdot \frac{2}{3}(\cos \frac{5}{18}\pi + i \sin \frac{5}{18}\pi)$
34. $10(\cos \frac{1}{6}\pi + i \sin \frac{1}{6}\pi) \cdot \frac{2}{5}(\cos \frac{1}{12}\pi + i \sin \frac{1}{12}\pi)$
35. $4(\cos \frac{5}{9}\pi + i \sin \frac{5}{9}\pi) \cdot (\cos \frac{7}{36}\pi + i \sin \frac{7}{36}\pi)$
36. $\frac{3}{4}(\cos \frac{7}{18}\pi + i \sin \frac{7}{18}\pi) \cdot \frac{8}{3}(\cos \frac{4}{9}\pi + i \sin \frac{4}{9}\pi)$
37. $5(\cos 75° + i \sin 75°) \cdot \frac{4}{5}(\cos 255° + i \sin 255°)$
38. $2(\cos 145° + i \sin 145°) \cdot 3(\cos 95° + i \sin 95°)$

In Exercises 39 through 46, express the quotient in the form $a + bi$.

39. $6(\cos 70° + i \sin 70°) \div 3(\cos 40° + i \sin 40°)$
40. $4(\cos 65° + i \sin 65°) \div 2(\cos 20° + i \sin 20°)$
41. $8(\cos \frac{4}{3}\pi + i \sin \frac{4}{3}\pi) \div 2(\cos \frac{7}{12}\pi + i \sin \frac{7}{12}\pi)$
42. $2(\cos \frac{17}{12}\pi + i \sin \frac{17}{12}\pi) \div 6(\cos \frac{3}{4}\pi + i \sin \frac{3}{4}\pi)$
43. $5(\cos \frac{5}{18}\pi + i \sin \frac{5}{18}\pi) \div (\cos \frac{11}{18}\pi + i \sin \frac{11}{18}\pi)$
44. $6(\cos \frac{5}{18}\pi + i \sin \frac{5}{18}\pi) \div \frac{3}{2}(\cos \frac{7}{9}\pi + i \sin \frac{7}{9}\pi)$
45. $\frac{2}{3}(\cos 350° + i \sin 350°) \div (\cos 80° + i \sin 80°)$
46. $3(\cos 310° + i \sin 310°) \div \frac{1}{2}(\cos 85° + i \sin 85°)$

47. Prove that if z is a complex number such that $|z| = 1$, then

$$\frac{1}{z} = \bar{z}$$

(*Hint:* Show that $\cos(-\theta) + i \sin(-\theta)$ is both the conjugate and the reciprocal of $\cos \theta + i \sin \theta$.)

48. Prove that if $z = r(\cos \theta + i \sin \theta)$, then

$$z^2 = r^2(\cos 2\theta + i \sin 2\theta)$$

and

$$z^3 = r^3(\cos 3\theta + i \sin 3\theta)$$

8.6 ## Powers and Roots of Complex Numbers and De Moivre's Theorem

The polar form of complex numbers can be applied to calculate their powers and roots. The procedure is provided by a theorem called De Moivre's theorem, named for the mathematician Abraham de Moivre (1667–1754), who was born in France but lived most of his life in London. He was a friend of Sir Isaac Newton, one of the inventors of calculus. To lead up to the theorem, we consider the complex number $z = r(\cos \theta + i \sin \theta)$ and compute some positive-integer powers of z. Certainly

$$z^1 = r(\cos \theta + i \sin \theta) \tag{1}$$

From Theorem 1(i) of Section 8.5,

$$
\begin{aligned}
z^2 &= [r(\cos \theta + i \sin \theta][r(\cos \theta + i \sin \theta)] \\
&= r^2[\cos(\theta + \theta) + i \sin(\theta + \theta)] \\
&= r^2(\cos 2\theta + i \sin 2\theta) \\
z^3 &= [r(\cos \theta + i \sin \theta)][r(\cos \theta + i \sin \theta)]^2
\end{aligned}
\tag{2}
$$

Substituting from (2) in the right side, we get

$$
\begin{aligned}
z^3 &= [r(\cos \theta + i \sin \theta)][r^2(\cos 2\theta + i \sin 2\theta)] \\
&= r^3[\cos(\theta + 2\theta) + i \sin(\theta + 2\theta)] \\
&= r^3(\cos 3\theta + i \sin 3\theta)
\end{aligned}
\tag{3}
$$

Illustration 1

a) From (2),

$$[4(\cos 60° + i \sin 60°)]^2 = 4^2[\cos(2 \cdot 60°) + i \sin(2 \cdot 60°)]$$
$$= 16(\cos 120° + i \sin 120°)$$
$$= 16\left[-\frac{1}{2} + i\left(\frac{\sqrt{3}}{2}\right)\right]$$
$$= -8 + 8\sqrt{3}\, i$$

b) From (3),

$$[4(\cos 60° + i \sin 60°)]^3 = 4^3[\cos(3 \cdot 60°) + i \sin(3 \cdot 60°)]$$
$$= 64(\cos 180° + i \sin 180°)$$
$$= 64[-1 + i(0)]$$
$$= -64$$

■

Equations (1), (2), and (3) are the special cases of De Moivre's theorem when n is 1, 2, and 3.

De Moivre's Theorem for Positive Integers

If n is any positive integer, then

$$[r(\cos \theta + i \sin \theta)]^n = r^n(\cos n\theta + i \sin n\theta)$$

The proof of De Moivre's theorem requires mathematical induction, which is discussed in Section 11.2.

Example   1 Use De Moivre's theorem to find the following.

a) $(1 + i)^5$ b) $(\sqrt{3} - 1)^6$

Solution   a) We first write $1 + i$ in polar form. If $1 + i = a + bi$, then $a = 1$ and $b = 1$. Thus

$$r = \sqrt{a^2 + b^2} \qquad \cos \theta = \frac{a}{r} \qquad \sin \theta = \frac{b}{r}$$
$$= \sqrt{1^2 + 1^2}$$
$$= \sqrt{2} \qquad\qquad = \frac{1}{\sqrt{2}} \qquad = \frac{1}{\sqrt{2}}$$

Hence $\theta = \frac{1}{4}\pi$. Therefore

$$1 + i = \sqrt{2}(\cos \tfrac{1}{4}\pi + i \sin \tfrac{1}{4}\pi)$$

From De Moivre's theorem,

$$(1 + i)^5 = (\sqrt{2})^5(\cos \tfrac{5}{4}\pi + i \sin \tfrac{5}{4}\pi)$$

$$= 4\sqrt{2}\left[-\frac{1}{\sqrt{2}} + i\left(-\frac{1}{\sqrt{2}}\right)\right]$$

$$= -4 - 4i$$

b) If $\sqrt{3} - i = a + bi$, then $a = \sqrt{3}$ and $b = -1$. Hence

$$r = \sqrt{(\sqrt{3})^2 + (-1)^2} \qquad \cos \theta = \frac{a}{r} \qquad \sin \theta = \frac{b}{r}$$

$$= 2$$

$$= \frac{\sqrt{3}}{2} \qquad\qquad = -\frac{1}{2}$$

Thus $\theta = \tfrac{11}{6}\pi$. Therefore

$$\sqrt{3} - i = 2(\cos \tfrac{11}{6}\pi + i \sin \tfrac{11}{6}\pi)$$

From De Moivre's theorem,

$$(\sqrt{3} - i)^6 = 2^6[\cos (6 \cdot \tfrac{11}{6}\pi) + i \sin(6 \cdot \tfrac{11}{6}\pi)]$$
$$= 64(\cos 11\pi + i \sin 11\pi)$$
$$= 64(-1 + i \cdot 0)$$
$$= -64 \qquad\qquad\qquad\blacksquare$$

We wish to define zero and negative-integer exponents of complex numbers in such a way that De Moivre's theorem holds for these integers. If the theorem is to be valid for $n = 0$, then if $z = r(\cos \theta + i \sin \theta)$,

$$z^0 = r^0(\cos 0 \cdot \theta + i \sin 0 \cdot \theta)$$
$$= 1(\cos 0 + i \sin 0)$$
$$= 1(1 + i \cdot 0)$$
$$= 1$$

If $z = r(\cos \theta + i \sin \theta)$ and the theorem is to be valid for $-n$ where $n > 0$, then

$$z^{-n} = r^{-n}[\cos(-n\theta) + i \sin(-n\theta)]$$

Because $\cos(-n\theta) = \cos n\theta$ and $\sin(-n\theta) = -\sin n\theta$, this equation becomes

$$z^{-n} = \frac{\cos n\theta - i \sin n\theta}{r^n}$$

We multiply the numerator and denominator by the conjugate of the numerator, and we have

$$z^{-n} = \frac{(\cos n\theta - i \sin n\theta)(\cos n\theta + i \sin n\theta)}{r^n(\cos n\theta + i \sin n\theta)}$$

$$= \frac{\cos^2 n\theta - i^2 \sin^2 n\theta}{r^n(\cos n\theta + i \sin n\theta)}$$

$$= \frac{\cos^2 n\theta + \sin^2 n\theta}{r^n(\cos n\theta + i \sin n\theta)}$$

$$= \frac{1}{r^n(\cos n\theta + i \sin n\theta)}$$

$$= \frac{1}{z^n}$$

Thus in order for De Moivre's theorem to hold for zero and negative-integer exponents, we make the following definition, which is consistent with the definition of zero and negative-integer exponents of real numbers.

Definition

Zero and Negative-Integer Exponents of Complex Numbers
If z is a complex number other than $0 + 0i$, and n is a positive integer,
  (i) $z^0 = 1$

  (ii) $z^{-n} = \dfrac{1}{z^n}$

From De Moivre's theorem for positive integers and the above definition, we have De Moivre's theorem for all integers.

De Moivre's Theorem for All Integers

If k is any integer, then

$$[r(\cos \theta + i \sin \theta)]^k = r^k(\cos k\theta + i \sin k\theta)$$

Example   2 Use De Moivre's theorem to find the following.
  a) $(-1 + \sqrt{3}i)^{-4}$ b) $(-1 - i)^{-14}$

Solution   a) If $-1 + \sqrt{3}i = a + bi$, then $a = -1$ and $b = \sqrt{3}$. Hence

$$r = \sqrt{(-1)^2 + (\sqrt{3})^2} \qquad \cos \theta = \frac{a}{r} \qquad \sin \theta = \frac{b}{r}$$
$$= 2 \qquad\qquad\qquad = -\frac{1}{2} \qquad\qquad = \frac{\sqrt{3}}{2}$$

Thus $\theta = \frac{2}{3}\pi$. Therefore

$$-1 + \sqrt{3}i = 2(\cos \tfrac{2}{3}\pi + i \sin \tfrac{2}{3}\pi)$$

From De Moivre's theorem

$$(-1 + \sqrt{3}i)^{-4} = 2^{-4}[\cos(-4 \cdot \tfrac{2}{3}\pi) + i\sin(-4 \cdot \tfrac{2}{3}\pi)]$$

$$= \tfrac{1}{16}[\cos(-\tfrac{8}{3}\pi) + i\sin(-\tfrac{8}{3}\pi)]$$

$$= \frac{1}{16}\left[-\frac{1}{2} + i\left(-\frac{\sqrt{3}}{2}\right)\right]$$

$$= -\frac{1}{32} - \frac{\sqrt{3}}{32}i$$

b) If $-1 - i = a + bi$, then $a = -1$ and $b = -1$. Thus

$$r = \sqrt{(-1)^2 + (-1)^2} \qquad \cos\theta = \frac{a}{r} \qquad\qquad \sin\theta = \frac{b}{r}$$

$$= \sqrt{2} \qquad\qquad\qquad\qquad = -\frac{1}{\sqrt{2}} \qquad\qquad\quad = -\frac{1}{\sqrt{2}}$$

Therefore $\theta = \tfrac{5}{4}\pi$. Hence

$$-1 - i = \sqrt{2}(\cos\tfrac{5}{4}\pi + i\sin\tfrac{5}{4}\pi)$$

From De Moivre's theorem,

$$(-1 - i)^{-14} = (\sqrt{2})^{-14}\left[\cos\left(-14 \cdot \frac{5}{4}\pi\right) + i\sin\left(-14 \cdot \frac{5}{4}\pi\right)\right]$$

$$= \frac{1}{2^7}\left[\cos\left(-\frac{35}{2}\pi\right) + i\sin\left(-\frac{35}{2}\pi\right)\right]$$

$$= \frac{1}{128}(0 + i \cdot 1)$$

$$= \frac{1}{128}i$$

■

For the treatment of roots of complex numbers, we first define an nth root.

Definition

nth Root of a Complex Number
If z is a nonzero complex number, and n is a positive integer, the number w is said to be an **nth root** of z if

$$w^n = z$$

To find a formula for computing nth roots of complex numbers, assume that $w^n = z$ and let

$$z = r(\cos\theta + i\sin\theta) \qquad \text{and} \qquad w = s(\cos\phi + i\sin\phi)$$

We wish to obtain formulas for s and ϕ in terms of r and θ. Because $w^n = z$,

$$[s(\cos \phi + i \sin \phi)]^n = r(\cos \theta + i \sin \theta)$$

Applying De Moivre's theorem to the left side, we have

$$s^n(\cos n\phi + i \sin n\phi) = r(\cos \theta + i \sin \theta) \tag{4}$$

This equation is an equality of two complex numbers. Therefore the absolute values of these complex numbers are equal; thus

$$s^n = r$$
$$s = r^{1/n}$$

Because $s^n = r$, then from (4),

$$\cos n\phi + i \sin n\phi = \cos \theta + i \sin \theta$$

From this equation and the definition of equality of two complex numbers,

$$\cos n\phi = \cos \theta \quad \text{and} \quad \sin n\phi = \sin \theta$$

Because the period of both the sine and cosine is 2π, these equations are valid if and only if

$$n\phi = \theta + k \cdot 2\pi \quad k \in J$$

Therefore

$$\phi = \frac{\theta + k \cdot 2\pi}{n} \quad k \in J$$

With this value of ϕ and $s = r^{1/n}$, and because $w = s(\cos \phi + i \sin \phi)$, we obtain

$$w = r^{1/n}\left[\cos\left(\frac{\theta + k \cdot 2\pi}{n}\right) + i \sin\left(\frac{\theta + k \cdot 2\pi}{n}\right)\right] \quad k \in J$$

If in this equation k is replaced successively by $0, 1, \ldots, n - 1$, we obtain n distinct values of w which are all nth roots of z. Observe that if k is replaced by an integer greater than $n - 1$ or less than 0, we get no new value of w. For instance, if $k = n$, the right side is

$$r^{1/n}\left[\cos\left(\frac{\theta + n \cdot 2\pi}{n}\right) + i \sin\left(\frac{\theta + n \cdot 2\pi}{n}\right)\right] = r^{1/n}\left[\cos\left(\frac{\theta}{n} + 2\pi\right) + i \sin\left(\frac{\theta}{n} + 2\pi\right)\right]$$

which gives the same value as $k = 0$. We have proved the following theorem.

nth Roots of Complex Numbers

If z is a nonzero complex number, where

$$z = r(\cos \theta + i \sin \theta)$$

and n is a positive integer, then z has exactly n distinct nth roots given by

$$r^{1/n}\left[\cos\left(\frac{\theta + k \cdot 2\pi}{n}\right) + i \sin\left(\frac{\theta + k \cdot 2\pi}{n}\right)\right] \qquad k \in J$$

where k is $0, 1, \ldots, n - 1$.

If θ is measured in degrees, then the formula of the above theorem is written as

$$r^{1/n}\left[\cos\left(\frac{\theta + k \cdot 360°}{n}\right) + i \sin\left(\frac{\theta + k \cdot 360°}{n}\right)\right] \qquad k \in J$$

The absolute value of each of the nth roots of the complex number z is

$$\sqrt{(r^{1/n})^2\left[\cos^2\left(\frac{\theta + k \cdot 2\pi}{n}\right) + \sin^2\left(\frac{\theta + k \cdot 2\pi}{n}\right)\right]} = \sqrt{(r^{1/n})^2(1)}$$

$$= r^{1/n} \qquad \text{because} \qquad r > 0$$

Thus the geometric representations of the nth roots lie on a circle with center at the origin and radius $\sqrt[n]{r}$. Furthermore, because the difference of the arguments of successive nth roots is $\dfrac{2\pi}{n}$, they are equally spaced around the circle.

Illustration 2

To find the four fourth roots of $-8 + 8\sqrt{3}i$, we first write this number in polar form. If $-8 + 8\sqrt{3}i = a + bi$, then $a = -8$ and $b = 8\sqrt{3}$. Therefore

$$r = \sqrt{(-8)^2 + (8\sqrt{3})^2} \qquad \cos \theta = \frac{a}{r} \qquad \sin \theta = \frac{b}{r}$$

$$= \sqrt{64 + 64(3)}$$

$$= 8\sqrt{4} \qquad\qquad\qquad = \frac{-8}{16} \qquad\qquad = \frac{8\sqrt{3}}{16}$$

$$= 16 \qquad\qquad\qquad\qquad = -\frac{1}{2} \qquad\qquad = \frac{\sqrt{3}}{2}$$

Hence $\theta = 120°$. Thus

$$-8 + 8\sqrt{3}i = 16(\cos 120° + i \sin 120°)$$

From the above theorem, the four fourth roots are given by

$$16^{1/4}\left[\cos\left(\frac{120° + k \cdot 360°}{4}\right) + i \sin\left(\frac{120° + k \cdot 360°}{4}\right)\right]$$

$$\Leftrightarrow \qquad 2[\cos(30° + k \cdot 90°) + i \sin(30° + k \cdot 90°)]$$

where k is 0, 1, 2, 3. Replacing k successively by 0, 1, 2, and 3, we obtain

$$2(\cos 30° + i \sin 30°) = \sqrt{3} + i$$
$$2(\cos 120° + i \sin 120°) = -1 + \sqrt{3}i$$
$$2(\cos 210° + i \sin 210°) = -\sqrt{3} - i$$
$$2(\cos 300° + i \sin 300°) = 1 - \sqrt{3}i$$

Observe that if k is replaced by any other integer, one of these four complex numbers is obtained. For instance, if $k = 4$, we get

$$2(\cos 390° + i \sin 390°) = \sqrt{3} + i$$

and if $k = -1$, we get

$$2[\cos(-60°) + i \sin(-60°)] = 1 - \sqrt{3}i$$

The geometric representations of these roots appear in Figure 1. All four lie on the circle having center at the origin and radius 2, and they are equally spaced around the circle. Note that the four fourth roots occur in pairs in which each member of a pair is the negative (additive inverse) of the other member: $\sqrt{3} + i$ and $-\sqrt{3} - i$; $-1 + \sqrt{3}i$ and $1 - \sqrt{3}i$.

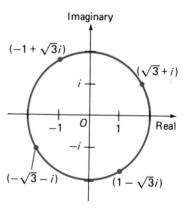

FIGURE 1

Example   3 Find the six sixth roots of 1 and show their geometric representations.

Solution   If $1 = a + bi$, then $a = 1$ and $b = 0$. Therefore

$$r = \sqrt{1^2 + 0^2} \qquad \cos \theta = \frac{a}{r} \qquad \sin \theta = \frac{b}{r}$$
$$= 1 \qquad\qquad = 1 \qquad\qquad = 0$$

Thus $\theta = 0$. Therefore

$$1 = 1(\cos 0 + i \sin 0)$$

The six sixth roots of 1 are given by

$$1^{1/6}\left[\cos\left(\frac{0 + k \cdot 2\pi}{6}\right) + i \sin\left(\frac{0 + k \cdot 2\pi}{6}\right)\right]$$

$$\Leftrightarrow \qquad \cos\frac{k}{3}\pi + i \sin\frac{k}{3}\pi$$

where k is 0, 1, 2, 3, 4, 5. Replacing k successively by 0, 1, 2, 3, 4, and 5, we obtain

$$\cos 0 + i \sin 0 = 1$$

$$\cos\frac{1}{3}\pi + i \sin\frac{1}{3}\pi = \frac{1}{2} + \frac{\sqrt{3}}{2}i$$

$$\cos\frac{2}{3}\pi + i \sin\frac{2}{3}\pi = -\frac{1}{2} + \frac{\sqrt{3}}{2}i$$

$$\cos\pi + i \sin\pi = -1$$

$$\cos\frac{4}{3}\pi + i \sin\frac{4}{3}\pi = -\frac{1}{2} - \frac{\sqrt{3}}{2}i$$

$$\cos\frac{5}{3}\pi + i \sin\frac{5}{3}\pi = \frac{1}{2} - \frac{\sqrt{3}}{2}i$$

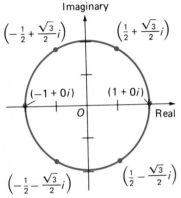

FIGURE 2

Figure 2 shows the geometric representations of the six roots. They all lie on the unit circle and are equally spaced around it.

The complex roots of 1 are called **roots of unity.** Because

$$1 = 1(\cos 0 + i \sin 0)$$

the n nth roots of unity are given by

$$\cos\frac{k \cdot 2\pi}{n} + i \sin\frac{k \cdot 2\pi}{n}$$

where k is 0, 1, 2, . . . , $n - 1$.

Illustration   3

The four fourth roots of unity are given by

$$\cos\frac{k \cdot 2\pi}{4} + i \sin\frac{k \cdot 2\pi}{4}$$

where k is 0, 1, 2, 3. Replacing k successively by 0, 1, 2, and 3, we get

$$\cos 0 + i \sin 0 = 1$$
$$\cos\tfrac{1}{2}\pi + i \sin\tfrac{1}{2}\pi = i$$
$$\cos\pi + i \sin\pi = -1$$
$$\cos\tfrac{3}{2}\pi + i \sin\tfrac{3}{2}\pi = -i$$

The roots of unity can be used to obtain the roots of other real numbers as shown in the following illustration and example.

Illustration 4

We use the four fourth roots of unity to find the four fourth roots of 81. A real fourth root of 81 is 3. If each of the four fourth roots of unity found in Illustration 3 is multiplied by 3, we obtain

$$3(1) = 3 \qquad 3(i) = 3i \qquad 3(-1) = -3 \qquad 3(-i) = -3i$$

Thus the four fourth roots of 81 are 3, -3, $3i$, and $-3i$.   ∎

Example 4   Use the six sixth roots of unity found in Example 3 to obtain the six sixth roots of 64.

Solution   A real sixth root of 64 is 2. If each of the six sixth roots of unity is multiplied by 2, we get the six sixth roots of 64:

$$2(1) = 2$$

$$2\left(\frac{1}{2} + \frac{\sqrt{3}}{2}i\right) = 1 + \sqrt{3}i$$

$$2\left(-\frac{1}{2} + \frac{\sqrt{3}}{2}i\right) = -1 + \sqrt{3}i$$

$$2(-1) = -2$$

$$2\left(-\frac{1}{2} - \frac{\sqrt{3}}{2}i\right) = -1 - \sqrt{3}i$$

$$2\left(\frac{1}{2} - \frac{\sqrt{3}}{2}i\right) = 1 - \sqrt{3}i$$

∎

Example 5   Express each of the three cube roots of $3 + 4i$ in the form $a + bi$. Approximate the values of a and b to two decimal places.

Solution   If $3 + 4i = a + bi$, then $a = 3$ and $b = 4$. Thus

$$r = \sqrt{3^2 + 4^2} \qquad \cos \theta = \frac{a}{r} \qquad \sin \theta = \frac{b}{r} \qquad \tan \theta = \frac{b}{a}$$

$$= 5 \qquad\qquad = \frac{3}{5} \qquad\qquad = \frac{4}{5} \qquad\qquad = \frac{4}{3}$$

Hence $\theta = \arctan \frac{4}{3}$. Therefore

$$3 + 4i = 5[\cos(\arctan \tfrac{4}{3}) + i \sin(\arctan \tfrac{4}{3})]$$

Thus the three cube roots of $3 + 4i$ are given by

$$5^{1/3}\left[\cos\left(\frac{\arctan \frac{4}{3} + k \cdot 2\pi}{3}\right) + i \sin\left(\frac{\arctan \frac{4}{3} + k \cdot 2\pi}{3}\right)\right]$$

where k is 0, 1, 2. Replacing k successively by 0, 1, and 2, and using a calculator to compute the approximate values, we obtain

$$5^{1/3}[\cos \tfrac{1}{3}(\text{arc tan } \tfrac{4}{3}) + i \sin \tfrac{1}{3}(\text{arc tan } \tfrac{4}{3})] \approx 1.710[0.953 + i(0.304)]$$
$$\approx 1.63 + 0.52i$$
$$5^{1/3}[\cos \tfrac{1}{3}(\text{arc tan } \tfrac{4}{3} + 2\pi) + i \sin \tfrac{1}{3}(\text{arc tan } \tfrac{4}{3} + 2\pi)] \approx 1.710[-0.740 + i(0.673)]$$
$$\approx -1.26 + 1.15i$$
$$5^{1/3}[\cos \tfrac{1}{3}(\text{arc tan } \tfrac{4}{3} + 4\pi) + i \sin \tfrac{1}{3}(\text{arc tan } \tfrac{4}{3} + 4\pi)] \approx 1.710[-0.213 + i(-0.977)]$$
$$\approx -0.36 - 1.67i \qquad \blacksquare$$

EXERCISES 8.6

In Exercises 1 through 20, use De Moivre's theorem to find the indicated power. Write the answer in cartesian form.

1. $[2(\cos 30° + i \sin 30°)]^2$
2. $[4(\cos 120° + i \sin 120°)]^2$
3. $[4(\cos 120° + i \sin 120°)]^3$
4. $[2(\cos 30° + i \sin 30°)]^3$
5. $(\cos \tfrac{1}{4}\pi + i \sin \tfrac{1}{4}\pi)^4$
6. $[3(\cos \tfrac{3}{4}\pi + i \sin \tfrac{3}{4}\pi)]^5$
7. $[2(\cos 48° + i \sin 48°)]^5$
8. $[(\cos 50° + i \sin 50°)]^6$
9. $(1 + i)^3$
10. $(-1 + i)^3$
11. $(-1 - \sqrt{3}i)^6$
12. $(\sqrt{3} - i)^6$
13. $(-4 + 4i)^5$
14. $\left(\dfrac{1}{4} + \dfrac{\sqrt{3}}{4}i\right)^5$
15. $\left(\dfrac{1}{2} - \dfrac{\sqrt{3}}{2}i\right)^{-3}$
16. $(-3 - 3i)^{-4}$
17. $\left(\dfrac{\sqrt{2}}{2} + \dfrac{\sqrt{2}}{2}i\right)^{-8}$
18. $(2 + 2\sqrt{3}i)^{-5}$
19. $(-3\sqrt{3} + 3i)^{30}$
20. $(4\sqrt{2} - 4\sqrt{2}i)^{40}$

In Exercises 21 through 30, find the indicated roots, write them in cartesian form, and show their geometric representations.

21. The two square roots of $-4i$
22. The two square roots of $25i$
23. The three cube roots of 125
24. The three cube roots of -64
25. The three cube roots of $4 - 4\sqrt{3}i$
26. The three cube roots of $-\dfrac{\sqrt{3}}{2} + \dfrac{1}{2}i$

27. The four fourth roots of $-\dfrac{1}{2} + \dfrac{\sqrt{3}}{2}i$
28. The four fourth roots of $-8 - 8\sqrt{3}i$
29. The five fifth roots of $-16\sqrt{3} - 16i$
30. The six sixth roots of $32\sqrt{3} + 32i$
31. (a) Find the three cube roots of unity. (b) Use the result of part (a) to find the three cube roots of 27.
32. (a) Find the five fifth roots of unity. (b) Use the result of part (a) to find the five fifth roots of 32.
33. (a) Find the eight eighth roots of unity. (b) Use the result of part (a) to find the eight eighth roots of 256.
34. (a) Find the seven seventh roots of unity. (b) Use the result of part (a) to find the seven seventh roots of 10,000,000.

In Exercises 35 through 42, solve the equation for all complex roots.

35. $x^4 + 81 = 0$
36. $x^4 + 16 = 0$
37. $x^3 + i = 0$
38. $x^3 - i = 0$
39. $x^3 - 8i = 0$
40. $x^3 + 27i = 0$
41. $x^5 + 1 = 0$
42. $x^6 + 1 = 0$

In Exercises 43 through 46, express each of the indicated roots in the form $a + bi$. Approximate the values of a and b to two decimal places.

43. The three cube roots of $5 + 2i$
44. The three cube roots of $4 + 7i$
45. The five fifth roots of $-4 + 3i$
46. The five fifth roots of $2 - i$

8.7 Graphs of Equations in Polar Coordinates (Supplementary)

In Section 8.4 we stated that the graph of a polar equation consists of those points, and only those points, having at least one pair of polar coordinates

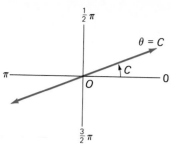

FIGURE 1

that satisfy the equation. In this section we show how to obtain a sketch of such a graph.

The equation

$$\theta = C$$

where C is a constant, is satisfied by all points having polar coordinates (r, C) whatever the value of r. Therefore the graph of this equation is a line containing the pole and making an angle of radian measure C with the polar axis. See Figure 1. The same line is given by the equation

$$\theta = C \pm k\pi$$

where k is any integer.

Illustration   1

a)   The graph of the equation

$$\theta = \frac{1}{4}\pi$$

appears in Figure 2. It is the line containing the pole and making an angle of radian measure $\frac{1}{4}\pi$ with the polar axis. The same line is given by the equations

$$\theta = \frac{5}{4}\pi \qquad \theta = \frac{9}{4}\pi \qquad \theta = -\frac{3}{4}\pi \qquad \theta = -\frac{7}{4}\pi$$

and so on.

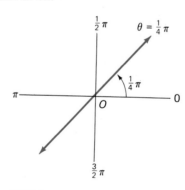

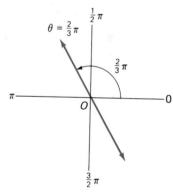

FIGURE 2   **FIGURE 3**

b)   The graph of the equation

$$\theta = \frac{2}{3}\pi$$

is shown in Figure 3. It is the line passing through the pole and making an angle of radian measure $\frac{2}{3}\pi$ with the polar axis. Other equations of this line are

$$\theta = \frac{5}{3}\pi \qquad \theta = \frac{8}{3}\pi \qquad \theta = -\frac{1}{3}\pi \qquad \theta = -\frac{4}{3}\pi$$

and so on.

■

In general, the polar form of an equation of a line is not as simple as the cartesian form. However, if the line is parallel to either the polar axis or the $\frac{1}{2}\pi$ axis, the equation is fairly simple.

Suppose a line is parallel to the polar axis and contains the point B whose cartesian coordinates are $(0, b)$, so that the polar coordinates of B are $(b, \frac{1}{2}\pi)$. A cartesian equation of this line is $y = b$. If we replace y by $r \sin \theta$, we have

$$r \sin \theta = b$$

which is a polar equation of any line parallel to the polar axis. If b is positive, the line is above the polar axis. If b is negative, it is below the polar axis.

Illustration 2

In Figure 4 we have a sketch of the graph of the equation

$$r \sin \theta = 3$$

and in Figure 5 there is a sketch of the graph of the equation

$$r \sin \theta = -3$$

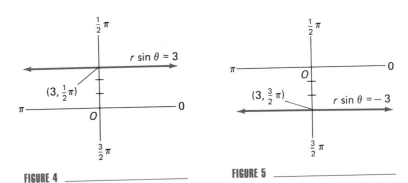

FIGURE 4 _____   **FIGURE 5** _____ ■

Now consider a line parallel to the $\frac{1}{2}\pi$ axis or, equivalently, perpendicular to the polar axis. If it goes through the point A whose cartesian coordinates are $(a, 0)$ and polar coordinates are $(a, 0)$, a cartesian equation is $x = a$. Replacing x by $r \cos \theta$, we obtain

$$r \cos \theta = a$$

which is an equation of any line perpendicular to the polar axis. If a is positive, the line is to the right of the $\frac{1}{2}\pi$ axis. If a is negative, it is to the left of the $\frac{1}{2}\pi$ axis.

Illustration 3

Figure 6 shows a sketch of the graph of the equation

$$r \cos \theta = 3$$

and Figure 7 shows a sketch of the graph of the equation

$$r \cos \theta = -3$$

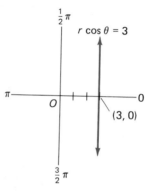

FIGURE 6 _____   FIGURE 7 _____   ■

The graph of the equation

$$r = C$$

where C is any constant, is a circle whose center is at the pole and radius is $|C|$. The same circle is given by the equation

$$r = -C$$

Illustration 4

In Figure 8 there is a sketch of the graph of the equation

$$r = 4$$

It is a circle with center at the pole and radius 4. The same circle is given by the equation

$$r = -4$$

FIGURE 8 _____   although the use of such an equation is uncommon.   ■

As was the case with the line, the general polar equation of a circle is not as simple as the cartesian form. However, there are further special cases of an equation of a circle that are worth considering in polar form.

If a circle contains the origin (the pole) and has its center at the point having cartesian coordinates (a, b), then a cartesian equation of the circle is

$$x^2 + y^2 - 2ax - 2by = 0$$

A polar equation of this circle is

$$(r \cos \theta)^2 + (r \sin \theta)^2 - 2a(r \cos \theta) - 2b(r \sin \theta) = 0$$
$$r^2(\cos^2 \theta + \sin^2 \theta) - 2ar \cos \theta - 2br \sin \theta = 0$$
$$r^2 - 2ar \cos \theta - 2br \sin \theta = 0$$
$$r(r - 2a \cos \theta - 2b \sin \theta) = 0$$
$$r = 0 \qquad r - 2a \cos \theta - 2b \sin \theta = 0$$

Because the graph of $r = 0$ is the pole and the pole is on the graph of $r - 2a \cos \theta - 2b \sin \theta = 0$, a polar equation of the circle is

$$r = 2a \cos \theta + 2b \sin \theta$$

When $b = 0$ in this equation, we have

$$\boxed{r = 2a \cos \theta}$$

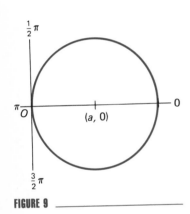

FIGURE 9

This is a polar equation of the circle of radius $|a|$ units, tangent to the $\frac{1}{2}\pi$ axis, and with its center on the polar axis or its extension. If $a > 0$, the circle is to the right of the pole as in Figure 9, and if $a < 0$, the circle is to the left of the pole.

If $a = 0$ in the equation $r = 2a \cos \theta + 2b \sin \theta$, we have

$$\boxed{r = 2b \sin \theta}$$

This is a polar equation of the circle of radius $|b|$ units, tangent to the polar axis, and with its center on the $\frac{1}{2}\pi$ axis or its extension. If $b > 0$, the circle is above the pole, and if $b < 0$, the circle is below the pole.

Example   1   Draw a sketch of the graph of each of the following equations.

a) $r = 5 \cos \theta$ b) $r = -6 \sin \theta$

Solution   a) The equation

$$r = 5 \cos \theta$$

is of the form $r = 2a \cos \theta$ with $a = \frac{5}{2}$. Thus the graph is a circle with center at the point having polar coordinates $(\frac{5}{2}, 0)$ and tangent to the $\frac{1}{2}\pi$ axis. A sketch of the graph appears in Figure 10.

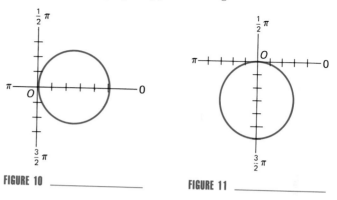

FIGURE 10 _____   FIGURE 11 _____

b) The equation

$$r = -6 \sin \theta$$

is of the form $r = 2b \sin \theta$ with $b = -3$. The graph is the circle with center at the point having polar coordinates $(3, \frac{3}{2}\pi)$ and tangent to the polar axis. Figure 11 shows a sketch of the graph. ■

To draw a sketch of the graph of a polar equation it is helpful to consider properties of symmetry of the graph. In Section 3.2 we stated that two points P and Q are symmetric with respect to a line if and only if the line is the perpendicular bisector of the line segment PQ, and that two points P and Q are symmetric with respect to a third point if and only if the third point is the midpoint of the line segment PQ. Therefore the points $(2, \frac{1}{3}\pi)$ and $(2, \frac{2}{3}\pi)$ are symmetric with respect to the $\frac{1}{2}\pi$ axis, and the points $(2, \frac{1}{3}\pi)$ and $(2, -\frac{2}{3}\pi)$ are symmetric with respect to the pole. We also stated in Section 3.2 that the graph of an equation is symmetric with respect to a line l if and only if for every point P on the graph there is a point Q, also on the graph, such that P and Q are symmetric with respect to l. Similarly, the graph of an equation is symmetric with respect to a point R if and only if for every point P on the graph there is a point S, also on the graph, such that P and S are symmetric with respect to R. We have three theorems giving tests for symmetry of graphs of polar equations.

Theorem 1   If for an equation in polar coordinates an equivalent equation is obtained when (r, θ) is replaced by either $(r, -\theta + 2n\pi)$ or $(-r, \pi - \theta + 2n\pi)$, where n is any integer, the graph of the equation is symmetric with respect to the polar axis.

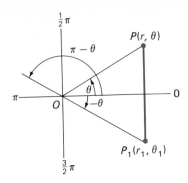

FIGURE 12

Proof If the point $P(r, \theta)$ is a point on the graph of an equation, then the graph is symmetric with respect to the polar axis if there is a point $P_1(r_1, \theta_1)$ on the graph such that the polar axis is the perpendicular bisector of the line segment P_1P (see Figure 12). So if $r_1 = r$, then θ_1 must equal $-\theta + 2n\pi$, where n is an integer. And if $r_1 = -r$, then θ_1 must be $\pi - \theta + 2n\pi$. ∎

Theorem 2

If for an equation in polar coordinates an equivalent equation is obtained when (r, θ) is replaced by either $(r, \pi - \theta + 2n\pi)$ or $(-r, -\theta + 2n\pi)$, where n is any integer, the graph of the equation is symmetric with respect to the $\frac{1}{2}\pi$ axis.

Theorem 3

If for an equation in polar coordinates an equivalent equation is obtained when (r, θ) is replaced by either $(-r, \theta + 2n\pi)$ or $(r, \pi + \theta + 2n\pi)$, where n is any integer, the graph of the equation is symmetric with respect to the pole.

The proofs of Theorems 2 and 3 are similar to the proof of Theorem 1 and are omitted.

Illustration 5

For the graph of the equation

$$r = 4 \cos 2\theta$$

we test for symmetry with respect to the polar axis, the $\frac{1}{2}\pi$ axis, and the pole.

Using Theorem 1 to test for symmetry with respect to the polar axis, we replace (r, θ) by $(r, -\theta)$ and obtain $r = 4 \cos(-2\theta)$, which is equivalent to $r = 4 \cos 2\theta$. So the graph is symmetric with respect to the polar axis.

Using Theorem 2 to test for symmetry with respect to the $\frac{1}{2}\pi$ axis, we replace (r, θ) by $(r, \pi - \theta)$ and get $r = 4 \cos(2(\pi - \theta))$ or, equivalently, $r = 4 \cos(2\pi - 2\theta)$, which is equivalent to the equation $r = 4 \cos 2\theta$. Therefore the graph is symmetric with respect to the $\frac{1}{2}\pi$ axis.

To test for symmetry with respect to the pole, we replace (r, θ) by $(-r, \theta)$ and obtain the equation $-r = 4 \cos 2\theta$, which is not equivalent to the given equation. But we must also determine whether the other set of coordinates works. We replace (r, θ) by $(r, \pi + \theta)$ and obtain $r = 4 \cos 2(\pi + \theta)$ or, equivalently, $r = 4 \cos(2\pi + 2\theta)$, which is equivalent to the equation $r = 4 \cos 2\theta$. Therefore the graph is symmetric with respect to the pole. ∎

When drawing a sketch of a graph, it is desirable to determine whether the pole is on the graph. This is done by substituting 0 for r and solving for θ.

TABLE 1

θ	r
0	-1
$\frac{1}{6}\pi$	$1 - \sqrt{3}$
$\frac{1}{3}\pi$	0
$\frac{1}{2}\pi$	1
$\frac{2}{3}\pi$	2
$\frac{5}{6}\pi$	$1 + \sqrt{3}$
π	3

Example 2 Draw a sketch of the graph of the equation

$$r = 1 - 2 \cos \theta$$

Solution Replacing (r, θ) by $(r, -\theta)$, we obtain an equivalent equation. Therefore the graph is symmetric with respect to the polar axis.

Table 1 gives the coordinates of some points on the graph. From these points we draw half of the graph; the remainder is drawn from its symmetry with respect to the polar axis.

If $r = 0$, we obtain $\cos \theta = \frac{1}{2}$, and if $0 \leq \theta \leq \pi$, then $\theta = \frac{1}{3}\pi$. Thus the point $(0, \frac{1}{3}\pi)$ is on the graph.

A sketch of the graph appears in Figure 13. ∎

The curve in Example 2 is called a *limaçon*. The graph of an equation of the form

$$r = a \pm b \cos \theta \qquad \text{or} \qquad r = a \pm b \sin \theta$$

is a **limaçon.** There are four types of limaçon, and the particular type depends on the ratio a/b, where a and b are positive. We show these four types obtained from the equation

$$r = a + b \cos \theta \qquad a > 0 \text{ and } b > 0$$

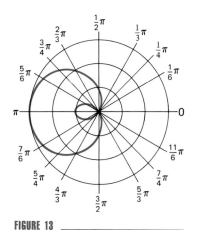

FIGURE 13

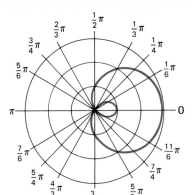

$0 < \dfrac{a}{b} < 1$

(a) Limaçon with a loop

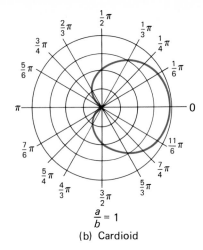

$\dfrac{a}{b} = 1$

(b) Cardioid

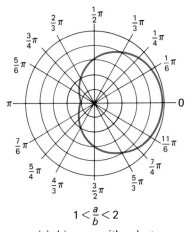

$1 < \dfrac{a}{b} < 2$

(c) Limaçon with a dent

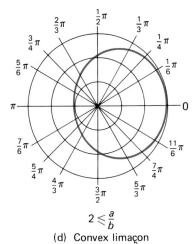

$2 \le \dfrac{a}{b}$

(d) Convex limaçon

FIGURE 14

---

1. $0 < \dfrac{a}{b} < 1$ **Limaçon with a loop.** See Figure 14(a).

2. $\dfrac{a}{b} = 1$ **Cardioid** (heart-shaped). See Figure 14(b).

3. $1 < \dfrac{a}{b} < 2$ **Limaçon with a dent.** See Figure 14(c).

4. $2 \le \dfrac{a}{b}$ **Convex limaçon** (no dent). See Figure 14(d).

If you study calculus, where horizontal and tangent lines of polar curves are discussed, the reason that limaçons of type 3 have a dent and those of type 4 have no dent will be apparent.

The limaçons obtained from the equation

$$r = a + b \sin \theta \qquad a > 0 \text{ and } b > 0$$

have the $\frac{1}{2}\pi$ axis as the axis of symmetry. If a limaçon has the equation

$$r = a - b \cos \theta \qquad a > 0 \text{ and } b > 0$$

the limaçon points in the direction of π, and if it has the equation

$$r = a - b \sin \theta \qquad a > 0 \text{ and } b > 0$$

it points in the direction of $\frac{3}{2}\pi$.

Example 3 Draw a sketch of the graph of each of the following limaçons: (a) $r = 3 + 2 \sin \theta$; (b) $r = 2 + 2 \cos \theta$; (c) $r = 2 - \sin \theta$.

Solution a) The equation $r = 3 + 2 \sin \theta$ is of the form of $r = a + b \sin \theta$ with $a = 3$ and $b = 2$. Because $a/b = 3/2$, and $1 < \frac{3}{2} < 2$, the graph is a limaçon with a dent. It is symmetric with respect to the $\frac{1}{2}\pi$ axis. Table 2 gives the coordinates of some of the points on the graph. A sketch of the graph, shown in Figure 15, is drawn by plotting the points whose coordinates are given in Table 2 and using the symmetry property.

TABLE 2

θ	r
0	3
$\frac{1}{6}\pi$	4
$\frac{1}{3}\pi$	$3 + \sqrt{3}$
$\frac{1}{2}\pi$	5
π	3
$\frac{7}{6}\pi$	2
$\frac{4}{3}\pi$	$3 - \sqrt{3}$
$\frac{3}{2}\pi$	1

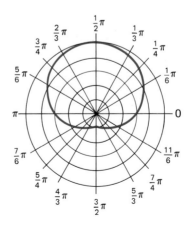

FIGURE 15

TABLE 3

θ	r
0	4
$\frac{1}{6}\pi$	$2 + \sqrt{3}$
$\frac{1}{3}\pi$	3
$\frac{1}{2}\pi$	2
$\frac{2}{3}\pi$	1
$\frac{5}{6}\pi$	$2 - \sqrt{3}$
π	0

TABLE 4

θ	r
0	2
$\frac{1}{6}\pi$	$\frac{3}{2}$
$\frac{1}{3}\pi$	$2 - \frac{1}{2}\sqrt{3}$
$\frac{1}{2}\pi$	1
π	2
$\frac{7}{6}\pi$	$\frac{5}{2}$
$\frac{4}{3}\pi$	$2 + \frac{1}{2}\sqrt{3}$
$\frac{3}{2}\pi$	3

b) The equation $r = 2 + 2\cos\theta$ is of the form of $r = a + b\cos\theta$ with $a = 2$ and $b = 2$. Because $a/b = 1$, the graph is a cardioid. It is symmetric with respect to the polar axis. The coordinates of some of the points on the graph are given in Table 3, and the sketch of the graph, appearing in Figure 16, is drawn by plotting these points and using the symmetry property.

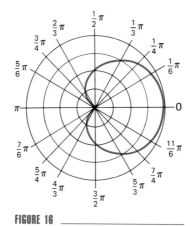

FIGURE 16

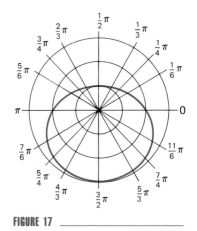

FIGURE 17

c) The equation $r = 2 - \sin\theta$ is of the form of $r = a - b\sin\theta$ with $a = 2$ and $b = 1$. Because $a/b = 2$, the graph is a convex limaçon. It is symmetric with respect to the $\frac{1}{2}\pi$ axis and points in the direction of $\frac{3}{2}\pi$. Figure 17 shows a sketch of the graph obtained by plotting the points whose coordinates are given in Table 4 and using the symmetry property. ∎

The graph of an equation of the form

$$r = a\cos n\theta \qquad \text{or} \qquad r = a\sin n\theta$$

is a **rose,** having n leaves if n is odd and $2n$ leaves if n is even.

Example **4**   Draw a sketch of the four-leafed rose

$$r = 4\cos 2\theta$$

Solution   In Illustration 5 we proved that the graph is symmetric with respect to the polar axis, the $\frac{1}{2}\pi$ axis, and the pole. Substituting 0 for r in the given equation we get

$$\cos 2\theta = 0$$

TABLE 5

θ	r
0	4
$\frac{1}{12}\pi$	$2\sqrt{3}$
$\frac{1}{6}\pi$	2
$\frac{1}{4}\pi$	0
$\frac{1}{3}\pi$	-2
$\frac{5}{12}\pi$	$-2\sqrt{3}$
$\frac{1}{2}\pi$	-4

from which we obtain, for $0 \le \theta < 2\pi$,

$$\theta = \tfrac{1}{4}\pi \qquad \theta = \tfrac{3}{4}\pi \qquad \theta = \tfrac{5}{4}\pi \qquad \theta = \tfrac{7}{4}\pi$$

Table 5 gives values of r for some values of θ from 0 to $\frac{1}{2}\pi$. From these values and the symmetry properties we draw a sketch of the graph, shown in Figure 18. ∎

Observe that if in the equations for a rose we take $n = 1$, we get

$$r = a \cos \theta \qquad \text{or} \qquad r = a \sin \theta$$

which are equations for a circle. Thus a circle can be considered as a one-leafed rose.

Other polar curves that occur frequently are *lemniscates* (see Exercises 29 through 32) and *spirals* (see Exercises 25 through 28). The curve in the next example is called a *spiral of Archimedes*.

Example 5   Draw a sketch of the graph of

$$r = \theta \qquad \theta \ge 0$$

Solution   When $\theta = n\pi$, where n is any integer, the graph intersects the polar axis or its extension, and when $\theta = \frac{1}{2}n\pi$, where n is any odd integer, the graph intersects the $\frac{1}{2}\pi$ axis or its extension. When $r = 0$, $\theta = 0$. A sketch of the graph appears in Figure 19.

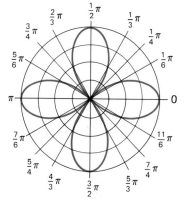

FIGURE 18

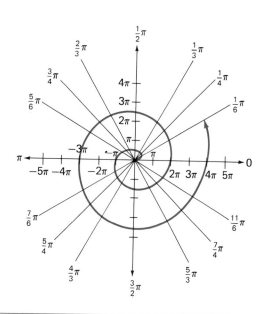

FIGURE 19

EXERCISES   8.7

In each exercise, sketch the graph of the equation.

1. (a) $\theta = \frac{1}{3}\pi$; (b) $r = \frac{1}{3}\pi$ 2. (a) $\theta = \frac{3}{4}\pi$; (b) $r = \frac{3}{4}\pi$
3. (a) $\theta = 2$; (b) $r = 2$ 4. $\theta = -3$; (b) $r = -3$
5. (a) $r\cos\theta = 4$; (b) $r = 4\cos\theta$
6. (a) $r\sin\theta = 2$; (b) $r = 2\sin\theta$
7. (a) $r\sin\theta = -4$; (b) $r = -4\sin\theta$
8. (a) $r\cos\theta = -5$; (b) $r = -5\cos\theta$
9. $r = 4 - 4\cos\theta$ 10. $r = 3 - 3\sin\theta$
11. $r = 2 + 2\sin\theta$ 12. $r = 3 + 3\cos\theta$
13. $r = 2 - 3\sin\theta$ 14. $r = 4 - 3\sin\theta$
15. $r = 3 - 2\cos\theta$ 16. $r = 3 - 4\cos\theta$
17. $r = 4 + 2\sin\theta$ 18. $r = 6 + 2\cos\theta$
19. $r = 2\sin 3\theta$ 20. $r = 4\sin 5\theta$ 21. $r = 2\cos 4\theta$
22. $r = 3\cos 2\theta$ 23. $r = 4\sin 2\theta$ 24. $r = 3\cos 3\theta$

25. $r = e^\theta$ (logarithmic spiral)
26. $r = e^{\theta/3}$ (logarithmic spiral)
27. $r = \dfrac{1}{\theta}$ (reciprocal spiral)
28. $r = 2\theta$ (spiral of Archimedes)
29. $r^2 = 9\sin 2\theta$ (lemniscate)
30. $r^2 = 16\cos 2\theta$ (lemniscate)
31. $r^2 = -25\cos 2\theta$ (lemniscate)
32. $r^2 = -4\sin 2\theta$ (lemniscate)
33. $r = 2\sin\theta\tan\theta$ (cissoid)
34. $(r - 2)^2 = 8\theta$ (parabolic spiral)
35. $r = 2\sec\theta - 1$ (conchoid of Nicomedes)
36. $r = 2\csc\theta + 3$ (conchoid of Nicomedes)
37. $r = |\sin 2\theta|$ 38. $r = 2|\cos\theta|$

REVIEW EXERCISES FOR CHAPTER 8

In Exercises 1 through 4, find the measure of the indicated side, but do not use a calculator or tables for the trigonometric function values. Express the results to the number of significant digits justified by the given information.

1. $\alpha = 30°$, $\gamma = 45°$, $c = 5.3$; find a.
2. $\beta = 135°$, $a = 29$, $c = 14$; find b.
3. $\alpha = 120°$, $b = 35$, $c = 46$; find a.
4. $\beta = 120°$, $\gamma = 30°$, $b = 7.8$; find c.

In Exercises 5 through 10, solve the triangle. Express the results to the number of significant digits justified by the given information.

5. $\alpha = 43.2°$, $\beta = 61.4°$, $b = 26.8$
6. $a = 60.4$, $b = 72.3$, $c = 54.7$
7. $\gamma = 105.3°$, $a = 21.6$, $b = 32.4$
8. $\alpha = 114°$, $\gamma = 32°$, $a = 85$
9. $a = 518.2$, $b = 439.7$, $c = 630.4$
10. $\alpha = 29.42°$, $b = 7134$, $c = 6024$

In Exercises 11 through 14, find the area of the triangle of the indicated exercise.

11. Exercise 5   12. Exercise 8
13. Exercise 9   14. Exercise 10

In Exercises 15 through 18 because the measures of two sides of a triangle and the angle opposite one of them are given, we have the ambiguous case. Determine the number of triangles that satisfy the given set of conditions and solve each triangle. Express the results to the number of significant digits justified by the given information.

15. $\alpha = 54.4°$, $a = 112$, $b = 131$
16. $\gamma = 32.4°$, $b = 50.3$, $c = 25.1$
17. $\beta = 39.7°$, $b = 12.8$, $c = 10.8$
18. $\beta = 42.5°$, $a = 12.4$, $b = 10.1$

In Exercises 19 through 22, $\mathbf{A} = \langle 4, -6\rangle$, $\mathbf{B} = \langle 1, 7\rangle$, and $\mathbf{C} = \langle 9, -5\rangle$.

19. Find the indicated vector and illustrate geometrically: (a) $\mathbf{A} + \mathbf{B}$; (b) $\mathbf{A} - \mathbf{C}$.
20. Find the indicated vector and illustrate geometrically: (a) $\mathbf{A} + (\mathbf{B} - \mathbf{C})$; (b) $(\mathbf{A} + \mathbf{B}) - \mathbf{C}$.
21. Find: (a) $|\mathbf{A}|$; (b) $|\mathbf{B} + \mathbf{C}|$; (c) $|\mathbf{B}| + |\mathbf{C}|$.
22. Find: (a) $|\mathbf{B}|$; (b) $|\mathbf{A} - \mathbf{C}|$; (c) $|\mathbf{A}| - |\mathbf{C}|$.

In Exercises 23 and 24, find the components of the vector having the given magnitude and direction angle.

23. (a) 12; $\frac{1}{6}\pi$; (b) 36; 112°
24. (a) 25; $\frac{3}{4}\pi$; (b) 130; 335.2°

In Exercises 25 and 26, plot the point having the given set of polar coordinates; then give two other sets of polar coordinates of the same point, one with the same value of r and one with an r having opposite sign.

25. (a) $(2, \frac{3}{4}\pi)$; (b) $(-3, \frac{7}{6}\pi)$
26. (a) $(4, -\frac{1}{3}\pi)$; (b) $(-1, \frac{1}{4}\pi)$

In Exercises 27 and 28, find the rectangular cartesian coordinates of the point whose polar coordinates are given.

27. (a) $(1, \frac{1}{2}\pi)$; (b) $(2, -\frac{1}{3}\pi)$; (c) $(4, \frac{5}{4}\pi)$; (d) $(-3, \frac{1}{6}\pi)$
28. (a) $(5, \pi)$; (b) $(-2, \frac{5}{6}\pi)$; (c) $(-\sqrt{2}, \frac{1}{4}\pi)$; (d) $(1, \frac{4}{3}\pi)$

In Exercises 29 and 30, find a set of polar coordinates of the point whose rectangular cartesian coordinates are given. Take $r > 0$ and $0 \leq \theta < 2\pi$.

29. (a) $(-4, 4)$; (b) $(1, -\sqrt{3})$; (c) $(0, 6)$;
    (d) $(-2\sqrt{3}, -2)$
30. (a) $(-4, 0)$; (b) $(\sqrt{3}, 1)$; (c) $(-2, -2)$;
    (d) $(3, -3\sqrt{3})$

In Exercises 31 through 34, find a polar equation of the graph having the given cartesian equation.

31. $4x^2 - 9y^2 = 36$
32. $2xy = 1$
33. $x^2 + y^2 - 9x + 8y = 0$
34. $y^4 = x^2(a^2 - y^2)$

In Exercises 35 through 38, find a cartesian equation of the graph having the given polar equation.

35. $r^2 \sin 2\theta = 4$
36. $r(1 - \cos \theta) = 2$
37. $r^2 = \sin^2 \theta$
38. $r = a \tan^2 \theta$

In Exercises 39 and 40, show the geometric representation of the complex number as a point in the complex plane.

39. (a) $3 + 7i$; (b) $6 - 3i$; (c) $-1 + 5i$; (d) $-4i$
40. (a) $-3 + 2i$; (b) $-7 - i$; (c) -4; (d) $3i$

In Exercises 41 and 42, write the expression without absolute-value bars.

41. (a) $|4 - 3i|$; (b) $|-6 + 2i|$
42. (a) $|5 + 12i|$; (b) $|-6i|$

In Exercises 43 and 44, express the complex number in standard polar form.

43. (a) $-3 + \sqrt{3}i$; (b) $2\sqrt{3} + 2i$; (c) $-4i$; (d) $-1 - i$
44. (a) $\frac{1}{2} + \frac{1}{2}i$; (b) $\sqrt{3} - i$; (c) $6i$; (d) -6

In Exercises 45 through 48, show the geometric representation of the complex number in the complex plane, and write it in cartesian form.

45. $2(\cos \frac{1}{6}\pi + i \sin \frac{1}{6}\pi)$
46. $6(\cos \frac{2}{3}\pi + i \sin \frac{2}{3}\pi)$
47. $4(\cos 135° + i \sin 135°)$
48. $8(\cos 330° + i \sin 330°)$

In Exercises 49 and 50, express the product in the form $a + bi$.

49. $4(\cos 50° + i \sin 50°) \cdot \frac{3}{4}(\cos 70° + i \sin 70°)$
50. $5(\cos \frac{7}{12}\pi + i \sin \frac{7}{12}\pi) \cdot 2(\cos \frac{2}{3}\pi + i \sin \frac{2}{3}\pi)$

In Exercises 51 and 52, express the quotient in the form $a + bi$.

51. $10(\cos \frac{4}{9}\pi + i \sin \frac{4}{9}\pi) \div 5(\cos \frac{43}{36}\pi + i \sin \frac{43}{36}\pi)$
52. $2(\cos 200° + i \sin 200°) \div 3(\cos 50° + i \sin 50°)$

In Exercises 53 through 56, express the power in the form $a + bi$.

53. $(\sqrt{3} + i)^3$
54. $(4 - 4i)^4$
55. $(-2 + 2i)^{-6}$
56. $\left(\frac{1}{2} + \frac{\sqrt{3}}{2}i\right)^{-5}$

In Exercises 57 through 62, find the indicated roots, write them in cartesian form, and show their geometric representations.

57. The three cube roots of 8
58. The three cube roots of -27
59. The four fourth roots of $2 + i$
60. The three cube roots of $8i$
61. The five fifth roots of $2 - 5i$
62. The four fourth roots of $\dfrac{\sqrt{3}}{2} - \dfrac{1}{2}i$

63. A tree is situated on a hill and at a point P, 23 m down the hill from the tree, the angle subtended by the tree is 18.5°. If the height of the tree is 36.5 m, at what angle is the hill inclined with the horizontal?

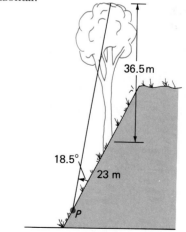

36.5 m

18.5°

23 m

P

64. A triangular lot has sides of lengths 242 ft, 160 ft, and 184 ft. If the cost per square foot of land is appraised at $40, what is the appraised value of the lot?

65. A pilot going from city A to city B must avoid a particular mountain range. The pilot first flies a course of 52° for a distance of 160 mi and then

alters the course to 105° and arrives at B after flying another 108 mi. (a) What is the direct distance from A to B? (b) What is the bearing from A to B?

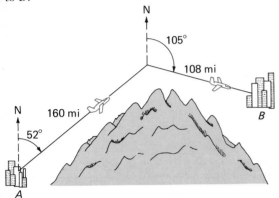

66. A hill is inclined at an angle of 16.2° with the horizontal, and a tunnel running through the side of the hill is inclined at an angle of 11.3° with the horizontal. From a point 256 ft down the tunnel, what is the vertical distance to the surface of the hill?

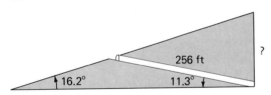

67. A vacant lot in the form of a parallelogram is situated on the corner of two streets that intersect at an angle of 98.3°, and the street frontages of the lot are 76.7 ft and 91.4 ft. If instead of going around the street sides of the lot, a girl decides to cross the lot along a diagonal from one corner to another, what distance does she save?

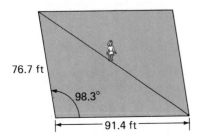

68. To determine the distance between two points P and Q on opposite sides of a building, a third

point R is chosen such that the distance from P to R is 120 m and the distance from Q to R is 140 m. If the angle formed by the line segments PR and QR is 72.3°, what is the distance from P to Q?

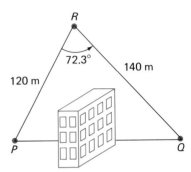

69. Find the components of the vector having a magnitude of 12 and a direction angle of $\frac{1}{6}\pi$.

70. Two forces of magnitudes 50 lb and 70 lb make an angle of 60° with each other and are applied to an object at the same point. Find (a) the magnitude of the resultant force and (b) to the nearest degree, the angle it makes with the force of 50 lb.

71. Determine the angle between two forces of 112 lb and 136 lb applied to an object at the same point if the resultant force has a magnitude of 168 lb.

72. The compass heading of an airplane is 107° and its air speed is 210 mi/hr. If there is a wind blowing from the west at 36 mi/hr, what is (a) the plane's ground speed and (b) its course?

In Exercises 73 and 74, solve the equation for all complex roots.

73. $x^6 + 1{,}000{,}000 = 0$ 74. $x^5 - 243 = 0$

Exercises 75 through 86 pertain to Supplementary Section 8.7. In Exercises 75 through 84, draw a sketch of the graph of the equation.

75. (a) $\theta = \frac{1}{4}\pi$; (b) $r = 4$
76. (a) $r\cos\theta = 3$; (b) $r = 3\cos\theta$
77. $r = 3\sin 2\theta$ 78. $r = 3 + 2\cos\theta$
79. $r = 2(1 - \cos\theta)$ 80. $r = 1 - 2\sin\theta$
81. $r = \sqrt{|\cos\theta|}$ 82. $r = 2\sin 3\theta$
83. $r^2 = -\sin 2\theta$ 84. $r^2 = 16\cos\theta$
85. Draw a sketch of the graph of (a) $r\theta = 3$ (reciprocal spiral) and (b) $3r = \theta$ (spiral of Archimedes).
86. Show that the equations $r = 1 + \sin\theta$ and $r = \sin\theta - 1$ have the same graph.

C H A P T E R

9

Systems of Equations and Inequalities, and Matrices

Introduction

In Chapter 2 you learned how to solve equations and inequalities involving a single variable. In this chapter we are concerned with equations and inequalities in two or more variables. A set of such equations or inequalities forms a system. In Sections 9.1 and 9.2 we solve systems of linear equations, whereas in Section 9.3 we solve systems involving quadratic equations. Systems of linear inequalities are treated in Section 9.4, and in the same section there is an introduction to linear programming.

We apply the Gaussian reduction method to solve systems of linear equations in Section 9.5. This technique paves the way for the treatment of matrices in the remaining sections. We discuss determinants and Cramer's rule in Section 9.6. Matrix algebra is presented in Supplementary Section 9.7. In Supplementary Section 9.8 we introduce the multiplicative inverse of a square matrix and use it to solve systems of linear equations.

9.1 Systems of Linear Equations in Two Variables

Many applications of mathematics lead to more than one equation in several variables. The resulting equations are called a **system** of equations. The solution set of a system of equations consists of all solutions that are common to the equations in the system.

In Section 3.3 we proved that the graph of an equation of the form

$$ax + by = c$$

is a line and that all ordered pairs (x, y) satisfying the equation are coordinates of points on the line. A system of two linear equations in two variables x and y can be written as

$$\begin{cases} a_1x + b_1y = c_1 \\ a_2x + b_2y = c_2 \end{cases}$$

where a_1, b_1, c_1, a_2, b_2, and c_2 are real numbers. The left brace is used to indicate that the two equations form a system. If an ordered pair (x, y) is to satisfy a system of two linear equations, the corresponding point (x, y) must lie on the two lines that are the graphs of the equations.

Illustration 1

A particular system of two linear equations is

$$\begin{cases} 2x + y = 3 \\ 5x + 3y = 10 \end{cases}$$

The solution set of each of the equations in the system is an infinite set of ordered pairs of real numbers, and the graphs of these sets are lines. Recall that to draw a

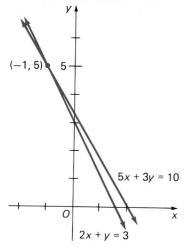

FIGURE 1

sketch of a line we need to find two points on the line; usually we plot the points where the line intersects the coordinate axes. On the line $2x + y = 3$ we have the points $(\frac{3}{2}, 0)$ and $(0, 3)$, while on the line $5x + 3y = 10$ we have the points $(2, 0)$ and $(0, \frac{10}{3})$. Figure 1 shows on the same coordinate system sketches of the two lines. From the figure it is apparent that the two lines intersect at exactly one point. This point, $(-1, 5)$ can be verified by substituting into the equations as follows:

$$2(-1) + 5 = 3$$
$$5(-1) + 3(5) = 10$$

The only ordered pair that is common to the solution sets of the two equations is $(-1, 5)$. Hence the solution set of the system is $\{(-1, 5)\}$. ∎

Illustration 2

Consider the system

$$\begin{cases} 6x - 3y = 5 \\ 2x - y = 4 \end{cases}$$

As may be seen in Figure 2, the lines having these equations appear to be parallel. It can easily be proved that the lines are indeed parallel by writing each of the equations in the slope-intercept form $y = mx + b$. Solving for y in each equation, we have

$$\begin{array}{ll} 6x - 3y = 5 & 2x - y = 4 \\ -3y = -6x + 5 & -y = -2x + 4 \\ y = 2x - \frac{5}{3} & y = 2x - 4 \end{array}$$

For each equation the slope $m = 2$. The two lines are not the same because the y intercepts, $-\frac{5}{3}$ and -4, are not equal. Therefore the lines are parallel and have no point in common. The solution set of the system is $\varnothing$. ∎

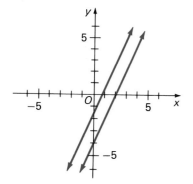

FIGURE 2

Illustration 3

For the system

$$\begin{cases} 3x + 2y = 4 \\ 6x + 4y = 8 \end{cases}$$

the graphs of the two equations are the same line. See Figure 3. This fact is evident when the equations are written in the slope-intercept form. Solving each of the equations for y, we have

$$\begin{array}{ll} 3x + 2y = 4 & 6x + 4y = 8 \\ 2y = -3x + 4 & 4y = -6x + 8 \\ y = -\frac{3}{2}x + 2 & y = -\frac{3}{2}x + 2 \end{array}$$ ∎

We have seen that for a system of two linear equations in two variables, three possibilities arise:

Possibility 1: The intersection of the two solution sets contains exactly one ordered pair, as in Illustration 1. The graphs intersect in exactly one point. The equations are said to be **consistent** and **independent**.

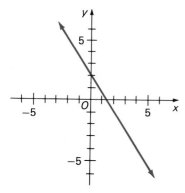

FIGURE 3

Possibility 2: The intersection of the two solution sets is the empty set, as in Illustration 2. The graphs are distinct parallel lines. The equations are said to be **inconsistent.**

Possibility 3: The solution sets of the two equations are equal, as in Illustration 3. The graphs are the same line. The equations are said to be **dependent.**

When two linear equations in two variables are consistent and independent (as in Illustration 1), a solution obtained from the graphs is generally only an approximation because reading numbers from the graphs depends on measurement. To obtain exact solutions of systems of linear equations we must use algebraic methods. These methods consist of replacing the given system by an equivalent system, one that has exactly the same solution set.

If any equation in a given system is replaced by an equivalent equation, the resulting system is equivalent to the given system. Furthermore, if any two equations of a given system are interchanged, the resulting system is equivalent to the given system.

One method for finding the solution set of a system of two linear equations in two variables is called the **substitution method.** For any ordered pair in the solution set of a system of equations, the variables in one equation represent the same numbers as the corresponding variables in the other equation. Therefore, if we replace one of the variables in one of the equations by its equal from the other equation, we have an equivalent system. The following example shows the procedure.

Example 1 Use the substitution method to find the solution set of the system in Illustration 1:

$$\begin{cases} 2x + y = 3 \\ 5x + 3y = 10 \end{cases}$$

Solution We solve the first equation for y, and get the equivalent system

$$\begin{cases} y = 3 - 2x \\ 5x + 3y = 10 \end{cases}$$

We replace y in the second equation by its equal, $3 - 2x$, from the first equation. We then have the equivalent system

$$\begin{cases} y = 3 - 2x \\ 5x + 3(3 - 2x) = 10 \end{cases}$$

Simplifying the second equation, we have

$$\begin{cases} y = 3 - 2x \\ -x + 9 = 10 \end{cases}$$

Solving the second equation for x, we get

$$\begin{cases} y = 3 - 2x \\ x = -1 \end{cases}$$

Finally, we substitute the value of x from the second equation into the first equation and we have

$$\begin{cases} y = 5 \\ x = -1 \end{cases}$$

This system is equivalent to the given one. Hence the solution set is $\{(-1, 5)\}$. ∎

Another approach to solving a system of two linear equations is called the **elimination method.** In this method we replace one of the equations in the system by an equation obtained in the following way: Multiply each equation by a nonzero real number, and add the resulting equations. An equivalent system is obtained. (It is understood that "to multiply an equation by a number" means to multiply each side of the equation by that number, and "to add two equations" means to add the corresponding sides of the equations.) We choose the multipliers in such a way that adding the resulting equations eliminates one of the variables. The elimination method is demonstrated in the following illustration.

Illustration   4

We apply the elimination method to find the solution set of the system of equations in Example 1:

$$\begin{cases} 2x + \ y = 3 \\ 5x + 3y = 10 \end{cases}$$

Remember that our goal is to eliminate one of the variables. Observe that the coefficient of y is 1 in the first equation and 3 in the second equation. To obtain an equation not involving y, we therefore replace the second equation by the sum of the second equation and -3 times the first. We begin by multiplying the first equation by -3 and writing the equivalent system

$$\begin{cases} -6x - 3y = -9 \\ \ \ 5x + 3y = 10 \end{cases}$$

Adding the two equations gives the following computation:

$$\begin{array}{rcl} -6x - 3y &=& -9 \\ 5x + 3y &=& 10 \\ \hline -x &=& 1 \end{array}$$

With this equation and the first equation in the given system, we can write the following equivalent system.

$$\begin{cases} 2x + y = 3 \\ -x = 1 \end{cases}$$

If we now multiply both sides of the second equation by -1, we have the equivalent system

$$\begin{cases} 2x + y = 3 \\ x = -1 \end{cases}$$

We next substitute -1 for x in the first equation to obtain

$$\begin{cases} 2(-1) + y = 3 \\ x = -1 \end{cases}$$

$$\Leftrightarrow \qquad \begin{cases} y = 5 \\ x = -1 \end{cases}$$

Therefore the solution set is $\{(-1, 5)\}$, which agrees with the result of Example 1. ∎

The next two illustrations show what happens when the elimination method is used if the two equations in the given system are inconsistent or dependent.

Illustration 5

The system of Illustration 2 is

$$\begin{cases} 6x - 3y = 5 \\ 2x - y = 4 \end{cases}$$

We now write the equivalent system obtained by multiplying the second equation by -3. We have

$$\begin{cases} 6x - 3y = 5 \\ -6x + 3y = -12 \end{cases}$$

This system is equivalent to the following one, obtained by replacing the second equation by the sum of the two equations:

$$\begin{cases} 6x - 3y = 5 \\ 0 = -7 \end{cases}$$

The solution set of this latter system is the empty set $\varnothing$, because there is no ordered pair (x, y) for which the second equation is a true statement. Hence the solution set of the given system is $\varnothing$. The two equations are inconsistent. ∎

Illustration 6

The system of Illustration 3 is

$$\begin{cases} 3x + 2y = 4 \\ 6x + 4y = 8 \end{cases}$$

Multiplying the second equation by $-\frac{1}{2}$, we have the equivalent system

$$\begin{cases} 3x + 2y = 4 \\ -3x - 2y = -4 \end{cases}$$

Replacing the second equation by the sum of these two equations, we have the equivalent system

$$\begin{cases} 3x + 2y = 4 \\ 0 = 0 \end{cases}$$

The second equation of this latter system is an identity; that is, it is a true statement for any ordered pair (x, y). Therefore the solution set of the system is the same as the solution set of the first equation and it can be written as $\{(x, y)|3x + 2y = 4\}$. The equations are dependent. Another way of indicating the solution set arises by solving the equation $3x + 2y = 4$ for x in terms of y and getting

$$x = \frac{4}{3} - \frac{2}{3}y$$

Then the solution set is the set of ordered pairs $\{(\frac{4}{3} - \frac{2}{3}y, y)\}$. Alternatively, by letting y be any real number t, x is $\frac{4}{3} - \frac{2}{3}t$ and the solution set can be written as $\{(\frac{4}{3} - \frac{2}{3}t, t)\}$. ∎

Example 2 Find the solution set of the system

$$\begin{cases} \dfrac{4}{x} + \dfrac{3}{y} = 4 \\ \dfrac{2}{x} - \dfrac{6}{y} = -3 \end{cases} \qquad \text{(I)}$$

Solution The system can be written as

$$\begin{cases} 4\left(\dfrac{1}{x}\right) + 3\left(\dfrac{1}{y}\right) = 4 \\ 2\left(\dfrac{1}{x}\right) - 6\left(\dfrac{1}{y}\right) = -3 \end{cases}$$

Thus the system is linear in $\dfrac{1}{x}$ and $\dfrac{1}{y}$. If we make the substitutions $u = \dfrac{1}{x}$ and $v = \dfrac{1}{y}$, the system is equivalent to

$$\begin{cases} 4u + 3v = 4 \\ 2u - 6v = -3 \end{cases} \qquad \text{(II)}$$

To solve this system we first multiply the second equation by -2 and obtain

$$\begin{cases} 4u + 3v = 4 \\ -4u + 12v = 6 \end{cases}$$

We now replace the second equation by the sum of the two equations. We get

$$\begin{cases} 4u + 3v = 4 \\ \phantom{4u + {}}15v = 10 \end{cases}$$

Solving the second equation for v, we have

$$\begin{cases} 4u + 3v = 4 \\ v = \frac{2}{3} \end{cases}$$

We substitute $\frac{2}{3}$ for v in the first equation and obtain

$$\begin{cases} 4u + 3(\frac{2}{3}) = 4 \\ \phantom{4u + 3(\frac{2}{3})}v = \frac{2}{3} \end{cases}$$

$$\Leftrightarrow \quad \begin{cases} 4u + 2 = 4 \\ v = \frac{2}{3} \end{cases}$$

$$\Leftrightarrow \quad \begin{cases} 4u = 2 \\ v = \frac{2}{3} \end{cases}$$

$$\Leftrightarrow \quad \begin{cases} u = \frac{1}{2} \\ v = \frac{2}{3} \end{cases}$$

Therefore the solution of system (II) is $u = \frac{1}{2}$ and $v = \frac{2}{3}$. To obtain the solution of system (I) we replace u and v by $\dfrac{1}{x}$ and $\dfrac{1}{y}$, respectively. Thus

$$\frac{1}{x} = \frac{1}{2} \qquad \frac{1}{y} = \frac{2}{3}$$

$$x = 2 \qquad y = \frac{3}{2}$$

Hence the solution set of system (I) is $\{(2, \frac{3}{2})\}$. ∎

In Example 2 the variables are treated as $\dfrac{1}{x}$ and $\dfrac{1}{y}$. If, instead, each of the equations in system (I) is multiplied by xy (the LCD), we obtain the system

$$\begin{cases} 4y + 3x = 4xy \\ 2y - 6x = -3xy \end{cases} \tag{III}$$

Because $4xy$ and $-3xy$ are second-degree terms, the equations in system (III) are of the second degree. To solve this system for x and y requires a more complicated procedure than that used in Example 2. Observe that $(0, 0)$ is a solution of system (III), while the equations of system (I) are not defined when $x = 0$ or $y = 0$.

In our previous discussions of word problems it was necessary to represent each of the unknown numbers by using only one variable. In this section and the next we use systems of equations to solve word problems by using a different variable to represent each of the unknown numbers. You will see that some problems that can be solved by applying a system of linear equations in two or more variables can also be solved by using a single linear equation in one variable. By assigning a separate variable to represent each unknown number, it is usually easier to form an equation. However, it is necessary to have as many independent equations as there are variables.

Example   3   A college rowing team can row 2 mi downstream in 8 min, but it takes 12 min for the team to row the same distance upstream. How fast can the team row in still water, and what is the rate of the current?

Solution   Let x represent the number of miles per hour in the rate of rowing in still water, and let y represent the number of miles per hour in the rate of the current. Then the effective rate of the boat going downstream is the sum of these rates, that is $(x + y)$ miles per hour. The effective rate of the boat going upstream is the rate in still water minus the rate of the current, that is, $(x - y)$ miles per hour. Table 1 gives expressions involving x and y for the number of miles in the distance each way.

TABLE 1

	Number of Hours in Time	×	Number of Miles per Hour in Effective Rate	=	Number of Miles in Distance
Downstream	$\frac{2}{15}$		$x + y$		$\frac{2}{15}(x + y)$
Upstream	$\frac{1}{5}$		$x - y$		$\frac{1}{5}(x - y)$

The number of miles in the distance traveled each way is 2, but it can also be represented by either of the entries in the last column of the table. Therefore we have the system of equations

$$\begin{cases} \frac{2}{15}(x + y) = 2 \\ \frac{1}{5}(x - y) = 2 \end{cases}$$

$$\Leftrightarrow \qquad \begin{cases} x + y = 15 \\ x - y = 10 \end{cases}$$

We replace the first equation by the sum of the two equations and we replace the second equation by their difference. We have then the equivalent system

$$\begin{cases} 2x = 25 \\ 2y = 5 \end{cases}$$

$$\Leftrightarrow \qquad \begin{cases} x = \frac{25}{2} \\ y = \frac{5}{2} \end{cases}$$

Thus the team can row $\frac{25}{2}$ mi/hr in still water, and the rate of the current is $\frac{5}{2}$ mi/hr.

Check Because $\frac{25}{2} + \frac{5}{2} = 15$, the effective rate downstream is 15 mi/hr. Furthermore, $\frac{2}{15}(15) = 2$; thus in 8 min the team can row 2 mi down the river. Because $\frac{25}{2} - \frac{5}{2} = 10$, the effective rate upstream is 10 mi/hr. Therefore in 12 min the team can row 2 mi up the river. ■

EXERCISES 9.1

In Exercises 1 through 10, draw a sketch of the graph of the system of equations. Classify the equations as (i) consistent and independent, (ii) inconsistent, or (iii) dependent. If the equations are consistent and independent, determine the solution set of the system from the graphs.

1. $\begin{cases} x - y = 8 \\ 2x + y = 1 \end{cases}$
2. $\begin{cases} y = 8 + 2x \\ 6x + 3y = 0 \end{cases}$

3. $\begin{cases} 2x + y = 6 \\ 8x = 6y + 9 \end{cases}$
4. $\begin{cases} 9x - 3y = 7 \\ y = 3x - \frac{5}{2} \end{cases}$

5. $\begin{cases} y = 2x - 4 \\ 6x - 3y - 12 = 0 \end{cases}$
6. $\begin{cases} 2x - 3y = -1 \\ 5x - 4y = 8 \end{cases}$

7. $\begin{cases} 4x - 2y - 7 = 0 \\ x = \frac{1}{2}y + 5 \end{cases}$
8. $\begin{cases} 3x - y = 1 \\ 6x + 5y = 2 \end{cases}$

9. $\begin{cases} 2x + 6y = -11 \\ 4x - 3y = -2 \end{cases}$
10. $\begin{cases} y = 3x - 5 \\ 6x - 2y = 10 \end{cases}$

In Exercises 11 through 26, find the solution set of the system by using either the substitution method or the elimination method.

11. $\begin{cases} 5x - 2y - 5 = 0 \\ 3x + y - 3 = 0 \end{cases}$
12. $\begin{cases} 3x + 4y - 4 = 0 \\ 5x + 2y - 8 = 0 \end{cases}$

13. $\begin{cases} 4x + 3y + 6 = 0 \\ 3x - 2y - 4 = 0 \end{cases}$
14. $\begin{cases} 8x - 3y = 5 \\ 5x - 2y = 4 \end{cases}$

15. $\begin{cases} 5x + 3y = 3 \\ x + 9y = 2 \end{cases}$
16. $\begin{cases} 5x + 6y = -5 \\ 15x - 3y = 13 \end{cases}$

17. $\begin{cases} 3x + 4y - 4 = 0 \\ 6x - 2y - 3 = 0 \end{cases}$

18. $\begin{cases} 18x + 3y - 10 = 0 \\ 2x - 2y - 5 = 0 \end{cases}$

19. $\begin{cases} 8x + 5y = 3 \\ 7x + 3y = -7 \end{cases}$

20. $\begin{cases} 2x - 5y = -21 \\ 5x + 3y = -6 \end{cases}$

21. $\begin{cases} \dfrac{x}{3} + \dfrac{y}{2} = 1 \\[2mm] \dfrac{x}{4} - \dfrac{y}{3} = -1 \end{cases}$
22. $\begin{cases} \dfrac{x}{2} - \dfrac{y}{6} = 1 \\[2mm] \dfrac{x}{3} + \dfrac{y}{2} = -1 \end{cases}$

23. $\begin{cases} \dfrac{6}{x} + \dfrac{3}{y} = -2 \\[2mm] \dfrac{4}{x} + \dfrac{7}{y} = -2 \end{cases}$
24. $\begin{cases} \dfrac{2}{x} + \dfrac{3}{y} = 2 \\[2mm] \dfrac{4}{x} - \dfrac{3}{y} = 1 \end{cases}$

25. $\begin{cases} \dfrac{3}{x} - \dfrac{2}{y} = 14 \\[2mm] \dfrac{6}{x} + \dfrac{3}{y} = -7 \end{cases}$
26. $\begin{cases} \dfrac{1}{x} - \dfrac{10}{y} = 6 \\[2mm] \dfrac{2}{x} + \dfrac{5}{y} = 2 \end{cases}$

In Exercises 27 through 30, determine whether the equations are consistent or inconsistent. To show they are inconsistent, solve a system of two of the equations and show that no member of the solution set satisfies the third equation. To show they are consistent, find the solution set.

27. $\begin{cases} 4x - y = 1 \\ 2x + y = 5 \\ 5x - 2y = -3 \end{cases}$ 28. $\begin{cases} 3x + 4y = 4 \\ 2x - y = 10 \\ x + 3y = -2 \end{cases}$

29. $\begin{cases} 2x - 5y = 4 \\ 3x - 2y = -5 \\ -3x + 4y = 1 \end{cases}$

30. $\begin{cases} 2x + y = 10 \\ 3x - 4y = -5 \\ 4x - 3y = 0 \end{cases}$

31. Three pounds of tea and 8 lb of coffee cost $39.70, and 5 lb of tea and 6 lb of coffee cost $47.10. What is the cost per pound of tea, and what is the cost per pound of coffee?

32. The cost of sending a telegram is based on a flat rate for the first 10 words and a fixed charge for each additional word. If a telegram of 15 words costs $11.65 and a telegram of 19 words costs $14.57, what is the flat rate, and what is the fixed charge for each additional word?

33. A group of women decided to contribute equal amounts toward obtaining a speaker for a book review. If there were 10 more women, each would have paid $2 less. However, if there were 5 less women, each would have paid $2 more. How many women were in the group and how much was the speaker paid?

34. A woman has a certain amount of money invested. If she had $6000 more invested at a rate 1 percent lower, she would have the same yearly income from the investment. Furthermore, if she had $4500 less invested at a rate 1 percent higher, her yearly income from the investment would also be the same. How much does she have invested, and at what rate is it invested?

35. A chemist has two acid solutions. One contains 15 percent acid and the other contains 6 percent acid. How many cubic centimeters of each solution should be used to obtain 400 cm^3 of a solution that is 9 percent acid?

36. A tank contains a mixture of insect spray and water in which there are 5 gal of insect spray and 25 gal of water. A second tank also contains 5 gal of spray but only 15 gal of water. If it is desired to have 7.5 gal of a mixture of which 20 percent is spray, how many gallons should be taken from each tank?

37. A boat goes downstream a distance of 5 mi in 15 min, but it takes 20 min for the return trip. Find the rate of the boat in still water and the rate of the current.

38. An airplane travels a distance of 1980 mi in 6 hr with the wind, but it takes 7 hr 20 min to make the return trip. Find the rate of the airplane in still air and the rate of the wind.

39. It takes a man 23 min longer to jog 5 mi than it takes his son to run the same distance. However, if the man doubles his rate, he can run the distance in 1 min less than his son. What is the man's rate of jogging and what is the son's rate of running?

40. The distance between two automobiles is 140 km. If the cars are driven toward each other, they will meet in 48 min. However, if they are driven in the same direction they will meet in 4 hr. What is the rate at which each car is driven?

41. If a girl works for 8 min and her brother works for 15 min, they can wash the front windows of their house. Also, if the girl works for 12 min and her brother works for 10 min, they can wash the same windows. How long will it take each person alone to wash the windows? (*Hint:* If it takes t minutes for a person to do a job alone, then in 1 min, the person can do $\dfrac{1}{t}$ of the job.)

42. A painter and his son can paint a room together in 8 hr. If the father works alone for 3 hr and then is joined by his son, the two together can complete the job in 6 hr more. How long will it take each person alone to paint the room? (See the hint for Exercise 41.)

43. If either 4 is added to the denominator of a fraction or 2 is subtracted from the numerator of the fraction, the resulting fraction is equivalent to $\frac{1}{2}$. What is the fraction?

44. If the numerator and denominator of a fraction are both increased by 5, the resulting fraction is equivalent to $\frac{2}{3}$. However, if the numerator and denominator are both decreased by 5, the resulting fraction is equivalent to $\frac{3}{7}$. What is the fraction?

9.2 Systems of Linear Equations in Three Variables

So far the linear (first-degree) equations we have discussed have contained at most two variables. In this section we introduce systems of linear equations in three variables.

Consider the equation

$$2x - y + 4z = 10$$

for which the replacement set of each of the three variables x, y, and z is the set R of real numbers. This equation is linear in the three variables. A solution of a linear equation in the three variables x, y, and z is the ordered triple of real numbers (r, s, t) such that if x is replaced by r, y by s, and z by t, the resulting statement is true. The set of all solutions is the solution set of the equation.

Illustration   1

For the equation

$$2x - y + 4z = 10$$

the ordered triple $(3, 4, 2)$ is a solution because

$$2(3) - 4 + 4(2) = 10$$

Some other ordered triples that satisfy this equation are $(-1, 8, 5)$, $(2, -6, 0)$, $(1, 0, 2)$, $(5, 0, 0)$, $(0, -6, 1)$, $(8, 2, -1)$, and $(7, 2, -\frac{1}{2})$. It appears that the solution set is infinite. ∎

The graph of an equation in three variables is a set of points represented by ordered triples of real numbers. Such points appear in a three-dimensional coordinate system, which we do not discuss in this book. You should, however, be aware that the graph of a linear equation in three variables is a plane.

Suppose that we have the following system of linear equations in the variables x, y, and z:

$$\begin{cases} a_1x + b_1y + c_1z = d_1 \\ a_2x + b_2y + c_2z = d_2 \\ a_3x + b_3y + c_3z = d_3 \end{cases}$$

The solution set of this system is the intersection of the solution sets of the three equations. Because the graph of each equation is a plane, the solution set can be interpreted geometrically as the intersection of three planes. When this intersection consists of a single point as in Figure 1, the equations of the system are said to be consistent and independent. As we proceed with the discussion, we shall show other possible relative positions of three planes.

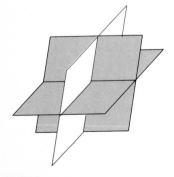

FIGURE 1

Algebraic methods for finding the solution set of a system of three linear equations in three variables are analogous to those used to solve linear systems in two variables. The following example shows the substitution method.

Example 1 Find the solution set of the system

$$\begin{cases} x - y - 4z = 3 \\ 2x - 3y + 2z = 0 \\ 2x - y + 2z = 2 \end{cases}$$

Solution We solve the first equation for x and obtain

$$\begin{cases} x = y + 4z + 3 \\ 2x - 3y + 2z = 0 \\ 2x - y + 2z = 2 \end{cases}$$

We now substitute the value of x from the first equation into the other two equations, and we obtain the equivalent system

$$\begin{cases} x = y + 4z + 3 \\ 2(y + 4z + 3) - 3y + 2z = 0 \\ 2(y + 4z + 3) - y + 2z = 2 \end{cases}$$

$$\Leftrightarrow \quad \begin{cases} x = y + 4z + 3 \\ -y + 10z = -6 \\ y + 10z = -4 \end{cases}$$

We next solve the second equation for y and get

$$\begin{cases} x = y + 4z + 3 \\ y = 10z + 6 \\ y + 10z = -4 \end{cases}$$

Substituting the value of y from the second equation into the third gives the equivalent system

$$\begin{cases} x = y + 4z + 3 \\ y = 10z + 6 \\ (10z + 6) + 10z = -4 \end{cases}$$

$$\Leftrightarrow \quad \begin{cases} x = y + 4z + 3 \\ y = 10z + 6 \\ 20z = -10 \end{cases}$$

$$\Leftrightarrow \quad \begin{cases} x = y + 4z + 3 \\ y = 10z + 6 \\ z = -\frac{1}{2} \end{cases}$$

Substituting the value of z from the third equation into the second equation, we obtain

$$\begin{cases} x = y + 4z + 3 \\ y = 1 \\ z = -\frac{1}{2} \end{cases}$$

Substituting the values of y and z from the second and third equations into the first equation, we get

$$\begin{cases} x = 2 \\ y = 1 \\ z = -\frac{1}{2} \end{cases}$$

This latter system is equivalent to the given system. Hence the solution set of the given system is $\{(2, 1, -\frac{1}{2})\}$.

    The solution can be checked by substituting into each of the given equations. Doing this, we have

$$\begin{cases} 2 - 1 + 2 = 3 \\ 4 - 3 - 1 = 0 \\ 4 - 1 - 1 = 2 \end{cases}$$

The equations of the given system are consistent and independent.   ■

    In the next example we use the elimination method to solve a system of three linear equations.

Example 2   Find the solution set of the system

$$\begin{cases} 4x - 2y - 3z = 8 \\ 5x + 3y - 4z = 4 \\ 6x - 4y - 5z = 12 \end{cases} \tag{I}$$

Solution   We first obtain an equivalent system in which the second and third equations do not involve the variable x. To eliminate x between the first two equations, we multiply the first equation by 5 and the second by -4 and add them:

$$\begin{array}{r} 20x - 10y - 15z = 40 \\ -20x - 12y + 16z = -16 \\ \hline -22y + z = 24 \end{array}$$

To eliminate x between the first and third equations of the given system, we multiply the first equation by 6 and the third by -4 and add them. Following is the computation:

$$24x - 12y - 18z = 48$$
$$-24x + 16y + 20z = -48$$
$$\overline{4y + 2z = 0}$$

The following system (II) is equivalent to the given system (I). Its first equation is the same as the first equation in (I), and its second and third equations are those obtained above.

$$\begin{cases} 4x - 2y - 3z = 8 \\ - 22y + z = 24 \\ 4y + 2z = 0 \end{cases} \qquad \text{(II)}$$

Because the coefficient of y is -22 in the second equation and 4 in the third equation, we can obtain an equation not involving y by computing the sum of 2 times the second equation and 11 times the third equation:

$$-44y + 2z = 48$$
$$44y + 22z = 0$$
$$\overline{24z = 48}$$

Dividing both sides of this equation by 24, we have

$$z = 2$$

We now replace the third equation of system (II) by this equation. We have the following equivalent system:

$$\begin{cases} 4x - 2y - 3z = 8 \\ - 22y + z = 24 \\ z = 2 \end{cases} \qquad \text{(III)}$$

Substituting the value of z from the third equation into the second, we obtain

$$-22y + 2 = 24$$
$$-22y = 22$$
$$y = -1$$

Replacing the second equation of system (III) by this equation, we have the equivalent system

$$\begin{cases} 4x - 2y - 3z = 8 \\ y = -1 \\ z = 2 \end{cases} \qquad \text{(IV)}$$

Substituting the values of y and z from the second and third equations into the first, we get

$$4x - 2(-1) - 3(2) = 8$$
$$4x + 2 - 6 = 8$$
$$4x = 12$$
$$x = 3$$

We replace the first equation of system (IV) by this equation and obtain the equivalent system

$$\begin{cases} x = 3 \\ y = -1 \\ z = 2 \end{cases} \qquad \text{(V)}$$

Because systems (V) and (I) are equivalent, the required solution set is $\{(3, -1, 2)\}$. ∎

The solution set in the above example could have been found by working with other combinations of equations. The procedure we followed is a methodical one. It relied on obtaining system (III), which is in what is called **triangular form.** With the system in triangular form we readily found first the value of y and then the value of x by substituting known values of the variables back into the equations. This process is called **back substitution.** When a system is in triangular form, it is a simple matter to use back substitution to obtain an equivalent system like (V), which is the eventual goal.

In the next illustration we have a system of three linear equations in three variables where the equations are *inconsistent.* We show what happens when we try to solve such a system.

Illustration 2

Suppose that we have the system

$$\begin{cases} 2x + y - z = 2 \\ x + 2y + 4z = 1 \\ 5x + y - 7z = 4 \end{cases} \qquad \text{(VI)}$$

We can replace this system by an equivalent one in which two of the equations do not involve x. We eliminate x between the first two equations by adding the first equation and -2 times the second. We eliminate x between the first and third equations by finding the sum of 5 times the first and -2 times the third. The computation is as follows:

$$
\begin{array}{r}
2x + y - z = 2 \\
-2x - 4y - 8z = -2 \\
\hline
-3y - 9z = 0
\end{array}
\qquad
\begin{array}{r}
10x + 5y - 5z = 10 \\
-10x - 2y + 14z = -8 \\
\hline
3y + 9z = 2
\end{array}
$$

The following system, containing the first equation of system (VI) and the above two equations, is equivalent to (VI):

$$\begin{cases} 2x + y - z = 2 \\ -3y - 9z = 0 \\ 3y + 9z = 2 \end{cases}$$

Replacing the third equation by the sum of the second equation and the third

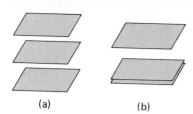

(a)   (b)

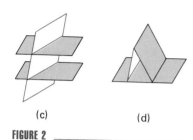

(c)   (d)

FIGURE 2

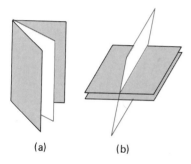

(a)   (b)

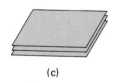

(c)

FIGURE 3

equation, we have the equivalent system

$$\begin{cases} 2x + y - z = 2 \\ - 3y - 9z = 0 \\ 0 = 2 \end{cases}$$

We see that the solution set of this system is the empty set $\varnothing$, because there is no ordered triple (x, y, z) for which the third equation is a true statement. Therefore the solution set of the given system (VI) is $\varnothing$, and the three equations are inconsistent. ∎

In Section 9.1 we showed that when a system of two linear equations in two variables is inconsistent, the graphs of the two equations are parallel lines. For a system of three inconsistent linear equations in three variables, the graphs of the three equations are planes that have no common intersection. The various possibilities are shown in Figure 2(a)–(d). In (a) the three planes are parallel. In (b) two of the planes are the same plane, and the third plane is parallel to it. In (c) two of the planes are parallel, the intersection of each of these planes with the third plane is a line, and the lines are parallel. In (d) no two planes are parallel, but two of the planes intersect in a line that is parallel to the third plane.

The graphs of three dependent linear equations in three variables are either three planes having a line in common or else they are the same plane. The various possibilities are shown in Figure 3(a)–(c). In (a) the graphs are three distinct planes having a line in common. In (b) two of the planes are identical, and the third plane intersects them in a line. In (c) the three planes are identical.

In the next example, we have a word problem involving three unknown quantities, each represented by a different variable. A system of three equations is obtained, but the equations turn out to be dependent. However, the word problem has a finite number of solutions.

Example 3 A group of 14 people spent $28 for admission tickets to Cinema One. The theatre charges $2.50 for adults, $1.50 for students, and $1 for children. If the same people had attended Cinema Two, which charges $4 for adults, $2 for students, and $1 for children, they would have spent $42 for admission tickets. How many adults, how many students, and how many children were in the group?

Solution The unknown quantities are the number of adults, the number of students, and the number of children in the group. We represent these numbers by a, s, and c, respectively. Then, because there are 14 people in the group,

$$a + s + c = 14$$

Because Cinema One charges $2.50 for adults, $1.50 for students, and $1 for children, and the total for admission tickets to Cinema One was $28, we have the equation

$$2.5a + 1.5s + c = 28$$

or, equivalently, if we multiply on each side of the equation by 2 to eliminate the decimals,

$$5a + 3s + 2c = 56$$

Because Cinema Two charges \$4 for adults, \$2 for students, and \$1 for children, and the total for admission tickets to Cinema Two was \$42, we have the equation

$$4a + 2s + c = 42$$

We have then the system of equations

$$\begin{cases} a + s + c = 14 \\ 5a + 3s + 2c = 56 \\ 4a + 2s + c = 42 \end{cases} \tag{VII}$$

We eliminate a between the first two equations by finding the sum of 5 times the first and -1 times the second. We eliminate a between the first and third equations by adding 4 times the first and -1 times the third. The computation is as follows:

$$\begin{aligned} 5a + 5s + 5c &= 70 \\ -5a - 3s - 2c &= -56 \\ \hline 2s + 3c &= 14 \end{aligned} \qquad \begin{aligned} 4a + 4s + 4c &= 56 \\ -4a - 2s - c &= -42 \\ \hline 2s + 3c &= 14 \end{aligned}$$

The following system, containing the first equation of system (VII) and the above two equations, is equivalent to (VII):

$$\begin{cases} a + s + c = 14 \\ 2s + 3c = 14 \\ 2s + 3c = 14 \end{cases}$$

Replacing the third equation by the sum of 1 times the second equation and -1 times the third equation, we have the equivalent system

$$\begin{cases} a + s + c = 14 \\ 2s + 3c = 14 \\ 0 = 0 \end{cases}$$

The third equation of this system is an identity because it is a true statement for any ordered triple (a, s, c). In particular, it is true for the ordered triple $(0, 0, t)$; so we replace the third equation by the equation $c = t$ and we have the equivalent system

$$\begin{cases} a + s + c = 14 \\ 2s + 3c = 14 \\ c = t \end{cases}$$

Substituting the value of c from the third equation into the second, and solving for s, we have

$$\begin{cases} a + s + c = 14 \\ \qquad s = 7 - \frac{3}{2}t \\ \qquad c = t \end{cases}$$

We now substitute into the first equation the values of s and c, and we have

$$\begin{cases} a + (7 - \frac{3}{2}t) + t = 14 \\ \qquad\qquad s = 7 - \frac{3}{2}t \\ \qquad\qquad c = t \end{cases}$$

$$\Leftrightarrow \qquad \begin{cases} a = 7 + \frac{1}{2}t \\ s = 7 - \frac{3}{2}t \\ c = t \end{cases}$$

This system is equivalent to system (VII), and so the original equations are dependent. It follows that a solution of (VII) is an ordered triple of the form $(7 + \frac{1}{2}t, 7 - \frac{3}{2}t, t)$.

Because a, s, and c must represent nonnegative integers, each of the numbers t, $7 - \frac{3}{2}t$, and $7 + \frac{1}{2}t$ must be nonnegative integers. If $t = 0$, $7 - \frac{3}{2}t = 7$, and $7 + \frac{1}{2}t = 7$. Hence $(7, 7, 0)$ is a solution. If $t = 1$, $7 - \frac{3}{2}t = \frac{11}{2}$, and $7 + \frac{1}{2}t = \frac{15}{2}$; thus $t = 1$ does not give a solution to the problem. If $t = 2$, $7 - \frac{3}{2}t = 4$, and $7 + \frac{1}{2}t = 8$. Therefore $(8, 4, 2)$ is a solution. If $t = 3$, both $7 - \frac{3}{2}t$ and $7 + \frac{1}{2}t$ are not integers and so $t = 3$ does not give a solution. If $t = 4$, $7 - \frac{3}{2}t = 1$, and $7 + \frac{1}{2}t = 9$. Hence $(9, 1, 4)$ is a solution. If t is an integer greater than 4, $7 - \frac{3}{2}t$ is a negative number. Therefore the solution set is $\{(7, 7, 0), (8, 4, 2), (9, 1, 4)\}$. Thus there are three possible combinations of people in the group: seven adults, seven students, and no children; eight adults, four students, and two children; or nine adults, one student, and four children.

Check If seven adults, seven students, and no children are in the group, the number of dollars in the total cost of admission tickets to Cinema One is $(2.5)(7) + (1.5)(7) = 28$, and the number of dollars in the total cost of admission tickets to Cinema Two is $(4)(7) + (2)(7) = 42$.

If eight adults, four students, and two children are in the group, the number of dollars in the total cost of admission tickets to Cinema One is $(2.5)(8) + (1.5)(4) + (1)(2) = 28$, and the number of dollars in the total cost of admission tickets to Cinema Two is $(4)(8) + (2)(4) + (1)(2) = 42$.

If nine adults, one student, and four children are in the group, the number of dollars in the total cost of admission tickets to Cinema One is $(2.5)(9) + (1.5)(1) + (1)(4) = 28$, and the number of dollars in the total cost of admission tickets to Cinema Two is $(4)(9) + (2)(1) + (1)(4) = 42$.

EXERCISES 9.2

In Exercises 1 through 14, find the solution set of the system. If the equations are either inconsistent or dependent, then so indicate.

1. $\begin{cases} 4x + 3y + z = 15 \\ x - y - 2z = 2 \\ 2x - 2y + z = 4 \end{cases}$

2. $\begin{cases} 2x + 3y + z = 8 \\ 5x + 2y - 3z = -13 \\ x - 2y + 5z = 15 \end{cases}$

3. $\begin{cases} x - y + 3z = 2 \\ 2x + 2y - z = 5 \\ 5x + 2z = 7 \end{cases}$

4. $\begin{cases} 3x + 2y - z = 4 \\ 3x + y + 3z = -2 \\ 6x - 3y - 2z = -6 \end{cases}$

5. $\begin{cases} x + \frac{1}{3}(y - z) = -1 \\ y - \frac{1}{2}(z - 2x) = 1 \\ z - \frac{1}{4}(2x - y) = -2 \end{cases}$

6. $\begin{cases} 3x - 2y + 4z = 4 \\ 7x + 5y - z = 9 \\ x + 9y - 9z = 1 \end{cases}$

7. $\begin{cases} 2x - 3y - 5z = 4 \\ x + 7y + 6z = -7 \\ 7x - 2y - 9z = 6 \end{cases}$

8. $\begin{cases} 3x - 5y + 2z = -2 \\ 2x + 3z = -3 \\ 4y - 3z = 8 \end{cases}$

9. $\begin{cases} x - y = 2 \\ 3y + z = 1 \\ x - 2z = 7 \end{cases}$

10. $\begin{cases} 3x - 2y = 1 \\ z - y = 5 \\ z - 2x = 5 \end{cases}$

11. $\begin{cases} x - y + 5z = 2 \\ 4x - 3y + 5z = 3 \\ 3x - 2y + 4z = 1 \end{cases}$

12. $\begin{cases} 5x - 4y + 5z = 6 \\ 6x + y - 2x = 4 \\ 4x - 9y + 12z = 5 \end{cases}$

13. $\begin{cases} \dfrac{1}{x} + \dfrac{1}{y} - \dfrac{1}{z} = 5 \\ \dfrac{3}{x} - \dfrac{1}{y} + \dfrac{2}{z} = 12 \\ \dfrac{1}{x} + \dfrac{2}{y} + \dfrac{1}{z} = 9 \end{cases}$

14. $\begin{cases} \dfrac{3}{x} - \dfrac{3}{y} + \dfrac{1}{z} = -1 \\ \dfrac{2}{x} + \dfrac{1}{y} - \dfrac{4}{z} = 0 \\ \dfrac{1}{x} + \dfrac{4}{y} + \dfrac{1}{z} = 5 \end{cases}$

In Exercises 15 through 18, show that the system has an infinite number of solutions in the solution set and use the variable t to express the solution set as an infinite set of ordered triples. Then assign the values 0, 1, 2, -1, and -2 to t and find five ordered triples in the solution set.

15. $\begin{cases} 5x - 4y + 3z = 1 \\ 3x - 5y + 7z = 11 \end{cases}$

16. $\begin{cases} 3x - 2y - 5z = 5 \\ 2x - y - 4z = 3 \end{cases}$

17. $\begin{cases} x + y - z = 4 \\ x - 2y - z = -5 \end{cases}$

18. $\begin{cases} x - y - 2z = -7 \\ x + y - 3z = -6 \end{cases}$

In Exercises 19 through 22, determine whether the equations are consistent or inconsistent. To show they are inconsistent, solve a system of three of the equations and show that no member of the solution set satisfies the fourth equation. To show they are consistent, find the solution set.

19. $\begin{cases} 2x + y + 3z = -4 \\ x - 4y - 2z = 3 \\ 4x - 2y + z = 4 \\ 5x + 3y + 4z = 5 \end{cases}$

20. $\begin{cases} 2x + 3y - z = 5 \\ 4x - 3y + 3z = 5 \\ 3x + y + 4z = 2 \\ x - 2y + z = 1 \end{cases}$

21. $\begin{cases} 2x + 4y + 3z = 5 \\ x - 4y - 2z = 7 \\ 4x - 3y + 5z = 2 \\ 3x + 2y + 4z = 8 \end{cases}$

22. $\begin{cases} x - y - 4z = 0 \\ x - y + 2z = 6 \\ 3x + y - 5z = -1 \\ x - 2y + z = 7 \end{cases}$

In Exercises 23 and 24, find an equation of the circle containing the given points. (*Hint:* Recall that an equation of a circle is of the form $x^2 + y^2 + Dx + Ey + F = 0$.)

23. $(-2, 8)$, $(2, 6)$, and $(-7, 3)$

24. $(5, 4)$, $(-2, 3)$, and $(-4, 1)$

25. A parabola has its axis parallel to the y axis and contains the points $(1, -1)$, $(2, 3)$, and $(3, 15)$. Find its equation. (*Hint:* An equation of such a parabola is of the form $y = ax^2 + bx + c$.)

26. A parabola has its axis parallel to the x axis and contains the points $(3, 0)$, $(1, -4)$, and $(-1, 6)$. Find its equation. (*Hint:* An equation of such a parabola is of the form $x = ay^2 + by + c$.)

27. Part of $25,000 is invested at 10 percent, another part is invested at 12 percent, and a third part is invested at 16 percent. The total yearly income from these three investments is $3200. Furthermore, the income from the 16 percent investment yields the same amount as the sum of the incomes from the other two investments. How much is invested at each rate?

28. If t degrees Celsius is the temperature at which water boils at a height of h feet above sea level, then $h = a + bt + ct^2$. Given that water boils at 100° Celsius at sea level, 95° Celsius at a height of 7400 ft above sea level, and 90° Celsius at a height of 14,550 ft above sea level, find a, b, and c.

29. In 20 oz of one alloy there are 6 oz of copper, 4 oz of zinc, and 10 oz of lead. In 20 oz of a second alloy there are 12 oz of copper, 5 oz of zinc, and 3 oz of lead, while in 20 oz of a third alloy there are 8 oz of copper, 6 oz of zinc, and 6 oz of lead. How many ounces of each alloy should be

combined to make a new alloy containing 34 oz of copper, 17 oz of zinc, and 19 oz of lead?

30. A total of $50,000 is invested in three different securities for which the annual dividends are computed at 8 percent, 10 percent, and 12 percent. If the total annual income from the three securities is $5320, and the income from the 12 percent security is $1080 more than that from the 10 percent security, what is the amount invested in each security?

31. On a store counter, there was a supply of three sizes of Christmas cards. The large cards cost $1 each; the medium cards cost 80 cents each; and the small cards cost 60 cents each. A woman purchased ten cards, which consisted of one-fourth of the available large cards, one-third of the available medium cards, and one-half of the available small cards. The total cost of her cards was $8.20. If there were 21 cards remaining on the counter after her purchase, how many of each kind of card did she buy?

32. Food A has 560 calories per pound and 80 units of vitamins per pound, food B has 240 calories per pound and 400 units of vitamins per pound, and food C has 480 calories per pound and 160 units of vitamins per pound. It is desired to have a 10-lb mixture of foods A, B, and C to contain a total of 2000 calories and 1200 units of vitamins. Show that these requirements lead to an inconsistent system of equations, and therefore the mixture is not possible.

33. Suppose in Exercise 32, instead of a 10-lb mixture, it is desired to have a 5-lb mixture of foods A, B, and C to contain a total of 2000 calories and 1200 units of vitamins. (a) Show that these conditions lead to a dependent system of equations. (b) If x pounds of A, y pounds of B, and z pounds of C are to be used to make up the 5 lb, what are the restrictions on x, y, and z?

34. A brochure promoting exhibitions at an art gallery is sent each month to the persons whose names appear on a mailing list, and employees A, B, and C assist in the preparation of the mailing. When all three work together, it takes them 2 hr 55 min to complete the job. Last month employee C was away and employees A and B together took 5 hr to prepare the mailing. For this month's mailing each employee started at a different time: employee A began at 9 A.M.; employee B joined A at 10 A.M.; and employee C joined A and B at 10:54 A.M. They finished the work at 1 P.M. How long does it take each employee working alone to prepare the mailing? (See the hint for Exercise 41 in Exercises 9.1.)

9.3 Systems Involving Quadratic Equations

In Sections 9.1 and 9.2 our discussion of systems of equations was confined to linear systems. However, a number of applications lead to nonlinear systems as illustrated in Exercises 25 through 36. The word problems in these exercises use concepts presented previously, but the resulting systems involve at least one quadratic equation. In this section we discuss methods of solving such systems of two equations in two variables.

We consider first a system that contains a linear equation and a quadratic equation. In this case the system can be solved by the substitution method. The linear equation can be solved for one variable in terms of the other, and the resulting expression can be substituted into the quadratic equation, as shown in the following example.

Example **1** Find the solution set of the system

$$\begin{cases} y^2 = 4x \\ x + y = 3 \end{cases} \tag{I}$$

Solution We solve the second equation for x and obtain the equivalent system

$$\begin{cases} y^2 = 4x \\ x = 3 - y \end{cases}$$

Replacing x in the first equation by its equal from the second, we have the equivalent system

$$\begin{cases} y^2 = 4(3 - y) \\ x = 3 - y \end{cases}$$

$$\Leftrightarrow \quad \begin{cases} y^2 + 4y - 12 = 0 \\ \qquad\qquad x = 3 - y \end{cases} \tag{II}$$

We now solve the first equation.

$$(y - 2)(y + 6) = 0$$
$$y - 2 = 0 \qquad y + 6 = 0$$
$$y = 2 \qquad\quad y = -6$$

Because the first equation of system (II) is equivalent to the two equations $y = 2$ and $y = -6$, system (II) is equivalent to the two systems

$$\begin{cases} y = 2 \\ x = 3 - y \end{cases} \quad \text{and} \quad \begin{cases} y = -6 \\ x = 3 - y \end{cases}$$

In each of the latter two systems we substitute into the second equation the value of y from the first, and we have

$$\begin{cases} y = 2 \\ x = 1 \end{cases} \quad \text{and} \quad \begin{cases} y = -6 \\ x = 9 \end{cases}$$

These two systems are equivalent to system (I). Thus the solution set of (I) is $\{(1, 2), (9, -6)\}$. ∎

Figure 1 shows sketches of the graphs of the two equations of system (I) on the same coordinate system. The graph of the first equation is a parabola, and the graph of the second is a line. The graphs intersect at the points $(1, 2)$ and $(9, -6)$.

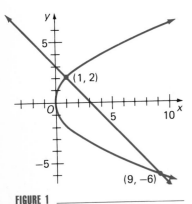

FIGURE 1

Example 2 Find the solution set of the system

$$\begin{cases} x^2 + y^2 = 25 \\ 3x + 4y = 25 \end{cases} \tag{III}$$

Draw sketches of the graphs of the two equations on the same coordinate system.

Solution Solving the second equation for y and replacing y in the first equation by the resulting expression, we have the equivalent system

$$\begin{cases} x^2 + \left(\dfrac{25 - 3x}{4}\right)^2 = 25 \\ \\ \qquad\qquad y = \dfrac{25 - 3x}{4} \end{cases} \qquad\text{(IV)}$$

We solve the first equation by first multiplying each side by 16.

$$16x^2 + (625 - 150x + 9x^2) = 400$$
$$25x^2 - 150x + 225 = 0$$
$$x^2 - 6x + 9 = 0$$
$$(x - 3)^2 = 0$$

Hence the roots of this quadratic equation are 3 and 3; that is, 3 is a double root. Therefore system (IV) is equivalent to the system

$$\begin{cases} x = 3 \\ y = \dfrac{25 - 3x}{4} \end{cases}$$

Substituting 3 for x in the second equation, we obtain

$$\begin{cases} x = 3 \\ y = 4 \end{cases}$$

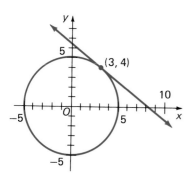

FIGURE 2

Thus the solution set of the given system (III) is $\{(3, 4)\}$.

    Sketches of the graphs of the two equations of system (III) are shown in Figure 2. We see that the line is tangent to the circle at the point (3, 4); this is the geometric significance of the double root. The point of tangency can be considered as two intersections of the line and the circle. ∎

Example 3 Find the solution set of the system

$$\begin{cases} x^2 + y^2 = 2 \\ \quad x - y = 4 \end{cases} \qquad\text{(V)}$$

Draw sketches of the graphs of the two equations on the same coordinate system.

Solution Solving the second equation for x and replacing x in the first equation by the resulting expression, we have the equivalent system

$$\begin{cases} (y + 4)^2 + y^2 = 2 \\ \qquad\qquad x = y + 4 \end{cases} \qquad\text{(VI)}$$

We solve the first equation for y.

$$y^2 + 8y + 16 + y^2 = 2$$

$$2y^2 + 8y + 14 = 0$$

$$y^2 + 4y + 7 = 0$$

$$y = \frac{-4 \pm \sqrt{4^2 - 4(1)(7)}}{2(1)}$$

$$= \frac{-4 \pm \sqrt{-12}}{2}$$

$$= \frac{-4 \pm 2i\sqrt{3}}{2}$$

$$= -2 \pm i\sqrt{3}$$

Hence the first equation of system (VI) is equivalent to the two equations $y = -2 + i\sqrt{3}$ and $y = -2 - i\sqrt{3}$. Therefore, (VI) is equivalent to the two systems

$$\begin{cases} y = -2 + i\sqrt{3} \\ x = y + 4 \end{cases} \quad \text{and} \quad \begin{cases} y = -2 - i\sqrt{3} \\ x = y + 4 \end{cases}$$

In each of these systems we substitute the value of y from the first equation into the second and we have

$$\begin{cases} y = -2 + i\sqrt{3} \\ x = 2 + i\sqrt{3} \end{cases} \quad \text{and} \quad \begin{cases} y = -2 - i\sqrt{3} \\ x = 2 - i\sqrt{3} \end{cases}$$

These two systems are equivalent to the given system (V). Thus the solution set of (V) is

$$\{(2 + i\sqrt{3}, -2 + i\sqrt{3}), (2 - i\sqrt{3}, -2 - i\sqrt{3})\}$$

Sketches of the graphs of the two equations of (V) appear in Figure 3. The line and the circle do not intersect. ∎

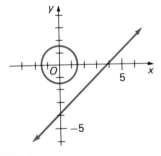

FIGURE 3

As in Example 3, when the solutions of a system are ordered pairs of imaginary numbers, there are no points of intersection of the graphs that correspond to the solutions. Remember that the coordinates of points in the real plane are real numbers.

In Section 9.1 we introduced the elimination method to solve a system of linear equations. We apply this method in the following example involving a system of two quadratic equations.

Example 4   Find the solution set of the system

$$\begin{cases} 2x^2 - 3y^2 = 6 \\ 6x^2 + y^2 = 58 \end{cases} \tag{VII}$$

Draw sketches of the graphs of the two equations on the same coordinate system.

Solution   We wish to replace the given system by one that has an equation containing only one variable. We can eliminate y between the two equations by adding the first equation and 3 times the second as follows:

$$
\begin{array}{rcl}
2x^2 - 3y^2 &=& 6 \\
18x^2 + 3y^2 &=& 174 \\
\hline
20x^2 &=& 180
\end{array}
$$

The following system, involving the first equation of system (VII) and the above equation, is equivalent to (VII):

$$
\begin{cases}
2x^2 - 3y^2 = 6 \\
\quad\quad 20x^2 = 180
\end{cases}
$$

$$
\Leftrightarrow \quad
\begin{cases}
2x^2 - 3y^2 = 6 \\
\quad\quad x^2 = 9
\end{cases}
$$

The second equation of this system is equivalent to the two equations $x = 3$ and $x = -3$. Therefore this system is equivalent to the two systems

$$
\begin{cases}
2x^2 - 3y^2 = 6 \\
\quad\quad x = 3
\end{cases}
\quad \text{and} \quad
\begin{cases}
2x^2 - 3y^2 = 6 \\
\quad\quad x = -3
\end{cases}
$$

In each of these two systems we substitute into the first equation the value of x from the second, and we have

$$
\begin{cases}
2(3)^2 - 3y^2 = 6 \\
\quad\quad x = 3
\end{cases}
\quad \text{and} \quad
\begin{cases}
2(-3)^2 - 3y^2 = 6 \\
\quad\quad x = -3
\end{cases}
$$

$$
\Leftrightarrow \quad
\begin{cases}
18 - 3y^2 = 6 \\
\quad\quad x = 3
\end{cases}
\quad \text{and} \quad
\begin{cases}
18 - 3y^2 = 6 \\
\quad\quad x = -3
\end{cases}
$$

$$
\Leftrightarrow \quad
\begin{cases}
y^2 = 4 \\
x = 3
\end{cases}
\quad \text{and} \quad
\begin{cases}
y^2 = 4 \\
x = -3
\end{cases}
$$

The first equation in each of these two systems is equivalent to the two equations $y = 2$ and $y = -2$. Hence the two systems are equivalent to the four systems

$$
\begin{cases}
y = 2 \\
x = 3
\end{cases}
\quad
\begin{cases}
y = -2 \\
x = 3
\end{cases}
\quad
\begin{cases}
y = 2 \\
x = -3
\end{cases}
\quad
\begin{cases}
y = -2 \\
x = -3
\end{cases}
$$

These four systems are equivalent to the given system (VII). Therefore the solution set of (VII) is $\{(3, 2), (3, -2), (-3, 2), (-3, -2)\}$.

Sketches of the graphs of the two equations of system (VII) are shown in Figure 4. The graph of the first equation is a hyperbola and the graph of the second is an ellipse. The two graphs intersect at the four points $(3, 2)$, $(3, -2)$, $(-3, 2)$, and $(-3, -2)$. ■

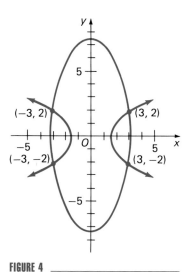

FIGURE 4

In the next example there is a system of two quadratic equations in which the second equation involves three second-degree terms. Because the right side of the second equation is zero and the left side can be factored, the second equation is equivalent to two linear equations. Therefore the given system is equivalent to two systems, each consisting of a quadratic equation and a linear equation.

Example 5 Find the solution set of the system

$$\begin{cases} x^2 + y^2 = 16 \\ 2x^2 - 3xy + y^2 = 0 \end{cases}$$

Draw sketches of the graphs of the two equations on the same coordinate system.

Solution We factor the left side of the second equation, and we obtain the equivalent system

$$\begin{cases} x^2 + y^2 = 16 \\ (2x - y)(x - y) = 0 \end{cases}$$

Because the second equation is equivalent to the two equations $2x - y = 0$ and $x - y = 0$, the given system is equivalent to the two systems

$$\begin{cases} x^2 + y^2 = 16 \\ 2x - y = 0 \end{cases} \quad \text{and} \quad \begin{cases} x^2 + y^2 = 16 \\ x - y = 0 \end{cases}$$

$$\Leftrightarrow \qquad \begin{cases} x^2 + y^2 = 16 \\ y = 2x \end{cases} \quad \text{and} \quad \begin{cases} x^2 + y^2 = 16 \\ y = x \end{cases}$$

In each of the latter two systems we substitute into the first equation the value of y from the second, and we have

$$\begin{cases} x^2 + 4x^2 = 16 \\ y = 2x \end{cases} \quad \text{and} \quad \begin{cases} x^2 + x^2 = 16 \\ y = x \end{cases}$$

$$\Leftrightarrow \qquad \begin{cases} x^2 = \frac{16}{5} \\ y = 2x \end{cases} \quad \text{and} \quad \begin{cases} x^2 = 8 \\ y = x \end{cases}$$

The equation $x^2 = \frac{16}{5}$ is equivalent to the two equations $x = \frac{4}{5}\sqrt{5}$ and $x = -\frac{4}{5}\sqrt{5}$. The equation $x^2 = 8$ is equivalent to the two equations $x = 2\sqrt{2}$ and $x = -2\sqrt{2}$. Therefore the preceding two systems are equivalent to the four systems

$$\begin{cases} x = \frac{4}{5}\sqrt{5} \\ y = 2x \end{cases} \quad \begin{cases} x = -\frac{4}{5}\sqrt{5} \\ y = 2x \end{cases} \quad \begin{cases} x = 2\sqrt{2} \\ y = x \end{cases} \quad \begin{cases} x = -2\sqrt{2} \\ y = x \end{cases}$$

In each of these latter systems we substitute the value of x from the first

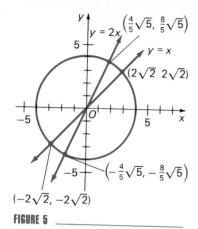

FIGURE 5

equation into the second equation, and we have

$$\begin{cases} x = \frac{4}{5}\sqrt{5} \\ y = \frac{8}{5}\sqrt{5} \end{cases} \quad \begin{cases} x = -\frac{4}{5}\sqrt{5} \\ y = -\frac{8}{5}\sqrt{5} \end{cases} \quad \begin{cases} x = 2\sqrt{2} \\ y = 2\sqrt{2} \end{cases} \quad \begin{cases} x = -2\sqrt{2} \\ y = -2\sqrt{2} \end{cases}$$

Thus the solution set of the given system is

$$\{(\tfrac{4}{5}\sqrt{5}, \tfrac{8}{5}\sqrt{5}), (-\tfrac{4}{5}\sqrt{5}, -\tfrac{8}{5}\sqrt{5}), (2\sqrt{2}, 2\sqrt{2}), (-2\sqrt{2}, -2\sqrt{2})\}$$

Figure 5 shows sketches of the graphs of the two equations. The graph of the equation $x^2 + y^2 = 16$ is a circle and the graph of the equation $2x^2 - 3xy + y^2 = 0$ consists of the two lines having the equations $y = 2x$ and $y = x$. Each line intersects the circle at two points. With $\sqrt{5} \approx 2.2$ and $\sqrt{2} \approx 1.4$, these points are $(1.8, 3.6)$, $(-1.8, -3.6)$, $(2.8, 2.8)$, and $(-2.8, -2.8)$. ■

The system in the next example consists of two quadratic equations in which all the terms containing variables are of the second degree; that is, there are no first-degree terms. If one of the equations is replaced by a combination of the two equations in which no constant term appears, the system that results can be solved by the method used in Example 5.

Example 6 Find the solution set of the system

$$\begin{cases} 4x^2 + xy + y^2 = 6 \\ 2x^2 - xy + y^2 = 8 \end{cases} \tag{VIII}$$

Solution We can obtain an equation having 0 on the right side by adding 4 times the first equation and -3 times the second as follows:

$$\begin{array}{rcl} 16x^2 + 4xy + 4y^2 &=& 24 \\ -6x^2 + 3xy - 3y^2 &=& -24 \\ \hline 10x^2 + 7xy + y^2 &=& 0 \end{array}$$

The following system, involving the first equation of system (VIII) and the above equation, is equivalent to (VIII):

$$\begin{cases} 4x^2 + xy + y^2 = 6 \\ 10x^2 + 7xy + y^2 = 0 \end{cases}$$

Factoring the left member of the second equation, we obtain

$$\begin{cases} 4x^2 + xy + y^2 = 6 \\ (5x + y)(2x + y) = 0 \end{cases}$$

The second equation is equivalent to the two equations $5x + y = 0$ and

$2x + y = 0$. Hence this system is equivalent to the two systems

$$\begin{cases} 4x^2 + xy + y^2 = 6 \\ \qquad\qquad y = -5x \end{cases} \quad \text{and} \quad \begin{cases} 4x^2 + xy + y^2 = 6 \\ \qquad\qquad y = -2x \end{cases}$$

In each of these two systems, if we substitute into the first equation the value of y from the second equation, we have

$$\begin{cases} 4x^2 + x(-5x) + (-5x)^2 = 6 \\ \qquad\qquad\qquad\qquad y = -5x \end{cases} \quad \text{and} \quad \begin{cases} 4x^2 + x(-2x) + (-2x)^2 = 6 \\ \qquad\qquad\qquad\qquad y = -2x \end{cases}$$

$$\Leftrightarrow \quad \begin{cases} 4x^2 - 5x^2 + 25x^2 = 6 \\ \qquad\qquad\qquad y = -5x \end{cases} \quad \text{and} \quad \begin{cases} 4x^2 - 2x^2 + 4x^2 = 6 \\ \qquad\qquad\qquad y = -2x \end{cases}$$

$$\Leftrightarrow \quad \begin{cases} x^2 = \frac{1}{4} \\ y = -5x \end{cases} \quad \text{and} \quad \begin{cases} x^2 = 1 \\ y = -2x \end{cases}$$

The equation $x^2 = \frac{1}{4}$ is equivalent to the two equations $x = \frac{1}{2}$ and $x = -\frac{1}{2}$, and the equation $x^2 = 1$ is equivalent to the two equations $x = 1$ and $x = -1$. Thus the preceding two systems are equivalent to the four systems

$$\begin{cases} x = \frac{1}{2} \\ y = -5x \end{cases} \quad \begin{cases} x = -\frac{1}{2} \\ y = -5x \end{cases} \quad \begin{cases} x = 1 \\ y = -2x \end{cases} \quad \begin{cases} x = -1 \\ y = -2x \end{cases}$$

If in each of these equations we substitute in the second the value of x from the first, we have the equivalent systems

$$\begin{cases} x = \frac{1}{2} \\ y = -\frac{5}{2} \end{cases} \quad \begin{cases} x = -\frac{1}{2} \\ y = \frac{5}{2} \end{cases} \quad \begin{cases} x = 1 \\ y = -2 \end{cases} \quad \begin{cases} x = -1 \\ y = 2 \end{cases}$$

These four systems are equivalent to the given system. Therefore the solution set is $\{(\frac{1}{2}, -\frac{5}{2}), (-\frac{1}{2}, \frac{5}{2}), (1, -2), (-1, 2)\}$. ∎

EXERCISES 9.3

In Exercises 1 through 24, find the solution set of the system. In Exercises 1 through 12, draw sketches of the graphs of the equations.

1. $\begin{cases} x^2 + y^2 = 25 \\ x - y + 1 = 0 \end{cases}$

2. $\begin{cases} x^2 + y^2 = 25 \\ x - 2y = -2 \end{cases}$

3. $\begin{cases} x^2 - y = 1 \\ x - 2y = -1 \end{cases}$

4. $\begin{cases} x^2 - y^2 = 9 \\ 2x + y = 6 \end{cases}$

5. $\begin{cases} x^2 - y^2 = 9 \\ x + y - 5 = 0 \end{cases}$

6. $\begin{cases} 4x^2 + y^2 = 25 \\ 2x + y + 1 = 0 \end{cases}$

7. $\begin{cases} x^2 - y - 4 = 0 \\ x - y - 3 = 0 \end{cases}$

8. $\begin{cases} 4x^2 + y - 3 = 0 \\ 8x + y - 7 = 0 \end{cases}$

9. $\begin{cases} x^2 - 2y^2 = 2 \\ x + 2y = 2 \end{cases}$

10. $\begin{cases} 4x^2 + y^2 = 17 \\ x^2 + y = 5 \end{cases}$

11. $\begin{cases} x^2 - y^2 = 15 \\ xy = 4 \end{cases}$

12. $\begin{cases} x^2 + y^2 = 25 \\ xy = 12 \end{cases}$

13. $\begin{cases} x^2 + y^2 = 4 \\ x^2 + 2y = 4 \end{cases}$

14. $\begin{cases} x^2 + xy + y^2 = 3 \\ x + y + 1 = 0 \end{cases}$

15. $\begin{cases} x^2 + y^2 = 25 \\ x^2 + 4y^2 = 64 \end{cases}$

16. $\begin{cases} 3x^2 + 2y^2 = 59 \\ 2x^2 + y^2 = 34 \end{cases}$

17. $\begin{cases} x^2 - y^2 = 9 \\ y^2 - 2x = 6 \end{cases}$

18. $\begin{cases} x^2 + y^2 = 16 \\ 9x^2 - 4y^2 = 36 \end{cases}$

19. $\begin{cases} x^2 - xy + 4 = 0 \\ 2x^2 - 2xy + y^2 = 8 \end{cases}$

20. $\begin{cases} 2x^2 - xy - y^2 = 0 \\ xy = 9 \end{cases}$

21. $\begin{cases} 10x^2 - xy + 4y^2 = 28 \\ 2x^2 - 3xy - 2y^2 = 0 \end{cases}$

22. $\begin{cases} 4x^2 - 5xy + 3y^2 = 24 \\ 2x^2 - 3xy + 2y^2 = 16 \end{cases}$

23. $\begin{cases} \dfrac{7}{x^2} - \dfrac{8}{y^2} = 5 \\[2mm] \dfrac{3}{x^2} - \dfrac{4}{y^2} = 2 \end{cases}$

24. $\begin{cases} \dfrac{3}{x^2} + \dfrac{1}{y^2} = 7 \\[2mm] \dfrac{5}{x^2} - \dfrac{2}{y^2} = -3 \end{cases}$

25. The sum of the reciprocals of two numbers is $\frac{4}{15}$ and their product is 60. What are the numbers?

26. The sum of the squares of two numbers is $\frac{5}{18}$ and the sum of 6 times the smaller number and 4 times the larger number is 3. What are the numbers?

27. The length of the hypotenuse of a right triangle is 37 cm and its area is 210 cm^2. Find the lengths of the legs of the triangle.

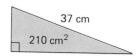

28. Determine the dimensions of a rectangle of area 60 in^2 that is inscribed in a circle of radius 6.5 in.

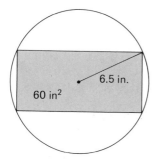

29. A rectangular lot has a perimeter of 40 m and an area of 96 m^2. What are its dimensions?

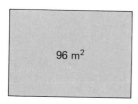

30. Find an equation of the common chord of the two circles

$$x^2 + y^2 - 4x - 1 = 0$$

and

$$x^2 + y^2 - 2y - 9 = 0$$

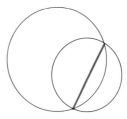

31. A group of students planned a field trip and agreed to contribute equal amounts toward the transportation costs of $150. Later five more students decided to go on the trip and the transportation cost for each student was reduced by $1.50. Find the number of students who actually made the trip and the amount each paid for transportation.

32. An investment yields an annual interest of $1500. If $500 more is invested and the rate is 2 percent less, the annual interest is $1300. What is the amount of the investment and the rate of interest?

33. A piece of tin is in the form of a rectangle whose area is 486 cm^2. A square of side 3 cm is cut from each corner, and an open box is made by turning up the ends and sides. If the volume of the box is 504 cm^3, what are the dimensions of the piece of tin?

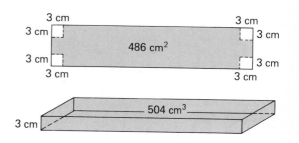

34. A closed rectangular box, having a square base, has a total surface area of 16 ft². If a main diagonal of the box has a length of 3 ft, what are the dimensions of the box?

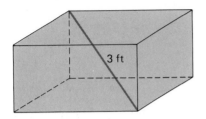

35. An open rectangular box, with a square base, has a surface area of 128 ft². If the cost per square foot of material for the sides was $1 and for the bottom was $1.20, and the total cost of the material was $131.20, what are the dimensions of the box?

36. A cyclist traveled a certain distance at her usual speed. If her speed had been 2 mi/hr faster, she would have traveled the distance in 1 hr less time. If her speed had been 2 mi/hr slower, she would have taken 2 hr longer. Find the distance traveled and her usual speed.

9.4   Systems of Linear Inequalities and Introduction to Linear Programming

Systems of linear inequalities are important in economics, business, statistics, science, engineering, and other fields. With electronic computers performing most of the computation, large numbers of inequalities with many unknowns are usually involved. In this section we briefly discuss how to solve systems of linear inequalities. We then give an introduction to linear programming, a related approach to decision-making problems.

Statements of the form

$$Ax + By + C > 0 \qquad Ax + By + C < 0$$
$$Ax + By + C \ge 0 \qquad Ax + By + C \le 0$$

where A, B, and C are constants, and A and B are not both zero, are inequalities of the first degree in two variables. By the graph of such an inequality, we mean the set of all points (x, y) in a rectangular cartesian coordinate system for which (x, y) is an ordered pair satisfying the inequality.

Every line in a plane divides the plane into two regions, one on each side of the line. Each of these regions is called a **half plane.** The graphs of inequalities of the forms

$$Ax + By + C > 0 \qquad \text{and} \qquad Ax + By + C < 0$$

are half planes. We shall show this for the particular inequalities

$$2x - y - 4 > 0 \qquad \text{and} \qquad 2x - y - 4 < 0$$

Let L be the line having the equation $2x - y - 4 = 0$. If we solve this equation for y, we obtain $y = 2x - 4$. If (x, y) is any point in the plane, exactly one of the following statements holds:

$$y = 2x - 4 \qquad y > 2x - 4 \qquad y < 2x - 4$$

Now, $y > 2x - 4$ if and only if the point (x, y) is above the point $(x, 2x - 4)$ on line L; see Figure 1. Furthermore, $y < 2x - 4$ if and only if the point (x, y) is below the point $(x, 2x - 4)$ on L; see Figure 2. Therefore the line L

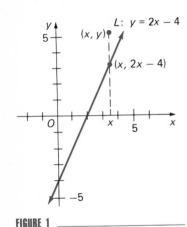

FIGURE 1

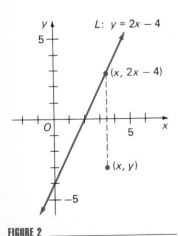

FIGURE 2

divides the plane into two regions. One region is the half plane above L, which is the graph of the inequality $y > 2x - 4$, and the other region is the half plane below L, which is the graph of the inequality $y < 2x - 4$. A similar discussion holds for any line L having an equation of the form $Ax + By + C = 0$ where $B \neq 0$.

If $B = 0$, an equation of line L is $Ax + C = 0$, and L is a vertical line whose equation can be written as $x = -\dfrac{C}{A}$. In particular, consider the line having the equation $x = 4$. Then if (x, y) is any point in the plane, exactly one of the following statements is true:

$$x = 4 \qquad x > 4 \qquad x < 4$$

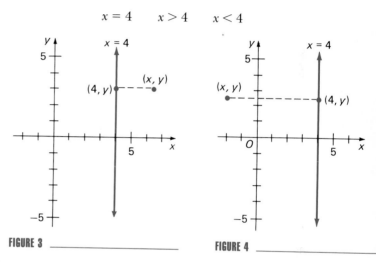

FIGURE 3 _____   FIGURE 4 _____

The point (x, y) is to the right of the point $(4, y)$ if and only if $x > 4$. See Figure 3, showing the graph of the inequality $x > 4$ as the half plane lying to the right of the line $x = 4$. Similarly, the graph of $x < 4$ is the half plane lying to the left of the line $x = 4$ because $x < 4$ if and only if the point (x, y) is to the left of the point $(4, y)$. See Figure 4. This discussion can be extended to any line having an equation of the form $Ax + C = 0$.

By generalizing the above arguments to any line, we can prove the following theorem.

Theorem 1

(i) The graph of $y > mx + b$ is the half plane lying above the line $y = mx + b$.

(ii) The graph of $y < mx + b$ is the half plane lying below the line $y = mx + b$.

(iii) The graph of $x > a$ is the half plane lying to the right of the line $x = a$.

(iv) The graph of $x < a$ is the half plane lying to the left of the line $x = a$.

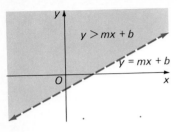

FIGURE 5 ————————

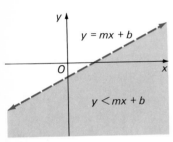

FIGURE 6 ————————

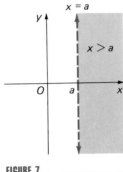

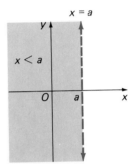

FIGURE 7 ————————   **FIGURE 8** ————————

Figures 5 and 6 show sketches of the graphs of the inequalities $y > mx + b$ and $y < mx + b$, respectively. Observe that the appropriate half plane is shaded. The graph of the line $y = mx + b$ is indicated by a dashed line to show that it is not part of the graph.

Sketches of the graphs of $x > a$ and $x < a$ appear in Figures 7 and 8, respectively.

Example **1**   Draw a sketch of the graph of the inequality

$$2x - 4y + 5 > 0$$

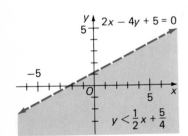

FIGURE 9 ————————

Solution   The given inequality is equivalent to

$$-4y > -2x - 5$$
$$y < \tfrac{1}{2}x + \tfrac{5}{4}$$

The graph of this inequality is the half plane below the line having the equation $y = \tfrac{1}{2}x + \tfrac{5}{4}$. A sketch of this graph is the shaded half plane appearing in Figure 9.   ▪

A **closed half plane** is a half plane together with the line bounding it, and it is the graph of an inequality of the form

$$Ax + By + C \geq 0 \qquad \text{or} \qquad Ax + By + C \leq 0$$

Illustration **1**

The inequality

$$4x + 5y - 20 \geq 0$$

is equivalent to

$$5y \geq -4x + 20$$
$$y \geq -\tfrac{4}{5}x + 4$$

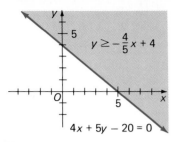

$$y \geq -\tfrac{4}{5}x + 4$$

$$4x + 5y - 20 = 0$$

FIGURE 10

Therefore the graph of this inequality is the closed half plane consisting of the line $y = -\tfrac{4}{5}x + 4$ and the half plane above it. A sketch of the graph is shown in Figure 10.   ■

Two intersecting lines divide the points of the plane into four regions. Each of these regions is the intersection of two half planes and is defined by a system of two linear inequalities.

Illustration   2

The system of inequalities

$$\begin{cases} 6x - y - 5 > 0 \\ 4x + 3y - 7 > 0 \end{cases}$$

$$\Leftrightarrow \qquad \begin{cases} y < 6x - 5 \\ y > -\tfrac{4}{3}x + \tfrac{7}{3} \end{cases}$$

defines the region that is the intersection of the half plane below the line $y = 6x - 5$ and the half plane above the line $y = -\tfrac{4}{3}x + \tfrac{7}{3}$. The region is shaded in Figure 11.

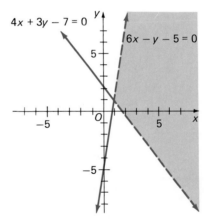

$$4x + 3y - 7 = 0$$

$$6x - y - 5 = 0$$

FIGURE 11   ■

The next example involves a region defined by a system of five linear inequalities.

Example   2
On a particular airline flight a passenger is allowed to check a piece of luggage, without additional cost, provided the sum of the three dimensions (length, width, and height) does not exceed 62 in. Suppose a piece of luggage has a height of 20 in. and the length is greater than the width, but not more than twice the width.

a) If the length is x inches and the width is y inches what is the system of inequalities involving x and y?

b) The region of permissible values of x and y is the graph of the system in part (a). Draw a sketch of this graph.

Solution   a) Because both x and y must be positive, we have the inequalities $x > 0$ and $y > 0$. Furthermore, x must be greater than y and less than or equal to $2y$; therefore we have the inequalities $x > y$ and $x \leq 2y$. Because $x + y + 20$ must be less than or equal to 62, we also have the inequality $x + y \leq 42$. Therefore we have the following system of inequalities:

$$\begin{cases} x > 0 \\ y > 0 \\ x > y \\ x \leq 2y \\ x + y \leq 42 \end{cases}$$

b) Let us designate the lines associated with the five inequalities by L_1, L_2, L_3, L_4, and L_5, respectively. These lines have the equations

$$L_1\colon x = 0 \qquad L_2\colon \quad y = 0 \qquad L_3\colon x = y$$
$$L_4\colon x = 2y \qquad L_5\colon x + y = 42$$

Lines L_1, L_2, L_3, and L_4 intersect at the origin. The x intercept of L_5 is 42 and the y intercept is also 42. If we solve simultaneously the equations for L_3 and L_5, we obtain the point of intersection $P(21, 21)$. The point of intersection of L_4 and L_5 is $Q(28, 14)$. Figure 12 shows the five lines.

The third inequality, $x > y$, defines the half plane below line L_3, and the fourth inequality, $x \leq 2y$, defines the closed half plane above line L_4. The inequality $y \leq 42 - x$, which is equivalent to the fifth inequality, defines the closed half plane below L_5. The region, the coordinates of whose points satisfy all five inequalities, consists of the interior of the triangle OPQ and the points on the line segments OQ and QP, excluding the origin and the point P. This region is shaded in Figure 12. ∎

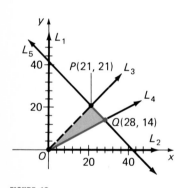

FIGURE 12 _____

The procedure demonstrated in the next example can be extended to an arbitrary number of linear inequalities.

Example   **3**   Find the region, if there is one, the coordinates of whose points satisfy each of the inequalities in the following system:

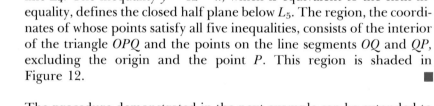

$$\begin{cases} x \geq 1 \\ y \geq 0 \\ 4x - 3y - 12 \leq 0 \\ 2x - y - 4 \leq 0 \\ x + y - 7 \leq 0 \end{cases}$$

Solution   Let lines L_1, L_2, L_3, L_4, and L_5 have the following equations, which are obtained from the inequalities of the system:

$$L_1: x = 1$$
$$L_2: y = 0$$
$$L_3: 4x - 3y - 12 = 0$$
$$L_4: 2x - y - 4 = 0$$
$$L_5: x + y - 7 = 0$$

We now determine whether there are any points whose coordinates satisfy all five of the inequalities. Solving simultaneously the equations for L_1 and L_2, we obtain the point of intersection $P(1, 0)$. We check to see whether the coordinates of this point satisfy the third, fourth, and fifth inequalities of the system; we see that they do. The coordinates of P then satisfy each of the five inequalities; therefore P is in the required region. We find the point of intersection of L_1 and L_3 to be $Q(1, -\frac{8}{3})$. The coordinates of Q satisfy the fifth inequality but not the second and fourth. Hence the point Q is not in the region.

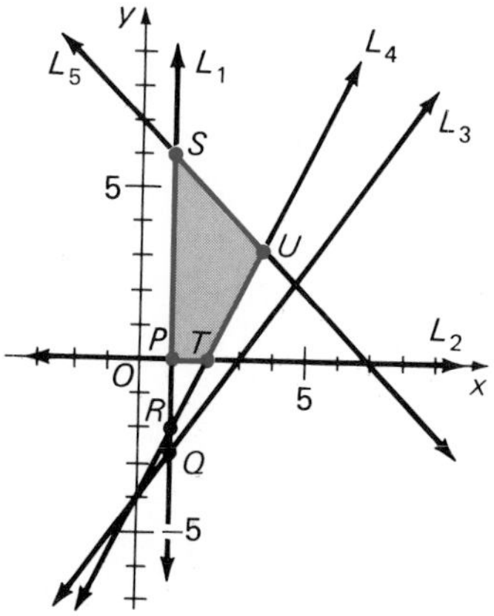

Continuing, we see that L_1 and L_4 intersect at the point $R(1, -2)$, and the coordinates of R satisfy the third and fifth inequalities but not the second. Lines L_1 and L_5 intersect at the point $S(1, 6)$ and the coordinates of this point satisfy the second, third, and fourth inequalities. Thus the point S is in the required region. Taking the equations in pairs, we have ten points of intersection. We see that the point of intersection of L_2 and L_4, which is $T(2, 0)$, and the point of intersection of L_4 and L_5, which is $U(\frac{11}{3}, \frac{10}{3})$, are in the required region. The remaining four points are not. Therefore we have four points, P, S, T, and U, whose coordinates satisfy all five inequalities. We plot these four points and the five lines in Figure 13. The five inequalities define half planes, and the interior and sides of the quadrilateral $PSUT$ give us all the points whose coordinates satisfy the given system.   ■

In the discussion that follows we refer to a *convex* region. A region is said to be **convex** if and only if for every pair of points P and Q in the region, the line segment PQ lies entirely in the region.

FIGURE 13 _____

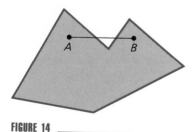

FIGURE 14 _____

Illustration   3

The shaded region in Figure 13 is convex. The region in Figure 14 is not convex because not every point of the line segment AB shown in the figure is in the region.   ■

Suppose a merchant has a fixed amount of floor space available and a certain amount of money to invest in two different kinds of merchandise. Knowing the cost and profit per unit of each kind of merchandise and the amount of floor space required for each unit, the dealer would be interested in determining the number of units of each kind he should stock in order for his profit to be a maximum. Or suppose three customers, each having specific requirements, are to be serviced with goods from two warehouses whose capacities are fixed. Knowing the costs per unit of servicing each customer from each warehouse, it is desired to determine the amount of goods to be shipped to each customer from each warehouse in order for the total cost of servicing to be a minimum. These are two decision problems that can be solved by methods referred to as **mathematical programming.** The independent variables in such decision problems are called **primary variables,** and they are subject to restrictions called **constraints.** The problem involves maximizing or minimizing an algebraic expression involving the primary variables. When the constraints can be expressed as a system of linear inequalities, and the algebraic expression is linear, we have a problem in **linear programming.**

We can solve a problem in linear programming involving two primary variables by geometric methods. We shall demonstrate this geometric method by an example because this will appeal to your intuition and aid you in understanding more complicated techniques. The geometric method involves first finding the set of all **feasible solutions to the problem.** By a feasible solution, we mean one that satisfies all the constraints. Each of the constraints is an inequality, and the region common to the graphs of these inequalities is the **set of all feasible solutions.** In the case of two primary variables this region is a polygon. To determine which feasible solution is the optimum (the "best") solution, we make use of the following theorem, which we state without proof.

Theorem 2

> If a, b, and c are constants, and
>
> $$z = ax + by + c$$
>
> where (x, y) corresponds to a point in a closed convex polygonal region, then the values of x and y that maximize and minimize z occur at vertices of the polygon.

Illustration 4

Suppose

$$z = 9x - 3y + 10$$

where x and y satisfy the five inequalities of the system in Example 3. The set of all feasible solutions to this system of inequalities consists of the points that are either in the interior or on the sides of the quadrilateral $PSUT$ in Figure 13. By Theorem 2 the maximum and minimum values of z must each occur at a vertex. We compute the value of z at each vertex.

At $(1, 0)$, $z = 9(1) - 3(0) + 10$ At $(1, 6)$, $z = 9(1) - 3(6) + 10$
             $= 19$ $= 1$

At $(2, 0)$, $z = 9(2) - 3(0) + 10$ At $(\frac{11}{3}, \frac{10}{3})$, $z = 9(\frac{11}{3}) - 3(\frac{10}{3}) + 10$
             $= 28$ $= 33$

The maximum value of z occurs at $(\frac{11}{3}, \frac{10}{3})$, and it is 33. The minimum value of z occurs at $(1, 6)$, and it is 1.   ∎

Example   4   A company manufactures two products, A and B, and each of these products must be processed on two different machines. Product A requires 1 min of work time per unit on machine 1 and 4 min of work time per unit on machine 2. Product B requires 2 min of work time per unit on machine 1 and 3 min of work time per unit on machine 2. Each day 100 min are available on machine 1 and 200 min are available on machine 2. To satisfy certain customers, the company must produce at least 6 units per day of product A and at least 12 units per day of product B. If the profit of each unit of product A is \$50 and the profit of each unit of product B is \$60, how many units of each product should be produced daily in order to maximize the company's profits?

Solution   Let x represent the number of units of product A to be produced daily, and let y represent the number of units of product B to be produced daily. If P dollars is the company's daily profit, then

$$P = 50x + 60y$$

We wish to maximize P subject to the following constraints:

$$x + 2y \leq 100 \qquad 4x + 3y \leq 200 \qquad x \geq 6 \qquad y \geq 12$$

The set of all feasible solutions to this system of inequalities consists of the points that are either in the interior or on the sides of the quadrilateral shown in Figure 15. The vertices of this quadrilateral are at the points $(6, 12)$, $(41, 12)$, $(20, 40)$, and $(6, 47)$. The maximum solution must occur at one of the vertices. We compute the value of P at each vertex.

At $(6, 12)$, $P = 50(6) + 60(12)$ At $(41, 12)$, $P = 50(41) + 60(12)$
              $= 1020$ $= 2770$

At $(20, 40)$, $P = 50(20) + 60(40)$ At $(6, 47)$, $P = 50(6) + 60(47)$
               $= 3400$ $= 3120$

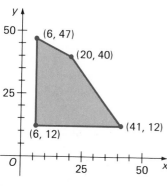

FIGURE 15

Therefore the maximum value of P occurs at $(20, 40)$. The company should manufacture 20 units of product A and 40 units of product B daily to realize a maximum daily profit of \$3400.   ∎

EXERCISES 9.4

In Exercises 1 through 12, draw a sketch of the graph of the inequality.

1. (a) $x > 2$; (b) $x \leq -3$
2. (a) $x \geq -4$; (b) $x < 7$
3. (a) $y \geq -5$; (b) $y < 6$
4. (a) $y > -1$; (b) $y \leq 8$
5. $y < 4x - 2$
6. $y \geq 2x - 3$
7. $2x - 7 \geq 0$
8. $y + 8 < 0$
9. $3x - 6y + 4 \leq 0$
10. $2y - 8x + 5 > 0$
11. $5x + 6 > 2y$
12. $9x + 3y \geq 7$

In Exercises 13 through 16, define the region by an inequality.

13. The half plane below the line $3x + 2y - 12 = 0$.
14. The half plane above the line $5x - 4y - 20 = 0$.
15. The closed half plane bounded by $2x - 7y + 3 = 0$ and containing the point $(-3, 2)$.
16. The closed half plane bounded by $4x - y + 8 = 0$ and containing the point $(5, -3)$.

In Exercises 17 through 26, draw a sketch of the region (if any) defined by the system of inequalities.

17. $\begin{cases} y - 2x \leq 4 \\ 2x + y \leq 4 \end{cases}$

18. $\begin{cases} 4x + y \geq 6 \\ 6x - 2y \leq 7 \end{cases}$

19. $\begin{cases} 2x + y \geq 4 \\ y - 2x \geq 4 \end{cases}$

20. $\begin{cases} 4x + 3y - 7 > 0 \\ 6x - y - 5 < 0 \end{cases}$

21. $\begin{cases} 3 + y \leq x \\ x < y - 4 \end{cases}$

22. $\begin{cases} -1 < x - y \\ x - y \leq 2 \end{cases}$

23. $\begin{cases} x - y - 1 \leq 0 \\ x + y + 1 \geq 0 \\ x - y + 1 \geq 0 \\ x + y - 1 \leq 0 \end{cases}$

24. $\begin{cases} x - 2y < 4 \\ 11 - 6x < y \\ 4x + 5y < 29 \end{cases}$

25. $\begin{cases} x \geq 0 \\ y \geq 0 \\ y \leq 3 \\ x + y - 5 \leq 0 \\ 2x + y - 8 \leq 0 \end{cases}$

26. $\begin{cases} x \geq 0 \\ y \geq 0 \\ y - x + 1 \geq 0 \\ x + y - 5 \leq 0 \\ x + 3y - 8 \leq 0 \end{cases}$

In Exercises 27 through 29, maximize z subject to the constraints.

27. $z = 2x + 3.5y$; $x \geq 0$; $y \geq 0$; $x + y \leq 6$; $3x + 5y \leq 20$; $x - y \geq 1$
28. $z = 4x + 5y$; $x \geq 0$; $y \geq 0$; $2x + y \leq 5$; $2x + y \geq 3$
29. $z = 6x + 5y$; $x \geq 1$; $y \geq 0$; $16 - x - 3y \geq 0$; $2x + y - 20 \leq 0$

In Exercises 30 through 32, minimize z subject to the constraints.

30. $z = 4x + 3y$; $x \geq 0$; $y \geq 0$; $x + 2y - 6 \leq 0$; $3x + y - 8 \leq 0$
31. $z = 5x + 3y$; $y \leq 7$; $2x - y + 3 \geq 0$; $2x + y - 7 \geq 0$; $x - y + 1 \leq 0$
32. $z = 6x + 3y$; $x + 2y - 2 \geq 0$; $x + y - 2 \leq 0$; $2x + y - 2 \geq 0$

33. Suppose, on the flight in Example 2, that a passenger has two pieces of luggage and that she is allowed to check the two pieces without additional cost provided their total dimensions do not exceed 106 in. The sum of the three dimensions of the larger piece is 58 in. Furthermore, the smaller piece has a height of 12 in. and a length greater than the width but not more than 10 in. greater. (a) If the smaller piece of luggage has a length of x inches and a width of y inches, what is the system of inequalities involving x and y? (b) Draw a sketch of the graph of the system in part (a) that shows the region of permissible values of x and y.

34. A mixture, weighing not more than 78 oz of food A and food B is to be made so that it contains at least 12 oz of a certain nutrient. Food A contains 0.4 oz of nutrient per unit and the weight of 1 unit is 3 oz. In food B there is 0.8 oz of nutrient per unit and the weight of 1 unit is 4 oz. (a) If there are x units of food A and y units of food B in the mixture, what is the system of inequalities involving x and y? (b) Draw a sketch of the graph of the system in part (a) that shows the region of permissible values of x and y.

35. A charity organization wishes to raise at least $12,000 from a movie premiere to be held at a theatre with 800 seats. The ticket prices are to be $20 and $15, with at least 500 tickets to be sold at $20. (a) If x tickets at $20 and y tickets at $15 are sold, what is the system of inequalities involving x and y? (b) Draw a sketch of the graph of the system in part (a) that shows the region of permissible values of x and y.

36. A wholesaler has 24,000 ft² of storage space available and $200,000 that can be spent for merchandise of types A and B. Each unit of type A costs $40 and requires 6 ft² of storage space; and each unit of type B costs $80 and requires 8 ft² of stor-

age space. If the wholesaler expects a profit of $20 per unit on type A and $45 per unit on type B, how many units of each should be bought and stocked in order to maximize the profit?

37. A company manufactures two products, A and B, and it requires three different machines to process each product. Product A requires 10 hr of time on machine M_1, 6 hr of time on machine M_2, and 12 hr of time on machine M_3. Product B requires 10 hr of time on M_1, 12 hr on M_2, and 4 hr on M_3. If the profit of each unit of product A is $400 and the profit of each unit of product B is $720, how many units of each product should be produced in each two-week period if there are 240 hr of time available on each machine and the company wishes to maximize the profit?

38. A distributor of video recorders has two warehouses that supply three different retailers. To deliver a recorder to retailer R_1 costs $27 from warehouse W_1 and $36 from warehouse W_2. It costs $9 to deliver a recorder from W_1 to retailer R_2 and $6 to deliver one from W_2 to R_2. For retailer R_3, it costs $15 if the recorder comes from W_1 and $30 if it comes from W_2. Suppose that R_1 orders three recorders, R_2 orders four recorders, and two recorders are ordered from R_3. If the distributor has five recorders in stock in W_1 and four in W_2, how many recorders should be shipped from each warehouse to each retailer in order for the distributor to minimize the delivery costs?

39. A mixture of food A and food B is to be made so that it contains at least 45 oz of nutrient N_1 and 40 oz of nutrient N_2. The cost per pound of A is $4 and each pound of A contains 1 oz of N_1 and 2 oz of N_2. Food B costs $8 per pound and each pound of B contains 1.5 oz of N_1 and 0.5 oz of N_2. If the weight of the mixture must not exceed 40 lb, how many pounds of each food should be used so that the total cost is a minimum?

9.5 Solutions of Systems of Linear Equations by Matrices

In Section 9.2 we showed how the elimination method is used to put a system of three linear equations in triangular form, from which we then obtain the solution of the system by back substitution. This procedure leads to the solution of a system of linear equations by *matrices*. The matrix method is of special importance because most computer solutions of linear systems depend on matrices.

In Example 2 of Section 9.2 we had the system of equations

$$\begin{cases} 4x - 2y - 3z = 8 \\ 5x + 3y - 4z = 4 \\ 6x - 4y - 5z = 12 \end{cases} \quad \text{(I)}$$

We solved this system by obtaining the equivalent system in triangular form,

$$\begin{cases} 4x - 2y - 3z = 8 \\ - 22y + z = 24 \\ z = 2 \end{cases}$$

By using a zero as the coefficient of a variable that does not appear in an equation, this latter system can be written as

$$\begin{cases} 4x - 2y - 3z = 8 \\ 0x - 22y + z = 24 \\ 0x + 0y + z = 2 \end{cases} \quad \text{(II)}$$

The procedure used to obtain system (II) involves operations on the equations of system (I) and a series of equivalent systems until a system in triangular form is found. These operations cause changes in the coefficients of the variables and changes in the constant terms in the right members of the equations. Thus we are concerned essentially with these numbers (the coefficients and the constant terms) that appear in each of the equivalent systems. Therefore, to simplify the calculations, we introduce a notation for recording the coefficients and constant terms so that the variables do not have to be written. For system (I) the coefficients of the variables are listed in the following way:

$$\begin{bmatrix} 4 & -2 & -3 \\ 5 & 3 & -4 \\ 6 & -4 & -5 \end{bmatrix} \tag{III}$$

If we also list the constant terms appearing on the right-hand sides of the equations of system (I), we have

$$\left[\begin{array}{ccc|c} 4 & -2 & -3 & 8 \\ 5 & 3 & -4 & 4 \\ 6 & -4 & -5 & 12 \end{array}\right] \tag{IV}$$

where a vertical line is used to separate the coefficients from the constant terms. The numbers involved in system (II) are listed as

$$\left[\begin{array}{ccc|c} 4 & -2 & -3 & 8 \\ 0 & -22 & 1 & 24 \\ 0 & 0 & 1 & 2 \end{array}\right] \tag{V}$$

Each of the arrays (III), (IV), and (V) is called a **matrix.** The numbers in the matrix are called **elements.** The elements that appear next to each other horizontally form a **row,** and those that appear vertically form a **column.** Hence in matrix (IV) the elements in the first row are 4, -2, -3, and 8, and the elements in the second column are -2, 3, and -4. If there are m rows and n columns in a matrix, then its order is $m \times n$ (read "m by n"). Notice that the number of rows is stated first. If $m = n$, we have a **square matrix of order n.** Matrices (IV) and (V) are of order 3×4, and matrix (III) is a square matrix of order 3. The **main diagonal** of a matrix contains the elements on the diagonal line starting at the upper left-hand corner. In matrix (III) the elements on the main diagonal are 4, 3, and -5, while in matrix (V) they are 4, -22, and 1.

Suppose that we have a system of linear equations where the constants are on the right side and the terms involving the variables, appearing in the same order in each equation, are on the left side. System (I) is such a system. The matrix whose only elements are the coefficients of the variables, listed as they appear in the equations, is called the **coefficient matrix.** Thus matrix (III) is the coefficient matrix of system (I). The matrix obtained from the coefficient matrix by listing the constants in the additional column on the right, where the coefficients are separated from the constants by a vertical line, is called the **augmented matrix.** For system (I) the augmented matrix is matrix (IV).

To solve a system of linear equations by using matrices, we start with the augmented matrix and perform operations on the rows to obtain a matrix of an equivalent system. The process is continued until we obtain a matrix of a system in triangular form, that is, a matrix that has zeros everywhere below the main diagonal. Such a matrix is said to be in **echelon form.** For example, (II) is the system in triangular form that is equivalent to system (I). Furthermore, matrix (V) is the augmented matrix of system (II) and is in echelon form.

The rules for performing operations on matrices are called **elementary row operations,** and they are given in the following theorem.

Theorem 1

> If we have an augmented matrix of a system of linear equations, each of the following operations produces a matrix of an equivalent system of linear equations:
>
> (i) Interchanging any two rows.
> (ii) Multiplying each element of a row by the same nonzero number.
> (iii) Replacing a given row by a new row whose elements are the sum of k_1 times the elements of the given row and k_2 times the corresponding elements of any other row, where k_1 and k_2 are real numbers, and $k_1 \neq 0$.

The proof of Theorem 1 utilizes the corresponding operations on the equations of the system and is omitted. Observe that the theorem involves *operations on rows only.* Do not make the mistake of applying the operations to columns, as doing so does not produce a matrix of an equivalent system.

In part (ii) of Theorem 1, it is understood that to multiply a row by a number means to multiply each element of the row by the number. In a similar manner, when applying part (iii) by adding one row to another row, we are actually adding corresponding elements of the two rows.

The following illustration shows how Theorem 1 is used to solve system (I). You should compare the computation with that of Example 2 in Section 9.2, which involves the same system.

Illustration   1

System (I) is

$$\begin{cases} 4x - 2y - 3z = 8 \\ 5x + 3y - 4z = 4 \\ 6x - 4y - 5z = 12 \end{cases}$$

and the augmented matrix is

$$\begin{bmatrix} 4 & -2 & -3 & | & 8 \\ 5 & 3 & -4 & | & 4 \\ 6 & -4 & -5 & | & 12 \end{bmatrix}$$

To obtain 0 as the first element in the second row, we replace the second row by the sum of 5 times the first row and -4 times the second row; and to obtain 0 as the first element in the third row, we replace the third row by the sum of 6 times the first row and -4 times the third row. Thus we have the matrix

$$\begin{bmatrix} 4 & -2 & -3 & | & 8 \\ 0 & -22 & 1 & | & 24 \\ 0 & 4 & 2 & | & 0 \end{bmatrix}$$

From this matrix we obtain 0 as the second element in the third row by replacing the third row by the sum of 2 times the second row and 11 times the third row. Doing this gives

$$\begin{bmatrix} 4 & -2 & -3 & | & 8 \\ 0 & -22 & 1 & | & 24 \\ 0 & 0 & 24 & | & 48 \end{bmatrix}$$

Multiplying the third row by $\frac{1}{24}$, we get

$$\begin{bmatrix} 4 & -2 & -3 & | & 8 \\ 0 & -22 & 1 & | & 24 \\ 0 & 0 & 1 & | & 2 \end{bmatrix}$$

which is matrix (V) and is in echelon form; it is the augmented matrix for system (II), which is in triangular form. The solution set is then found by back substitution as in Example 2 of Section 9.2. ∎

The procedure demonstrated in Illustration 1 for solving a system of linear equations by matrices is called the **Gaussian reduction method,** named after the German mathematician and scientist Karl Friedrich Gauss (1777–1855). The method requires reducing a matrix to echelon form. In linear algebra, a more advanced course, a formal step-by-step process for doing this is given. For the fairly simple systems in this text, the echelon form can be obtained by a sequence of elementary row operations determined by observation and trial and error.

In the following example a system of four linear equations in four variables is solved by the Gaussian reduction method. By a solution of a system of four equations in w, x, y, and z, we mean an ordered four-tuple (r, s, t, u) such that each of the equations is satisfied if w, x, y, and z are replaced by r, s, t, and u, respectively.

Example   1   Use the Gaussian reduction method to find the solution set of the system

$$\begin{cases} w + x + y + z = 5 \\ \quad\quad 3z + x = w - 2 \\ \quad 2x + 2y = 3w + 2 \\ \quad\quad\quad y = w + z \end{cases}$$

Solution   We replace each equation by an equivalent one in which the terms involving variables are on the left side and the constant terms are on the right side. We have the following equivalent system in which the equations are written so that terms containing the same variable are in a vertical column.

$$\begin{cases} w + x + y + z = 5 \\ -w + x \quad\quad + 3z = -2 \\ -3w + 2x + 2y \quad\quad = 2 \\ -w \quad\quad + y - z = 0 \end{cases}$$

The augmented matrix of this system is

$$\left[\begin{array}{cccc|c} 1 & 1 & 1 & 1 & 5 \\ -1 & 1 & 0 & 3 & -2 \\ -3 & 2 & 2 & 0 & 2 \\ -1 & 0 & 1 & -1 & 0 \end{array}\right]$$

Initially we obtain zeros as the first elements in the second, third, and fourth rows. We replace the second row by the sum of the first and second rows; we replace the third row by the sum of 3 times the first row and the third row; we replace the fourth row by the sum of the first and fourth rows. We then have the matrix

$$\left[\begin{array}{cccc|c} 1 & 1 & 1 & 1 & 5 \\ 0 & 2 & 1 & 4 & 3 \\ 0 & 5 & 5 & 3 & 17 \\ 0 & 1 & 2 & 0 & 5 \end{array}\right]$$

We now interchange the second and fourth rows because then we will have 1 as an element in the second row and second column; this will make it easier to obtain zeros in the third and fourth rows of the second column. We then have the matrix

$$\left[\begin{array}{cccc|c} 1 & 1 & 1 & 1 & 5 \\ 0 & 1 & 2 & 0 & 5 \\ 0 & 5 & 5 & 3 & 17 \\ 0 & 2 & 1 & 4 & 3 \end{array}\right]$$

Replacing the third row by the sum of the third row and -5 times the second row, and replacing the fourth row by the sum of the fourth row and -2 times the second row, we obtain the matrix

$$\begin{bmatrix} 1 & 1 & 1 & 1 & | & 5 \\ 0 & 1 & 2 & 0 & | & 5 \\ 0 & 0 & -5 & 3 & | & -8 \\ 0 & 0 & -3 & 4 & | & -7 \end{bmatrix}$$

Replacing the fourth row by the sum of 3 times the third row and -5 times the fourth row, we get the matrix

$$\begin{bmatrix} 1 & 1 & 1 & 1 & | & 5 \\ 0 & 1 & 2 & 0 & | & 5 \\ 0 & 0 & -5 & 3 & | & -8 \\ 0 & 0 & 0 & -11 & | & 11 \end{bmatrix}$$

We now multiply the fourth row by $-\frac{1}{11}$ and obtain

$$\begin{bmatrix} 1 & 1 & 1 & 1 & | & 5 \\ 0 & 1 & 2 & 0 & | & 5 \\ 0 & 0 & -5 & 3 & | & -8 \\ 0 & 0 & 0 & 1 & | & -1 \end{bmatrix}$$

This matrix is in echelon form and is the augmented matrix of the system in triangular form,

$$\begin{cases} w + x + y + z = 5 \\ \phantom{w + {}} x + \phantom{y + {}} 2y = 5 \\ \phantom{w + x + {}} -5y + 3z = -8 \\ \phantom{w + x + y + {}} z = -1 \end{cases}$$

We back substitute into the third equation the value of z from the fourth equation and obtain $y = 1$; then, back substituting 1 for y in the second equation, we have $x = 3$. By back substituting in the first equation 3 for x, 1 for y, and -1 for z, we obtain $w = 2$. We have then the following system, which is equivalent to the given system:

$$\begin{cases} w = 2 \\ x = 3 \\ y = 1 \\ z = -1 \end{cases}$$

Thus the solution set of the given system is $\{(2, 3, 1, -1)\}$. ∎

Example   2   Use the Gaussian reduction method to find the solution set of the system

$$\begin{cases} 2x + y - 3z = 0 \\ 3x + 2y - 4z = 2 \\ x - y - 3z = -6 \end{cases}$$

Solution   The augmented matrix of the system is

$$\begin{bmatrix} 2 & 1 & -3 & | & 0 \\ 3 & 2 & -4 & | & 2 \\ 1 & -1 & -3 & | & -6 \end{bmatrix}$$

Replacing the second row by the sum of 3 times the first row and -2 times the second row, and replacing the third row by the sum of the first row and -2 times the third row, we obtain

$$\begin{bmatrix} 2 & 1 & -3 & | & 0 \\ 0 & -1 & -1 & | & -4 \\ 0 & 3 & 3 & | & 12 \end{bmatrix}$$

We now replace the third row by the sum of 3 times the second row and the third row and get

$$\begin{bmatrix} 2 & 1 & -3 & | & 0 \\ 0 & -1 & -1 & | & -4 \\ 0 & 0 & 0 & | & 0 \end{bmatrix}$$

This matrix is in echelon form and is the augmented matrix of the system

$$\begin{cases} 2x + y - 3z = 0 \\ -y - z = -4 \\ 0 = 0 \end{cases}$$

From the third equation we observe that the equations of the system are dependent. The third equation is an identity for any ordered triple (x, y, z) and in particular for $(0, 0, t)$. Thus we replace the equation $0 = 0$ by the equation $z = t$, and we have the equivalent system in triangular form

$$\begin{cases} 2x + y - 3z = 0 \\ -y - z = -4 \\ z = t \end{cases}$$

We now back substitute the value of z from the third equation into the second and solve for y. We then have the equivalent system

$$\begin{cases} 2x + y - 3z = 0 \\ \qquad\quad y = 4 - t \\ \qquad\qquad z = t \end{cases}$$

Back substituting the values of y and z from the second and third equations into the first gives

$$\begin{cases} 2x + (4 - t) - 3t = 0 \\ \qquad\qquad\quad y = 4 - t \\ \qquad\qquad\qquad z = t \end{cases}$$

$$\Leftrightarrow \quad \begin{cases} x = 2t - 2 \\ y = 4 - t \\ z = t \end{cases}$$

Therefore the solution set of the given system is $\{(2t - 2, 4 - t, t)\}$.

In Example 2 observe that when the Gaussian reduction method is used to solve a system of dependent equations, the matrix in echelon form has all zeros in the last row. ∎

EXERCISES 9.5

In Exercises 1 through 6, use the Gaussian reduction method to find the solution set of the system of the indicated exercise of Exercises 9.1.

1. Exercise 11
2. Exercise 12
3. Exercise 15
4. Exercise 16
5. Exercise 21
6. Exercise 22

In Exercises 7 through 10, use the Gaussian reduction method to find the solution set of the system.

7. $\begin{cases} 4x - 3y = 5 \\ 6x + 2y = 1 \end{cases}$

8. $\begin{cases} 4x + 3y = 2 \\ 5x + 4y = 1 \end{cases}$

9. $\begin{cases} \dfrac{x}{2} - \dfrac{y}{6} = 1 \\ \dfrac{x}{4} + \dfrac{y}{3} = 3 \end{cases}$

10. $\begin{cases} \dfrac{x}{3} + \dfrac{y}{2} = 0 \\ \dfrac{x}{6} + \dfrac{y}{8} = \dfrac{1}{2} \end{cases}$

In Exercises 11 through 16, use the Gaussian reduction method to find the solution set of the system of the indicated exercise of Exercises 9.2.

11. Exercise 1
12. Exercise 2
13. Exercise 5
14. Exercise 6
15. Exercise 7
16. Exercise 10

In Exercises 17 through 26, use the Gaussian reduction method to find the solution set of the system.

17. $\begin{cases} x - 2y - z = 3 \\ x + y + z = 4 \\ x - 3y - z = 4 \end{cases}$

18. $\begin{cases} x + y - 2z = 5 \\ 3x + 2y = 4 \\ 2x + z = 2 \end{cases}$

19. $\begin{cases} \dfrac{x}{2} + \dfrac{y}{3} + \dfrac{z}{4} = 2 \\ \dfrac{x}{4} - \dfrac{2y}{3} - \dfrac{z}{4} = \dfrac{1}{2} \\ \dfrac{x}{6} + \dfrac{y}{9} + \dfrac{z}{2} = 4 \end{cases}$

20. $\begin{cases} \dfrac{x}{3} + \dfrac{y}{2} - \dfrac{z}{2} = 0 \\ \dfrac{x}{6} + \dfrac{3y}{4} + \dfrac{z}{3} = \dfrac{5}{8} \\ \dfrac{3x}{2} + \dfrac{y}{4} + \dfrac{z}{6} = -\dfrac{1}{8} \end{cases}$

21. $\begin{cases} 2x + 3y - 4z = 4 \\ x + 2y - 5z = 6 \\ 4x + 5y - 2z = 0 \end{cases}$

22. $\begin{cases} w + 3x + y + z = 3 \\ -w - 3y - z = 0 \\ -2w + 3x - 4y + z = 0 \\ -w - 6x - 2y + 2z = -4 \end{cases}$

23. $\begin{cases} 2w + x - 3y - 3z = 4 \\ \quad\quad x + 2y + \; z = -3 \\ 2w - x \quad\quad\; + 3z = -3 \\ 5w \quad\quad\; - y + \; z = -6 \end{cases}$

25. $\begin{cases} 2w + 3x - 4y - \; z = 3 \\ 3w + \; x + \; y + 2z = 1 \\ \; w - 2x + 3y - \; z = 0 \\ \; w - 2x - \; y - 9z = 5 \end{cases}$

24. $\begin{cases} 2w - 4x + \; y - \; 2z = 3 \\ 3w + \; x + 2y + \; 3z = 12 \\ \; w - 4x + 2y - \; 6z = 1 \\ 5w + \; x \quad\quad\; + 11z = 16 \end{cases}$

26. $\begin{cases} 4w + \; x - 2y + \; z = 4 \\ 2w + 3x + 4y - 3z = -2 \\ 3w - 2x + 5y + 3z = 2 \\ \; w + 4x + 3y + \; z = 5 \end{cases}$

9.6 Determinants and Cramer's Rule

Associated with each square matrix is a number called the *determinant*. Determinants can be used to solve systems of linear equations by formulas known as *Cramer's rule*, named after the Swiss mathematician Gabriel Cramer (1704–1752). We begin by defining a second-order determinant. We shall then obtain Cramer's rule for a system of two linear equations in two variables.

Definition

> **Second-Order Determinant**
>
> If H is the square matrix of order two
>
> $$\begin{bmatrix} a_1 & b_1 \\ a_2 & b_2 \end{bmatrix}$$
>
> then the **determinant** of H, denoted by either det H or $\begin{vmatrix} a_1 & b_1 \\ a_2 & b_2 \end{vmatrix}$, is defined by
>
> $$\begin{vmatrix} a_1 & b_1 \\ a_2 & b_2 \end{vmatrix} = a_1 b_2 - a_2 b_1$$
>
> and is of the **second order.**

Observe that the notation for the determinant of a matrix consists of the elements of the matrix with vertical lines in place of brackets. We emphasize that the determinant of a matrix is a number. It can be remembered as a difference of products of elements in the diagonals.

Illustration 1

If

$$H = \begin{bmatrix} 3 & -2 \\ 4 & -1 \end{bmatrix}$$

we compute det H as follows:

$$\begin{vmatrix} 3 & -2 \\ 4 & -1 \end{vmatrix} = 3(-1) - 4(-2)$$

$$= 5 \qquad \blacksquare$$

Example   **1**   Compute the value of each determinant.

a) $\begin{vmatrix} 4 & -8 \\ -3 & 10 \end{vmatrix}$ b) $\begin{vmatrix} 7 & 2 \\ 5 & 0 \end{vmatrix}$

Solution

a) $\begin{vmatrix} 4 & -8 \\ -3 & 10 \end{vmatrix} = 4(10) - (-3)(-8)$

$$= 40 - 24$$

$$= 16$$

b) $\begin{vmatrix} 7 & 2 \\ 5 & 0 \end{vmatrix} = 7(0) - (5)(2)$

$$= 0 - 10$$

$$= -10 \qquad \blacksquare$$

Consider now the system

$$\begin{cases} a_1 x + b_1 y = c_1 \\ a_2 x + b_2 y = c_2 \end{cases} \qquad (I)$$

where in each equation at least one of the coefficients of the variables is nonzero. Without loss of generality, we assume in the first equation that the coefficient of the first variable x is nonzero. (If the coefficient of x were zero, then the coefficient of y would be nonzero, and we could consider y as the first variable.) To solve system (I) we use the Gaussian reduction method to find an equivalent system in triangular form.

The augmented matrix of system (I) is

$$\begin{bmatrix} a_1 & b_1 & \bigm| & c_1 \\ a_2 & b_2 & \bigm| & c_2 \end{bmatrix}$$

We obtain a matrix of an equivalent system by replacing the second row by the sum of $-a_2$ times the first row and a_1 times the second row. (Here we are applying Theorem 1(iii) of Section 9.5, where it is necessary that $a_1 \neq 0$.) We then have

$$\begin{bmatrix} a_1 & b_1 & \bigm| & c_1 \\ 0 & (a_1 b_2 - a_2 b_1) & \bigm| & (a_1 c_2 - a_2 c_1) \end{bmatrix}$$

which is the augmented matrix of the system

$$\begin{cases} a_1x + b_1y = c_1 \\ (a_1b_2 - a_2b_1)y = a_1c_2 - a_2c_1 \end{cases}$$

In the second equation of this system the coefficient of y and the right side can be written as determinants; that is, the system is

$$\begin{cases} a_1x + b_1y = c_1 \\ \begin{vmatrix} a_1 & b_1 \\ a_2 & b_2 \end{vmatrix} y = \begin{vmatrix} a_1 & c_1 \\ a_2 & c_2 \end{vmatrix} \end{cases} \qquad \text{(II)}$$

If

$$\begin{vmatrix} a_1 & b_1 \\ a_2 & b_2 \end{vmatrix} \neq 0$$

we can solve the second equation for y, and we have

$$y = \frac{\begin{vmatrix} a_1 & c_1 \\ a_2 & c_2 \end{vmatrix}}{\begin{vmatrix} a_1 & b_1 \\ a_2 & b_2 \end{vmatrix}} \qquad \text{if} \qquad \begin{vmatrix} a_1 & b_1 \\ a_2 & b_2 \end{vmatrix} \neq 0$$

We can use the same procedure to solve for x and obtain

$$x = \frac{\begin{vmatrix} c_1 & b_1 \\ c_2 & b_2 \end{vmatrix}}{\begin{vmatrix} a_1 & b_1 \\ a_2 & b_2 \end{vmatrix}} \qquad \text{if} \qquad \begin{vmatrix} a_1 & b_1 \\ a_2 & b_2 \end{vmatrix} \neq 0$$

These formulas for x and y are known as **Cramer's rule** for the solution of a system of two linear equations in two variables. Observe that in each of the denominators we have the determinant of the coefficient matrix of the given system. Let us denote this determinant by D.

$$D = \begin{vmatrix} a_1 & b_1 \\ a_2 & b_2 \end{vmatrix}$$

In the numerator for the value of x we have the determinant obtained from D by replacing the coefficients, a_1 and a_2, of x by the numbers c_1 and c_2, respectively. We denote this determinant by D_x. In the numerator, for the value of y we have the determinant obtained from D by replacing the coefficients, b_1 and b_2, of y by c_1 and c_2, respectively. We denote this determinant by D_y.

We summarize the results of the preceding discussion in the following formal statement of Cramer's rule.

Cramer's Rule for a System of Two Linear Equations

Suppose we have the system of two linear equations in two variables x and y

$$\begin{cases} a_1x + b_1y = c_1 \\ a_2x + b_2y = c_2 \end{cases}$$

where either $a_1 \neq 0$ or $b_1 \neq 0$, and either $a_2 \neq 0$ or $b_2 \neq 0$. Let

$$D = \begin{vmatrix} a_1 & b_1 \\ a_2 & b_2 \end{vmatrix} \qquad D_x = \begin{vmatrix} c_1 & b_1 \\ c_2 & b_2 \end{vmatrix} \qquad D_y = \begin{vmatrix} a_1 & c_1 \\ a_2 & c_2 \end{vmatrix}$$

If $D \neq 0$, then the system has a unique solution given by

$$x = \frac{D_x}{D} \qquad \text{and} \qquad y = \frac{D_y}{D}$$

Example 2 Use Cramer's rule to find the solution set of the system

$$\begin{cases} 4x + 3y = 6 \\ 2x - 5y = 16 \end{cases}$$

Solution We first compute D, D_x, and D_y.

$$D = \begin{vmatrix} 4 & 3 \\ 2 & -5 \end{vmatrix} \qquad D_x = \begin{vmatrix} 6 & 3 \\ 16 & -5 \end{vmatrix} \qquad D_y = \begin{vmatrix} 4 & 6 \\ 2 & 16 \end{vmatrix}$$

$$= -20 - 6 \qquad\quad = -30 - 48 \qquad\quad = 64 - 12$$

$$= -26 \qquad\qquad = -78 \qquad\qquad = 52$$

Therefore, from Cramer's rule,

$$x = \frac{D_x}{D} \qquad\quad y = \frac{D_y}{D}$$

$$= \frac{-78}{-26} \qquad\quad = \frac{52}{-26}$$

$$= 3 \qquad\qquad = -2$$

Thus the solution set is $\{(3, -2)\}$. ∎

Note that Cramer's rule requires that $D \neq 0$. If $D = 0$, then system (II) is

$$\begin{cases} a_1x + b_1y = c_1 \\ \qquad\qquad 0 = D_y \end{cases} \qquad\qquad \text{(III)}$$

If $D_y \neq 0$, there is no ordered pair (x, y) for which the second equation of system (III) is a true statement; thus the equations of system (I) are inconsistent. If $D_y = 0$, then system (III) is

$$\begin{cases} a_1x + b_1y = c_1 \\ \qquad\quad 0 = 0 \end{cases}$$

Therefore the equations of system (I) are dependent. The two results are summarized in the following theorem.

Theorem 1

For the determinants defined in Cramer's rule for a system of two linear equations,

  (i) if $D = 0$ and $D_y \neq 0$, then the equations are inconsistent;
(ii) if $D = 0$ and $D_y = 0$, then the equations are dependent.

In parts (i) and (ii), D_y can be replaced by D_x.

Example 3 Apply Cramer's rule to each of the following systems.

a) $\begin{cases} 6x - 3y = 5 \\ 2x - y = 4 \end{cases}$ b) $\begin{cases} 3x + 2y = 4 \\ 6x + 4y = 8 \end{cases}$

Solution a) We compute D, D_x, and D_y.

$$D = \begin{vmatrix} 6 & -3 \\ 2 & -1 \end{vmatrix} \qquad D_x = \begin{vmatrix} 5 & -3 \\ 4 & -1 \end{vmatrix} \qquad D_y = \begin{vmatrix} 6 & 5 \\ 2 & 4 \end{vmatrix}$$

$$ = -6 + 6 \qquad\quad = -5 + 12 \qquad\quad = 24 - 10$$

$$ = 0 \qquad\qquad\quad\; = 7 \qquad\qquad\quad\;\; = 14$$

Because $D = 0$ and $D_y \neq 0$, it follows from part (i) of Theorem 1 that the equations of the given system are inconsistent.

b) We compute D, D_x, and D_y.

$$D = \begin{vmatrix} 3 & 2 \\ 6 & 4 \end{vmatrix} \qquad D_x = \begin{vmatrix} 4 & 2 \\ 8 & 4 \end{vmatrix} \qquad D_y = \begin{vmatrix} 3 & 4 \\ 6 & 8 \end{vmatrix}$$

$$ = 12 - 12 \qquad\quad = 16 - 16 \qquad\quad = 24 - 24$$

$$ = 0 \qquad\qquad\quad\; = 0 \qquad\qquad\quad\;\; = 0$$

From part (ii) of Theorem 1, we conclude that the equations of the given system are dependent. ∎

Observe that both systems in Example 3 are taken from Section 9.1, the first from Illustrations 2 and 5 of that section and the second from Illustrations 3 and 6.

We can extend Cramer's rule to a system of three linear equations in three variables. However, before we do that we discuss a *third-order determinant*.

Definition

> **Third-Order Determinant**
> If H is the square matrix of order three
>
> $$\begin{bmatrix} a_1 & b_1 & c_1 \\ a_2 & b_2 & c_2 \\ a_3 & b_3 & c_3 \end{bmatrix}$$
>
> then the **determinant** of H, denoted by either det H or $\begin{vmatrix} a_1 & b_1 & c_1 \\ a_2 & b_2 & c_2 \\ a_3 & b_3 & c_3 \end{vmatrix}$, is defined by
>
> $$\begin{vmatrix} a_1 & b_1 & c_1 \\ a_2 & b_2 & c_2 \\ a_3 & b_3 & c_3 \end{vmatrix} = a_1b_2c_3 - a_1b_3c_2 + a_3b_1c_2 - a_2b_1c_3 + a_2b_3c_1 - a_3b_2c_1$$
>
> and is of the **third order.**

In practice it is not necessary to remember the complicated formula in the above definition. To obtain a better method for computing a third-order determinant, we first define the *minor* and the *cofactor* of an element in a matrix.

Definition

> **Minor**
> Let e be an element of a square matrix H, and let G be the matrix obtained by deleting the row and column in which e appears. Then the **minor** of e is det G.

Illustration 2

Suppose that H is the matrix

$$\begin{bmatrix} 2 & 3 & -1 \\ -4 & 1 & -2 \\ 5 & 0 & -3 \end{bmatrix}$$

Consider the elements of the first row: 2, 3, and -1. The minor of 2 is

$$\begin{vmatrix} 1 & -2 \\ 0 & -3 \end{vmatrix} = (1)(-3) - (0)(-2)$$

$$= -3$$

The minor of 3 is

$$\begin{vmatrix} -4 & -2 \\ 5 & -3 \end{vmatrix} = (-4)(-3) - (5)(-2)$$

$$= 22$$

The minor of -1 is

$$\begin{vmatrix} -4 & 1 \\ 5 & 0 \end{vmatrix} = (-4)(0) - (5)(1)$$

$$= -5 \qquad \blacksquare$$

Definition

> **Cofactor**
>
> Let e be the element in row i and column j of a square matrix H, and let M be the minor of e. Then the **cofactor** K of the element e is defined by
>
> $$K = (-1)^{i+j}M$$

Illustration 3

Suppose that H is the matrix of Illustration 2. Consider the elements of the first row: 2, 3, and -1. The cofactor of 2 is

$$(-1)^{1+1}\begin{vmatrix} 1 & -2 \\ 0 & -3 \end{vmatrix} = (-1)^2(-3)$$

$$= -3$$

The cofactor of 3 is

$$(-1)^{1+2}\begin{vmatrix} -4 & -2 \\ 5 & -3 \end{vmatrix} = (-1)^3(22)$$

$$= -22$$

The cofactor of -1 is

$$(-1)^{1+3}\begin{vmatrix} -4 & 1 \\ 5 & 0 \end{vmatrix} = (-1)^4(-5)$$

$$= -5 \qquad \blacksquare$$

    A mnemonic device for determining whether $(-1)^{i+j}$ is $+1$ or -1 is the so-called *checkerboard rule*, which alternates $+$ and $-$ signs over the determinant starting with a $+$ in the upper left corner as follows:

$$\begin{vmatrix} + & - & + \\ - & + & - \\ + & - & + \end{vmatrix}$$

Illustration 4

Suppose that H is the matrix

$$\begin{bmatrix} a_1 & b_1 & c_1 \\ a_2 & b_2 & c_2 \\ a_3 & b_3 & c_3 \end{bmatrix}$$

The product of a_1 and its cofactor is

$$a_1 \cdot (-1)^{1+1} \begin{vmatrix} b_2 & c_2 \\ b_3 & c_3 \end{vmatrix} = a_1 \begin{vmatrix} b_2 & c_2 \\ b_3 & c_3 \end{vmatrix}$$

The product of b_1 and its cofactor is

$$b_1 \cdot (-1)^{1+2} \begin{vmatrix} a_2 & c_2 \\ a_3 & c_3 \end{vmatrix} = -b_1 \begin{vmatrix} a_2 & c_2 \\ a_3 & c_3 \end{vmatrix}$$

The product of c_1 and its cofactor is

$$c_1 \cdot (-1)^{1+3} \begin{vmatrix} a_2 & b_2 \\ a_3 & b_3 \end{vmatrix} = c_1 \begin{vmatrix} a_2 & b_2 \\ a_3 & b_3 \end{vmatrix} \qquad \blacksquare$$

If we now group pairs of terms in the equation for a third-order determinant, we find that

$$\begin{vmatrix} a_1 & b_1 & c_1 \\ a_2 & b_2 & c_2 \\ a_3 & b_3 & c_3 \end{vmatrix} = a_1(b_2c_3 - b_3c_2) - b_1(a_2c_3 - a_3c_2) + c_1(a_2b_3 - a_3b_2)$$

Each of the expressions in parentheses on the right side of this equation is a second-order determinant. Thus, with determinant notation, the equation can be written

$$\begin{vmatrix} a_1 & b_1 & c_1 \\ a_2 & b_2 & c_2 \\ a_3 & b_3 & c_3 \end{vmatrix} = a_1 \begin{vmatrix} b_2 & c_2 \\ b_3 & c_3 \end{vmatrix} - b_1 \begin{vmatrix} a_2 & c_2 \\ a_3 & c_3 \end{vmatrix} + c_1 \begin{vmatrix} a_2 & b_2 \\ a_3 & b_3 \end{vmatrix} \qquad (1)$$

By referring to Illustration 4 we notice that the terms on the right side of Equation (1) are the products of the elements of the first row and their cofactors. Equation (1) is a special case of the following theorem, the proof of which is omitted. However, in Exercises 43 and 44, you are asked to verify the theorem for two other special cases.

Theorem 2

Let H be a square matrix of order three, and let e_1, e_2, and e_3 be the elements in any row or column of H. Furthermore, let K_1, K_2, and K_3 be the cofactors of e_1, e_2, and e_3, respectively. Then

$$\det H = e_1K_1 + e_2K_2 + e_3K_3$$

When Theorem 2 is used, we say that the determinant is evaluated by the cofactors of the elements of the row, or column, in which the elements e_1, e_2, and e_3 appear.

Example   **4**   If H is the matrix of Illustration 2, evaluate det H in two ways: (a) by the cofactors of the elements of the first row; (b) by the cofactors of the elements of the second column.

Solution

a) $$\begin{vmatrix} 2 & 3 & -1 \\ -4 & 1 & -2 \\ 5 & 0 & -3 \end{vmatrix} = 2 \cdot (-1)^{1+1} \begin{vmatrix} 1 & -2 \\ 0 & -3 \end{vmatrix} + 3 \cdot (-1)^{1+2} \begin{vmatrix} -4 & -2 \\ 5 & -3 \end{vmatrix} + (-1)(-1)^{1+3} \begin{vmatrix} -4 & 1 \\ 5 & 0 \end{vmatrix}$$

$$= 2(1)(-3 - 0) + 3(-1)(12 + 10) + (-1)(1)(0 - 5)$$
$$= 2(-3) - 3(22) - 1(-5)$$
$$= -67$$

b) $$\begin{vmatrix} 2 & 3 & -1 \\ -4 & 1 & -2 \\ 5 & 0 & -3 \end{vmatrix} = 3 \cdot (-1)^{1+2} \begin{vmatrix} -4 & -2 \\ 5 & -3 \end{vmatrix} + 1 \cdot (-1)^{2+2} \begin{vmatrix} 2 & -1 \\ 5 & -3 \end{vmatrix} + 0 \cdot (-1)^{3+2} \begin{vmatrix} 2 & -1 \\ -4 & -2 \end{vmatrix}$$

$$= 3(-1)(12 + 10) + 1(1)(-6 + 5) + 0$$
$$= -3(22) + 1(-1)$$
$$= -67 \qquad \blacksquare$$

Observe in Example 4 that because 0 is an element in the second column, less computation is involved in part (b) than in part (a). It is always best to evaluate the determinant by the cofactors of the row, or column, containing the most zeros.

We have introduced determinants in order to apply them to solve systems of linear equations. Our treatment is a brief one, and we do not give a formal definition of a determinant of order higher than three. However, the determinant of a square matrix of order n can be evaluated by a theorem similar to Theorem 2. For instance, the determinant of a square matrix of order four can be evaluated by the cofactors of the elements of any row, or column, where each of the corresponding minors is a determinant of order three.

The German mathematician Gottfried Wilhelm Leibniz (1646–1716) is often said to be the inventor of determinants. In fact, however, the Japanese mathematician Seki Kowa (1642–1708) used them first.

We are now ready to state Cramer's rule for a system of three linear equations in three variables. The proof is analogous to that for a system of two linear equations, but its details are omitted.

Cramer's Rule for a System of Three Linear Equations

Suppose we have the system of three linear equations in three variables x, y, and z

$$\begin{cases} a_1x + b_1y + c_1z = d_1 \\ a_2x + b_2y + c_2z = d_2 \\ a_3x + b_3y + c_3z = d_3 \end{cases}$$

where in each equation at least one of the coefficients of the variables is nonzero. Let

$$D = \begin{vmatrix} a_1 & b_1 & c_1 \\ a_2 & b_2 & c_2 \\ a_3 & b_3 & c_3 \end{vmatrix}$$

$$D_x = \begin{vmatrix} d_1 & b_1 & c_1 \\ d_2 & b_2 & c_2 \\ d_3 & b_3 & c_3 \end{vmatrix} \qquad D_y = \begin{vmatrix} a_1 & d_1 & c_1 \\ a_2 & d_2 & c_2 \\ a_3 & d_3 & c_3 \end{vmatrix} \qquad D_z = \begin{vmatrix} a_1 & b_1 & d_1 \\ a_2 & b_2 & d_2 \\ a_3 & b_3 & d_3 \end{vmatrix}$$

If $D \neq 0$, then the system has a unique solution given by

$$x = \frac{D_x}{D} \qquad y = \frac{D_y}{D} \qquad z = \frac{D_z}{D}$$

Observe that in each of the denominators for x, y, and z we have the determinant of the coefficient matrix of the given system. In each of the numerators is the determinant of the matrix obtained from the coefficient matrix by substituting the numbers d_1, d_2, and d_3 for the coefficients of the variable for which we are solving.

Example 5 Use Cramer's rule to find the solution set of the system

$$\begin{cases} 4x + 3y + z = 1 \\ 3x - 2y - 3z = -2 \\ 5x + 4y + 2z = 3 \end{cases}$$

Solution Because the computation of a determinant is simplified when some elements are 0, we first replace the given system by an equivalent one that has some 0 coefficients. If we replace the second equation by the sum of 3 times the first equation and the second equation, and we replace the third equation by the sum of -2 times the first equation and the third equation, we have the equivalent system

$$\begin{cases} 4x + 3y + z = 1 \\ 15x + 7y + 0z = 1 \\ -3x - 2y + 0z = 1 \end{cases}$$

We first evaluate D, the determinant of the coefficient matrix, by the cofactors of the elements of the third column.

$$\begin{vmatrix} 4 & 3 & 1 \\ 15 & 7 & 0 \\ -3 & -2 & 0 \end{vmatrix} = 1 \begin{vmatrix} 15 & 7 \\ -3 & -2 \end{vmatrix}$$

$$= 1(-30 + 21)$$

$$= -9$$

From Cramer's rule,

$$x = \frac{D_x}{-9} \qquad y = \frac{D_y}{-9} \qquad z = \frac{D_z}{-9}$$

We now evaluate D_x and D_y. In each case the computation is performed by using the cofactors of the elements of the third column.

$$D_x = \begin{vmatrix} 1 & 3 & 1 \\ 1 & 7 & 0 \\ 1 & -2 & 0 \end{vmatrix} \qquad D_y = \begin{vmatrix} 4 & 1 & 1 \\ 15 & 1 & 0 \\ -3 & 1 & 0 \end{vmatrix}$$

$$= 1 \begin{vmatrix} 1 & 7 \\ 1 & -2 \end{vmatrix} \qquad = 1 \begin{vmatrix} 15 & 1 \\ -3 & 1 \end{vmatrix}$$

$$= 1(-2 - 7) \qquad = 1(15 + 3)$$

$$= -9 \qquad\qquad = 18$$

Therefore

$$x = \frac{-9}{-9} \qquad y = \frac{18}{-9}$$

$$= 1 \qquad\quad = -2$$

Instead of computing z by using D_z, we substitute the values obtained for x and y in one of the original equations. Substituting into the first equation of the given system, we have

$$4(1) + 3(-2) + z = 1$$

$$z = 3$$

Therefore the solution set is $\{(1, -2, 3)\}$. The solution can be checked by verifying that the second and third equations of the given system are satisfied when $x = 1$, $y = -2$, and $z = 3$.   ■

Cramer's rule can be extended to systems of n linear equations in n variables. For any such system, if the determinant of the coefficient matrix is not zero, there is a unique solution. A discussion of systems for which $D = 0$ requires concepts of linear algebra. When there are many equations, it is tedious to apply Cramer's rule because of the many determinants that are needed. Of course, computers can be used to solve such systems. If a computer is not used, the Gaussian reduction method is then more practical for systems of more than three equations.

EXERCISES 9.6

In Exercises 1 through 8, evaluate the given determinant.

1. (a) $\begin{vmatrix} 3 & -1 \\ 2 & 5 \end{vmatrix}$; (b) $\begin{vmatrix} 6 & 5 \\ -2 & 1 \end{vmatrix}$

2. (a) $\begin{vmatrix} -2 & 4 \\ 3 & 7 \end{vmatrix}$; (b) $\begin{vmatrix} 8 & 5 \\ 5 & 8 \end{vmatrix}$

3. (a) $\begin{vmatrix} -4 & 8 \\ -3 & 7 \end{vmatrix}$; (b) $\begin{vmatrix} -1 & 3 \\ 9 & 0 \end{vmatrix}$

4. (a) $\begin{vmatrix} 2 & 6 \\ 0 & -1 \end{vmatrix}$; (b) $\begin{vmatrix} 3 & 8 \\ -7 & -6 \end{vmatrix}$

5. $\begin{vmatrix} -1 & 6 & 4 \\ -1 & 2 & 1 \\ -2 & 7 & 4 \end{vmatrix}$

6. $\begin{vmatrix} 4 & -3 & 3 \\ 6 & -4 & 5 \\ -3 & 3 & -2 \end{vmatrix}$

7. $\begin{vmatrix} 1 & 2 & 0 \\ 2 & -1 & 3 \\ 1 & 5 & -4 \end{vmatrix}$

8. $\begin{vmatrix} -6 & -3 & 2 \\ 3 & 2 & 0 \\ 4 & 2 & -1 \end{vmatrix}$

In Exercises 9 through 16, solve for x.

9. $\begin{vmatrix} x & 2 \\ -1 & 5 \end{vmatrix} = 7$

10. $\begin{vmatrix} 4 & -2 \\ x & 3 \end{vmatrix} = 6$

11. $\begin{vmatrix} x & -2 \\ 5 & x \end{vmatrix} = 14$

12. $\begin{vmatrix} x^2 & -3 \\ x & 1 \end{vmatrix} = 4$

13. $\begin{vmatrix} x & 0 & 0 \\ 3 & -1 & 5 \\ 2 & 2 & -6 \end{vmatrix} = 8$

14. $\begin{vmatrix} x & 1 & 0 \\ 3 & -2 & -2 \\ -4 & 4 & 5 \end{vmatrix} = 5$

15. $\begin{vmatrix} x & 2 & 5 \\ 1 & 0 & 4 \\ x^2 & -1 & 2 \end{vmatrix} = 3$

16. $\begin{vmatrix} x & 0 & -1 \\ 3 & x & 4 \\ 2 & 2 & 1 \end{vmatrix} = 10$

In Exercises 17 through 42, find the solution set of the system by Cramer's rule. If the equations of the system are either inconsistent or dependent, so indicate.

17. Exercise 11 of Exercises 9.1
18. Exercise 12 of Exercises 9.1
19. Exercise 15 of Exercises 9.1
20. Exercise 16 of Exercises 9.1
21. Exercise 21 of Exercises 9.1
22. Exercise 22 of Exercises 9.1

23. $\begin{cases} 2x - 3y = 4 \\ 3x - y = 1 \end{cases}$

24. $\begin{cases} x + y = 6 \\ 3x - 5y = -16 \end{cases}$

25. $\begin{cases} 2x + 6y = 3 \\ 8x - 3y = -6 \end{cases}$

26. $\begin{cases} 6x + y = 6 \\ 2x - 2y = -5 \end{cases}$

27. Exercise 1 of Exercises 9.2
28. Exercise 2 of Exercises 9.2
29. Exercise 5 of Exercises 9.2
30. Exercise 6 of Exercises 9.2
31. Exercise 7 of Exercises 9.2
32. Exercise 12 of Exercises 9.2

33. $\begin{cases} 2x - 2y + z = 0 \\ x + 5y - 7z = 3 \\ x - y - 3z = -7 \end{cases}$

34. $\begin{cases} x + 4y + z = 5 \\ 2x + y - 4z = 0 \\ 3x - 3y + z = -1 \end{cases}$

35. $\begin{cases} 4x + 3y + 5z = 0 \\ 2x - 4y - 3z = 0 \\ 6x - 2y + z = 0 \end{cases}$

36. $\begin{cases} 3x + y + 2z = 2 \\ 3x + 2y + 4z = 3 \\ 3x - y - 2z = 0 \end{cases}$

37. $\begin{cases} 6x + y - 2z = 0 \\ 2x + 3y - z = 0 \\ 4x + 2y - 3z = 2 \end{cases}$

38. $\begin{cases} x - y - z = 0 \\ 8x - 2y + z = 0 \\ x + 3y + 5z = 0 \end{cases}$

39. $\begin{cases} 2x - y + 3z = 5 \\ 3x - 2y + 7z = 3 \\ 3x - y + 2z = 10 \end{cases}$

40. $\begin{cases} 5x - 6y = 1 \\ 2x + 3z = 3 \\ 4y + 6z = 5 \end{cases}$

41. $\begin{cases} 4x - 3z = 6 \\ 3y + 4z = 5 \\ 5z - 2x = 4 \end{cases}$

42. $\begin{cases} 5x + 3y - z = 5 \\ 3x - 4z = 8 \\ 4x + 7y = 7 \end{cases}$

43. Use the equation in the definition of a third-order determinant to verify Theorem 2 where $\det H$ is the sum of the products of the elements of the third row and their cofactors.

44. Use the equation in the definition of a third-order determinant to verify Theorem 2 where $\det H$ is the sum of the products of the elements of the second column and their cofactors.

In Exercises 45 and 46, evaluate the fourth-order determinant by using an extension of Theorem 2; that is, if e_1, e_2, e_3, and e_4 are the elements of any row or column and K_1, K_2, K_3, and K_4 are the cofactors of e_1, e_2, e_3, and e_4, then the value of the determinant is $e_1K_1 + e_2K_2 + e_3K_3 + e_4K_4$.

45. $\begin{vmatrix} 2 & 1 & -1 & 5 \\ -1 & -3 & 0 & -2 \\ 5 & 8 & 2 & 6 \\ 0 & 9 & 0 & 5 \end{vmatrix}$

46. $\begin{vmatrix} 3 & -1 & 2 & -3 \\ 0 & 1 & -1 & 3 \\ -1 & 2 & 0 & -4 \\ 0 & 6 & 4 & 3 \end{vmatrix}$

In Exercises 47 and 48, find the solution set of the system by extending Cramer's rule to a system of four linear equations in four variables.

47. $\begin{cases} 4w + 2x + 3y = 5 \\ 2w + 3x + z = 2 \\ 3w + 5x + 2z = 3 \\ 4x + 3y + 6z = 1 \end{cases}$ 48. $\begin{cases} 4w - 2x + y - z = 3 \\ 2w - 4x + y + z = -1 \\ 4w - 3x - 2z = 5 \\ 3w + x - y = 2 \end{cases}$

9.7   Properties of Matrices (Supplementary)

In Section 9.5 we showed how matrices can be used to solve a system of linear equations. There are many other applications of matrices in a variety of fields including both the social and physical sciences. In this section and the next we present some of the properties of matrices that are used in these applications.

We defined a matrix of order $m \times n$ as one having m rows and n columns. If $m = n$, we have a square matrix of order n. If a matrix has only one column it is called a **column matrix,** and if it has only one row it is called a **row matrix.**

Illustration 1

The following matrices have the order indicated below the matrix:

$$\begin{bmatrix} 5 & 4 \\ -2 & 7 \\ 0 & -4 \end{bmatrix} \quad \begin{bmatrix} 9 & -1 & 6 \\ 2 & 10 & -8 \end{bmatrix} \quad \begin{bmatrix} 7 \\ -3 \\ 0 \\ 4 \end{bmatrix} \quad \begin{bmatrix} 0 & 2 & -5 \end{bmatrix} \quad \begin{bmatrix} -4 & 1 & 8 \\ 0 & -2 & 11 \\ 6 & -3 & 0 \end{bmatrix}$$

3×2 2×3 4×1 1×3 3×3

                         Column matrix   Row matrix   Square matrix ∎

Two matrices are said to be **equal** if and only if their orders are equal and the corresponding elements are equal. That is,

$$\begin{bmatrix} r & s \\ t & u \end{bmatrix} = \begin{bmatrix} a & b \\ c & d \end{bmatrix}$$

if and only if

$$r = a \quad s = b \quad t = c \quad u = d$$

Illustration 2

If

$$A = \begin{bmatrix} 4 & -3 & 0 \\ 1 & 3 & 2 \end{bmatrix} \quad B = \begin{bmatrix} 1 & 3 & 2 \\ 4 & -3 & 0 \end{bmatrix} \quad C = \begin{bmatrix} 2 \times 2 & -3 & 0 \\ 3 - 2 & \frac{6}{2} & 2 \\ & & 1 \end{bmatrix}$$

then $A \neq B$ and $B \neq C$, but $A = C$. ∎

We now define *addition* of two matrices that have the same number of rows and columns.

Definition

> **Sum of Two Matrices**
> If A and B are two matrices having the same order, the **sum** of A and B, denoted by $A + B$, is the matrix for which each of its elements is the sum of the corresponding elements of A and B.

Illustration 3

If

$$A = \begin{bmatrix} 5 & -2 & 1 \\ 3 & 0 & -4 \end{bmatrix} \quad \text{and} \quad B = \begin{bmatrix} 4 & 7 & 6 \\ -1 & 8 & -3 \end{bmatrix}$$

then

$$A + B = \begin{bmatrix} 5 + 4 & -2 + 7 & 1 + 6 \\ 3 + (-1) & 0 + 8 & -4 + (-3) \end{bmatrix}$$

$$= \begin{bmatrix} 9 & 5 & 7 \\ 2 & 8 & -7 \end{bmatrix} \qquad \blacksquare$$

Observe that addition of matrices having different orders is not defined.

Associated with the operation of addition of matrices is an *identity element* called a **zero matrix** and denoted by **0.** It is one whose elements are all zeros and can be of any order.

Illustration 4

The following matrices are zero matrices:

$$\begin{bmatrix} 0 & 0 \\ 0 & 0 \\ 0 & 0 \end{bmatrix} \quad \begin{bmatrix} 0 & 0 & 0 & 0 \end{bmatrix} \quad \begin{bmatrix} 0 \\ 0 \end{bmatrix} \qquad \blacksquare$$

Theorem 1

> If H is any matrix and **0** is the zero matrix of the same order as H, then
>
> $$H + \mathbf{0} = H \qquad \text{and} \qquad \mathbf{0} + H = H$$

The proof of Theorem 1 follows from the definitions of the zero matrix and matrix addition. For instance, if H is a matrix of order 2×3, then

$$\begin{bmatrix} a_1 & b_1 & c_1 \\ a_2 & b_2 & c_2 \end{bmatrix} + \begin{bmatrix} 0 & 0 & 0 \\ 0 & 0 & 0 \end{bmatrix} = \begin{bmatrix} a_1 & b_1 & c_1 \\ a_2 & b_2 & c_2 \end{bmatrix}$$

If H is any matrix, then the **negative** (or **additive inverse**) of H, denoted by $-H$, is the matrix for which each of the elements is the negative of the corresponding element of H.

Illustration 5

If

$$H = \begin{bmatrix} 5 & -2 & 1 & 3 \\ 4 & 0 & -4 & 6 \end{bmatrix} \quad \text{then} \quad -H = \begin{bmatrix} -5 & 2 & -1 & -3 \\ -4 & 0 & 4 & -6 \end{bmatrix}$$ ■

Theorem 2

> The sum of a matrix H and the negative of H is a zero matrix; that is, if H is any matrix, then
>
> $$H + (-H) = \mathbf{0}$$

The proof of Theorem 2 follows from the definition of the negative of a matrix.

Recall that subtraction of real numbers was defined in terms of addition; that is $a - b = a + (-b)$. A similar definition is given for *subtraction of matrices.*

Definition

> **Difference of Two Matrices**
> If A and B are matrices having the same order, then the **difference** of A and B, denoted by $A - B$, is defined by
>
> $$A - B = A + (-B)$$

Illustration 6

$$\begin{bmatrix} 7 & -3 \\ 2 & 1 \\ -6 & 0 \end{bmatrix} - \begin{bmatrix} 4 & -2 \\ -5 & 0 \\ 3 & -8 \end{bmatrix} = \begin{bmatrix} 7 & -3 \\ 2 & 1 \\ -6 & 0 \end{bmatrix} + \begin{bmatrix} -4 & 2 \\ 5 & 0 \\ -3 & 8 \end{bmatrix}$$

$$= \begin{bmatrix} 3 & -1 \\ 7 & 1 \\ -9 & 8 \end{bmatrix}$$ ■

Matrix addition is commutative and associative, as stated in the next theorem.

Theorem 3

If A, B, and C are matrices of the same order, then

(i) $A + B = B + A$

(ii) $A + (B + C) = (A + B) + C$

The proof of Theorem 3 follows from the definition of matrix addition and the commutativity and associativity of addition of real numbers.

Example   **1**   Verify Theorem 3(ii) for the following matrices:

$$A = \begin{bmatrix} 2 & -1 & 0 \\ 3 & 4 & -2 \end{bmatrix} \qquad B = \begin{bmatrix} -3 & 5 & 1 \\ -6 & 0 & 2 \end{bmatrix} \qquad C = \begin{bmatrix} 1 & -4 & -5 \\ -3 & -1 & 7 \end{bmatrix}$$

Solution   We first compute $B + C$.

$$B + C = \begin{bmatrix} -3 & 5 & 1 \\ -6 & 0 & 2 \end{bmatrix} + \begin{bmatrix} 1 & -4 & -5 \\ -3 & -1 & 7 \end{bmatrix}$$

$$= \begin{bmatrix} -3 + 1 & 5 + (-4) & 1 + (-5) \\ -6 + (-3) & 0 + (-1) & 2 + 7 \end{bmatrix}$$

$$= \begin{bmatrix} -2 & 1 & -4 \\ -9 & -1 & 9 \end{bmatrix}$$

Therefore

$$A + (B + C) = \begin{bmatrix} 2 & -1 & 0 \\ 3 & 4 & -2 \end{bmatrix} + \begin{bmatrix} -2 & 1 & -4 \\ -9 & -1 & 9 \end{bmatrix}$$

$$= \begin{bmatrix} 2 + (-2) & -1 + 1 & 0 + (-4) \\ 3 + (-9) & 4 + (-1) & -2 + 9 \end{bmatrix}$$

$$= \begin{bmatrix} 0 & 0 & -4 \\ -6 & 3 & 7 \end{bmatrix}$$

We now determine $A + B$.

$$A + B = \begin{bmatrix} 2 & -1 & 0 \\ 3 & 4 & -2 \end{bmatrix} + \begin{bmatrix} -3 & 5 & 1 \\ -6 & 0 & 2 \end{bmatrix}$$

$$= \begin{bmatrix} 2 + (-3) & -1 + 5 & 0 + 1 \\ 3 + (-6) & 4 + 0 & -2 + 2 \end{bmatrix}$$

$$= \begin{bmatrix} -1 & 4 & 1 \\ -3 & 4 & 0 \end{bmatrix}$$

Thus

$$(A + B) + C = \begin{bmatrix} -1 & 4 & 1 \\ -3 & 4 & 0 \end{bmatrix} + \begin{bmatrix} 1 & -4 & -5 \\ -3 & -1 & 7 \end{bmatrix}$$

$$= \begin{bmatrix} -1 + 1 & 4 + (-4) & 1 + (-5) \\ -3 + (-3) & 4 + (-1) & 0 + 7 \end{bmatrix}$$

$$= \begin{bmatrix} 0 & 0 & -4 \\ -6 & 3 & 7 \end{bmatrix}$$

By comparing the results, we observe that for these three matrices

$$A + (B + C) = (A + B) + C \qquad \blacksquare$$

We now define *multiplication* of a matrix by a *scalar* where, for our purposes, a **scalar** denotes a real number.

Definition

> **Product of a Scalar and a Matrix**
> The **product** of a scalar k and a matrix H, denoted by kH, is the matrix in which each element is obtained by multiplying k by the corresponding element of H.

Illustration 7

$$2 \begin{bmatrix} 4 & -1 & 6 \\ -3 & 0 & \dfrac{1}{2} \end{bmatrix} = \begin{bmatrix} 8 & -2 & 12 \\ -6 & 0 & 1 \end{bmatrix}$$

and

$$-3 \begin{bmatrix} -5 & -3 \\ 2 & 1 \\ 0 & -4 \end{bmatrix} = \begin{bmatrix} 15 & 9 \\ -6 & -3 \\ 0 & 12 \end{bmatrix} \qquad \blacksquare$$

The definition of the product of two matrices may be peculiar to you. However, there are applications that warrant such a definition. Before giving the formal definition, we consider some special cases and begin with the following illustration involving a row matrix and a column matrix.

Illustration 8

Let

$$R = \begin{bmatrix} 2 & 3 & -5 \end{bmatrix} \quad \text{and} \quad C = \begin{bmatrix} 1 \\ 4 \\ 2 \end{bmatrix}$$

The product of R and C, denoted by RC, is the matrix containing a single element that is the sum of the products formed by multiplying each element in R by the corresponding element in C:

$$RC = \begin{bmatrix} 2 & 3 & -5 \end{bmatrix} \begin{bmatrix} 1 \\ 4 \\ 2 \end{bmatrix}$$
$$= \begin{bmatrix} (2)(1) + (3)(4) + (-5)(2) \end{bmatrix}$$
$$= \begin{bmatrix} 2 + 12 - 10 \end{bmatrix}$$
$$= \begin{bmatrix} 4 \end{bmatrix} \qquad \blacksquare$$

Observe that the method of matrix multiplication in Illustration 8 allows us to write the linear equation

$$ax + by + cz = d$$

in matrix form as

$$\begin{bmatrix} a & b & c \end{bmatrix} \begin{bmatrix} x \\ y \\ z \end{bmatrix} = \begin{bmatrix} d \end{bmatrix}$$

Illustration 9

Consider the product of the matrices C and D, where

$$C = \begin{bmatrix} 3 & 0 & -4 \\ -2 & 2 & -1 \end{bmatrix} \quad \text{and} \quad D = \begin{bmatrix} 2 & -3 \\ 4 & -1 \\ 1 & 5 \end{bmatrix}$$

To obtain the element in row 1, column 1 of CD we multiply the elements in row 1 of C by the corresponding elements in column 1 of D and add the products as follows:

$$(3)(2) + (0)(4) + (-4)(1) = 2$$

For the element in row 1, column 2 of CD we multiply the elements in row 1 of C by the corresponding elements in column 2 of D and add the products:

$$(3)(-3) + (0)(-1) + (-4)(5) = -29$$

For the element in row 2, column 1 of CD we add the products obtained by multiplying the elements in row 2 of C by the corresponding elements in column 1 of D:

$$(-2)(2) + (2)(4) + (-1)(1) = 3$$

For the element in row 2, column 2 of CD we add the products obtained by multiplying the elements in row 2 of C by the corresponding elements in column 2 of D:

$$(-2)(-3) + (2)(-1) + (-1)(5) = -1$$

We have therefore obtained the following product:

$$\begin{bmatrix} 3 & 0 & -4 \\ -2 & 2 & -1 \end{bmatrix} \begin{bmatrix} 2 & -3 \\ 4 & -1 \\ 1 & 5 \end{bmatrix} = \begin{bmatrix} 2 & -29 \\ 3 & -1 \end{bmatrix}$$ ∎

We now give the formal definition of the product of two matrices A and B. It is necessary that the number of columns of the first matrix A be the same as the number of rows of the second matrix B.

Definition

> **Product of Two Matrices**
> Suppose that A is a matrix of order $m \times p$ and B is a matrix of order $p \times n$. Then the **product** of A and B, denoted by AB, is the $m \times n$ matrix for which the element in the ith row and jth column is the sum of the products formed by multiplying each element in the ith row of A by the corresponding element in the jth column of B.

Illustration   10

We apply the definition to find the product DC for the matrices of Illustration 9.

$$DC = \begin{bmatrix} 2 & -3 \\ 4 & -1 \\ 1 & 5 \end{bmatrix} \begin{bmatrix} 3 & 0 & -4 \\ -2 & 2 & -1 \end{bmatrix}$$

$$= \begin{bmatrix} (2)(3) + (-3)(-2) & (2)(0) + (-3)(2) & (2)(-4) + (-3)(-1) \\ (4)(3) + (-1)(-2) & (4)(0) + (-1)(2) & (4)(-4) + (-1)(-1) \\ (1)(3) + (5)(-2) & (1)(0) + (5)(2) & (1)(-4) + (5)(-1) \end{bmatrix}$$

$$= \begin{bmatrix} 12 & -6 & -5 \\ 14 & -2 & -15 \\ -7 & 10 & -9 \end{bmatrix}$$ ∎

Again we stress that in order to obtain the product of matrices A and B, the number of columns in A must be the same as the number of rows in B; otherwise the product is not defined. When AB is defined, this product has as many rows as A and as many columns as B. Refer back to Illustrations 8, 9, and 10 and observe the following:

Illustration   8: 1×3 matrix times 3×1 matrix yields 1×1 matrix

Illustration   9: 2×3 matrix times 3×2 matrix yields 2×2 matrix

Illustration 10: 3×2 matrix times 2×3 matrix yields 3×3 matrix

Another observation from Illustrations 9 and 10 is that CD is not equal to DC. That is, matrix multiplication is *not commutative*. However, the following theorem states that matrix multiplication is associative and distributive when the products and sums exist.

Theorem 4

> If A, B, and C are matrices such that the following products and sums exist, then
>
> (i) $A(BC) = (AB)C$
> (ii) $A(B + C) = AB + AC$
> (iii) $(B + C)A = BA + CA$

The proof of Theorem 4 is omitted. However, in the next example we verify part (i) for particular matrices. You are asked to verify parts (ii) and (iii) for particular matrices in Exercises 36 and 37.

Example 2 Let

$$A = \begin{bmatrix} -3 & 0 \\ 2 & 1 \\ 3 & -2 \end{bmatrix} \qquad B = \begin{bmatrix} 2 & -1 \\ 1 & 3 \end{bmatrix} \qquad C = \begin{bmatrix} 4 & 3 \\ -1 & -2 \end{bmatrix}$$

Solution

$$BC = \begin{bmatrix} 2 & -1 \\ 1 & 3 \end{bmatrix} \begin{bmatrix} 4 & 3 \\ -1 & -2 \end{bmatrix}$$

$$= \begin{bmatrix} (2)(4) + (-1)(-1) & (2)(3) + (-1)(-2) \\ (1)(4) + (3)(-1) & (1)(3) + (3)(-2) \end{bmatrix}$$

$$= \begin{bmatrix} 9 & 8 \\ 1 & -3 \end{bmatrix}$$

$$A(BC) = \begin{bmatrix} -3 & 0 \\ 2 & 1 \\ 3 & -2 \end{bmatrix} \begin{bmatrix} 9 & 8 \\ 1 & -3 \end{bmatrix}$$

$$= \begin{bmatrix} (-3)(9) + (0)(1) & (-3)(8) + (0)(-3) \\ (2)(9) + (1)(1) & (2)(8) + (1)(-3) \\ (3)(9) + (-2)(1) & (3)(8) + (-2)(-3) \end{bmatrix}$$

$$= \begin{bmatrix} -27 & -24 \\ 19 & 13 \\ 25 & 30 \end{bmatrix}$$

We now compute $(AB)C$.

$$AB = \begin{bmatrix} -3 & 0 \\ 2 & 1 \\ 3 & -2 \end{bmatrix} \begin{bmatrix} 2 & -1 \\ 1 & 3 \end{bmatrix}$$

$$= \begin{bmatrix} (-3)(2) + (0)(1) & (-3)(-1) + (0)(3) \\ (2)(2) + (1)(1) & (2)(-1) + (1)(3) \\ (3)(2) + (-2)(1) & (3)(-1) + (-2)(3) \end{bmatrix}$$

$$= \begin{bmatrix} -6 & 3 \\ 5 & 1 \\ 4 & -9 \end{bmatrix}$$

$$(AB)C = \begin{bmatrix} -6 & 3 \\ 5 & 1 \\ 4 & -9 \end{bmatrix} \begin{bmatrix} 4 & 3 \\ -1 & -2 \end{bmatrix}$$

$$= \begin{bmatrix} (-6)(4) + 3(-1) & (-6)(3) + (3)(-2) \\ (5)(4) + (1)(-1) & (5)(3) + (1)(-2) \\ (4)(4) + (-9)(-1) & (4)(3) + (-9)(-2) \end{bmatrix}$$

$$= \begin{bmatrix} -27 & -24 \\ 19 & 13 \\ 25 & 30 \end{bmatrix}$$

Hence $A(BC) = (AB)C$.   ■

The method of matrix multiplication can be used to write a system of linear equations in matrix form. For instance, the system

$$\begin{cases} a_1x + b_1y + c_1z = d_1 \\ a_2x + b_2y + c_2z = d_2 \\ a_3x + b_3y + c_3z = d_3 \end{cases}$$

can be written as

$$\begin{bmatrix} a_1 & b_1 & c_1 \\ a_2 & b_2 & c_2 \\ a_3 & b_3 & c_3 \end{bmatrix} \begin{bmatrix} x \\ y \\ z \end{bmatrix} = \begin{bmatrix} d_1 \\ d_2 \\ d_3 \end{bmatrix}$$

In Section 9.8 we show how this matrix form can be used to solve a system of linear equations.

We now summarize the facts we learned in this section about matrix addition and multiplication.

1. The zero matrix is the identity element for matrix addition.

2. The matrix $-H$ is the additive inverse of the matrix H.

3. Matrix addition is commutative and associative.

4. Matrix multiplication is *not* commutative.

5. Matrix multiplication is associative if the products exist.

6. Matrix multiplication is distributive with respect to addition if the products and sums exist.

We leave the discussions of the multiplicative identity element and multiplicative inverses of matrices for the next section.

EXERCISES 9.7

1. Given the matrices

$$A = \begin{bmatrix} 3 & -2 & 7 & 0 \\ 4 & 5 & -1 & 8 \end{bmatrix} \quad B = \begin{bmatrix} -6 \\ 2 \\ 5 \end{bmatrix}$$

what is each of the following: (a) the order of A; (b) the order of B; (c) the element in the first row and second column of A; (d) the negative of A; (e) the negative of B; (f) the product of 3 and A; (g) the product of -2 and B?

2. Given the matrices

$$A = \begin{bmatrix} 5 & -1 & -3 & 0 \\ 1 & 2 & -1 & -4 \\ 0 & -6 & 2 & 3 \\ 4 & -1 & 7 & 2 \end{bmatrix} \quad B = \begin{bmatrix} 2 & -1 & 5 & -6 \end{bmatrix}$$

what is each of the following: (a) the order of A; (b) the order of B; (c) the element in the third row and second column of A; (d) the negative of A; (e) the negative of B; (f) the product of 2 and A; (b) the product of -3 and B?

3. Given the matrices

$$A = \begin{bmatrix} 7 & 2 & -1 \\ -3 & -4 & 0 \\ -2 & 1 & -5 \\ 0 & -4 & -1 \end{bmatrix} \quad B = \begin{bmatrix} -3 & 5 \\ 1 & -2 \end{bmatrix}$$

what is each of the following: (a) the order of A; (b) the order of B; (c) the element in the second row and third column of A; (d) the zero matrix of the same order as A; (e) the zero matrix of the same order as B; (f) the product of -4 and A; (g) the product of 6 and B?

4. Given the matrices

$$A = \begin{bmatrix} 9 & -4 & 10 \\ 2 & -5 & 1 \\ -7 & -2 & -3 \end{bmatrix} \quad B = \begin{bmatrix} 0 & -6 \\ -2 & 3 \\ -3 & 1 \end{bmatrix}$$

what is each of the following: (a) the order of A; (b) the order of B; (c) the element in the third row and first column of A; (d) the zero matrix of the same order as A; (e) the zero matrix of the same order as B; (f) the product of -1 and A; (b) the negative of B?

In Exercises 5 through 10, perform the indicated operations.

5. $\begin{bmatrix} 3 & -5 \\ -1 & 2 \end{bmatrix} + \begin{bmatrix} -4 & 1 \\ -8 & 6 \end{bmatrix}$

6. $\begin{bmatrix} 4 & -1 \\ -5 & 6 \\ 0 & -2 \end{bmatrix} + \begin{bmatrix} -5 & -1 \\ 3 & -6 \\ -2 & -2 \end{bmatrix}$

7. $\begin{bmatrix} -8 & 3 & -1 \\ 9 & -2 & -5 \end{bmatrix} - \begin{bmatrix} 3 & -2 & 5 \\ -4 & -7 & 3 \end{bmatrix}$

8. $\begin{bmatrix} -4 & -3 & 5 \\ 2 & -4 & 1 \\ 0 & 6 & -7 \end{bmatrix} - \begin{bmatrix} -1 & 4 & -2 \\ 5 & 4 & 0 \\ -7 & 6 & -4 \end{bmatrix}$

9. $\begin{bmatrix} 2 & 6 & -2 \\ 1 & -1 & 3 \\ -5 & -7 & 2 \end{bmatrix} + \begin{bmatrix} -7 & -6 & 5 \\ 4 & -8 & -3 \\ 0 & 2 & -3 \end{bmatrix}$

10. $\begin{bmatrix} 9 & -10 \\ -6 & 2 \end{bmatrix} - \begin{bmatrix} -3 & -6 \\ 8 & -1 \end{bmatrix}$

11. Find a, b, c, and d so that

$$\begin{bmatrix} 7 & -2 \\ -4 & 3 \end{bmatrix} - \begin{bmatrix} a & b \\ c & d \end{bmatrix} = \begin{bmatrix} -6 & 5 \\ -1 & 4 \end{bmatrix}$$

12. Find a, b, c, d, e, and f so that

$$\begin{bmatrix} a & b \\ c & d \\ e & f \end{bmatrix} + \begin{bmatrix} -2 & -1 \\ 5 & 7 \\ -5 & 0 \end{bmatrix} = \begin{bmatrix} 6 & -1 \\ -5 & 3 \\ -8 & -3 \end{bmatrix}$$

In Exercises 13 through 28, find the product.

13. $4\begin{bmatrix} -1 & 2 \\ 3 & 5 \\ -6 & 1 \end{bmatrix}$

14. $-5\begin{bmatrix} -7 & 0 \\ -4 & 6 \end{bmatrix}$

15. $-2\begin{bmatrix} -3 & 6 & -1 \\ 4 & -2 & 0 \\ -5 & 1 & -8 \end{bmatrix}$

16. $3\begin{bmatrix} 4 & 3 & -3 \\ 1 & -2 & 7 \end{bmatrix}$

17. $\begin{bmatrix} 2 & 3 \\ -1 & 5 \end{bmatrix}\begin{bmatrix} 2 & -1 \\ 0 & 3 \end{bmatrix}$

18. $\begin{bmatrix} 4 & -5 \\ 7 & 3 \end{bmatrix}\begin{bmatrix} 5 & -1 \\ -2 & 7 \end{bmatrix}$

19. $\begin{bmatrix} 1 & 2 & 3 \\ 4 & 5 & 7 \end{bmatrix}\begin{bmatrix} 1 & -1 \\ 2 & 0 \\ -1 & 1 \end{bmatrix}$

20. $\begin{bmatrix} 2 & 1 \end{bmatrix}\begin{bmatrix} 3 \\ 2 \end{bmatrix}$

21. $\begin{bmatrix} 1 & 2 & -3 \end{bmatrix}\begin{bmatrix} 2 \\ 1 \\ -1 \end{bmatrix}$

22. $\begin{bmatrix} 3 \\ 2 \end{bmatrix}\begin{bmatrix} 2 & 1 \end{bmatrix}$

23. $\begin{bmatrix} 2 \\ 1 \\ -1 \end{bmatrix}\begin{bmatrix} 1 & 2 & -3 \end{bmatrix}$

24. $\begin{bmatrix} -1 & 3 & -2 \end{bmatrix}\begin{bmatrix} 2 & 0 \\ 1 & 4 \\ -1 & -2 \end{bmatrix}$

25. $\begin{bmatrix} 4 & -1 \\ 0 & 2 \\ 5 & 1 \end{bmatrix}\begin{bmatrix} -2 & 3 \\ 0 & -3 \end{bmatrix}$

26. $\begin{bmatrix} 2 & 0 & -2 \\ 3 & -1 & 0 \end{bmatrix}\begin{bmatrix} 5 & 2 & -1 \\ 0 & -3 & 1 \\ -2 & 6 & 0 \end{bmatrix}$

27. $\begin{bmatrix} 1 & -4 & 0 & -1 \\ 2 & 0 & 3 & -2 \end{bmatrix}\begin{bmatrix} 1 & 0 & -2 \\ -1 & 6 & 0 \\ 2 & 1 & 3 \\ 0 & 2 & -4 \end{bmatrix}$

28. $\begin{bmatrix} 1 & -1 & 0 & 3 \\ 2 & 0 & -3 & -2 \\ 0 & -4 & 5 & 2 \end{bmatrix}\begin{bmatrix} -2 & 6 \\ 0 & 1 \\ -1 & 3 \\ 2 & -3 \end{bmatrix}$

In Exercises 29 through 40, verify the statement where

$$A = \begin{bmatrix} -2 & 3 \\ 2 & -1 \end{bmatrix} \quad B = \begin{bmatrix} 2 & -1 \\ 3 & -2 \end{bmatrix} \quad C = \begin{bmatrix} -3 & 0 \\ 1 & -2 \end{bmatrix}$$

$$D = \begin{bmatrix} -1 & 4 \\ 0 & 0 \end{bmatrix} \quad I = \begin{bmatrix} 1 & 0 \\ 0 & 1 \end{bmatrix}$$

29. $A + B = B + A$
30. $B + C = C + B$
31. $A + (B + C) = (A + B) + C$
32. $A + \mathbf{0} = A$ and $A + (-A) = \mathbf{0}$
33. $A\mathbf{0} = \mathbf{0}$ and $\mathbf{0}A = \mathbf{0}$
34. $AI = A$ and $IA = A$
35. $AB \neq BA$
36. $A(B + C) = AB + AC$
37. $(B + C)A = BA + CA$
38. $A(BC) = (AB)C$
39. $DA = DB$ even though $D \neq \mathbf{0}$ and $A \neq B$
40. $(B + C)(B - C) \neq B^2 - C^2$, where $B^2 = BB$ and $C^2 = CC$

In Exercises 41 through 44, find the matrix X that satisfies the equation.

41. $X - 3\begin{bmatrix} 2 & -1 \\ 3 & 1 \end{bmatrix} = \begin{bmatrix} -2 & 5 \\ 1 & 4 \end{bmatrix}$

42. $2\begin{bmatrix} -4 & -1 & 3 & 0 \\ 2 & -5 & -2 & 6 \end{bmatrix} + 3X = \begin{bmatrix} -5 & 10 & 3 & -6 \\ 7 & -7 & 11 & 0 \end{bmatrix}$

43. $-2\begin{bmatrix} 5 & -1 \\ 6 & 0 \\ -2 & -5 \end{bmatrix} - 4X = \begin{bmatrix} -2 & 6 \\ 4 & 4 \\ 0 & 2 \end{bmatrix}$

44. $-3X - 4\begin{bmatrix} 6 \\ 3 \\ -1 \\ 2 \end{bmatrix} = \begin{bmatrix} -9 \\ -6 \\ 7 \\ 1 \end{bmatrix}$

9.8   # Solutions of Linear Systems by Matrix Inverses (Supplementary)

In this section we introduce the *multiplicative identity element* and the *multiplicative inverse* of a square matrix. While matrix inverses have various applications, we confine their use here to solving systems of linear equations. In our discussion we need the concept of the *transpose* of a matrix.

Definition

> **Transpose of a Matrix**
> The **transpose** of a matrix H, denoted by H^t, is the matrix for which the ith row of H^t is the ith column of H and the jth column of H^t is the jth row of H; that is, corresponding rows and columns are interchanged.

Illustration 1

If

$$A = \begin{bmatrix} 3 & 1 \\ 4 & -2 \\ 0 & 5 \end{bmatrix} \quad \text{and} \quad B = \begin{bmatrix} -1 & 5 \\ -6 & 2 \end{bmatrix}$$

then

$$A^t = \begin{bmatrix} 3 & 4 & 0 \\ 1 & -2 & 5 \end{bmatrix} \quad \text{and} \quad B^t = \begin{bmatrix} -1 & -6 \\ 5 & 2 \end{bmatrix} \quad\blacksquare$$

In Section 9.6 we used cofactors of elements of square matrices of order three to evaluate third-order determinants. To refer to cofactors of elements of square matrices of order two, we define such a cofactor as follows: if e is the element in row i and column j then the **cofactor** of e is $(-1)^{i+j}$ times the element obtained by deleting the row and column in which e appears.

Illustration 2

In the matrix

$$\begin{bmatrix} -4 & 3 \\ -2 & 5 \end{bmatrix}$$

the cofactor of -4 is $(-1)^{1+1}(5) = 5$; the cofactor of 3 is $(-1)^{1+2}(-2) = 2$; the cofactor of -2 is $(-1)^{2+1}(3) = -3$; the cofactor of 5 is $(-1)^{2+2}(-4) = -4$. $\blacksquare$

The number 1 is the multiplicative identity for real numbers such that for any real number a

$$a \cdot 1 = a$$

Does the set of matrices of a given order have a multiplicative identity? The answer, in general, is no. However, there is a multiplicative identity for the set of all square matrices of a given order.

Definition

> **Multiplicative Identity**
> The **multiplicative identity** for the set of square matrices of order n, denoted by I_n, is the square matrix of order n whose elements on the main diagonal are ones and whose other elements are all zeros.

Illustration 3

The multiplicative identities for the sets of square matrices of orders 2, 3, and 4, are, respectively, I_2, I_3, and I_4, where

$$I_2 = \begin{bmatrix} 1 & 0 \\ 0 & 1 \end{bmatrix} \qquad I_3 = \begin{bmatrix} 1 & 0 & 0 \\ 0 & 1 & 0 \\ 0 & 0 & 1 \end{bmatrix} \qquad I_4 = \begin{bmatrix} 1 & 0 & 0 & 0 \\ 0 & 1 & 0 & 0 \\ 0 & 0 & 1 & 0 \\ 0 & 0 & 0 & 1 \end{bmatrix} \qquad \blacksquare$$

Illustration 4

Let H be any square matrix of order three:

$$H = \begin{bmatrix} a_1 & b_1 & c_1 \\ a_2 & b_2 & c_2 \\ a_3 & b_3 & c_3 \end{bmatrix}$$

Then

$$HI_3 = \begin{bmatrix} a_1 & b_1 & c_1 \\ a_2 & b_2 & c_2 \\ a_3 & b_3 & c_3 \end{bmatrix} \begin{bmatrix} 1 & 0 & 0 \\ 0 & 1 & 0 \\ 0 & 0 & 1 \end{bmatrix} \qquad I_3H = \begin{bmatrix} 1 & 0 & 0 \\ 0 & 1 & 0 \\ 0 & 0 & 1 \end{bmatrix} \begin{bmatrix} a_1 & b_1 & c_1 \\ a_2 & b_2 & c_2 \\ a_3 & b_3 & c_3 \end{bmatrix}$$

$$= \begin{bmatrix} a_1 & b_1 & c_1 \\ a_2 & b_2 & c_2 \\ a_3 & b_3 & c_3 \end{bmatrix} \qquad = \begin{bmatrix} a_1 & b_1 & c_1 \\ a_2 & b_2 & c_2 \\ a_3 & b_3 & c_3 \end{bmatrix}$$

$$= H \qquad\qquad\qquad\qquad = H \qquad \blacksquare$$

In Illustration 4 we have proved the following theorem for square matrices of order three. A similar proof can be given for square matrices of any order.

Theorem 1

> If H is a square matrix of order n and if I_n is the multiplicative identity of order n, then
>
> $$HI_n = H \quad \text{and} \quad I_nH = H$$

We know that for every real number a, except 0, there exists a real number a^{-1}, called the multiplicative inverse of a, such that

$$a \cdot a^{-1} = 1 \quad \text{and} \quad a^{-1} \cdot a = 1$$

If a square matrix H of order n is to have a multiplicative inverse H^{-1}, it is necessary that

$$HH^{-1} = I_n \quad \text{and} \quad H^{-1}H = I_n$$

We now give a theorem that states that if $\det H \neq 0$, there is such a matrix H^{-1}. The theorem gives a method for computing H^{-1} when it exists. The proof is omitted.

Theorem 2

> Suppose that H is a square matrix of order n for which $\det H \neq 0$. If M is the matrix obtained by replacing each element e in H by its cofactor and
>
> $$H^{-1} = \frac{1}{\det H} M^t$$
>
> then
>
> $$HH^{-1} = I_n \quad \text{and} \quad H^{-1}H = I_n$$

If a matrix H has an inverse (that is, if $\det H \neq 0$), H is said to be **nonsingular.** If a matrix H does not have an inverse (that is, if $\det H = 0$), then H is said to be **singular.**

Example   1 If

$$H = \begin{bmatrix} 2 & -4 \\ 3 & -5 \end{bmatrix}$$

find H^{-1} and show that $HH^{-1} = I_2$ and $H^{-1}H = I_2$.

Solution

$$\det H = 2(-5) - 3(-4)$$
$$= 2$$

Because $\det H \neq 0$, H^{-1} exists and

$$H^{-1} = \frac{1}{\det H} \begin{bmatrix} A_1 & B_1 \\ A_2 & B_2 \end{bmatrix}^t$$

$$= \frac{1}{2} \begin{bmatrix} -5 & -3 \\ 4 & 2 \end{bmatrix}^t$$

$$= \frac{1}{2} \begin{bmatrix} -5 & 4 \\ -3 & 2 \end{bmatrix}$$

$$= \begin{bmatrix} -\dfrac{5}{2} & 2 \\ -\dfrac{3}{2} & 1 \end{bmatrix}$$

Therefore

$$HH^{-1} = \begin{bmatrix} 2 & -4 \\ 3 & -5 \end{bmatrix} \begin{bmatrix} -\dfrac{5}{2} & 2 \\ -\dfrac{3}{2} & 1 \end{bmatrix} \qquad\qquad H^{-1}H = \begin{bmatrix} -\dfrac{5}{2} & 2 \\ -\dfrac{3}{2} & 1 \end{bmatrix} \begin{bmatrix} 2 & -4 \\ 3 & -5 \end{bmatrix}$$

$$= \begin{bmatrix} -5 + 6 & 4 - 4 \\ \dfrac{-15}{2} + \dfrac{15}{2} & 6 - 5 \end{bmatrix} \qquad\qquad = \begin{bmatrix} -5 + 6 & 10 - 10 \\ -3 + 3 & 6 - 5 \end{bmatrix}$$

$$= \begin{bmatrix} 1 & 0 \\ 0 & 1 \end{bmatrix} \qquad\qquad\qquad\qquad = \begin{bmatrix} 1 & 0 \\ 0 & 1 \end{bmatrix}$$

$$= I_2 \qquad\qquad\qquad\qquad\qquad = I_2 \qquad\qquad\blacksquare$$

Example 2   If

$$H = \begin{bmatrix} 1 & 2 & 1 \\ 1 & 3 & 0 \\ 4 & 0 & 2 \end{bmatrix}$$

find H^{-1} and show that $HH^{-1} = I_3$.

Solution   We first compute det H by expanding by elements of the third column.

$$\det H = 1 \begin{vmatrix} 1 & 3 \\ 4 & 0 \end{vmatrix} + 2 \begin{vmatrix} 1 & 2 \\ 1 & 3 \end{vmatrix}$$

$$= -12 + 2(1)$$

$$= -10$$

Because det $H \neq 0$, H^{-1} exists and

$$H^{-1} = \frac{1}{-10} \begin{bmatrix} A_1 & B_1 & C_1 \\ A_2 & B_2 & C_2 \\ A_3 & B_3 & C_3 \end{bmatrix}^t$$

We now compute the cofactors of the elements of H.

$$A_1 = \begin{vmatrix} 3 & 0 \\ 0 & 2 \end{vmatrix} \qquad B_1 = (-1)\begin{vmatrix} 1 & 0 \\ 4 & 2 \end{vmatrix} \qquad C_1 = \begin{vmatrix} 1 & 3 \\ 4 & 0 \end{vmatrix}$$
$$= 6 \qquad\qquad = -2 \qquad\qquad = -12$$

$$A_2 = (-1)\begin{vmatrix} 2 & 1 \\ 0 & 2 \end{vmatrix} \qquad B_2 = \begin{vmatrix} 1 & 1 \\ 4 & 2 \end{vmatrix} \qquad C_2 = (-1)\begin{vmatrix} 1 & 2 \\ 4 & 0 \end{vmatrix}$$
$$= -4 \qquad\qquad = -2 \qquad\qquad = 8$$

$$A_3 = \begin{vmatrix} 2 & 1 \\ 3 & 0 \end{vmatrix} \qquad B_3 = (-1)\begin{vmatrix} 1 & 1 \\ 1 & 0 \end{vmatrix} \qquad C_3 = \begin{vmatrix} 1 & 2 \\ 1 & 3 \end{vmatrix}$$
$$= -3 \qquad\qquad = 1 \qquad\qquad = 1$$

Therefore

$$H^{-1} = -\frac{1}{10} \begin{bmatrix} 6 & -2 & -12 \\ -4 & -2 & 8 \\ -3 & 1 & 1 \end{bmatrix}^t$$

$$= -\frac{1}{10} \begin{bmatrix} 6 & -4 & -3 \\ -2 & -2 & 1 \\ -12 & 8 & 1 \end{bmatrix}$$

$$= \begin{bmatrix} -\dfrac{3}{5} & \dfrac{2}{5} & \dfrac{3}{10} \\[2mm] \dfrac{1}{5} & \dfrac{1}{5} & -\dfrac{1}{10} \\[2mm] \dfrac{6}{5} & -\dfrac{4}{5} & -\dfrac{1}{10} \end{bmatrix}$$

Thus

$$HH^{-1} = \begin{bmatrix} 1 & 2 & 1 \\ 1 & 3 & 0 \\ 4 & 0 & 2 \end{bmatrix} \begin{bmatrix} -\dfrac{3}{5} & \dfrac{2}{5} & \dfrac{3}{10} \\[2mm] \dfrac{1}{5} & \dfrac{1}{5} & -\dfrac{1}{10} \\[2mm] \dfrac{6}{5} & -\dfrac{4}{5} & -\dfrac{1}{10} \end{bmatrix}$$

$$= \begin{bmatrix} -\dfrac{3}{5} + \dfrac{2}{5} + \dfrac{6}{5} & \dfrac{2}{5} + \dfrac{2}{5} - \dfrac{4}{5} & \dfrac{3}{10} - \dfrac{2}{10} - \dfrac{1}{10} \\[2mm] -\dfrac{3}{5} + \dfrac{3}{5} + 0 & \dfrac{2}{5} + \dfrac{3}{5} + 0 & \dfrac{3}{10} - \dfrac{3}{10} + 0 \\[2mm] -\dfrac{12}{5} + 0 + \dfrac{12}{5} & \dfrac{8}{5} + 0 - \dfrac{8}{5} & \dfrac{12}{10} + 0 - \dfrac{2}{10} \end{bmatrix}$$

$$= \begin{bmatrix} 1 & 0 & 0 \\ 0 & 1 & 0 \\ 0 & 0 & 1 \end{bmatrix}$$

$$= I_3$$ ∎

It is rather cumbersome to use Theorem 2 to compute the inverse of a matrix having an order greater than three. There are other systematic and more efficient methods, but they involve concepts that we do not discuss in this book.

In the following example we show how the inverse of a matrix is used to solve a system of linear equations.

Example 3 Use the inverse of a matrix to solve the following system of equations:

$$\begin{cases} x + 2y + \ z = 2 \\ x + 3y = 1 \\ 4x + 2z = -4 \end{cases}$$

Solution If

$$H = \begin{bmatrix} 1 & 2 & 1 \\ 1 & 3 & 0 \\ 4 & 0 & 2 \end{bmatrix}$$

then, in matrix form, the system is

$$H\begin{bmatrix} x \\ y \\ z \end{bmatrix} = \begin{bmatrix} 2 \\ 1 \\ -4 \end{bmatrix}$$

Because H is the matrix of Example 2, we know that H^{-1} exists. Therefore we can multiply both sides of this equation on the left by H^{-1} and we obtain

$$\begin{bmatrix} x \\ y \\ z \end{bmatrix} = H^{-1}\begin{bmatrix} 2 \\ 1 \\ -4 \end{bmatrix}$$

Using the result of Example 2 for H^{-1}, we have

$$\begin{bmatrix} x \\ y \\ z \end{bmatrix} = \begin{bmatrix} -\dfrac{3}{5} & \dfrac{2}{5} & \dfrac{3}{10} \\ \dfrac{1}{5} & \dfrac{1}{5} & -\dfrac{1}{10} \\ \dfrac{6}{5} & -\dfrac{4}{5} & -\dfrac{1}{10} \end{bmatrix}\begin{bmatrix} 2 \\ 1 \\ -4 \end{bmatrix}$$

$$= \begin{bmatrix} -\dfrac{6}{5} + \dfrac{2}{5} - \dfrac{6}{5} \\ \dfrac{2}{5} + \dfrac{1}{5} + \dfrac{2}{5} \\ \dfrac{12}{5} - \dfrac{4}{5} + \dfrac{2}{5} \end{bmatrix}$$

$$= \begin{bmatrix} -2 \\ 1 \\ 2 \end{bmatrix}$$

Therefore $x = -2$, $y = 1$, and $z = 2$. ■

EXERCISES 9.8

In Exercises 1 through 4, write the transpose of the matrix.

1. $\begin{bmatrix} 2 & -4 & 1 \\ -3 & 5 & 0 \end{bmatrix}$

2. $\begin{bmatrix} 9 & -3 \\ -7 & 2 \end{bmatrix}$

3. $\begin{bmatrix} -5 & 0 & 2 \\ 3 & -6 & 8 \\ 0 & -1 & 4 \end{bmatrix}$

4. $\begin{bmatrix} 4 & -3 & 6 & 1 \\ 0 & -2 & -5 & -1 \\ 7 & 0 & 3 & 4 \\ 2 & -6 & 0 & 5 \end{bmatrix}$

In Exercises 5 through 14, the given matrix is H. Find H^{-1}, if it exists, and show that $HH^{-1} = I_n$. If H is singular (that is, H^{-1} does not exist), then state this fact.

5. $\begin{bmatrix} 5 & 2 \\ 2 & 1 \end{bmatrix}$

6. $\begin{bmatrix} 2 & -1 \\ -5 & 3 \end{bmatrix}$

7. $\begin{bmatrix} -2 & 3 \\ 4 & -6 \end{bmatrix}$

8. $\begin{bmatrix} 7 & -1 \\ -5 & 1 \end{bmatrix}$

9. $\begin{bmatrix} -4 & 1 \\ -5 & 2 \end{bmatrix}$

10. $\begin{bmatrix} 4 & -1 & 2 \\ -2 & 0 & -2 \\ 3 & 1 & 5 \end{bmatrix}$

11. $\begin{bmatrix} 3 & 3 & 1 \\ 1 & 4 & 1 \\ 2 & 3 & 1 \end{bmatrix}$

12. $\begin{bmatrix} 0 & 1 & 2 \\ -3 & 4 & 0 \\ 0 & -2 & -1 \end{bmatrix}$

13. $\begin{bmatrix} 1 & -2 & 1 \\ 2 & 2 & -1 \\ 1 & 1 & 0 \end{bmatrix}$

14. $\begin{bmatrix} 1 & 0 & 0 \\ 2 & 1 & 0 \\ 4 & 0 & 1 \end{bmatrix}$

In Exercises 15 and 16, find H^{-1} for the given matrix H without using Theorem 2. (*Hint:* Let $H^{-1} = \begin{bmatrix} a & b \\ c & d \end{bmatrix}$, and from the fact that $HH^{-1} = I_2$, obtain four equations in a, b, c, and d.)

15. $\begin{bmatrix} 3 & -4 \\ -2 & 3 \end{bmatrix}$

16. $\begin{bmatrix} -3 & 1 \\ 2 & -2 \end{bmatrix}$

In Exercises 17 through 26, verify the equality if

$$M = \begin{bmatrix} 3 & 1 \\ 7 & 3 \end{bmatrix} \quad \text{and} \quad N = \begin{bmatrix} -2 & 1 \\ -3 & 2 \end{bmatrix}$$

17. $(MN)^{-1} = N^{-1}M^{-1}$

18. $(NM)^{-1} = M^{-1}N^{-1}$

19. $(M^{-1})^{-1} = M$

20. $(N^{-1})^{-1} = N$

21. $(M^{-1})^t = (M^t)^{-1}$

22. $(N^{-1})^t = (N^t)^{-1}$

23. $(M^{-1})^2 = (M^2)^{-1}$

24. $(N^{-1})^2 = (N^2)^{-1}$

25. $\det(M^{-1}) = \dfrac{1}{\det M}$

26. $\det(N^{-1}) = \dfrac{1}{\det N}$

In Exercises 27 through 34, use the inverse of a matrix to solve the system. In Exercises 27 and 28, use the result that

$$\begin{bmatrix} 2 & -1 \\ -5 & 3 \end{bmatrix}^{-1} = \begin{bmatrix} 3 & 1 \\ 5 & 2 \end{bmatrix}$$

and in Exercises 31 and 32, use the result that

$$\begin{bmatrix} 1 & 0 & 2 \\ 0 & -1 & 1 \\ 1 & 3 & 0 \end{bmatrix}^{-1} = \begin{bmatrix} 3 & -6 & -2 \\ -1 & 2 & 1 \\ -1 & 3 & 1 \end{bmatrix}$$

27. $\begin{cases} 2x - y = -2 \\ -5x + 3y = 7 \end{cases}$

28. $\begin{cases} 2x - y = 1 \\ -5x + 3y = -5 \end{cases}$

29. $\begin{cases} 3x + 2y = -1 \\ 4x + 3y = 0 \end{cases}$

30. $\begin{cases} 3x + 2y = 3 \\ 6x - 6y = 1 \end{cases}$

31. $\begin{cases} x + 2z = 5 \\ -y + z = 3 \\ x + 3y = 0 \end{cases}$

32. $\begin{cases} x + 2z = 12 \\ -y + z = 7 \\ x + 3y = -4 \end{cases}$

33. $\begin{cases} 4x + 2y + 3z = 4 \\ 5x - y + 4z = 12 \\ x - 3y + 2z = 0 \end{cases}$

34. $\begin{cases} 2x + y + 3z = 5 \\ x + y + z = 0 \\ 4x + y - 2z = -15 \end{cases}$

REVIEW EXERCISES FOR CHAPTER 9

In Exercises 1 through 4, draw a sketch of the graph of the system. Classify the equations as (i) consistent and independent, (ii) inconsistent, or (iii) dependent. If the equations are consistent and independent, determine the solution set of the system from the graphs.

1. $\begin{cases} 4x + 3y = 6 \\ 2x + y = 4 \end{cases}$

2. $\begin{cases} 3x - 2y = 4 \\ 9x - 6y = 8 \end{cases}$

3. $\begin{cases} 2y = 4x - 6 \\ 6x = 3y + 9 \end{cases}$

4. $\begin{cases} 4x + 2y = 5 \\ 8x - 2y = 1 \end{cases}$

In Exercises 5 through 14, find the solution set of the system. If the equations are either inconsistent or dependent, then so indicate.

5. $\begin{cases} 2x + y + 1 = 0 \\ 3x + 2y + 4 = 0 \end{cases}$

6. $\begin{cases} 3x + 4y - 6 = 0 \\ x - 2y + 8 = 0 \end{cases}$

7. $\begin{cases} 2x - 5y = 7 \\ 6x - 15y = 14 \end{cases}$

8. $\begin{cases} 3x - 2y + 7 = 0 \\ 2x - 3y + 8 = 0 \end{cases}$

9. $\begin{cases} \dfrac{4}{x} - \dfrac{7}{y} = 4 \\ \dfrac{12}{x} + \dfrac{3}{y} = 4 \end{cases}$

10. $\begin{cases} \dfrac{3}{x} - \dfrac{2}{y} = 8 \\ \dfrac{9}{x} + \dfrac{4}{y} = -6 \end{cases}$

11. $\begin{cases} x + 2y + 2z = 1 \\ x - 3y - 2z = 4 \\ 6x + y - z = 21 \end{cases}$

12. $\begin{cases} 6x + 4y + 5z = 14 \\ 4x - 3y - z = 2 \\ 14x - 10y - 9z = 10 \end{cases}$

13. $\begin{cases} 4x - 3y - 6z = 7 \\ 2x - y - 4z = 3 \\ 3x - 2y - 5z = 5 \end{cases}$

14. $\begin{cases} 3x + 3y - 5z = 4 \\ 6x + 2y - 3z = 7 \\ 3x - 5y + 9z = 5 \end{cases}$

In Exercises 15 through 22, find the solution set of the

system. In Exercises 15 through 20, draw sketches of the graphs of the equations and obtain approximate values for the coordinates of the points of intersection.

15. $\begin{cases} x^2 + y^2 = 9 \\ 3x - 2y = 6 \end{cases}$

16. $\begin{cases} 7x + 3y = 9 \\ x^2 + 2y = 1 \end{cases}$

17. $\begin{cases} 3x + 2y = -5 \\ x^2 - 3y = 3 \end{cases}$

18. $\begin{cases} x^2 + y^2 = 50 \\ 3x - 4y = 0 \end{cases}$

19. $\begin{cases} 3x^2 + 2y^2 = 7 \\ 5x^2 - y^2 = 3 \end{cases}$

20. $\begin{cases} x^2 + y^2 = 25 \\ 4x^2 - xy - 3y^2 = 0 \end{cases}$

21. $\begin{cases} 2x^2 - xy + y^2 = 8 \\ 4x^2 + xy + y^2 = 6 \end{cases}$

22. $\begin{cases} 4x^2 - 3y^2 = -8 \\ y^2 + 2xy = 8 \end{cases}$

In Exercises 23 through 28, draw a sketch of the graph of the inequality.

23. (a) $x > 4$; (b) $y \le -2$ 24. (a) $x \ge -6$; (b) $y < 5$
25. $3x - 4y \ge 12$ 26. $5x + 3y \le 15$
27. $8x + 5y < 40$ 28. $7x - 4y > 28$

In Exercises 29 through 32, draw a sketch of the region (if any) defined by the system of inequalities.

29. $\begin{cases} 5x + 3y \le 1 \\ 3x - 2y > 12 \end{cases}$

30. $\begin{cases} 3x + 2y \le 12 \\ 2y - 4 > x \\ x \ge 0 \end{cases}$

31. $\begin{cases} x - 2y < 5 \\ 4x + 5y < 20 \\ 6x + y > 4 \end{cases}$

32. $\begin{cases} 2x + y \le 8 \\ x + y \le 4 \\ x \ge 1 \\ y \ge -2 \\ y \le 1 \end{cases}$

33. Let $z = x + 4y$; minimize z subject to the following constraints: $x + 4y \le 8$; $x - y \le 4$; $x \ge 2$.

34. Let $z = 2x + y$; maximize z subject to the following constraints: $x \ge 0$; $y \ge 0$; $x \ge y - 1$; $3x + 2y \le 17$; $x + 4y \ge 9$.

In Exercises 35 through 40, use the Gaussian reduction method to find the solution set of the system.

35. $\begin{cases} 5x - 7y = 4 \\ 2x + 3y = 19 \end{cases}$

36. $\begin{cases} 2x + 3y = 0 \\ 3x + 4y = 2 \end{cases}$

37. $\begin{cases} 3x - 5y + 2z = 4 \\ 4x + 2y + 7z = 1 \\ 5x - 9y - 3z = -11 \end{cases}$

38. $\begin{cases} 3x - 2y - 2z = 3 \\ 6x + 4y + 3z = 3 \\ 3x - 6y + z = -2 \end{cases}$

39. $\begin{cases} w + x + y + 3z = 3 \\ 3w + x - y = 0 \\ 2w - 2x - y + 6z = 4 \\ 4w - x - 2y - 3z = 0 \end{cases}$

40. $\begin{cases} 2w + 2x - z = 3 \\ w + 2x - 2z = 2 \\ w + y + z = 0 \\ w - x - 2z = -4 \end{cases}$

In Exercises 41 through 44, evaluate the determinant.

41. $\begin{vmatrix} 5 & -7 \\ 3 & -4 \end{vmatrix}$

42. $\begin{vmatrix} 6 & 3 \\ -2 & -4 \end{vmatrix}$

43. $\begin{vmatrix} 2 & 3 & -1 \\ 3 & 1 & 5 \\ 2 & 3 & 4 \end{vmatrix}$

44. $\begin{vmatrix} 1 & 1 & -1 \\ 0 & 1 & 1 \\ -4 & 2 & -3 \end{vmatrix}$

In Exercises 45 through 48, solve for x.

45. $\begin{vmatrix} x & 3 \\ 4 & 2 \end{vmatrix} = 8$

46. $\begin{vmatrix} x & 1 \\ 6 & x \end{vmatrix} = 10$

47. $\begin{vmatrix} 0 & x & 0 \\ -2 & 4 & -1 \\ 3 & 7 & x \end{vmatrix} = -1$

48. $\begin{vmatrix} x & 2 & 0 \\ 4 & -1 & x \\ 2 & 3 & 1 \end{vmatrix} = -14$

In Exercises 49 through 56, use Cramer's rule to solve the system.

49. $\begin{cases} 6x + 3y = 1 \\ 3x - 2y = 4 \end{cases}$

50. $\begin{cases} 2x + 5y = 4 \\ 3x + 4y = -1 \end{cases}$

51. $\begin{cases} \dfrac{x}{3} + \dfrac{y}{2} = \dfrac{5}{6} \\ \dfrac{x}{2} + \dfrac{y}{3} = \dfrac{2}{3} \end{cases}$

52. $\begin{cases} x + \dfrac{y}{2} = \dfrac{5}{4} \\ \dfrac{5x}{3} - y = -\dfrac{2}{3} \end{cases}$

53. $\begin{cases} x + y - z = 0 \\ 2x - y + 3z = -1 \\ 2y - 3z = 1 \end{cases}$

54. $\begin{cases} x + 2z = 2 \\ 3x + y = 5 \\ 2y - 3z = -5 \end{cases}$

55. $\begin{cases} 2x + y = 0 \\ 4x - 3z = -2 \\ 2y + 3z = 2 \end{cases}$

56. $\begin{cases} 4x + 2y - 3z = 10 \\ 2x - 3y - 4z = 8 \\ 6x - 5y - 2z = 6 \end{cases}$

57. A man placed his savings in two investments. The interest rate on investment A is 10 percent and on investment B it is 12 percent. The annual income from the two investments is \$3760. If investment A were at 12 percent and investment B at 10 percent, his annual income would be \$3720. What is the total amount of his savings?

58. At a supermarket that sells fruit by the pound, one person bought 3 lb of oranges and 6 lb of grapefruit for a total cost of \$6, while a second

person paid $6.40 for 5 lb of oranges and 4 lb of grapefruit. What is the price per pound of the oranges and the grapefruit?

59. A chemist has a solution that is 50 percent acid. By adding water the solution is reduced to one containing 40 percent acid. By adding 500 cm^3 more water the solution then contains only 35 percent acid. Determine (a) how many cubic centimeters of the 50 percent acid solution the chemist had originally and (b) how many cubic centimeters of the 35 percent acid solution the chemist had finally.

60. An investment yields an annual interest of $750. If $500 more is invested and the rate is 1 percent less, the annual interest is $650. What is the amount of the investment and the rate of interest?

61. Workers A and B can complete a particular job if they work together for 12 days. If A works alone for 20 days and then B completes the job alone in 6 more days, how long does it take each worker to do the job alone? (See the hint for Exercise 41 in Exercises 9.1.)

62. A pilot makes a check flight in an airplane. The pilot flies a distance of 80 km against the wind in 10 min, and then flies back the same distance with the wind in 8 min. If the plane's rate in still air is the same in both directions, what is the rate of the wind and what is the plane's rate in still air?

63. Find an equation of the circle containing the points $(-2, -1)$, $(5, 0)$, and $(2, 1)$. (See the hint for Exercises 23 and 24 in Exercises 9.2.)

64. A parabola has its axis parallel to the y axis and contains the points $(2, 3)$, $(1, 0)$, and $(5, 36)$. Find its equation. (See the hint for Exercise 25 in Exercises 9.2.)

65. A woman bought 100 stamps in denominations of 20, 30, and 42 cents, and the number of 30-cent stamps purchased was ten less than the combined total of the other two denominations. If she paid $30 for the stamps, how many stamps of each denomination did she buy?

66. A woman has a certain amount of money invested at a particular rate of interest. If she had $2000 more invested at a rate 2 percent lower, she would receive the same annual interest. If she had $2000 less invested at a rate 3 percent higher, she also would receive the same annual interest. How much does she have invested and at what rate?

67. A company makes two types of lamp shades. It takes twice as much time to make one type A shade as it does to make a type B shade, but if all shades were of type B the company could make 1250 per day. However, because of the availability of materials, the total daily output of both types cannot be greater than 800. Furthermore the daily output of type A cannot exceed 400 and the daily output of type B cannot be greater than 700. If the profit of each type A shade is $8 and the profit of each type B shade is $6, how many shades of each type should be made per day to maximize the profit?

68. The area of a right triangle is 84 cm^2 and the length of the hypotenuse is 25 cm. Find the lengths of the legs of the triangle.

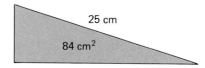

69. An open box is constructed from a piece of cardboard, having an area of 120 in^2, by cutting a square of side 2 in. from each corner and turning up the ends and sides. If the volume of the box is 96 in^3, what are the dimensions of the original piece of cardboard?

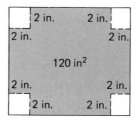 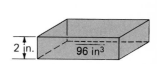

Exercises 70 through 78 pertain to Supplementary Section 9.7. In Exercises 70 through 76, perform the indicated operations.

70. $\begin{bmatrix} 3 & -4 \\ -7 & 1 \end{bmatrix} + \begin{bmatrix} -6 & 7 \\ -3 & 5 \end{bmatrix}$

71. $2\begin{bmatrix} -4 & 0 \\ 1 & -3 \\ 8 & -2 \end{bmatrix} - 3\begin{bmatrix} -2 & 2 \\ 3 & 5 \\ -1 & -3 \end{bmatrix}$

72. $-4\begin{bmatrix} 2 & -1 & 0 \\ 3 & 1 & -2 \end{bmatrix} + 3\begin{bmatrix} -2 & 1 & -1 \\ -3 & 0 & 2 \end{bmatrix}$

73. $\begin{bmatrix} 1 & 1 \\ 0 & 1 \end{bmatrix}\begin{bmatrix} 1 & 1 \\ -1 & 1 \end{bmatrix}$

74. $\begin{bmatrix} 1 & -1 \end{bmatrix}\begin{bmatrix} 1 \\ -1 \end{bmatrix}$

75. $\begin{bmatrix} 1 & 2 \\ -1 & 3 \end{bmatrix}\begin{bmatrix} 2 & 1 & 3 \\ 0 & -1 & 0 \end{bmatrix}$

76. $\begin{bmatrix} -1 & 2 & 0 \end{bmatrix}\begin{bmatrix} -2 & 1 & 0 \\ 0 & -1 & 1 \\ 3 & 0 & 2 \end{bmatrix}$

In Exercises 77 and 78, find the matrix X that satisfies the equation.

77. $2X + 3\begin{bmatrix} 2 & -4 \\ 1 & 0 \end{bmatrix} = \begin{bmatrix} 4 & -6 \\ 3 & 4 \end{bmatrix}$

78. $4\begin{bmatrix} 1 & -2 \\ 0 & -1 \\ 3 & 2 \end{bmatrix} - X = \begin{bmatrix} 1 & -7 \\ -5 & -4 \\ 9 & 7 \end{bmatrix}$

Exercises 79 through 84 pertain to Supplementary Section 9.8. In Exercises 79 and 80, find the transpose of the matrix.

79. $\begin{bmatrix} 2 & -1 \\ 3 & 4 \end{bmatrix}$

80. $\begin{bmatrix} 3 & -1 & 0 \\ -2 & 1 & 2 \\ 0 & -3 & 4 \end{bmatrix}$

In Exercises 81 and 82, the matrix is H. Find H^{-1}, if it exists, and show that $HH^{-1} = I$.

81. $\begin{bmatrix} 1 & 3 & 0 \\ 0 & 1 & 1 \\ 2 & 0 & 0 \end{bmatrix}$

82. $\begin{bmatrix} -2 & 1 \\ 4 & -3 \end{bmatrix}$

In Exercises 83 and 84, solve the system by using the inverse of a matrix.

83. $\begin{cases} x + y - z = 0 \\ 2x - y + 3z = -1 \\ 2y - 3z = 1 \end{cases}$

84. $\begin{cases} 2x + 5y = 4 \\ 3x + 4y = -1 \end{cases}$

Polynomial Functions and Polynomial Equations

Introduction

The main objective of this chapter is to learn methods for finding the zeros of a polynomial function P or, equivalently, the roots of the corresponding polynomial equation $P(x) = 0$. Some of the tools that we need are the remainder and factor theorems and synthetic division, which form the subject matter of Section 10.1. Then in Section 10.2 we show how to find the rational zeros of polynomial functions with real coefficients. We develop a systematic procedure for determining the exact or approximate value of all real zeros of polynomial functions with real coefficients in Section 10.3. In Section 10.4 we discuss complex zeros of polynomial functions. Supplementary Section 10.5 gives a treatment of partial fractions, which are important in computational work in more advanced courses such as calculus.

10.1 The Remainder Theorem, the Factor Theorem, and Synthetic Division

In Section 4.2 we stated that a function f is called a polynomial function if $f(x)$ is a polynomial of degree n, that is, if

$$f(x) = a_n x^n + a_{n-1} x^{n-1} + a_{n-2} x^{n-2} + \cdots + a_1 x + a_0$$

where $a_n, a_{n-1}, \ldots, a_0$ are real numbers with $a_n \neq 0$ and n is a nonnegative integer. For instance, the function defined by

$$f(x) = 2x^4 - 5x^3 + 7x - 1$$

is a polynomial function of degree 4. In this chapter we shall be dividing polynomials by linear expressions of the form $x - r$, where r is a real number. The following illustration reviews such a division, first discussed in Section 1.3.

Illustration   1

Consider the polynomial $P(x) = 2x^3 - 5x^2 + 6x - 3$ and divide $P(x)$ by $x - 2$.

$$
\begin{array}{r}
2x^2 - x + 4 \\
x - 2 \overline{)\, 2x^3 - 5x^2 + 6x - 3} \\
\underline{2x^3 - 4x^2} \\
-x^2 + 6x \\
\underline{-x^2 + 2x} \\
4x - 3 \\
\underline{4x - 8} \\
5
\end{array}
$$

Hence the quotient is $2x^2 - x + 4$, and the remainder is 5. Therefore we can write

$$2x^3 - 5x^2 + 6x - 3 = (x - 2)(2x^2 - x + 4) + 5 \qquad \blacksquare$$

Illustration 1 gives a special case of the following theorem that we state without proof.

Theorem 1

> If $P(x)$ is a polynomial and r is a real number, then, when $P(x)$ is divided by $x - r$, we obtain as the quotient a unique polynomial $Q(x)$ and as the remainder a real number R, such that for all values of x
>
> $$P(x) = (x - r)Q(x) + R$$

Illustration 2

In Illustration 1, $P(x) = 2x^3 - 5x^2 + 6x - 3$. Thus

$$P(2) = 2(2)^3 - 5(2)^2 + 6(2) - 3$$
$$= 5$$

Observe that this value of $P(2)$ is the same number as the remainder obtained in Illustration 1 when $P(x)$ was divided by $x - 2$. ∎

In Illustration 2 we have a special case of the following theorem, known as the *remainder theorem*.

Remainder Theorem

> If $P(x)$ is a polynomial and r is a real number, then if $P(x)$ is divided by $x - r$, the remainder is $P(r)$.

Proof From Theorem 1, it follows that when $P(x)$ is divided by $x - r$, we obtain a polynomial $Q(x)$ as the quotient and a real number R as the remainder such that for all values of x

$$P(x) = (x - r)Q(x) + R$$

Because this equation is an identity, it is satisfied when $x = r$. Thus

$$P(r) = (r - r)Q(r) + R$$
$$= 0 + R$$
$$= R$$

and the theorem is proved. ∎

Example   1   Use long division to find the remainder when $2x^3 - 3x^2 - 4x - 17$ is divided by $x - 3$. Then find the remainder by the remainder theorem.

Solution Let $P(x) = 2x^3 - 3x^2 - 4x - 17$. We use long division to divide $P(x)$ by $x - 3$.

$$
\begin{array}{r}
2x^2 + 3x\ + 5 \\
x - 3\overline{)\ 2x^3 - 3x^2 - 4x - 17} \\
\underline{2x^3 - 6x^2} \\
3x^2 - 4x \\
\underline{3x^2 - 9x} \\
5x - 17 \\
\underline{5x - 15} \\
-\ 2
\end{array}
$$

The remainder is -2.

By the remainder theorem, when $P(x)$ is divided by $x - 3$, the remainder is $P(3)$. Because

$$P(x) = 2x^3 - 3x^2 - 4x - 17$$

we have

$$
\begin{aligned}
P(3) &= 2(3)^3 - 3(3)^2 - 4(3) - 17 \\
&= 54 - 27 - 12 - 17 \\
&= -2
\end{aligned}
$$

which agrees with the remainder obtained by long division. ∎

A consequence of the remainder theorem is the *factor theorem*. It enables us to determine whether a specific expression of the form $x - r$ is a factor of a given polynomial.

Factor Theorem

> If $P(x)$ is a polynomial and r is a real number, then $P(x)$ has $x - r$ as a factor if and only if $P(r) = Q$.

Proof Because the statement of the theorem has an *if and only if* qualification, there are two parts to be proved—Part 1: $x - r$ is a factor of $P(x)$ if $P(r) = 0$; Part 2: $x - r$ is a factor of $P(x)$ only if $P(r) = 0$.

Proof of Part 1: From Theorem 1 and the remainder theorem it follows that for the polynomial $P(x)$ and the real number r there exists a unique polynomial $Q(x)$ such that

$$P(x) = (x - r)Q(x) + P(r)$$

If $P(r) = 0$, then

$$P(x) = (x - r)Q(x)$$

Therefore $x - r$ is a factor of $P(x)$.

Proof of Part 2: We wish to prove that if $x - r$ is a factor of $P(x)$, then $P(r) = 0$. If $x - r$ is a factor of $P(x)$, then, when $P(x)$ is divided by $x - r$, the remainder must be zero. Thus, from the remainder theorem, it follows that $P(r) = 0$. ■

Example **2** Show that $x - 4$ is a factor of $2x^3 - 6x^2 - 5x - 12$.

Solution If $P(x) = 2x^3 - 6x^2 - 5x - 12$, then

$$
\begin{aligned}
P(4) &= 2(4)^3 - 6(4)^2 - 5(4) - 12 \\
&= 2(64) - 6(16) - 20 - 12 \\
&= 128 - 96 - 32 \\
&= 0
\end{aligned}
$$

Therefore, by the factor theorem, $x - 4$ is a factor of $P(x)$. ■

Example **3** Determine whether $x + 1$ is a factor of
$$5x^4 + x^3 - 4x^2 - 6x - 10$$

Solution Let $P(x) = 5x^4 + x^3 - 4x^2 - 6x - 10$. By the factor theorem, $x + 1$ or, equivalently, $x - (-1)$ is a factor of $P(x)$ only if $P(-1) = 0$.

$$
\begin{aligned}
P(-1) &= 5(-1)^4 + (-1)^3 - 4(-1)^2 - 6(-1) - 10 \\
&= 5(1) - 1 - 4(1) + 6 - 10 \\
&= -4
\end{aligned}
$$

Because $P(-1) \neq 0$, we conclude that $x + 1$ is not a factor of $P(x)$. ■

Example **4** Find a value of k so that $x + 3$ is a factor of $3x^3 + kx^2 - 7x + 6$.

Solution Let $P(x) = 3x^3 + kx^2 - 7x + 6$. By the factor theorem, $x + 3$ is a factor of $P(x)$ if $P(-3) = 0$. Equating $P(-3)$ to zero, we have

$$
\begin{aligned}
3(-3)^3 + k(-3)^2 - 7(-3) + 6 &= 0 \\
-81 + 9k + 21 + 6 &= 0 \\
9k &= 54 \\
k &= 6
\end{aligned}
$$

Thus $x + 3$ is a factor of $P(x)$ if $k = 6$. ■

Applications of the remainder theorem and the factor theorem involve the division of a polynomial by linear expressions of the form $x - r$. To simplify the computation of such divisions, we use a procedure called *synthetic division*, which we now explain.

In Illustration 1 we used long division to divide $2x^3 - 5x^2 + 6x - 3$ by $x - 2$. The computation is as follows:

$$
\begin{array}{r}
2x^2 - x + 4 \\
x - 2 \overline{\smash{)}\; 2x^3 - 5x^2 + 6x - 3} \\
2x^3 - 4x^2 \\
\hline
-x^2 + 6x \\
-x^2 + 2x \\
\hline
4x - 3 \\
4x - 8 \\
\hline
5
\end{array}
$$

The writing can be shortened by omitting the powers of x and recording only the coefficients. By doing this, the computation takes the following form:

$$
\begin{array}{r}
2 -1 4 \\
1 - 2 \overline{\smash{)}\; 2 -5 6 -3} \\
2 -4 \\
\hline
-1 6 \\
-1 2 \\
\hline
4 -3 \\
4 -8 \\
\hline
5
\end{array}
$$

In the divisor, $x - 2$, the coefficient of x is 1. Thus the coefficient of the first term in each remainder is the same as that of the succeeding term of the quotient. Furthermore, the first term of the next partial product is the same as the coefficient of the first term in each remainder. Hence we can omit the terms of the quotient as well as the first terms of the partial products. With these terms omitted, we have

$$
\begin{array}{r}
\underline{1 - 2} \; \vert \; 2 -5 6 -3 \\
-4 \\
\hline
-1 6 \\
2 \\
\hline
4 -3 \\
-8 \\
\hline
5
\end{array}
$$

In synthetic division the divisor is a polynomial of the form $x - r$, and so the first coefficient in the divisor is always 1; thus we delete the coefficient 1. We can also move the numbers up so that they are arranged in three lines; doing this, we have

$$
\begin{array}{r}
\underline{-2} \; \vert \; 2 -5 6 -3 \\
-4 2 -8 \\
\hline
-1 4 5
\end{array}
$$

We now write 2, the first coefficient in the dividend, in the first position in the bottom row, and we have

$$\underline{-2\ |}\ \begin{array}{rrrr} 2 & -5 & 6 & -3 \\ & -4 & 2 & -8 \\ \hline 2 & -1 & 4 & 5 \end{array}$$

We notice that the first three numbers in the bottom row are the coefficients 2, -1, and 4 of the quotient; the last number in the bottom row is 5, and 5 is the remainder. The numbers in the second row are obtained by multiplying the number in the bottom row of the preceding column by -2, and the numbers in the bottom row are found by subtracting the numbers in the second row from those of the top row. If the multiplier, -2, is replaced by 2, the numbers in the second row can then be added to those of the top row to obtain the numbers in the bottom row. We make this change and the work appears as follows:

$$\underline{2\ |}\ \begin{array}{rrrr} 2 & -5 & 6 & -3 \\ & 4 & -2 & 8 \\ \hline 2 & -1 & 4 & 5 \end{array}$$

This arrangement of the computation is the **synthetic division** of the polynomial $2x^3 - 5x^2 + 6x - 3$ by $x - 2$, with the quotient $2x^2 - x + 4$ and the remainder 5.

In general, the following steps give the procedure for synthetic division of a polynomial $P(x)$ by $x - r$. As you read Illustrations 3 and 4, refer back to these steps.

1. Write $P(x)$ in the form $a_nx^n + a_{n-1}x^{n-1} + a_{n-2}x^{n-2} + \ldots + a_1x + a_0$ and insert a zero coefficient for any missing term.

2. Write the coefficients of $P(x)$ in order in a horizontal row.

3. Bring down the first coefficient a_n of $P(x)$ to the bottom row.

4. Multiply a_n by r, and write the product in the second row below the coefficient a_{n-1}; then add the product to a_{n-1} and write the sum in the bottom row.

5. Multiply this sum by r and write the product in the second row below the coefficient a_{n-2}; add the product to a_{n-2} and write the sum in the bottom row.

6. Continue the process of steps 4 and 5 as long as possible.

7. The last number in the bottom row is the remainder and the preceding numbers are the coefficients of the successive terms of the quotient which is a polynomial of degree one less than that of $P(x)$.

Illustration 3

We use synthetic division to divide $3x^3 - 7x^2 + x + 5$ by $x - 1$. The coefficients of the dividend are 3, -7, 1, and 5. The colored arrows indicate the order in which the numbers are found.

$$\begin{array}{r|rrrr} 1 & 3 & -7 & 1 & 5 \\ & & 3 & -4 & -3 \\ \hline & 3 & -4 & -3 & 2 \end{array}$$

Because the bottom row consists of the numbers 3, -4, -3, and 2, the quotient is $3x^2 - 4x - 3$ and the remainder is 2. Therefore

$$3x^3 - 7x^2 + x + 5 = (x - 1)(3x^2 - 4x - 3) + 2 \qquad \blacksquare$$

Illustration   4

We use synthetic division to find the quotient and remainder when $x^4 - 7x^2 + 2x - 6$ is divided by $x + 3$. Because we are dividing by $x + 3$ or, equivalently, $x - (-3)$, r is -3. The coefficients of $x^4 - 7x^2 + 2x - 6$ are 1, 0, -7, 2, and -6 (we insert a zero for the coefficient of the missing term involving x^3). The computation has the following form:

$$\begin{array}{r|rrrrr} -3 & 1 & 0 & -7 & 2 & -6 \\ & & -3 & 9 & -6 & 12 \\ \hline & 1 & -3 & 2 & -4 & 6 \end{array}$$

Therefore the quotient is $x^3 - 3x^2 + 2x - 4$, and the remainder is 6. Thus

$$x^4 - 7x^2 + 2x - 6 = (x + 3)(x^3 - 3x^2 + 2x - 4) + 6$$

Example   5   Use synthetic division to find the quotient and remainder when

$$x^5 - 3x^4 + 4x + 5$$

is divided by $x - 2$.

Solution   The coefficients of $x^5 - 3x^4 + 4x + 5$ are 1, -3, 0, 0, 4, and 5, where the two zeros represent the coefficients of the missing terms involving x^3 and x^2. Following is the computation by synthetic division:

$$\begin{array}{r|rrrrrr} 2 & 1 & -3 & 0 & 0 & 4 & 5 \\ & & 2 & -2 & -4 & -8 & -8 \\ \hline & 1 & -1 & -2 & -4 & -4 & -3 \end{array}$$

The quotient is $x^4 - x^3 - 2x^2 - 4x - 4$, and the remainder is -3. $\qquad \blacksquare$

The remainder theorem states that for a given polynomial $P(x)$, the value of $P(r)$ is the remainder when $P(x)$ is divided by $x - r$. Because synthetic division provides a fast way of obtaining this remainder, it is usually easier to compute $P(r)$ by synthetic division than by direct substitution.

Example 6 If

$$P(x) = 2x^5 + 4x^4 - 10x^3 - 20x - 10$$

find $P(0)$, $P(-1)$, $P(3)$, and $P(-4)$.

Solution We obtain $P(0)$ and $P(-1)$ by direct substitution.

$$P(0) = -10 \qquad P(-1) = -2 + 4 + 10 + 20 - 10$$
$$= 22$$

We obtain $P(3)$ and $P(-4)$ by synthetic division.

$$
\begin{array}{r|rrrrrr}
3 & 2 & 4 & -10 & 0 & -20 & -10 \\
  &   & 6 & 30 & 60 & 180 & 480 \\
\hline
  & 2 & 10 & 20 & 60 & 160 & 470
\end{array}
$$

$$
\begin{array}{r|rrrrrr}
-4 & 2 & 4 & -10 & 0 & -20 & -10 \\
   &   & -8 & 16 & -24 & 96 & -304 \\
\hline
   & 2 & -4 & 6 & -24 & 76 & -314
\end{array}
$$

Thus $P(3) = 470$ and $P(-4) = -314$.   ■

Example 7 Use synthetic division to determine whether the linear expression is a factor of $P(x)$.

a) $x - 2$; $P(x) = 4x^3 - 7x^2 + x - 2$
b) $x + 3$; $P(x) = 2x^4 + 5x^3 + 11x + 6$

Solution a) We use synthetic division to find $P(2)$.

$$
\begin{array}{r|rrrr}
2 & 4 & -7 & 1 & -2 \\
  &   & 8 & 2 & 6 \\
\hline
  & 4 & 1 & 3 & 4
\end{array}
$$

Therefore $P(2) = 4$. Because $P(2) \neq 0$, we conclude from the factor theorem that $x - 2$ is not a factor of $P(x)$.

b) We compute $P(-3)$ by synthetic division.

$$
\begin{array}{r|rrrrr}
-3 & 2 & 5 & 0 & 11 & 6 \\
   &   & -6 & 3 & -9 & -6 \\
\hline
   & 2 & -1 & 3 & 2 & 0
\end{array}
$$

Because $P(-3) = 0$, it follows from the factor theorem that $x + 3$ is a factor of $P(x)$. Thus

$$2x^4 + 5x^3 + 11x + 6 = (x + 3)(2x^3 - x^2 + 3x + 2)$$ ■

Example 8 Use synthetic division to show that $3x - 1$ is a factor of

$$6x^3 + x^2 - 4x + 1$$

Solution Because $3x - 1 = 3(x - \frac{1}{3})$, $3x - 1$ will be a factor if $x - \frac{1}{3}$ is a factor. We compute $P(\frac{1}{3})$ where $P(x) = 6x^3 + x^2 - 4x + 1$ by synthetic division.

$$\frac{1}{3} \underline{\big|\ 6 \quad 1 \quad -4 \quad 1}$$
$$\phantom{\frac{1}{3}\big|\ 6}\ \ 2 \quad\ 1 \quad -1$$
$$\overline{\phantom{\frac{1}{3}\big|}\ 6 \quad 3 \quad -3 \quad\ 0}$$

Because $P(\frac{1}{3}) = 0$, it follows from the factor theorem that $x - \frac{1}{3}$ is a factor of $P(x)$. Furthermore,

$$\begin{aligned}P(x) &= (x - \tfrac{1}{3})(6x^2 + 3x - 3) \\ &= (x - \tfrac{1}{3})[3(2x^2 + x - 1)] \\ &= (3x - 1)(2x^2 + x - 1)\end{aligned}$$

Therefore $3x - 1$ is a factor of $P(x)$. ∎

EXERCISES 10.1

In Exercises 1 through 8, use long division to find the remainder when the polynomial is divided by the linear expression. Then find the remainder by the remainder theorem.

1. $(3x^2 - 4x + 5) \div (x - 3)$
2. $(4x^2 + 7x - 5) \div (x - 1)$
3. $(3x^4 + 7x^3 + x^2 + x + 9) \div (x + 1)$
4. $(x^3 - 4x^2 + 5) \div (x + 3)$
5. $(x^3 + 9) \div (x + 2)$ 6. $(x^4 - 8) \div (x - 2)$
7. $(3x^5 - 7x^4 - 5x^3 - 4x^2 + 1) \div (x - 3)$
8. $(8x^5 + 7x^2 - 3) \div (x - \frac{1}{2})$

In Exercises 9 through 14, use the factor theorem to answer the question.

9. Is $x - 3$ a factor of $2x^3 - 6x^2 - 5x + 15$?
10. Is $x + 3$ a factor of $3x^3 - x^2 - 22x - 24$?
11. Is $x + 2$ a factor of $x^4 + 2x^3 - 12x^2 - 11x + 6$?
12. Is $x - 2$ a factor of $x^7 - 128$?
13. Is $x + 3$ a factor of $x^5 + 243$?
14. Is $x - a$ a factor of $x^8 + a^8$?
15. Find a value of k so that $x + 2$ is a factor of $3x^3 + 5x^2 + kx - 10$.
16. Find a value of k so that $x - 5$ is a factor of $kx^3 - 17x^2 - 4kx + 5$.
17. Find values of k so that $x - 4$ is a factor of $x^3 - k^2x^2 - 8kx - 16$.

18. Find values of k so that $x + 1$ is a factor of $5x^3 + k^2x^2 + 2kx - 3$.

In Exercises 19 through 28, use synthetic division to find the quotient and remainder.

19. $(2x^3 - x^2 + 3x + 12) \div (x - 4)$
20. $(y^3 + 4y^2 + 3y - 6) \div (y - 2)$
21. $(2x^4 + 5x^3 - 2x - 1) \div (x + 4)$
22. $(x^3 + 4x^2 - 7) \div (x + 3)$
23. $(3z^5 + z^4 - 4z^2 + 7) \div (z - 2)$
24. $(4x^6 + 21x^5 - 26x^3 + 27x) \div (x + 5)$
25. $(6x^3 - x^2 + 2x + 2) \div (x + \frac{1}{3})$
26. $(8x^3 - 6x^2 + 5x - 3) \div (x - \frac{1}{4})$
27. $(x^7 - 1) \div (x - 1)$
28. $(x^7 + 1) \div (x + 1)$

In Exercises 29 through 34, use synthetic division.

29. If $P(x) = 4x^3 - 5x^2 - 4$, find $P(2)$ and $P(-3)$.
30. If $P(x) = 3x^3 + 4x^2 - 9$, find $P(-2)$ and $P(1)$.
31. If $P(x) = x^4 + 3x^3 - 5x^2 + 9$, find $P(-4)$ and $P(3)$.
32. If $P(x) = 2x^4 - 7x^3 - 15x + 1$, find $P(4)$ and $P(-2)$.
33. If $P(x) = 6x^3 - x^2 - 7x + 2$, find $P(-\frac{1}{3})$ and $P(\frac{3}{2})$.
34. If $P(x) = x^3 + 2x + 4$, find $P(-1.3)$ and $P(2.1)$.

In Exercises 35 through 38, show by synthetic division that the linear expression is a factor of $P(x)$.

35. $x + 3$; $P(x) = 4x^3 + 9x^2 - 8x + 3$
36. $x + 5$; $P(x) = 2x^3 + 9x^2 - 3x + 10$
37. $2x - 1$; $P(x) = 6x^3 - 7x^2 + 4x - 1$
    [*Hint:* $2x - 1 = 2(x - \frac{1}{2})$.]
38. $3x + 2$; $P(x) = 12x^3 + 5x^2 - 11x - 6$
    [*Hint:* $3x + 2 = 3(x + \frac{2}{3})$.]

In Exercises 39 through 42, use synthetic division to determine whether the linear expression is a factor of $P(x)$.

39. $x - 4$; $P(x) = 2x^4 - 7x^3 - 14x + 8$
40. $x - 3$; $P(x) = x^4 - 6x^2 - 5x - 12$

41. $2x + 3$; $P(x) = 4x^3 - 4x^2 - 11x + 6$
42. $3x - 1$; $P(x) = 9x^3 + 3x^2 - 5x - 1$
43. (a) Is $x - 3$ a factor of $x^{50} - 3^{50}$? (b) Is $x - 3$ a factor of $x^{50} + 3^{50}$? (c) Is $x + 3$ a factor of $x^{49} - 3^{49}$? (d) Is $x + 3$ a factor of $x^{49} + 3^{49}$?
44. (a) Is $x + 3$ a factor of $x^{50} - 3^{50}$? (b) Is $x + 3$ a factor of $x^{50} + 3^{50}$? (c) Is $x - 3$ a factor of $x^{49} - 3^{49}$? (d) Is $x - 3$ a factor of $x^{49} + 3^{49}$?
45. For what integer values of n is $x - y$ a factor of $x^n - y^n$?
46. For what integer values of n is $x + y$ a factor of $x^n - y^n$?
47. For what integer values of n is $x + y$ a factor of $x^n + y^n$?

10.2   Rational Zeros of Polynomial Functions

Recall from Section 4.4 that the number r is a zero of polynomial function P if $P(r) = 0$. That is, the zeros of P are the roots of the polynomial equation $P(x) = 0$.

Illustration   1

If $P(x) = x^2 - 5x + 6$, the zeros of P are found by solving the equation

$$x^2 - 5x + 6 = 0$$
$$(x - 2)(x - 3) = 0$$
$$x = 2 \quad x = 3$$

Therefore the zeros of P are 2 and 3.   ∎

For a polynomial function of degree three or four, the general method for obtaining the zeros is complicated. Furthermore, for the zeros of a polynomial function of degree greater than four, there is no general formula in terms of a finite number of operations on the coefficients. However, in this section we shall show how the remainder and factor theorems and synthetic division can be applied to find the rational zeros of polynomial functions.

We shall make use of two theorems proved in Section 10.4, where complex zeros (including both real and imaginary numbers) of polynomial functions are treated. The first of these theorems states that if

$$P(x) = a_n x^n + a_{n-1}x^{n-1} + a_{n-2}x^{n-2} + \ldots + a_1 x + a_0 \qquad n \geq 1, a_n \neq 0$$

and if $r_1, r_2, \ldots, r_n$ are complex zeros of P, then

$$P(x) = a_n(x - r_1)(x - r_2) \ldots (x - r_n) \tag{1}$$

If in this equation a factor $x - r_i$ occurs k times, then r_i is called a **zero of multiplicity k.** If a zero of multiplicity k is counted as k zeros, then it follows from Equation (1) that a polynomial function P, for which $P(x)$ is of degree $n \geq 1$, has *at least n zeros*, some of which may be repeated. Theorem 2 proved in Section 10.4 states that such a polynomial has *exactly n zeros*.

Illustration 2

The function P defined by

$$P(x) = (x - 4)^3(x + 1)^2(x - 3)$$

is of degree six and P has six zeros; they are 4, 4, 4, -1, -1, and 3. The number 4 is a zero of multiplicity three, and -1 is a zero of multiplicity two. ■

   For each theorem relating to the zeros of a polynomial function, we have a statement regarding the roots of a polynomial equation. For instance, from the theorem which states that a polynomial of degree n has exactly n zeros, we have the fact that a polynomial equation of degree n has exactly n roots.

Illustration 3

The polynomial equation corresponding to the polynomial function of Illustration 2 is

$$(x - 4)^3(x + 1)^2(x - 3) = 0$$

This is an equation of the sixth degree, and the six roots are 4, 4, 4, -1, -1, and 3. ■

Example 1 Show that 3 is a zero of multiplicity two of the polynomial function defined by

$$P(x) = 2x^4 - 11x^3 + 11x^2 + 15x - 9$$

and find the other two zeros.

Solution To show that 3 is a zero of multiplicity two of $P(x)$, we show that $(x - 3)^2$ is a factor of $P(x)$. We use synthetic division to divide $P(x)$ by $x - 3$.

$$
\begin{array}{r|rrrrr}
3 & 2 & -11 & 11 & 15 & -9 \\
  &   & 6 & -15 & -12 & 9 \\
\hline
  & 2 & -5 & -4 & 3 & 0
\end{array}
$$

Hence

$$2x^4 - 11x^3 + 11x^2 + 15x - 9 = (x - 3)(2x^3 - 5x^2 - 4x + 3) \qquad (2)$$

We now divide $2x^3 - 5x^2 - 4x + 3$ by $x - 3$.

$$
\begin{array}{r|rrrr}
3 & 2 & -5 & -4 & 3 \\
  &   & 6 & 3 & -3 \\
\hline
  & 2 & 1 & -1 & 0
\end{array}
$$

Therefore

$$2x^3 - 5x^2 - 4x + 3 = (x - 3)(2x^2 + x - 1) \tag{3}$$

Substituting from Equation (3) into Equation (2), we obtain

$$2x^4 - 11x^3 + 11x^2 + 15x - 9 = (x - 3)^2(2x^2 + x - 1)$$

The quadratic factor can now be factored into two linear factors, and we have

$$2x^4 - 11x^3 + 11x^2 + 15x - 9 = (x - 3)^2(2x - 1)(x + 1)$$

Because $2x - 1 = 2(x - \frac{1}{2})$, it follows that

$$2x^4 - 11x^3 + 11x^2 + 15x - 9 = 2(x - 3)^2(x - \tfrac{1}{2})(x + 1)$$

Thus the zeros of the given polynomial function are 3, 3, $\frac{1}{2}$, and -1. ■

If the coefficients in the equation defining $P(x)$ are integers, then the rational zeros of P can be found by applying the following theorem.

Theorem 1

> Suppose that
>
> $$P(x) = a_n x^n + a_{n-1} x^{n-1} + a_{n-2} x^{n-2} + \ldots + a_1 x + a_0$$
>
> where $a_0, a_1, \ldots, a_n$ are integers. If $\dfrac{p}{q}$, in lowest terms, is a rational number and a zero of P, then p is an integer factor of a_0 and q is an integer factor of a_n.

Before proving Theorem 1 we give an illustration showing its application.

Illustration 4

Suppose that

$$P(x) = 3x^3 - 2x^2 - 7x - 2$$

From Theorem 1, we know that any rational zero $\dfrac{p}{q}$ of P must be such that p is an integer factor of -2 and q is an integer factor of 3. Therefore the possible values of p are 1, -1, 2, and -2; and the possible values of q are 1, -1, 3, and -3. Thus the set of possible rational zeros of P is

$$\left\{ 1, -1, 2, -2, \frac{1}{3}, -\frac{1}{3}, \frac{2}{3}, -\frac{2}{3} \right\}$$

Because the degree of P is three, not more than three members of this set can be zeros. We apply synthetic division to find out whether any of them are actual zeros.

$$
\underline{1}\ \begin{array}{rrrr} 3 & -2 & -7 & -2 \\ & 3 & 1 & -6 \\ \hline 3 & 1 & -6 & -8 \end{array}
\qquad
\underline{-1}\ \begin{array}{rrrr} 3 & -2 & -7 & -2 \\ & -3 & 5 & 2 \\ \hline 3 & -5 & -2 & 0 \end{array}
$$

Because $P(-1) = 0$, it follows that -1 is a zero of P. Furthermore, $x + 1$ is a factor of $P(x)$, and

$$P(x) = (x + 1)(3x^2 - 5x - 2)$$

The other two zeros of P are found by equating the quadratic factor to zero and solving the equation.

$$3x^2 - 5x - 2 = 0$$
$$(x - 2)(3x + 1) = 0$$

$$x - 2 = 0 \qquad 3x + 1 = 0$$
$$x = 2 \qquad x = -\tfrac{1}{3}$$

Thus the three zeros of P are -1, 2, and $-\tfrac{1}{3}$.
Observe that we have also factored $P(x)$.

$$P(x) = (x + 1)(x - 2)(3x + 1) \qquad \blacksquare$$

Proof of Theorem 1 Because $\dfrac{p}{q}$ is a zero of P, it is a solution of the equation

$$a_n x^n + a_{n-1}x^{n-1} + a_{n-2}x^{n-2} + \ldots + a_1 x + a_0 = 0$$

Therefore

$$a_n\left(\frac{p}{q}\right)^n + a_{n-1}\left(\frac{p}{q}\right)^{n-1} + a_{n-2}\left(\frac{p}{q}\right)^{n-2} + \ldots + a_1\left(\frac{p}{q}\right) + a_0 = 0$$

Multiplying on each side of this equation by q^n, we obtain

$$a_n p^n + a_{n-1}p^{n-1}q + a_{n-2}p^{n-2}q^2 + \ldots + a_1 p q^{n-1} + a_0 q^n = 0 \qquad (4)$$

We now add $-a_0 q^n$ on each side and factor p from each term on the resulting left side. Thus we have the equivalent equation

$$p(a_n p^{n-1} + a_{n-1}p^{n-2}q + a_{n-2}p^{n-3}q^2 + \ldots + a_1 q^{n-1}) = -a_0 q^n$$

Because $a_i (i$ is $1, 2, \ldots, n)$, p, and q are integers, and the sum and product of integers are integers, the expression in parentheses on the left side is an integer. If we represent this integer by t, we have the equation

$$pt = -a_0 q^n$$

The left side is an integer having p as a factor. Therefore p must be a factor of the right side, $-a_0 q^n$. Because $\dfrac{p}{q}$ is in lowest terms, p has no factor in common with q. Thus p must be a factor of a_0.

Equation (4) is also equivalent to the equation

$$q(a_{n-1}p^{n-1} + a_{n-2}p^{n-2}q + \ldots + a_1pq^{n-2} + a_0q^{n-1}) = -a_np^n$$

Now the left side is an integer having q as a factor; hence q must be a factor of the right member, $-a_np^n$. Because q has no factor in common with p, it follows that q must be a factor of a_n. ∎

Observe that Theorem 1 does not guarantee that a polynomial function with integer coefficients has a rational zero; however, the theorem provides the means of locating the numbers that could be rational zeros. We can then use synthetic division to determine whether any of the possible zeros are indeed actual zeros.

Illustration 5

If

$$P(x) = 4x^3 + 14x^2 + 10x - 3$$

then, by Theorem 1, any rational zero $\dfrac{p}{q}$ of P must be such that p is an integer factor of -3 and q is an integer factor of 4. Thus the possible values of p are 1, -1, 3, and -3; and the possible values of q are 1, -1, 2, -2, 4, and -4. Hence the set of possible rational zeros of P is

$$\left\{ 1, -1, 3, -3, \frac{1}{2}, -\frac{1}{2}, \frac{3}{2}, -\frac{3}{2}, \frac{1}{4}, -\frac{1}{4}, \frac{3}{4}, -\frac{3}{4} \right\}$$

Because the polynomial is of the third degree, not more than three of the possibilities can be zeros. We now use synthetic division to ascertain which of them, if any, are zeros.

```
 1 | 4   14   10   -3        -1 | 4    14    10   -3
   |     4   18   28            |      -4  -10    0
   ----------------------       -------------------------
     4   18   28   25             4    10     0   -3


 3 | 4   14   10   -3        -3 | 4    14    10   -3
   |    12   78  264            |     -12   -6  -12
   ----------------------       -------------------------
     4   26   88  261             4     2     4  -15


 1                            1
 2 | 4   14   10   -3        -2 | 4    14    10   -3
   |     2    8    9            |      -2   -6   -2
   ----------------------       -------------------------
     4   16   18    6             4    12     4   -5
```

So far, we have seen that $P(1) = 25$, $P(-1) = -3$, $P(3) = 261$, $P(-3) = -15$, $P\left(\dfrac{1}{2}\right) = 6$, and $P\left(-\dfrac{1}{2}\right) = -5$. We continue.

```
 3                            3
 2 | 4   14   10   -3        -2 | 4    14    10   -3
   |     6   30   60            |      -6  -12    3
   ----------------------       -------------------------
     4   20   40   57             4     8    -2    0
```

Because $P\left(-\dfrac{3}{2}\right) = 0$, it follows that $-\dfrac{3}{2}$ is a zero of P. Furthermore, $x + \dfrac{3}{2}$ is a factor of $P(x)$, and

$$P(x) = \left(x + \frac{3}{2}\right)(4x^2 + 8x - 2)$$

$$= 2\left(x + \frac{3}{2}\right)(2x^2 + 4x - 1)$$

The other two zeros of P can be found by setting the quadratic factor equal to zero and solving the equation.

$$2x^2 + 4x - 1 = 0$$

$$x = \frac{-4 \pm \sqrt{16 + 8}}{4}$$

$$= \frac{-4 \pm 2\sqrt{6}}{4}$$

$$= \frac{-2 \pm \sqrt{6}}{2}$$

Therefore the three zeros of P are $-\dfrac{3}{2}$, $\dfrac{-2 + \sqrt{6}}{2}$, and $\dfrac{-2 - \sqrt{6}}{2}$.   ∎

A special case of Theorem 1 occurs when a_n, the coefficient of x^n, is 1. Then

$$P(x) = x^n + a_{n-1}x^{n-1} + a_{n-2}x^{n-2} + \ \ldots \ + a_1x + a_0$$

where $a_0, a_1, \ldots, a_{n-1}$ are integers. For such a polynomial any rational zero of P must be an integer and, furthermore, must be an integer factor of a_0. This follows from the fact that if $\dfrac{p}{q}$ is a rational zero of P, then p must be a factor of a_0 and q must be a factor of 1, the coefficient of x^n.

Example   **2**   Find all the rational zeros of the polynomial function defined by

$$P(x) = x^5 + 4x^4 - 4x^3 - 34x^2 - 45x - 18$$

Solution   The possible rational zeros are integer factors of -18. These numbers are 1, -1, 2, -2, 3, -3, 6, -6, 9, -9, 18, and -18. We use synthetic division to determine which of them, if any, are zeros.

$$
\begin{array}{r|rrrrrr}
1 & 1 & 4 & -4 & -34 & -45 & -18 \\
  &   & 1 & 5 & 1 & -33 & -78 \\
\hline
  & 1 & 5 & 1 & -33 & -78 & -96 \\
\end{array}
$$

$$
\begin{array}{r|rrrrrr}
-1 & 1 & 4 & -4 & -34 & -45 & -18 \\
   &   & -1 & -3 & 7 & 27 & 18 \\
\hline
   & 1 & 3 & -7 & -27 & -18 & 0 \\
\end{array}
$$

Because $P(-1) = 0$, it follows that -1 is a zero of P, and

$$P(x) = (x + 1)(x^4 + 3x^3 - 7x^2 - 27x - 18)$$

Any other rational zeros of P must be zeros of the second factor. Because -1 is a possible zero of the second factor (and if it is, then -1 is a multiple zero of P), we divide that factor by $x + 1$ to determine whether the remainder is zero.

$$
\begin{array}{r|rrrr}
-1 & 1 & 3 & -7 & -27 & -18 \\
   &   & -1 & -2 & 9 & 18 \\
\hline
   & 1 & 2 & -9 & -18 & 0
\end{array}
$$

Therefore

$$x^4 + 3x^3 - 7x^2 - 27x - 18 = (x + 1)(x^3 + 2x^2 - 9x - 18)$$

so that

$$P(x) = (x + 1)(x + 1)(x^3 + 2x^2 - 9x - 18)$$

Continuing, we check for further rational zeros of P by considering the polynomial $x^3 + 2x^2 - 9x - 18$.

$$
\begin{array}{r|rrrr}
-1 & 1 & 2 & -9 & -18 \\
   &   & -1 & -1 & 10 \\
\hline
   & 1 & 1 & -10 & -8
\end{array}
\qquad
\begin{array}{r|rrrr}
2 & 1 & 2 & -9 & -18 \\
  &   & 2 & 8 & -2 \\
\hline
  & 1 & 4 & -1 & -20
\end{array}
$$

$$
\begin{array}{r|rrrr}
-2 & 1 & 2 & -9 & -18 \\
   &   & -2 & 0 & 18 \\
\hline
   & 1 & 0 & -9 & 0
\end{array}
$$

Thus

$$x^3 + 2x^2 - 9x - 18 = (x + 2)(x^2 - 9)$$

so

$$P(x) = (x + 1)(x + 1)(x + 2)(x^2 - 9)$$

Factoring $x^2 - 9$, we obtain

$$P(x) = (x + 1)^2(x + 2)(x - 3)(x + 3)$$

Hence the zeros of P are -1, -1, -2, 3, and -3.   ∎

In searching for the zeros of a polynomial function, the procedure can often be shortened if *upper and lower bounds* of the zeros can be determined.

Definition

Bounds for Zeros of a Polynomial Function
If P is a polynomial function, an **upper bound** of the real zeros of P is any number that is greater than or equal to the largest real zero. A **lower bound** of the real zeros of P is any number that is less than or equal to the smallest real zero.

Illustration 6

In Example 2 the zeros of the given function are -3, -2, -1, -1, and 3. The number 4 is an upper bound of these zeros. Another upper bound is 3. Actually, any number that is greater than or equal to 3 is an upper bound of these zeros. The number -3 is a lower bound of these zeros; the numbers $-\frac{7}{2}$ and -10 are also lower bounds. Any number that is less than or equal to -3 will serve as a lower bound.

■

The following theorem states that upper and lower bounds of the real zeros of a polynomial function P can be determined by observing the behavior of the signs of the numbers in the bottom row of the synthetic division of P.

Theorem 2

Suppose that

$$P(x) = a_n x^n + a_{n-1} x^{n-1} + a_{n-2} x^{n-2} + \ldots + a_1 x + a_0$$

and $a_n > 0$. In the synthetic division of $P(x)$ by $x - r$,

(i) if $r > 0$, and there are no negative numbers in the bottom row, then r is an upper bound of the real zeros of P;

(ii) if $r < 0$, and the signs of the numbers in the bottom row are alternately plus and minus (where zero can be appropriately denoted by either $+0$ or -0), then r is a lower bound of the real zeros of P.

Proof   To prove part (i), we consider the quotient $Q(x)$ and the remainder R when $P(x)$ is divided by $x - r$, and we have

$$P(x) = (x - r)Q(x) + R$$

By hypothesis, the numbers in the bottom row of the synthetic division of $P(x)$ by $x - r$ are nonnegative. Therefore R and the coefficients in $Q(x)$ are positive or zero. Furthermore, we are given that the coefficient a_n of x^{n-1} in $Q(x)$ is positive.

To show that r is an upper bound of the real zeros of P, we shall show that if s is any positive number for which $s > r$, then $P(s) \neq 0$. If s is substituted for x in the equation $P(x) = (x - r)Q(x) + R$, we have

$$P(s) = (s - r)Q(s) + R$$

Because $s > r$, then $s - r > 0$. Furthermore, $s > 0$ and hence $Q(s) > 0$ since the first coefficient in $Q(x)$ is positive and all the other coefficients are nonnegative. Also $R \geq 0$. Thus from the equation defining $P(s)$, it follows that $P(s) > 0$. Therefore $P(s) \neq 0$, and hence r is an upper bound of the real zeros of P.

The proof of part (ii) is similar if we consider

$$P(-x) = (-x - r)Q(-x) + R$$

However, here we must discuss separately the case when n is odd and the case when n is even. This proof is omitted. ∎

Illustration 7

Suppose that $P(x) = 2x^3 - x^2 + 4x - 2$. Then, if we use synthetic division to divide $P(x)$ by $x - 1$, we have

$$\underline{1}\,\begin{array}{|rrrr} 2 & -1 & 4 & -2 \\ & 2 & 1 & 5 \\ \hline 2 & 1 & 5 & 3 \end{array}$$

Because all the numbers in the bottom row are positive, it follows from Theorem 2(i) that 1 is an upper bound of the real zeros of P. If we divide $P(x)$ by $x + 1$, we have

$$\underline{-1}\,\begin{array}{|rrrr} 2 & -1 & 4 & -2 \\ & -2 & 3 & -7 \\ \hline 2 & -3 & 7 & -9 \end{array}$$

Because the signs of the numbers in the bottom row are alternately plus and minus, it follows from Theorem 2(ii) that -1 is a lower bound of the real zeros of P. ∎

We can apply Theorem 1 to find the rational roots of a polynomial equation if the equation is equivalent to one in which the coefficients are integers. If all but two of the roots of such a polynomial equation are rational, then any irrational or imaginary roots can be found by the quadratic formula; this procedure was demonstrated in Illustration 5.

Example 3   Find the rational roots of the equation

$$\tfrac{5}{3}x^4 - \tfrac{2}{3}x^3 - \tfrac{11}{6}x^2 + 2x = -\tfrac{3}{2}$$

If possible, find any irrational or imaginary roots.

Solution   We first multiply each side of the given equation by 6 so that the coefficients are integers, because to apply Theorem 1 this must be the case. We obtain the equivalent equation

$$10x^4 - 4x^3 - 11x^2 + 12x + 9 = 0$$

Let $P(x) = 10x^4 - 4x^3 - 11x^2 + 12x + 9$. If $\dfrac{p}{q}$ is a rational root of the equation $P(x) = 0$, then p must be an integer factor of 9 and q must be an integer factor of 10. Therefore the possible values of p are $1, -1, 3, -3, 9$, and -9;

the possible values of q are $1, -1, 2, -2, 5, -5, 10,$ and -10. The set of possible rational roots of the equation is therefore

$$\{1, -1, 3, -3, 9, -9, \tfrac{1}{2}, -\tfrac{1}{2}, \tfrac{3}{2}, -\tfrac{3}{2}, \tfrac{9}{2}, -\tfrac{9}{2}, \tfrac{1}{5}, -\tfrac{1}{5}, \tfrac{3}{5}, -\tfrac{3}{5}, \tfrac{9}{5}, -\tfrac{9}{5}, \tfrac{1}{10},$$
$$-\tfrac{1}{10}, \tfrac{3}{10}, -\tfrac{3}{10}, \tfrac{9}{10}, -\tfrac{9}{10}\}$$

We apply synthetic division to test these possible roots one by one.

$$
\begin{array}{r|rrrrr}
1 & 10 & -4 & -11 & 12 & 9 \\
  &   & 10 & 6 & -5 & 7 \\
\hline
  & 10 & 6 & -5 & 7 & 16
\end{array}
\qquad
\begin{array}{r|rrrrr}
-1 & 10 & -4 & -11 & 12 & 9 \\
   &   & -10 & 14 & -3 & -9 \\
\hline
   & 10 & -14 & 3 & 9 & 0
\end{array}
$$

Therefore $P(-1) = 0$, and so -1 is a root of the equation $P(x) = 0$; furthermore, $x + 1$ is a factor of $P(x)$. We now apply synthetic division to the quotient $Q(x) = 10x^3 - 14x^2 + 3x + 9$. We test -1 by dividing $Q(x)$ by $x + 1$.

$$
\begin{array}{r|rrrr}
-1 & 10 & -14 & 3 & 9 \\
   &   & -10 & 24 & -27 \\
\hline
   & 10 & -24 & 27 & -18
\end{array}
$$

Because the signs of the numbers in the bottom row are alternately plus and minus, it follows from Theorem 2(ii) that -1 is a lower bound of the real zeros of Q (and hence of P). Thus -9, $-\tfrac{9}{2}$, -3, $-\tfrac{9}{5}$, and $-\tfrac{3}{2}$ are eliminated as possible roots of the given equation. We test 3.

$$
\begin{array}{r|rrrr}
3 & 10 & -14 & 3 & 9 \\
  &   & 30 & 48 & 153 \\
\hline
  & 10 & 16 & 51 & 162
\end{array}
$$

Because all the numbers in the bottom row are positive, it follows from Theorem 2(i) that 3 is an upper bound of the real zeros of Q (and hence of P). Therefore, we eliminate 9 and $\tfrac{9}{2}$ as possible roots of the given equation. We now test $\tfrac{1}{2}$.

$$
\begin{array}{r|rrrr}
\tfrac{1}{2} & 10 & -14 & 3 & 9 \\
             &   & 5 & -\tfrac{9}{2} & \\
\hline
             & 10 & -9 & \text{stop} &
\end{array}
$$

The notation "stop" is indicated because once a fraction has appeared in the second row, each successive entry in the bottom row will also be a fraction. Thus the last number in the bottom row cannot be zero. A similar situation occurs when we test $-\tfrac{1}{2}$, $\tfrac{3}{2}$, $\pm\tfrac{1}{5}$, and $\tfrac{3}{5}$. We now test $-\tfrac{3}{5}$.

$$
\begin{array}{r|rrrr}
-\tfrac{3}{5} & 10 & -14 & 3 & 9 \\
              &   & -6 & 12 & -9 \\
\hline
              & 10 & -20 & 15 & 0
\end{array}
$$

Thus $Q(-\tfrac{3}{5}) = 0$; consequently, $x + \tfrac{3}{5}$ is a factor of $Q(x)$ and therefore also a factor of $P(x)$. Hence $-\tfrac{3}{5}$ is a root of the equation $P(x) = 0$. The equation $P(x) = 0$ can be written as

$$(x + 1)(x + \tfrac{3}{5})(10x^2 - 20x + 15) = 0$$

or, equivalently (dividing both sides by 5),

$$(x + 1)(x + \tfrac{3}{5})(2x^2 - 4x + 3) = 0$$

Because the third factor is a quadratic polynomial, we can find the other two roots by solving the equation

$$2x^2 - 4x + 3 = 0$$

$$x = \frac{4 \pm \sqrt{16 - 24}}{4}$$

$$= \frac{4 \pm \sqrt{-8}}{4}$$

$$= \frac{4 \pm 2i\sqrt{2}}{4}$$

$$= \frac{2 \pm i\sqrt{2}}{2}$$

The given equation, then, has the rational roots -1 and $-\tfrac{3}{5}$ and the two imaginary roots $\tfrac{1}{2}(2 \pm i\sqrt{2})$.   ■

Example   4   Prove that the equation

$$2x^3 - 2x^2 - 4x + 1 = 0$$

has no rational roots.

Solution   Let $P(x) = 2x^3 - 2x^2 - 4x + 1$. If $\dfrac{p}{q}$ is a rational root of the equation $P(x) = 0$, then p must be an integer factor of 1 and q must be an integer factor of 2. The possible values of p are therefore 1 and -1; the possible values of q are 1, -1, 2, and -2. Thus the set of possible rational roots of the equation is $\{1, -1, \tfrac{1}{2}, -\tfrac{1}{2}\}$. We test each of these possible roots by synthetic division.

$$
\begin{array}{r|rrrr}
1 & 2 & -2 & -4 & 1 \\
  &   & 2 & 0 & -4 \\
\hline
  & 2 & 0 & -4 & -3
\end{array}
\qquad
\begin{array}{r|rrrr}
-1 & 2 & -2 & -4 & 1 \\
   &   & -2 & 4 & 0 \\
\hline
   & 2 & -4 & 0 & 1
\end{array}
$$

$$
\begin{array}{r|rrrr}
\tfrac{1}{2} & 2 & -2 & -4 & 1 \\
             &   & 1 & -\tfrac{1}{2} & \\
\hline
             & 2 & -1 & \text{stop} &
\end{array}
\qquad
\begin{array}{r|rrrr}
-\tfrac{1}{2} & 2 & -2 & -4 & 1 \\
              &   & -1 & \tfrac{3}{2} & \\
\hline
              & 2 & -3 & \text{stop} &
\end{array}
$$

Therefore $P(1) \neq 0$, $P(-1) \neq 0$, $P(\tfrac{1}{2}) \neq 0$, and $P(-\tfrac{1}{2}) \neq 0$. Hence the equation has no rational roots.   ■

Example   5   From a square piece of tin that measures 12 in. on a side, a rectangular box is to be made by cutting out small squares of the same size

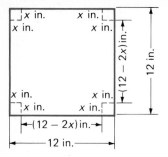

FIGURE 1 _____

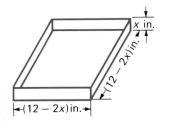

FIGURE 2 _____

from the four corners and turning up the sides. If the volume of the box is to be 108 in³, what should be the length of the side of the square to be cut out?

Solution Let x inches be the length of the side of the square to be cut out. Figure 1 shows the original piece of tin and Figure 2 shows the resulting box, having a depth of x inches and a square base of side $(12 - 2x)$ inches.

Because the volume of the box is to be 108 in³,

$$x(12 - 2x)^2 = 108$$
$$x[2(6 - x)]^2 = 108$$
$$4x(6 - x)^2 = 108$$
$$x(36 - 12x + x^2) = 27$$
$$x^3 - 12x^2 + 36x - 27 = 0$$

The possible rational roots of this equation are factors of -27, which are 1, -1, 3, -3, 9, -9, 27, and -27. We use synthetic division to determine which of them, if any, are roots.

$$
\begin{array}{r|rrrr}
1 & 1 & -12 & 36 & -27 \\
  &   & 1 & -11 & 25 \\
\hline
  & 1 & -11 & 25 & 2
\end{array}
\qquad
\begin{array}{r|rrrr}
3 & 1 & -12 & 36 & -27 \\
  &   & 3 & -27 & 27 \\
\hline
  & 1 & -9 & 9 & 0
\end{array}
$$

Therefore 3 is a root of the equation and the left side can be factored to obtain

$$(x - 3)(x^2 - 9x + 9) = 0$$

The other two roots of the equation can be found by solving the quadratic equation

$$x^2 - 9x + 9 = 0$$

$$x = \frac{9 \pm \sqrt{81 - 36}}{2}$$

$$= \frac{9 \pm \sqrt{45}}{2}$$

$$= \frac{9 \pm 3\sqrt{5}}{2}$$

$$x = \frac{9 + 3\sqrt{5}}{2} \qquad x = \frac{9 - 3\sqrt{5}}{2}$$

$$\approx 7.85 \qquad\qquad \approx 1.15$$

Because the base of the box is $(12 - 2x)$ inches, and $12 - 2x$ is negative when $x = 7.85$, we reject this value of x as a solution of the problem. However, $x = 1.15$ is a possible solution. Thus the length of the side of the square to be cut out may be either 3 in. or 1.15 in. ∎

Exercises   10.2

In Exercises 1 through 8, find the zeros of the polynomial function defined by the equation. State the multiplicity of each zero.

1. $P(x) = (x - 4)^2(x^2 - 4)$ 2. $P(x) = x^3(x^2 - 5)$
3. $P(x) = x^2(x + 1)^2(x^2 - 3)$ 4. $P(x) = (x^2 - 25)^2$
5. $P(x) = (x + 7)^3(x^2 - 7)^2$
6. $P(x) = (x^2 - 2)^2(x^2 - 4)(2x + 1)$
7. $P(x) = (3x + 4)^3(4x^2 - 9)^2(4x^2 + 12x + 9)$
8. $P(x) = (x^2 - 9)^2(5x^2 - 17x + 6)^2$
9. Show that -2 and 3 are zeros of the polynomial function defined by

$$P(x) = x^4 - 4x^3 - 7x^2 + 22x + 24$$

    and find the other two zeros.
10. Show that 5 and -1 are zeros of the polynomial function defined by

$$P(x) = x^4 + x^3 - 31x^2 - x + 30$$

    and find the other two zeros.
11. Show that -4 is a zero of multiplicity two of the polynomial function defined by

$$P(x) = x^4 + 9x^3 + 23x^2 + 8x - 16$$

    and find the other two zeros.
12. Show that 3 is a zero of multiplicity two of the polynomial function defined by

$$P(x) = x^4 - 3x^3 - 11x^2 + 39x - 18$$

    and find the other two zeros.
13. Given that -2 is a root of the equation

$$5x^3 + 3x^2 - 12x + 4 = 0$$

    find the other two roots.
14. Given that $\frac{4}{3}$ is a root of the equation

$$3x^3 - 16x^2 + 28x - 16 = 0$$

    find the other two roots.
15. Given that $\frac{1}{2}$ and $-\frac{2}{3}$ are roots of the equation

$$6x^4 + 25x^3 + 8x^2 - 7x - 2 = 0$$

    find the other two roots.
16. Given that $\sqrt{3}$ and $-\sqrt{3}$ are roots of the equation $x^4 + 3x^3 - 5x^2 - 9x + 6 = 0$, find the other two roots.

In Exercises 17 through 28, find all the rational zeros of the polynomial function. If possible, find any irrational or imaginary zeros.

17. $P(x) = x^3 - 3x^2 - x + 3$
18. $P(x) = x^3 - 4x^2 + x + 6$
19. $P(x) = x^3 - 7x - 6$ 20. $P(x) = x^3 - x^2 - 8x + 12$
21. $P(x) = x^4 + 3x^3 - 12x^2 - 13x - 15$
22. $P(x) = x^4 - 3x^3 + x^2 + 7x - 30$
23. $P(x) = 3x^3 + 8x^2 - 1$ 24. $P(x) = 4x^3 - 31x + 15$
25. $P(x) = 6x^4 - 37x^3 + 63x^2 - 33x + 5$
26. $P(x) = 8x^4 + 6x^3 - 13x^2 - x + 3$
27. $P(x) = x^4 - 2x^3 - 9x^2 + 20x - 4$
28. $P(x) = 2x^4 - x^3 + 2x^2 - 7x + 3$

In Exercises 29 through 40, find all the rational roots of the equation. If possible, find any irrational or imaginary roots.

29. $x^3 + 2x^2 - 7x + 4 = 0$
30. $x^3 - 3x^2 - 10x + 24 = 0$
31. $2x^3 - 13x^2 + 27x - 18 = 0$
32. $x^3 - 8x - 8 = 0$
33. $x^5 + 2x^4 - 13x^3 - 14x^2 + 24x = 0$
34. $9x^4 - 3x^3 + 7x^2 - 3x - 2 = 0$
35. $12x^4 - 5x^3 - 38x^2 + 15x + 6 = 0$
36. $2x^3 - \frac{25}{2}x^2 + \frac{7}{2}x - 3 = 0$
37. $x^3 + \frac{17}{3}x^2 - \frac{5}{3}x + 2 = 0$
38. $3x^4 + x^3 + 12x^2 - 5x - 3 = 0$
39. $\frac{1}{14}x^4 + \frac{1}{7}x^3 - \frac{1}{2}x^2 - \frac{1}{2}x + 1 = 0$
40. $18x^6 + 3x^5 - 25x^4 - 41x^3 - 15x^2 = 0$

In Exercises 41 through 44, prove that the equation has no rational roots.

41. $x^3 - 9x - 6 = 0$ 42. $2x^3 + 6x^2 - 3 = 0$
43. $3x^4 - x^3 + 4x^2 + 2x - 2 = 0$
44. $x^4 - x^3 - 4x^2 - 16 = 0$
45. A rectangular box is to be made from a piece of cardboard 6 cm wide and 14 cm long by cutting out squares of the same size from the four corners and turning up the sides. If the volume of the box is to be 40 cm³, what should the length of the side of the square to be cut out be?

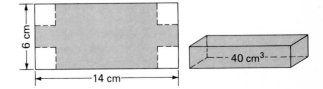

46. A slice of thickness 1 cm is cut off from one side of a cube. If the volume of the remaining figure is 180 cm^3, how long is the edge of the original cube?

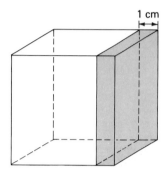

1 cm

47. A right circular cone is inscribed in a sphere and 32 times its volume is equal to 9 times the volume of the sphere. If the radius of the sphere is 2 ft, what is the altitude of the cone? There are two possible answers. The formula for the volume of a cone is $V = \frac{1}{3}\pi r^2 h$ and the formula for the volume of a sphere is $V = \frac{4}{3}\pi r^3$.

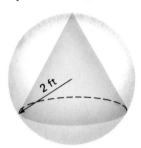

2 ft

48. A rectangular box has dimensions 12 in., 4 in., and 4 in. If the first two dimensions are decreased and the other dimension is increased by the same amount, a second box is formed, and its volume is five-eighths of the volume of the first box. Determine the dimensions of the second box.

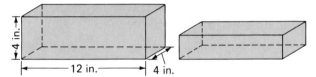

4 in.

12 in.

4 in.

10.3 Real Roots of Polynomial Equations

In Example 4 of Section 10.2 we showed that the equation

$$2x^3 - 2x^2 - 4x + 1 = 0 \tag{1}$$

has no rational roots. However, because the equation is of the third degree, it has three roots. These roots must therefore be either irrational or imaginary. In Section 10.4 we show that the imaginary roots of a polynomial equation with real coefficients must occur in pairs. Since this means there must always be an even number of imaginary roots, Equation (1) has either three real roots, all of which are irrational, or two imaginary roots and one real irrational root.

In the discussion that follows we shall need the concept of *variation in sign* of a polynomial. If the terms of a polynomial with real coefficients are written in descending powers of the variable (the terms involving zero coefficients are omitted), then a **variation in sign** occurs if two successive coefficients are opposite in sign. For example, if

$$P(x) = 2x^3 - 2x^2 - 4x + 1$$

the coefficients have, successively, the signs $+, -, -, +$; thus there are two variations in sign.

Illustration 1

If $Q(x) = x^4 - 6x^2 - 2x - 1$, then $Q(x)$ has one variation in sign. Furthermore, $Q(-x) = x^4 - 6x^2 + 2x - 1$, and $Q(-x)$ has three variations in sign. Moreover, if $R(x) = x^3 + 2x + 5$, then $R(x)$ has no variations in sign. ■

Descartes' Rule of Signs

> If $P(x)$ is a polynomial having real coefficients, the number of positive roots of the equation $P(x) = 0$ either is equal to the number of variations in sign of $P(x)$ or is less than this number by an even natural number. Furthermore, the number of negative roots of the equation is equal to the number of variations in sign of $P(-x)$ or is less than this number by an even natural number.

Descartes' rule of signs is named for the French mathematician René Descartes, mentioned in Section 3.1 as the originator of analytic geometry. Its proof is beyond the scope of this book.

Illustration 2

We apply Descartes' rule of signs to Equation (1). If

$$P(x) = 2x^3 - 2x^2 - 4x + 1$$

$P(x)$ has two variations in sign. By Descartes' rule of signs, the number of positive roots of Equation (1) is either two or zero. Furthermore, $P(-x) = -2x^3 - 2x^2 + 4x + 1$ has one variation in sign. Thus Equation (1) has one negative root. ■

The three roots of Equation (1) can now be described more fully. From Illustration 2 and the discussion earlier in this section, either two are positive irrational numbers and one is a negative irrational number, or else two are imaginary numbers and one is a negative irrational number.

Observe that Descartes' rule of signs states that if a polynomial $P(x)$ has an odd number of variations in sign, then the equation $P(x) = 0$ has an odd number of positive roots; thus we are certain of at least one positive root in such a case. Similarly, if $P(-x)$ has an odd number of variations in sign, then there is at least one negative root of the equation $P(x) = 0$.

Example 1

From Descartes' rule of signs, determine information concerning the number of positive, negative, and imaginary roots of each of the following equations.

a) $x^3 + 6x - 2 = 0$

b) $x^4 + 2x^2 - 5 = 0$

c) $6x^4 - x^3 + 2x - 3 = 0$

Solution a) Let $P(x) = x^3 + 6x - 2$. Because $P(x)$ has one variation in sign, there is one positive root of the equation. $P(-x) = -x^3 - 6x - 2$. Because $P(-x)$ has no variations in sign, there are no negative roots of the equation. Because the equation is of the third degree, it has three roots, and hence there are two imaginary roots.

b) Let $Q(x) = x^4 + 2x^2 - 5$. There is one positive root of the equation because $Q(x)$ has one variation in sign. $Q(-x) = x^4 + 2x^2 - 5$. Thus $Q(-x)$ has one variation in sign; consequently, there is one negative root of the equation. The equation has four roots, and hence there are two imaginary roots.

c) Let $R(x) = 6x^4 - x^3 + 2x - 3$. Because $R(x)$ has three variations in sign, the number of positive roots of the equation is either three or one. $R(-x) = 6x^4 + x^3 - 2x - 3$, and $R(-x)$ has one variation in sign. Therefore the equation has one negative root. Thus the equation has either three positive roots, one negative root, and no imaginary roots; or else it has one positive root, one negative root, and two imaginary roots. ■

The following theorem gives information about the real roots of a polynomial equation.

Theorem 1

> Suppose $P(x)$ is a polynomial and a and b are real numbers such that $a < b$. Then if $P(a)$ and $P(b)$ are opposite in sign, there is a real number c between a and b such that $P(c) = 0$.

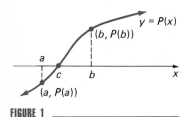

FIGURE 1

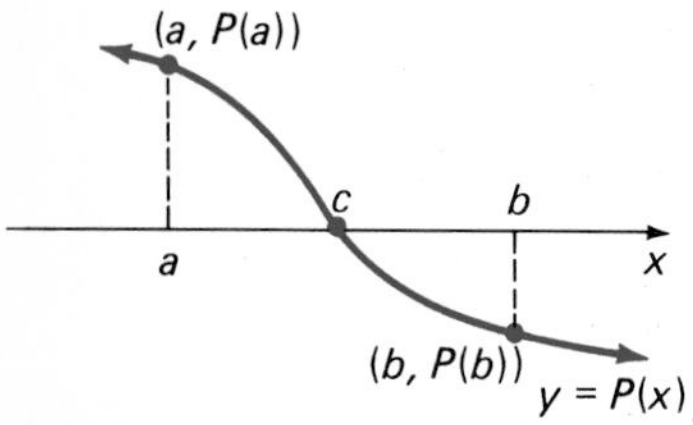

FIGURE 2

The proof of Theorem 1 is omitted. However, because the graph of a polynomial function is a continuous unbroken curve, the following argument should make the theorem seem reasonable: If $P(a)$ and $P(b)$ are opposite in sign, then the points $(a, P(a))$ and $(b, P(b))$ are on opposite sides of the x axis; thus the graph of $y = P(x)$ must intersect the x axis in at least one point $(c, 0)$ where c is between a and b. We show this situation in Figures 1 and 2. In Figure 1 there is a portion of the graph of a polynomial function P from the point $(a, P(a))$ to $(b, P(b))$ where $P(a) < 0$ and $P(b) > 0$. The graph intersects the x axis at the point $(c, 0)$ where $a < c < b$. Figure 2 shows the case when $P(a) > 0$ and $P(b) < 0$.

Illustration 3

We have learned that the three roots of Equation (1) are such that either two are positive irrational numbers and one is a negative irrational number, or two are imaginary numbers and one is a negative irrational number. With

$P(x) = 2x^3 - 2x^2 - 4x + 1$, we compute $P(1)$ and $P(2)$ by synthetic division.

$$
\begin{array}{r|rrrr}
1 & 2 & -2 & -4 & 1 \\
  &   & 2 & 0 & -4 \\
\hline
  & 2 & 0 & -4 & -3
\end{array}
\qquad
\begin{array}{r|rrrr}
2 & 2 & -2 & -4 & 1 \\
  &   & 4 & 4 & 0 \\
\hline
  & 2 & 2 & 0 & 1
\end{array}
$$

Therefore $P(1) = -3$ and $P(2) = 1$. Because $P(1)$ and $P(2)$ are opposite in sign, it follows from Theorem 1 that there is a real number c between 1 and 2 such that $P(c) = 0$; therefore Equation (1) has a positive root between 1 and 2. Thus we are certain that the equation has two positive irrational roots and one negative irrational root.

We can apply Theorem 1 to locate integers between which the other two roots lie. Because $P(0) = 1$, it follows that $P(0)$ and $P(1)$ are opposite in sign, and hence the equation has a positive root between 0 and 1. We compute $P(-1)$ and $P(-2)$ by synthetic division.

$$
\begin{array}{r|rrrr}
-1 & 2 & -2 & -4 & 1 \\
   &   & -2 & 4 & 0 \\
\hline
   & 2 & -4 & 0 & 1
\end{array}
\qquad
\begin{array}{r|rrrr}
-2 & 2 & -2 & -4 &   1 \\
   &   & -4 & 12 & -16 \\
\hline
   & 2 & -6 & 8 & -15
\end{array}
$$

Thus $P(-1)$ and $P(-2)$ are opposite in sign, and consequently a negative root is between -2 and -1.   ■

Example   2   Determine all the information you can concerning the number of positive, negative, and imaginary roots of each of the following equations.

a) $3x^4 + x^2 + 7x + 1 = 0$ b) $x^5 + 5x^2 - 4 = 0$

Solution   a) Let $P(x) = 3x^4 + x^2 + 7x + 1$. Because $P(x)$ has no variations in sign, there is no positive root. $P(-x) = 3x^4 + x^2 - 7x + 1$. Because $P(-x)$ has two variations in sign, there are either two or no negative roots. Furthermore, $P(0) = 1$ and $P(-1) = -2$. Because $P(0)$ and $P(-1)$ have opposite signs, it follows from Theorem 1 that there is a number c between 0 and -1 such that $P(c) = 0$; therefore the number c is a negative root of the equation. Thus the equation has two negative roots and two imaginary roots.

b) Let $Q(x) = x^5 + 5x^2 - 4$. Because $Q(x)$ has one variation in sign, there is one positive root. $Q(-x) = -x^5 + 5x^2 - 4$, and $Q(-x)$ has two variations in sign. Therefore there are either two negative roots or no negative roots. We compute $Q(-1)$ by synthetic division.

$$
\begin{array}{r|rrrrrr}
-1 & 1 & 0 & 0 & 5 & 0 & -4 \\
   &   & -1 & 1 & -1 & -4 & 4 \\
\hline
   & 1 & -1 & 1 & 4 & -4 & 0
\end{array}
$$

Therefore $Q(-1) = 0$ and thus -1 is a root of the equation. Hence the equation has one positive root, two negative roots, and two imaginary roots.   ■

We have seen that some polynomial equations with integer coefficients have no rational roots. Equation (1) is an example. It has three irrational roots, and in Illustration 3 we located them between -2 and -1, between 0 and 1, and between 1 and 2. The irrational roots can be approximated to any degree of accuracy by numerical methods and the application of computer programs. One technique that is often used is *Newton's method of approximation*, which involves concepts of calculus.

There is also a graphical way of approximating roots of polynomial equations. It involves approximating the x intercepts of the graph of the corresponding polynomial function. Obviously this procedure is not very accurate. An elementary method makes repeated use of Theorem 1, and the following example demonstrates it. With a calculator the procedure can be extended to an accuracy of several decimal places.

Example   **3**   Find the approximate value, to the nearest hundredth, of the smallest positive root of Equation (1).

Solution   The equation is

$$2x^3 - 2x^2 - 4x + 1 = 0$$

We know from Illustration 3 that the smallest positive root is between 0 and 1. Let $P(x) = 2x^3 - 2x^2 - 4x + 1$. The interval between 0 and 1 is divided into ten equal subintervals to give the numbers $0, 0.1, 0.2, 0.3, \ldots, 0.9, 1$. We use synthetic division to find $P(x)$ at each of these numbers until there is a change in sign. Of course, a calculator can also be used for the computation.

$$
\begin{array}{r|rrrr}
0.1 & 2 & -2 & -4 & 1 \\
& & 0.2 & -0.18 & -0.418 \\
\hline
& 2 & -1.8 & -4.18 & 0.582
\end{array}
\qquad
\begin{array}{r|rrrr}
0.2 & 2 & -2 & -4 & 1 \\
& & 0.4 & -0.32 & -0.864 \\
\hline
& 2 & -1.6 & -4.32 & 0.136
\end{array}
$$

$$
\begin{array}{r|rrrr}
0.3 & 2 & -2 & -4 & 1 \\
& & 0.6 & -0.42 & -1.326 \\
\hline
& 2 & -1.4 & -4.42 & -0.326
\end{array}
$$

Because $P(0.2) = 0.136$ and $P(0.3) = -0.326$, the root lies between 0.2 and 0.3. We now divide the interval between 0.2 and 0.3 into ten equal subintervals to give the numbers $0.2, 0.21, 0.22, 0.23, \ldots, 0.29, 0.3$. We use synthetic division to find $P(x)$ at each of these numbers until there is a change of sign.

$$
\begin{array}{r|rrrr}
0.22 & 2 & -2 & -4 & 1 \\
& & 0.44 & -0.3432 & -0.9555 \\
\hline
& 2 & -1.56 & -4.3432 & 0.0445
\end{array}
$$

$$
\begin{array}{r|rrrr}
0.23 & 2 & -2 & -4 & 1 \\
& & 0.46 & -0.3542 & -1.0015 \\
\hline
& 2 & -1.54 & -4.3542 & -0.0015
\end{array}
$$

Thus $P(0.22) = 0.0445$ and $P(0.23) = -0.0015$, and hence the root lies between 0.22 and 0.23. To obtain the root accurate to the nearest hundredth, we find $P(0.225)$.

$$\underline{0.225 \,\big|\; 2 \quad -2 \qquad -4 \qquad\quad 1}$$
$$ \quad\;\; 0.45 \quad -0.349 \quad -0.9785$$
$$\overline{ 2 \quad -1.55 \quad -4.349 \qquad 0.0215}$$

Therefore $P(0.225) = 0.0215$. It follows that the root is greater than 0.225, and hence it is closer to 0.23 than to 0.22. Thus the smallest positive root to the nearest hundredth is 0.23. ■

We conclude this section by giving an example showing how to prove that certain numbers are irrational.

Example   4 Prove that $\sqrt{3}$ is irrational.

Solution   Let $x = \sqrt{3}$. Then $x^2 = 3$ or, equivalently,

$$x^2 - 3 = 0$$

If $P(x) = x^2 - 3$, then $P(x)$ has one variation in sign. Therefore, by Descartes' rule of signs, there is one positive root of the equation. From Theorem 1 of Section 10.2, the only possible positive rational roots are 1 and 3. Obviously neither is a root of the equation. Thus the positive root is an irrational number and it is $\sqrt{3}$. ■

EXERCISES   10.3

In Exercises 1 through 12, use Descartes' rule of signs to determine information concerning the number of positive, negative, and imaginary roots of the equation.

1. $x^3 - 4x^2 - 2 = 0$ 2. $5x^3 - 3x - 7 = 0$
3. $4x^3 + 6x^2 - 3x + 5 = 0$ 4. $2x^3 + x + 1 = 0$
5. $3x^3 - 4x^2 + 2x - 5 = 0$ 6. $x^3 + 2x^2 + 3x + 1 = 0$
7. $x^4 + 7x^3 + x - 8 = 0$
8. $3x^4 - 2x^3 + 8x^2 - x - 7 = 0$
9. $6x^4 + 8x^2 + x = 0$ 10. $5x^4 - 3x^3 - 2 = 0$
11. $2x^5 - 6x^4 - x^2 + 4x - 1 = 0$
12. $x^6 + 3x^4 + 2x^3 - 4x^2 + 2x - 5 = 0$

In Exercises 13 through 20, determine all the information you can concerning the number of positive, negative, and imaginary roots of the equation. Find any rational roots and locate any irrational roots between two consecutive integers.

13. $x^3 - 6x + 3 = 0$ 14. $x^3 + 3x - 20 = 0$
15. $x^4 + x^2 - 1 = 0$ 16. $x^3 + 3x^2 - 2x - 5 = 0$

17. $4x^4 - 3x^3 + 2x - 5 = 0$
18. $2x^4 - 14x^3 + 24x^2 + x - 4 = 0$
19. $3x^4 - 21x^3 + 36x^2 + 2x - 8 = 0$
20. $x^4 + 2x^3 - 9x^2 - 8x + 14 = 0$

In Exercises 21 through 26, use the method of Example 3 to find the approximate value, to the nearest hundredth, of the indicated root.

21. $x^3 - 4x - 8 = 0$; the positive root
22. $x^3 - 2x + 7 = 0$; the negative root
23. $x^4 - 10x + 5 = 0$; the smallest positive root
24. $x^4 - 10x + 5 = 0$; the largest positive root
25. $2x^4 - 2x^3 + x^2 + 3x - 4 = 0$; the negative root
26. $x^4 + x^3 - 3x^2 - x - 4 = 0$; the positive root

In Exercises 27 through 32, prove that the number is irrational.

27. $\sqrt{5}$ 28. $-2\sqrt{7}$ 29. $\sqrt[3]{10}$
30. $\sqrt[4]{8}$ 31. $2 - \sqrt{5}$ 32. $3 + 2\sqrt{3}$

33. A spherical solid having radius r units and specific gravity k will sink in water to a depth of x units, where x is a positive root of the equation

$$x^3 - 3rx^2 + 4kr^3 = 0$$

Determine to the nearest tenth of an inch the depth to which a spherical buoy, of radius 10 in. and specific gravity 0.1, will sink.

34. Use the equation of Exercise 33 to find, to the nearest tenth of an inch, the depth to which a spherical ball will sink if the radius is 3 in. and the specific gravity is 0.5.

35. An open metal pan, having a volume of 1 gal (231 in^3), is to be made by cutting out squares of

the same size from the corners of a rectangular piece of metal 14 in. by 18 in. and turning up the sides. Determine to the nearest hundredth of an inch the length of a side of the squares to be cut from each corner. There are two possible answers.

36. A closed rectangular safe is to be made of lead of uniform thickness on the top and bottom and sides. The inside dimensions are to be 4 ft by 4 ft by 6 ft and the volume of the lead to be used in the construction is to be 450 ft^3. Find the thickness of the sides to the nearest hundredth of a foot.

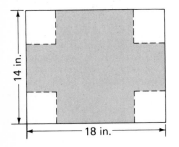

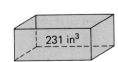

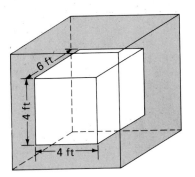

10.4   Complex Zeros of Polynomial Functions

So far our main concern in this chapter has been finding real zeros of polynomials with real coefficients. We now extend our discussion to include complex zeros as well as polynomial functions having complex coefficients. Such functions arise in more advanced courses and have applications in many fields, especially engineering and physics. Because the properties of complex numbers form a part of our presentation you may wish to review Section 1.8 at this time.

In Section 10.1 we stated the factor theorem when the number r is a real number and the polynomial $P(x)$ has real coefficients. The theorem is also valid when r is complex and $P(x)$ has complex coefficients.

Example   **1**   Show by synthetic division that $-2i$ is a zero of the polynomial function defined by

$$P(x) = x^3 - x^2 + 4x - 4$$

Then write $P(x)$ as the product of two polynomials having complex coefficients.

Solution   We compute $P(-2i)$ by synthetic division.

$$
\underline{-2i}\ \begin{array}{rrrr} 1 & -1 & 4 & -4 \\ & -2i & -4+2i & 4 \\ \hline 1 & -1-2i & 2i & 0 \end{array}
$$

Because $P(-2i) = 0$, $-2i$ is a zero of P. Furthermore, by the factor theorem, $x + 2i$ is a factor of $P(x)$ and

$$
P(x) = (x + 2i)[x^2 - (1 + 2i)x + 2i] \qquad \blacksquare
$$

In Section 10.2 we stated that it is difficult to find the zeros of a polynomial function except in special cases. However, there is a theorem, called the *fundamental theorem of algebra,* which guarantees that every polynomial function of nonzero degree has at least one complex zero.

The Fundamental Theorem of Algebra

> Every polynomial function of degree greater than zero, with complex coefficients, has at least one complex zero.

There are many proofs of this theorem but all involve concepts beyond the level of this book. The theorem was first proved in 1799 by the German mathematician Karl Friedrich Gauss (1777–1855) in his doctoral dissertation.

The fundamental theorem of algebra and the factor theorem are used to prove the next theorem. Recall that this theorem, whose proof was delayed until now, was applied in Section 10.2.

Theorem 1

> If $P(x)$ is the polynomial with complex coefficients defined by
>
> $$P(x) = a_n x^n + a_{n-1} x^{n-1} + a_{n-2} x^{n-2} + \ldots + a_1 x + a_0$$
>
> where $n \geq 1$, then
>
> $$P(x) = a_n(x - r_1)(x - r_2) \cdot \ldots \cdot (x - r_n) \qquad a_n \neq 0 \qquad (1)$$
>
> where each r_i $(i = 1, 2, \ldots, n)$ is a complex zero of P.

Proof   From the fundamental theorem of algebra, the function P has at least one complex zero, r_1. That is, there exists a complex number r_1 such that $P(r_1) = 0$. Therefore, by the factor theorem, $x - r_1$ is a factor of $P(x)$. Thus

$$
P(x) = (x - r_1)Q_1(x) \qquad (2)
$$

where $Q_1(x)$ is the quotient obtained when $P(x)$ is divided by $x - r_1$, and $Q_1(x)$ is of degree $n - 1$. From the fundamental theorem of algebra, if

$n - 1 \geq 1$, there exists a complex number r_2 such that $Q_1(r_2) = 0$. Then, by the factor theorem,

$$Q_1(x) = (x - r_2)Q_2(x) \tag{3}$$

where $Q_2(x)$ is the quotient obtained when $Q_1(x)$ is divided by $x - r_2$. Substituting from Equation (3) into Equation (2), we get

$$P(x) = (x - r_1)(x - r_2)Q_2(x)$$

Because $Q_1(r_2) = 0$, it follows from Equation (2) that $P(r_2) = 0$, and hence r_2 is a complex zero of P. We continue this procedure until the factoring has been performed n times; then we have

$$P(x) = (x - r_1)(x - r_2) \cdot \ \ldots \ \cdot (x - r_n)Q_n(x)$$

where each r_i $(i = 1, 2, \ldots, n)$ is a complex zero of P. Because there are n factors of the form $x - r_i$, the polynomial $Q_n(x)$ must be a constant and that constant must be the coefficient of x^n in the expansion. Thus $Q_n(x) = a_n$, and therefore

$$P(x) = a_n(x - r_1)(x - r_2) \cdot \ \ldots \ \cdot (x - r_n)$$

where each r_i is a complex zero of P. ■

In Section 10.2 you learned that if in Equation (1) a factor $x - r_i$ occurs exactly k times, then r_i is called a zero of multiplicity k. If such a zero is counted as k zeros, then it follows from Theorem 1 that a polynomial function P, for which $P(x)$ is of degree $n \geq 1$, has *at least n zeros*, some of which may be repeated. However, we can prove that such a polynomial function has *exactly n zeros*, and this fact is stated in the next theorem, which we also applied in Section 10.2.

Theorem 2   If $P(x)$ is a polynomial of degree $n \geq 1$, with complex coefficients, then P has exactly n complex zeros.

Proof   From Theorem 1, P has at least n complex zeros. If we now show that P cannot have more than n zeros, the theorem will be proved.

Equation (1) is

$$P(x) = a_n(x - r_1)(x - r_2) \cdot \ \ldots \ \cdot (x - r_n) \qquad a_n \neq 0$$

where each r_i (i is $1, 2, \ldots, n$) is a complex zero of P. Let r be any number other than $r_1, r_2, \ldots, r_n$. Because Equation (1) is an identity,

$$P(r) = a_n(r - r_1)(r - r_2) \cdot \ \ldots \ \cdot (r - r_n) \qquad a_n \neq 0$$

Because $r \neq r_i$, none of the factors $r - r_i$ is zero; therefore the right side of this equation is not zero. Thus $P(r) \neq 0$; so r is not a zero of P. Hence P has exactly n complex zeros. ■

Example   2   Find the zeros of the polynomial function P defined by

$$P(x) = x^3 - 4x^2 + 5x - 6$$

Solution   Any rational zero of P must be a factor of -6. Thus the set of possible rational zeros is $\{1, -1, 2, -2, 3, -3, 6, -6\}$. We use synthetic division to determine which, if any, are zeros.

$$
\begin{array}{r|rrrr}
1 & 1 & -4 & 5 & -6 \\
  &   & 1 & -3 & 2 \\
\hline
  & 1 & -3 & 2 & -4
\end{array}
\qquad
\begin{array}{r|rrrr}
-1 & 1 & -4 & 5 & -6 \\
   &   & -1 & 5 & -10 \\
\hline
   & 1 & -5 & 10 & -16
\end{array}
$$

$$
\begin{array}{r|rrrr}
2 & 1 & -4 & 5 & -6 \\
  &   & 2 & -4 & 2 \\
\hline
  & 1 & -2 & 1 & -4
\end{array}
\qquad
\begin{array}{r|rrrr}
3 & 1 & -4 & 5 & -6 \\
  &   & 3 & -3 & 6 \\
\hline
  & 1 & -1 & 2 & 0
\end{array}
$$

Because $P(3) = 0$, then 3 is a zero of P and

$$P(x) = (x - 3)(x^2 - x + 2)$$

To determine the other two zeros, we solve the quadratic equation

$$x^2 - x + 2 = 0$$

$$x = \frac{1 \pm \sqrt{1 - 8}}{2}$$

$$= \frac{1 \pm i\sqrt{7}}{2}$$

Therefore the three zeros are 3, $\frac{1}{2}(1 + i\sqrt{7})$, and $\frac{1}{2}(1 - i\sqrt{7})$. ∎

The polynomial function with real coefficients in Example 2 has two zeros, $\frac{1}{2}(1 + i\sqrt{7})$ and $\frac{1}{2}(1 - i\sqrt{7})$, that are conjugates of each other. Furthermore, from the quadratic formula, it follows that if a quadratic function having real coefficients has a complex zero, then the other zero is its conjugate. These two situations are special cases of a theorem which states that if a polynomial function P with real coefficients has a complex zero, then the conjugate is also a zero of P. To prove this fact we need the following properties of conjugates of complex numbers. The notation $\bar{z}$ is used to denote the conjugate of the complex number z; that is,

$$\text{if} \quad z = a + bi \quad \text{then} \quad \bar{z} = a - bi$$

Properties of Complex Conjugates

If z_1, z_2, and z are complex numbers, then

(i) $\bar{z}_1 + \bar{z}_2 = \overline{z_1 + z_2}$,

(ii) $\bar{z}_1 \cdot \bar{z}_2 = \overline{z_1 \cdot z_2}$,

(iii) $\overline{z^n} = \bar{z}^n$, for all positive integers n.

The proof of property (i) is given below, and the proof of property (ii) is left as an exercise. See Exercise 39. To prove property (iii) we need to use mathematical induction (discussed in Section 11.2) and property (ii).

Proof of property (i) Let $z_1 = a + bi$ and $z_2 = c + di$. Then $\overline{z}_1 = a - bi$ and $\overline{z}_2 = c - di$.

$$z_1 + z_2 = (a + bi) + (c + di)$$
$$= (a + c) + (b + d)i$$

Thus

$$\overline{z_1 + z_2} = (a + c) - (b + d)i$$

However,

$$\overline{z}_1 + \overline{z}_2 = (a - bi) + (c - di)$$
$$= (a + c) - (b + d)i$$

Thus we conclude

$$\overline{z}_1 + \overline{z}_2 = \overline{z_1 + z_2}$$ ∎

Properties (i) and (ii) can be extended to more than two complex numbers. That is, if $z_1, z_2, \ldots, z_n$ are complex numbers, then

$$\overline{z}_1 + \overline{z}_2 + \ldots + \overline{z}_n = \overline{z_1 + z_2 + \ldots + z_n} \tag{4}$$

and

$$\overline{z}_1 \cdot \overline{z}_2 \cdot \ldots \cdot \overline{z}_n = \overline{z_1 \cdot z_2 \cdot \ldots \cdot z_n}$$

These formulas can be proved by mathematical induction.

We are now ready to prove the following theorem, which we used in Section 10.3.

Theorem 3

> If $P(x)$ is a polynomial with real coefficients and if z is a complex zero of P, then the conjugate $\overline{z}$ is also a zero of P.

Proof Let

$$P(x) = a_n x^n + a_{n-1} x^{n-1} + a_{n-2} x^{n-2} + \ldots + a_1 x + a_0$$

where all the coefficients a_i are real numbers. Because z is a zero of P, $P(z) = 0$; thus

$$a_n z^n + a_{n-1} z^{n-1} + a_{n-2} z^{n-2} + \ldots + a_1 z + a_0 = 0$$

Hence

$$\overline{a_n z^n + a_{n-1} z^{n-1} + a_{n-2} z^{n-2} + \ldots + a_1 z + a_0} = \overline{0}$$

From Equation (4) (the conjugate of the sum of complex numbers equals the sum of the conjugates of the numbers) and the fact that $\overline{0} = 0$, we have from the above equation

$$\overline{a_n z^n} + \overline{a_{n-1} z^{n-1}} + \overline{a_{n-2} z^{n-2}} + \ldots + \overline{a_1 z} + \overline{a_0} = 0$$

Applying property (ii) of conjugates in each term of the left side of this equation, we have

$$\overline{a_n}\,\overline{z^n} + \overline{a_{n-1}}\,\overline{z^{n-1}} + \overline{a_{n-2}}\,\overline{z^{n-2}} + \ldots + \overline{a_1}\,\overline{z} + \overline{a_0} = 0$$

Because each a_j is a real number, $\overline{a_j} = a_j (j$ is $0, 1, 2, \ldots, n)$; furthermore, from property (iii) of conjugates, $\overline{z^j} = \overline{z}^j$. Thus the above equation can be written as

$$a_n \overline{z}^n + a_{n-1} \overline{z}^{n-1} + a_{n-2} \overline{z}^{n-2} + \ldots + a_1 \overline{z} + a_0 = 0$$

The left side of this equation is $P(\overline{z})$. Therefore $P(\overline{z}) = 0$; hence $\overline{z}$ is a zero of P.   ■

Example   **3**   Find a polynomial $P(x)$ of the fourth degree that has real coefficients if P has $1 - i$ and $-2i$ as zeros.

Solution   From Theorem 3, if $1 - i$ and $-2i$ are zeros of P, then their conjugates $1 + i$ and $2i$ are also zeros. Therefore

$$\begin{aligned} P(x) &= [x - (1 - i)][x - (1 + i)][x - (-2i)][x - 2i] \\ &= (x^2 - 2x + 2)(x^2 + 4) \\ &= x^4 - 2x^3 + 6x^2 - 8x + 8 \end{aligned}$$ ■

Example   **4**   Given that i is a root of the equation

$$2x^4 - 5x^3 + 3x^2 - 5x + 1 = 0$$

find the solution set of the equation.

Solution   Because i is a root of the given equation, its conjugate, $-i$, is also a root. We use synthetic division to divide the polynomial $2x^4 - 5x^3 + 3x^2 - 5x + 1$ by $x - i$ and then we divide the quotient by $x - (-i)$.

$$\begin{array}{r|rrrrr} i & 2 & -5 & 3 & -5 & 1 \\ & & 2i & -2 - 5i & 5 + i & -1 \\ \hline -i & 2 & -5 + 2i & 1 - 5i & i & -1 \\ & & -2i & 5i & -i & \\ \hline & 2 & -5 & 1 & 0 & \end{array}$$

Therefore the given equation can be written as

$$(x - i)(x + i)(2x^2 - 5x + 1) = 0$$

Equating each factor to zero, we obtain

$$x - i = 0 \qquad x + i = 0 \qquad 2x^2 - 5x + 1 = 0$$

$$x = i \qquad\qquad x = -i$$

$$x = \frac{5 \pm \sqrt{25 - 8}}{4}$$

$$= \frac{5 \pm \sqrt{17}}{4}$$

Thus the solution set of the given equation is

$$\{i, -i, \tfrac{1}{4}(5 + \sqrt{17}), \tfrac{1}{4}(5 - \sqrt{17})\}$$ ∎

Following are two interesting theorems that are consequences of Theorem 3.

Theorem 4

> If $P(x)$ is a polynomial with real coefficients and the degree of P is an odd number, then P has at least one real zero.

Proof Assume P has no real zeros. Then because the degree of P is an odd number, it has an odd number of imaginary zeros. However, the number of imaginary zeros of P must be even, because by Theorem 3, for each imaginary zero, its conjugate must also be a zero. Therefore we have a contradiction. Thus our assumption that P has no real zeros is false. Hence P has at least one real zero. ∎

Theorem 5

> If $P(x)$ is a polynomial with real coefficients, then $P(x)$ can be expressed as a product of linear or quadratic polynomials with real coefficients.

Proof If $P(x)$ is of degree n, then by Theorem 2, P has exactly n complex zeros. Denote these zeros by $c_1, c_2, \ldots, c_n$. Thus

$$P(x) = a_n(x - c_1)(x - c_2) \cdot \ldots \cdot (x - c_n)$$

For every c_i that is a real number, $x - c_i$ is a linear factor of $P(x)$. Suppose z_i is an imaginary zero of P. Then by Theorem 3, $\bar{z}_i$ is also a zero of P and is one of the complex zeros $c_1, c_2, \ldots, c_n$. Therefore $x - z_i$ and $x - \bar{z}_i$ are factors of $P(x)$ and

$$(x - z_i)(x - \bar{z}_i) = x^2 - (z_i + \bar{z}_i)x + z_i \cdot \bar{z}_i$$

Let $z_i = a + bi$ and $\bar{z}_i = a - bi$. Then

$$z_i + \bar{z}_i = (a + bi) + (a - bi) \qquad z_i \cdot \bar{z}_i = (a + bi)(a - bi)$$

$$= 2a \qquad\qquad\qquad = a^2 - b^2 i^2$$

$$\qquad\qquad\qquad\qquad\qquad = a^2 + b^2$$

Hence

$$(x - z_i)(x - \overline{z}_i) = x^2 - 2ax + (a^2 + b^2)$$

Because $-2a$ and $a^2 + b^2$ are real numbers, $(x - z_i)(x - \overline{z}_i)$ is a quadratic polynomial with real coefficients.

Therefore we conclude that $P(x)$ can be expressed as a product of linear or quadratic polynomials with real coefficients. ∎

The following two illustrations verify Theorem 5 for two particular polynomials.

Illustration 1

Let $P(x) = x^3 - 4x^2 + 5x - 6$. From Example 2

$$P(x) = (x - 3)(x^2 - x + 2)$$

which is the product of a linear and a quadratic polynomial. ∎

Illustration 2

Let $P(x) = 2x^4 - 5x^3 + 3x^2 - 5x + 1$. From Example 4

$$\begin{aligned} P(x) &= (x - i)(x + i)(2x^2 - 5x + 1) \\ &= (x^2 + 1)(2x^2 - 5x + 1) \end{aligned}$$

which is the product of two quadratic polynomials. ∎

EXERCISES 10.4

In Exercises 1 through 4, show by synthetic division that the complex number is a zero of P, and find the other zeros.

1. $-3i$; $P(x) = x^4 - 3x^3 + 11x^2 - 27x + 18$
2. $4i$; $P(x) = x^4 - x^3 + 15x^2 - 16x - 16$
3. $2 + i$; $P(x) = x^3 - 7x^2 + 17x - 15$
4. $-1 - i$; $P(x) = 2x^3 + 3x^2 + 2x - 2$

In Exercises 5 through 12, find the zeros of the polynomial function.

5. $P(x) = x^3 - 4x^2 + 6x - 4$
6. $P(x) = x^3 + 2x^2 - 2x + 3$
7. $P(x) = 2x^4 - 7x^3 + 21x^2 + 17x - 13$
8. $P(x) = x^4 - 2x^3 + 2x^2 + 2x - 3$
9. $P(x) = x^5 - 2x^4 + 8x^3 - 16x^2 + 16x - 32$
10. $P(x) = 3x^5 - 18x^4 + 38x^3 - 36x^2 + 24x - 16$
11. $P(x) = 9x^4 - 42x^3 + 79x^2 - 40x + 6$
12. $P(x) = 6x^4 + 5x^3 + 4x^2 - 2x - 1$

In Exercises 13 through 18, find the solution set of the equation.

13. $x^3 + 5x^2 + 5x + 4 = 0$
14. $x^3 + 3x^2 + 8x + 24 = 0$
15. $x^4 - 3x^3 + 4x^2 - 6x + 4 = 0$
16. $x^4 + x^3 - 5x^2 + x - 6 = 0$
17. $4x^4 + 12x^3 + 17x^2 + 10x + 2 = 0$
18. $6x^4 + 13x^3 + 24x^2 - 8 = 0$

In Exercises 19 through 26, find a polynomial $P(x)$ of the stated degree with real coefficients for which the numbers are zeros of P.

19. Second degree; $4 + 3i$ is a zero of P.
20. Second degree; $3 - i$ is a zero of P.
21. Third degree; $2 - i\sqrt{5}$ and -4 are zeros of P.
22. Third degree; $5 + i\sqrt{3}$ and 2 are zeros of P.
23. Fourth degree; $-3i$ is a zero of multiplicity two of P.
24. Fourth degree; $2 + i$ and $1 - i\sqrt{2}$ are zeros of P.
25. Fifth degree; 3, $3 + i\sqrt{2}$, and $-i\sqrt{2}$ are zeros of P.
26. Fifth degree; $3 - i$ (multiplicity two) and 1 are zeros of P.

In Exercises 27 through 30, find the solution set of the equation if the number is a root.

27. $3x^4 - 2x^3 + 2x^2 - 8x - 40 = 0$; $2i$ is a root.
28. $5x^4 - 2x^3 + 46x^2 - 18x + 9 = 0$; $-3i$ is a root.
29. $2x^4 + 6x^3 + 33x^2 - 36x + 20 = 0$; $-2 - 4i$ is a root.
30. $3x^4 + 4x^3 + 9x^2 - 6x + 4 = 0$; $-1 + i\sqrt{3}$ is a root.

In Exercises 31 through 38, use the result of the indicated exercise to express the polynomial of that exercise as a product of linear or quadratic polynomials.

31. Exercise 5   32. Exercise 6   33. Exercise 7
34. Exercise 8   35. Exercise 9   36. Exercise 10
37. Exercise 11   38. Exercise 12

39. Prove that if z_1 and z_2 are complex numbers, then $\overline{z_1} \cdot \overline{z_2} = \overline{z_1 \cdot z_2}$.

40. Prove that $2 - i$ is a root of the equation

$$x^2 - 2x + 1 + 2i = 0$$

but that its conjugate is not a root. Does this contradict Theorem 3?

10.5 Partial Fractions

In Section 1.6 we discussed how to combine two or more rational expressions into one rational expression by addition or subtraction. For example,

$$\frac{3}{x + 2} + \frac{4}{x - 3} = \frac{7x - 1}{(x + 2)(x - 3)}$$

In more advanced mathematics courses, such as calculus, it is sometimes necessary to express a single rational expression as a sum of two or more simpler quotients called **partial fractions.** We now discuss how to decompose a rational expression into partial fractions.

Consider a rational function H defined by

$$H(x) = \frac{P(x)}{Q(x)}$$

where $P(x)$ and $Q(x)$ are polynomials. We shall assume that we have a **proper fraction,** that is, one for which the degree of $P(x)$ is less than the degree of $Q(x)$. If we have a rational function for which the degree of the numerator is not less than the degree of the denominator, then we have an **improper fraction,** and in that case we divide the numerator by the denominator until a proper fraction is obtained. For instance,

$$\frac{x^4 - 10x^2 + 3x + 1}{x^2 - 4} = x^2 - 6 + \frac{3x - 23}{x^2 - 4}$$

In general, then, we are concerned with a method of decomposing a proper fraction of the form $P(x)/Q(x)$ into two or more partial fractions. The denominators of the partial fractions are obtained by factoring $Q(x)$ into a product of linear and quadratic factors. Sometimes it may be difficult to find these factors. However, recall Theorem 5 from Section 10.4, which states that a polynomial with real coefficients can be expressed as a product of linear and quadratic polynomials with real coefficients.

After $Q(x)$ has been factored into products of linear and quadratic factors, the method of determining the partial fractions depends on the nature of these factors. We consider various cases separately. The results of advanced algebra, which are not proved here, provide us with the form of the partial fractions in each case.

Case 1 The factors of $Q(x)$ are all linear, and none is repeated. That is,

$$Q(x) = (a_1 x + b_1)(a_2 x + b_2 \cdot \ldots \cdot a_n x + b_n)$$

where no two of the factors are identical. In this case we write

$$\frac{P(x)}{Q(x)} = \frac{A_1}{a_1 x + b_1} + \frac{A_2}{a_2 x + b_2} + \cdots + \frac{A_n}{a_n x + b_n}$$

where $A_1, A_2, \ldots, A_n$ are constants to be determined. Observe that this equation is an identity because it is true for each value of x for which a denominator is not zero. The following illustration shows a method for determining the values of A_i.

Illustration 1

To decompose the fraction

$$\frac{7x - 1}{x^2 - x - 6}$$

into partial fractions, we factor the denominator and obtain

$$\frac{7x - 1}{x^2 - x - 6} = \frac{7x - 1}{(x + 2)(x - 3)}$$

Therefore we have

$$\frac{7x - 1}{(x + 2)(x - 3)} = \frac{A}{x + 2} + \frac{B}{x - 3} \tag{1}$$

Equation (1) is an identity for all x except -2 and 3. By multiplying on both sides of the equation by the LCD, we obtain

$$7x - 1 = A(x - 3) + B(x + 2)$$

This equation is an identity. It is true for all values of x including -2 and 3. We now find the constants A and B. Substituting 3 for x in the preceding equation, we get

$$20 = 5B \quad \Leftrightarrow \quad B = 4$$

Substituting -2 for x in the same equation, we obtain

$$-15 = -5A \quad \Leftrightarrow \quad A = 3$$

With these values for A and B, we have from Equation (1)

$$\frac{7x - 1}{(x + 2)(x - 3)} = \frac{3}{x + 2} + \frac{4}{x - 3}$$

Observe that this equation is equivalent to the one at the beginning of this section. ■

Example 1 Decompose the fraction

$$\frac{x-1}{x^3 - x^2 - 2x}$$

into partial fractions.

Solution We factor the denominator and have

$$\frac{x-1}{x^3 - x^2 - 2x} = \frac{x-1}{x(x-2)(x+1)}$$

Thus we have

$$\frac{x-1}{x(x-2)(x+1)} = \frac{A}{x} + \frac{B}{x-2} + \frac{C}{x+1} \tag{2}$$

Equation (2) is an identity for all x except 0, 2, and -1. We multiply on both sides of the equation by the LCD and get

$$x - 1 = A(x-2)(x+1) + Bx(x+1) + Cx(x-2)$$

This equation is an identity that is true for all values of x including 0, 2, and -1. To find the constants, we first substitute 0 for x and obtain

$$-1 = -2A \qquad \Leftrightarrow \qquad A = \tfrac{1}{2}$$

Substituting 2 for x, we get

$$1 = 6B \qquad \Leftrightarrow \qquad B = \tfrac{1}{6}$$

Substituting -1 for x, we obtain

$$-2 = 3C \qquad \Leftrightarrow \qquad C = -\tfrac{2}{3}$$

With these values for A, B, and C we have from (2)

$$\frac{x-1}{x(x-2)(x+1)} = \frac{\tfrac{1}{2}}{x} + \frac{\tfrac{1}{6}}{x-2} + \frac{-\tfrac{2}{3}}{x+1}$$

or, equivalently,

$$\frac{x-1}{x(x-2)(x+1)} = \frac{1}{2x} + \frac{1}{6(x-2)} - \frac{2}{3(x+1)} \qquad \blacksquare$$

Case 2 The factors of $Q(x)$ are all linear, and some are repeated.

Suppose that $(ax + b)^p$ occurs as a factor of $Q(x)$. Then $ax + b$ is said to be a p-fold factor of $Q(x)$, and corresponding to this factor there will be the sum of p partial fractions

$$\frac{A_1}{ax + b} + \frac{A_2}{(ax + b)^2} + \ldots + \frac{A_{p-1}}{(ax + b)^{p-1}} + \frac{A_p}{(ax + b)^p}$$

where $A_1, A_2 \ldots, A_p$ are constants to be determined. Example 2 illustrates this case and the method of determining each A_i.

Example 2 Decompose the fraction

$$\frac{x^4 + x^2 + 16x - 12}{x^3(x - 2)^2}$$

into partial fractions.

Solution We write the given fraction as a sum of partial fractions as follows:

$$\frac{x^4 + x^2 + 16x - 12}{x^3(x - 2)^2} = \frac{A}{x} + \frac{B}{x^2} + \frac{C}{x^3} + \frac{D}{x - 2} + \frac{E}{(x - 2)^2} \qquad (3)$$

Multiplying on both sides of Equation (3) by the LCD, we get

$$x^4 + x^2 + 16x - 12 = Ax^2(x - 2)^2 + Bx(x - 2)^2 + C(x - 2)^2$$
$$+ Dx^3(x - 2) + Ex^3 \qquad (4)$$

We substitute 0 for x in this equation and obtain

$$-12 = 4C \qquad \Leftrightarrow \qquad C = -3$$

Substituting 2 for x in Equation (4), we get

$$40 = 8E \qquad \Leftrightarrow \qquad E = 5$$

With these values for C and E in (4) and expanding the powers of the binomials, we have

$$x^4 + x^2 + 16x - 12 = Ax^2(x^2 - 4x + 4) + Bx(x^2 - 4x + 4)$$
$$- 3(x^2 - 4x + 4) + Dx^3(x - 2) + 5x^3$$
$$x^4 + x^2 + 16x - 12 = (A + D)x^4 + (-4A + B - 2D + 5)x^3$$
$$+ (4A - 4B - 3)x^2 + (4B + 12)x - 12$$

Because this equation is an identity, the coefficients on the left must equal the corresponding coefficients on the right. Therefore we have the following system of equations:

$$\begin{cases} A + D = 1 \\ -4A + B - 2D + 5 = 0 \\ 4A - 4B - 3 = 1 \\ 4B + 12 = 16 \end{cases}$$

From the fourth equation, $B = 1$. Replacing B by 1 in the third equation and solving for A, we get $A = 2$. With $A = 2$ in the first equation, we obtain $D = -1$. We use the second equation as a check:

$$-4A + B - 2D + 5 = -4(2) + 1 - 2(-1) + 5$$
$$= 0$$

Thus the values of the constants are as follows:

$$A = 2 \qquad B = 1 \qquad C = -3 \qquad D = -1 \qquad E = 5$$

With these values we have from (3)

$$\frac{x^4 + x^2 + 16x - 12}{x^3(x-2)^2} = \frac{2}{x} + \frac{1}{x^2} - \frac{3}{x^3} - \frac{1}{x-2} + \frac{5}{(x-2)^2} \qquad \blacksquare$$

Case 3 The factors of $Q(x)$ are linear and quadratic, and none of the quadratic factors is repeated.

Corresponding to the quadratic factor $ax^2 + bx + c$ in the denominator is the partial fraction of the form

$$\frac{Ax + B}{ax^2 + bx + c}$$

Example 3 Decompose the fraction

$$\frac{x^2 - x - 5}{x^3 + x^2 - 2}$$

into partial fractions.

Solution We attempt to factor the denominator by using synthetic division to divide $x^3 + x^2 - 2$ by linear expressions of the form $x - r$, where r is an integer factor of -2. The division by $x - 1$ is as follows:

$$\underline{1}\begin{array}{|rrrr} 1 & 1 & 0 & -2 \\ & 1 & 2 & 2 \\ \hline 1 & 2 & 2 & 0 \end{array}$$

Therefore $x^3 + x^2 - 2 = (x - 1)(x^2 + 2x + 2)$. The given fraction is written as a sum of partial fractions in the following way:

$$\frac{x^2 - x - 5}{(x - 1)(x^2 + 2x + 2)} = \frac{Ax + B}{x^2 + 2x + 2} + \frac{C}{x - 1} \qquad (5)$$

Multiplying on both sides by the LCD, we have

$$x^2 - x - 5 = (Ax + B)(x - 1) + C(x^2 + 2x + 2) \qquad (6)$$

We compute C by substituting 1 for x in Equation (6), and we get

$$-5 = 5C \quad \Leftrightarrow \quad C = -1$$

We replace C by -1 in (6) and multiply on the right side to obtain

$$x^2 - x - 5 = (A - 1)x^2 + (B - A - 2)x + (-B - 2)$$

Equating coefficients of like powers of x gives the system

$$\begin{cases} A - 1 = 1 \\ B - A - 2 = -1 \\ -B - 2 = -5 \end{cases}$$

Therefore

$$A = 2 \qquad B = 3$$

Substituting the values of A, B, and C in Equation (5), we obtain

$$\frac{x^2 - x - 5}{(x - 1)(x^2 + 2x + 2)} = \frac{2x + 3}{x^2 + 2x + 2} - \frac{1}{x - 1}$$ ∎

Case 4   The factors of $Q(x)$ are linear and quadratic, and some of the quadratic factors are repeated.

If $ax^2 + bx + c$ is a p-fold quadratic factor of $Q(x)$, then, corresponding to this factor $(ax^2 + bx + c)^p$, we have the sum of the following p partial fractions:

$$\frac{A_1x + B_1}{ax^2 + bx + c} + \frac{A_2x + B_2}{(ax^2 + bx + c)^2} + \cdots + \frac{A_px + B_p}{(ax^2 + bx + c)^p}$$

Illustration 2

If the denominator contains the factor $(x^2 - 5x + 2)^3$, we have, corresponding to this factor, the sum of partial fractions

$$\frac{Ax + B}{x^2 - 5x + 2} + \frac{Cx + D}{(x^2 - 5x + 2)^2} + \frac{Ex + F}{(x^2 - 5x + 2)^3}$$ ∎

Example 4   Decompose the fraction

$$\frac{3x^4 - 12x^3 + 4x^2 + 11x + 4}{x(x^2 - 3x - 2)^2}$$

into partial fractions.

Solution   The given fraction is written as a sum of partial fractions as follows:

$$\frac{3x^4 - 12x^3 + 4x^2 + 11x + 4}{x(x^2 - 3x - 2)^2} = \frac{Ax + B}{x^2 - 3x - 2} + \frac{Cx + D}{(x^2 - 3x - 2)^2} + \frac{E}{x} \quad (7)$$

We multiply on both sides by the LCD and get

$$3x^4 - 12x^3 + 4x^2 + 11x + 4 = x(Ax + B)(x^2 - 3x - 2) + x(Cx + D) + E(x^2 - 3x - 2)^2 \quad (8)$$

Substituting 0 for x in this equation, we obtain

$$4 = 4E \quad \Leftrightarrow \quad E = 1$$

With $E = 1$ in Equation (8) and multiplying the polynomials, we get

$$3x^4 - 12x^3 + 4x^2 + 11x + 4$$
$$= Ax^4 - 3Ax^3 - 2Ax^2 + Bx^3 - 3Bx^2 - 2Bx + Cx^2 + Dx + x^4 + 9x^2 + 4 - 6x^3 - 4x^2 + 12x$$
$$= (A + 1)x^4 + (-3A + B - 6)x^3 + (-2A - 3B + C + 5)x^2 + (-2B + D + 12)x + 4$$

We equate the coefficients of corresponding powers of x and obtain the system

$$\begin{cases} A + 1 = 3 \\ -3A + B - 6 = -12 \\ -2A - 3B + C + 5 = 4 \\ -2B + D + 12 = 11 \end{cases}$$

We solve this system to obtain

$$A = 2 \quad B = 0 \quad C = 3 \quad D = -1 \quad E = 1$$

Substituting these values in Equation (7), we have

$$\frac{3x^4 - 12x^3 + 4x^2 + 11x + 4}{x(x^2 - 3x - 2)^2} = \frac{2x}{x^2 - 3x - 2} + \frac{3x - 1}{(x^2 - 3x - 2)^2} + \frac{1}{x} \quad \blacksquare$$

Observe that in each of the examples, the number of constants to be determined is equal to the degree of the denominator of the original fraction being decomposed into partial fractions. This situation applies to all cases.

EXERCISES 10.5

In Exercises 1 through 10, decompose the fraction into partial fractions.

1. $\dfrac{12}{x^2 - 4}$

2. $\dfrac{1}{2x^2 - x}$

3. $\dfrac{x - 1}{x^2 + x}$

4. $\dfrac{x + 15}{x^2 - 9}$

5. $\dfrac{x + 5}{x^2 - 4x + 3}$

6. $\dfrac{3x}{x^2 + x - 2}$

7. $\dfrac{x + 12}{3x^2 - 5x - 2}$

8. $\dfrac{3x - 7}{4x^2 + 3x - 1}$

9. $\dfrac{3x^2 + 3x - 12}{6x^3 + 5x^2 - 6x}$

10. $\dfrac{2x^2 - 11x - 9}{x^3 - 2x^2 - 3x}$

In Exercises 11 through 14, express the improper fraction as the sum of a polynomial and partial fractions.

11. $\dfrac{2x^3 + 4}{x^2 - 4}$

12. $\dfrac{x^3 + 5}{x^2 - 1}$

13. $\dfrac{4x^3 - 8x^2 - 10x + 30}{2x^2 + x - 6}$

14. $\dfrac{6x^3 + x^2 - 5x - 7}{3x^2 - x - 2}$

In Exercises 15 through 38, decompose the fraction into partial fractions.

15. $\dfrac{3x^2 + 13x - 10}{x^3 - 2x^2}$

16. $\dfrac{x^2 + x + 1}{x^4 - x^3}$

17. $\dfrac{x^2 - 11x + 6}{(x + 2)(x^2 - 4x + 4)}$

18. $\dfrac{x^2 + 11}{(x - 5)(x^2 + 2x + 1)}$

19. $\dfrac{3x + 16}{(2x^2 - x - 1)^2}$

20. $\dfrac{9x^3 - 8x^2 - 4x + 48}{(x^2 - 4)^2}$

21. $\dfrac{x^3 + 6x - 4}{(x - 2)^3}$

22. $\dfrac{x^2 + 2}{(x - 3)^3}$

23. $\dfrac{3x^2 - x + 4}{x^3 + x^2 + x}$

24. $\dfrac{3x + 8}{x^3 + 4x}$

25. $\dfrac{3x^2 + 2x - 4}{x^3 - 8}$

26. $\dfrac{x^2 - 6x + 2}{x^3 + 1}$

27. $\dfrac{2x^2 - 7x + 1}{x^3 - x^2 + x - 1}$

28. $\dfrac{3x^2 - 9x + 8}{x^3 + x^2 + 3x + 3}$

29. $\dfrac{11x^2 + 11x + 8}{2x^3 + 8x^2 + 3x + 12}$

30. $\dfrac{3x^2 + 2x + 3}{x^4 + x^3 + x^2 + x}$

31. $\dfrac{3x^2 - 4x}{(x^2 + 1)(x^2 - x - 1)}$

32. $\dfrac{4x - 3}{x^4 + 2x^3 + 3x^2}$

33. $\dfrac{x + 6}{x^4 + 2x^3 + 3x^2}$

34. $\dfrac{3x^4 + 4}{x^4 + 4x^2 + 4}$

35. $\dfrac{x^3 - x^2}{x^4 + 2x^2 + 1}$

36. $\dfrac{x^4 + 2x^2 - 2x - 4}{(x^2 + 3)^3}$

37. $\dfrac{x^4 + x^3 - 5x^2 - 14x - 1}{x^5 - x^4 + 4x^3 - 4x^2 + 4x - 4}$

38. $\dfrac{11x - 28}{x^5 + 2x^4 + 2x^3 + 4x^2 + x + 2}$

In Exercises 1 and 2, use the remainder theorem to find the remainder for the indicated division.

1. $(3x^3 + 4x^2 - 3x - 5) \div (x + 2)$
2. $(2x^4 - 5x^2 - 2x + 1) \div (x - 1)$

In Exercises 3 and 4, use the factor theorem to answer the question.

3. Is $x - 3$ a factor of $x^3 + 2x^2 - 12x - 9$?
4. Is $x + 4$ a factor of $2x^3 + 9x^2 + 6x + 8$?
5. Find a value of k so that $x - 3$ is a factor of $2kx^3 - 5x^2 + 3kx$.
6. Find a value of k so that $x + 2$ is a factor of $2x^4 + 2kx^3 - x^2 - 3kx - 8$.

In Exercises 7 through 10, use synthetic division to find the quotient and remainder.

7. $(2x^4 + 7x^3 - 4x + 5) \div (x + 3)$
8. $(x^3 - 6x^2 + 8x - 5) \div (x - 4)$
9. $(x^6 - 64) \div (x - 2)$ 10. $(x^5 + 243) \div (x + 3)$

In Exercises 11 and 12, find the indicated function values by synthetic division.

11. $P(x) = 2x^4 - 8x^2 - 10x - 3$; (a) $P(-2)$ and (b) $P(3)$.
12. $P(x) = 3x^4 + 10x^3 - 6x^2 + 1$; (a) $P(-4)$ and (b) $P(-\frac{1}{3})$.

In Exercises 13 through 16, find the zeros of the polynomial function P defined by the equation. State the multiplicity of each zero.

13. $P(x) = (x^2 + 2x - 3)(2x^2 + x - 15)$
14. $P(x) = (x^2 - 1)(x^2 - 4)(x^2 + x - 2)$
15. $P(x) = (x^2 - 9)(x^2 - 4)^2(6x^2 + x - 15)$
16. $P(x) = (x - 5)^3(x^2 - 36)(x^2 + 2x - 1)^2$

In Exercises 17 and 18, show that the numbers are zeros of the polynomial function, and find the other two zeros.

17. $P(x) = x^4 + x^3 - 8x^2 + 8$; -1 and 2
18. $P(x) = 2x^4 + 5x^3 - 6x^2 - 7x + 6$; 1 and -3

In Exercises 19 through 22, find all the rational zeros of the polynomial function. If possible, find any irrational or imaginary zeros.

19. $P(x) = x^4 - 3x^3 - 8x^2 + 26x - 12$
20. $P(x) = x^3 + 5x^2 + 5x - 3$
21. $P(x) = 3x^3 - 11x^2 + 9x - 2$
22. $P(x) = 2x^4 + x^3 - 17x^2 - 4x + 6$

In Exercises 23 through 26, find all the rational roots of

the equation. If possible, find any irrational or imaginary roots.

23. $x^3 + x^2 - 15x + 9 = 0$
24. $x^4 - 3x^3 - 10x^2 + 28x - 16 = 0$
25. $6x^4 - 25x^3 - 3x^2 + 5x + 1 = 0$
26. $3x^3 - x^2 + 16x + 12 = 0$
27. Find the solution set of the equation
$$x^4 + 2x^3 - 4x^2 - 4x + 4 = 0$$
given that $\sqrt{2}$ and $-\sqrt{2}$ are roots.
28. Find the solution set of the equation
$$x^4 + x^3 - 4x^2 - 3x + 3 = 0$$
given that $\sqrt{3}$ and $-\sqrt{3}$ are roots.

In Exercises 29 through 32, prove that the equation has no rational roots. Then use Descartes' rule of signs to determine information concerning the number of positive, negative, and imaginary roots.

29. $x^3 - 7x^2 + x + 3 = 0$ 30. $x^3 - 3x^2 - 5 = 0$
31. $x^4 - 6x - 9 = 0$ 32. $x^4 + 2x^3 + 6x - 3 = 0$

In Exercises 33 through 36, determine all the information you can concerning the number of positive, negative, and imaginary roots of the equation. Find any rational roots and locate any irrational roots between two consecutive integers.

33. $x^3 - 3x^2 + 6x - 24 = 0$ 34. $x^3 + 3x^2 - 3x - 2 = 0$
35. $x^4 - 2x^3 - 13x^2 + 33x - 14 = 0$
36. $3x^4 + 10x^3 - 11x^2 - 4x + 2 = 0$

In Exercises 37 and 38, use the method of Example 3 in Section 10.3 to find the approximate value, to the nearest hundredth, of the indicated root.

37. $x^3 - 3x^2 - 6x + 2 = 0$; the negative root
38. $x^3 + 9x^2 + 15x - 21 = 0$; the positive root

In Exercises 39 through 42, find the solution set of the equation.

39. $4x^3 - 11x^2 + 26x - 15 = 0$
40. $3x^3 - x^2 + 16x + 12 = 0$
41. $6x^4 - 25x^3 - 3x^2 + 5x + 1 = 0$
42. $2x^4 - 9x^3 + 17x^2 - 3x - 7 = 0$

In Exercises 43 through 46, (a) find the zeros of the polynomial function and (b) use the result of part (a) to express the polynomial as a product of linear or quadratic factors.

43. $P(x) = 2x^3 + 3x^2 + 6x - 4$

44. $P(x) = 3x^4 + x^3 - 3x - 1$

45. $P(x) = x^4 + 2x^3 + 6x^2 + 8x + 8$; $2i$ is a zero

46. $P(x) = 2x^4 - 2x^3 + 3x^2 - 2x + 1$; $-i$ is a zero

47. Find the solution set of the equation

$$x^4 + 3x^3 + 3x^2 - 2 = 0$$

given that $-1 + i$ is a root.

48. Find the solution set of the equation

$$2x^4 - x^3 + 33x^2 - 16x + 16 = 0$$

given that $4i$ is a root.

In Exercises 49 through 52, find a polynomial $P(x)$ of the stated degree with real coefficients for which the numbers are zeros of P.

49. Fourth degree; $1 - i$ and $1 + i\sqrt{3}$

50. Fifth degree; $2 - i\sqrt{2}$ (multiplicity two) and -1

51. Sixth degree; $3i$ (multiplicity two) and $2 - i\sqrt{2}$

52. Sixth degree; $2 + 3i$, $2 + i$, and $1 + 3i$

53. The dimensions of a rectangular box are 3 in., 4 in., and 5 in. The volume of the box is doubled if each dimension is increased by the same number of inches. Determine this number.

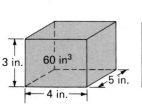
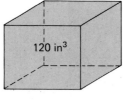

54. The volume of a rectangular box is 504 cm³, and the numbers of centimeters in the dimensions of the box are three consecutive integers. What are the dimensions?

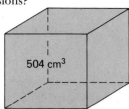

55. The area of a right triangle is 6 cm². Find the lengths of the sides of the triangle if the length of

one of the sides is 2 cm shorter than the length of the hypotenuse.

56. Determine to the nearest tenth of an inch the depth to which a wooden spherical ball will sink if the radius of the ball is 2 in. and the specific gravity is 0.6. (*Hint:* Use the equation of Exercise 33 in Exercises 10.3.)

57. An open rectangular box is to be made of wood of uniform thickness on the sides and bottom. The inside dimensions are to be 5 ft in length, 3 ft in width, and 2 ft in height. The wood weighs 40 lb/ft³. Determine to the nearest hundredth of an inch what the thickness of the wood should be if the weight of the empty box is to be 160 lb.

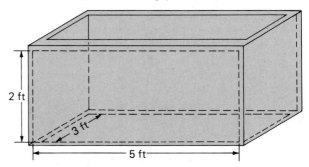

58. A hollow spherical container, whose capacity is 800 cm³, has an outer radius of 6 cm. Find the thickness, to the nearest hundredth of a centimeter, of the wall of the container. The formula for the volume of a sphere is $V = \frac{4}{3}\pi r^3$.

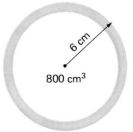

Exercises 59 through 64 pertain to Supplementary Section 10.5. In these exercises decompose the given fraction into partial fractions.

59. $\dfrac{2x^2 - 15x - 32}{(x - 1)(x^2 + 6x + 8)}$

60. $\dfrac{x^3 - 6x^2 + 8x - 1}{x^2 - 6x + 9}$

61. $\dfrac{x^2 + 2x - 2}{x^3 - x^2 - x - 2}$

62. $\dfrac{x^2 - 3x + 6}{2x^3 - 3x^2 - 2x + 3}$

63. $\dfrac{x^4 + x^3 + x - 4}{x^4 + 2x^2 + 1}$

64. $\dfrac{x^4 - x^2 + 9}{x(x^2 + 3x + 3)^2}$

C H A P T E R

11

Selected Topics in Algebra

Introduction

The subject matter of this chapter is diversified and the treatment of the various topics is brief. For more detailed discussions you need to refer to more advanced texts.

The introduction to sequences and series in Section 11.1 is based on the function concept. In Section 11.1 we also present the sigma notation for writing a summation. This notation is applied extensively in Section 11.2 on mathematical induction, which is used to prove that certain formulas and theorems are true for any positive integer. In particular, we prove by mathematical induction some properties of arithmetic and geometric sequences and series in Sections 11.3 and 11.4. In Section 11.5 we discuss infinite geometric series. If you study calculus you will realize that such series are just one of many types of infinite series that constitute a major part of the course.

General counting procedures as well as permutations and combinations are treated in Section 11.6, where we apply them to situations that arise in everyday life. In Section 11.7 we give an introduction to the theory of probability. In Section 11.8 the binomial theorem is presented. Our proof of the theorem utilizes the idea of the binomial coefficients as combinations.

11.1 Sequences, Series, and Sigma Notation

Sequences of numbers are often encountered in mathematics. For instance, the numbers

$$2, 4, 6, 8, 10$$

form a sequence. This sequence is called *finite* because there is a last number. If a set of numbers forming a sequence does not have a last number, the sequence is said to be *infinite*. For example,

$$1, \tfrac{1}{2}, \tfrac{1}{3}, \tfrac{1}{4}, \tfrac{1}{5}, \ldots$$

is an infinite sequence because the three dots with no number following indicate that there is no last number.

Before defining a sequence we define a **sequence function** as a function whose domain is a subset of positive integers. A **finite sequence function** is one whose domain is the set $\{1, 2, 3, \ldots, n\}$ of the first n positive integers. An **infinite sequence function** is one whose domain is the set of all positive integers. The numbers in the range of a sequence function are called **elements.** A **sequence** consists of the elements of a sequence function, listed in order.

Illustration   1

Let f be the function defined by

$$f(n) = 2n \qquad n \in \{1, 2, 3, 4, 5\}$$

Then f is a finite sequence function, and

$$f(1) = 2 \qquad f(2) = 4 \qquad f(3) = 6 \qquad f(4) = 8 \qquad f(5) = 10$$

The elements of the sequence defined by f are then 2, 4, 6, 8, and 10, and the sequence is

$$2, 4, 6, 8, 10$$

The ordered pairs in f are $(1, 2)$, $(2, 4)$, $(3, 6)$, $(4, 8)$, and $(5, 10)$.   ■

Illustration   2

Let g be the function defined by

$$g(n) = \frac{1}{n} \qquad n \in \{1, 2, 3, \ldots\}$$

The function g is an infinite sequence function. The elements of g are

$$g(1) = \frac{1}{1} \qquad g(2) = \frac{1}{2} \qquad g(3) = \frac{1}{3} \qquad g(4) = \frac{1}{4} \qquad g(5) = \frac{1}{5}$$

and so on. The sequence defined by g is therefore

$$1, \frac{1}{2}, \frac{1}{3}, \frac{1}{4}, \frac{1}{5}, \ldots$$

Some of the ordered pairs in g are $(1, 1)$, $(2, \frac{1}{2})$, $(3, \frac{1}{3})$, $(4, \frac{1}{4})$, $(5, \frac{1}{5})$, and $(6, \frac{1}{6})$.   ■

Illustration   3

Let h be the function defined by

$$h(n) = \frac{2n - 1}{n^2} \qquad n \in \{1, 2, 3, \ldots\}$$

The function h is an infinite sequence function. The elements of h are

$$h(1) = \frac{1}{1} \qquad h(2) = \frac{3}{4} \qquad h(3) = \frac{5}{9} \qquad h(4) = \frac{7}{16}$$

and so on. The sequence defined by h is therefore

$$1, \frac{3}{4}, \frac{5}{9}, \frac{7}{16}, \ldots$$

Some of the ordered pairs in h are $(1, 1)$, $(2, \frac{3}{4})$, $(3, \frac{5}{9})$, $(4, \frac{7}{16})$, and $(5, \frac{9}{25})$.   ■

We denote the first element of a sequence by a_1, the second element by a_2, the third element by a_3, and so on. The nth element is a_n, called the **general element** of the sequence.

For the sequence of Illustration 1, $a_n = f(n)$; that is, $a_n = 2n$. Therefore $a_1 = 2$, $a_2 = 4$, $a_3 = 6$, $a_4 = 8$, and $a_5 = 10$. For the sequence of Illustration 3, $a_n = h(n)$; that is,

$$a_n = \frac{2n - 1}{n^2}$$

Hence $a_1 = \frac{1}{1}$, $a_2 = \frac{3}{4}$, $a_3 = \frac{5}{9}$, and so on. Sometimes we state the general element of a sequence when we list the elements in order. Thus for the elements of the sequence of Illustration 3 we would write

$$\frac{1}{1}, \frac{3}{4}, \frac{5}{9}, \frac{7}{16}, \ldots, \frac{2n - 1}{n^2}, \ldots$$

A sequence

$$a_1, a_2, a_3, \ldots, a_n, \ldots$$

is said to be **equal** to a sequence

$$b_1, b_2, b_3, \ldots, b_n, \ldots$$

if and only if $a_i = b_i$ for every positive integer i. Remember that a sequence consists of an ordering of the elements of a sequence function. Therefore it is possible for two sequences to have the same elements and be unequal. For instance, the sequence for which $a_n = \dfrac{1}{n}$ has as its elements the reciprocals of the positive integers:

$$1, \frac{1}{2}, \frac{1}{3}, \frac{1}{4}, \ldots, \frac{1}{n}, \ldots$$

The sequence for which

$$a_n = \begin{cases} 1 & \text{if } n \text{ is odd} \\ \dfrac{2}{n + 2} & \text{if } n \text{ is even} \end{cases}$$

has as its elements

$$a_1 = 1 \qquad a_2 = \frac{1}{2} \qquad a_3 = 1 \qquad a_4 = \frac{1}{3} \qquad a_5 = 1 \qquad a_6 = \frac{1}{4}$$

and so on. The sequence is

$$1, \frac{1}{2}, 1, \frac{1}{3}, 1, \frac{1}{4}, \ldots$$

The elements of the sequences

$$1, \frac{1}{2}, \frac{1}{3}, \frac{1}{4}, \ldots, \frac{1}{n}, \ldots \qquad \text{and} \qquad 1, \frac{1}{2}, 1, \frac{1}{3}, 1, \frac{1}{4}, \ldots$$

are the same, but the sequences are not equal.

You should realize that several elements of a sequence do not determine a unique general element, as shown in the following illustration.

Illustration   4

a)   The sequence for which $a_n = 2n$ is

$$2, 4, 6, 8, 10, 12, \ldots, 2n, \ldots$$

b)   The sequence for which

$$a_n = \begin{cases} 2n & \text{if } n \text{ is odd} \\ 2a_{n-1} & \text{if } n \text{ is even} \end{cases}$$

is

$$2, 4, 6, 12, 10, 20, \ldots, a_n, \ldots$$

c)   The sequence for which $a_n = 2n + (n-1)(n-2)(n-3)$ is

$$2, 4, 6, 14, 34, 72, \ldots, 2n + (n-1)(n-2)(n-3), \ldots \qquad \blacksquare$$

Observe that all three sequences in Illustration 4 have 2, 4, and 6 as their first three elements, but each sequence, and thus its general element, is different. To determine a sequence uniquely, we must have a method for obtaining any element. One way is to find an equation that defines the general element, but that is not always possible. For instance, the sequence of prime numbers can be written as

$$2, 3, 5, 7, 11, 13, 17, 19, \ldots, a_n, \ldots$$

where a_n is the nth prime number. We cannot write an equation that defines a_n. However, a_n can be determined (theoretically) for every positive integer n.

Example   1   Write the first five elements of the sequence for each of the following general elements.

a) $a_n = \dfrac{n+2}{n(n+1)}$ b) $a_n = (-1)^n \dfrac{1}{3^{n-1}}$ c) $a_n = (-1)^{n-1}x^{2n+1}$

Solution

a) $a_1 = \dfrac{3}{1 \cdot 2}$ $a_2 = \dfrac{4}{2 \cdot 3}$ $a_3 = \dfrac{5}{3 \cdot 4}$ $a_4 = \dfrac{6}{4 \cdot 5}$ $a_5 = \dfrac{7}{5 \cdot 6}$

   $= \dfrac{3}{2}$ $= \dfrac{2}{3}$ $= \dfrac{5}{12}$ $= \dfrac{3}{10}$ $= \dfrac{7}{30}$

Hence the first five elements are

$$\tfrac{3}{2}, \tfrac{2}{3}, \tfrac{5}{12}, \tfrac{3}{10}, \tfrac{7}{30}$$

b) $a_1 = (-1)^1 \dfrac{1}{3^0}$; $a_2 = (-1)^2 \dfrac{1}{3^1}$; $a_3 = (-1)^3 \dfrac{1}{3^2}$; $a_4 = (-1)^4 \dfrac{1}{3^3}$;

$a_5 = (-1)^5 \dfrac{1}{3^4}$. Therefore the first five elements are

$$-1, \tfrac{1}{3}, -\tfrac{1}{9}, \tfrac{1}{27}, -\tfrac{1}{81}$$

c) $a_1 = (-1)^0 x^3$; $a_2 = (-1)^1 x^5$; $a_3 = (-1)^2 x^7$; $a_4 = (-1)^3 x^9$; $a_5 = (-1)^4 x^{11}$. Thus the first five elements are

$$x^3, -x^5, x^7, -x^9, x^{11}$$ ■

Example **2** Write the first 12 elements of the sequence for which

$$a_n = \begin{cases} n & \text{if } n \text{ is odd} \\ n & \text{if } n \text{ is even and not exactly divisible by 4} \\ \tfrac{1}{2}(n + a_{n-2}) & \text{if } n \text{ is even and exactly divisible by 4} \end{cases}$$

Solution $a_1 = 1$; $a_2 = 2$; $a_3 = 3$; $a_4 = \tfrac{1}{2}(4 + 2)$; $a_5 = 5$; $a_6 = 6$; $a_7 = 7$; $a_8 = \tfrac{1}{2}(8 + 6)$; $a_9 = 9$; $a_{10} = 10$; $a_{11} = 11$; $a_{12} = \tfrac{1}{2}(12 + 10)$. Therefore the first twelve elements are

$$1, 2, 3, 3, 5, 6, 7, 7, 9, 10, 11, 11$$ ■

We introduce the **sigma notation** to facilitate writing the sum of the elements of a sequence. This notation involves the use of the symbol Σ, the capital sigma of the Greek alphabet, which corresponds to the letter S. The symbol i, called the **index of summation,** is also involved. Prior to the formal definition of the sigma notation, we give some examples of it in the following illustration.

Illustration 5

a) $\displaystyle\sum_{i=1}^{5} i^2 = 1^2 + 2^2 + 3^2 + 4^2 + 5^2$

Observe that $i = 1$ appears under the sigma symbol. This indicates that the first term on the right-hand side is the value of i^2 when $i = 1$. The next four terms are the values of i^2 when i is 2, 3, 4, and 5. We stop there because the number 5 appears above the sigma symbol.

b) $\displaystyle\sum_{i=-2}^{3} (4i + 1) = [4(-2) + 1] + [4(-1) + 1] + [4(0) + 1] + [4(1) + 1]$

$$+ [4(2) + 1] + [4(3) + 1]$$

$$= (-7) + (-3) + 1 + 5 + 9 + 13$$

Observe that $i = -2$ appears under the sigma symbol and 3 appears above it. Therefore the first term on the right-hand side is the value of $4i + 1$ when

$i = -2$, and the remaining terms are the values of $4i + 1$ when i is $-1, 0, 1, 2$, and 3.

c) $\displaystyle\sum_{k=3}^{10} \frac{1}{k} = \frac{1}{3} + \frac{1}{4} + \frac{1}{5} + \frac{1}{6} + \frac{1}{7} + \frac{1}{8} + \frac{1}{9} + \frac{1}{10}$

d) $\displaystyle\sum_{j=1}^{n} j^3 = 1^3 + 2^3 + 3^3 + \ldots + n^3$ ■

The sigma notation can be defined by the equation

$$\sum_{i=m}^{n} F(i) = F(m) + F(m + 1) + F(m + 2) + \ldots + F(n)$$

where m and n are integers and $m \leq n$. The right-hand side of this equation consists of the sum of $n - m + 1$ terms, the first of which is obtained by replacing i by m in $F(i)$, the second by replacing i by $m + 1$ in $F(i)$, and so on until the last term is obtained by replacing i by n in $F(i)$. The number m is called the **lower limit** of the sum and n is called the **upper limit.** The symbol i is a "dummy" symbol because any other symbol could be used without changing the right-hand side. For example,

$$\sum_{i=4}^{6} i^3 = 4^3 + 5^3 + 6^3$$

can be written also as

$$\sum_{j=4}^{6} j^3 = 4^3 + 5^3 + 6^3$$

Sometimes the terms of a sum involve subscripts. For instance, the sum

$$a_1 + a_2 + a_3 + \ldots + a_n$$

can be written with sigma notation as

$$\sum_{i=1}^{n} a_i$$

Example 3 Write the following sums with sigma notation.

a) $2 + 4 + 6 + 8$ b) $1 + 3 + 5 + 7 + 9$

c) $-3a_3 + 4a_4 - 5a_5 + 6a_6 - 7a_7 + 8a_8$

d) $x^2 - x^4 + x^6 - x^8 + x^{10} - x^{12} + x^{14}$

Solution   a) The numbers 2, 4, 6, and 8 are the first four positive even integers, and they can be written as $2i$, where i is 1, 2, 3, and 4. Thus

$$2 + 4 + 6 + 8 = \sum_{i=1}^{4} 2i$$

b) Observe that 1, 3, 5, 7, and 9 are the first five positive odd integers, and they can be written as $2i - 1$, where i is 1, 2, 3, 4, and 5. Thus

$$1 + 3 + 5 + 7 + 9 = \sum_{i=1}^{5} (2i - 1)$$

c) Notice in the given summation that the odd-numbered terms are preceded by a minus sign and the even-numbered terms are preceded by a plus sign. So that the odd-numbered terms contain an odd power of -1 and the even-numbered terms contain an even power of -1, we write the factor $(-1)^i$ in the sigma notation. Thus we have

$$-3a_3 + 4a_4 - 5a_5 + 6a_6 - 7a_7 + 8a_8 = \sum_{i=3}^{8} (-1)^i i a_i$$

d) In the given summation the odd-numbered terms are preceded by a plus sign and the even-numbered terms are preceded by a minus sign. Therefore the odd-numbered terms require an even power of -1 and the even-numbered terms require an odd power of -1; so we write the factor $(-1)^{i-1}$ in the sigma notation. We have

$$x^2 - x^4 + x^6 - x^8 + x^{10} - x^{12} + x^{14} = \sum_{i=1}^{7} (-1)^{i-1} x^{2i} \qquad \blacksquare$$

Observe that we can write a sum an unlimited number of ways with sigma notation. For instance, in Example 3 the sum in part (b) can also be written as

$$\sum_{i=2}^{6} (2i - 3) \qquad \sum_{i=0}^{4} (2i + 1) \qquad \sum_{i=3}^{7} (2i - 5)$$

and so on.

The sum of the elements of a sequence is a **series**. For example, associated with the sequence 3, 6, 9, 12, 15, 18 is the series

$$3 + 6 + 9 + 12 + 15 + 18 = \sum_{i=1}^{6} 3i$$

Associated with the sequence

$$a_1, a_2, a_3, \ldots, a_n$$

is the series

$$a_1 + a_2 + a_3 + \ldots + a_n = \sum_{i=1}^{n} a_i$$

The terms in a series are the same as the corresponding elements in the associated sequence. The **general term** of a series is the general element of the associated sequence.

Example 4   Write the following series with sigma notation.

a) $1 + \frac{1}{2} + \frac{1}{3} + \frac{1}{4} + \frac{1}{5} + \frac{1}{6}$ b) $-1 + 5 - 9 + 13 - 17$

c) $\frac{1}{3}x^3 - \frac{1}{9}x^5 + \frac{1}{27}x^7 - \frac{1}{81}x^9$

Solution   a) The first term can also be written as $\frac{1}{1}$. Thus each term is a fraction whose numerator is 1 and whose denominator is the number of the term. Therefore

$$1 + \frac{1}{2} + \frac{1}{3} + \frac{1}{4} + \frac{1}{5} + \frac{1}{6} = \sum_{i=1}^{6} \frac{1}{i}$$

b) Except for the first term, each of the numbers 1, 5, 9, 13, and 17 is 4 more than the preceding one. This suggests $4i$ in the sigma notation. Because the first term is 1, we need to write $4i - 3$ in order to obtain 1 when $i = 1$. Then observe that $4i - 3 = 5$ when $i = 2$, $4i - 3 = 9$ when $i = 3$, and so on. Because the odd-numbered terms in the given series are preceded by a minus sign and the even-numbered terms are preceded by a plus sign, we need a factor of $(-1)^i$. Thus we have

$$-1 + 5 - 9 + 13 - 17 = \sum_{i=1}^{5} (-1)^i (4i - 3)$$

c) The exponents of x are the odd integers 3, 5, 7, and 9, which can be written as $2i + 1$, where i is 1, 2, 3, and 4. Thus in the sigma notation we must have a factor of x^{2i+1}. Because the numerical coefficients are the reciprocals of successive powers of 3, there will also be a factor of $\frac{1}{3^i}$.

Furthermore, since the odd-numbered terms are preceded by a plus sign and the even-numbered terms are preceded by a minus sign, we need a factor of $(-1)^{i-1}$. Hence we have

$$\frac{1}{3}x^3 - \frac{1}{9}x^5 + \frac{1}{27}x^7 - \frac{1}{81}x^9 = \sum_{i=1}^{4} (-1)^{i-1} \frac{1}{3^i} x^{2i+1}$$ ■

EXERCISES 11.1

In Exercises 1 through 16, write the first eight elements of the sequence whose general element is given.

1. $a_n = 2n + 3$

2. $a_n = \dfrac{3n - 1}{2}$

3. $a_n = \dfrac{n^2 + 1}{n}$

4. $a_n = \dfrac{1}{n^2 + 2}$

5. $a_n = (-1)^{n-1} \dfrac{n + 1}{2n - 1}$

6. $a_n = (-1)^{n+1} \dfrac{n}{2^n}$

7. $a_n = (-1)^n \dfrac{2^n}{1 + 2^n}$

8. $a_n = \dfrac{(-1)^{n-1}}{n(n+1)}$

9. $a_n = \dfrac{(-1)^{n+1}}{n+2} x^n$

10. $a_n = \dfrac{(-1)^n}{n^2} x^{2n-1}$

11. $a_n = n + (-1)^n n$

12. $a_n = \dfrac{1}{2n} - \dfrac{1}{3n}$

13. $a_n = \begin{cases} \dfrac{2}{n+1} & \text{if } n \text{ is odd} \\ 2 & \text{if } n \text{ is even} \end{cases}$

14. $a_n = \begin{cases} 1 & \text{if } n \text{ is odd} \\ \dfrac{4}{(n+2)^2} & \text{if } n \text{ is even} \end{cases}$

15. $a_n = \begin{cases} \dfrac{n+1}{2} & \text{if } n \text{ is odd} \\ a_{n-1} & \text{if } n \text{ is even} \end{cases}$

16. $a_n = \begin{cases} n & \text{if } n \text{ is odd} \\ \dfrac{1}{2}n & \text{if } n \text{ is even and not} \\ & \text{exactly divisible by } 4 \\ \dfrac{1}{2}(a_{n-2} + a_{n-1}) & \text{if } n \text{ is even and exactly} \\ & \text{divisible by } 4 \end{cases}$

In Exercises 17 through 30, write the series. In Exercises 17 through 24, find the sum of the series.

17. $\displaystyle\sum_{i=1}^{5} (4i - 3)$

18. $\displaystyle\sum_{i=1}^{7} (i + 1)^2$

19. $\displaystyle\sum_{j=2}^{6} \dfrac{j}{j-1}$

20. $\displaystyle\sum_{k=1}^{4} \dfrac{(-1)^{k+1}}{k}$

21. $\displaystyle\sum_{i=1}^{100} 5$

22. $\displaystyle\sum_{i=1}^{8} \dfrac{3i - 6}{2}$

23. $\displaystyle\sum_{k=0}^{5} \dfrac{1}{2^k}$

24. $\displaystyle\sum_{j=0}^{3} \dfrac{(-1)^j}{2^j + 1}$

25. $\displaystyle\sum_{i=1}^{8} (-1)^{i-1} x^{2i-1}$

26. $\displaystyle\sum_{i=1}^{5} 2^i x^{3i}$

27. $\displaystyle\sum_{i=0}^{7} (-1)^i (i + 1) a_i$

28. $\displaystyle\sum_{i=0}^{6} a_i x^{2i+1}$

29. $\displaystyle\sum_{i=1}^{n} f(x_{i-1})$

30. $\displaystyle\sum_{i=0}^{n} f(x_{i+1}) h$

In Exercises 31 through 38, write the series with sigma notation (there is no unique solution).

31. $1 + 3 + 5 + 7 + 9 + 11$

32. $2 + 4 + 6 + 8 + 10$

33. $4 - 7 + 10 - 13 + 16$

34. $1 + \dfrac{1}{3} + \dfrac{1}{9} + \dfrac{1}{27} + \dfrac{1}{81} + \dfrac{1}{243}$

35. $1 + \dfrac{3}{4} + \dfrac{5}{9} + \dfrac{7}{16} + \dfrac{9}{25} + \dfrac{11}{36}$

36. $1 - \dfrac{1}{4} + \dfrac{1}{16} - \dfrac{1}{64}$

37. $\dfrac{1}{2} - \dfrac{x^2}{4} + \dfrac{x^4}{6} - \dfrac{x^6}{8} + \dfrac{x^8}{10}$

38. $x - \dfrac{1}{2}x^3 + \dfrac{1}{3}x^5 - \dfrac{1}{4}x^7 + \dfrac{1}{5}x^9$

39. Write the general element and the first six elements of three different sequences, each having as the first three elements 1, 3, and 5.

40. Write the general element of a sequence whose first three elements are 2, 4, and 6, and whose fourth element is x, where x can be any real number.

11.2 Mathematical Induction

The formula for the sum of the first n positive odd integers is

$$1 + 3 + 5 + 7 + \ldots + (2n - 1) = n^2$$

$$\Leftrightarrow \qquad \sum_{i=1}^{n} (2i - 1) = n^2$$

Let us verify this formula for some values of n.

$$n = 1: \qquad\qquad\qquad 1 = 1 \ \Leftrightarrow 1 = 1^2$$
$$n = 2: \qquad\qquad\qquad 1 + 3 = 4 \ \Leftrightarrow 1 + 3 = 2^2$$
$$n = 3: \qquad\qquad 1 + 3 + 5 = 9 \ \Leftrightarrow 1 + 3 + 5 = 3^2$$
$$n = 4: \qquad\qquad 1 + 3 + 5 + 7 = 16 \Leftrightarrow 1 + 3 + 5 + 7 = 4^2$$
$$n = 5: \qquad 1 + 3 + 5 + 7 + 9 = 25 \Leftrightarrow 1 + 3 + 5 + 7 + 9 = 5^2$$

From these calculations we are convinced that the formula holds for values of n from 1 to 5. We could continue on with more values of n and we would observe that the formula is still true. Of course, such a procedure is *not a proof* of the formula for all positive-integer values of n. However, there is a proof involving a technique called **mathematical induction.** We now state the *principle of mathematical induction* that provides the basis for this method of proof, which is presented in Illustration 1.

Principle of Mathematical Induction

A statement P_n, involving the positive integer n, is true for all positive integer values of n if the following two conditions are satisfied:

(i) P_1 is true; that is, the statement is true for $n = 1$.
(ii) If k is an arbitrary positive integer for which P_k is true, then P_{k+1} is also true; that is, whenever the statement is true for $n = k$, it is also true for $n = k + 1$, where k is an arbitrary positive integer.

We do not prove the principle of mathematical induction. Its use involves the following reasoning:

From (i) the statement P_n is true for $n = 1$. Because P_n is true for $n = 1$, it follows from (ii) that P_n is true for $n = 1 + 1$ or 2. Because P_n is true for $n = 2$, it is true for $n = 2 + 1$ or 3. Because P_n is true for $n = 3$, it is true for $n = 3 + 1$ or 4. And so on. Thus P_n is true for all positive-integer values of n.

Illustration 1:

We shall use mathematical induction to prove that

$$1 + 3 + 5 + 7 + \ldots + (2n - 1) = n^2$$

for all positive-integer values of n. The proof consists of two parts, verifying the two conditions of the principle of mathematical induction, and a conclusion.

Part 1   We first verify that the formula is true for $n = 1$. If $n = 1$, the formula becomes

$$1 = 1^2$$

which is true.

Part 2 We now show that if the formula is true for $n = k$, then it is also true for $n = k + 1$, where k is an arbitrary positive integer. That is, we assume

$$1 + 3 + 5 + \ldots + (2k - 1) = k^2 \qquad (1)$$

We wish to prove that if Equation (1) is true, then

$$1 + 3 + 5 + \ldots + (2k - 1) + [2(k + 1) - 1] = (k + 1)^2 \qquad (2)$$

is also true. The last term on the left side of Equation (2) can be written as $2k + 1$. We add $2(k + 1) - 1$ to the left side of Equation (1) and its equivalent $2k + 1$ to the right side, and we obtain

$$1 + 3 + 5 + \ldots + (2k - 1) + [2(k + 1) - 1] = k^2 + (2k + 1)$$
$$1 + 3 + 5 + \ldots + (2k - 1) + [2(k + 1) - 1] = (k + 1)^2$$

which is Equation (2).

Conclusion In part 1 we proved that the formula is true for $n = 1$. In part 2 we proved that when the formula is true for $n = k$, it is also true for $n = k + 1$. Therefore, by the principle of mathematical induction, the formula holds for all positive-integer values of n. ■

In the following example, we prove that the sum of the first n positive even integers is given by the formula

$$2 + 4 + 6 + \ldots + 2n = n(n + 1)$$

Example 1 Use mathematical induction to prove

$$\sum_{i=1}^{n} 2i = n(n + 1)$$

Solution

Part 1 First the formula is verified for $n = 1$. When $n = 1$, the left side is

$$\sum_{i=1}^{1} 2i = 2$$

and the right side is

$$1(1 + 1) = 2$$

Therefore the formula is true when $n = 1$.

Part 2 We assume that the formula is true when $n = k$, where k is any positive integer:

$$\sum_{i=1}^{k} 2i = k(k + 1) \qquad (3)$$

With this assumption we wish to prove that the formula is also true when $n = k + 1$. Thus we wish to prove

$$\sum_{i=1}^{k+1} 2i = (k + 1)[(k + 1) + 1]$$

$$\Leftrightarrow \qquad \sum_{i=1}^{k+1} 2i = (k + 1)(k + 2) \tag{4}$$

When $n = k + 1$, we have

$$\sum_{i=1}^{k+1} 2i = 2 + 4 + 6 + \ldots + 2k + 2(k + 1)$$

$$= \sum_{i=1}^{k} 2i + (2k + 2)$$

$$= k(k + 1) + (2k + 2) \qquad \text{(by applying Equation (3))}$$

$$= k^2 + k + 2k + 2$$

$$= k^2 + 3k + 2$$

$$= (k + 1)(k + 2)$$

which is Equation (4).

Conclusion   We have proved that the formula is true when $n = 1$; and we have also proved that when the formula is true for $n = k$, it is also true for $n = k + 1$. Therefore, by the principle of mathematical induction, the formula is true when n is any positive integer.   ■

In the next example we prove the formula that gives the sum of the squares of the first n positive integers.

Example   2   Use mathematical induction to prove

$$\sum_{i=1}^{n} i^2 = \frac{n(n + 1)(2n + 1)}{6}$$

Solution

Part 1   We first verify the formula for $n = 1$. With this value of n the left side is

$$\sum_{i=1}^{1} i^2 = 1^2$$

$$= 1$$

and the right side is

$$\frac{1(1 + 1)(2 + 1)}{6} = \frac{1 \cdot 2 \cdot 3}{6}$$

$$= 1$$

Thus the formula is true when $n = 1$.

Part 2 We assume that the formula is true when $n = k$, where k is any positive integer; or

$$\sum_{i=1}^{k} i^2 = \frac{k(k + 1)(2k + 1)}{6} \tag{5}$$

With this assumption we wish to prove that the formula is also true when $n = k + 1$; that is, we wish to prove

$$\sum_{i=1}^{k+1} i^2 = \frac{(k + 1)[(k + 1) + 1][2(k + 1) + 1]}{6} \tag{6}$$

When $n = k + 1$, we have

$$\sum_{i=1}^{k+1} i^2 = 1^2 + 2^2 + 3^2 + \ldots + k^2 + (k + 1)^2$$

$$= \sum_{i=1}^{k} i^2 + (k + 1)^2$$

$$= \frac{k(k + 1)(2k + 1)}{6} + (k + 1)^2 \qquad \text{(by applying Equation (5))}$$

$$= \frac{k(k + 1)(2k + 1) + 6(k + 1)^2}{6}$$

$$= \frac{(k + 1)[k(2k + 1) + 6(k + 1)]}{6}$$

$$= \frac{(k + 1)(2k^2 + 7k + 6)}{6}$$

$$= \frac{(k + 1)(k + 2)(2k + 3)}{6}$$

$$= \frac{(k + 1)[(k + 1) + 1][2(k + 1) + 1]}{6}$$

which is Equation (6).

Conclusion We have proved that the formula is true when $n = 1$; and we have also proved that when the formula is true for $n = k$, it is also true for

$n = k + 1$. Therefore, by the principle of mathematical induction, the formula is true when n is any positive integer.   ∎

Example   3   Use mathematical induction to prove that

$$2^n \geq 2n$$

for all positive-integer values of n.

Solution

Part 1   We verify the inequality for $n = 1$. When $n = 1$, the left side is 2 and the right side is 2. Because $2 \geq 2$, the inequality is true when $n = 1$.

Part 2   We assume that the inequality is true when $n = k$, where k is any positive integer; that is, we assume

$$2^k \geq 2k \qquad (7)$$

With this assumption we wish to prove that the inequality is true when $n = k + 1$; that is, we wish to prove

$$2^{k+1} \geq 2(k + 1) \qquad (8)$$

On both sides of Inequality (7) we multiply by 2 and obtain

$$2 \cdot 2^k \geq 2 \cdot 2k$$
$$2^{k+1} \geq 2k + 2k \qquad (9)$$

Because $k \geq 1$, then $2k \geq 2$. Thus

$$2k + 2k \geq 2k + 2 \qquad (10)$$

From Inequalities (9) and (10)

$$2^{k+1} \geq 2k + 2$$
$$2^{k+1} \geq 2(k + 1)$$

which is Inequality (8).

Conclusion   Because the inequality is true when $n = 1$, and we have proved that when it is true for $n = k$, it is also true for $n = k + 1$, it follows from the principle of mathematical induction that the inequality is true when n is any positive integer.   ∎

In Example 3 we proved that the nonstrict inequality

$$2^n \geq 2n$$

is valid for all positive-integer values of n. By a similar procedure, as indicated in the following illustration, we can prove the corresponding strict inequality

$$2^n > 2n \qquad \text{where} \qquad n > 2 \qquad (11)$$

and n is a positive integer. Because 3 is the smallest value of n for which the inequality is true, Part 1 of the proof by mathematical induction is to verify that P_3 (rather than P_1) is true. In this case we are applying an extension of the principle of mathematical induction.

Illustration 2

To prove Inequality (11) by mathematical induction, part 1 is as follows:

We verify (11) for $n = 3$. When $n = 3$, the left side of the inequality is 8 and the right side is 6. Because $8 > 6$, (11) is true when $n = 3$.

Part 2 of the proof of (11) is the same as that in part 2 in the solution of Example 3 except that the inequality symbol $\geq$ is replaced by $>$.

Then the conclusion states that because the inequality is true when $n = 3$, and in part 2 we proved that when it is true for $n = k$, it is also true for $n = k + 1$, it follows from the principle of mathematical induction that Inequality (11) is true when n is a positive integer and $n > 2$. ∎

Example 4   Prove the following statement by mathematical induction: $x - y$ is a factor of $x^n - y^n$ for all positive-integer values of n.

Solution

Part 1   When $n = 1$, $x^n - y^n$ becomes $x - y$, which certainly has $x - y$ as a factor.

Part 2   We assume that $x - y$ is a factor of $x^k - y^k$, where k is any positive integer; and with this assumption we wish to prove that $x - y$ is also a factor of $x^{k+1} - y^{k+1}$. If we subtract and add xy^k to $x^{k+1} - y^{k+1}$, we obtain

$$x^{k+1} - y^{k+1} = x^{k+1} - xy^k + xy^k - y^{k+1}$$

$$\Leftrightarrow \qquad x^{k+1} - y^{k+1} = x(x^k - y^k) + y^k(x - y) \qquad (12)$$

We have assumed that $x - y$ is a factor of $x^k - y^k$; furthermore, $x - y$ is a factor of $y^k(x - y)$. Hence $x - y$ is a factor of each of the two terms in the right member of (12). Therefore $x - y$ is a factor of $x^{k+1} - y^{k+1}$.

Conclusion   We have shown that the statement is true when $n = 1$ and we have proved that when the statement is true for $n = k$, it is also true for $n = k + 1$. Therefore, by the principle of mathematical induction, the statement is true for all positive-integer values of n. ∎

As stated in Chapter 1, certain laws of exponents can be proved by mathematical induction. In Example 5 we give such a proof that utilizes the following definition of positive-integer exponents:

Let a be any real number. Then

$$a^1 = a \qquad (13)$$

If k is any positive integer such that a^k is defined, let

$$a^{k+1} = a^k \cdot a \qquad (14)$$

Example   5   Prove that if m and n are positive integers and a is a real number, then

$$a^m \cdot a^n = a^{m+n} \qquad (15)$$

Solution   Let m be an arbitrary positive integer. We wish to prove that Equation (15) is true for all positive-integer values of n.

Part 1   We verify that (15) is true when $n = 1$. From (13),

$$a^m \cdot a^1 = a^m \cdot a$$

Applying (14) on the right side of this equation, we have

$$a^m \cdot a^1 = a^{m+1}$$

Hence (15) is true when $n = 1$.

Part 2   We assume that (15) is true when $n = k$, where k is any positive integer:

$$a^m \cdot a^k = a^{m+k} \qquad (16)$$

With this assumption we wish to prove that

$$a^m \cdot a^{k+1} = a^{m+(k+1)}$$

To prove this, we start with the left side and replace a^{k+1} by $a^k \cdot a$, which follows from (14). Thus

$$
\begin{aligned}
a^m \cdot a^{k+1} &= a^m \cdot (a^k \cdot a) \\
&= (a^m \cdot a^k) \cdot a && \text{(from the associative law for multiplication} \\
&= a^{m+k} \cdot a && \text{(from (16))} \\
&= a^{(m+k)+1} && \text{(from (14))} \\
&= a^{m+(k+1)} && \text{(from the associative law for addition)}
\end{aligned}
$$

which is what we wished to prove.

Conclusion   From Part 1 we know that (15) is true when $n = 1$. From part 2 we know that when (15) is true for $n = k$, it is also true for $n = k + 1$, where k is any positive integer. Therefore, by the principle of mathematical induction, (15) is true when n is any positive integer.   ∎

EXERCISES 11.2

In Exercises 1 through 14, use mathematical induction to prove that the formula is true for all positive-integer values of n.

1. $\displaystyle\sum_{i=1}^{n} i = \frac{n(n+1)}{2}$

2. $\displaystyle\sum_{i=1}^{n} 4i = 2n(n+1)$

3. $\displaystyle\sum_{i=1}^{n} (3i - 2) = \frac{n(3n-1)}{2}$

4. $\displaystyle\sum_{i=1}^{n} (3i - 1) = \frac{n(3n+1)}{2}$

5. $\displaystyle\sum_{i=1}^{n} \frac{i(i+1)}{2} = \frac{n(n+1)(n+2)}{6}$

6. $\displaystyle\sum_{i=1}^{n} (2i - 1)^2 = \frac{n(2n-1)(2n+1)}{3}$

7. $\displaystyle\sum_{i=1}^{n} 2^i = 2(2^n - 1)$

8. $\displaystyle\sum_{i=1}^{n} 3^i = \frac{3}{2}(3^n - 1)$

9. $\displaystyle\sum_{i=1}^{n} i^3 = \frac{n^2(n+1)^2}{4}$

10. $\displaystyle\sum_{i=1}^{n} (2i - 1)^3 = n^2(2n^2 - 1)$

11. $\displaystyle\sum_{i=1}^{n} \frac{1}{i(i+1)} = \frac{n}{n+1}$

12. $\displaystyle\sum_{i=1}^{n} \frac{1}{(2i-1)(2i+1)} = \frac{n}{2n+1}$

13. $\displaystyle\sum_{i=1}^{n} \frac{1}{(3i-1)(3i+2)} = \frac{n}{2(3n+2)}$

14. $\displaystyle\sum_{i=1}^{n} ar^{i-1} = \frac{a - ar^n}{1 - r}$

In Exercises 15 through 18, use mathematical induction to prove that the inequality is true for all positive-integer values of n.

15. $3^n \geq 3n$

16. $2^n > n$

17. $a^n > 1$, if a is a real number and $a > 1$.

18. $0 < a^n < 1$, if a is a real number and $0 < a < 1$.

In Exercises 19 and 20, use mathematical induction to prove that the inequality is true for the indicated positive-integer values of n.

19. $2^n > n^2$, if $n > 4$

20. $3^n > 2^n + 10n$, if $n > 3$

In Exercises 21 through 26, use mathematical induction to prove that the statement is true for all positive-integer values of n.

21. 2 is a factor of $n^2 + n$.

22. 2 is a factor of $n^2 - n + 2$.

23. 6 is a factor of $n^3 + 3n^2 + 2n$.

24. 3 is a factor of $4^n - 1$.

25. $x + y$ is a factor of $x^{2n} - y^{2n}$.

26. $x + y$ is a factor of $x^{2n-1} + y^{2n-1}$.

In Exercises 27 through 31, use mathematical induction to prove the statement.

27. If m and n are positive integers and a is a real number, then

$$(a^n)^m = a^{nm}$$

28. If n is a positive integer, and a and b are real numbers, then

$$(ab)^n = a^n b^n$$

29. If n is a positive integer, a and b are real numbers, and $b \neq 0$, then

$$\left(\frac{a}{b}\right)^n = \frac{a^n}{b^n}$$

30. If P dollars is invested at an annual interest rate of $100i$ percent compounded m times per year, and if A_n dollars is the amount of the investment at the end of n interest periods, then

$$A_n = P\left(1 + \frac{i}{m}\right)^n$$

31. If $n \geq 3$, the sum of the interior angles of an n-sided polygon is $(n-2)180°$. (*Hint:* Choose one vertex and form $n - 2$ triangles by drawing $n - 3$ lines through the selected vertex and each of $n - 3$ other vertices.)

11.3   Arithmetic Sequences and Series

Each element, except the first, in the sequence

$$2, 5, 8, 11, 14, 17, 20$$

is 3 more than the one preceding. Here we have an example of an *arithmetic sequence.*

Definition

> **Arithmetic Sequence**
>
> An **arithmetic sequence** is a sequence for which any element, except the first, can be obtained by adding a constant to the preceding element.

An arithmetic sequence is sometimes called an **arithmetic progression.** The constant addend in an arithmetic sequence is called the **common difference** and is denoted by d. We can ascertain whether a given sequence is an arithmetic sequence by subtracting each element from the succeeding one.

Illustration 1

For the sequence

$$9, 5, 1, -3, -7, -11$$

we observe that $5 - 9 = -4$; $1 - 5 = -4$; $-3 - 1 = -4$; $-7 - (-3) = -4$; and $-11 - (-7) = -4$. Therefore we have an arithmetic sequence where the common difference d is -4. ∎

In an arithmetic sequence the number of elements is denoted by N, the first element is denoted by a_1, and the last element is denoted by a_N. In the sequence of Illustration 1, $N = 6$, $a_1 = 9$, and $a_6 = -11$.

The definition of an arithmetic sequence can be stated symbolically by giving the value of the first element a_1, the number of elements N, and the formula

$$a_{n+1} = a_n + d$$

from which every element after the first can be obtained from the preceding one. This formula is called a **recursive formula.** Recursive formulas are used in computer programming because the repeated application of a single formula is often involved.

Illustration 2

If $a_1 = 4$, $N = 8$, and

$$a_{n+1} = a_n + 3$$

the arithmetic sequence is

$$4, 7, 10, 13, 16, 19, 22, 25$$ ∎

From the recursive formula we can write the general arithmetic sequence, for which the first element is a_1, the common difference is d, and the number of elements is N. We start with the element a_1 and each successive element is obtained from the preceding one by adding d to it. Hence we have

$$a_1, \, a_1 + d, \, a_1 + 2d, \, a_1 + 3d, \, a_1 + 4d, \ldots, a_N$$

Refer to the first five elements and observe that each element is a_1 plus a multiple of d, where the coefficient of d is one less than the number of the element. Intuitively, it appears that $a_N = a_1 + (N - 1)d$. We state this formally and prove it by mathematical induction.

Theorem 1

> The Nth element of an arithmetic sequence is given by
>
> $$a_N = a_1 + (N - 1)d$$

Proof

Part 1   We first show that the formula is true if $N = 1$ by substituting 1 for N.

$$a_1 = a_1 + (1 - 1)d$$
$$= a_1$$

Part 2   We now assume that the formula is true if $N = k$, that is,

$$a_k = a_1 + (k - 1)d$$

We wish to show that the formula is true if $N = k + 1$, that is,

$$a_{k+1} = a_1 + [(k + 1) - 1]d$$

By the definition of an arithmetic sequence

$$a_{k+1} = a_k + d$$

Replacing a_k by $a_1 + (k - 1)d$, we have

$$a_{k+1} = [a_1 + (k - 1)d] + d$$
$$= a_1 + kd - d + d$$
$$= a_1 + [(k + 1) - 1]d$$

which is what we wished to show.

Conclusion We have proved that the formula is true when $N = 1$, and we have also proved that when the formula is true for $N = k$, it is also true for $N = k + 1$. Therefore, by the principle of mathematical induction, it is valid for all natural numbers. ∎

Example   1 Find the thirtieth element of the arithmetic sequence for which the first element is 5 and the second element is 9.

Solution   Let a_{30} be the thirtieth element of the arithmetic sequence

$$5, 9, \ldots, a_{30}$$

Then $d = 9 - 5$, or 4, $a_1 = 5$, and $N = 30$. From Theorem 1,

$$a_{30} = a_1 + (30 - 1)d$$
$$= 5 + 29 \cdot 4$$
$$= 121$$

∎

Example   2 Find the first element of an arithmetic sequence whose common difference is -5 and whose eighteenth element is -21.

Solution   From Theorem 1, with $N = 18$,

$$a_{18} = a_1 + (18 - 1)d$$

Substituting -21 for a_{18} and -5 for d, we have

$$-21 = a_1 + 17(-5)$$
$$a_1 = 64$$

∎

Example   3 If the twelfth element of an arithmetic sequence is -21 and the twenty-fifth element is 18, what is the fourth element?

Solution   We can consider the arithmetic sequence consisting of the first 25 elements:

$$a_1, \ldots, a_4, \ldots, a_{12}, \ldots, a_{25}$$

From Theorem 1, with $N = 25$ and $a_{25} = 18$,

$$18 = a_1 + (25 - 1)d$$
$$18 = a_1 + 24d$$

The first 12 elements of the sequence also form an arithmetic sequence, with $N = 12$ and $a_{12} = -21$. From Theorem 1, with these values of N and a_{12},

$$-21 = a_1 + (12 - 1)d$$
$$-21 = a_1 + 11d$$

We have then the following system of equations:

$$\begin{cases} 18 = a_1 + 24d \\ -21 = a_1 + 11d \end{cases}$$

If we replace the second equation by the difference of the two equations we obtain the equivalent system

$$\begin{cases} 18 = a_1 + 24d \\ 39 = 13d \end{cases}$$

$$\Leftrightarrow \quad \begin{cases} a_1 = -54 \\ d = 3 \end{cases}$$

Again from Theorem 1, with $N = 4$, $a_1 = -54$, and $d = 3$,

$$a_4 = -54 + (4 - 1)(3)$$
$$= -45 \qquad \blacksquare$$

Definition

> **Arithmetic Means**
>
> If $a, c_1, c_2, \ldots, c_k, b$ is an arithmetic sequence, then the numbers $c_1, c_2, \ldots, c_k$ are the k arithmetic means between a and b.

Illustration 3

Because

$$2, 5, 8, 11, 14, 17, 20$$

is an arithmetic sequence, it follows that 5, 8, 11, 14, and 17 are the five arithmetic means between 2 and 20. $\blacksquare$

Example 4 Insert three arithmetic means between 11 and 14.

Solution   If $c_1, c_2,$ and c_3 are the three arithmetic means, then

$$11, c_1, c_2, c_3, 14$$

is an arithmetic sequence. With $N = 5$ in Theorem 1,

$$a_5 = a_1 + (5 - 1)d$$

Because $a_1 = 11$ and $a_5 = 14$,

$$14 = 11 + 4d$$
$$d = \tfrac{3}{4}$$

Thus

$$c_1 = 11 + \tfrac{3}{4} \qquad c_2 = 11\tfrac{3}{4} + \tfrac{3}{4} \qquad c_3 = 12\tfrac{1}{2} + \tfrac{3}{4}$$
$$= 11\tfrac{3}{4} \qquad\qquad = 12\tfrac{1}{2} \qquad\qquad = 13\tfrac{1}{4}$$

The three arithmetic means are therefore $11\tfrac{3}{4}$, $12\tfrac{1}{2}$, and $13\tfrac{1}{4}$. $\blacksquare$

Illustration   4

To insert one arithmetic mean between the numbers x and y, let M be the arithmetic mean, and we have the arithmetic sequence

$$x, M, y$$

The common difference can be represented by either $M - x$ or $y - M$. Therefore

$$M - x = y - M$$
$$2M = x + y$$
$$M = \frac{x + y}{2}$$

∎

The number M obtained in Illustration 4 is called *the arithmetic mean (or average) of* the numbers x and y. We can generalize this concept and refer to the *arithmetic mean* of a set of numbers.

Definition

Arithmetic Mean

(i) **The arithmetic mean** (or **average**) of the numbers x and y is the number

$$\frac{x + y}{2}$$

(ii) The **arithmetic mean** (or **average**) of a set of numbers $x_1, x_2, x_3, \ldots, x_n$ is the number

$$\frac{x_1 + x_2 + x_3 + \ldots + x_n}{n}$$

Example   5   On five separate examinations a student received the following test scores: 78, 89, 62, 75, and 84. Find the arithmetic mean of these scores.

Solution   If M is the arithmetic mean, we have from the definition

$$M = \frac{78 + 89 + 62 + 75 + 84}{5}$$
$$= \frac{388}{5}$$
$$= 77.6$$

∎

An **arithmetic series** is the indicated sum of the elements of an arithmetic sequence.

Illustration 5

The arithmetic series associated with the arithmetic sequence of Illustration 2 is

$$4 + 7 + 10 + 13 + 16 + 19 + 22 + 25$$

This arithmetic series can be written with the sigma notation as

$$\sum_{i=1}^{8} (3i + 1)$$ ■

The arithmetic series associated with the general arithmetic sequence is

$$a_1 + (a_1 + d) + (a_1 + 2d) + \ldots + [a_1 + (N - 1)d]$$

If we denote this sum by S_N, we have

$$S_N = a_1 + (a_1 + d) + (a_1 + 2d) + \ldots + [a_1 + (N - 1)d]$$

The series on the right can be written in the reverse order with the Nth term being written as a_N, the $(N - 1)$th term being written as $a_N - d$, and so on, until the first term is written as $a_N - (N - 1)d$. Therefore

$$S_N = a_N + (a_N - d) + (a_N - 2d) + \ldots + [a_N - (N - 1)d]$$

If we add term-by-term the two equations defining S_N, we obtain

$$S_N + S_N = (a_1 + a_N) + (a_1 + a_N) + (a_1 + a_N) + \ldots + (a_1 + a_N)$$

where on the right side the term $a_1 + a_N$ occurs N times. Hence

$$2S_N = N(a_1 + a_N)$$

$$S_N = \frac{N}{2}(a_1 + a_N)$$

If we substitute into this equation the value of a_N from the formula of Theorem 1, we obtain

$$S_N = \frac{N}{2}(a_1 + [a_1 + (N - 1)d])$$

$$S_N = \frac{N}{2}[2a_1 + (N - 1)d]$$

We have proved the following theorem.

Theorem 2

If $a_1, a_2, a_3, \ldots, a_N$ is an arithmetic sequence with common difference d, and

$$S_N = a_1 + a_2 + a_3 + \ldots + a_N$$

then

$$S_N = \frac{N}{2}(a_1 + a_N)$$

and

$$S_N = \frac{N}{2}[2a_1 + (N - 1)d]$$

The first formula in Theorem 2 can be written as

$$S_N = N\left(\frac{a_1 + a_N}{2}\right)$$

Thus S_N is the product of the number of terms and the arithmetic mean of the first and last terms.

Example   6 Find the sum of the positive even integers less than 100.

Solution The positive even integers less than 100 form the arithmetic sequence

$$2, 4, 6, \ldots, 96, 98$$

We wish to find the sum of the associated arithmetic series, which is

$$2 + 4 + 6 + \ldots + 96 + 98$$

For this series $a_1 = 2$, $d = 2$, $N = 49$, $a_{49} = 98$. From Theorem 2,

$$S_{49} = \tfrac{49}{2}(a_1 + a_{49})$$
$$= \tfrac{49}{2}(2 + 98)$$
$$= 2450 \qquad \blacksquare$$

Example   7 The seller of a certain piece of real estate received the following two offers from prospective purchasers:

Offer 1: The payment for the first year is $24,000, and for nine years thereafter there is an annual increase of $1800 in the payments.

Offer 2: The payment for the first six months is $12,000, and for the second six months it is $12,450. For nine years thereafter there is a semiannual increase of $450 in the payments.

Which offer will give the seller more money over a ten-year period and how much more?

Solution   According to offer 1, the number of dollars received by the seller over a period of ten years is the sum of the following arithmetic series of ten terms:

$$24,000 + 25,800 + 27,600 + \ldots + a_{10}$$

Let S_{10} be this sum. Then from the second formula of Theorem 2,

$$S_{10} = \tfrac{10}{2}[2a_1 + (10 - 1)d]$$

Because $a_1 = 24,000$ and $d = 1800$,

$$S_{10} = 5[2(24,000) + 9(1800)]$$
$$= 321,000$$

According to offer 2, the number of dollars received by the seller over a period of ten years is the sum of the following arithmetic series of twenty terms:

$$12,000 + 12,450 + 12,900 + \ldots + a_{20}$$

Denoting this sum by S_{20} and from Theorem 2 with $a_1 = 12,000$ and $d = 450$, we have

$$S_{20} = \tfrac{20}{2}[2a_1 + (20 - 1)d]$$
$$= 10[2(12,000) + 19(450)]$$
$$= 325,500$$

Therefore offer 2 will give the seller $4500 more money over period. ∎

EXERCISES   11.3

In Exercises 1 through 8, write the first five elements of an arithmetic sequence whose first element is a and whose common difference is d.

1. $a = 5; d = 3$
2. $a = -3; d = 2$
3. $a = 10; d = -4$
4. $a = 16; d = -5$
5. $a = -5; d = -7$
6. $a = 20; d = 10$
7. $a = x; d = 2y$
8. $a = u + v; d = -3v$

In Exercises 9 through 16, determine if the elements form an arithmetic sequence. If they do, write the next two elements of the arithmetic sequence.

9. $3, -1, -5, -9$
10. $12, 7, 2, -3$
11. $2, -6, 10, -14$
12. $-1, -\frac{1}{3}, \frac{1}{3}, 1$

13. $\frac{1}{3}, \frac{1}{4}, \frac{1}{6}, \frac{1}{12}$
14. $\frac{1}{2}, \frac{3}{4}, \frac{7}{8}, \frac{15}{16}$
15. $x, 2x + y, 3x + 2y$
16. $s, t, 2t - s$
17. Find the twelfth element of an arithmetic sequence whose first element is 2 and whose second element is 5.
18. Find the tenth element of an arithmetic sequence whose first element is 8 and whose third element is 2.
19. Find the first element of an arithmetic sequence whose eighth element is 2 and whose common difference is -2.
20. The ninth element of an arithmetic sequence is 28 and the twenty-first element is 100. What is the fifteenth element?

21. The first three elements of an arithmetic sequence are 20, 16, and 12. Which element is -96?

22. In the arithmetic sequence whose first three elements are $\frac{1}{6}$, $\frac{1}{4}$, and $\frac{1}{3}$, which element is 4?

23. Insert four arithmetic means between 5 and 6.

24. Insert seven arithmetic means between 3 and 9.

25. Find the arithmetic mean of the following set of test scores: 72, 53, 85, 74, 62, and 83.

In Exercises 26 through 31, find the sum of the arithmetic series.

26. $\displaystyle\sum_{i=1}^{8}(3i-1)$

27. $\displaystyle\sum_{i=1}^{18}\frac{2i-1}{3}$

28. $\displaystyle\sum_{i=2}^{12}(8-2i)$

29. $\displaystyle\sum_{k=1}^{50}(2k-1)$

30. $\displaystyle\sum_{j=1}^{20}(5j-1)$

31. $\displaystyle\sum_{i=3}^{12}(\tfrac{1}{2}i-5)$

32. Find the sum of all the positive integers less than 100.

33. Find the sum of all the positive even integers consisting of two digits.

34. Find the sum of all the integer multiples of 8 between 9 and 199.

35. The sum of $1000 is distributed among four people so that each person after the first receives $20 less than the preceding person. How much does each person receive?

36. A student's grade on the first of 12 quizzes in her algebra course was 45. However, on each successive quiz her score was 5 more than on the preceding one. What was the arithmetic mean (average) of the twelve scores?

37. In a display window a grocer wishes to place boxes of detergent in pyramid form so that the bottom row contains 15 boxes, the next row contains 14 boxes, the next row contains 13 boxes, and so on, with one box on top. How many boxes of detergent are necessary for the pyramid?

38. To dig a well a company charges $80 for the first foot, $100 for the second foot, $120 for the third foot, and so on; the cost of each foot is $20 more than the cost of the preceding foot. What is the depth of a well that costs $23,400 to dig?

39. A contractor who does not meet the deadline on the construction of a building is fined $800 per day for each of the first ten days of extra time, and for each additional day thereafter the fine is increased by $160 each day. If the contractor is fined $20,160, by how many extra days was the construction time delayed?

40. It is desired to pile some logs in layers so that the top layer contains one log, the next layer contains two logs, the next layer contains three logs, and so on; each layer containing one more log than the layer on top of it. There are 190 logs. Can all the logs be used in such a grouping, and if so, how many logs are in the bottom layer?

11.4 Geometric Sequences and Series

The sequence

$$1, 2, 4, 8, 16, 32, 64, 128$$

is an example of a *geometric sequence*. Each element, except the first, can be obtained by multiplying the preceding element by 2.

Definition

Geometric Sequence

A **geometric sequence** is a sequence such that any element after the first can be obtained by multiplying the preceding element by a constant.

A geometric sequence is also called a **geometric progression.**

The constant multiplier in a geometric sequence is called the **common ratio** and is denoted by r. We may compute r by dividing any term by the preceding one.

Illustration   1

For the sequence

$$1, 2, 4, 8, 16, 32, 64, 128$$

we have

$$\frac{2}{1} = 2 \qquad \frac{4}{2} = 2 \qquad \frac{8}{4} = 2 \qquad \frac{16}{8} = 2 \qquad \frac{32}{16} = 2 \qquad \frac{64}{32} = 2 \qquad \frac{128}{64} = 2$$

The common ratio $r = 2$.   ■

As with an arithmetic sequence, the number of elements in a geometric sequence is denoted by N, the first element is denoted by a_1, and the last element is denoted by a_N. In the sequence of Illustration 1, $N = 8$, $a_1 = 1$, and $a_8 = 128$.

A geometric sequence can be defined by giving the values of a_1 and N and a recursive formula

$$a_{n+1} = a_n r$$

from which every element after the first can be obtained from the preceding one.

Illustration   2

Consider the geometric sequence for which $a_1 = 128$, $N = 5$, and $a_{n+1} = a_n(-\frac{1}{4})$. Then

$$a_2 = 128\left(-\frac{1}{4}\right) \qquad a_3 = -32\left(-\frac{1}{4}\right) \qquad a_4 = 8\left(-\frac{1}{4}\right) \qquad a_5 = -2\left(-\frac{1}{4}\right)$$

$$= -32 \qquad\qquad = 8 \qquad\qquad = -2 \qquad\qquad = \frac{1}{2}$$

Therefore the sequence is

$$128, -32, 8, -2, \frac{1}{2}$$

■

The general geometric sequence, for which the first element is a_1, the common ratio is r, and the number of elements is N, can be obtained by applying the recursive formula. Starting with the element a_1, we obtain each successive element by multiplying the preceding one by r. Doing this, we have

$$a_1, a_1 r, a_1 r^2, a_1 r^3, a_1 r^4, \ldots, a_N$$

In the first five elements we observe that each element is the product of a_1 and a power of r, where the exponent of r is one less than the number of the element. Therefore our intuition suggests that the Nth (last) element is $a_N = a_1 r^{N-1}$.

Theorem 1

> The Nth element of a geometric sequence is given by
> $$a_N = a_1 r^{N-1}.$$

The proof of this theorem is by mathematical induction and is left as an exercise. See Exercise 43.

Example   1 Find the tenth element of the geometric sequence for which the first element is $\frac{5}{2}$ and the second element is -5.

Solution   Let a_{10} be the tenth element of the geometric sequence

$$\tfrac{5}{2}, -5, \ldots, a_{10}$$

Then
$$r = (-5) \div \tfrac{5}{2}$$
$$= (-5)\tfrac{2}{5}$$
$$= -2$$

From Theorem 1 with $a_1 = \frac{5}{2}$ and $N = 10$,
$$a_{10} = a_1 r^{10-1}$$
$$= \tfrac{5}{2}(-2)^9$$
$$= -5(2)^8$$
$$= -1280$$

∎

Example   2 A city has a population of 100,000. If the population increases 10 percent every 5 years, what will the population be at the end of 40 years?

Solution   The population at the end of 5 years will be

$$100{,}000 + 0.10(100{,}000) = (1.10)(100{,}000)$$

The population at the end of each successive 5-year period is 1.10 times the population at the end of the preceding 5-year period. Hence we have the geometric sequence of nine elements

$$100{,}000, \ (1.10)(100{,}000), \ (1.10)^2(100{,}000), \ldots, a_9$$

where a_9 is the population at the end of 40 years. From Theorem 1 with

$N = 9$, $a_1 = 100,000$, and $r = 1.10$,

$$a_9 = a_1 r^{9-1}$$
$$= 100,000(1.10)^8$$

From a calculator $(1.10)^8 \approx 2.144$. Thus

$$a_9 \approx (2.144)10^5$$

Hence, to four significant digits, the population at the end of 40 years will be 214,400. ■

Definition

> **Geometric Means**
> If $a, c_1, c_2, \ldots, c_k, b$ is a geometric sequence, then the numbers c_1, $c_2, \ldots, c_k$ are a set of k geometric means between a and b.

Illustration 3

The sequence

$$2, 6, 18, 54, 162$$

is a geometric sequence with $r = 3$. From the definition, the numbers 6, 18, and 54 form a set of three geometric means between 2 and 162.

Because the sequence

$$2, -6, 18, -54, 162$$

is also a geometric sequence ($r = -3$), the numbers -6, 18, and -54 form another set of three geometric means between 2 and 162. ■

Example 3 Insert two geometric means between 1000 and 64.

Solution We consider the geometric sequence

$$1000, c_1, c_2, 64$$

where c_1 and c_2 are the required geometric means. In this geometric sequence $N = 4$, $a_1 = 1000$, and $a_4 = 64$. From Theorem 1,

$$a_4 = a_1 r^{4-1}$$

Therefore

$$64 = 1000 r^3$$
$$r^3 = \tfrac{8}{125}$$
$$r = \sqrt[3]{\tfrac{8}{125}}$$
$$r = \tfrac{2}{5}$$

Thus the geometric sequence is 1000, 400, 160, 64, and so 400 and 160 are two geometric means between 1000 and 64. ■

If m is a geometric mean between two numbers x and y, then

$$x, m, y$$

is a geometric sequence. Therefore

$$\frac{m}{x} = \frac{y}{m}$$
$$m^2 = xy$$

This equation implies that either both x and y are positive or both x and y are negative. Furthermore, the equation has two solutions: $m = \sqrt{xy}$ and $m = -\sqrt{xy}$. Because we want the geometric mean to be between the numbers x and y, we choose for the value of m the number having the same sign as x and y. Therefore we have the following definition.

Definition

> **Geometric Mean**
> The **geometric mean** between the numbers x and y is
> $$\sqrt{xy} \quad \text{if } x \text{ and } y \text{ are positive}$$
> $$-\sqrt{xy} \quad \text{if } x \text{ and } y \text{ are negative}$$

Example   4   Find the geometric mean between each of the following sets of numbers.

a) 4 and 9   b) $-\frac{3}{10}$ and $-\frac{5}{6}$.

Solution   In each part let m be the geometric mean. We compute m by applying the definition.

a) $m = \sqrt{4 \cdot 9}$ b) $m = -\sqrt{(-\frac{3}{10})(-\frac{5}{6})}$

$\quad = \sqrt{36}$ $\quad = -\sqrt{\frac{15}{60}}$

$\quad = 6$ $\quad = -\sqrt{\frac{1}{4}}$

$\qquad\qquad\qquad\qquad = -\frac{1}{2}$ ∎

By a generalization of the definition of the geometric mean between two numbers, we define the **geometric mean** of a set of numbers x_1, $x_2, x_3, \ldots, x_n$ to be the number $\sqrt[n]{x_1 x_2 x_3 \ldots x_n}$.

Illustration   4

The geometric mean of the numbers 4, 10, and 25 is

$$\sqrt[3]{(4)(10)(25)} = \sqrt[3]{1000}$$
$$= 10 \quad ∎$$

With any geometric sequence, there is an associated **geometric series,** which is the indicated sum of the elements of the geometric sequence.

Illustration 5

The geometric sequence of Illustration 2 is

$$128, -32, 8, -2, \frac{1}{2}$$

Associated with this sequence is the geometric series

$$128 - 32 + 8 - 2 + \frac{1}{2}$$

which can be written with sigma notation as

$$\sum_{i=1}^{5} 128\left(-\frac{1}{4}\right)^{i-1}$$ ■

Let S_N be the sum of N terms of a geometric series. Then

$$S_N = a_1 + a_1 r + a_1 r^2 + a_1 r^3 + \ldots + a_1 r^{N-2} + a_1 r^{N-1}$$

If we multiply both members of this equation by r, we have

$$r S_N = a_1 r + a_1 r^2 + a_1 r^3 + a_1 r^4 + \ldots + a_1 r^{N-1} + a_1 r^N$$

The sum of the first equation and -1 times the second gives

$$S_n - r S_N = a_1 - a_1 r^N$$
$$(1 - r) S_N = a_1 - a_1 r^N$$

If $1 - r \neq 0$, we can divide each member of this equation by $1 - r$ and obtain

$$S_N = \frac{a_1 - a_1 r^N}{1 - r} \qquad \text{if } r \neq 1$$

$$S_N = \frac{a_1(1 - r^N)}{1 - r} \qquad \text{if } r \neq 1$$

A formula for S_N in terms of a_1, r, and a_N is found by expressing $a_1 r^N$ as $r(a_1 r^{N-1})$. Doing this, we have

$$S_N = \frac{a_1 - r(a_1 r^{N-1})}{1 - r} \qquad \text{if } r \neq 1$$

From Theorem 1, $a_1 r^{N-1} = a_N$. Thus

$$S_N = \frac{a_1 - r a_N}{1 - r} \qquad \text{if } r \neq 1$$

We have proved the following theorem.

Theorem 2   If $a_1, a_2, a_3, \ldots, a_N$ is a geometric sequence with common ratio r, and

$$S_N = a_1 + a_2 + a_3 + \ldots + a_N$$

then

(i) $S_N = \dfrac{a_1(1 - r^N)}{1 - r}$ if $r \neq 1$

and

(ii) $S_N = \dfrac{a_1 - ra_N}{1 - r}$ if $r \neq 1$

If $r = 1$, we have the trivial geometric series

$$a_1 + a_1 + a_1 + \ldots + a_1 \qquad (N \text{ terms of } a_1)$$

For this series, $S_N = Na_1$.

Example   5   Find the sum of the geometric series

$$\sum_{i=1}^{5} 2(\tfrac{1}{3})^{i-1}$$

Solution   For the given series, $a_1 = 2$, $r = \frac{1}{3}$, and $N = 5$. Thus, from Theorem 2(i),

$$S_5 = \frac{a_1(1 - r^5)}{1 - r}$$

$$= \frac{2[1 - (\tfrac{1}{3})^5]}{1 - \tfrac{1}{3}}$$

$$= \frac{2(1 - \tfrac{1}{243})}{\tfrac{2}{3}}$$

$$= 3 - \tfrac{1}{81}$$

$$= 2\tfrac{80}{81} \qquad\blacksquare$$

Example   6   Find the sum of the geometric series associated with the sequence of Example 1.

Solution   The sequence of Example 1, consisting of 10 elements, is

$$\tfrac{5}{2}, -5, 10, \ldots, -1280$$

Therefore the associated geometric series is

$$\tfrac{5}{2} - 5 + 10 - \ldots - 1280 \qquad (10 \text{ terms})$$

From Theorem 2(ii) with $a_{10} = -1280$, $r = -2$, and $a_1 = \tfrac{5}{2}$,

$$
\begin{aligned}
S_{10} &= \frac{a_1 - r a_{10}}{1 - r} \\
&= \frac{\tfrac{5}{2} - (-2)(-1280)}{1 - (-2)} \\
&= \frac{\tfrac{5}{2} - 2560}{3} \\
&= \frac{5 - 5120}{6} \\
&= -852\tfrac{1}{2}
\end{aligned}
$$

■

In the next example we use Theorem 3 of Section 5.1, which states that if P dollars is invested at an interest rate of $100i$ percent compounded m times per year and if A_n dollars is the amount of the investment at the end of n interest periods, then

$$A_n = P\left(1 + \frac{i}{m}\right)^n$$

Example   7   In order to create a sinking fund that will provide capital to purchase some new equipment, a company deposits $25,000 into an account on January 1 every year for 10 years. If the account earns 12 percent interest, compounded annually, how much is in the sinking fund immediately after the tenth deposit is made?

Solution   Immediately after the tenth deposit is made, the tenth payment has earned no interest; the ninth payment has earned interest for 1 year; the eighth payment has earned interest for 2 years; and so on; and the first payment has earned interest for 9 years. To find the number of dollars in the fund immediately after the tenth payment, we apply the formula for A_n with $P = 25,000$, $i = 0.12$, and $m = 1$ to find the dollar amount of each payment. The results are as follows:

10th payment: 25,000	(no interest)
9th payment: $25,000(1.12)^1$	(interest for 1 year; $n = 1$)
8th payment: $25,000(1.12)^2$	(interest for 2 years; $n = 2$)
$\vdots$	
1st payment: $25,000(1.12)^9$	(interest for 9 years; $n = 9$)

If x dollars is the total amount in the sinking fund immediately after the tenth deposit is made, then

$$x = 25{,}000 + 25{,}000(1.12)^1 + 25{,}000(1.12)^2 + \ldots + 25{,}000(1.12)^9$$

We observe that x is the sum of a geometric series for which $N = 10$, $r = 1.12$, and $a_1 = 25{,}000$. From Theorem 2(i),

$$x = \frac{25{,}000[1 - (1.12)^{10}]}{1 - 1.12}$$

From a calculator, $(1.12)^{10} \approx 3.106$. Thus

$$x \approx \frac{25{,}000(1 - 3.106)}{-0.12}$$

$$\approx (4.39)10^5$$

Therefore the amount in the sinking fund immediately after the tenth deposit is made is \$439,000, to three significant digits.   ■

EXERCISES   11.4

In Exercises 1 through 8, write the first five elements of a geometric sequence whose first element is a and whose common ratio is r.

1. $a = 5$; $r = 3$
2. $a = 3$; $r = 2$
3. $a = 8$; $r = -\frac{1}{2}$
4. $a = 2$; $r = \sqrt{2}$
5. $a = -\frac{9}{16}$; $r = -\frac{2}{3}$
6. $a = -81$; $r = \frac{1}{3}$
7. $a = \dfrac{x}{y}$; $r = -\dfrac{y}{x}$
8. $a = \dfrac{s}{t}$; $r = \dfrac{1}{u}$

In Exercises 9 through 16, determine whether the elements form a geometric sequence. If they do, write the next two elements.

9. $1, 3, 9$
10. $2, -4, 8$
11. $\sqrt{2}, \sqrt{6}, 3\sqrt{2}$
12. $\frac{1}{2}, \frac{1}{3}, \frac{1}{4}$
13. $3.33, 2.22, 1.11$
14. $3^{-2}, 3^0, 3^2$
15. $-6, 2, -\frac{2}{3}$
16. $\sqrt[3]{3}, \sqrt[6]{3}, 1$

17. Find the third element of a geometric sequence whose fifth element is 81 and whose ninth element is 16.

18. If the first element of a geometric sequence is $\frac{1}{8}$ and the eighth element is -16, what is the sixth element?

19. Find the common ratio of a geometric sequence whose third element is -2 and whose sixth element is 54.

20. The first element of a geometric sequence is 1 and the common ratio is 3; determine the smallest four-digit numeral that represents an element of this geometric sequence.

21. In the geometric sequence whose first element is 0.0003 and whose common ratio is 10, which element is 3,000,000?

22. In the geometric sequence whose first three elements are 27, -18, and 12, which element is $-\frac{512}{729}$?

23. Insert five geometric means between 1 and 64.
24. Insert three geometric means between 162 and 2.
25. Insert two geometric means between $\sqrt{3}$ and 3.
26. Find the geometric mean between 16 and 25.
27. Find the geometric mean between $-\frac{2}{3}$ and -6.
28. Find the geometric mean of the numbers $\frac{1}{3}$, 4, and 6.
29. Find the geometric mean of the numbers 9, 21, and 49.

30. If p and q are unequal positive numbers, show that their geometric mean is less than their arithmetic mean.

In Exercises 31 through 36, find the sum of the geometric series.

31. $\displaystyle\sum_{i=1}^{8} 2^i$

32. $\displaystyle\sum_{i=2}^{7} (-3)^i$

33. $\displaystyle\sum_{j=3}^{9} 5\left(\frac{1}{3}\right)^{j-3}$

34. $\displaystyle\sum_{k=1}^{6} \left(\frac{2}{3}\right)^k$

35. $\displaystyle\sum_{i=1}^{10} (1.02)^i$

36. $\displaystyle\sum_{i=1}^{12} (1.02)^{1-i}$

37. Find a sequence of four numbers, the first of which is 6 and the fourth of which is 16, if the first three numbers form an arithmetic sequence and the last three numbers form a geometric sequence.

38. Three numbers form an arithmetic sequence having a common difference of 4. If the first number is increased by 2, the second number by 3, and the third number by 5, the resulting numbers form a geometric sequence. Find the original numbers.

39. From a barrel filled with 1 gal of wine, 1 pint is withdrawn and then the barrel is filled with water. If this procedure is followed six times, what fractional part of the original contents is in the barrel?

40. If a town having a population of 5000 in 1971 has a 20 percent increase every 5 years, what is its expected population in the year 2001?

41. Payments of $1000 are deposited into a sinking fund every 6 months and the account earns 10 percent interest, compounded semiannually. How much is in the fund immediately after the twentieth payment is made?

42. Three numbers whose sum is 3 form an arithmetic sequence and their squares form a geometric sequence. What are the numbers?

43. Prove Theorem 1 by mathematical induction.

11.5 Infinite Geometric Series

Our discussion of series has so far been restricted to those associated with finite sequences. The series associated with the infinite sequence

$$a_1, a_2, a_3, \ldots, a_n \ldots$$

is denoted by

$$a_1 + a_2 + a_3 + \ldots + a_n + \ldots$$

and is called an **infinite series.** But what is the meaning of such an expression? That is, what do we mean by the "sum" of an infinite number of terms, and under what circumstances does such a sum exist? The answers to these questions depend on the concept of *limit,* which is studied in calculus. However, for some particular infinite series we can give an intuitive idea of the concept of sum.

Suppose a piece of string of length 2 ft is cut in half. One of these halves of length 1 ft is set aside and the other piece is cut in half again. One of the resulting pieces of length $\frac{1}{2}$ ft is set aside and the other piece is cut in half so that two pieces, each of length $\frac{1}{4}$ ft, are obtained. One of the pieces of length $\frac{1}{4}$ ft is set aside and then the other piece is cut in half; so two pieces, each of length $\frac{1}{8}$ ft, are obtained. Again one of the pieces is set aside and the other is cut in half. If this procedure is continued indefinitely, the number of feet in the sum of the lengths of the pieces set aside can be considered as the infinite series

$$1 + \frac{1}{2} + \frac{1}{4} + \frac{1}{8} + \frac{1}{16} + \ldots + \frac{1}{2^{n-1}} + \ldots \tag{1}$$

Series (1) is an infinite geometric series with $r = \frac{1}{2}$. Because we started with a piece of string 2 ft in length, your intuition indicates that the sum of infinite series (1) should be 2. We can demonstrate this situation by applying our knowledge of geometric series. From Theorem 2 of Section 11.4, if S_N is the sum of N terms of a geometric series,

$$S_N = \frac{a_1(1 - r^N)}{1 - r}$$

Therefore, for a finite number N of terms of series (1) with $a_1 = 1$ and $r = \frac{1}{2}$,

$$S_N = \frac{(1)[1 - (\frac{1}{2})^N]}{1 - \frac{1}{2}}$$

$$= \frac{1 - (\frac{1}{2})^N}{\frac{1}{2}}$$

$$\Leftrightarrow \qquad S_N = 2[1 - (\tfrac{1}{2})^N]$$

Applying this formula to successive values of N, we obtain

$$
\begin{aligned}
S_1 &= 2[1 - (\tfrac{1}{2})^1] & S_2 &= 2[1 - (\tfrac{1}{2})^2] & S_3 &= 2[1 - (\tfrac{1}{2})^3] \\
&= 2 - 2(\tfrac{1}{2}) & &= 2 - 2(\tfrac{1}{4}) & &= 2 - 2(\tfrac{1}{8}) \\
&= 2 - 1 & &= 2 - \tfrac{1}{2} & &= 2 - \tfrac{1}{4}
\end{aligned}
$$

$$
\begin{aligned}
S_4 &= 2[1 - (\tfrac{1}{2})^4] & S_5 &= 2[1 - (\tfrac{1}{2})^5] & S_6 &= 2[1 - (\tfrac{1}{2})^6] \\
&= 2 - 2(\tfrac{1}{16}) & &= 2 - 2(\tfrac{1}{32}) & &= 2 - 2(\tfrac{1}{64}) \\
&= 2 - \tfrac{1}{8} & &= 2 - \tfrac{1}{16} & &= 2 - \tfrac{1}{32}
\end{aligned}
$$

$$\vdots$$

$$
\begin{aligned}
S_{10} &= 2[1 - (\tfrac{1}{2})^{10}] \\
&= 2 - 2(\tfrac{1}{1024}) \\
&= 2 - \tfrac{1}{512}
\end{aligned}
$$

and so on. We intuitively see that we can make the value of S_N as close to 2 as we please by taking N large enough. In other words, we can make the difference between 2 and S_N as small as we please by taking N sufficiently large. Therefore we state that S_N approaches 2 as N increases without bound, and we write

$$S_N \to 2 \qquad \text{as} \quad N \to +\infty \tag{2}$$

Consider now the general infinite geometric series

$$a_1 + a_1 r + a_1 r^2 + a_1 r^3 + \ldots + a_1 r^{n-1} + \ldots \qquad |r| < 1$$

The sum of the first N terms of this series is given by

$$S_N = \frac{a_1}{1-r}(1 - r^N) \tag{3}$$

Let us consider what happens to r^N as N increases without bound, when $|r| < 1$. For instance,

$$\left(\frac{1}{2}\right)^1 = \frac{1}{2}, \left(\frac{1}{2}\right)^2 = \frac{1}{4}, \left(\frac{1}{2}\right)^3 = \frac{1}{8}, \ldots, \left(\frac{1}{2}\right)^{10} = \frac{1}{1024}, \ldots$$

$$\left(\frac{1}{3}\right)^1 = \frac{1}{3}, \left(\frac{1}{3}\right)^2 = \frac{1}{9}, \left(\frac{1}{3}\right)^3 = \frac{1}{27}, \ldots, \left(\frac{1}{3}\right)^{10} = \frac{1}{59,049}, \ldots$$

$$\left(\frac{2}{3}\right)^1 = \frac{2}{3}, \left(\frac{2}{3}\right)^2 = \frac{4}{9}, \left(\frac{2}{3}\right)^3 = \frac{8}{27}, \ldots, \left(\frac{2}{3}\right)^{10} = \frac{1024}{59,049}, \ldots$$

More generally, for any r for which $|r| < 1$, when N increases without bound, $|r^N|$ gets smaller and smaller; that is,

$$r^N \to 0 \qquad \text{as} \quad N \to +\infty$$

Therefore, from (3),

$$S_N \to \frac{a_1}{1-r} \qquad \text{as} \quad N \to +\infty$$

This statement leads us to make the following definition.

Definition

> **Sum of an Infinite Geometric Series**
> The **sum** S of an infinite geometric series, for which $|r| < 1$, is given by
>
> $$S = \frac{a_1}{1-r}$$

Illustration 1

For series (1), $a_1 = 1$ and $r = \frac{1}{2}$. Therefore, if S is the sum of this series, we have from the definition

$$S = \frac{a_1}{1-r}$$

$$= \frac{1}{1-\frac{1}{2}}$$

$$= 2$$

This result agrees with statement (2).   ∎

Observe that the sum of an infinite geometric series is defined only when $|r| < 1$. If we have an infinite geometric series for which $|r| > 1$, then as N increases without bound, r^N will not approach a finite number; so by referring to formula (3) it is apparent that the infinite series does not have a sum. If $r = 1$, then $S_N = Na_1$, and as N increases without bound, $|S_N|$ also increases without bound.

Example   1   Determine the sum of the infinite geometric series

$$6 + 4 + \tfrac{8}{3} + \ \ldots \ + 6(\tfrac{2}{3})^{n-1} + \ \ldots$$

Solution   From the definition with $a_1 = 6$ and $r = \tfrac{2}{3}$,

$$\begin{aligned}
S &= \frac{a_1}{1 - r} \\
  &= \frac{6}{1 - \tfrac{2}{3}} \\
  &= 18
\end{aligned}$$ ∎

An important application of infinite geometric series is to express a given nonterminating repeating decimal as a fraction. We can thus show that a decimal numeral represents a rational number. To indicate a nonterminating repeating decimal, we write a bar over the repeated digits. Then $0.33\overline{3}$ indicates $0.3333 \ldots$ and $4.0242\overline{24}$ indicates $4.024242424 \ldots$.

Illustration   2

The nonterminating repeating decimal $0.33\overline{3}$ can be written as

$$0.3 + 0.03 + 0.003 + 0.0003 + \ \ldots$$

$$\Leftrightarrow \qquad \frac{3}{10} + \frac{3}{100} + \frac{3}{1000} + \frac{3}{10{,}000} + \ \ldots \ + \frac{3}{10^n} + \ \ldots$$

which is an infinite geometric series with $a_1 = \tfrac{3}{10}$ and $r = \tfrac{1}{10}$. If S is the sum of this series,

$$\begin{aligned}
S &= \frac{a_1}{1 - r} \\
  &= \frac{\dfrac{3}{10}}{1 - \dfrac{1}{10}} \\
  &= \frac{1}{3}
\end{aligned}$$

Therefore the nonterminating repeating decimal $0.33\overline{3}$ and the fraction $\tfrac{1}{3}$ are representations for the same rational number.   ∎

Example   2   Express the nonterminating repeating decimal $4.0242\overline{424}$ as a fraction.

Solution   The given decimal can be written as

$$4 + \left[\frac{24}{1000} + \frac{24}{100,000} + \frac{24}{10,000,000} + \ldots + \frac{24}{10^{2n+1}} + \ldots\right]$$

The series in brackets is an infinite geometric series with $a_1 = \frac{24}{1000}$ and $r = \frac{1}{100}$. If S is the sum of this series, then

$$S = \frac{a_1}{1 - r}$$

$$= \frac{\frac{24}{1000}}{1 - \frac{1}{100}}$$

$$= \frac{\frac{24}{1000}}{\frac{99}{100}}$$

$$= \frac{24}{990}$$

$$= \frac{4}{165}$$

Thus

$$4.0242\overline{424} = 4 + \frac{4}{165}$$

$$= \frac{664}{165}$$

■

Example   3   A ball is dropped from a height of 36 m, and each time it strikes the ground it rebounds to a height of two-thirds of the distance from which it fell. Find the total distance traveled by the ball before it comes to rest.

Solution   Let d meters be the total distance traveled by the ball. To obtain d, we must add the distances it falls as well as the distances it rebounds. Thus

$$d = 36 + [(36)(\tfrac{2}{3}) + (36)(\tfrac{2}{3}) + (36)(\tfrac{2}{3})^2 + (36)(\tfrac{2}{3})^2 + \ldots]$$
$$= 36 + 2[(36)(\tfrac{2}{3}) + (36)(\tfrac{2}{3})^2 + \ldots + (36)(\tfrac{2}{3})^n + \ldots]$$

The series in brackets is an infinite geometric series with $a_1 = (36)(\tfrac{2}{3})$ and $r = \tfrac{2}{3}$. If S is the sum of this series, then

$$S = \frac{a_1}{1 - r}$$

$$= \frac{24}{1 - \frac{2}{3}}$$

$$= 72$$

Therefore

$$d = 36 + 2(72)$$
$$= 180$$

Thus the ball travels 180 m before coming to rest. ■

EXERCISES 11.5

In Exercises 1 through 10, find the sum of the infinite geometric series.

1. $16 + 4 + 1 + \ldots$

2. $\frac{1}{3} + \frac{1}{9} + \frac{1}{27} + \ldots$

3. $60 - 6 + 0.6 - \ldots$

4. $4 - 1.6 + 0.64 - \ldots$

5. $\frac{2}{3} + \frac{1}{9} + \frac{1}{54} + \ldots$

6. $1 + (1.04)^{-1} + (1.04)^{-2} + \ldots$

7. $\frac{4}{3} - 1 + \frac{3}{4} - \ldots$

8. $-2 - \frac{1}{2} - \frac{1}{8} - \ldots$

9. $3 + \sqrt{3} + 1 + \ldots$

10. $(2 + \sqrt{3}) + 1 + (2 - \sqrt{3}) + \ldots$

In Exercises 11 through 22, express the nonterminating repeating decimal as a fraction.

11. $0.66\overline{6}$

12. $0.27\overline{27}$

13. $0.81\overline{81}$

14. $0.252\overline{252}$

15. $2.99\overline{9}$

16. $3.1416\overline{1416}$

17. $1.234\overline{234}$

18. $7.99\overline{9}$

19. $0.4653\overline{4653}$

20. $2.045\overline{045}$

21. $3.2254\overline{44}$

22. $6.507\overline{11}$

23. Express the nonterminating repeating decimal $2.46\overline{46}$ as a fraction by two methods: (a) Consider $2.46\overline{46}$ as $2 + 0.46 + 0.0046 + 0.000046 + \ldots$ and (b) consider $2.46\overline{46}$ as $2.4 + 0.064 + 0.00064 + 0.0000064 + \ldots$.

24. Express the nonterminating repeating decimal $5.169\overline{69}$ as a fraction by two methods: (a) Consider $5.169\overline{69}$ as $5.1 + 0.069 + 0.00069 + 0.0000069 + \ldots$ and (b) consider $5.169\overline{69}$ as $5.16 + 0.0096 + 0.000096 + 0.00000096 + \ldots$.

25. A ball is dropped from a height of 12 m. Each time it strikes the ground it bounces back to a height of three-fourths of the distance from which it fell. Determine the total distance traveled by the ball before it comes to rest.

26. What is the total distance traveled by a tennis ball before coming to rest if it is dropped from a height of 100 m and if, after each fall, it rebounds $\frac{11}{20}$ of the distance from which it fell?

27. The path of each swing, after the first, of a pendulum bob is 0.93 as long as the path of the previous swing from one side to the other side. If the path of the first swing is 28 cm long, and air resistance eventually brings the pendulum to rest, how far does the bob travel before it comes to rest?

28. An equilateral triangle has sides of length 4 units; therefore its perimeter is 12 units. Another equilateral triangle is constructed by drawing line segments through the midpoints of the sides of the first triangle. This triangle has sides of length 2 units, and its perimeter is 6 units. If this procedure can be repeated an unlimited number of times, what is the total perimeter of all the triangles that are formed?

29. After a woman riding a bicycle removes her feet from the pedals, the front wheel rotates 100 times during the first 10 sec. Then in each succeeding 10-sec time period the wheel rotates $\frac{4}{5}$ as many times as it did the previous period. Determine the number of rotations of the wheel before the bicycle stops.

30. Find an infinite geometric series whose sum is 6 and such that each term is four times the sum of all the terms that follow it.

11.6   Counting, Permutations, and Combinations

We now consider the number of ways a set of events can occur under certain conditions. For instance, how many positive integers of two different digits can be formed from the integers 1, 2, 3, and 4? In how many ways can a chairman and a secretary be chosen from a committee of five people? If there are seven doors providing access to a building, in how many different ways can a person enter the building by one door and leave by a different one? The answers to these questions can be obtained by applying the following axiom.

Fundamental Principle of Counting

If one event can occur in m different ways, and if, after it has happened in one of these ways, a second event can occur in n different ways, then both events can occur, in the order stated, in $m \cdot n$ different ways.

Illustration   1

To determine how many positive integers of two different digits can be formed from the integers 1, 2, 3, and 4, we apply the fundamental principle of counting with $m = 4$ (the number of ways the tens' digit can be selected from among the four digits) and $n = 3$ (the number of ways the units' digit can be selected from among the three remaining digits after the tens' digit has been chosen). Hence there are $4 \cdot 3$, or 12, positive integers. To enumerate the twelve positive integers, we use a **tree diagram** as shown in Figure 1. The twelve positive integers are 12, 13, 14, 21, 23, 24, 31, 32, 34, 41, 42, and 43.   ■

Illustration   2

Suppose that we wish to determine how many ways a chairman and a secretary can be chosen from a committee of five people. The number of ways that the chairman can be selected from among the five people is five and then, after the chairman has been selected, there are four ways in which the secretary can be chosen from the remaining four people. Hence, from the fundamental principle of counting, the two positions can be filled in $5 \cdot 4$, or 20, different ways.   ■

Example   1 If there are seven doors providing access to a building, in how many ways can a person enter the building by one door and leave by a different door?

Solution The entry door can be chosen in any one of seven ways, and then the exit door can be selected from among the six remaining doors. Therefore, from the fundamental principle of counting, the number of ways that a person can enter the building by one door and leave by a different door is

$$7 \cdot 6 = 42$$ ■

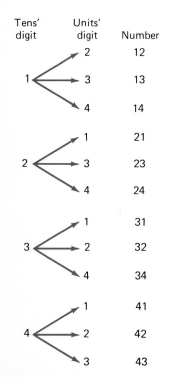

Tens' Units'
digit digit   Number

1	2	12
	3	13
	4	14
2	1	21
	3	23
	4	24
3	1	31
	2	32
	4	34
4	1	41
	2	42
	3	43

FIGURE 1

Hundreds' digit	Tens' digit	Units' digit	Number

FIGURE 2 _____

Illustration 3

In Illustration 1, we enumerated, by the use of a tree diagram, the twelve positive integers of two different digits that can be formed from the integers 1, 2, 3, and 4. Suppose now that we wish to find how many positive integers of three different digits can be formed with the same integers 1, 2, 3, and 4. In this case we have a tree diagram similar to that of Figure 1 but at the ends of each of the branches there are two additional branches. Such a tree diagram is shown in Figure 2. We see that there are 24 possible positive integers. Observe that 24 is the product of the number of ways, four, that the hundreds' digit can be chosen, the number of ways, three, that the tens' digit can be selected (after determining the hundreds' digit), and the number of possible choices, two, for the units' digit (after the tens' digit is chosen); that is, $24 = 4 \cdot 3 \cdot 2$. ∎

Illustration 3 shows an extension of the fundamental principle of counting to three events. In a similar manner it can be extended to any finite number of events.

Illustration 4

To determine how many different arrangements of six distinct books each can be made on a shelf with space for six books, we apply the fundamental principle of counting to six events. In the first space we may place any one of the six books. In the next space we may place any one of the five remaining books; in the next space we may place any one of the four remaining books; and so on. Therefore the number of different arrangements is

$$6 \cdot 5 \cdot 4 \cdot 3 \cdot 2 \cdot 1 = 720$$ ∎

When applying the fundamental principle of counting, it is sometimes convenient to write a horizontal mark for each of the events that is to occur. Then, in the order in which the events occur, insert above the corresponding mark the number of ways that the event can happen. The number of ways the set of events can occur in the required order is equal to the product of the numbers above the horizontal marks. If there are any special conditions affecting the order of events, these conditions should be considered first. The following example demonstrates the procedure.

Example 2 How many different arrangements, each consisting of four different letters, can be formed from the letters of the word "personal" if each arrangement is to begin and end with a vowel?

Solution There are three vowels and five consonants and four positions to be filled. We write four horizontal marks: __ __ __ __. In the first position we may choose any one of the three vowels and for the fourth position we may choose any one of the two remaining vowels. The second position can then be filled by any one of the six remaining letters, and the third position by any one of the five remaining letters. We write the numbers of choices over the horizontal marks and multiply, and we have

$$\underline{3} \cdot \underline{6} \cdot \underline{5} \cdot \underline{2} = 180$$

Therefore there are 180 different arrangements.   ■

Example   3   Suppose that three of the books in Illustration 4 are mathematics books and three are physics books.

a) How many different arrangements of the six books can be made on the shelf if books on the same subject are to be kept together?

b) How many different ways can the books be arranged on a shelf if the three mathematics books are to be kept together but the three physics books can be placed anywhere?

Solution   a) If the mathematics books are on the left, then the first book may be chosen in any one of three ways; the second book can be chosen from either of the two remaining books; and the third book must be the remaining mathematics book. The fourth book can be any one of the three physics books; the fifth book can be selected from either of the two remaining physics books; and the sixth book must be the remaining physics book. Therefore, if the mathematics books are on the left, the number of arrangements is

$$\underline{3} \cdot \underline{2} \cdot \underline{1} \cdot \underline{3} \cdot \underline{2} \cdot \underline{1} = 36$$

There are also 36 arrangements of the books for which the physics books are on the left. Therefore the total number of arrangements of the six books, for which books on the same subject are kept together, is

$$2 \cdot 36 = 72$$

b) We first consider the three mathematics books as a single element and each of the physics books as a single element. Then, we are arranging four elements in a row on a shelf. The first space can be occupied by any one of the four elements, the second space can be occupied by any one of the remaining three elements, and so on. The number of ways of arranging the four elements in order is

$$\underline{4} \cdot \underline{3} \cdot \underline{2} \cdot \underline{1} = 24$$

However, for each way of arranging the four elements, the three mathematics books can be arranged in order $3 \cdot 2 \cdot 1$, or 6, ways. Thus the number of ways the six books can be arranged in the desired order is

$$6 \cdot 24 = 144$$ ■

We now introduce the *factorial notation* that we need for the discussion of *permutations* and *combinations*. To denote the product of the first n positive integers, we write $n!$, read *"n factorial."* Thus

$$n! = n(n-1)(n-2)\ldots 3 \cdot 2 \cdot 1$$

Illustration   5

$$1! = 1 \qquad 2! = 2 \cdot 1 \qquad 3! = 3 \cdot 2 \cdot 1 \qquad 4! = 4 \cdot 3 \cdot 2 \cdot 1$$
$$ = 2 = 6 = 24$$

and so on.   ∎

Because

$$n! = n(n-1)(n-2)\ldots 3 \cdot 2 \cdot 1$$

and

$$(n-1)! = (n-1)(n-2)\ldots 3 \cdot 2 \cdot 1$$

it follows that

$$n! = n(n-1)!$$

In particular $5! = 5 \cdot 4!$ and $26! = 26 \cdot 25!$. If we substitute 1 for n in this formula, we obtain

$$1! = 1(1-1)!$$
$$1! = 1 \cdot 0!$$

Therefore, we define

$$0! = 1$$

In Illustration 1 we determined that from the integers 1, 2, 3, and 4, we can form the following twelve positive integers consisting of two different digits:

$$12 \quad 13 \quad 14 \quad 21 \quad 23 \quad 24 \quad 31 \quad 32 \quad 34 \quad 41 \quad 42 \quad 43$$

Each of these orderings is called a *permutation* of the elements of the set $\{1, 2, 3, 4\}$ taken two at a time.

Definition

> **Permutation**
> Let S be a set containing n elements, and suppose r is a positive integer such that $r \le n$. Then a **permutation** of r elements of S is an arrangement in a definite order, without repetitions, of r elements of S.

We use the notation $_nP_r$ to denote the number of permutations of n elements taken r at a time. Other symbols for this number are $P(n, r)$, $P_{n,r}$, and P^n_r.

We wish to find a formula for computing $_nP_r$. The number of arrangements in a definite order of n distinct elements taken r at a time is the number of ways that r positions can be filled from n elements. The first position can be chosen in n ways. Then there are $n - 1$ ways of choosing the second position, $n - 2$ ways of choosing the third position, and so on. The rth position can be selected from any of the remaining $n - (r - 1)$ (or, equivalently, $n - r + 1$) elements. Hence, from the fundamental principle of counting, we have

$$_nP_r = n(n - 1)(n - 2) \ldots (n - r + 1)$$

There is another formula for $_nP_r$ that involves the factorial notation. If we multiply the right side of the above formula by $\dfrac{(n - r)!}{(n - r)!}$, we obtain

$$_nP_r = \frac{n(n - 1)(n - 2) \ldots (n - r + 1)[(n - r)!]}{(n - r)!}$$

$$= \frac{n!}{(n - r)!}$$

We have proved the following theorem.

Theorem 1

> The number of permutations of n elements taken r at a time is given by either of the following formulas:
> (i) $_nP_r = n(n - 1)(n - 2) \ldots (n - r + 1)$
> (ii) $_nP_r = \dfrac{n!}{(n - r)!}$

Illustration 6

From the integers 1, 2, 3, and 4, the number of positive integers of two different digits that can be formed is $_4P_2$ and the number of positive integers of three different digits that can be formed is $_4P_3$. We compute $_4P_2$ and $_4P_3$ from Theorem 1(i).

$$_4P_2 = 4 \cdot 3 \qquad _4P_3 = 4 \cdot 3 \cdot 2$$
$$= 12 \qquad\qquad = 24$$

These results agree with those of Illustrations 1 and 3, respectively.   ■

Example   4   A bus has seven vacant seats. If three additional passengers enter the bus, in how many different ways can they be seated?

Solution   Each seating arrangement of the three passengers is a different arrangement of three seats from a set of seven seats. Therefore the number of ways the passengers can be seated is

$$_7P_3 = 7 \cdot 6 \cdot 5$$
$$= 210$$ ■

To determine the number of permutations of n elements taken n (or all) at a time, we use Theorem 1(i) with $r = n$ and then $n - r + 1$ is $n - n + 1$, or 1; thus

$$_nP_n = n(n - 1)(n - 2) \ldots 3 \cdot 2 \cdot 1$$

⇔

$$_nP_n = n!$$

Example   5   In how many ways can four boys and four girls be seated in a row containing eight seats if (a) a person may sit in any seat, and (b) boys and girls must sit in alternate seats?

Solution   a) The number of ways that eight people may be arranged in order eight at a time is

$$_8P_8 = 8 \cdot 7 \cdot 6 \cdot 5 \cdot 4 \cdot 3 \cdot 2 \cdot 1$$
$$= 40{,}320$$

  b) If a boy is in the first seat on the left, then the first seat can be occupied by any one of the four boys, the second seat can be occupied by any one of the four girls, the third seat can be occupied by any one of the three remaining boys, the fourth seat can be occupied by any one of the three remaining girls, and so on. Hence, if a boy is in the first seat on the left, the number of ways the eight people can be seated is

$$\underline{4} \cdot \underline{4} \cdot \underline{3} \cdot \underline{3} \cdot \underline{2} \cdot \underline{2} \cdot \underline{1} \cdot \underline{1} = 576$$

There are also 576 arrangements for which a girl is in the first seat on the left. Hence the total number of ways that four boys and four girls can be seated in a row containing eight seats, if boys and girls must alternate, is

$$2 \cdot 576 = 1152$$ ■

Illustration 7

The number of permutations of the six distinct letters of the word PAINTS, taken six at a time is 6!, or 720. Now consider the letters of the word DEGREE. For this word, we also have six letters, but because there are three E's the letters are not distinct. Hence the number of *distinguishable* permutations of the six letters of the word DEGREE taken six at a time is not 720. To demonstrate this concept, we consider reordering the second, fifth, and sixth letters in each of the words DEGREE and PAINTS, without changing the order of the other letters. For the word DEGREE the three E's are distinguished by using subscripts and we write

$$D \quad E_1 \quad G \quad R \quad E_2 \quad E_3$$

Rearranging the second, fifth, and sixth letters in each of the words gives the following permutations:

$$
\begin{array}{cccccc cccccc}
D & E_1 & G & R & E_2 & E_3 & \quad P & A & I & N & T & S \\
D & E_1 & G & R & E_3 & E_2 & \quad P & A & I & N & S & T \\
D & E_2 & G & R & E_1 & E_3 & \quad P & T & I & N & A & S \\
D & E_2 & G & R & E_3 & E_1 & \quad P & T & I & N & S & A \\
D & E_3 & G & R & E_1 & E_2 & \quad P & S & I & N & A & T \\
D & E_3 & G & R & E_2 & E_1 & \quad P & S & I & N & T & A \\
\end{array}
$$

Observe that corresponding to the one permutation DEGREE there are 6, or 3!, permutations for the letters of the word PAINTS. Such a situation exists for each distinguishable permutation of the letters of the word DEGREE. Let P be the number of distinguishable permutations of the six letters

$$D, \quad E, \quad G, \quad R, \quad E, \quad E$$

Then, because for each of these permutations there are 3! ways in which the E's can be rearranged without changing the order of the other letters, we have

$$3! \cdot P = 6!$$

$$P = \frac{6!}{3!}$$

Hence $P = 120$. ■

Illustration 8

To determine the number of different nine-digit numerals that can be formed from the digits 6, 6, 6, 6, 5, 5, 5, 4, and 3, we first consider one such numeral, for instance,

$$665566543$$

With this ordering of the nine digits, there are 4! permutations of the digits 6 and 3! permutations of the digits 5 that have no effect on the numeral. Therefore there are $4! \cdot 3!$ arrangements of the digits in the numeral 665566543 that do not result in a distinguishable permutation of the given nine digits. Hence if P is the number of distinguishable permutations of the nine digits,

$$4! \cdot 3! \cdot P = 9!$$

because 9! is the number of permutations of nine distinct elements taken all at a time. Thus

$$P = \frac{9!}{4! \cdot 3!}$$

$$= \frac{9 \cdot 8 \cdot 7 \cdot 6 \cdot 5}{3 \cdot 2 \cdot 1} \cdot \frac{4!}{4!}$$

$$= 2520$$

Therefore, there are 2520 different nine-digit numerals that can be formed from the given digits. ∎

By the procedure used in Illustrations 7 and 8, we can prove the following theorem. The details of the proof are omitted.

Theorem 2

If we are given n elements, of which exactly m_1 are alike of one kind, exactly m_2 are alike of a second kind, . . . , and exactly m_k are alike of a kth kind, and if $n = m_1 + m_2 + \ldots + m_k$, then the number of distinguishable permutations that can be made of the n elements taking them all at one time is

$$\frac{n!}{m_1! \cdot m_2! \cdot \ldots \cdot m_k!}$$

Example   6 How many different signals, each consisting of eight flags hung one above the other, can be formed from a set of three indistinguishable red flags, two indistinguishable blue flags, two indistinguishable white flags, and one black flag?

Solution We have a set of eight elements in which three are alike of one kind (the red flags), two are alike of a second kind (the blue flags), two are alike of a third kind (the white flags), and one is of a fourth kind (the black flag). If P is the number of different signals (the number of distinguishable permutations of these eight elements taken all at one time), it follows from Theorem 2 that

$$P = \frac{8!}{3! \cdot 2! \cdot 2! \cdot 1!}$$

$$= \frac{8 \cdot 7 \cdot 6 \cdot 5 \cdot 4}{2 \cdot 2 \cdot 1} \cdot \frac{3!}{3!}$$

$$= 1680$$

∎

So far we have been concerned with ordered arrangements of elements of a set. Our discussion now pertains to subsets of a set without regard to the relative order of the elements in the subsets. For example, if a committee of four people is to be selected from a group of eight, we are concerned only with the four people on the committee and not with the order in which they are chosen. Or if a student has ten posters to pin up on the walls of her room, but there is space for only seven, then the specific seven posters that she decides to hang are of importance, but the order in which she selects them is not important. In each of these situations, we are referring to a *combination* of elements of a set.

Definition

> **Combination**
> Let S be a set containing n elements, and suppose r is a positive integer such that $r \leq n$. Then a **combination** of r elements of S is a subset of S containing r distinct elements.

The notation we use for the number of combinations of n elements taken r at a time is $_nC_r$. Other symbols used for this number are $C(n, r)$, $C_{n,r}$, C_r^n, and $\binom{n}{r}$.

Illustration 9

Suppose that $S = \{a, b, c, d\}$.

a) There is only one subset of the four letters taken all at a time. Hence

$$_4C_4 = 1$$

b) There are four subsets of the four letters taken three at a time. They are

$$\{a, b, c\} \quad \{a, b, d\} \quad \{a, c, d\} \quad \{b, c, d\}$$

Therefore,

$$_4C_3 = 4$$

c) There are six subsets of the four letters taken two at a time, and they are

$$\{a, b\} \quad \{a, c\} \quad \{a, d\} \quad \{b, c\} \quad \{b, d\} \quad \{c, d\}$$

Thus

$$_4C_2 = 6$$

d) The subsets of the four letters taken one at a time are

$$\{a\} \quad \{b\} \quad \{c\} \quad \{d\}$$

Therefore,

$$_4C_1 = 4$$

∎

Observe that the distinction between permutations and combinations is that changing the order of a set of elements gives a different permutation but the same combination.

To obtain a formula for computing $_nC_r$, we must find the total number of subsets of r elements each that can be obtained from a set of n elements. The elements of each of the subsets can be arranged in order $r!$ different ways. Therefore, for each combination of n elements taken r at a time, there are $r!$ permutations; thus, for the total of all the possible combinations, the number of different permutations is $r! \cdot {_nC_r}$. However, these are all the possible permutations of n elements taken r at a time, and therefore

$$r! \cdot {_nC_r} = {_nP_r}$$

$$_nC_r = \frac{_nP_r}{r!}$$

We have proved the following theorem.

Theorem 3

> The number of combinations of n elements taken r at a time is given by
>
> $$_nC_r = \frac{_nP_r}{r!}$$

Illustration   10

a) $\quad _4C_4 = \dfrac{_4P_4}{4!}$ b) $\quad _4C_3 = \dfrac{_4P_3}{3!}$ c) $\quad _4C_2 = \dfrac{_4P_2}{2!}$ d) $\quad _4C_1 = \dfrac{_4P_1}{1!}$

$\qquad\quad = \dfrac{4!}{4!}$ $\qquad\qquad = \dfrac{4 \cdot 3 \cdot 2}{3 \cdot 2 \cdot 1}$ $\qquad\qquad = \dfrac{4 \cdot 3}{2 \cdot 1}$ $\qquad\qquad = \dfrac{4}{1}$

$\qquad\quad = 1$ $\qquad\qquad\quad = 4$ $\qquad\qquad\qquad = 6$ $\qquad\qquad\quad = 4$

These results agree with those of Illustration 9.   ■

Example   7 A football conference consists of eight teams. If each team plays every other team, how many conference games are played?

Solution   The number of conference games that are played is the number of two-element subsets of an eight-element set. This number is

$$_8C_2 = \frac{_8P_2}{2!}$$

$$= \frac{8 \cdot 7}{1 \cdot 2}$$

$$= 28$$ ■

Theorem 1(ii) gives the following formula for computing $_nP_r$:

$$_nP_r = \frac{n!}{(n-r)!}$$

Substituting from this formula into the formula of Theorem 3, we obtain

$$_nC_r = \frac{n!}{(n-r)!r!} \qquad (1)$$

Now, if in (1) r is replaced by $n - r$, we get

$$_nC_{n-r} = \frac{n!}{[n-(n-r)]!(n-r)!}$$

$$_nC_{n-r} = \frac{n!}{r!(n-r)!} \qquad (2)$$

From Equations (1) and (2), $_nC_r = {_nC_{n-r}}$. We have proved the following theorem.

Theorem 4

$$_nC_r = {_nC_{n-r}}$$

Theorem 4 seems reasonable if you realize that each time a subset of r elements is chosen from a set of n elements, another set of $n - r$ elements remains unchosen. For instance, $_6C_4 = {_6C_2}$, $_{12}C_9 = {_{12}C_3}$, $_{100}C_{95} = {_{100}C_5}$, and so on.

Example 8 A student has ten posters to pin up on the walls of her room, but there is space for only seven. In how many ways can she choose the posters to be pinned up?

Solution The number of ways she can choose the posters is the number of combinations of ten elements taken seven at a time, which is $_{10}C_7$. From Theorem 4, $_{10}C_7 = {_{10}C_3}$; the significance of this equality is that each time she selects seven posters to be used she is also selecting the three posters not to be used. Therefore,

$$_{10}C_7 = {_{10}C_3}$$
$$= \frac{10 \cdot 9 \cdot 8}{3 \cdot 2 \cdot 1}$$
$$= 120$$

Example   9   How many committees of five can be formed from eight sophomores and four freshmen if each committee is to consist of three sophomores and two freshmen?

Solution   The number of ways of choosing three sophomores from eight is $_8C_3$, and the number of ways of choosing two freshmen from four is $_4C_2$. From the fundamental principle of counting, the number of committees consisting of three sophomores and two freshmen is

$$_8C_3 \cdot {}_4C_2 = \frac{8 \cdot 7 \cdot 6}{3 \cdot 2 \cdot 1} \cdot \frac{4 \cdot 3}{2 \cdot 1}$$
$$= 56 \cdot 6$$
$$= 336 \qquad \blacksquare$$

Example   10   From six history books and eight economics books, in how many ways can a person select two history books and three economics books and arrange them on a shelf?

Solution   The number of ways that two history books can be selected from six is $_6C_2$, and the number of ways that three economics books can be selected from eight is $_8C_3$. Then, after the five books are selected, the number of ways that they can be arranged on a shelf is $_5P_5$. Hence if x is the number of ways that a person can select the books and arrange them on a shelf,

$$x = {}_6C_2 \cdot {}_8C_3 \cdot {}_5P_5$$
$$= \frac{6 \cdot 5}{2 \cdot 1} \cdot \frac{8 \cdot 7 \cdot 6}{3 \cdot 2 \cdot 1} \cdot 5 \cdot 4 \cdot 3 \cdot 2 \cdot 1$$
$$= 100,800 \qquad \blacksquare$$

EXERCISES   11.6

In Exercises 1 and 2, simplify the rational expression.

1. (a) $\dfrac{6!}{9!}$; (b) $\dfrac{12!}{8!6!}$; (c) $\dfrac{3! + 4!}{7!}$

2. (a) $\dfrac{51!}{49!}$; (b) $\dfrac{2!5!}{7!}$; (c) $\dfrac{8!}{6! + 7!}$

In Exercises 3 through 6, find the number.

3. (a) $_5P_3$; (b) $_6P_6$; (c) $_7P_1$
4. (a) $_8P_4$; (b) $_5P_5$; (c) $_{20}P_2$
5. (a) $_7C_3$; (b) $_{50}C_{48}$; (c) $_{11}C_{11}$
6. (a) $_6C_2$; (b) $_{30}C_{26}$; (c) $_{15}C_{15}$

7. How many positive integers of three digits can be formed from the digits 1, 3, 5, 7, and 9 if repetitions (a) are not permitted, and (b) are permitted?

8. How many positive integers of four digits can be formed from the digits 1, 2, 3, 4, 5, and 6 if repetitions (a) are not permitted, and (b) are permitted?

9. How many positive integers of four different digits can be formed from the digits of Exercise 8 if (a) the integers are to be even, and (b) the integers are to be greater than 3000?

10. How many positive integers of three different digits can be formed from the digits of Exercise 8 if (a) the integers are to be odd, and (b) the integers are to be less than 400?

11. (a) If two cubical dice, one white and one black, are thrown, in how many ways can they fall? (b) In how many ways can three different-colored cubical dice fall?

12. (a) In a contest with twelve entries, in how many ways can a jury award first, second, and third prizes? (b) In a club of twelve members, in how many ways can the offices of president, secretary, and treasurer be filled if no person is to hold more than one office?

13. If there are six airlines flying between Los Angeles and San Francisco and four bus lines going between San Francisco and Santa Rosa, in how many ways can a person go by plane from Los Angeles to San Francisco, and then by bus from San Francisco to Santa Rosa, and return by the same modes of transport on each leg of the trip, without using the same company twice?

14. If a bookshelf has space for five books and there are ten different books available, how many different arrangements can be made of the five books on the shelf?

15. In how many ways can nine books be arranged on a shelf if four particular books are to be kept together?

16. In how many ways can four French books, two Spanish books, and three German books be arranged on a shelf so that all books in the same language are together?

17. In how many ways can a baseball coach prepare a batting order (a) if the coach has already selected the team of nine members, and (b) if the coach still has to select the team from among twelve players?

18. In how many ways can a baseball coach assign positions to a team of nine men if only two men are qualified to be pitcher, and only three other men are qualified to be catcher, but all of the men are qualified to play any other position?

19. How many different committees of three persons each can be chosen from a group of twelve persons?

20. How many different hands of five cards each can be dealt from a standard deck of 52 cards?

21. If a student is to answer any six questions on a test containing nine questions, in how many different ways can the student choose the questions?

22. If, in Exercise 21, the first three questions must be answered, in how many different ways can the student choose the six questions?

23. By using seven points, no three of which are collinear, how many different lines are determined?

24. How many triangles are determined from the points in Exercise 23?

25. A committee of five is to be chosen from seven Republicans and four Democrats. In how many ways can the committee be chosen if it is to contain three Republicans and two Democrats?

26. A bag contains four red balls, six white balls, and five blue balls. In how many ways can six balls be chosen if there are to be two balls of each color?

27. For the bag of Exercise 26, in how many ways can six balls be chosen if there are to be exactly four white balls?

28. How many different arrangements, each consisting of five different letters, can be formed from the letters of the word EQUATION for each of the following situations: (a) each arrangement begins and ends with a consonant; (b) in each arrangement vowels and consonants alternate?

29. How many different arrangements can be formed from the eight letters of the word EQUATION for each of the following situations: (a) in each arrangement all the vowels are together; (b) in each arrangement the letter q is immediately followed by the letter u?

30. From the letters of each of the following words, how many different permutations can be formed if the letters are taken all at a time: (a) COLLEGE; (b) MISSISSIPPI?

31. From the letters of each of the following words, how many different permutations can be formed if the letters are taken all at a time: (a) BOOKWORM; (b) TENNESSEE?

32. How many four-digit positive integers can be made without repeating any digits?

33. How many of the positive integers in Exercise 32 are odd?

34. How many different eight-digit numerals can be formed from the digits 2, 2, 2, 4, 4, 6, 6, and 8?

35. How many different ten-digit numerals can be formed from the digits 3, 3, 3, 3, 1, 1, 1, 7, 7, and 5?

In Exercises 36 through 38, note that in a standard deck of cards there are 13 spades, 13 hearts, 13 diamonds, and 13 clubs.

36. In how many different ways can a bridge hand of thirteen cards be dealt from a standard deck?

37. In how many different ways can a bridge hand containing six spades, four hearts, two diamonds, and one club be dealt from a standard deck?

38. In how many different ways can a bridge hand of thirteen cards, containing exactly eight spades, be dealt from a standard deck?

39. A classroom has only eight vacant seats, three in the front row and five in the back row. In how many ways can five additional students be seated if two of them insist on sitting in the front row?

40. In how many ways can five students be seated in a row of eight seats if a certain two students insist on sitting next to each other?

41. In how many ways can five students be seated in a row of eight seats if a certain two students refuse to sit next to each other?

42. There are ten desks in a row. In how many ways can five students be seated at consecutive desks? (*Hint:* First determine the number of ways that the five consecutive desks can be chosen.)

43. If, in Exercise 42, the five students are seated for a test and there is to be exactly one empty desk between each two students, in how many ways can the students be seated in the row?

44. In a display window a grocer wishes to put a row of fifteen cans of soup consisting of five identical cans of tomato soup, four identical cans of mushroom soup, three identical cans of celery soup, and three identical cans of vegetable soup. How many different displays are possible?

45. How many displays in Exercise 44 have a can of tomato soup at each end?

46. How many displays in Exercise 44 have a can of the same kind of soup at each end?

11.7   Probability

The theory of probability was founded in the seventeenth century by the two French mathematicians Blaise Pascal (1623–1662) and Pierre de Fermat (1601–1665). It began with problems related to games of chance such as throwing dice, tossing coins, and drawing lottery numbers. It now has applications in the social, physical, and life sciences. For instance, it is used to make predictions in problems related to insurance, annuities, statistics, quality control, mechanics, heredity, biometrics, and experimental methods in general. In this section we give a very brief introduction to the subject and we use simple examples related to games of chance.

We restrict our discussion to experiments having a finite number of mutually exclusive possible outcomes. For example, if a single cubical die is thrown, there are six equally likely outcomes: The die lands so that 1, 2, 3, 4, 5, or 6 faces up. If a coin is tossed, there are two possible outcomes: The coin lands heads up (H) or tails up (T). The set of all possible outcomes of a given experiment is called the **sample space** of the experiment.

Illustration 1

a) If S is the sample space of the experiment of throwing a single die, then

$$S = \{1, 2, 3, 4, 5, 6\}$$

b) If S is the sample space of the experiment of tossing a single coin, then

$$S = \{H, T\}$$

c) If S is the sample space of the experiment of tossing both a dime and a quarter, then

$$S = \{HH, HT, TH, TT\}$$ ∎

A subset of the sample space of an experiment is called an **event** associated with the experiment.

Illustration 2

a) For the experiment of Illustration 1(a), if E is the event of throwing a 5, $E = \{5\}$. If E is the event of throwing an even number, $E = \{2, 4, 6\}$.

b) For the experiment of Illustration 1(b), if E is the event of tossing a head, $E = \{H\}$. If E is the event of tossing either a head or a tail, $E = \{H, T\}$.

c) For the experiment of Illustration 1(c), if E is the event of tossing at least one tail, $E = \{HT, TH, TT\}$. If E is the event of tossing exactly one head, $E = \{HT, TH\}$. If E is the event of tossing no heads, $E = \{TT\}$. ∎

Definition

> **Probability of an Event**
> If S is the sample space of an experiment and E is an event associated with the experiment, the **probability** of E, denoted by $P(E)$, is defined by
>
> $$P(E) = \frac{n(E)}{n(S)}$$
>
> where $n(E)$ and $n(S)$ are the number of elements in E and S, respectively.

Because E is a subset of S, $0 \leq n(E) \leq n(S)$. Therefore

$$0 \leq P(E) \leq 1$$

Furthermore, if E has no elements, that is, $n(E) = 0$, then $P(E) = 0$; the event will never happen. If $E = S$, then $n(E) = n(S)$ and $P(E) = 1$; the event is certain to happen.

Example 1 Determine the probability of each of the following events: (a) obtaining a 5 on a throw of a single die; (b) obtaining a head on a toss of a coin; (c) obtaining at least one tail when tossing both a dime and a quarter.

Solution a) Since $S = \{1, 2, 3, 4, 5, 6\}$ and $E = \{5\}$, $n(S) = 6$ and $n(E) = 1$.

Therefore

$$P(E) = \frac{n(E)}{n(S)}$$

$$= \frac{1}{6}$$

b) Because $S = \{H, T\}$, and $E = \{H\}$, $n(S) = 2$ and $n(E) = 1$. Hence

$$P(E) = \frac{n(E)}{n(S)}$$

$$= \frac{1}{2}$$

c) Because $S = \{HH, HT, TH, TT\}$ and $E = \{HT, TH, TT\}$, $n(S) = 4$ and $n(E) = 3$. Thus

$$P(E) = \frac{n(E)}{n(S)}$$

$$= \frac{3}{4}$$ ■

From part (a) of Example 1, the probability of obtaining a 5 on a throw of a single die is $\frac{1}{6}$. This means that if we throw a single die many times, we will obtain a 5 for approximately one-sixth of the throws. For instance, with sixty throws, we will obtain a 5 approximately ten times. This does not mean that it will necessarily occur exactly ten times. However, as the number of throws increases, then the number of 5s that are obtained gets closer and closer to the product of $\frac{1}{6}$ and the number of throws. Similar comments pertain to parts (b) and (c) of Example 1.

Example 2 If two dice are thrown, what is the probability of obtaining a sum of (a) 7 and (b) 5?

Solution a) We distinguish between the two dice by referring to one as "the first die" and to the other as "the second die." We shall denote the outcome of a throw of the two dice by an ordered pair (x, y) where x is the number on the first die and y is the number on the second die. Thus both x and y can have six possible values. Hence there are $6 \cdot 6$, or 36, possible outcomes of the experiment. Therefore, if S is the sample space, $n(S) = 36$. If E is the event of obtaining a sum of 7, then

$$E = \{(1, 6), (2, 5), (3, 4), (4, 3), (5, 2), (6, 1)\}$$

and $n(E) = 6$. Therefore, if $P(E)$ is the probability of obtaining a sum of 7,

$$P(E) = \frac{n(E)}{n(S)}$$

$$= \frac{6}{36}$$

$$= \frac{1}{6}$$

b) If E is the event of obtaining a sum of 5, then

$$E = \{(1, 4), (2, 3), (3, 2), (4, 1)\}$$

and $n(E) = 4$. Thus, if $P(E)$ is the probability of obtaining a sum of 5,

$$P(E) = \frac{n(E)}{n(S)}$$

$$= \frac{4}{36}$$

$$= \frac{1}{9}$$ ■

Example   **3**   Determine the probability of each of the following events: (a) drawing a spade from a deck of 52 playing cards; (b) drawing four spades in succession from a deck of 52 playing cards if after each card is drawn it is not replaced in the deck.

Solution   a) Because there are 52 possible outcomes when a card is drawn, then if S is the sample space, $n(S) = 52$. Because the number of spades in a deck is 13, $n(E) = 13$. Therefore, if $P(E)$ is the probability of drawing a spade,

$$P(E) = \frac{n(E)}{n(S)}$$

$$= \frac{13}{52}$$

$$= \frac{1}{4}$$

b) The experiment consists of drawing four cards from a deck of 52. Thus if S is the sample space of the experiment, $n(S) = {}_{52}C_4$. Because there are 13 spades in the deck, the number of four-card hands containing exactly four spades is ${}_{13}C_4$. Therefore $n(E) = {}_{13}C_4$. Hence, if

$P(E)$ is the probability of drawing four spades in succession,

$$P(E) = \frac{n(E)}{n(S)}$$

$$= \frac{_{13}C_4}{_{52}C_4}$$

$$= \frac{13 \cdot 12 \cdot 11 \cdot 10}{1 \cdot 2 \cdot 3 \cdot 4} \div \frac{52 \cdot 51 \cdot 50 \cdot 49}{1 \cdot 2 \cdot 3 \cdot 4}$$

$$= \frac{13 \cdot 12 \cdot 11 \cdot 10}{52 \cdot 51 \cdot 50 \cdot 49}$$

$$\approx 0.0026$$

Example   4   If a French book, a Spanish book, a German book, a Russian book, and an English book are placed at random on a shelf with space for five books, what is the probability that the Russian and English books will be next to each other?

Solution   If S is the sample space of the experiment of placing five books in order on a shelf, $n(S) = {_5}P_5$. We now wish to compute $n(E)$ where E is the event that the Russian and English books are together. Let us consider the Russian and English books as a single element and each of the other three books as a single element. Thus we are arranging four elements in a row on a shelf. The number of ways of arranging four elements in order is $4 \cdot 3 \cdot 2 \cdot 1$. However, for each way of arranging the four elements, the Russian and English books can be arranged in order two ways. Thus $n(E) = 2 \cdot 4 \cdot 3 \cdot 2 \cdot 1$. Therefore, if $P(E)$ is the required probability,

$$P(E) = \frac{n(E)}{n(S)}$$

$$= \frac{2 \cdot 4 \cdot 3 \cdot 2 \cdot 1}{{_5}P_5}$$

$$= \frac{2 \cdot 4 \cdot 3 \cdot 2 \cdot 1}{5 \cdot 4 \cdot 3 \cdot 2 \cdot 1}$$

$$= \frac{2}{5}$$

Suppose now that E_1 and E_2 are two mutually exclusive events associated with an experiment having S as a sample space. By mutually exclusive events we mean that E_1 and E_2 have no elements in common; that is,

$E_1 \cap E_2 = \emptyset$. Thus $n(E_1 \cup E_2) = n(E_1) + n(E_2)$. Then if $E = E_1 \cup E_2$,

$$P(E) = \frac{n(E)}{n(S)}$$

$$= \frac{n(E_1 \cup E_2)}{n(S)}$$

$$= \frac{n(E_1) + n(E_2)}{n(S)}$$

$$= \frac{n(E_1)}{n(S)} + \frac{n(E_2)}{n(S)}$$

$$= P(E_1) + P(E_2)$$

Therefore the probability of E is the sum of the probabilities of E_1 and E_2. We extend this fact to any number of mutually exclusive events in the next theorem. The theorem can be proved by mathematical induction.

Theorem 1

If $E_1, E_2, \ldots, E_n$ are mutually exclusive events associated with a particular experiment and if

$$E = E_1 \cup E_2 \cup \ldots \cup E_n$$

then

$$P(E) = P(E_1) + P(E_2) + \ldots + P(E_n)$$

Example   5   What is the probability of obtaining a sum of 7 or 5 if two dice are thrown?

Solution   Let E_1 be the event of obtaining a 7 and E_2 be the event of obtaining a 5. From Example 3, $P(E_1) = \frac{1}{6}$ and $P(E_2) = \frac{1}{9}$. We wish to find $P(E)$ where $E = E_1 \cup E_2$. From Theorem 1

$$P(E) = P(E_1) + P(E_2)$$

$$= \tfrac{1}{6} + \tfrac{1}{9}$$

$$= \tfrac{3}{18} + \tfrac{2}{18}$$

$$= \tfrac{5}{18}$$ ∎

EXERCISES   11.7

1. On a throw of a single die, determine the probability of each of the following events: (a) obtaining a 3; (b) obtaining an odd number.
2. On a throw of a single die, determine the probability of each of the following events: (a) obtaining either a 1 or a 6; (b) obtaining a number less than 4.
3. On a throw of two dice, determine the probability of each of the following events: (a) obtaining a sum of 3; (b) obtaining a sum of 7.

4. On a throw of two dice, determine the probability of each of the following events: (a) obtaining a sum of 4; (b) obtaining a sum of 11.

5. On a throw of two dice, what is the probability of obtaining a sum that is at most 4?

6. On a throw of two dice, what is the probability of obtaining a sum that is at least 3?

7. If two coins are tossed, what is the probability of each of the following events: (a) obtaining no heads; (b) obtaining at least one head?

8. If two coins are tossed, what is the probability of each of the following events: (a) obtaining exactly one tail; (b) obtaining at most one tail?

9. If a single card is drawn from a deck of 52 playing cards, what is the probability of each of the following events: (a) obtaining an ace; (b) obtaining a spade; (c) obtaining an ace or a spade?

10. If a single card is drawn from a deck of 52 playing cards, what is the probability of each of the following events: (a) obtaining a king or a queen; (b) obtaining a heart; (c) obtaining a king, a queen, or a heart?

11. A bag contains five red balls, three blue balls, and two green balls. One ball is drawn from the bag. Determine the probability of each of the following events: (a) the ball is red; (b) the ball is blue; (c) the ball is either blue or green.

12. Assume that two balls are drawn simultaneously from the bag of Exercise 11. Determine the probability of each of the following events: (a) one ball is blue and one is green; (b) at least one ball is green; (c) neither ball is blue.

13. Assume that two balls are drawn simultaneously from the bag of Exercise 11. Determine the probability of each of the following events: (a) both balls are red; (b) neither ball is red; (c) at least one ball is blue.

14. Assume that three balls are drawn simultaneously from the bag of Exercise 11. Determine the probability of each of the following events: (a) all three balls are red; (b) all three balls are blue; (c) no ball is red.

15. A number of two different digits is to be formed from the digits 1, 2, 3, 4, and 5. Determine the probability of each of the following events: (a) the number is even; (b) the number is less than 31; (c) the number is greater than 34.

16. A number of two different digits is to be formed from the digits of Exercise 15. Determine the

probability of each of the following events: (a) the number is odd; (b) the number is greater than 35; (c) the number is less than 42.

17. A number of three different digits is to be formed from the digits of Exercise 15. What is the probability that it is less than 325?

18. A number of three different digits is to be formed from the digits of Exercise 15. What is the probability that it is greater than 235?

19. If six boys and six girls are seated at random in a row containing twelve seats, what is the probability that boys and girls sit in alternate seats?

20. If six boys and six girls are seated at random at a round table, what is the probability that boys and girls sit in alternate seats?

21. If ten students are assigned seats at random in a row of ten chairs, what is the probability that a certain two students will be seated next to each other?

22. If ten students are assigned seats at random in a row of ten chairs, what is the probability that a certain three students will be seated together?

23. A committee of five is to be chosen at random from a group of four Republicans and five Democrats. What is the probability that the committee will contain exactly two Republicans and three Democrats?

24. What is the probability that the committee of Exercise 23 will contain at least three Democrats?

25. What is the probability that the committee of Exercise 23 will contain all Democrats?

26. A committee of five is to be chosen at random from a group of ten students. What is the probability that a certain two students will both be on the committee?

27. What is the probability that a certain two students will both not be on the committee of Exercise 26?

28. In the game of craps, a player throws two dice and wins on the first throw if a sum of 7 or 11 is obtained. What is the probability of winning on the first throw?

29. Five cards are drawn in succession from a deck of 52 playing cards, and after each card is drawn it is not replaced in the deck. What is the probability that all five cards are of the same suit?

30. Three mathematics books and three physics books are to be placed at random on a shelf with space for six books. What is the probability that all the books on the same subject are together?

31. On a true-false test containing three questions a student guesses each answer. What is the probability that (a) all three answers are correct; (b) exactly two answers are correct; (c) at least two answers are correct; (d) at most one answer is correct?

32. Let E be an event of an experiment having a sample space S and $\bar{E}$ be the set of elements in S that are not in E. Prove that $P(\bar{E}) = 1 - P(E)$.

33. On a true-false test containing four questions a student guesses each answer. Use the result of Exercise 32 to determine the probability that at least one answer is correct.

34. Use the result of Exercise 32 to determine the probability of obtaining at least one queen if five cards are drawn in succession from a deck of 52 playing cards and after each card is drawn it is not replaced in the deck.

11.8 | The Binomial Theorem

A power of a binomial is a special kind of series called a **binomial expansion.** Let us consider the binomial expansion of

$$(a + b)^n$$

for specific values of n.

$$
\begin{aligned}
n = 1: \quad & (a + b)^1 = a + b \\
n = 2: \quad & (a + b)^2 = a^2 + 2ab + b^2 \\
n = 3: \quad & (a + b)^3 = a^3 + 3a^2b + 3ab^2 + b^3 \\
n = 4: \quad & (a + b)^4 = a^4 + 4a^3b + 6a^2b^2 + 4ab^3 + b^4 \\
n = 5: \quad & (a + b)^5 = a^5 + 5a^4b + 10a^3b^2 + 10a^2b^3 + 5ab^4 + b^5 \\
n = 6: \quad & (a + b)^6 = a^6 + 6a^5b + 15a^4b^2 + 20a^3b^3 + 15a^2b^4 + 6ab^5 + b^6
\end{aligned}
$$

Each equation after the first is obtained by multiplying on both sides of the previous equation by $a + b$. On the right side of each equation the first term can be written with a factor b^0 and the last term with a factor a^0. Therefore each term contains nonnegative integer powers of a and b. We also note the following properties of each of the six expansions:

1. There are $n + 1$ terms in the expansion.

2. The sum of the exponents of a and b in any term is n: the exponent of a decreases by 1 and the exponent of b increases by 1 from each term to the next term.

3.   (i) The first term in the expansion is

$$a^n = {}_nC_0a^n$$

     (ii) The second term is

$$\frac{n}{1}a^{n-1}b = {}_nC_1a^{n-1}b$$

(iii) The third term is

$$\frac{n(n-1)}{2 \cdot 1} a^{n-2} b^2 = {}_nC_2 a^{n-2} b^2$$

(iv) The fourth term is

$$\frac{n(n-1)(n-2)}{3 \cdot 2 \cdot 1} a^{n-3} b^3 = {}_nC_3 a^{n-3} b^3$$

(v) The fifth term is

$$\frac{n(n-1)(n-2)(n-3)}{4 \cdot 3 \cdot 2 \cdot 1} a^{n-4} b^4 = {}_nC_4 a^{n-4} b^4$$

(vi) The term involving b^r is

$$\frac{n(n-1)(n-2) \dots (n-r+1)}{r!} a^{n-r} b^r = {}_nC_r a^{n-r} b^r$$

(vii) The last term is

$$b^n = {}_nC_n b^n$$

Illustration   1

The binomial expansion of $(a + b)^n$ when $n = 6$ is

$$(a + b)^6 = a^6 + 6a^5 b + 15a^4 b^2 + 20a^3 b^3 + 15a^2 b^4 + 6ab^5 + b^6$$

We show that the preceding properties apply to this expansion.

1.   There are seven terms in the expansion.

2.   The sum of the exponents of a and b in any term is 6; the exponent of a decreases by 1 and the exponent of b increases by 1 from each term to the next term.

3.   (i) The first term in the expansion is

$$a^6 = {}_6C_0 a^6$$

(ii) The second term is

$$\frac{6}{1} a^{6-1} b = {}_6C_1 a^5 b$$

(iii) The third term is

$$\frac{6 \cdot 5}{2 \cdot 1} a^{6-2} b^2 = {}_6C_2 a^4 b^2$$

(iv) The fourth term is

$$\frac{6 \cdot 5 \cdot 4}{3 \cdot 2 \cdot 1} a^{6-3} b^3 = {}_6C_3 a^3 b^3$$

(v) The fifth term is

$$\frac{6 \cdot 5 \cdot 4 \cdot 3}{4 \cdot 3 \cdot 2 \cdot 1} a^{6-4} b^4 = {}_6C_4 a^2 b^4$$

(vi) The sixth term is

$$\frac{6 \cdot 5 \cdot 4 \cdot 3 \cdot 2}{5 \cdot 4 \cdot 3 \cdot 2 \cdot 1} a^{6-5} b^5 = {}_6C_5 ab^5$$

(vii) The last term is

$$b^6 = {}_6C_6 b^6 \qquad\qquad \blacksquare$$

Illustration 1 and the discussion preceding it suggest a similar expression for the expansion of $(a + b)^n$, where n is any positive integer. We state this in the binomial theorem.

The Binomial Theorem

> If n is any positive integer, then
>
> $$(a + b)^n = {}_nC_0 a^n + {}_nC_1 a^{n-1} b + {}_nC_2 a^{n-2} b^2$$
> $$+ \ldots + {}_nC_r a^{n-r} b^r + \ldots + {}_nC_n b^n$$

Proof: If n is a positive integer, then $(a + b)^n$ represents the product of n equal factors of $(a + b)$; that is,

$$(a + b)^n = (a + b)(a + b)(a + b) \ldots (a + b) \qquad n \text{ factors of } (a + b)$$

When performing the multiplication, we select either a or b from each of the n factors and multiply these n numbers; we form every such product by considering all possible choices of either a or b from each of the n factors. The term involving $a^{n-r} b^r$ is a sum of terms, each of which is a product of r factors of b (one factor from each of r factors $(a + b)$) and $(n - r)$ factors of a (one factor from each of the remaining $(n - r)$ factors of $(a + b)$). The number of ways that the r factors of b can be chosen from n factors of $(a + b)$ is ${}_nC_r$. Therefore there are ${}_nC_r$ terms involving $a^{n-r} b^r$. Thus the coefficient of $a^{n-r} b^r$ is ${}_nC_r$. Because r can be any integer from 0 through n, inclusive, the formula follows.

The binomial theorem can also be proved by mathematical induction. However, this proof is longer than the one presented. $\blacksquare$

Example **1** Expand and simplify by the binomial theorem $(x^2 + 3y)^5$.

Solution Applying the binomial theorem where a is x^2, b is $3y$, and n is 5, we have

$$(x^2 + 3y)^5 = {}_5C_0(x^2)^5 + {}_5C_1(x^2)^4(3y)^1 + {}_5C_2(x^2)^3(3y)^2$$
$$+ {}_5C_3(x^2)^2(3y)^3 + {}_5C_4(x^2)^1(3y)^4 + {}_5C_5(3y)^5$$

$$= 1 \cdot x^{10} + \frac{5}{1}x^8(3y) + \frac{5 \cdot 4}{2 \cdot 1}x^6(9y^2) + \frac{5 \cdot 4 \cdot 3}{3 \cdot 2 \cdot 1}x^4(27y^3)$$

$$+ \frac{5 \cdot 4 \cdot 3 \cdot 2}{4 \cdot 3 \cdot 2 \cdot 1}x^2(81y^4) + 1(243y^5)$$

$$= x^{10} + 15x^8y + 90x^6y^2 + 270x^4y^3 + 405x^2y^4 + 243y^5 \qquad \blacksquare$$

Because ${}_nP_r = n(n - 1)(n - 2) \ldots (n - r + 1)$, we have

$$_nC_r = \frac{{}_nP_r}{r!}$$

$$= \frac{n(n - 1)(n - 2) \ldots (n - r + 1)}{r!}$$

Therefore the binomial expansion for $(a + b)^n$ can be written as

$$(a + b)^n = a^n + \frac{n}{1!}a^{n-1}b + \frac{n(n - 1)}{2!}a^{n-2}b^2 + \frac{n(n - 1)(n - 2)}{3!}a^{n-3}b^3$$

$$+ \ldots + \frac{n(n - 1)(n - 2) \ldots (n - r + 1)}{r!}a^{n-r}b^r + \ldots + b^n$$

This formula is used in the next example.

Example 2 Expand and simplify $\left(2\sqrt{t} - \dfrac{1}{t}\right)^4$.

Solution We use the above formula where a is $2\sqrt{t}$, b is $-\dfrac{1}{t}$, and n is 4.

$$\left(2\sqrt{t} - \frac{1}{t}\right)^4 = (2\sqrt{t})^4 + \frac{4}{1!}(2\sqrt{t})^3\left(-\frac{1}{t}\right)^1 + \frac{4 \cdot 3}{2!}(2\sqrt{t})^2\left(-\frac{1}{t}\right)^2$$

$$+ \frac{4 \cdot 3 \cdot 2}{3!}(2\sqrt{t})^1\left(-\frac{1}{t}\right)^3 + \frac{4 \cdot 3 \cdot 2 \cdot 1}{4!}\left(-\frac{1}{t}\right)^4$$

$$= 16t^2 - 32t^{1/2} + \frac{24}{t} - \frac{8}{t^{5/2}} + \frac{1}{t^4} \qquad \blacksquare$$

From the binomial theorem, the term involving b^r in the expansion of $(a + b)^n$ is the $(r + 1)$st term, which is

$$_nC_r a^{n-r} b^r$$

The rth term in the expansion of $(a + b)^n$ is obtained from this expression by replacing r by $r - 1$. Thus

> The rth term in $(a + b)^n$ is $_nC_{r-1} a^{n-r+1} b^{r-1}$

Observe that the exponent of b is one less than the number of the term, and the sum of the exponents of a and b is n.

Example 3 Find the seventh term of the expansion of $(2u^3 - \frac{1}{4}v^4)^{10}$.

Solution Applying the formula for the rth term where r is 7, n is 10, a is $2u^3$, and b is $-\frac{1}{4}v^4$, we have

$$_{10}C_6(2u^3)^4\left(-\frac{1}{4}v^4\right)^6 = \frac{10 \cdot 9 \cdot 8 \cdot 7 \cdot 6 \cdot 5}{6 \cdot 5 \cdot 4 \cdot 3 \cdot 2 \cdot 1}(2^4 u^{12})\left(\frac{1}{2^{12}}v^{24}\right)$$

$$= \frac{210}{2^8}u^{12}v^{24}$$

$$= \frac{105}{128}u^{12}v^{24}$$

■

Example 4 Find the term involving x^3 in the expansion of $(x - 3x^{-1})^9$.

Solution From the formula for the rth term, where a is x, b is $-3x^{-1}$, and n is 9, the rth term has the factors

$$x^{10-r}(-3x^{-1})^{r-1} = (-3)^{r-1}x^{11-2r}$$

The term involving x^3 is the one for which the exponent of x is 3; hence we solve the equation

$$11 - 2r = 3$$

$$r = 4$$

Thus the fourth term is the desired term. It is

$$_9C_3 x^6(-3x^{-1})^3 = \frac{9 \cdot 8 \cdot 7}{3 \cdot 2 \cdot 1}(-27)x^3$$

$$= -2268x^3$$

■

There is an interesting pattern for the coefficients in the binomial expansion. The coefficients can be written in the following triangular arrangement:

$$
\begin{array}{ccccccc}
 & & & 1 & 1 & & & \\
 & & 1 & 2 & 1 & & \\
 & 1 & 3 & 3 & 1 & & \\
 1 & 4 & 6 & 4 & 1 & & \\
1 & 5 & 10 & 10 & 5 & 1 &
\end{array}
$$

$$
\begin{array}{ccccccc}
\cdot & \cdot & \cdot & \cdot & \cdot & \cdot & \cdot \\
\cdot & \cdot & \cdot & \cdot & \cdot & \cdot & \cdot \\
\cdot & \cdot & \cdot & \cdot & \cdot & \cdot & \cdot
\end{array}
$$

The first row contains the coefficients for $(a + b)^1$; the second row contains the coefficients in the expansion of $(a + b)^2$; in the third row are the coefficients for $(a + b)^3$; and so on. Observe that each row begins and ends with the number 1. Also observe that each of the other numbers is the sum of the two numbers in the previous row, one to the left and one to the right of the number. For instance, the numbers in the fifth row are the coefficients for $(a + b)^5$. The first and last numbers are 1. The second number 5 is the sum of 1 and 4; then 10 is the sum of 4 and 6; 10 is the sum of 6 and 4; and 5 is the sum of 4 and 1. This triangular array is called **Pascal's triangle,** named in honor of the French mathematician Blaise Pascal (1623–1662), who used it in his work with permutations, combinations, and probability, although it was known before his time.

Example   5   Expand $(a + b)^7$ by first finding the coefficients from Pascal's triangle.

Solution   We write the fifth row, and then obtain the sixth and seventh rows by the procedure described above.

$$
\begin{array}{ccccccccccccc}
 & & 1 & & 5 & & 10 & & 10 & & 5 & & 1 \\
 & 1 & & 6 & & 15 & & 20 & & 15 & & 6 & & 1 \\
1 & & 7 & & 21 & & 35 & & 35 & & 21 & & 7 & & 1
\end{array}
$$

The coefficients for $(a + b)^7$ are in the seventh row. Therefore

$$(a + b)^7 = a^7 + 7a^6b + 21a^5b^2 + 35a^4b^3 + 35a^3b^4 + 21a^2b^5 + 7ab^6 + b^7 \quad \blacksquare$$

When n is small, the use of Pascal's triangle for the coefficients of $(a + b)^n$ is advantageous. However, if n is large or a specific term is desired, you will want to use the binomial theorem or the formula for the rth term.

EXERCISES 11.8

In Exercises 1 through 12, expand the power of the binomial.

1. $(a + b)^8$
2. $(a + b)^4$
3. $(x - y)^9$
4. $(x - y)^{10}$
5. $(x + 3y)^5$
6. $(2x - y)^6$
7. $(4 - ab)^6$
8. $(2t + s^2)^7$
9. $(3e^x - 2e^{-x})^5$
10. $(2u^{-1} - 3u^2)^5$
11. $(a^{1/2} + b^{1/2})^8$
12. $(xy^{-1} - x^{-1}y)^7$

In Exercises 13 through 18, find the first four terms in the expansion of the power of the binomial and simplify each term.

13. $(2x^2 + y^2)^{12}$
14. $(a^3 - 2b^2)^{14}$
15. $(u^{-1} - 3v^2)^{11}$
16. $(e^{x/2} - e^{-x/2})^{20}$
17. $(a^{1/3} - b^{1/3})^9$
18. $(\frac{2}{5}a^{2/3} + b^{3/2})^{11}$

19. Find the seventh term of the expansion of $(a + b)^{14}$.
20. Find the sixth term of the expansion of $(\frac{1}{2}a - b)^{13}$.
21. Find the sixth term of the expansion of $(2x - 3)^9$.
22. Find the tenth term of the expansion of $(\sqrt{t} - t^{-1/2})^{15}$.
23. Find the middle term of the expansion of $(1 - x^3y^{-2})^{12}$.

24. Find the middle term of the expansion of $(\frac{1}{3}y + \sqrt{y})^{10}$.
25. Find the term involving a^6 in the expansion of $(\frac{1}{2} + a)^{12}$.
26. Find the term involving x^{12} in the expansion of $(x^2 - \frac{1}{2})^{11}$.
27. Find the term that does not contain x in the expansion of $(x^2 - 2x^{-2})^{10}$.
28. Find the term involving t^{-4} in the expansion of $(\frac{1}{5}t^2 - t^{-1})^{13}$.

In Exercises 29 through 32, write the binomial expansion by first finding the coefficients from Pascal's triangle.

29. $(a + b)^9$
30. $(x - y)^8$
31. $(r - t)^{12}$
32. $(u + v)^{11}$

In Exercises 33 through 36, find to four significant digits the value of the power by using the indicated binomial expansion. Compare the result with that obtained from a calculator.

33. $(1.02)^8 = (1 + 0.02)^8$
34. $(0.98)^6 = (1 - 0.02)^6$
35. $(99)^5 = (100 - 1)^5$
36. $(101)^7 = (100 + 1)^7$

REVIEW EXERCISES FOR CHAPTER 11

In Exercises 1 and 2, write the first six elements of the sequence whose general element is given.

1. $a_n = (-1)^{n-1}\dfrac{2n - 1}{3^n}$

2. $a_n = (-1)^n \dfrac{n^2 + 1}{2n} x^{2n-1}$

In Exercises 3 and 4, write the series and find the sum of the series.

3. $\displaystyle\sum_{i=2}^{7} \frac{i + 1}{i - 1}$

4. $\displaystyle\sum_{k=1}^{4} (-1)^{k+1} \frac{3k}{2^k}$

In Exercises 5 and 6, write the series with sigma notation (there is no unique solution).

5. $\frac{1}{2}x^2 - \frac{1}{4}x^4 + \frac{1}{8}x^6 - \frac{1}{16}x^8$
6. $-x + \frac{1}{4}x^3 - \frac{1}{7}x^5 + \frac{1}{10}x^7 - \frac{1}{13}x^9 + \frac{1}{16}x^{10}$

In Exercises 7 through 10, find the sum of the series.

7. $\displaystyle\sum_{j=1}^{10} 3\left(\frac{1}{2}\right)^j$

8. $\displaystyle\sum_{k=1}^{30} (3k + 1)$

9. $\displaystyle\sum_{i=1}^{20} \left(\frac{1}{3}i + 3\right)$

10. $\displaystyle\sum_{i=1}^{10} (1.01)^i$

In Exercises 11 and 12, find the number.

11. (a) $\dfrac{12!}{4!9!}$; (b) $\dfrac{4! + 5!}{3! + 5!}$
12. (a) $\dfrac{3!7!}{9!}$; (b) $\dfrac{4! + 5!}{6!}$

13. Find x so that the numbers $\frac{1}{16}$, $\frac{1}{4}$, and x form an arithmetic sequence.
14. Find x so that the numbers 25, x, and 9 form a geometric sequence.
15. Find the first element of a geometric sequence whose fourth element is -3 and whose eighth element is -243.

16. Find the sum of the positive odd integers between 10 and 100.
17. In the arithmetic sequence whose first three elements are -8, -5, and -2, which element is 52?
18. Insert three arithmetic means between $\frac{1}{2}$ and $\frac{2}{3}$.
19. Insert five geometric means between 192 and 3.
20. Find the sum of the infinite geometric series $\frac{1}{3} + \frac{1}{9} + \frac{1}{27} + \ldots$.
21. Find the sum of the infinite geometric series $0.4 + 0.02 + 0.001 + \ldots$.
22. Find the thirtieth element of the arithmetic sequence whose seventeenth element is 7 and whose forty-seventh element is 31.
23. (a) How many numbers between 100 and 500 are divisible by 8? (b) What is their sum?
24. Show that the reciprocals of the elements of a geometric sequence also form a geometric sequence.

In Exercises 25 and 26, express the nonterminating repeating decimal as a fraction.

25. $0.727\overline{272}$

26. $4.6636\overline{363}$

In Exercises 27 through 30, use mathematical induction to prove that the formula is true for all positive-integer values of n.

27. $\displaystyle\sum_{i=1}^{n} (4i + 1) = n(2n + 3)$

28. $\displaystyle\sum_{i=1}^{n} 4^i = \frac{4(4^n - 1)}{3}$

29. $\displaystyle\sum_{i=1}^{n} \frac{1}{(3i - 2)(3i + 1)} = \frac{n}{3n + 1}$

30. $\displaystyle\sum_{i=1}^{n} i(2i + 1) = \frac{4n^3 + 9n^2 + 5n}{6}$

In Exercises 31 and 32, find the number.

31. (a) $_{10}P_4$; (b) $_7C_3$; (c) $_{16}C_{14}$
32. (a) $_6P_3$; (b) $_8C_4$; (c) $_{20}C_{17}$

In Exercises 33 and 34, expand the power of the binomial.

33. $(2x - y)^7$

34. $(a + 3b)^8$

In Exercises 35 and 36, write and simplify the first four terms in the expansion of the binomial.

35. $(x + 2x^{-1})^{20}$

36. $(4w^{-1} - \frac{1}{2}w^2)^{15}$

37. Find the tenth term of the expansion of $(t^{1/2} - t^{-1/2})^{15}$.
38. Find the middle term of the expansion of $(x^2 + 3y)^6$.
39. Find the term involving z^{12} in the expansion of $(z^2 - \frac{1}{2})^{11}$.
40. Find the term involving u^{40} in the expansion of $\left(2u^4 - \dfrac{1}{2u}\right)^{15}$.
41. Write the binomial expansion of $(x - y)^9$ by first finding the coefficients from Pascal's triangle.
42. Find to four decimal places the value of $(0.95)^6$ by using the binomial expansion of $(1 - 0.05)^6$. Compare the result with that obtained from a calculator.
43. A pile of logs has 30 logs in the bottom layer, 29 logs in the next to bottom layer, and so on, and the top layer contains 5 logs; each layer except the last contains one less log than the layer beneath it. How many logs are in the pile?
44. In a certain culture the number of bacteria increases 20 percent every 30 min. There are 1000 bacteria present initially. Find a formula for determining the number of bacteria in the culture at the end of t hours. How many bacteria are in the culture at the end of 5 hours?
45. (a) How many ancestors, to the nearest thousand, did you have 20 generations ago under the assumption that each ancestor appears only once in your family tree? (b) What is the total number of ancestors, to the nearest thousand, in all 20 generations?
46. A man borrows $20,000 and places a mortgage on his home. He agrees that at the end of each year for 10 years he will repay $2000 of the principal together with interest at the rate of 15 percent per year on the amount outstanding during the year. What is the total amount to be paid in 10 years?
47. Three numbers whose sum is 35 form a geometric sequence. If 1 is subtracted from the first number, 2 is subtracted from the second number, and 8 is subtracted from the third number, the resulting differences form an arithmetic sequence. What are the numbers?
48. The path of each swing, after the first, of a pendulum bob is 80 percent as long as the path of the previous swing from one side to the other side. If the path of the first swing is 18 in. long,

and air resistance eventually brings the pendulum to rest, how far does the bob travel before it comes to rest?

49. A sheet of paper is torn in half and then each half is again torn in half, and so on until the tearing-in-half is done 30 times. Then each of the pieces of paper is placed in a pile with one piece on top of another. If the original sheet of paper has a thickness of 0.01 cm, what will be the height of the pile to the nearest kilometer?

50. In how many ways can six students be seated in a row of six desks?

51. For a baseball team of nine players, how many batting orders are possible?

52. If there are twelve teams in an athletic conference, how many games must be played if each team is to play every other team?

53. There are ten questions on an examination and a student is required to answer any five of them. How many different sets of five questions can be selected?

54. A committee of four is to be formed from eight sophomores and six freshmen. How many different committees can be chosen if the committee is to contain two sophomores and two freshmen?

55. From the digits 0, 1, 2, 3, 4, 5, 6, 7, 8, 9, how many three-digit numerals can be formed?

56. A basketball team consists of twelve players, of which only four play the center position. How many starting line-ups of five players are possible?

57. How many of the numerals in Exercise 55 are odd?

58. In how many ways can the six students of Exercise 50 be seated in the row of six desks if a certain two students (a) insist on sitting next to each other, and (b) refuse to sit next to each other?

59. (a) From the letters of the word PEOPLE, how many different permutations can be formed by taking all the letters at a time? (b) How many of these permutations begin and end with a vowel?

60. For an audition a tenor is to sing three operatic arias from a selection of ten. (a) In how many

ways can he choose the three arias? (b) In how many different orders can he present three arias?

61. On a throw of a single die, determine the probability of each of the following events: (a) obtaining a 3 or a 6; (b) obtaining a number greater than 3.

62. On a throw of two dice, determine the probability of each of the following events: (a) obtaining a sum of 6; (b) obtaining a sum of at least 4.

63. If two coins are tossed, what is the probability of each of the following events: (a) obtaining at least one tail; (b) obtaining at most one head?

64. If three coins are tossed, what is the probability of each of the following events: (a) obtaining all heads; (b) obtaining exactly one head?

65. If three coins are tossed, what is the probability of each of the following events: (a) obtaining no heads; (b) obtaining at least two heads?

66. If a single card is drawn from a deck of 52 playing cards, what is the probability of each of the following events: (a) obtaining a queen; (b) obtaining a diamond or a club; (c) obtaining a queen, a diamond, or a club?

67. Three French books and four Spanish books are to be placed at random on a shelf with space for seven books. What is the probability that all the books in the same language will be together?

68. A committee of four is to be chosen at random from a group of six men and eight women. What is the probability that the committee will contain two men and two women?

69. Five boys and five girls are seated at random in ten chairs. Show that the probability that the boys and girls will be seated alternately is the same whether they are seated in a row or in a circle.

70. Do Exercise 69 if there are n boys and n girls.

71. Prove by mathematical induction that the following inequality is true for all positive-integer values of n: $2^{n+3} < (n + 3)!$

72. Prove by mathematical induction that $x - y$ is a factor of $x^{2n} - y^{2n}$ for all positive-integer values of n.

Appendix

Properties of Real Numbers

A **binary** operation on the set of real numbers is a rule that assigns to any two real numbers a and b, taken in a definite order, a number c. **Addition** is a binary operation on R because addition assigns to the real numbers a and b a number, denoted by $a + b$, called the **sum** of a and b. The numbers a and b are called **addends** (or **terms**). **Multiplication** is also a binary operation on R because multiplication assigns to the real numbers a and b a number denoted by ab, called the **product** of a and b. The numbers a and b are called **factors.** The following seven axioms give laws governing the operations of addition and multiplication on the set R. In the illustrations that follow the axioms, we use elements of the set of natural numbers, and we assume that the sum and product of natural numbers are known.

Axiom 1

> **Closure Laws**
> If a and b are real numbers, then $a + b$ and ab are unique real numbers.

Axiom 1 guarantees that whenever the operations of addition and multiplication are performed on two real numbers, the sum and product are real numbers. The axiom is called *closure laws* because a set is said to be **closed** with respect to an operation if, whenever the operation is performed on elements of the set, an element of the set is obtained.

Illustration 1

a) The set $\{1, 2, 3, 4\}$ is not closed with respect to addition because whenever the operation of addition is performed on elements of this set we do not necessarily obtain an element of the set. For instance, $2 + 3 = 5$, but 5 is not an element of the set.

b) The set of even natural numbers is closed with respect to both addition and multiplication because whenever either addition or multiplication is performed on two even natural numbers, the sum and product are even natural numbers. For instance, 6 and 8 are even natural numbers; $6 + 8 = 14$, and $6 \cdot 8 = 48$; 14 and 48 are even natural numbers. ∎

Axiom 2

Commutative Laws

If a and b are real numbers,

$$a + b = b + a \qquad \text{and} \qquad ab = ba$$

Axiom 2 states that the sum and product of two real numbers are not affected by the order of the numbers.

Illustration 2

a) By the commutative law for addition, $4 + 9 = 9 + 4$.

b) By the commutative law for multiplication, $5 \cdot 7 = 7 \cdot 5$. ∎

Axiom 3

Associative Laws

If a, b, and c are real numbers,

$$a + (b + c) = (a + b) + c \qquad \text{and} \qquad a(bc) = (ab)c$$

Although addition and multiplication are binary operations, the expressions $a + b + c$ and abc are meaningful because the grouping symbols (parentheses and brackets) can be inserted in any possible way without affecting the results. From Axiom 3 it follows that the sum of three real numbers can be obtained by grouping the addends in either of two ways and that the product of three real numbers can be found by grouping the factors in either of two ways.

Illustration 3

a)
$$2 + 3 + 4 = 2 + (3 + 4) \qquad 2 + 3 + 4 = (2 + 3) + 4$$
$$= 2 + 7 \qquad\qquad\qquad\quad = 5 + 4$$
$$= 9 \qquad\qquad\qquad\qquad = 9$$

b)
$$3 \cdot 6 \cdot 5 = 3 \cdot (6 \cdot 5) \qquad 3 \cdot 5 \cdot 6 = (3 \cdot 5) \cdot 6$$
$$= 3 \cdot 30 \qquad\qquad\qquad = 15 \cdot 6$$
$$= 90 \qquad\qquad\qquad\quad = 90$$

∎

Axiom 4

Distributive Law
If a, b, and c are real numbers,
$$a(b + c) = ab + ac$$

Illustration   4

a)   When evaluating $4(7 + 2)$, the parentheses are used to indicate that we should first perform the operation of addition of $7 + 2$ and then the sum is multiplied by 4. Therefore we have
$$4(7 + 2) = 4 \cdot 9$$
$$= 36$$

b)   When computing $(4 \cdot 7) + (4 \cdot 2)$, the parentheses indicate that each of the multiplications is performed and then the sum of the two products is obtained. We have then
$$(4 \cdot 7) + (4 \cdot 2) = 28 + 8$$
$$= 36$$

Observe that from parts (a) and (b) we can conclude that
$$4(7 + 2) = (4 \cdot 7) + (4 \cdot 2)$$

and this equality is a special case of Axiom 4.   ∎

Axiom 5

Existence of Identity Elements
There exist two distinct real numbers 0 and 1, called the **additive identity** and **multiplicative identity,** respectively, such that for any real number a,
$$a + 0 = a \quad \text{and} \quad a \cdot 1 = a$$

Illustration   5
$$8 + 0 = 8 \quad \text{and} \quad 8 \cdot 1 = 8$$ ∎

Axiom 6

Existence of Additive Inverse
For every real number a, there exists a real number, called the **opposite of a (or additive inverse of a)**, denoted by $-a$, such that
$$a + (-a) = 0$$

Illustration   6

The opposite of 4 is denoted by -4 and
$$4 + (-4) = 0$$ ∎

Axiom 7

Existence of Multiplicative Inverse

For every real number a, except 0, there exists a real number, called the **reciprocal of a** (or **multiplicative inverse of a**), denoted by $\dfrac{1}{a}$, such that

$$a \cdot \frac{1}{a} = 1$$

Illustration 7

The reciprocal of 9 is $\dfrac{1}{9}$ and

$$9 \cdot \frac{1}{9} = 1$$

Axioms 1 through 7 are called **field axioms,** and if these axioms are satisfied by a set of elements, then the set is called a **field** under the two binary operations involved. Thus the set R is a field under addition and multiplication. For the set J of integers, each of Axioms 1 through 6 is satisfied, but Axiom 7 is not satisfied; for instance, the integer 2 has no multiplicative inverse in J. Therefore the set of integers is not a field under addition and multiplication.

The field axioms do not imply any order of the real numbers. That is, by means of the field axioms alone we cannot state that 2 is greater than 1, 3 is greater than 2, and so on. However, we introduce an axiom giving the concept of a real number being positive. It is called the *order axiom* because the notion of a positive number is used in Section 1.1 to define what is meant by one real number being *greater than* or *less than* another.

Axiom 8

Order Axiom

In the set of real numbers there exists a subset called the **positive numbers** such that

(i) if a is a real number, exactly one of the following three statements holds:

$$a = 0 \qquad a \text{ is positive} \qquad -a \text{ is positive}$$

(ii) The sum of two positive numbers is positive.

(iii) The product of two positive numbers is positive.

Because the set R of real numbers satisfies the order axiom and the field axioms, we say that R is an **ordered field.**

The opposites of the elements of the set of positive numbers form the set of *negative numbers,* as given in the following definition.

Definition

> **Negative Number**
> The real number a is **negative** if and only if $-a$ is positive.

We use the terminology *negative* when referring to a negative number. For instance, -5 can be read "negative 5," as well as "the opposite of 5." The symbolism $-(-5)$ is read "the opposite of negative 5." In summary, then, we are using the symbol $-$ in two different ways: (1) to denote the opposite of a real number; and (2) to denote a negative number. If x can be any real number, $-x$ denotes the opposite of x. Observe that $-x$ is not necessarily a negative number; if x is -3, for instance, then $-x = 3$.

From Axiom 8 and the preceding definition, it follows that a real number is either positive, negative, or zero.

Illustration 8

In each of the following, there is a statement that is an immediate consequence of one of the field axioms; the axiom that applies is indicated.

a) $5 + (-5) = 0$; additive inverse axiom (Axiom 6)

b) $1 \cdot y = y$; identity element for multiplication (Axiom 5)

c) $x + (y + 2) = (x + y) + 2$; associative law for addition (Axiom 3)

d) $3 + (x + 4) = (x + 4) + 3$; commutative law for addition (Axiom 2)

e) $2(5x) = (2 \cdot 5)x$; associative law for multiplication (Axiom 3)

f) $8(u + 12) = 8u + 8 \cdot 12$; distributive law (Axiom 4) ∎

EXERCISES   A.1

In Exercises 1 through 24, the equality follows immediately from one of the field axioms. Indicate which axiom applies. Assume that each variable is a real number.

1. $4 \cdot 5 = 5 \cdot 4$
2. $(6 + 2) + 4 = 6 + (2 + 4)$
3. $8 + 0 = 8$
4. $1 \cdot y = y$
5. $(5 + 2) + 4 = (2 + 5) + 4$
6. $17 + 41 = 41 + 17$

7. $3(xy) = (3x)y$
8. $(7a)b = b(7a)$
9. $\pi + (-\pi) = 0$
10. $x + (y + x) = (y + x) + x$
11. $7 + (8 + 11) = 7 + (11 + 8)$
12. $b + (-b) = 0$
13. $4 \cdot \frac{1}{4} = 1$
14. $x + 0 = x$
15. $3(a + b) = (a + b)3$
16. $11 + (y + 7) = (y + 7) + 11$

17. $a(b + 0) = ab$
18. $4(x + y) = 4x + 4y$
19. $0 \cdot 1 = 0$
20. $0 + 0 = 0$
21. $w + x(y + z) = w + (xy + xz)$
22. $(r + s)u + t = (s + r)u + t$
23. $(r + s) + (t + u) = r + [s + (t + u)]$
24. $(w + x) + (y + z) = [(w + x) + y] + z$

In Exercises 25 through 36, state whether or not the set is closed under the indicated operation and give an example to illustrate your answer.

25. The set N of natural numbers; addition
26. The set of negative numbers; addition
27. The set of odd natural numbers; multiplication
28. The set of odd natural numbers; addition
29. $\{0\}$; addition
30. $\{0\}$; multiplication
31. $\{1\}$; multiplication
32. $\{1\}$; addition
33. $\{1, 2\}$; addition
34. $\{1, 2\}$; multiplication
35. $\{0, 1\}$; multiplication
36. $\{0, 1\}$; addition

A.2 Linear Interpolation

To obtain an approximation of a trigonometric function value between two entries in either Table VI or Table VII we can apply the method of linear interpolation. This method is demonstrated in the following illustration.

Illustration 1

Suppose we wish to determine an approximation for sin 0.583 by using Table VI. Because sin x is increasing when $0 < x < 1.57$ (that is, x is in the first quadrant),

$$\sin 0.580 < \sin 0.583 < \sin 0.590$$

From Table VI, we find

$$\sin 0.580 \approx 0.5480 \quad \text{and} \quad \sin 0.590 \approx 0.5564$$

Figure 1 shows a portion of the graph of the sine function from the point $P(0.580, 0.5480)$ to the point $Q(0.590, 0.5564)$. In the figure, the units on the axes are magnified and the portion of the graph is distorted in order to demonstrate the procedure. In a more accurate figure, the line segment from P to Q would be much closer to the graph than it is in Figure 1.

The point R on the graph has an abscissa (x coordinate) of 0.583 and the ordinate (y coordinate) of R is sin 0.583. An approximation of the ordinate of R is the ordinate of the point S on the line segment from P to Q that has an abscissa of 0.583. The ordinate of S is represented by

$$0.5480 + d$$

where d units is the length of the line segment TS shown in Figure 1. To compute d, we use a property of similar triangles which states that the lengths of corresponding sides are proportional. Because triangle PTS is similar to triangle PUQ, we have

$$\frac{d}{0.0084} = \frac{0.003}{0.010}$$

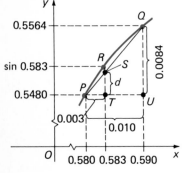

FIGURE 1

Therefore

$$d = \frac{3}{10}(0.0084)$$
$$= 0.0025$$

Thus

$$\sin 0.583 \approx 0.5480 + 0.0025$$
$$= 0.5505 \qquad \blacksquare$$

When performing the calculations involved in linear interpolation it is not necessary to draw a portion of the graph and show the similar triangles as we did in Illustration 1. Instead, arrange the computation as shown in the following example.

Example   1 Determine an approximation for $\tan 0.348$.

Solution Because $\tan x$ is increasing when $0 < x < 1.57$,

$$\tan 0.340 < \tan 0.348 < \tan 0.350$$

We use Table VI to find approximations for $\tan 0.340$ and $\tan 0.350$.

$$0.010 \left\{ 0.008 \left\{ \begin{array}{l} \tan 0.340 \approx 0.3537 \\ \tan 0.348 \approx ? \\ \tan 0.350 \approx 0.3650 \end{array} \right\} d \right\} 0.0113$$

Then

$$\frac{d}{0.0113} = \frac{0.008}{0.010}$$

$$d = \frac{8}{10}(0.0113)$$

$$d = 0.0090$$

Therefore

$$\tan 0.348 \approx 0.3537 + 0.0090$$
$$= 0.3627 \qquad \blacksquare$$

In the next example, where linear interpolation is used to approximate a cosine function value, the number d is subtracted from a value found from the table because the cosine function is decreasing in the first quadrant. Table VII is used because the angle is measured in degrees.

Example   2 Determine an approximation for $\cos 66.42°$.

Solution Because $\cos \theta$ is decreasing when $0° < \theta < 90°$,

$$\cos 66.50° < \cos 66.42° < \cos 66.40°$$

From Table VII we find approximations for $\cos 66.40°$ and $\cos 66.50°$.

$$0.10 \left\{ 0.02 \left\{ \begin{array}{l} \cos 66.40° \approx 0.4003 \\ \cos 66.42° \approx \, ? \\ \cos 66.50° \approx 0.3987 \end{array} \right\} d \right\} 0.0016$$

Hence

$$\frac{d}{0.0016} = \frac{0.02}{0.10}$$

$$d = \frac{2}{10} (0.0016)$$

$$d = 0.0003$$

Therefore

$$\cos 66.42° \approx 0.4003 - 0.0003$$
$$= 0.4000 \qquad \blacksquare$$

Example 3 Determine an approximation for $\sin 27°16'$.

Solution Because $\sin \theta$ is increasing when $0° < \theta < 90°$,

$$\sin 27°12' < \sin 27°16' < \sin 27°18'$$

We use Table VII for approximations of $\sin 27°12'$ and $\sin 27°18'$.

$$6 \left\{ 4 \left\{ \begin{array}{l} \sin 27°12' \approx 0.4571 \\ \sin 27°16' \approx \, ? \\ \sin 27°18' \approx 0.4586 \end{array} \right\} d \right\} 0.0015$$

Thus

$$\frac{d}{0.0015} = \frac{4}{6}$$

$$d = \frac{2}{3} (0.0015)$$

$$d = 0.0010$$

Therefore

$$\sin 27°16' \approx 0.4571 + 0.0010$$
$$= 0.4581 \qquad \blacksquare$$

We can use linear interpolation to find an approximation of the number or angle having a trigonometric function value that is not a table entry. The next two examples show the procedure.

Example 4 Find an approximation for the real number x given that $\cos x = 0.8856$ and $0 < x < 1.57$.

Solution We use Table VI and observe that 0.8856 does not appear in the column headed $\cos x$. However, the numbers 0.8870 and 0.8823 do appear, and they are values of $\cos 0.480$ and $\cos 0.490$, respectively.

$$0.010 \left\{ d \begin{cases} \cos 0.480 \approx 0.8870 \\ \cos x \qquad \approx 0.8856 \end{cases} 0.0014 \\ \cos 0.490 \approx 0.8823 \right\} 0.0047$$

Then

$$\frac{d}{0.010} = \frac{0.0014}{0.0047}$$

$$d = \frac{14}{47}(0.010)$$

$$d = 0.003$$

Therefore

$$x \approx 0.480 + 0.003$$
$$= 0.483 \qquad \blacksquare$$

Example 5 Find an approximation for the acute angle θ in degree measure if $\tan \theta = 0.2743$.

Solution Because θ is measured in degrees we use Table VII. The number 0.2743 does not appear in the column headed $\tan \theta$; however, from the table we observe that $\tan 15.30° \approx 0.2736$ and $\tan 15.40° \approx 0.2754$.

$$0.10 \left\{ d \begin{cases} \cos 15.30° \approx 0.2736 \\ \tan \theta \qquad \approx 0.2743 \end{cases} 0.0007 \\ \tan 15.40° \approx 0.2754 \right\} 0.0018$$

Therefore

$$\frac{d}{0.10} = \frac{0.0007}{0.0018}$$

$$d = \frac{7}{18}(0.10)$$

$$d = 0.04$$

Thus

$$\theta \approx 15.30° + 0.04°$$
$$= 15.34° \qquad \blacksquare$$

EXERCISES A.2

In Exercises 1 through 20, use linear interpolation with Table VI or VII to find an approximation of the function value.

1. sin 0.476
2. tan 1.073
3. cos 0.214
4. cos 0.818
5. tan 0.853
6. cot 0.532
7. cot 0.519
8. sin 1.465
9. sec 1.327
10. csc 0.459
11. cos 37.53°
12. tan 15.36°
13. tan 56.71°
14. sin 72.84°
15. sin 46.82°
16. cos 41.67°
17. csc 80.05°
18. cot 24.93°
19. cot 12.48°
20. sec 86.12°

In Exercises 21 through 30, use linear interpolation with Table VI to find an approximation for the real number x having the function value if $0 < x < 1.57$.

21. cos x = 0.8986
22. sin x = 0.3639
23. tan x = 1.464
24. cos x = 0.4203
25. sin x = 0.6553
26. tan x = 1.620
27. cot x = 2.587
28. sec x = 3.245
29. csc x = 1.183
30. cot x = 0.7951

In Exercises 31 through 40, use linear interpolation with Table VII to find in degree measure an approximation for the acute angle θ having the function value.

31. sin θ = 0.8601
32. tan θ = 0.8412
33. cot θ = 1.3068
34. cos θ = 0.4993
35. cos θ = 0.3740
36. sin θ = 0.1239
37. tan θ = 0.5527
38. cot θ = 3.4986
39. sec θ = 3.9885
40. csc θ = 1.3750

Tables

TABLE I **The Greek Alphabet**

α	alpha	ι	iota	ρ	rho
β	beta	κ	kappa	σ	sigma
γ	gamma	λ	lambda	τ	tau
δ	delta	μ	mu	υ	upsilon
ϵ	epsilon	ν	nu	ϕ	phi
ζ	zeta	ξ	xi	χ	chi
η	eta	o	omicron	ψ	psi
θ	theta	π	pi	ω	omega

TABLES

TABLE II   Powers and Roots

n	n^2	$\sqrt{n}$	n^3	$\sqrt[3]{n}$	n	n^2	$\sqrt{n}$	n^3	$\sqrt[3]{n}$
1	1	1.000	1	1.000	51	2,601	7.141	132,651	3.708
2	4	1.414	8	1.260	52	2,704	7.211	140,608	3.732
3	9	1.732	27	1.442	53	2,809	7.280	148,877	3.756
4	16	2.000	64	1.587	54	2,916	7.348	157,464	3.780
5	25	2.236	125	1.710	55	3,025	7.416	166,375	3.803
6	36	2.449	216	1.817	56	3,136	7.483	175,616	3.826
7	49	2.646	343	1.913	57	3,249	7.550	185,193	3.848
8	64	2.828	512	2.000	58	3,364	7.616	195,112	3.871
9	81	3.000	729	2.080	59	3,481	7.681	205,379	3.893
10	100	3.162	1,000	2.154	60	3,600	7.746	216,000	3.915
11	121	3.317	1,331	2.224	61	3,721	7.810	226,981	3.936
12	144	3.464	1,728	2.289	62	3,844	7.874	238,328	3.958
13	169	3.606	2,197	2.351	63	3,969	7.937	250,047	3.979
14	196	3.742	2,744	2.410	64	4,096	8.000	262,144	4.000
15	225	3.873	3,375	2.466	65	4,225	8.062	274,625	4.021
16	256	4.000	4,096	2.520	66	4,356	8.124	287,496	4.041
17	289	4.123	4,913	2.571	67	4,489	8.185	300,763	4.062
18	324	4.243	5,832	2.621	68	4,624	8.246	314,432	4.082
19	361	4.359	6,859	2.668	69	4,761	8.307	328,509	4.102
20	400	4.472	8,000	2.714	70	4,900	8.367	343,000	4.121
21	441	4.583	9,261	2.759	71	5,041	8.426	357,911	4.141
22	484	4.690	10,648	2.802	72	5,184	8.485	373,248	4.160
23	529	4.796	12,167	2.844	73	5,329	8.544	389,017	4.179
24	576	4.899	13,824	2.884	74	5,476	8.602	405,224	4.198
25	625	5.000	15,625	2.924	75	5,625	8.660	421,875	4.217
26	676	5.099	17,576	2.962	76	5,776	8.718	438,976	4.236
27	729	5.196	19,683	3.000	77	5,929	8.775	456,533	4.254
28	784	5.291	21,952	3.037	78	6,084	8.832	474,552	4.273
29	841	5.385	24,389	3.072	79	6,241	8.888	493,039	4.291
30	900	5.477	27,000	3.107	80	6,400	8.944	512,000	4.309
31	961	5.568	29,791	3.141	81	6,561	9.000	531,441	4.327
32	1,024	5.657	32,768	3.175	82	6,724	9.055	551,368	4.344
33	1,089	5.745	35,937	3.208	83	6,889	9.110	571,787	4.362
34	1,156	5.831	39,304	3.240	84	7,056	9.165	592,704	4.380
35	1,255	5.916	42,875	3.271	85	7,225	9.220	614,125	4.397
36	1,296	6.000	46,656	3.302	86	7,396	9.274	636,056	4.414
37	1,369	6.083	50,653	3.332	87	7,569	9.327	658,503	4.431
38	1,444	6.164	54,872	3.362	88	7,744	9.381	681,472	4.448
39	1,521	6.245	59,319	3.391	89	7,921	9.434	704,969	4.465
40	1,600	6.325	64,000	3.420	90	8,100	9.487	729,000	4.481
41	1,681	6.403	68,921	3.448	91	8,281	9.539	753,571	4.498
42	1,764	6.481	74,088	3.476	92	8,464	9.592	778,688	4.514
43	1,849	6.557	79,507	3.503	93	8,649	9.643	804,357	4.531
44	1,936	6.633	85,184	3.530	94	8,836	9.695	830,584	4.547
45	2,025	6.708	91,125	3.557	95	9,025	9.747	857,375	4.563
46	2,116	6.782	97,336	3.583	96	9,216	9.798	884,736	4.579
47	2,209	6.856	103,823	3.609	97	9,409	9.849	912,673	4.595
48	2,304	6.928	110,592	3.634	98	9,604	9.899	941,192	4.610
49	2,401	7.000	117,649	3.659	99	9,801	9.950	970,299	4.626
50	2,500	7.071	125,000	3.684	100	10,000	10.000	1,000,000	4.642

TABLE III   Common Logarithms

N	0	1	2	3	4	5	6	7	8	9
10	0000	0043	0086	0128	0170	0212	0253	0294	0334	0374
11	0414	0453	0492	0531	0569	0607	0645	0682	0719	0755
12	0792	0828	0864	0899	0934	0969	1004	1038	1072	1106
13	1139	1173	1206	1239	1271	1303	1335	1367	1399	1430
14	1461	1492	1523	1553	1584	1614	1644	1673	1703	1732
15	1761	1790	1818	1847	1875	1903	1931	1959	1987	2014
16	2041	2068	2095	2122	2148	2175	2201	2227	2253	2279
17	2304	2330	2355	2380	2405	2430	2455	2480	2504	2529
18	2553	2577	2601	2625	2648	2672	2695	2718	2742	2765
19	2788	2810	2833	2856	2878	2900	2923	2945	2967	2989
20	3010	3032	3054	3075	3096	3118	3139	3160	3181	3201
21	3222	3243	3263	3284	3304	3324	3345	3365	3385	3404
22	3424	3444	3464	3483	3502	3522	3541	3560	3579	3598
23	3617	3636	3655	3674	3692	3711	3729	3747	3766	3784
24	3802	3820	3838	3856	3874	3892	3909	3927	3945	3962
25	3979	3997	4014	4031	4048	4065	4082	4099	4116	4133
26	4150	4166	4183	4200	4216	4232	4249	4265	4281	4298
27	4314	4330	4346	4362	4378	4393	4409	4425	4440	4456
28	4472	4487	4502	4518	4533	4548	4564	4579	4594	4609
29	4624	4639	4654	4669	4683	4698	4713	4728	4742	4757
30	4771	4786	4800	4814	4829	4843	4857	4871	4886	4900
31	4914	4928	4942	4955	4969	4983	4997	5011	5024	5038
32	5051	5065	5079	5092	5105	5119	5132	5145	5159	5172
33	5185	5198	5211	5224	5237	5250	5263	5276	5289	5302
34	5315	5328	5340	5353	5366	5378	5391	5403	5416	5428
35	5441	5453	5465	5478	5490	5502	5514	5527	5539	5551
36	5563	5575	5587	5599	5611	5623	5635	5647	5658	5670
37	5682	5694	5705	5717	5729	5740	5752	5763	5775	5786
38	5798	5809	5821	5832	5843	5855	5866	5877	5888	5899
39	5911	5922	5933	5944	5955	5966	5977	5988	5999	6010
40	6021	6031	6042	6053	6064	6075	6085	6096	6107	6117
41	6128	6138	6149	6160	6170	6180	6191	6201	6212	6222
42	6232	6243	6253	6263	6274	6284	6294	6304	6314	6325
43	6335	6345	6355	6365	6375	6385	6395	6405	6415	6425
44	6435	6444	6454	6464	6474	6484	6493	6503	6513	6522
45	6532	6542	6551	6561	6571	6580	6590	6599	6609	6618
46	6628	6637	6646	6656	6665	6675	6684	6693	6702	6712
47	6721	6730	6739	6749	6758	6767	6776	6785	6794	6803
48	6812	6821	6830	6839	6848	6857	6866	6875	6884	6893
49	6902	6911	6920	6928	6937	6946	6955	6964	6972	6981
50	6990	6998	7007	7016	7024	7033	7042	7050	7059	7067
51	7076	7084	7093	7101	7110	7118	7126	7135	7143	7152
52	7160	7168	7177	7185	7193	7202	7210	7218	7226	7235
53	7243	7251	7259	7267	7275	7284	7292	7300	7308	7316
54	7324	7332	7340	7348	7356	7364	7372	7380	7388	7396

TABLES

TABLE III (*Continued*) **Common Logarithms**

N	0	1	2	3	4	5	6	7	8	9
55	7404	7412	7419	7427	7435	7443	7451	7459	7466	7474
56	7482	7490	7497	7505	7513	7520	7528	7536	7543	7551
57	7559	7566	7574	7582	7589	7597	7604	7612	7619	7627
58	7634	7642	7649	7657	7664	7672	7679	7686	7694	7701
59	7709	7716	7723	7731	7738	7745	7752	7760	7767	7774
60	7782	7789	7796	7803	7810	7818	7825	7832	7839	7846
61	7853	7860	7868	7875	7882	7889	7896	7903	7910	7917
62	7924	7931	7938	7945	7952	7959	7966	7973	7980	7987
63	7993	8000	8007	8014	8021	8028	8035	8041	8048	8055
64	8062	8069	8075	8082	8089	8096	8102	8109	8116	8122
65	8129	8136	8142	8149	8156	8162	8169	8176	8182	8189
66	8195	8202	8209	8215	8222	8228	8235	8241	8248	8254
67	8261	8267	8274	8280	8287	8293	8299	8306	8312	8319
68	8325	8331	8338	8344	8351	8357	8363	8370	8376	8382
69	8388	8395	8401	8407	8414	8420	8426	8432	8439	8445
70	8451	8457	8463	8470	8476	8482	8488	8494	8500	8506
71	8513	8519	8525	8531	8537	8543	8549	8555	8561	8567
72	8573	8579	8585	8591	8597	8603	8609	8615	8621	8627
73	8633	8639	8645	8651	8657	8663	8669	8675	8681	8686
74	8692	8698	8704	8710	8716	8722	8727	8733	8739	8745
75	8751	8756	8762	8768	8774	8779	8785	8791	8797	8802
76	8808	8814	8820	8825	8831	8837	8842	8848	8854	8859
77	8865	8871	8876	8882	8887	8893	8899	8904	8910	8915
78	8921	8927	8932	8938	8943	8949	8954	8960	8965	8971
79	8976	8982	8987	8993	8998	9004	9009	9015	9020	9025
80	9031	9036	9042	9047	9053	9058	9063	9069	9074	9079
81	9085	9090	9096	9101	9106	9112	9117	9122	9128	9133
82	9138	9143	9149	9154	9159	9165	9170	9175	9180	9186
83	9191	9196	9201	9206	9212	9217	9222	9227	9232	9238
84	9243	9248	9253	9258	9263	9269	9274	9279	9284	9289
85	9294	9299	9304	9309	9315	9320	9325	9330	9335	9340
86	9345	9350	9355	9360	9365	9370	9375	9380	9385	9390
87	9395	9400	9405	9410	9415	9420	9425	9430	9435	9440
88	9445	9450	9455	9460	9465	9469	9474	9479	9484	9489
89	9494	9499	9504	9509	9513	9518	9523	9528	9533	9538
90	9542	9547	9552	9557	9562	9566	9571	9576	9581	9586
91	9590	9595	9600	9605	9609	9614	9619	9624	9628	9633
92	9638	9643	9647	9652	9657	9661	9666	9671	9675	9680
93	9685	9689	9694	9699	9703	9708	9713	9717	9722	9727
94	9731	9736	9741	9745	9750	9754	9759	9763	9768	9773
95	9777	9782	9786	9791	9795	9800	9805	9809	9814	9818
96	9823	9827	9832	9836	9841	9845	9850	9854	9859	9863
97	9868	9872	9877	9881	9886	9890	9894	9899	9903	9908
98	9912	9917	9921	9926	9930	9934	9939	9943	9948	9952
99	9956	9961	9965	9969	9974	9978	9983	9987	9991	9996

TABLES

TABLE IV   Natural Logarithms

N	0	1	2	3	4	5	6	7	8	9
1.0	0000	0100	0198	0296	0392	0488	0583	0677	0770	0862
1.1	0953	1044	1133	1222	1310	1398	1484	1570	1655	1740
1.2	1823	1906	1989	2070	2151	2231	2311	2390	2469	2546
1.3	2624	2700	2776	2852	2927	3001	3075	3148	3221	3293
1.4	3365	3436	3507	3577	3646	3716	3784	3853	3920	3988
1.5	4055	4121	4187	4253	4318	4383	4447	4511	4574	4637
1.6	4700	4762	4824	4886	4947	5008	5068	5128	5188	5247
1.7	5306	5365	5423	5481	5539	5596	5653	5710	5766	5822
1.8	5878	5933	5988	6043	6098	6152	6206	6259	6313	6366
1.9	6419	6471	6523	6575	6627	6678	6729	6780	6831	6881
2.0	6931	6981	7031	7080	7129	7178	7227	7275	7324	7372
2.1	7419	7467	7514	7561	7608	7655	7701	7747	7793	7839
2.2	7885	7930	7975	8020	8065	8109	8154	8198	8242	8286
2.3	8329	8372	8416	8459	8502	8544	8587	8629	8671	8713
2.4	8755	8796	8838	8879	8920	8961	9002	9042	9083	9123
2.5	9163	9203	9243	9282	9322	9361	9400	9439	9478	9517
2.6	9555	9594	9632	9670	9708	9746	9783	9821	9858	9895
2.7	9933	9969	*0006	*0043	*0080	*0116	*0152	*0188	*0225	*0260
2.8	1.0296	0332	0367	0403	0438	0473	0508	0543	0578	0613
2.9	0647	0682	0716	0750	0784	0818	0852	0886	0919	0953
3.0	1.0986	1019	1053	1086	1119	1151	1184	1217	1249	1282
3.1	1314	1346	1378	1410	1442	1474	1506	1537	1569	1600
3.2	1632	1663	1694	1725	1756	1787	1817	1848	1878	1909
3.3	1939	1969	2000	2030	2060	2090	2119	2149	2179	2208
3.4	2238	2267	2296	2326	2355	2384	2413	2442	2470	2499
3.5	1.2528	2556	2585	2613	2641	2669	2698	2726	2754	2782
3.6	2809	2837	2865	2892	2920	2947	2975	3002	3029	3056
3.7	3083	3110	3137	3164	3191	3218	3244	3271	3297	3324
3.8	3350	3376	3403	3429	3455	3481	3507	3533	3558	3584
3.9	3610	3635	3661	3686	3712	3737	3762	3788	3813	3838
4.0	1.3863	3888	3913	3938	3962	3987	4012	4036	4061	4085
4.1	4110	4134	4159	4183	4207	4231	4255	4279	4303	4327
4.2	4351	4375	4398	4422	4446	4469	4493	4516	4540	4563
4.3	4586	4609	4633	4656	4679	4702	4725	4748	4770	4793
4.4	4816	4839	4861	4884	4907	4929	4951	4974	4996	5019
4.5	1.5041	5063	5085	5107	5129	5151	5173	5195	5217	5239
4.6	5261	5282	5304	5326	5347	5369	5390	5412	5433	5454
4.7	5476	5497	5518	5539	5560	5581	5602	5623	5644	5665
4.8	5686	5707	5728	5748	5769	5790	5810	5831	5851	5872
4.9	5892	5913	5933	5953	5974	5994	6014	6034	6054	6074
5.0	1.6094	6114	6134	6154	6174	6194	6214	6233	6253	6273
5.1	6292	6312	6332	6351	6371	6390	6409	6429	6448	6467
5.2	6487	6506	6525	6544	6563	6582	6601	6620	6639	6658
5.3	6677	6696	6715	6734	6752	6771	6790	6808	6827	6845
5.4	6864	6882	6901	6919	6938	6956	6974	6993	7011	7029

TABLE IV (*Continued*)   **Natural Logarithms**

N	0	1	2	3	4	5	6	7	8	9
5.5	1.7047	7066	7084	7102	7120	7138	7156	7174	7192	7210
5.6	7228	7246	7263	7281	7299	7317	7334	7352	7370	7387
5.7	7405	7422	7440	7457	7475	7492	7509	7527	7544	7561
5.8	7579	7596	7613	7630	7647	7664	7681	7699	7716	7733
5.9	7750	7766	7783	7800	7817	7834	7851	7867	7884	7901
6.0	1.7918	7934	7951	7967	7984	8001	8017	8034	8050	8066
6.1	8083	8099	8116	8132	8148	8165	8181	8197	8213	8229
6.2	8245	8262	8278	8294	8310	8326	8342	8358	8374	8390
6.3	8405	8421	8437	8453	8469	8485	8500	8516	8532	8547
6.4	8563	8579	8594	8610	8625	8641	8656	8672	8687	8703
6.5	1.8718	8733	8749	8764	8779	8795	8810	8825	8840	8856
6.6	8871	8886	8901	8916	8931	8946	8961	8976	8991	9006
6.7	9021	9036	9051	9066	9081	9095	9110	9125	9140	9155
6.8	9169	9184	9199	9213	9228	9242	9257	9272	9286	9301
6.9	9315	9330	9344	9359	9373	9387	9402	9416	9430	9445
7.0	1.9459	9473	9488	9502	9516	9530	9544	9559	9573	9587
7.1	9601	9615	9629	9643	9657	9671	9685	9699	9713	9727
7.2	9741	9755	9769	9782	9796	9810	9824	9838	9851	9865
7.3	9879	9892	9906	9920	9933	9947	9961	9974	9988	*0001
7.4	2.0015	0028	0042	0055	0069	0082	0096	0109	0122	0136
7.5	2.0149	0162	0176	0189	0202	0215	0229	0242	0255	0268
7.6	0281	0295	0308	0321	0334	0347	0360	0373	0386	0399
7.7	0412	0425	0438	0451	0464	0477	0490	0503	0516	0528
7.8	0541	0554	0567	0580	0592	0605	0618	0630	0643	0656
7.9	0669	0681	0694	0707	0719	0732	0744	0757	0769	0782
8.0	2.0794	0807	0819	0832	0844	0857	0869	0882	0894	0906
8.1	0919	0931	0943	0956	0968	0980	0992	1005	1017	1029
8.2	1041	1054	1066	1078	1090	1102	1114	1126	1138	1150
8.3	1163	1175	1187	1199	1211	1223	1235	1247	1258	1270
8.4	1282	1294	1306	1318	1330	1342	1353	1365	1377	1389
8.5	2.1401	1412	1424	1436	1448	1459	1471	1483	1494	1506
8.6	1518	1529	1541	1552	1564	1576	1587	1599	1610	1622
8.7	1633	1645	1656	1668	1679	1691	1702	1713	1725	1736
8.8	1748	1759	1770	1782	1793	1804	1815	1827	1838	1849
8.9	1861	1872	1883	1894	1905	1917	1928	1939	1950	1961
9.0	2.1972	1983	1994	2006	2017	2028	2039	2050	2061	2072
9.1	2083	2094	2105	2116	2127	2138	2148	2159	2170	2181
9.2	2192	2203	2214	2225	2235	2246	2257	2268	2279	2289
9.3	2300	2311	2322	2332	2343	2354	2364	2375	2386	2396
9.4	2407	2418	2428	2439	2450	2460	2471	2481	2492	2502
9.5	2.2513	2523	2534	2544	2555	2565	2576	2586	2597	2607
9.6	2618	2628	2638	2649	2659	2670	2680	2690	2701	2711
9.7	2721	2732	2742	2752	2762	2773	2783	2793	2803	2814
9.8	2824	2834	2844	2854	2865	2875	2885	2895	2905	2915
9.9	2925	2935	2946	2956	2966	2976	2986	2996	3006	3016

Use ln 10 = 2.30259 to find logarithms of numbers greater than 10 or less than 1. *Example:* ln 220 = ln 2.2 + 2 ln 10 = 0.7885 + 2(2.30259) = 5.3937.

TABLE V Exponential Functions

x	e^x	e^{-x}	x	e^x	e^{-x}	x	e^x	e^{-x}	x	e^x	e^{-x}
0.00	1.0000	1.000000	**0.50**	1.6487	0.606531	**1.00**	2.7183	0.367879	**1.50**	4.4817	0.223130
0.01	1.0101	0.990050	0.51	1.6653	.600496	1.01	2.7456	.364219	1.51	4.5267	.220910
0.02	1.0202	.980199	0.52	1.6820	.594521	1.02	2.7732	.360595	1.52	4.5722	.218712
0.03	1.0305	.970446	0.53	1.6989	.588605	1.03	2.8011	.357007	1.53	4.6182	.216536
0.04	1.0408	.960789	0.54	1.7160	.582748	1.04	2.8292	.353455	1.54	4.6646	.214381
0.05	1.0513	0.951229	**0.55**	1.7333	0.576950	**1.05**	2.8577	0.349938	**1.55**	4.7115	0.212248
0.06	1.0618	.941765	0.56	1.7507	.571209	1.06	2.8864	.346456	1.56	4.7588	.210136
0.07	1.0725	.932394	0.57	1.7683	.565525	1.07	2.9154	.343009	1.57	4.8066	.208045
0.08	1.0833	.923116	0.58	1.7860	.559898	1.08	2.9447	.339596	1.58	4.8550	.205975
0.09	1.0942	.913931	0.59	1.8040	.554327	1.09	2.9743	.336216	1.59	4.9037	.203926
0.10	1.1052	0.904837	**0.60**	1.8221	0.548812	**1.10**	3.0042	0.332871	**1.60**	4.9530	0.201897
0.11	1.1163	.895834	0.61	1.8404	.543351	1.11	3.0344	.329559	1.61	5.0028	.199888
0.12	1.1275	.886920	0.62	1.8589	.537944	1.12	3.0649	.326280	1.62	5.0531	.197899
0.13	1.1388	.878095	0.63	1.8776	.532592	1.13	3.0957	.323033	1.63	5.1039	.195930
0.14	1.1503	.869358	0.64	1.8965	.527292	1.14	3.1268	.319819	1.64	5.1552	.193980
0.15	1.1618	0.860708	**0.65**	1.9155	0.522046	**1.15**	3.1582	0.316637	**1.65**	5.2070	0.192050
0.16	1.1735	.852144	0.66	1.9348	.516851	1.16	3.1899	.313486	1.66	5.2593	.190139
0.17	1.1853	.843665	0.67	1.9542	.511709	1.17	3.2220	.310367	1.67	5.3122	.188247
0.18	1.1972	.835270	0.68	1.9739	.506617	1.18	3.2544	.307279	1.68	5.3656	.186374
0.19	1.2092	.826959	0.69	1.9937	.501576	1.19	3.2871	.304221	1.69	5.4195	.184520
0.20	1.2214	0.818731	**0.70**	2.0138	0.496585	**1.20**	3.3201	0.301194	**1.70**	5.4739	0.182684
0.21	1.2337	.810584	0.71	2.0340	.491644	1.21	3.3535	.298197	1.71	5.5290	.180866
0.22	1.2461	.802519	0.72	2.0544	.486752	1.22	3.3872	.295230	1.72	5.5845	.179066
0.23	1.2586	.794534	0.73	2.0751	.481909	1.23	3.4212	.292293	1.73	5.6407	.177284
0.24	1.2712	.786628	0.74	2.0959	.477114	1.24	3.4556	.289384	1.74	5.6973	.175520
0.25	1.2840	0.778801	**0.75**	2.1170	0.472367	**1.25**	3.4903	0.286505	**1.75**	5.7546	0.173774
0.26	1.2969	.771052	0.76	2.1383	.467666	1.26	3.5254	.283654	1.76	5.8124	.172045
0.27	1.3100	.763379	0.77	2.1598	.463013	1.27	3.5609	.280832	1.77	5.8709	.170333
0.28	1.3231	.755784	0.78	2.1815	.458406	1.28	3.5966	.278037	1.78	5.9299	.168638
0.29	1.3364	.748264	0.79	2.2034	.453845	1.29	3.6328	.275271	1.79	5.9895	.166960
0.30	1.3499	0.740818	**0.80**	2.2255	0.449329	**1.30**	3.6693	0.272532	**1.80**	6.0496	0.165299
0.31	1.3634	.733447	0.81	2.2479	.444858	1.31	3.7062	.269820	1.81	6.1104	.163654
0.32	1.3771	.726149	0.82	2.2705	.440432	1.32	3.7434	.267135	1.82	6.1719	.162026
0.33	1.3910	.718924	0.83	2.2933	.436049	1.33	3.7810	.264477	1.83	6.2339	.160414
0.34	1.4049	.711770	0.84	2.3164	.431711	1.34	3.8190	.261846	1.84	6.2965	.158817
0.35	1.4191	0.704688	**0.85**	2.3396	0.427412	**1.35**	3.8574	0.259240	**1.85**	6.3598	0.157237
0.36	1.4333	.697676	0.86	2.3632	.423162	1.36	3.8962	.256661	1.86	6.4237	.155673
0.37	1.4477	.690734	0.87	2.3869	.418952	1.37	3.9354	.254107	1.87	6.4483	.154124
0.38	1.4623	.683861	0.88	2.4109	.414783	1.38	3.9749	.251579	1.88	6.5535	.152590
0.39	1.4770	.677057	0.89	2.4351	.410656	1.39	4.0149	.249075	1.89	6.6194	.151072
0.40	1.4918	0.670320	**0.90**	2.4596	0.406570	**1.40**	4.0552	0.246597	**1.90**	6.6859	0.149569
0.41	1.5068	.663650	0.91	2.4843	.402524	1.41	4.0960	.244143	1.91	6.7531	.148080
0.42	1.5220	.657047	0.92	2.5093	.398519	1.42	4.1371	.241714	1.92	6.8210	.146607
0.43	1.5373	.650509	0.93	2.5345	.394554	1.43	4.1787	.239309	1.93	6.8895	.145148
0.44	1.5527	.644036	0.94	2.5600	.390628	1.44	4.2207	.236928	1.94	6.9588	.143704
0.45	1.5683	0.637628	**0.95**	2.5857	0.386741	**1.45**	4.2631	0.234570	**1.95**	7.0287	0.142274
0.46	1.5841	.631284	0.96	2.6117	.382893	1.46	4.3060	.232236	1.96	7.0993	.140858
0.47	1.6000	.625002	0.97	2.6379	.379083	1.47	4.3492	.229925	1.97	7.1707	.139457
0.48	1.6161	.618783	0.98	2.6645	.375311	1.48	4.3929	.227638	1.98	7.2427	.138069
0.49	1.6323	.612626	0.99	2.6912	.371577	1.49	4.4371	.225373	1.99	7.3155	.136695
0.50	1.6487	0.606531	**1.00**	2.7183	0.367879	**1.50**	4.4817	0.223130	**2.00**	7.3891	0.135335

TABLES

TABLE V (*Continued*)   Exponential Functions

x	e^x	e^{-x}	x	e^x	e^{-x}	x	e^x	e^{-x}	x	e^x	e^{-x}
2.00	7.3891	0.135335	**2.50**	12.182	0.082085	**3.00**	20.086	0.049787	**3.50**	33.115	0.030197
2.01	7.4633	.133989	2.51	12.305	.081268	3.01	20.287	.049292	3.51	33.448	.029897
2.02	7.5383	.132655	2.52	12.429	.080460	3.02	20.491	.048801	3.52	33.784	.029599
2.03	7.6141	.131336	2.53	12.554	.079659	3.03	20.697	.048316	3.53	34.124	.029305
2.04	7.6906	.130029	2.54	12.680	.078866	3.04	20.905	.047835	3.54	34.467	.029013
2.05	7.7679	0.128735	**2.55**	12.807	0.078082	**3.05**	21.115	0.047359	**3.55**	34.813	0.028725
2.06	7.8460	.127454	2.56	12.936	.077305	3.06	21.328	.046888	3.56	35.163	.028439
2.07	7.9248	.126186	2.57	13.066	.076536	3.07	21.542	.046421	3.57	35.517	.028156
2.08	8.0045	.124930	2.58	13.197	.075774	3.08	21.758	.045959	3.58	35.874	.027876
2.09	8.0849	.123687	2.59	13.330	.075020	3.09	21.977	.045502	3.59	36.234	.027598
2.10	8.1662	0.122456	**2.60**	13.464	0.074274	**3.10**	22.198	0.045049	**3.60**	36.598	0.027324
2.11	8.2482	.121238	2.61	13.599	.073535	3.11	22.421	.044601	3.61	36.966	.027052
2.12	8.3311	.120032	2.62	13.736	.072803	3.12	22.646	.044157	3.62	37.338	.026783
2.13	8.4149	.118837	2.63	13.874	.072078	3.13	22.874	.043718	3.63	37.713	.026516
2.14	8.4994	.117655	2.64	14.013	.071361	3.14	23.104	.043283	3.64	38.092	.026252
2.15	8.5849	0.116484	**2.65**	14.154	0.070651	**3.15**	23.336	0.042852	**3.65**	38.475	0.025991
2.16	8.6711	.115325	2.66	14.296	.069948	3.16	23.571	.042426	3.66	38.861	.025733
2.17	8.7583	.114178	2.67	14.440	.069252	3.17	23.807	.042004	3.67	39.252	.025476
2.18	8.8463	.113042	2.68	14.585	.068563	3.18	24.047	.041586	3.68	39.646	.025223
2.19	8.9352	.111917	2.69	14.732	.067881	3.19	24.288	.041172	3.69	40.045	.024972
2.20	9.0250	0.110803	**2.70**	14.880	0.067206	**3.20**	24.533	0.040764	**3.70**	40.447	0.024724
2.21	9.1157	.109701	2.71	15.029	.066537	3.21	24.779	.040357	3.71	40.854	.024478
2.22	9.2073	.108609	2.72	15.180	.065875	3.22	25.028	.039955	3.72	41.264	.024234
2.23	9.2999	.107528	2.73	15.333	.065219	3.23	25.280	.039557	3.73	41.679	.023993
2.24	9.3933	.106459	2.74	15.487	.064570	3.24	25.534	.039164	3.74	42.098	.023754
2.25	9.4877	0.105399	**2.75**	15.643	0.063928	**3.25**	25.790	0.038774	**3.75**	42.521	0.023518
2.26	9.5831	.104350	2.76	15.800	.063292	3.26	26.050	.038388	3.76	42.948	.023284
2.27	9.6794	.103312	2.77	15.959	.062662	3.27	26.311	.038006	3.77	43.380	.023052
2.28	9.7767	.102284	2.78	16.119	.062039	3.28	26.576	.037628	3.78	43.816	.022823
2.29	9.8749	.101266	2.79	16.281	.061421	3.29	26.843	.037254	3.79	44.256	.022596
2.30	9.9742	0.100259	**2.80**	16.445	0.060810	**3.30**	27.113	0.036883	**3.80**	44.701	0.022371
2.31	10.074	.099261	2.81	16.610	.060205	3.31	27.385	.036516	3.81	45.150	.022148
2.32	10.176	.098274	2.82	16.777	.059606	3.32	27.660	.036153	3.82	45.604	.021928
2.33	10.278	.097296	2.83	16.945	.059013	3.33	27.938	.035793	3.83	46.063	.021710
2.34	10.381	.096328	2.84	17.116	.058426	3.34	28.219	.035437	3.84	46.525	.021494
2.35	10.486	0.095369	**2.85**	17.288	0.057844	**3.35**	28.503	0.035084	**3.85**	46.993	0.021280
2.36	10.591	.094420	2.86	17.462	.057269	3.36	28.789	.034735	3.86	47.465	.021068
2.37	10.697	.093481	2.87	17.637	.056699	3.37	29.079	.034390	3.87	47.942	.020858
2.38	10.805	.092551	2.88	17.814	.056135	3.38	29.371	.034047	3.88	48.424	.020651
2.39	10.913	.091630	2.89	17.993	.055576	3.39	29.666	.033709	3.89	48.911	.020445
2.40	11.023	0.090718	**2.90**	18.174	0.055023	**3.40**	29.964	0.033373	**3.90**	49.402	0.020242
2.41	11.134	.089815	2.91	18.357	.054476	3.41	30.265	.033041	3.91	49.899	.020041
2.42	11.246	.088922	2.92	18.541	.053934	3.42	30.569	.032712	3.92	50.400	.019840
2.43	11.359	.088037	2.93	18.728	.053397	3.43	30.877	.032387	3.93	50.907	.019644
2.44	11.473	.087161	2.94	18.916	.052866	3.44	31.187	.032065	3.94	51.419	.019448
2.45	11.588	0.086294	**2.95**	19.106	0.052340	**3.45**	31.500	0.031746	**3.95**	51.935	0.019255
2.46	11.705	.085435	2.96	19.298	.051819	3.46	31.817	.031430	3.96	52.457	.019063
2.47	11.822	.084585	2.97	19.492	.051303	3.47	32.137	.031117	3.97	52.985	.018873
2.48	11.941	.083743	2.98	19.688	.050793	3.48	32.460	.030807	3.98	53.517	.018686
2.49	12.061	.082910	2.99	19.886	.050287	3.49	32.786	.030501	3.99	54.055	.018500
2.50	12.182	0.082085	**3.00**	20.086	0.049787	**3.50**	33.115	0.030197	**4.00**	54.598	0.018316

(*Continued*)

TABLE V (*Continued*) **Exponential Functions**

x	e^x	e^{-x}	x	e^x	e^{-x}	x	e^x	e^{-x}	x	e^x	e^{-x}
4.00	54.598	0.018316	4.50	90.017	0.011109	5.00	148.41	0.006738	5.50	244.69	0.0040868
4.01	55.147	.018133	4.51	90.922	.010998	5.01	149.90	.006671	5.55	257.24	.0038875
4.02	55.701	.017953	4.52	91.836	.010889	5.02	151.41	.006605	5.60	270.43	.0036979
4.03	56.261	.017774	4.53	92.759	.010781	5.03	152.93	.006539	5.65	284.29	.0035175
4.04	56.826	.017597	4.54	93.691	.010673	5.04	154.47	.006474	5.70	298.87	.0033460
4.05	57.397	0.017422	4.55	94.632	0.010567	5.05	156.02	0.006409	5.75	314.19	0.0031828
4.06	57.974	.017249	4.56	95.583	.010462	5.06	157.59	.006346	5.80	330.30	.0030276
4.07	58.577	.017077	4.57	96.544	.010358	5.07	159.17	.006282	5.85	347.23	.0028799
4.08	59.145	.016907	4.58	97.514	.010255	5.08	160.77	.006220	5.90	365.04	.0027394
4.09	59.740	.016739	4.59	98.494	.010153	5.09	162.39	.006158	5.95	383.75	.0026058
4.10	60.340	0.016573	4.60	99.484	0.010052	5.10	164.02	0.006097	6.00	403.43	0.0024788
4.11	60.947	.016408	4.61	100.48	.009952	5.11	165.67	.006036	6.05	424.11	.0023579
4.12	61.559	.016245	4.62	101.49	.009853	5.12	167.34	.005976	6.10	445.86	.0022429
4.13	62.178	.016083	4.63	102.51	.009755	5.13	169.02	.005917	6.15	468.72	.0021335
4.14	62.803	.015923	4.64	103.54	.009658	5.14	170.72	.005858	6.20	492.75	.0020294
4.15	63.434	0.015764	4.65	104.58	0.009562	5.15	172.43	0.005799	6.25	518.01	0.0019305
4.16	64.072	.015608	4.66	105.64	.009466	5.16	174.16	.005742	6.30	544.57	.0018363
4.17	64.715	.015452	4.67	106.70	.009372	5.17	175.91	.005685	6.35	572.49	.0017467
4.18	65.366	.015299	4.68	107.77	.009279	5.18	177.68	.005628	6.40	601.85	.0016616
4.19	66.023	.015146	4.69	108.85	.009187	5.19	179.47	.005572	6.45	632.70	.0015805
4.20	66.686	0.014996	4.70	109.95	0.009095	5.20	181.27	0.005517	6.50	665.14	0.0015034
4.21	67.357	.014846	4.71	111.05	.009005	5.21	183.09	.005462	6.55	699.24	.0014301
4.22	68.033	.014699	4.72	112.17	.008915	5.22	184.93	.005407	6.60	735.10	.0013604
4.23	68.717	.014552	4.73	113.30	.008826	5.23	186.79	.005354	6.65	772.78	.0012940
4.24	69.408	.014408	4.74	114.43	.008739	5.24	188.67	.005300	6.70	812.41	.0012309
4.25	70.105	0.014264	4.75	115.58	0.008652	5.25	190.57	0.005248	6.75	854.06	0.0011709
4.26	70.810	.014122	4.76	116.75	.008566	5.26	192.48	.005195	6.80	897.85	.0011138
4.27	71.522	.013982	4.77	117.92	.008480	5.27	194.42	.005144	6.85	943.88	.0010595
4.28	72.240	.013843	4.78	119.10	.008396	5.28	196.37	.005092	6.90	992.27	.0010078
4.29	72.966	.013705	4.79	120.30	.008312	5.29	198.34	.005042	6.95	1043.1	.0009586
4.30	73.700	0.013569	4.80	121.51	0.008230	5.30	200.34	0.004992	7.00	1096.6	0.0009199
4.31	74.440	.013434	4.81	122.73	.008148	5.31	202.35	.004942	7.05	1152.9	.0008674
4.32	75.189	.013300	4.82	123.97	.008067	5.32	204.38	.004893	7.10	1212.0	.0008251
4.33	75.944	.013168	4.83	125.21	.007987	5.33	206.44	.004844	7.15	1274.1	.0007849
4.34	76.708	.013037	4.84	126.47	.007907	5.34	208.51	.004796	7.20	1339.4	.0007466
4.35	77.478	0.012907	4.85	127.74	0.007828	5.35	210.61	0.004748	7.25	1408.1	0.0007102
4.36	78.257	.012778	4.86	129.02	.007750	5.36	212.72	.004701	7.30	1480.3	.0006755
4.37	79.044	.012651	4.87	130.32	.007673	5.37	214.86	.004654	7.35	1556.2	.0006426
4.38	79.838	.012525	4.88	131.63	.007597	5.38	217.02	.004608	7.40	1636.0	.0006113
4.39	80.640	.012401	4.89	132.95	.007521	5.39	219.20	.004562	7.45	1719.9	.0005814
4.40	81.451	0.012277	4.90	134.29	0.007477	5.40	221.41	0.004517	7.50	1808.0	0.0005531
4.41	82.269	.012155	4.91	135.64	.007372	5.41	223.63	.004472	7.55	1900.7	.0005261
4.42	83.096	.012034	4.92	137.00	.007299	5.42	225.88	.004427	7.60	1998.2	.0005005
4.43	83.931	.011914	4.93	138.38	.007227	5.43	228.15	.004383	7.65	2100.6	.0004760
4.44	84.775	.011796	4.94	139.77	.007155	5.44	230.44	.004339	7.70	2208.3	.0004528
4.45	85.627	0.011679	4.95	141.17	0.007083	5.45	232.76	0.004296	7.75	2321.6	0.0004307
4.46	86.488	.011562	4.96	142.59	.007013	5.46	235.10	.004254	7.80	2440.6	.0004097
4.47	87.357	.011447	4.97	144.03	.006943	5.47	237.46	.004211	7.85	2565.7	.0003898
4.48	88.235	.011333	4.98	145.47	.006874	5.48	239.85	.004169	7.90	2697.3	.0003707
4.49	89.121	.011221	4.99	146.94	.006806	5.49	242.26	.004128	7.95	2835.6	.0003527
4.50	90.017	0.011109	5.00	148.41	0.006738	5.50	244.69	0.004087	8.00	2981.0	0.0003355

TABLE V (*Continued*)   **Exponential Functions**

x	e^x	e^{-x}	x	e^x	e^{-x}	x	e^x	e^{-x}	x	e^x	e^{-x}
8.00	2981.0	0.0003355	**8.50**	4914.8	0.0002036	**9.00**	8103.1	0.0001234	**9.50**	13360	0.0000749
8.05	3133.8	.0003191	8.55	5166.8	.0001935	9.05	8518.5	.0001174	9.55	14045	.0000712
8.10	3294.5	.0003035	8.60	5431.7	.0001841	9.10	8955.3	.0001117	9.60	14765	.0000677
8.15	3463.4	.0002887	8.65	5710.0	.0001751	9.15	9414.4	.0001062	9.65	15522	.0000644
8.20	3641.0	.0002747	8.70	6002.9	.0001666	9.20	9897.1	.0001010	9.70	16318	.0000613
8.25	3827.6	0.0002613	**8.75**	6310.7	0.0001585	**9.25**	10405	0.0000961	**9.75**	17154	0.0000583
8.30	4023.9	.0002485	8.80	6634.2	.0001507	9.30	10938	.0000914	9.80	18034	.0000555
8.35	4230.2	.0002364	8.85	6974.4	.0001434	9.35	11499	.0000870	9.85	18958	.0000527
8.40	4447.1	.0002249	8.90	7332.0	.0001364	9.40	12088	.0000827	9.90	19930	.0000502
8.45	4675.1	.0002139	8.95	7707.9	.0001297	9.45	12708	.0000787	9.95	20952	0.0000477
8.50	4914.8	0.0002036	**9.00**	8103.1	0.0001234	**9.50**	13360	0.0000749	**10.00**	22026	0.0000454

TABLE VI **Trigonometric Functions of Real Numbers and Angle Measurements in Radians**

θ radians or Real Number x	θ degrees	$\sin \theta$ or $\sin x$	$\cos \theta$ or $\cos x$	$\tan \theta$ or $\tan x$	$\csc \theta$ or $\csc x$	$\sec \theta$ or $\sec x$	$\cot \theta$ or $\cot x$
0.00	0° 00′	0.0000	1.000	0.0000	No value	1.000	No value
.01	0° 34′	.0100	1.000	.0100	100.0	1.000	100.0
.02	1° 09′	.0200	0.9998	.0200	50.00	1.000	49.99
.03	1° 43′	.0300	0.9996	.0300	33.34	1.000	33.32
.04	2° 18′	.0400	0.9992	.0400	25.01	1.001	24.99
0.05	2° 52′	0.0500	0.9988	0.0500	20.01	1.001	19.98
.06	3° 26′	.0600	.9982	.0601	16.68	1.002	16.65
.07	4° 01′	.0699	.9976	.0701	14.30	1.002	14.26
.08	4° 35′	.0799	.9968	.0802	12.51	1.003	12.47
.09	5° 09′	.0899	.9960	.0902	11.13	1.004	11.08
0.10	5° 44′	0.0998	0.9950	0.1003	10.02	1.005	9.967
.11	6° 18′	.1098	.9940	.1104	9.109	1.006	9.054
.12	6° 53′	.1197	.9928	.1206	8.353	1.007	8.293
.13	7° 27′	.1296	.9916	.1307	7.714	1.009	7.649
.14	8° 01′	.1395	.9902	.1409	7.166	1.010	7.096
0.15	8° 36′	0.1494	0.9888	0.1511	6.692	1.011	6.617
.16	9° 10′	.1593	.9872	.1614	6.277	1.013	6.197
.17	9° 44′	.1692	.9856	.1717	5.911	1.015	5.826
.18	10° 19′	.1790	.9838	.1820	5.586	1.016	5.495
.19	10° 53′	.1889	.9820	.1923	5.295	1.018	5.200
0.20	11° 28′	0.1987	0.9801	0.2027	5.033	1.020	4.933
.21	12° 02′	.2085	.9780	.2131	4.797	1.022	4.692
.22	12° 36′	.2182	.9759	.2236	4.582	1.025	4.472
.23	13° 11′	.2280	.9737	.2341	4.386	1.027	4.271
.24	13° 45′	.2377	.9713	.2447	4.207	1.030	4.086
0.25	14° 19′	0.2474	0.9689	0.2553	4.042	1.032	3.916
.26	14° 54′	.2571	.9664	.2660	3.890	1.035	3.759
.27	15° 28′	.2667	.9638	.2768	3.749	1.038	3.613
.28	16° 03′	.2764	.9611	.2876	3.619	1.041	3.478
.29	16° 37′	.2860	.9582	.2984	3.497	1.044	3.351
.030	17° 11′	0.2955	0.9553	0.3093	3.384	1.047	3.233
.31	17° 46′	.3051	.9523	.3203	3.278	1.050	3.122
.32	18° 20′	.3146	.9492	.3314	3.179	1.053	3.018
.33	18° 54′	.3240	.9460	.3425	3.086	1.057	2.920
.34	19° 29′	.3335	.9428	.3537	2.999	1.061	2.827
0.35	20° 03′	0.3429	0.9394	0.3650	2.916	1.065	2.740
.36	20° 38′	.3523	.9359	.3764	2.839	1.068	2.657
.37	21° 12′	.3616	.9323	.3879	2.765	1.073	2.578
.38	21° 46′	.3709	.9287	.3994	2.696	1.077	2.504
.39	22° 21′	.3802	.9249	.4111	2.630	1.081	2.433
0.40	22° 55′	0.3894	0.9211	0.4228	2.568	1.086	2.365
.41	23° 29′	.3986	.9171	.4346	2.509	1.090	2.301
.42	24° 04′	.4078	.9131	.4466	2.452	1.095	2.239
.43	24° 38′	.4169	.9090	.4586	2.399	1.100	2.180
.44	25° 13′	.4259	.9048	.4708	2.348	1.105	2.124

NOTE: Reprinted by permission of the publisher from *Trigonometry: An Analytic Approach*, Fourth Edition, by Irving Drooyan, Walter Hadel, and Charles C. Carico, Table II, pp. 320–323. Copyright © 1983 by Macmillan Publishing Company.

TABLE VI (*Continued*) **Trigonometric Functions of Real Numbers and Angle Measurements in Radians**

θ radians or Real Number x	θ degrees	$\sin \theta$ or $\sin x$	$\cos \theta$ or $\cos x$	$\tan \theta$ or $\tan x$	$\csc \theta$ or $\csc x$	$\sec \theta$ or $\sec x$	$\cot \theta$ or $\cot x$
0.45	25° 47′	0.4350	0.9004	0.4831	2.299	1.111	2.070
.46	26° 21′	.4439	.8961	.4954	2.253	1.116	2.018
.47	26° 56′	.4529	.8916	.5080	2.208	1.122	1.969
.48	27° 30′	.4618	.8870	.5206	2.166	1.127	1.921
.49	28° 04′	.4706	.8823	.5334	2.125	1.133	1.875
0.50	28° 39′	0.4794	0.8776	0.5463	2.086	1.139	1.830
.51	29° 13′	.4882	.8727	.5594	2.048	1.146	1.788
.52	29° 48′	.4969	.8678	.5726	2.013	1.152	1.747
.53	30° 22′	.5055	.8628	.5859	1.978	1.159	1.707
.54	30° 56′	.5141	.8577	.5994	1.945	1.166	1.668
0.55	31° 31′	0.5227	0.8525	0.6131	1.913	1.173	1.631
.56	32° 05′	.5312	.8473	.6269	1.883	1.180	1.595
.57	32° 40′	.5396	.8419	.6410	1.853	1.188	1.560
.58	33° 14′	.5480	.8365	.6552	1.825	1.196	1.526
.59	33° 48′	.5564	.8309	.6696	1.797	1.203	1.494
0.60	34° 23′	0.5646	0.8253	0.6841	1.771	1.212	1.462
.61	34° 57′	.5729	.8196	.6989	1.746	1.220	1.431
.62	35° 31′	.5810	.8139	.7139	1.721	1.229	1.401
.63	36° 06′	.5891	.8080	.7291	1.697	1.238	1.372
.64	36° 40′	.5972	.8021	.7445	1.674	1.247	1.343
.065	37° 15′	0.6052	0.7961	0.7602	1.652	1.256	1.315
.66	37° 49′	.6131	.7900	.7761	1.631	1.266	1.288
.67	38° 23′	.6210	.7838	.7923	1.610	1.276	1.262
.68	38° 58′	.6288	.7776	.8087	1.590	1.286	1.237
.69	39° 32′	.6365	.7712	.8253	1.571	1.297	1.212
.070	40° 06′	0.6442	0.7648	0.8423	1.552	1.307	1.187
.71	40° 41′	.6518	.7584	.8595	1.534	1.319	1.163
.72	41° 15′	.6594	.7518	.8771	1.517	1.330	1.140
.73	41° 50′	.6669	.7452	.8949	1.500	1.342	1.117
.74	42° 24′	.6743	.7385	.9131	1.483	1.354	1.095
0.75	42° 58′	0.6816	0.7317	0.9316	1.467	1.367	4.073
.76	43° 33′	.6889	.7248	.9505	1.452	1.380	1.052
.77	44° 07′	.6961	.7179	.9697	1.436	1.393	1.031
.78	44° 41′	.7033	.7109	.9893	1.422	1.407	1.011
.79	45° 16′	.7104	.7038	1.009	1.408	1.421	0.9908
0.80	45° 50′	0.7174	0.6967	1.030	1.394	1.435	0.9712
.81	46° 25′	.7243	.6895	1.050	1.381	1.450	.9520
.82	46° 59′	.7311	.6822	1.072	1.368	1.466	.9331
.83	47° 33′	.7379	.6749	1.093	1.355	1.482	.9146
.84	48° 08′	.7446	.6675	1.116	1.343	1.498	.8964
0.85	48° 42′	0.7513	0.6600	1.138	1.331	1.515	0.8785
.86	49° 16′	.7578	.6524	1.162	1.320	1.533	.8609
.87	49° 51′	.7643	.6448	1.185	1.308	1.551	.8437
.88	50° 25′	.7707	.6372	1.210	1.297	1.569	.8267
.89	51° 00′	.7771	.6294	1.235	1.287	1.589	.8100

(*Continued*)

TABLE VI (*Continued*) **Trigonometric Functions of Real Numbers and Angle Measurements in Radians**

θ radians or Real Number x	θ degrees	$\sin \theta$ or $\sin x$	$\cos \theta$ or $\cos x$	$\tan \theta$ or $\tan x$	$\csc \theta$ or $\csc x$	$\sec \theta$ or $\sec x$	$\cot \theta$ or $\cot x$
0.90	51° 34′	0.7833	0.6216	1.260	1.277	1.609	0.7936
.91	52° 08′	.7895	.6137	1.286	1.267	1.629	.7774
.92	52° 43′	.7956	.6058	1.313	1.257	1.651	.7615
.93	53° 17′	.8016	.5978	1.341	1.247	1.673	.7458
.94	53° 51′	.8076	.5898	1.369	1.238	1.696	.7303
0.95	54° 26′	0.8134	0.5817	1.398	1.229	1.719	0.7151
.96	55° 00′	.8192	.5735	1.428	1.221	1.744	.7001
.97	55° 35′	.8249	.5653	1.459	1.212	1.769	.6853
.98	56° 09′	.8305	.5570	1.491	1.204	1.795	.6707
.99	56° 43′	.8360	.5487	1.524	1.196	1.823	.6563
1.00	57° 18′	0.8415	0.5403	1.557	1.188	1.851	0.6421
1.01	57° 52′	.8468	.5319	1.592	1.181	1.880	.6281
1.02	58° 27′	.8521	.5234	1.628	1.174	1.911	.6142
1.03	59° 01′	.8573	.5148	1.665	1.166	1.942	.6005
1.04	59° 35′	.8624	.5062	1.704	1.160	1.975	.5870
1.05	60° 10′	0.8674	0.4976	1.743	1.153	2.010	0.5736
1.06	60° 44′	.8724	.4889	1.784	1.146	2.046	.5604
1.07	61° 18′	.8772	.4801	1.827	1.140	2.083	.5473
1.08	61° 53′	.8820	.4713	1.871	1.134	2.122	.5344
1.09	62° 27′	.8866	.4625	1.917	1.128	2.162	.5216
1.10	63° 02′	0.8912	0.4536	1.965	1.122	2.205	0.5090
1.11	63° 36′	.8957	.4447	2.014	1.116	2.249	.4964
1.12	64° 10′	.9001	.4357	2.066	1.111	2.295	.4840
1.13	64° 45′	.9044	.4267	2.120	1.106	2.344	.4718
1.14	65° 19′	.9086	.4176	2.176	1.101	2.395	.4596
1.15	65° 53′	0.9128	0.4085	2.234	1.096	2.448	0.4475
1.16	66° 28′	.9168	.3993	2.296	1.091	2.504	.4356
1.17	67° 02′	.9208	.3902	2.360	1.086	2.563	.4237
1.18	67° 37′	.9246	.3809	2.427	1.082	2.625	.4120
1.19	68° 11′	.9284	.3717	2.498	1.077	2.691	.4003
1.20	68° 45′	0.9320	0.3624	2.572	1.073	2.760	0.3888
1.21	69° 20′	.9356	.3530	2.650	1.069	2.833	.3773
1.22	69° 54′	.9391	.3436	2.733	1.065	2.910	.3659
1.23	70° 28′	.9425	.3342	2.820	1.061	2.992	.3546
1.24	71° 03′	.9458	.3248	2.912	1.057	3.079	.3434
1.25	71° 37′	0.9490	0.3153	3.010	1.054	3.171	0.3323
1.26	72° 12′	.9521	.3058	3.113	1.050	3.270	.3212
1.27	72° 46′	.9551	.2963	3.224	1.047	3.375	.3102
1.28	73° 20′	.9580	.2867	3.341	1.044	3.488	.2993
1.29	73° 55′	.9608	.2771	3.467	1.041	3.609	.2884
1.30	74° 29′	0.9636	0.2675	3.602	1.038	3.738	0.2776
1.31	75° 03′	.9662	.2579	3.747	1.035	3.878	.2669
1.32	75° 38′	.9687	.2482	3.903	1.032	4.029	.2562
1.33	76° 12′	.9711	.2385	4.072	1.030	4.193	.2456
1.34	76° 47′	.9735	.2288	4.256	1.027	4.372	.2350

TABLE VI (*Continued*) Trigonometric Functions of Real Numbers and
Angle Measurements in Radians

θ radians or Real Number x	θ degrees	$\sin \theta$ or $\sin x$	$\cos \theta$ or $\cos x$	$\tan \theta$ or $\tan x$	$\csc \theta$ or $\csc x$	$\sec \theta$ or $\sec x$	$\cot \theta$ or $\cot x$
1.35	77° 21′	0.9757	0.2190	4.455	1.025	4.566	0.2245
1.36	77° 55′	.9779	.2092	4.673	1.023	4.779	.2140
1.37	78° 30′	.9799	.1994	4.913	1.021	5.014	.2035
1.38	79° 04′	.9819	.1896	5.177	1.018	5.273	.1931
1.39	79° 38′	.9837	.1798	5.471	1.017	5.561	.1828
1.40	80° 13′	0.9854	0.1700	5.798	1.015	5.883	0.1725
1.41	80° 47′	.9871	.1601	6.165	1.013	6.246	.1622
1.42	81° 22′	.9887	.1502	6.581	1.011	6.657	.1519
1.43	81° 56′	.9901	.1403	7.055	1.010	7.126	.1417
1.44	82° 30′	.9915	.1304	7.602	1.009	7.667	.1315
1.45	83° 05′	0.9927	0.1205	8.238	1.007	8.299	0.1214
1.46	83° 39′	.9939	.1106	8.989	1.006	9.044	.1113
1.47	84° 13′	.9949	.1006	9.887	1.005	9.938	.1001
1.48	84° 48′	.9959	.0907	10.98	1.004	11.03	.0910
1.49	85° 22′	.9967	.0807	12.35	1.003	12.39	.0810
1.50	85° 57′	0.9975	0.0707	14.10	1.003	14.14	0.0709
1.51	86° 31′	.9982	.0608	16.43	1.002	16.46	.0609
1.52	87° 05′	.9987	.0508	19.67	1.001	19.69	.0508
1.53	87° 40′	.9992	.0408	24.50	1.001	24.52	.0408
1.54	88° 14′	.9995	.0308	32.46	1.000	32.48	.0308
1.55	88° 49′	0.9998	0.0208	48.08	1.000	48.09	0.0208
1.56	89° 23′	.9999	.0108	92.62	1.000	92.63	.0108
1.57	89° 57′	1.000	.0008	1256	1.000	1256	.0008

TABLE VII **Trigonometric Functions of Angle Measurements in Degrees**

θ deg	deg	min	$\sin\theta$	$\cos\theta$	$\tan\theta$	$\csc\theta$	$\sec\theta$	$\cot\theta$			
0.0	0	0	0.0000	1.0000	0.0000	No value	1.0000	No value	90	0	90.0
0.1	0	6	0.0017	1.0000	0.0017	572.96	1.0000	572.96	89	54	89.9
0.2	0	12	0.0035	1.0000	0.0035	286.48	1.0000	286.48	89	48	89.8
0.3	0	18	0.0052	1.0000	0.0052	190.99	1.0000	190.98	89	42	89.7
0.4	0	24	0.0070	1.0000	0.0070	143.24	1.0000	143.24	89	36	89.6
0.5	0	30	0.0087	1.0000	0.0087	114.59	1.0000	114.59	89	30	89.5
0.6	0	36	0.0105	0.9999	0.0105	95.495	1.0001	95.490	89	24	89.4
0.7	0	42	0.0122	0.9999	0.0122	81.853	1.0001	81.847	89	18	89.3
0.8	0	48	0.0140	0.9999	0.0140	71.622	1.0001	71.615	89	12	89.2
0.9	0	54	0.0157	0.9999	0.0157	63.665	1.0001	63.657	89	6	89.1
1.0	1	0	0.0175	0.9998	0.0175	57.299	1.0002	57.290	89	0	89.0
1.1	1	6	0.0192	0.9998	0.0192	52.090	1.0002	52.081	88	54	88.9
1.2	1	12	0.0209	0.9998	0.0209	47.750	1.0002	47.740	88	48	88.8
1.3	1	18	0.0227	0.9997	0.0227	44.077	1.0003	44.066	88	42	88.7
1.4	1	24	0.0244	0.9997	0.0244	40.930	1.0003	40.917	88	36	88.6
1.5	1	30	0.0262	0.9997	0.0262	38.202	1.0003	38.188	88	30	88.5
1.6	1	36	0.0279	0.9996	0.0279	35.815	1.0004	35.801	88	24	88.4
1.7	1	42	0.0297	0.9996	0.0297	33.708	1.0004	33.694	88	18	88.3
1.8	1	48	0.0314	0.9995	0.0314	31.836	1.0005	31.821	88	12	88.2
1.9	1	54	0.0332	0.9995	0.0332	30.161	1.0005	30.145	88	6	88.1
2.0	2	0	0.0349	0.9994	0.0349	28.654	1.0006	28.636	88	0	88.0
2.1	2	6	0.0366	0.9993	0.0367	27.290	1.0007	27.271	87	54	87.9
2.2	2	12	0.0384	0.9993	0.0384	26.050	1.0007	26.031	87	48	87.8
2.3	2	18	0.0401	0.9992	0.0402	24.918	1.0008	24.898	87	42	87.7
2.4	2	24	0.0419	0.9991	0.0419	23.880	1.0009	23.859	87	36	87.6
2.5	2	30	0.0436	0.9990	0.0437	22.926	1.0010	22.904	87	30	87.5
2.6	2	36	0.0454	0.9990	0.0454	22.044	1.0010	22.022	87	24	87.4
2.7	2	42	0.0471	0.9989	0.0472	21.229	1.0011	21.205	87	18	87.3
2.8	2	48	0.0488	0.9988	0.0489	20.471	1.0012	20.446	87	12	87.2
2.9	2	54	0.0506	0.9987	0.0507	19.766	1.0013	19.740	87	6	87.1
3.0	3	0	0.0523	0.9986	0.0524	19.107	1.0014	19.081	87	0	87.0
3.1	3	6	0.0541	0.9985	0.0542	18.492	1.0015	18.464	86	54	86.9
3.2	3	12	0.0558	0.9984	0.0559	17.914	1.0016	17.886	86	48	86.8
3.3	3	18	0.0576	0.9983	0.0577	17.372	1.0017	17.343	86	42	86.7
3.4	3	24	0.0593	0.9982	0.0594	16.862	1.0018	16.832	86	36	86.6
3.5	3	30	0.0610	0.9981	0.0612	16.380	1.0019	16.350	86	30	86.5
3.6	3	36	0.0628	0.9980	0.0629	15.926	1.0020	15.895	86	24	86.4
3.7	3	42	0.0645	0.9979	0.0647	15.496	1.0021	15.464	86	18	86.3
3.8	3	48	0.0663	0.9978	0.0664	15.089	1.0022	15.056	86	12	86.2
3.9	3	54	0.0680	0.9977	0.0682	14.703	1.0023	14.669	86	6	86.1
4.0	4	0	0.0698	0.9976	0.0699	14.336	1.0024	14.301	86	0	86.0
4.1	4	6	0.0715	0.9974	0.0717	13.987	1.0026	13.951	85	54	85.9
4.2	4	12	0.0732	0.9973	0.0734	13.654	1.0027	13.617	85	48	85.8
4.3	4	18	0.0750	0.9972	0.0752	13.337	1.0028	13.300	85	42	85.7
4.4	4	24	0.0767	0.9971	0.0769	13.035	1.0030	12.996	85	36	85.6
4.5	4	30	0.0785	0.9969	0.0787	12.746	1.0031	12.706	85	30	85.5
4.6	4	36	0.0802	0.9968	0.0805	12.469	1.0032	12.429	85	24	85.4
4.7	4	42	0.0819	0.9966	0.0822	12.204	1.0034	12.163	85	18	85.3
4.8	4	48	0.0837	0.9965	0.0840	11.951	1.0035	11.909	85	12	85.2
4.9	4	54	0.0854	0.9963	0.0857	11.707	1.0037	11.665	85	6	85.1
			$\cos\theta$	$\sin\theta$	$\cot\theta$	$\sec\theta$	$\csc\theta$	$\tan\theta$	deg	min	θ deg

NOTE: Reprinted by permission of the publisher from *Trigonometry: An Analytic Approach*, Fourth Edition, by Irving Drooyan, Walter Hadel, and Charles C. Carico, Table III, pp. 324–332. Copyright © 1983 by Macmillan Publishing Company.

TABLE VII (Continued)   Trigonometric Functions of Angle Measurements in Degrees

θ deg	deg	min	sin θ	cos θ	tan θ	csc θ	sec θ	cot θ			θ deg
5.0	5	0	0.0872	0.9962	0.0875	11.474	1.0038	11.430	85	0	85.0
5.1	5	6	0.0889	0.9960	0.0892	11.249	1.0040	11.205	84	54	84.9
5.2	5	12	0.0906	0.9959	0.0910	11.034	1.0041	10.988	84	48	84.8
5.3	5	18	0.0924	0.9957	0.0928	10.826	1.0043	10.780	84	42	84.7
5.4	5	24	0.0941	0.9956	0.0945	10.626	1.0045	10.579	84	36	84.6
5.5	5	30	0.0958	0.9954	0.0963	10.433	1.0046	10.385	84	30	84.5
5.6	5	36	0.0976	0.9952	0.0981	10.248	1.0048	10.199	84	24	84.4
5.7	5	42	0.0993	0.9951	0.0998	10.069	1.0050	10.019	84	18	84.3
5.8	5	48	0.1011	0.9949	0.1016	9.8955	1.0051	9.8448	84	12	84.2
5.9	5	54	0.1028	0.9947	0.1033	9.7283	1.0053	9.6768	84	6	84.1
6.0	6	0	0.1045	0.9945	0.1051	9.5668	1.0055	9.5144	84	0	84.0
6.1	6	6	0.1063	0.9943	0.1069	9.4105	1.0057	9.3573	83	54	83.9
6.2	6	12	0.1080	0.9942	0.1086	9.2593	1.0059	9.2052	83	48	83.8
6.3	6	18	0.1097	0.9940	0.1104	9.1129	1.0061	9.0579	83	42	83.7
6.4	6	24	0.1115	0.9938	0.1122	8.9711	1.0063	8.9152	83	36	83.6
6.5	6	30	0.1132	0.9936	0.1139	8.8337	1.0065	8.7769	83	30	83.5
6.6	6	36	0.1149	0.9934	0.1157	8.7004	1.0067	8.6428	83	24	83.4
6.7	6	42	0.1167	0.9932	0.1175	8.5711	1.0069	8.5126	83	18	83.3
6.8	6	48	0.1184	0.9930	0.1192	8.4457	1.0071	8.3863	83	12	83.2
6.9	6	54	0.1201	0.9928	0.1210	8.3238	1.0073	8.2636	83	6	83.1
7.0	7	0	0.1219	0.9925	0.1228	8.2055	1.0075	8.1444	83	0	83.0
7.1	7	6	0.1236	0.9923	0.1246	8.0905	1.0077	8.0285	82	54	82.9
7.2	7	12	0.1253	0.9921	0.1263	7.9787	1.0079	7.9158	82	48	82.8
7.3	7	18	0.1271	0.9919	0.1281	7.8700	1.0082	7.8062	82	42	82.7
7.4	7	24	0.1288	0.9917	0.1299	7.7642	1.0084	7.6996	82	36	82.6
7.5	7	30	0.1305	0.9914	0.1317	7.6613	1.0086	7.5958	82	30	82.5
7.6	7	36	0.1323	0.9912	0.1334	7.5611	1.0089	7.4947	82	24	82.4
7.7	7	42	0.1340	0.9910	0.1352	7.4635	1.0091	7.3962	82	18	82.3
7.8	7	48	0.1357	0.9907	0.1370	7.3684	1.0093	7.3002	82	12	82.2
7.9	7	54	0.1374	0.9905	0.1388	7.2757	1.0096	7.2066	82	6	82.1
8.0	8	0	0.1392	0.9903	0.1405	7.1853	1.0098	7.0264	82	0	82.0
8.1	8	6	0.1409	0.9900	0.1423	7.0972	1.0101	7.0264	81	54	81.9
8.2	8	12	0 1426	0.9898	0.1441	7.0112	1.0103	6.9395	81	48	81.8
8.3	8	18	0.1444	0.9895	0.1459	6.9273	1.0106	6.8548	81	42	81.7
8.4	8	24	0.1461	0.9893	0.1477	6.8454	1.0108	6.7720	81	36	81.6
8.5	8	30	0.1478	0.9890	0.1495	6.7655	1.0111	6.6912	81	30	81.5
8.6	8	36	0.1495	0.9888	0.1512	6.6874	1.0114	6.6122	81	24	81.4
8.7	8	42	0.1513	0.9885	0.1530	6.6111	1.0116	6.5350	81	18	81.3
8.8	8	48	0.1530	0.9882	0.1548	6.5366	1.0119	6.4596	81	12	81.2
8.9	8	54	0.1547	0.9880	0.1566	6.4637	1.0122	6.3859	81	6	81.1
9.0	9	0	0.1564	0.9877	0.1584	6.3925	1.0125	6.3138	81	0	81.0
9.1	9	6	0.1582	0.9874	0.1602	6.3228	1.0127	6.2432	80	54	80.9
9.2	9	12	0.1599	0.9871	0.1620	6.2547	1.0130	6.1742	80	48	80.8
9.3	9	18	0.1616	0.9869	0.1638	6.1880	1.0133	6.1066	80	42	80.7
9.4	9	24	0.1633	0.9866	0.1655	6.1227	1.0136	6.0405	80	36	80.6
9.5	9	30	0.1650	0.9863	0.1673	6.0589	1.0139	5.9758	80	30	80.5
9.6	9	36	0.1668	0.9860	0.1691	5.9963	1.0142	5.9124	80	24	80.4
9.7	9	42	0.1685	0.9857	0.1709	5.9351	1.0145	5.8502	80	18	80.3
9.8	9	48	0.1702	0.9854	0.1727	5.8751	1.0148	5.7894	80	12	80.2
9.9	9	54	0.1719	0.9851	0.1745	5.8164	1.0151	5.7297	80	6	80.1
			cos θ	sin θ	cot θ	sec θ	csc θ	tan θ	deg	min	θ deg

(Continued)

TABLE VII (*Continued*) **Trigonometric Functions of Angle Measurements in Degrees**

θ deg	deg	min	$\sin \theta$	$\cos \theta$	$\tan \theta$	$\csc \theta$	$\sec \theta$	$\cot \theta$			
10.0	10	0	0.1736	0.9848	0.1763	5.7588	1.0154	5.6713	80	0	80.0
10.1	10	6	0.1754	0.9845	0.1781	5.7023	1.0157	5.6140	79	54	79.9
10.2	10	12	0.1771	0.9842	0.1799	5.6470	1.0161	5.5578	79	48	79.8
10.3	10	18	0.1788	0.9839	0.1817	5.5928	1.0164	5.5027	79	42	79.7
10.4	10	24	0.1805	0.9836	0.1835	5.5396	1.0167	5.4486	79	36	79.6
10.5	10	30	0.1822	0.9833	0.1853	5.4874	1.0170	5.3955	79	30	79.5
10.6	10	36	0.1840	0.9829	0.1871	5.4362	1.0174	5.3435	79	24	79.4
10.7	10	42	0.1857	0.9826	0.1890	5.3860	1.0177	5.2924	79	18	79.3
10.8	10	48	0.1874	0.9823	0.1908	5.3367	1.0180	5.2422	79	12	79.2
10.9	10	54	0.1891	0.9820	0.1926	5.2883	1.0184	5.1929	79	6	79.1
11.0	11	0	0.1908	0.9816	0.1944	5.2408	1.0187	5.1446	79	0	79.0
11.1	11	6	0.1925	0.9813	0.1962	5.1942	1.0191	5.0970	78	54	78.9
11.2	11	12	0.1942	9.9810	0.1980	5.1484	1.0194	5.0504	78	48	78.8
11.3	11	18	0.1959	0.9806	0.1998	5.1034	1.0198	5.0045	78	42	78.7
11.4	11	24	0.1977	0.9803	0.2016	5.0593	1.0201	4.9595	78	36	78.6
11.5	11	30	0.1994	0.9799	0.2035	5.0159	1.0205	4.9152	78	30	78.5
11.6	11	36	0.2011	0.9796	0.2053	4.9732	1.0209	4.8716	78	24	78.4
11.7	11	42	0.2028	0.9792	0.2071	4.9313	1.0212	4.8288	78	18	78.3
11.8	11	48	0.2045	0.9789	0.2089	4.8901	1.0216	4.7867	78	12	78.2
11.9	11	54	0.2062	0.9785	0.2107	4.8496	1.0220	4.7453	78	6	78.1
12.0	12	0	0.2079	0.9781	0.2126	4.8097	1.0223	4.7046	78	0	78.0
12.1	12	6	0.2096	0.9778	0.2144	4.7706	1.0227	4.6646	77	54	77.9
12.2	12	12	0.2113	0.9774	0.2162	4.7321	1.0231	4.6252	77	48	77.8
12.3	12	18	0.2130	0.9770	0.2180	4.6942	1.0235	4.5864	77	42	77.7
12.4	12	24	0.2147	0.9767	0.2199	4.6569	1.0239	4.5483	77	36	77.6
12.5	12	30	0.2164	0.9763	0.2217	4.6202	1.0243	4.5107	77	30	77.5
12.6	12	36	0.2181	0.9759	0.2235	4.5841	1.0247	4.4737	77	24	77.4
12.7	12	42	0.2198	0.9755	0.2254	4.5486	1.0251	4.4374	77	18	77.3
12.8	12	48	0.2215	0.9751	0.2272	4.5137	1.0255	4.4015	77	12	77.2
12.9	12	54	0.2232	0.9748	0.2290	4.4793	1.0259	4.3662	77	6	77.1
13.0	13	0	0.2250	0.9744	0.2309	4.4454	1.0263	4.3315	77	0	77.0
13.1	13	6	0.2267	0.9740	0.2327	4.4121	1.0267	4.2972	76	54	76.9
13.2	13	12	0.2284	0.9736	0.2345	4.3792	1.0271	4.2635	76	48	76.8
13.3	13	18	0.2300	0.9732	0.2364	4.3469	1.0276	4.2303	76	42	76.7
13.4	13	24	0.2317	0.9728	0.2382	4.3150	1.0280	4.1976	76	36	76.6
13.5	13	30	0.2334	0.9724	0.2401	4.2837	1.0284	4.1653	76	30	76.5
13.6	13	36	0.2351	0.9720	0.2419	4.2528	1.0288	4.1335	76	24	76.4
13.7	13	42	0.2368	0.9715	0.2438	4.2223	1.0293	4.1022	76	18	76.3
13.8	13	48	0.2385	0.9711	0.2456	4.1923	1.0297	4.0713	76	12	76.2
13.9	13	54	0.2402	0.9707	0.2475	4.1627	1.0302	4.0408	76	6	76.1
14.0	14	0	0.2419	0.9703	0.2493	4.1336	1.0306	4.0108	76	0	76.0
14.1	14	6	0.2436	0.9699	0.2512	4.1048	1.0311	3.9812	75	54	75.9
14.2	14	12	0.2453	0.9694	0.2530	4.0765	1.0315	3.9520	75	48	75.8
14.3	14	18	0.2470	0.9690	0.2549	4.0486	1.0320	3.9232	75	42	75.7
14.4	14	24	0.2487	0.9686	0.2568	4.0211	1.0324	3.8947	75	36	75.6
14.5	14	30	0.2504	0.9681	0.2586	3.9939	1.0329	3.8667	75	30	75.5
14.6	14	36	0.2521	0.9677	0.2605	3.9672	1.0334	3.8391	75	24	75.4
14.7	14	42	0.2538	0.9673	0.2623	3.9408	1.0338	3.8118	75	18	75.3
14.8	14	48	0.2554	0.9668	0.2642	3.9147	1.0343	3.7849	75	12	75.2
14.9	14	54	0.2571	0.9664	0.2661	3.8890	1.0348	3.7583	75	6	75.1
			$\cos \theta$	$\sin \theta$	$\cot \theta$	$\sec \theta$	$\csc \theta$	$\tan \theta$	deg	min	θ deg

TABLE VII (*Continued*) **Trigonometric Functions of Angle Measurements in Degrees**

θ deg	deg	min	sin θ	cos θ	tan θ	csc θ	sec θ	cot θ			θ deg
15.0	15	0	0.2588	0.9659	0.2679	3.8637	1.0353	3.7321	75	0	75.0
15.1	15	6	0.2605	0.9655	0.2698	3.8387	1.0358	3.7062	74	54	74.9
15.2	15	12	0.2622	0.9650	0.2717	3.8140	1.0363	3.6806	74	48	74.8
15.3	15	18	0.2639	0.9646	0.2736	3.7897	1.0367	3.6554	74	42	74.7
15.4	15	24	0.2656	0.9641	0.2754	3.7657	1.0372	3.6305	74	36	74.6
15.5	15	30	0.2672	0.9636	0.2773	3.7420	1.0377	3.6059	74	30	74.5
15.6	15	36	0.2689	0.9632	0.2792	3.7186	1.0382	3.5816	74	24	74.4
15.7	15	42	0.2706	0.9627	0.2811	3.6955	1.0388	3.5576	74	18	74.3
15.8	15	48	0.2723	0.9622	0.2830	3.6727	1.0393	3.5339	74	12	74.2
15.9	15	54	0.2740	0.9617	0.2849	2.6502	1.0398	3.5105	74	6	74.1
16.0	16	0	0.2756	0.9613	0.2867	3.6280	1.0403	3.4874	74	0	74.0
16.1	16	6	0.2773	0.9608	0.2886	3.6060	1.0408	3.4646	73	54	73.9
16.2	16	12	0.2790	0.9603	0.2905	3.5843	1.0413	3.4420	73	48	73.8
16.3	16	18	0.2807	0.9598	0.2924	3.5629	1.0419	3.4197	73	42	73.7
16.4	16	24	0.2823	0.9593	0.2943	3.5418	1.0424	3.3977	73	36	73.6
16.5	16	30	0.2840	0.9588	0.2962	3.5209	1.0429	3.3759	73	30	73.5
16.6	16	36	0.2857	0.9583	0.2981	3.5003	1.0435	3.3544	73	24	73.4
16.7	16	42	0.2874	0.9578	0.3000	3.4800	1.0440	3.3332	73	18	73.3
16.8	16	48	0.2890	0.9573	0.3019	3.4598	1.0446	3.3122	73	12	73.2
16.9	16	54	0.2907	0.9568	0.3038	3.4399	1.0451	3.2914	73	6	73.1
17.0	17	0	0.2924	0.9563	0.3057	3.4203	1.0457	3.2709	73	0	73.0
17.1	17	6	0.2940	0.9558	0.3076	3.4009	1.0463	3.2506	72	54	72.9
17.2	17	12	0.2957	0.9553	0.3096	3.3817	1.0468	3.2305	72	48	72.8
17.3	17	18	0.2974	0.9548	0.3115	3.3628	1.0474	3.2106	72	42	72.7
17.4	17	24	0.2990	0.9542	0.3134	3.3440	1.0480	3.1910	72	36	72.6
17.5	17	30	0.3007	0.9537	0.3153	3.3255	1.0485	3.1716	72	30	72.5
17.6	17	36	0.3024	0.9532	0.3172	3.3072	1.0491	3.1524	72	24	72.4
17.7	17	42	0.3040	0.9527	0.3191	3.2891	1.0497	3.1334	72	18	72.3
17.8	17	48	0.3057	0.9521	0.3211	3.2712	1.0503	3.1146	72	12	72.2
17.9	17	54	0.3074	0.9516	0.3230	3.2536	1.0509	3.0961	72	6	72.1
18.0	18	0	0.3090	0.9511	0.3249	3.2361	1.0515	3.0777	72	0	72.0
18.1	18	6	0.3107	0.9505	0.3268	3.2188	1.0521	3.0595	71	54	71.9
18.2	18	12	0.3123	0.9500	0.3288	3.2017	1.0527	3.0415	71	48	71.8
18.3	18	18	0.3140	0.9494	0.3307	3.1848	1.0533	3.0237	71	42	71.7
18.4	18	24	0.3156	0.9489	0.3327	3.1681	1.0539	3.0061	71	36	71.6
18.5	18	30	0.3173	0.9483	0.3346	3.1515	1.0545	2.9887	71	30	71.5
18.6	18	36	0.3190	0.9478	0.3365	3.1352	1.0551	2.9714	71	24	71.4
18.7	18	42	0.3206	0.9472	0.3385	3.1190	1.0557	2.9544	71	18	71.3
18.8	18	48	0.3223	0.9466	0.3404	3.1030	1.0564	2.9375	71	12	71.2
18.9	18	54	0.3239	0.9461	0.3424	3.0872	1.0570	2.9208	71	6	71.1
19.0	19	0	0.3256	0.9455	0.3443	3.0716	1.0576	2.9042	71	0	71.0
19.1	19	6	0.3272	0.9449	0.3463	3.0561	1.0583	2.8878	70	54	70.9
19.2	19	12	0.3289	0.9444	0.3482	3.0407	1.0589	2.8716	70	48	70.8
19.3	19	18	0.3305	0.9438	0.3502	3.0256	1.0595	2.8556	70	42	70.7
19.4	19	24	0.3322	0.9432	0.3522	3.0106	1.0602	2.8397	70	36	70.6
19.5	19	30	0.3338	0.9426	0.3541	2.9957	1.0608	2.8239	70	30	70.5
19.6	19	36	0.3355	0.9421	0.3561	2.9811	1.0615	2.8083	70	24	70.4
19.7	19	42	0.3371	0.9415	0.3581	2.9665	1.0622	2.7929	70	18	70.3
19.8	19	48	0.3387	0.9409	0.3600	2.9521	1.0628	2.7776	70	12	70.2
19.9	19	54	0.3404	0.9403	0.3620	2.9379	1.0635	2.7625	70	6	70.1
			cos θ	sin θ	cot θ	sec θ	csc θ	tan θ	deg	min	θ deg

(*Continued*)

TABLE VII (*Continued*) **Trigonometric Functions of Angle Measurements in Degrees**

θ deg	deg	min	sin θ	cos θ	tan θ	csc θ	sec θ	cot θ			
20.0	20	0	0.3420	0.9397	0.3640	2.9238	1.0642	2.7475	**70**	**0**	**70.0**
20.1	20	6	0.3437	0.9391	0.3659	2.9099	1.0649	2.7326	**69**	**54**	**69.9**
20.2	20	12	0.3453	0.9385	0.3679	2.8960	1.0655	2.7179	**69**	**48**	**69.8**
20.3	20	18	0.3469	0.9379	0.3699	2.8824	1.0662	2.7034	**69**	**42**	**69.7**
20.4	20	24	0.3486	0.9373	0.3719	2.8688	1.0669	2.6889	**69**	**36**	**69.6**
20.5	20	30	0.3502	0.9367	0.3739	2.8555	1.0676	2.6746	**69**	**30**	**69.5**
20.6	20	36	0.3518	0.9361	0.3759	2.8422	1.0683	2.6605	**69**	**24**	**69.4**
20.7	20	42	0.3535	0.9354	0.3779	2.8291	1.0690	2.6464	**69**	**18**	**69.3**
20.8	20	48	0.3551	0.9348	0.3799	2.8161	1.0697	2.6325	**69**	**12**	**69.2**
20.9	20	54	0.3567	0.9342	0.3819	2.8032	1.0704	2.6187	**69**	**6**	**69.1**
21.0	21	0	0.3584	0.9336	0.3839	2.7904	1.0711	2.6051	**69**	**0**	**69.0**
21.1	21	6	0.3600	0.9330	0.3859	2.7778	1.0719	2.5916	**68**	**54**	**68.9**
21.2	21	12	0.3616	0.9323	0.3879	2.7653	1.0726	2.5782	**68**	**48**	**68.8**
21.3	21	18	0.3633	0.9317	0.3899	2.7529	1.0733	2.5649	**68**	**42**	**68.7**
21.4	21	24	0.3649	0.9311	0.3919	2.7407	1.0740	2.5517	**68**	**36**	**68.6**
21.5	21	30	0.3665	0.9304	0.3939	2.7285	1.0748	2.5386	**68**	**30**	**68.5**
21.6	21	36	0.3681	0.9298	0.3959	2.7165	1.0755	2.5257	**68**	**24**	**68.4**
21.7	21	42	0.3697	0.9291	0.3979	2.7046	1.0763	2.5129	**68**	**18**	**68.3**
21.8	21	48	0.3714	0.9285	0.4000	2.6927	1.0770	2.5002	**68**	**12**	**68.2**
21.9	21	54	0.3730	0.9278	0.4020	2.6811	1.0778	2.4876	**68**	**6**	**68.1**
22.0	22	0	0.3746	0.9272	0.4040	2.6695	1.0785	2.4751	**68**	**0**	**68.0**
22.1	22	6	0.3762	0.9265	0.4061	2.6580	1.0793	2.4627	**67**	**54**	**67.9**
22.2	22	12	0.3778	0.9259	0.4081	2.6466	1.0801	2.4504	**67**	**48**	**67.8**
22.3	22	18	0.3795	0.9252	0.4101	2.6354	1.0808	2.4383	**67**	**42**	**67.7**
22.4	22	24	0.3811	0.9245	0.4122	2.6242	1.0816	2.4262	**67**	**36**	**67.6**
22.5	22	30	0.3827	0.9239	0.4142	2.6131	1.0824	2.4142	**67**	**30**	**67.5**
22.6	22	36	0.3843	0.9232	0.4163	2.6022	1.0832	2.4023	**67**	**24**	**67.4**
22.7	22	42	0.3859	0.9225	0.4183	2.5913	1.0840	2.3906	**67**	**18**	**67.3**
22.8	22	48	0.3875	0.9219	0.4204	2.5805	1.0848	2.3789	**67**	**12**	**67.2**
22.9	22	54	0.3891	0.9212	0.4224	2.5699	1.0856	2.3673	**67**	**6**	**67.1**
23.0	23	0	0.3907	0.9205	0.4245	2.5593	1.0864	2.3559	**67**	**0**	**67.0**
23.1	23	6	0.3923	0.9198	0.4265	2.5488	1.0872	2.3445	**66**	**54**	**66.9**
23.2	23	12	0.3939	0.9191	0.4286	2.5384	1.0880	2.3332	**66**	**48**	**66.8**
23.3	23	18	0.3955	0.9184	0.4307	2.5282	1.0888	2.3220	**66**	**42**	**66.7**
23.4	23	24	0.3971	0.9178	0.4327	2.5180	1.0896	2.3109	**66**	**36**	**66.6**
23.5	23	30	0.3987	0.9171	0.4348	2.5078	1.0904	2.2998	**66**	**30**	**66.5**
23.6	23	36	0.4003	0.9164	0.4369	2.4978	1.0913	2.2889	**66**	**24**	**66.4**
23.7	23	42	0.4019	0.9157	0.4390	2.4879	1.0921	2.2781	**66**	**18**	**66.3**
23.8	23	48	0.4035	0.9150	0.4411	2.4780	1.0929	2.2673	**66**	**12**	**66.2**
23.9	23	54	0.4051	0.9143	0.4431	2.4683	1.0938	2.2566	**66**	**6**	**66.1**
24.0	24	0	0.4067	0.9135	0.4452	2.4586	1.0946	2.2460	**66**	**0**	**66.0**
24.1	24	6	0.4083	0.9128	0.4473	2.4490	1.0955	2.2355	**65**	**54**	**65.9**
24.2	24	12	0.4099	0.9121	0.4494	2.4395	1.0963	2.2251	**65**	**48**	**65.8**
24.3	24	18	0.4115	0.9114	0.4515	2.4301	1.0972	2.2148	**65**	**42**	**65.7**
24.4	24	24	0.4131	0.9107	0.4536	2.4207	1.0981	2.2045	**65**	**36**	**65.6**
24.5	24	30	0.4147	0.9100	0.4557	2.4114	1.0989	2.1943	**65**	**30**	**65.5**
24.6	24	36	0.4163	0.9092	0.4578	2.4022	1.0998	2.1842	**65**	**24**	**65.4**
24.7	24	42	0.4179	0.9085	0.4599	2.3931	1.1007	2.1742	**65**	**18**	**65.3**
24.8	24	48	0.4195	0.9078	0.4621	2.3841	1.1016	2.1642	**65**	**12**	**65.2**
24.9	24	54	0.4210	0.9070	0.4642	2.3751	1.1025	2.1543	**65**	**6**	**65.1**
			cos θ	sin θ	cot θ	sec θ	csc θ	tan θ	deg	min	θ deg

TABLE VII (*Continued*) Trigonometric Functions of Angle Measurements in Degrees

θ deg	deg	min	$\sin \theta$	$\cos \theta$	$\tan \theta$	$\csc \theta$	$\sec \theta$	$\cot \theta$			
25.0	25	0	0.4226	0.9063	0.4663	2.3662	1.1034	2.1445	65	0	65.0
25.1	25	6	0.4242	0.9056	0.4684	2.3574	1.1043	2.1348	64	54	64.9
25.2	25	12	0.4258	0.9048	0.4706	2.3486	1.1052	2.1251	64	48	64.8
25.3	25	18	0.4274	0.9041	0.4727	2.3400	1.1061	2.1155	64	42	64.7
25.4	25	24	0.4289	0.9033	0.4748	2.3314	1.1070	2.1060	64	36	64.6
25.5	25	30	0.4305	0.9026	0.4770	2.3228	1.1079	2.0965	64	30	64.5
25.6	25	36	0.4321	0.9018	0.4791	2.3144	1.1089	2.0872	64	24	64.4
25.7	25	42	0.4337	0.9011	0.4813	2.3060	1.1098	2.0778	64	18	64.3
25.8	25	48	0.4352	0.9003	0.4834	2.2976	1.1107	2.0686	64	12	64.2
25.9	25	54	0.4368	0.8996	0.4856	2.2894	1.1117	2.0594	64	6	64.1
26.0	26	0	0.4384	0.8988	0.4877	2.2812	1.1126	2.0503	64	0	64.0
26.1	26	6	0.4399	0.8980	0.4899	2.2730	1.1136	2.0413	63	54	63.9
26.2	26	12	0.4415	0.8973	0.4921	2.2650	1.1145	2.0323	63	48	63.8
26.3	26	18	0.4431	0.8965	0.4942	2.2570	1.1155	2.0233	63	42	63.7
26.4	26	24	0.4446	0.8957	0.4964	2.2490	1.1164	2.0145	63	36	63.6
26.5	26	30	0.4462	0.8949	0.4986	2.2412	1.1174	2.0057	63	30	63.5
26.6	26	36	0.4478	0.8942	0.5008	2.2333	1.1184	1.9970	63	24	63.4
26.7	26	42	0.4493	0.8934	0.5029	2.2256	1.1194	1.9883	63	18	63.3
26.8	26	48	0.4509	0.8926	0.5051	2.2179	1.1203	1.9797	63	12	63.2
26.9	26	54	0.4524	0.8918	0.5073	2.2103	1.1213	1.9711	63	6	63.1
27.0	27	0	0.4540	0.8910	0.5095	2.2027	1.1223	1.9626	63	0	63.0
27.1	27	6	0.4555	0.8902	0.5117	2.1952	1.1233	1.9542	62	54	62.9
27.2	27	12	0.4571	0.8894	0.5139	2.1877	1.1243	1.9458	62	48	62.8
27.3	27	18	0.4586	0.8886	0.5161	2.1803	1.1253	1.9375	62	42	62.7
27.4	27	24	0.4602	0.8878	0.5184	2.1730	1.1264	1.9292	62	36	62.6
27.5	27	30	0.4617	0.8870	0.5206	2.1657	1.1274	1.9210	62	30	62.5
27.6	27	36	0.4633	0.8862	0.5228	2.1584	1.1284	1.9128	62	24	62.4
27.7	27	42	0.4648	0.8854	0.5250	2.1513	1.1294	1.9047	62	18	62.3
27.8	27	48	0.4664	0.8846	0.5272	2.1441	1.1305	1.8967	62	12	62.2
27.9	27	54	0.4679	0.8838	0.5295	2.1371	1.1315	1.8887	62	6	62.1
28.0	28	0	0.4695	0.8829	0.5317	2.1301	1.1326	1.8807	62	0	62.0
28.1	28	6	0.4710	0.8821	0.5339	2.1231	1.1336	1.8728	61	54	61.9
28.2	28	12	0.4726	0.8813	0.5362	2.1162	1.1347	1.8650	61	48	61.8
28.3	28	18	0.4741	0.8805	0.5384	2.1093	1.1357	1.8572	61	42	61.7
28.4	28	24	0.4756	0.8796	0.5407	2.1025	1.1368	1.8495	61	36	61.6
28.5	28	30	0.4772	0.8788	0.5430	2.0957	1.1379	1.8418	61	30	61.5
28.6	28	36	0.4787	0.8780	0.5452	2.0890	1.1390	1.8341	61	24	61.4
28.7	28	42	0.4802	0.8771	0.5475	2.0824	1.1401	1.8265	61	18	61.3
28.8	28	48	0.4818	0.8763	0.5498	2.0758	1.1412	1.8190	61	12	61.2
28.9	28	54	0.4833	0.8755	0.5520	2.0692	1.1423	1.8115	61	6	61.1
29.0	29	0	0.4848	0.8746	0.5543	2.0627	1.1434	1.8040	61	0	61.0
29.1	29	6	0.4863	0.8738	0.5566	2.0562	1.1445	1.7966	60	54	60.9
29.2	29	12	0.4879	0.8729	0.5589	2.0598	1.1456	1.7893	60	48	60.8
29.3	29	18	0.4894	0.8721	0.5612	2.0434	1.1467	1.7820	60	42	60.7
29.4	29	24	0.4909	0.8712	0.5635	2.0371	1.1478	1.7747	60	36	60.6
29.5	29	30	0.4924	0.8704	0.5658	2.0308	1.1490	1.7675	60	30	60.5
29.6	29	36	0.4939	0.8695	0.5681	2.0245	1.1501	1.7603	60	24	60.4
29.7	29	42	0.4955	0.8686	0.5704	2.0183	1.1512	1.7532	60	18	60.3
29.8	29	48	0.4970	0.8678	0.5727	2.0122	1.1524	1.7461	60	12	60.2
29.9	29	54	0.4985	0.8669	0.5750	2.0061	1.1535	1.7391	60	6	60.1
			$\cos \theta$	$\sin \theta$	$\cot \theta$	$\sec \theta$	$\csc \theta$	$\tan \theta$	deg	min	θ deg

(Continued)

TABLE VII (*Continued*) **Trigonometric Functions of Angle Measurements in Degrees**

θ deg	deg	min	sin θ	cos θ	tan θ	csc θ	sec θ	cot θ			
30.0	30	0	0.5000	0.8660	0.5774	2.0000	1.1547	1.7321	**60**	**0**	**60.0**
30.1	30	6	0.5015	0.8652	0.5797	1.9940	1.1559	1.7251	59	54	59.9
30.2	30	12	0.5030	0.8643	0.5820	1.9880	1.1570	1.7182	59	48	59.8
30.3	30	18	0.5045	0.8634	0.5844	1.9821	1.1582	1.7113	59	42	59.7
30.4	30	24	0.5060	0.8625	0.5867	1.9762	1.1594	1.7045	59	36	59.6
30.5	30	30	0.5075	0.8616	0.5890	1.9703	1.1606	1.6977	59	30	59.5
30.6	30	36	0.5090	0.8607	0.5914	1.9645	1.1618	1.6909	59	24	59.4
30.7	30	42	0.5105	0.8599	0.5938	1.9587	1.1630	1.6842	59	18	59.3
30.8	30	48	0.5120	0.8590	0.5961	1.9530	1.1642	1.6775	59	12	59.2
30.9	30	54	0.5135	0.8581	0.5985	1.9473	1.1654	1.6709	59	6	59.1
31.0	31	0	0.5150	0.8572	0.6009	1.9416	1.1666	1.6643	59	0	59.0
31.1	31	6	0.5165	0.8563	0.6032	1.9360	1.1679	1.6577	58	54	58.9
31.2	31	12	0.5180	0.8554	0.6056	1.9304	1.1691	1.6512	58	48	58.8
31.3	31	18	0.5195	0.8545	0.6080	1.9249	1.1703	1.6447	58	42	58.7
31.4	31	24	0.5210	0.8536	0.6104	1.9194	1.1716	1.6383	58	36	58.6
31.5	31	30	0.5225	0.8526	0.6128	1.9139	1.1728	1.6319	58	30	58.5
31.6	31	36	0.5240	0.8517	0.6152	1.9084	1.1741	1.6255	58	24	58.4
31.7	31	42	0.5255	0.8508	0.6176	1.9031	1.1753	1.6191	58	18	58.3
31.8	31	48	0.5270	0.8499	0.6200	1.8977	1.1766	1.6128	58	12	58.2
31.9	31	54	0.5284	0.8490	0.6224	1.8924	1.1779	1.6066	58	6	58.1
32.0	32	0	0.5299	0.8480	0.6249	1.8871	1.1792	1.6003	58	0	58.0
32.1	32	6	0.5314	0.8471	0.6273	1.8818	1.1805	1.5941	57	54	57.9
32.2	32	12	0.5329	0.8462	0.6297	1.8766	1.1818	1.5880	57	48	57.8
32.3	32	18	0.5344	0.8453	0.6322	1.8714	1.1831	1.5818	57	42	57.7
32.4	32	24	0.5358	0.8443	0.6346	1.8663	1.1844	1.5757	57	36	57.6
32.5	32	30	0.5373	0.8434	0.6371	1.8612	1.1857	1.5697	57	30	57.5
32.6	32	36	0.5388	0.8425	0.6395	1.8561	1.1870	1.5637	57	24	57.4
32.7	32	42	0.5402	0.8415	0.6420	1.8510	1.1883	1.5577	57	18	57.3
32.8	32	48	0.5417	0.8406	0.6445	1.8460	1.1897	1.5517	57	12	57.2
32.9	32	54	0.5432	0.8396	0.6469	1.8410	1.1910	1.5458	57	6	57.1
33.0	33	0	0.5446	0.8387	0.6494	1.8361	1.1924	1.5399	57	0	57.0
33.1	33	6	0.5461	0.8377	0.6519	1.8312	1.1937	1.5340	56	54	56.9
33.2	33	12	0.5476	0.8368	0.6544	1.8263	1.1951	1.5282	56	48	56.8
33.3	33	18	0.5490	0.8358	0.6569	1.8214	1.1964	1.5224	56	42	56.7
33.4	33	24	0.5505	0.8348	0.6594	1.8166	1.1978	1.5166	56	36	56.6
33.5	33	30	0.5519	0.8339	0.6619	1.8118	1.1992	1.5108	56	30	56.5
33.6	33	36	0.5534	0.8329	0.6644	1.8070	1.2006	1.5051	56	24	56.4
33.7	33	42	0.5548	0.8320	0.6669	1.8023	1.2020	1.4994	56	18	56.3
33.8	33	48	0.5563	0.8310	0.6694	1.7976	1.2034	1.4938	56	12	56.2
33.9	33	54	0.5577	0.8300	0.6720	1.7929	1.2048	1.4882	56	6	56.1
34.0	34	0	0.5592	0.8290	0.6745	1.7883	1.2062	1.4826	56	0	56.0
34.1	34	6	0.5606	0.8281	0.6771	1.7837	1.2076	1.4770	55	54	55.9
34.2	34	12	0.5621	0.8271	0.6796	1.7791	1.2091	1.4715	55	48	55.8
34.3	34	18	0.5635	0.8261	0.6822	1.7745	1.2105	1.4659	55	42	55.7
34.4	34	24	0.5650	0.8251	0.6847	1.7700	1.2120	1.4605	55	36	55.6
34.5	34	30	0.5664	0.8241	0.6873	1.7655	1.2134	1.4550	55	30	55.5
34.6	34	36	0.5678	0.8231	0.6899	1.7610	1.2149	1.4496	55	24	55.4
34.7	34	42	0.5693	0.8221	0.6924	1.7566	1.2163	1.4442	55	18	55.3
34.8	34	48	0.5707	0.8211	0.6950	1.7522	1.2178	1.4388	55	12	55.2
34.9	34	54	0.5721	0.8202	0.6976	1.7478	1.2193	1.4335	55	6	55.1
			cos θ	sin θ	cot θ	sec θ	csc θ	tan θ	deg	min	θ deg

TABLE VII (*Continued*)   Trigonometric Functions of Angle Measurements in Degrees

θ deg	deg	min	sin θ	cos θ	tan θ	csc θ	sec θ	cot θ			
35.0	35	0	0.5736	0.8192	0.7002	1.7434	1.2208	1.4281	55	0	55.0
35.1	35	6	0.5750	0.8181	0.7028	1.7391	1.2223	1.4229	54	54	54.9
35.2	35	12	0.5764	0.8171	0.7054	1.7348	1.2238	1.4176	54	48	54.8
35.3	35	18	0.5779	0.8161	0.7080	1.7305	1.2253	1.4124	54	42	54.7
35.4	35	24	0.5793	0.8151	0.7107	1.7263	1.2268	1.4071	54	36	54.6
35.5	35	30	0.5807	0.8141	0.7133	1.7221	1.2283	1.4019	54	30	54.5
35.6	35	36	0.5821	0.8131	0.7159	1.7179	1.2299	1.3968	54	24	54.4
35.7	35	42	0.5835	0.8121	0.7186	1.7137	1.2314	1.3916	54	18	54.3
35.8	35	48	0.5850	0.8111	0.7212	1.7095	1.2329	1.3865	54	12	54.2
35.9	35	54	0.5864	0.8100	0.7239	1.7054	1.2345	1.3814	54	6	54.1
36.0	36	0	0.5878	0.8090	0.7265	1.7013	1.2361	1.3764	54	0	54.0
36.1	36	6	0.5892	0.8080	0.7292	1.6972	1.2376	1.3713	53	54	53.9
36.2	36	12	0.5906	0.8070	0.7319	1.6932	1.2392	1.3663	53	48	53.8
36.3	36	18	0.5920	0.8059	0.7346	1.6892	1.2408	1.3613	53	42	53.7
36.4	36	24	0.5934	0.8049	0.7373	1.6852	1.2424	1.3564	53	36	53.6
36.5	36	30	0.5948	0.8039	0.7400	1.6812	1.2440	1.3514	53	30	53.5
36.6	36	36	0.5962	0.8028	0.7427	1.6772	1.2456	1.3465	53	24	53.4
36.7	36	42	0.5976	0.8018	0.7454	1.6733	1.2472	1.3416	53	18	53.3
36.8	36	48	0.5990	0.8007	0.7481	1.6694	1.2489	1.3367	53	12	53.2
36.9	36	54	0.6004	0.7997	0.7508	1.6655	1.2505	1.3319	53	6	53.1
37.0	37	0	0.6018	0.7986	0.7536	1.6616	1.2521	1.3270	53	0	53.0
37.1	37	6	0.6032	0.7976	0.7563	1.6578	1.2538	1.3222	52	54	52.9
37.2	37	12	0.6046	0.7965	0.7590	1.6540	1.2554	1.3175	52	48	52.8
37.3	37	18	0.6060	0.7955	0.7618	1.6502	1.2571	1.3127	52	42	52.7
37.4	37	24	0.6074	0.7944	0.7646	1.6464	1.2588	1.3079	52	36	52.6
37.5	37	30	0.6088	0.7934	0.7673	1.6427	1.2605	1.3032	52	30	52.5
37.6	37	36	0.6101	0.7923	0.7701	1.6390	1.2622	1.2985	52	24	52.4
37.7	37	42	0.6115	0.7912	0.7729	1.6353	1.2639	1.2938	52	18	52.3
37.8	37	48	0.6129	0.7902	0.7757	1.6316	1.2656	1.2892	52	12	52.2
37.9	37	54	0.6143	0.7891	0.7785	1.6279	1.2673	1.2846	52	6	52.1
38.0	38	0	0.6157	0.7880	0.7813	1.6243	1.2690	1.2799	52	0	52.0
38.1	38	6	0.6170	0.7869	0.7841	1.6207	1.2708	1.2753	51	54	51.9
38.2	38	12	0.6184	0.7859	0.7869	1.6171	1.2725	1.2708	51	48	51.8
38.3	38	18	0.6198	0.7848	0.7898	1.6135	1.2742	1.2662	51	42	51.7
38.4	38	24	0.6211	0.7837	0.7926	1.6099	1.2760	1.2617	51	36	51.6
38.5	38	30	0.6225	0.7826	0.7954	1.6064	1.2778	1.2572	51	30	51.5
38.6	38	36	0.6239	0.7815	0.7983	1.6029	1.2796	1.2527	51	24	51.4
38.7	38	42	0.6252	0.7804	0.8012	1.5994	1.2813	1.2482	51	18	51.3
38.8	38	48	0.6266	0.7793	0.8040	1.5959	1.2831	1.2437	51	12	51.2
38.9	38	54	0.6280	0.7782	0.8069	1.5925	1.2849	1.2393	51	6	51.1
39.0	39	0	0.6293	0.7771	0.8098	1.5890	1.2868	1.2349	51	0	51.0
39.1	39	6	0.6307	0.7760	0.8127	1.5856	1.2886	1.2305	50	54	50.9
39.2	39	12	0.6320	0.7749	0.8156	1.5822	1.2904	1.2261	50	48	50.8
39.3	39	18	0.6334	0.7738	0.8185	1.5788	1.2923	1.2218	50	42	50.7
39.4	39	24	0.6347	0.7727	0.8214	1.5755	1.2941	1.2174	50	36	50.6
39.5	39	30	0.6361	0.7716	0.8243	1.5721	1.2960	1.2131	50	30	50.5
39.6	39	36	0.6374	0.7705	0.8273	1.5688	1.2978	1.2088	50	24	50.4
39.7	39	42	0.6388	0.7694	0.8302	1.5655	1.2997	1.2045	50	18	50.3
39.8	39	48	0.6401	0.7683	0.8332	1.5622	1.3016	1.2002	50	12	50.2
39.9	39	54	0.6414	0.7672	0.8361	1.5590	1.3035	1.1960	50	6	50.1
			cos θ	sin θ	cot θ	sec θ	csc θ	tan θ	deg	min	θ deg

(Continued)

TABLE VII (*Continued*)   Trigonometric Functions of Angle Measurements in Degrees

θ deg	deg	min	sin θ	cos θ	tan θ	csc θ	sec θ	cot θ			
40.0	40	0	0.6428	0.7660	0.8391	1.5557	1.3054	1.1918	50	0	50.0
40.1	40	6	0.6441	0.7649	0.8421	1.5525	1.3073	1.1875	49	54	49.9
40.2	40	12	0.6455	0.7638	0.8451	1.5493	1.3092	1.1833	49	48	49.8
40.3	40	18	0.6468	0.7627	0.8481	1.5461	1.3112	1.1792	49	42	49.7
40.4	40	24	0.6481	0.7615	0.8511	1.5429	1.3131	1.1750	49	36	49.6
40.5	40	30	0.6494	0.7604	0.8541	1.5398	1.3151	1.1708	49	30	49.5
40.6	40	36	0.6508	0.7593	0.8571	1.5366	1.3171	1.1667	49	24	49.4
40.7	40	42	0.6521	0.7581	0.8601	1.5335	1.3190	1.1626	49	18	49.3
40.8	40	48	0.6534	0.7570	0.8632	1.5304	1.3210	1.1585	49	12	49.2
40.9	40	54	0.6547	0.7559	0.8662	1.5273	1.3230	1.1544	49	6	49.1
41.0	41	0	0.6561	0.7547	0.8693	1.5243	1.3250	1.1504	49	0	49.0
41.1	41	6	0.6574	0.7536	0.8724	1.5212	1.3270	1.1463	48	54	48.9
41.2	41	12	0.6587	0.7524	0.8754	1.5182	1.3291	1.1423	48	48	48.8
41.3	41	18	0.6600	0.7513	0.8785	1.5151	1.3311	1.1383	48	42	48.7
41.4	41	24	0.6613	0.7501	0.8816	1.5121	1.3331	1.1343	48	36	48.6
41.5	41	30	0.6626	0.7490	0.8847	1.5092	1.3352	1.1303	48	30	48.5
41.6	41	36	0.6639	0.7478	0.8878	1.5062	1.3373	1.1263	48	24	48.4
41.7	41	42	0.6652	0.7466	0.8910	1.5032	1.3393	1.1224	48	18	48.3
41.8	41	48	0.6665	0.7455	0.8941	1.5003	1.3414	1.1184	48	12	48.2
41.9	41	54	0.6678	0.7443	0.8972	1.4974	1.3435	1.1145	48	6	48.1
42.0	42	0	0.6691	0.7431	0.9004	1.4945	1.3456	1.1106	48	0	48.0
42.1	42	6	0.6704	0.7420	0.9036	1.4916	1.3478	1.1067	47	54	47.9
42.2	42	12	0.6717	0.7408	0.9067	1.4887	1.3499	1.1028	47	48	47.8
42.3	42	18	0.6730	0.7396	0.9099	1.4859	1.3520	1.0990	47	42	47.7
42.4	42	24	0.6743	0.7385	0.9131	1.4830	1.3542	1.0951	47	36	47.6
42.5	42	30	0.6756	0.7373	0.9163	1.4802	1.3563	1.0913	47	30	47.5
42.6	42	36	0.6769	0.7361	0.9195	1.4774	1.3585	1.0875	47	24	47.4
42.7	42	42	0.6782	0.7349	0.9228	1.4746	1.3607	1.0837	47	18	47.3
42.8	42	48	0.6794	0.7337	0.9260	1.4718	1.3629	1.0799	47	12	47.2
42.9	42	54	0.6807	0.7325	0.9293	1.4690	1.3651	1.0761	47	6	47.1
43.0	43	0	0.6820	0.7314	0.9325	1.4663	1.3673	1.0724	47	0	47.0
43.1	43	6	0.6833	0.7302	0.9358	1.4635	1.3696	1.0686	46	54	46.9
43.2	43	12	0.6845	0.7290	0.9391	1.4608	1.3718	1.0649	46	48	46.8
43.3	43	18	0.6858	0.7278	0.9424	1.4581	1.3741	1.0612	46	42	46.7
43.4	43	24	0.6871	0.7266	0.9457	1.4554	1.3763	1.0575	46	36	46.6
43.5	43	30	0.6884	0.7254	0.9490	1.4527	1.3786	1.0538	46	30	46.5
43.6	43	36	0.6896	0.7242	0.9523	1.4501	1.3809	1.0501	46	24	46.4
43.7	43	42	0.6909	0.7230	0.9556	1.4474	1.3832	1.0464	46	18	46.3
43.8	43	48	0.6921	0.7218	0.9590	1.4448	1.3855	1.0428	46	12	46.2
43.9	43	54	0.6934	0.7206	0.9623	1.4422	1.3878	1.0392	46	6	46.1
44.0	44	0	0.6947	0.7193	0.9657	1.4396	1.3902	1.0355	46	0	46.0
44.1	44	6	0.6959	0.7181	0.9691	1.4370	1.3925	1.0319	45	54	45.9
44.2	44	12	0.6972	0.7169	0.9725	1.4344	1.3949	1.0283	45	48	45.8
44.3	44	18	0.6984	0.7157	0.9759	1.4318	1.3972	1.0247	45	42	45.7
44.4	44	24	0.6997	0.7145	0.9793	1.4293	1.3996	1.0212	45	36	45.6
44.5	44	30	0.7009	0.7133	0.9827	1.4267	1.4020	1.0176	45	30	45.5
44.6	44	36	0.7022	0.7120	0.9861	1.4242	1.4044	1.0141	45	24	45.4
44.7	44	42	0.7034	0.7108	0.9896	1.4217	1.4069	1.0105	45	18	45.3
44.8	44	48	0.7046	0.7096	0.9930	1.4192	1.4093	1.0070	45	12	45.2
44.9	44	54	0.7059	0.7083	0.9965	1.4167	1.4118	1.0035	45	6	45.1
45.0	45	0	0.7071	0.7071	1.0000	1.4142	1.4142	1.0000	45	0	45.0
			cos θ	sin θ	cot θ	sec θ	csc θ	tan θ	deg	min	θ deg

Answers to Odd-Numbered Exercises

CHAPTER 1

Exercises 1.1 (Page 13)

1. (a) $\in$; (b) $\notin$; (c) $\in$; (d) $\notin$ **3.** (a) $N \subseteq Q$; (b) $Q \subseteq R$;
(c) $N \subseteq J$; (d) $J \subseteq R$ **5.** (a) $\subseteq$; (b) $\not\subseteq$; (c) $\subseteq$; (d) $\subseteq$
7. (a) Q; (b) R; (c) J; (d) $\varnothing$ **9.** (a) $\{12, 571\}$;
(b) $\{12, \frac{5}{3}, 0, -38, 571, -\frac{1}{10}, 0.666 \ldots, 16.34\}$;
(c) $\{\sqrt{7}, -\sqrt{2}, \pi\}$; (d) $\{12, 0, -38, 571\}$
11. $\{-10, -7, -\sqrt{5}, -2, -\frac{7}{4}, -\frac{5}{3}, -1, 0, \frac{2}{3}, \frac{3}{4}, \sqrt{2}, 3, 5, 21\}$
13. (a) $\{x \mid -9 < x < 8\}$; (b) $\{y \mid -12 < y < -3\}$;
(c) $\{z \mid 4z - 5 < 0\}$ **15.** (a) $\{x \mid 2x + 4 \geq 0\}$;
(b) $\{r \mid 2 \leq r < 8\}$; (c) $\{a \mid -5 < a - 2 \leq 7\}$
17. (a) $(2, +\infty)$; (b) $(-4, 4]$
19. (a) $(2, 12)$; (b) $(-\infty, -4] \cup (4, +\infty)$
21. (a) $(2, 12)$; (b) $(-\infty, -4] \cup (4, +\infty)$
23. (a) $(-4, 0]$; (b) $(-\infty, 7]$ **25.** (a) $\{x \mid 2 < x < 7\}$;
(b) $\{x \mid -3 \leq x \leq 6\}$; (c) $\{x \mid -5 < x \leq 4\}$;
(d) $\{x \mid -10 \leq x < -2\}$ **27.** (a) $\{x \mid x \geq 3\}$;
(b) $\{x \mid x < 0\}$; (c) $\{x \mid x > -4\}$; (d) $\{x \mid x < 0\}$
$\cup \{x \mid x \geq 0\}$ **29.** (a) 7; (b) $\frac{3}{4}$; (c) $3 - \sqrt{3}$; (d) $3 - \sqrt{3}$
31. (a) 6; (b) 10; (c) 10; (d) 6
35.

Exercises 1.2 (Page 24)

1. (a) 125; (b) -32; (c) $\frac{1}{81}$; (d) $\frac{1}{1,000,000}$; (e) 1
3. (a) 6561; (b) 7776; (c) $\frac{49}{169}$; (d) $\frac{5184}{25}$ **5.** (a) $\frac{1}{625}$;
(b) $\frac{1}{36}$; (c) 8; (d) $-\frac{243}{32}$; (e) 1   **7.** (a) $\frac{1}{4}$; (b) 243; (c) $\frac{1}{100}$;
(d) 486   **9.** (a) $\frac{11}{18}$; (b) $\frac{1}{324}$; (c) $-\frac{2}{27}$; (d) $\frac{64}{17}$
11. (a) 200; (b) $\frac{9}{4}$; (c) $-\frac{64}{81}$; (d) 490,000   **13.** (a) $\frac{9}{7}$;

(b) $-\frac{17}{6}$; (c) $\frac{1255}{202}$ **15.** (a) x^7; (b) a^3; (c) y^{12}; (d) $\dfrac{1}{x^5}$

17. (a) $\dfrac{1}{x^{12}}$; (b) $\dfrac{1}{x^9 y^6}$; (c) $\dfrac{t^2}{s^{10}}$; (d) $\dfrac{c^2}{a^3 b}$ **19.** (a) $20x^8$;

(b) x^{15}; (c) $-6a^4 b^6$; (d) $-7x^{12}y^6$ **21.** (a) $72x^6 y^6$;

(b) $x^6 y^9$; (c) $\dfrac{81u^{16}}{256v^{12}}$; (d) $-\dfrac{r^{10}s^{15}t^{20}}{32u^{25}}$ **23.** (a) $-\dfrac{10}{3}r^2 t^2$;

(b) $-2x^3 y^2 z^3$ **25.** (a) $9x^4 y^{12}$; (b) $\dfrac{9c^{12}}{4a^2 b^6}$ **27.** (a) x^{4n};

(b) x^{3n^2}; (c) x^{2n}; (d) x^{4n+2}; (e) x^{n+1} **29.** 100
31. (a) $(4.55)10^4$; (b) $(1.30)10^{-3}$; (c) $(2.15)10^{-5}$;
(d) $(1.53)10^1$ **33.** (a) $(5.44)10^{15}$; (b) $(2.96)10^{-8}$;
(c) $(1.49)10^{33}$; (d) $(7.84)10^{-26}$ **35.** (a) $(5.260)10^1$;
(b) $(6.1)10^{-3}$; (c) $(1.72)10^5$; (d) $(1.720)10^5$
37. (a) $(3.960)10^{-2}$; (b) $(8.0022)10^{-6}$; (c) $(1.723)10^0$;
(d) $(4.260)10^2$ **39.** (a) 243.2; (b) 60,130;
(c) 0.008390; (d) 0.0000508   **41.** (a) $(4.81)10^9$;
(b) $(3.04)10^{-5}$ **43.** $(1.50)10^{11}$ **45.** $(6.58)10^{21}$

Exercises 1.3 (Page 33)

1. (a) Trinomial, 2; (b) Binomial, 1; (c) Monomial, 6;
(d) Trinomial, 4   **3.** (a) $10x^2 - 7x$; (b) $-30y^3 + 20y^2 - 5y + 15$ **5.** (a) $-17t^2 + 11st - 4s^2$; (b) $-5x + 77$
7. (a) $7x^3 + x^2 + 5x - 11$; (b) $x^3 - 15x^2 - x + 3$
9. (a) $14y^4 - 5y^3 - 4y^2 + 3y + 5$; (b) $-2y^4 + 5y^3 + 3y - 7$
11. (a) $-12x^4 + 6x^3 - 21x^2$; (b) $a^5 b^2 + \frac{5}{2}a^4 b^3 - \frac{1}{2}a^3 b^4$
13. (a) $6x^2 y^3 - 12xy^2 z^3 - 2x^2 y^2 z^2 - 2xyz^2$;
(b) $3x^{3n+1} - 12x^{3n} + 15x^{2n}$ **15.** (a) $2x^2 - 11x - 21$;
(b) $10s^2 + 8st - 24t^2$ **17.** (a) $-36a^2 - 88ab + 60b^2$;
(b) $32y^3 + 16y^2 + 2y$ **19.** $9x^3 + 12x^2 y + 7xy^2 + 20y^3$

21. $2x^4 - 8x^3 - 17x^2 + 38x - 15$ **23.** $-15b^4 + 14b^3 + 26b^2 - 19b + 14$ **25.** $12x^{4n} - 11x^{2n}y^n - 5y^{2n}$

29. (a) $x^2 + 9x + 20$; (b) $4x^2 + 12xy + 9y^2$

31. (a) $w^2 - 36$; (b) $12r^2 - 5rs - 2s^2$

33. (a) $t^4 + 4t^2 - 45$; (b) $16x^4 - 24x^2y^2 + 9y^4$

35. (a) $121u^2 - 64v^2$; (b) $35a^4 + 11a^2b^2 - 6b^4$

37. (a) $2 - 7x^2$; (b) $-7v^2 + 4uv$

39. (a) $-3abc^2 + 4b^2 - 2a^3bc$; (b) $8t^{2n} - 32t^{4n}$

41. $2x - 1$ **43.** $2a + 3$ **45.** $-3x + 5 + \dfrac{7}{x + 1}$

47. $5b^2 - 3b + 6$ **49.** $y - 3 + \dfrac{5y - 1}{y^2 + 3y - 2}$

51. $x^2 + 3x + 9$ **53.** $2x - 4y + \dfrac{9y^2}{2x + 3y}$

55. $a^5 + a^4b + a^3b^2 + a^2b^3 + ab^4 + b^5$

Exercises 1.4 (Page 41)

1. (a) $2 \cdot 5^2$; (b) $2^3 \cdot 3^2$; (c) $2 \cdot 3 \cdot 5^3$; (d) $2^2 \cdot 3^3 \cdot 11$

3. (a) $2^2 \cdot 3 \cdot 5^2$; (b) $2^2 \cdot 3^2 \cdot 7 \cdot 13$; (c) $3 \cdot 5^2 \cdot 7^2 \cdot 11$

5. $4x(2x + 1)$ **7.** $a^3(a^2 - 3a + 1)$ **9.** $6y(6x - y)$

11. $a^2b^3(a^2 - ab + b^3)$ **13.** $-4y^2z(3xyz + 7y + 5x^2z)$

15. $y^{2n}(y^n - y + 1)$ **17.** $(a + 8)(a - 8)$

19. $(x + 7y)(x - 7y)$ **21.** $(2s + 5r)(2s - 5r)$

23. $(3xy + 4w^3)(3xy - 4w^3)$ **25.** $(x^{2n} + y^{3n})(x^{2n} - y^{3n})$

27. $(x + 2)(x + 5)$ **29.** $(a - 4)(a - 6)$

31. $(t + 4)(t - 8)$ **33.** $(3x - y)(7x - y)$ **35.** $(x + 3)^2$

37. $(4x - 1)^2$ **39.** $(2a - 3b)^2$ **41.** $(5t + 3)(t - 2)$

43. $(6u - 5v)(3u + 4v)$ **45.** $(6x - 5y)(3x - 7y)$

47. $(x^{2n} + 2)(x^{2n} - 6)$ **49.** $(t + 2)(t^2 - 2t + 4)$

51. $(4x - y)(16x^2 + 4xy + y^2)$

53. $(a^2b - 3c)(a^4b^2 + 3a^2bc + 9c^2)$ **55.** $(a + 2b)(c + d)$

57. $(y^2 + 1)(4y - 1)$ **59.** $(2a^2 + 5)(5a - 2)$

61. $(y + w)(3x - z)$ **63.** $(2t^2 + 9s^2)(3s - t)$

65. $(2x - 3y + 4)(2x - 3y - 4)$

67. $(r + 5s + 3)(r + 5s - 3)$

69. $(x - 4y + 6a - b)(x - 4y - 6a + b)$

71. $(3a + 4b)(3a - 4b - 1)$

73. $(x - 4y)(1 - x^2 - 4xy - 16y^2)$

75. $(x + 2y - 1)(x^2 + 4xy + y^2 + x + 2y + 1)$

77. $(y^2 + 4)(y + 2)(y - 2)$

79. $(x^4 + 1)(x^2 + 1)(x + 1)(x - 1)$

81. $(a + 1)^2(a^2 - a + 1)^2$

83. $(2x - 1)(x + 1)(4x^2 + 2x + 1)(x^2 - x + 1)$

85. $(t^2 + 1)(t^4 + 1)$

87. $(2x - y)(2x + y)(4x^2 + 2xy + y^2)(4x^2 - 2xy + y^2)$

89. $(x - y)(ab + ac + bc)$

91. $(2a^n + b^n)(2a^n - b^n)(4a^{2n} - 2a^nb^n + b^{2n})$
$(4a^{2n} + 2a^nb^n + b^{2n})$

Exercises 1.5 (Page 50)

1. (a) $\dfrac{4}{a^4}$; (b) $\dfrac{y^2}{3x^3}$; (c) $-\dfrac{6t}{5s}$ **3.** (a) $\dfrac{2}{x + 2}$; (b) $\dfrac{x - y}{x}$

5. (a) $\dfrac{t + 4}{t + 3}$; (b) $\dfrac{3x + y}{x - 4y}$ **7.** (a) $-x - 4$; (b) $x^2 + 3x + 9$

9. (a) $\dfrac{3x}{2y^2}$; (b) $\dfrac{20a^2}{21c^2}$ **11.** (a) $\dfrac{-4t^3w}{9sv}$; (b) $\dfrac{x^2}{9z^3}$

13. $\dfrac{3(x - 4)}{5(x - 3)}$ **15.** $\dfrac{x^2 - xy - 6y^2}{x^2 + xy - 6y^2}$ **17.** $\dfrac{1}{b}$ **19.** $-\dfrac{x}{y}$

21. (a) z; (b) $\dfrac{20r^2sv^2}{9tw^3}$ **23.** $\dfrac{x + 7}{x - 3}$ **25.** $\dfrac{2x^2 + 2xy}{xy - y^2}$

27. $-\dfrac{1}{4}$ **29.** $-\dfrac{a^3 + ab^2}{a^2 + 2ab + b^2}$ **31.** $\dfrac{a^3 + 2a^2 - a + 1}{a^3}$

33. $\dfrac{-6xy}{2x^2 + 5xy + 2y^2}$ **35.** $\dfrac{2t^2 - 10t + 14}{3t^2 - 15t + 12}$

37. $\dfrac{18x - 8}{(x - 3)(x - 5)(x + 2)}$ **39.** $\dfrac{-4t}{s + t}$ **41.** $\dfrac{y - x}{y + x}$

43. $\dfrac{x + y}{x - y}$ **45.** -2 **47.** $-\dfrac{1}{x(x + h)}$

49. $-\dfrac{3}{(3x + 3h + 2)(3x + 2)}$ **51.** $-\dfrac{1}{xy(x + y)}$

53. $\dfrac{b^2 - a^2}{ab}$ **55.** $\dfrac{xy}{x + y}$

Exercises 1.6 (Page 59)

1. (a) 9; (b) 3; (c) 5; (d) 2 **3.** (a) 64; (b) 25; (c) $\frac{1}{6}$; (d) $\frac{8}{343}$ **5.** (a) $\frac{1}{12}$; (b) -1; (c) $\frac{1}{3}$; (d) 10 **7.** (a) 4;

(b) $\frac{8}{27}$; (c) $-\dfrac{16,807}{100,000}$ (d) -0.064 **9.** (a) $x^{1/6}$; (b) $a^{3/8}$;

(c) y^8; (d) $x^{1/4}$ **11.** (a) $\dfrac{1}{x^{1/4}}$; (b) $y^{1/4}$ **13.** (a) $\dfrac{2a^{5/6}}{3b}$;

(b) $\dfrac{s^{1/2}}{t^{2/3}}$ **15.** $\dfrac{2x^3}{3y^2}$ **17.** $\dfrac{z}{x^{1/2}y^{1/3}}$ **19.** (a) $x^{4/3} - x$;

(b) $a^{1/8} + a^{5/4}$ **21.** $\dfrac{(x - 1)^2}{x}$ **23.** (a) $y^{3/2}(2 - 3y)$;

(b) $x^{-4/3}(5 + 4x^3)$ **25.** (a) $b^{-3/5}(1 - 7b^{1/5} + 2b)$;

(b) $(x + 3)^{-1/3}[3 + 2(x + 3)^{5/3}]$ **27.** (a) a^{n+9}; (b) $x^{n/4}$

29. (a) $\dfrac{1}{y^n}$; (b) 2 **31.** $\dfrac{x + 2}{2(x + 1)^{3/2}}$ **33.** $\dfrac{5x^2(x + 3)}{(2x + 5)^{3/2}}$

35. $\dfrac{x^3(9x^2 + 4)}{(3x^2 + 1)^{3/2}}$ **37.** (a) $|x|^{3/2}y^2$; (b) $2s^2|t|^5$

39. (a) $9y^2|y - 2|$; (b) $4(x - 2)^2|2 - y|$

41. $4(u + 1)^2|u - 4|$

43. (a) $|x + 3| - |x - 3|$; (b) $x \geq 3$

Exercises 1.7 (Page 75)

1. (a) 9; (b) -0.1; (c) $\frac{4}{25}$; (d) $\frac{2}{5}$ **3.** (a) $\frac{6}{5}$; (b) 5; (c) -4;
(d) 3 **5.** (a) 24; (b) 2; (c) 6 **7.** (a) 3; (b) 3; (c) $\frac{1}{2}$
9. (a) $4x\sqrt{3}$; (b) $3x^2\sqrt[3]{2}$; (c) $2c^2\sqrt[3]{c^2}$
11. (a) b; (b) $-2x^5y^2\sqrt[3]{3y^2}$; (c) $2r^2ts^4\sqrt{10rt}$
13. (a) $10\sqrt{3}$; (b) $-24\sqrt[3]{2}$; (c) $-3s^2t^2\sqrt[3]{2s}$
15. (a) $30a^2bc\sqrt{2c}$; (b) $6x^2y^2\sqrt[3]{15x^2y}$

17. (a) $\dfrac{\sqrt{14}}{2}$; (b) $3\sqrt[3]{4}$ **19.** (a) $\dfrac{\sqrt{15}}{6}$; (b) $\dfrac{\sqrt[3]{100}}{2}$

21. (a) $\dfrac{3s\sqrt{2t}}{2t}$; (b) $\dfrac{\sqrt[4]{x^2y}}{3}$ **23.** (a) $\dfrac{2x^3y\sqrt{30yz}}{15z^4}$;

(b) $\dfrac{45w^2\sqrt[4]{72u^3v^2w^2}}{2v}$ **25.** $6 + 3\sqrt{3}$ **27.** $2\sqrt{2} - \sqrt{3}$

29. $\dfrac{13\sqrt{14} - 54}{10}$ **31.** $\dfrac{2\sqrt{x} - \sqrt{y}}{4x - y}$ **33.** $\dfrac{1}{\sqrt{x} + 3}$

35. $\dfrac{1}{\sqrt{x + 4} + 2}$ **37.** $\dfrac{2}{\sqrt{2(x + h) + 1} + \sqrt{2x + 1}}$

39. $\dfrac{\sqrt[3]{x^2} + \sqrt[3]{xy} + \sqrt[3]{y^2}}{x - y}$ **41.** (a) $-\dfrac{b}{a}$; (b) $\dfrac{c}{a}$

43. (a) $\dfrac{-b + \sqrt{b^2 - 4ac}}{2c}$; (b) $\dfrac{b^2 - 2ac}{a^2}$

Exercises 1.8 (Page 75)

1. (a) $5 + 0i$; (b) $0 + 7i$; (c) $3 + 5i$; (d) $3 - 5i$
3. (a) $8 - 5i$; (b) $-8 + 5i$; (c) $-6 + 6i$; (d) $\frac{1}{3} - \frac{3}{5}\sqrt{5}i$
5. (a) $12 + 3i$; (b) $1 - 2i$ **7.** (a) $-6 + 0i$; (b) $-5 + 8i$
9. $8 + 26i$ **11.** $-9 + \sqrt{5}i$ **13.** (a) $-15 + 0i$;
(b) $-4 + 0i$ **15.** $0 - 72i$ **17.** $8 - 18\sqrt{2} + 0i$
19. $53 + 0i$ **21.** $-24 + 5\sqrt{3}i$ **23.** $-18 + 18\sqrt{3}i$
25. $0 + 5i$ **27.** $-\frac{3}{13} - \frac{2}{13}i$ **29.** $\frac{4}{5} + \frac{7}{5}i$ **31.** $\frac{1}{2} - \frac{3}{4}i$
33. $\dfrac{3}{11} - \dfrac{\sqrt{2}}{11}i$ **35.** $\frac{5}{169} - \frac{12}{169}i$ **37.** (a) $-i$; (b) i;
(c) -1 **39.** (a) $-i$; (b) i; (c) -1 **41.** $8 - 6i$
43. $16 - 16i$ **45.** 0 **47.** 0

Review Exercises for Chapter 1 (Page 76)

1. (a) Q; (b) $\varnothing$; (c) N; (d) Q **3.** (a) $\{2, 15\}$;
(b) $\{-5, -4, 0, 2, 15\}$; (c) $\{-5, -4, -\frac{1}{3}, 0, \frac{3}{4}, 2, 15\}$;
(d) $\{-\sqrt{3}, \sqrt{2}, \frac{1}{2}\pi, 7\pi\}$ **5.** (a) $(5, +\infty)$; (b) $(-2, 3]$;
(c) $(-\infty, -4] \cup [4, +\infty)$; (d) $(-1, 6)$ **7.** (a) $\frac{1}{2}$; (b) π;
(c) $\sqrt{10} - 3$; (d) $5 - 2\sqrt{5}$ **9.** (a) $\frac{125}{729}$; (b) 81; (c) $\frac{125}{8}$;

(d) $-\frac{49}{8}$ **11.** (a) $\frac{1}{12}$; (b) $\frac{9}{2}$; (c) $\frac{3}{5}$ **13.** (a) x^{12}; (b) x^{32};
(c) $2x^4$; (d) $x^{3/8}$ **15.** (a) $-20x^6y^8$; (b) $\dfrac{16a^{12}}{81b^8}$; (c) $\dfrac{y^8}{x^6}$

17. (a) $60u^6v^5$; (b) $-216x^{11}y^4z^7$

19. (a) $\dfrac{9a^6b^4}{4}$; (b) $\dfrac{u^{28/3}}{t^7}$

21. (a) $x + 4y$; (b) $-4y^2 + 17y$
23. (a) $6x^3 + 2x^2 + 2x - 5$; (b) $2x^3 + 12x^2 - 20x + 11$
25. (a) $11y^4 - 3y^3 - 3y^2 + 7y + 7$;
(b) $-y^4 - 13y^3 + 3y^2 + 7y - 11$
27. (a) $-10y^5 + 6y^4 - 16y^3$; (b) $21x^2 - 23x - 20$
29. (a) $81t^2 - 16s^2$; (b) $3x^4 + 20x^3 - 8x^2 + 37x - 10$
31. (a) $5x^2 - 2$; (b) $y + 5$

33. (a) $w^2 - 3w + 6$; (b) $5x + 4y - \dfrac{3y^2}{2x - 3y}$

35. $a^3 - a^2b + ab^2 - b^3 + \dfrac{2b^4}{a + b}$

37. (a) $(x - 7)(x - 9)$; (b) $(3w + 7)(w + 2)$
39. (a) $(x + 8)(x - 8)$; (b) $(x - 4)(x^2 + 4x + 16)$
41. (a) $(x^2 + 8)(x^2 - 8)$; (b) $(x + 4)(x^2 - 4x + 16)$
43. (a) $(3x + 5y)(2x - 7y)$; (b) $9c(a + 2b)(a - 2b)$
45. (a) $(x - y)(2x + 3)$; (b) $(a + c)(b - d)$
47. (a) $y^2(x^2y^2 + 9w^3)(x^2y^2 - 9w^3)$;
(b) $(x^2 + 1)^2 (x + 1)^2(x - 1)^2$

49. (a) $\dfrac{7ac}{3b}$; (b) $\dfrac{y^2 - 3y - 4}{y^2 + 2y}$

51. (a) $\dfrac{16r^3}{15t^2u^3}$; (b) 1 **53.** (a) $\dfrac{13y + 3}{2y^2 + 3y - 9}$; (b) $\dfrac{2x^3}{x^2 - 9}$

55. (a) $\dfrac{1}{4 - x}$; (b) $\dfrac{x^2 - 15x - 9}{(3x + 4)(x - 2)(2x + 1)}$

57. $\dfrac{r^2 + rs + s^2}{r^2 + 2rs + s^2}$ **59.** $\dfrac{x - 2}{x - 1}$

61. $\dfrac{-2}{(2x + 2h + 1)(2x + 1)}$ **63.** $\dfrac{y - x}{x^2y^2}$

65. (a) $2y^2|xz^3|$; (b) $|y + 1|(x^2 + 4)$ **67.** (a) $\frac{9}{4}$; (b) 5;
(c) 3 **69.** (a) $5y\sqrt{3}$; (b) $4xy^2\sqrt[3]{3x}$; (c) $\sqrt[3]{5xy}$

71. (a) $\dfrac{\sqrt{22}}{11}$; (b) $\dfrac{\sqrt[3]{10st^2}}{2st}$; (c) $3\sqrt[4]{2xy^2}$ **73.** (a) $\dfrac{\sqrt{6}}{2}$;

(b) $\dfrac{1 - 2\sqrt{x} + x}{1 - x}$ **75.** (a) $\dfrac{1}{\sqrt{x} + 4}$; (b) $\dfrac{1}{\sqrt{x + 9} + 3}$

77. (a) $18 + i$; (b) $\frac{1}{4} - \frac{2}{3}i$ **79.** (a) $-21 + 0i$;
(b) $10 - 10i$ **81.** $-\frac{14}{25} + \frac{23}{25}i$ **83.** (a) i; (b) $-i$; (c) -1
85. (a) $(4.523)10^2$; (b) $(3.710)10^{-3}$; (c) 8620;
(d) 0.000509 **87.** (a) $(2.182)10^{10}$; (b) $(1.435)10^{-13}$

CHAPTER 2

Exercises 2.1 (Page 86)

1. $\{3\}$ **3.** $\{2\}$ **5.** $\{3\}$ **7.** $\varnothing$ **9.** $\{-\frac{13}{49}\}$ **11.** $\{2\}$
13. $\{2\}$ **15.** $\{2\}$ **17.** $\{-1\}$ **19.** $\{3\}$ **21.** $\{\frac{1}{2}\}$
23. $\{\frac{4}{3}\}$ **25.** $x = \frac{3}{4}b$ **27.** $y = a + 3b, a \neq 2b$

29. $x = 3a - 5b, a \neq -b$ **31.** $h = \dfrac{2A}{a+b}$

33. $r = \dfrac{E - IR}{I}$ **35.** $p = \dfrac{fq}{q-f}$ **37.** $r = \dfrac{a-S}{l-S}$

39. (a) $\{x | x \in R,\ x \neq 0\}$; (b) $\varnothing$; (c) $\varnothing$

Exercises 2.2 (Page 94)

1. $\frac{15}{2}, \frac{3}{2}$ **3.** 23, 14 **5.** 12 cm, 8 cm **7.** 423 adults,
387 students **9.** \$15,840, \$7,920, \$2,640 **11.** \$5000
13. 24 g of 80% gold alloy, 16 g of 55% gold alloy
15. $15\frac{5}{8}$ lb of \$4.10 tea, $9\frac{3}{8}$ lb of \$4.90 tea
17. $8\frac{1}{3}$ quarts **19.** 6 liters **21.** 2 hr **23.** 52 mi/hr
25. 420 ft **27.** $22\frac{2}{9}$ ft/sec, $21\frac{8}{9}$ ft/sec **29.** $2\frac{2}{9}$ hr
31. Newer press, $4\frac{1}{2}$ hr; older press, 9 hr **33.** $11\frac{1}{4}$ hr
35. 590 in English Composition, 650 in English
Fundamentals

Exercises 2.3 (Page 106)

1. $\{\pm 7\}$ **3.** $\{\pm\frac{2}{5}\sqrt{15}\}$ **5.** $\{0, \frac{1}{4}\}$ **7.** $\{3, 5\}$
9. $\{-\frac{3}{2}, \frac{1}{4}\}$ **11.** $\{-\frac{6}{7}\}$ **13.** $\{-\frac{1}{3}, 2\}$
15. (a) $x^2 + 6x + 9, (x+3)^2$; (b) $x^2 - 5x + \frac{25}{4}, (x - \frac{5}{2})^2$
17. (a) $x^2 - \frac{2}{3}x + \frac{1}{9}, (x - \frac{1}{3})^2$; (b) $x^2 + \frac{3}{5}x + \frac{9}{100}, (x + \frac{3}{10})^2$
19. $\{-4, -2\}$ **21.** $\{-\frac{3}{2}, 2\}$

23. $\left\{\dfrac{1 \pm \sqrt{5}}{2}\right\}$ **25.** $\{-\frac{2}{3} \pm \frac{\sqrt{2}}{3}i\}$ **27.** $\{-1, 4\}$

29. $\{1 \pm \sqrt{3}\}$ **31.** $\left\{\dfrac{2 \pm \sqrt{14}}{5}\right\}$ **33.** $\{\frac{1}{2} \pm \frac{1}{2}i\}$

35. $\{2 \pm \sqrt{3}i\}$ **37.** (a) 361, roots are real and
unequal; (b) 52, roots are real and unequal
39. (a) 0, roots are real and equal; (b) -31, roots are
imaginary and unequal **41.** $r = \sqrt{\dfrac{3V}{\pi h}}$

43. $v = \sqrt{\dfrac{rF}{kM}}$ **45.** $x = \dfrac{3 \pm \sqrt{9 + 40a^2}}{10a}, a \neq 0$

47. $x = y + 2 \pm 2\sqrt{y^2 + y + 1}$ **49.** 8, 10
51. 30 ft/sec and 25 ft/sec, or 25 ft/sec and 20 ft/sec
53. $\frac{25}{3}$ m^3 **55.** 10.8 m **57.** $x = \dfrac{-b \pm \sqrt{b^2 - 4a(c-y)}}{2a}$

Exercises 2.4 (Page 114)

1. $\{64\}$ **3.** $\varnothing$ **5.** $\{\frac{7}{2}\}$ **7.** $\{4\}$ **9.** $\{3\}$ **11.** $\{4\}$
13. $\varnothing$ **15.** $\{0, 3\}$ **17.** $\{7\}$ **19.** $\{-2\}$ **21.** $\{\frac{5}{3}, 3\}$
23. $\{-7\}$ **25.** $\{1, 2\}$ **27.** $\{-3, 0\}$ **29.** $\{\pm 1, \pm 2\}$

31. $\{\pm\sqrt{2}, \pm\sqrt{3}\}$ **33.** $\left\{\pm\dfrac{\sqrt{5}}{2}, \pm\dfrac{\sqrt{2}}{2}i\right\}$

35. $\{-5, \frac{1}{2}\}$ **37.** $\left\{2, 3, -1 \pm \sqrt{3}i, -\dfrac{3}{2} \pm \dfrac{3\sqrt{3}}{2}i\right\}$

39. $\{1\}$ **41.** $\{-5, -1, 3\}$ **43.** $\{\frac{1}{4}, \frac{1}{2}, 1, 2\}$ **45.** $\varnothing$

47. $1, -\dfrac{1}{2} \pm \dfrac{\sqrt{3}}{2}i$ **49.** $-2, 1 \pm \sqrt{3}i$ **51.** $\pm 3, \pm 3i$

53. 40 in., 30 in. **55.** 4 ft

Exercises 2.5 (Page 119)

1. $[5, +\infty)$ **3.** $(-\infty, \frac{7}{2})$ **5.** $(-5, +\infty)$ **7.** $(-\infty, \frac{17}{2}]$
9. $(-\infty, -\frac{17}{3})$ **11.** $(-8, +\infty)$ **13.** $(-\infty, 4]$
15. $(-\infty, 5)$ **17.** $[4, 8)$ **19.** $(-4, 1)$ **21.** $(1, 4)$
23. $[-6, -4]$ **25.** $(-\frac{5}{3}, \frac{4}{3}]$ **27.** $[-3, \frac{21}{5})$
29. $\{w | 0 \leq w \leq 7\}$ **31.** \$6000 **33.** At least 49
35. At most 20 g and at least 14 g

Exercises 2.6 (Page 126)

1. $(-\infty, -3) \cup (3, +\infty)$ **3.** $(-3, 4)$
5. $(-\infty, -\frac{1}{2}) \cup (\frac{7}{2}, +\infty)$ **7.** $[1, 3]$ **9.** $[-4, 1]$
11. $(-\infty, -4) \cup (2, +\infty)$ **13.** $[-2, \frac{3}{2}]$
15. Set R of real numbers **17.** $(-\infty, \frac{5}{3}) \cup (2, +\infty)$
19. $(-3, 1) \cup (4, +\infty)$ **21.** $(-\infty, -4] \cup [0, 4]$
23. $\left(\dfrac{1}{2} - \dfrac{\sqrt{5}}{2}, \dfrac{1}{2} + \dfrac{\sqrt{5}}{2}\right)$ **25.** Set R of real numbers
27. $(-\frac{5}{2}, 1)$ **29.** $(4, 24)$ **31.** $(-\infty, -\frac{23}{9}) \cup (-\frac{5}{3}, +\infty)$
33. $(-\infty, -4) \cup [-2, +\infty)$ **35.** $(-\infty, -2) \cup (1, 3)$
37. $(-\frac{1}{2}, \frac{1}{2}) \cup (3, +\infty)$ **39.** $(-4, -3] \cup (2, \frac{8}{3}]$
41. More than 10 and less than 70 **43.** If x feet
is the length of the side parallel to the river, then
$\frac{88}{3} \leq x \leq \frac{176}{5}$; and if y feet is the length of the other
side, then $100 \leq y \leq 120$

Exercises 2.7 (Page 132)

1. $\{1, 9\}$ **3.** $\{\frac{4}{3}, 4\}$ **5.** $\{-5, \frac{5}{2}\}$ **7.** $\{-1, 8\}$ **9.** $\{1, 3\}$

11. $\{\frac{9}{4}, 4\}$ **13.** $\left\{1, 2, \dfrac{-3 \pm \sqrt{17}}{2}\right\}$ **15.** $[-5, 5]$

17. $(-\infty, -6) \cup (8, +\infty)$ **19.** $[2, 8]$ **21.** $(-1, 8)$
23. $(-\infty, -1) \cup (8, +\infty)$ **25.** $(-\infty, -\frac{16}{3}) \cup (-\frac{8}{3}, +\infty)$
27. $[1, \frac{9}{5}]$ **29.** $(-\infty, \frac{1}{2}) \cup (1, +\infty)$ **37.** $(-3, -1) \cup (1, 3)$
39. $[-1, 2] \cup [3, 6]$ **41.** $(-\infty, -5] \cup [-3, 3] \cup [5, +\infty)$
43. $(-\infty, -1) \cup (2, +\infty)$
45. $(-\infty, -1) \cup (-1, -\frac{1}{3}] \cup [1, +\infty)$

Review Exercises for Chapter 2 (Page 133)

1. $\{3\}$ **3.** $\{\frac{4}{3}\}$ **5.** $\{3\}$ **7.** $\varnothing$ **9.** $\{\pm\frac{8}{7}\}$ **11.** $\{-2, 5\}$
13. $\{-\frac{6}{5}, \frac{1}{2}\}$ **15.** $\left\{\dfrac{2 \pm \sqrt{14}}{2}\right\}$ **17.** $\left\{\dfrac{1}{3} \pm \dfrac{\sqrt{5}}{3}i\right\}$
19. $\{-3, 5\}$ **21.** $\{-1, 4\}$ **23.** $\{-\frac{5}{3}, -2\}$ **25.** $\{2\}$
27. $\{-2\}$ **29.** $\{2, -1 \pm \sqrt{3}i\}$ **31.** $\{\pm\frac{1}{3}, \pm\frac{1}{2}\}$
33. $\left\{-2, -1, \dfrac{-3 \pm \sqrt{29}}{2}\right\}$ **35.** $\{6\}$ **37.** $\{-\frac{1}{27}, \frac{1}{8}\}$
39. $\{-6, 1\}$ **41.** $\{\frac{1}{2}, 2\}$ **43.** $x = -\dfrac{By + C}{A}$

45. $x = b \pm a$ **47.** $\left\{\dfrac{1}{2}, \dfrac{y - 1}{3}\right\}$
49. (a) $x^2 - 8x + 16 = (x - 4)^2$; (b) $y^2 + 3y + \frac{9}{4} = (y + \frac{3}{2})^2$;
(c) $x^2 + \frac{5}{3}x + \frac{25}{36} = (x + \frac{5}{6})^2$ **51.** (a) 0, roots are real
and equal; (b) 196, roots are real and unequal;
(c) -23, roots are imaginary and unequal
53. $(-\infty, 4]$ **55.** $(-\infty, -\frac{5}{4})$ **57.** $[12, +\infty)$
59. $(-2, 1)$ **61.** $(-4, 5]$ **63.** $(-\infty, -3] \cup [1, +\infty)$
65. $(-1, 6)$ **67.** $(-\infty, \frac{1}{6}) \cup (\frac{17}{6}, +\infty)$ **69.** $(-5, 9)$
71. $(-\infty, -1] \cup [\frac{5}{2}, +\infty)$ **73.** $(-\infty, 0)$
75. $(-\infty, -6) \cup (5, +\infty)$ **77.** $(-\frac{1}{2}, \frac{1}{2}) \cup (\frac{2}{3}, +\infty)$
79. $(-\infty, -4) \cup [-2, 3) \cup [9, +\infty)$
81. $(-2, -1) \cup (4, 5)$
83. $(-\infty, -4] \cup [-2, -1) \cup (-1, +\infty)$
87. \$22,500 at 16%, \$7500 at 12%
89. $\frac{4}{3}$ liters   **91.** 3 hr   **93.** 26 minutes and 40
seconds   **95.** 43.4 rods   **97.** 70 km/hr
99. At least 95

CHAPTER 3

Exercises 3.1 (Page 141)

1. (a) First quadrant; (b) third quadrant; (c) fourth
quadrant; (d) second quadrant   **3.** (a) $(1, 2)$;
(b) $(-1, -2)$; (c) $(-1, 2)$; (d) $(-2, 1)$ **5.** (a) $(2, -2)$;
(b) $(-2, 2)$; (c) $(-2, -2)$; (d) does not apply
7. (a) $(-1, 3)$; (b) $(1, -3)$; (c) $(1, 3)$; (d) $(-3, -1)$
9. (a) 7; (b) -7 **11.** (a) -4; (b) 4   **13.** (a) -10;
(b) 6   **15.** (b) 5; (c) $(-\frac{1}{2}, 5)$ **17.** (b) 13; (c) $(\frac{11}{2}, -1)$
19. (a) $(-\frac{3}{2}, 2)$ **21.** (a) $(\frac{5}{2}, \frac{3}{2})$ **23.** $|\overline{AB}| = 10$;
$|\overline{BC}| = \sqrt{17}$; $|\overline{CA}| = 13$ **25.** $\sqrt{26}$; $\frac{1}{2}\sqrt{89}$; $\frac{1}{2}\sqrt{53}$
27. $|\overline{AB}| = \sqrt{41}$; $|\overline{AC}| = \sqrt{41}$, $|\overline{BC}| = \sqrt{82}$, and
$|\overline{AB}|^2 + |\overline{AC}|^2 = |\overline{BC}|^2$; $\frac{41}{2}$ **33.** $17\sqrt{2}$ **35.** $(-8, 12)$
37. -2 or 8

Exercises 3.2 (Page 151)

1.

3.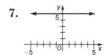

5.

7.

9.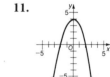

11.

13.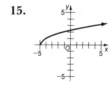

15.

17.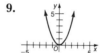

19.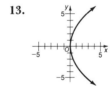

21.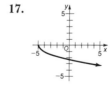

23.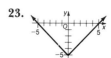

25.   **27.**   **29.**

31.   **33.** **35.**

37.

39. $(x - 4)^2 + (y + 3)^2 = 25$; $x^2 + y^2 - 8x + 6y = 0$
41. $(x + 5)^2 + (y + 12)^2 = 9$; $x^2 + y^2 + 10x + 24y + 160 = 0$ **43.** $x^2 + (y - 7)^2 = 1$; $x^2 + y^2 - 14y + 48 = 0$
45. $(x - 1)^2 + (y - 2)^2 = 13$ **47.** $(x - 5)^2 + (y + 1)^2 = 13$
49. $(3, 4)$; 4 **51.** $(-1, -5)$; $2\sqrt{2}$ **53.** $(0, -\frac{2}{3})$; $\frac{5}{3}$
55. Circle **57.** The empty set **59.** Point $(\frac{1}{2}, -\frac{3}{2})$

Exercises 3.3 (Page 161)

1. (a) $\frac{1}{5}$; (b) -1 **3.** (a) $-\frac{3}{4}$; (b) $-\frac{1}{7}$ **5.** (a) 0; (b) $\frac{1}{4}$
9. (a) $4x - y - 11 = 0$; (b) $11x - 4y - 9 = 0$
11. (a) $2x + 3y - 3 = 0$; (b) $6x - 3y + 8 = 0$
13. (a) $y = -7$; (b) $x = 2$ **15.** (a) $4x - 3y + 12 = 0$;
(b) $x - y = 0$ **17.** $2x - y + 7 = 0$ **19.** $5x + y - 14 = 0$
21. (a) $m = -\frac{1}{3}$, $b = 2$; (b) $m = 0$, $b = \frac{9}{4}$ **23.** (a) $m = \frac{7}{8}$,
$b = 0$; (b) $m = -\frac{1}{2}$, $b = 3$ **25.** $y = -5x + 8$
27. (a) $a = -2$, $b = 3$; (b) $a = -5$, $b = -4$
31. Slope of each line is $-\frac{3}{5}$ **33.** Slopes of lines are $\frac{2}{3}$
and $-\frac{3}{2}$ **35.** $\pm\frac{2}{3}$ **37.** (a) Collinear; (b) not collinear
39. (a) Not collinear; (b) collinear **41.** Slopes of two
sides are $-\frac{1}{2}$; slopes of other two sides are $\frac{3}{5}$
43. Area is 5 square units **45.** $9x + 4y - 19 = 0$;
$9x - 4y - 11 = 0$; $y = 1$ **47.** (a) $y = 25x + 3000$
49. (a) $600; (b) $y = 30x + 600$

Exercises 3.4 (Page 168)

1. (a) $(0, 0)$;
(b) $x = 0$;
(c) $(0, 1)$;
(d) $y = -1$;
(e) $(-2, 1)$, $(2, 1)$

(f)

3. (a) $(0, 0)$;
(b) $x = 0$;
(c) $(0, -4)$;
(d) $y = 4$;
(e) $(-8, -4)$, $(8, -4)$
(f)

5. (a) $(0, 0)$;
(b) $x = 0$;
(c) $(0, \frac{1}{4})$;
(d) $y = -\frac{1}{4}$;
(e) $(-\frac{1}{2}, \frac{1}{4})$, $(\frac{1}{2}, \frac{1}{4})$
(f)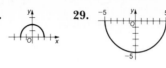

7. (a) $(0, 0)$;
(b) $y = 0$;
(c) $(3, 0)$;
(d) $x = -3$;
(e) $(3, -6)$, $(3, 6)$
(f)

9. (a) $(0, 0)$;
(b) $y = 0$;
(c) $(-2, 0)$;
(d) $x = 2$;
(e) $(-2, -4)$, $(-2, 4)$
(f)

11. (a) $(0, 0)$;
(b) $y = 0$;
(c) $(\frac{5}{4}, 0)$;
(d) $x = -\frac{5}{4}$;
(e) $(\frac{5}{4}, -\frac{5}{2})$, $(\frac{5}{4}, \frac{5}{2})$
(f)

13. (a) $(0, 0)$;
(b) $x = 0$;
(c) $(0, -\frac{2}{3})$;
(d) $y = \frac{2}{3}$;
(e) $(-\frac{4}{3}, -\frac{2}{3})$, $(\frac{4}{3}, -\frac{2}{3})$
(f)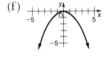

15. (a) $(0, 0)$;
(b) $y = 0$;
(c) $(\frac{9}{8}, 0)$;
(d) $x = -\frac{9}{8}$;
(e) $(\frac{9}{8}, -\frac{9}{4})$, $(\frac{9}{8}, \frac{9}{4})$
(f)

17. $x^2 = 16y$ **19.** $x^2 = -20y$ **21.** $y^2 = 8x$
23. $3y^2 = -20x$ **25.** $x^2 = 12y$
27. $3y^2 = -8x$ **29.** $x^2 = y$
31. $\frac{32}{45}$ in.
33. 16.6 m

Exercises 3.5 (Page 174)

1. $x'^2 + y'^2 = 13$; $x' = x + 3$, $y' = y + 2$
3. $x'^2 + y'^2 = \frac{1}{4}$; $x' = x + \frac{1}{2}$, $y' = y - 1$
5. (a) $(2, -4)$;
   (b) $x = 2$;
   (c) $(2, -2)$;
   (d) $(-2, -2)$, $(6, -2)$
   (e)

7. (a) $(-5, -3)$;
   (b) $y = -3$;
   (c) $(-\frac{13}{2}, -3)$;
   (d) $(-\frac{13}{2}, -6)$, $(-\frac{13}{2}, 0)$
   (e)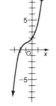

9. (a) $(3, 1)$;
   (b) $x = 3$;
   (c) $(3, 2)$;
   (d) $(1, 2)$, $(5, 2)$
   (e)

11. (a) $(9, -6)$;
   (b) $y = -6$;
   (c) $(8, -6)$;
   (d) $(8, -8)$, $(8, -4)$
   (e)

13. (a) $(0, -4)$;
   (b) $x = 0$;
   (c) $(0, -\frac{15}{4})$;
   (d) $(-\frac{1}{2}, -\frac{15}{4})$, $(\frac{1}{2}, -\frac{15}{4})$
   (e)

15. (a) $(-9, 3)$;
   (b) $y = 3$;
   (c) $(-\frac{35}{4}, 3)$;
   (d) $(-\frac{35}{4}, \frac{5}{2})$, $(-\frac{35}{4}, \frac{7}{2})$
   (e)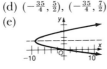

17. (a) $(2, -1)$;
   (b) $x = 2$;
   (c) $(2, -\frac{5}{4})$;
   (d) $(\frac{3}{2}, -\frac{5}{4})$, $(\frac{5}{2}, -\frac{5}{4})$
   (e)

19. (a) $(3, -2)$; (b) $y = -2$; (c) $(2\frac{7}{8}, -2)$;
(d) $(2\frac{7}{8}, -\frac{9}{4})$, $(2\frac{7}{8}, -\frac{7}{4})$ **21.** (a) $(4, 3)$; (b) $x = 4$;
(c) $(4, \frac{5}{2})$; (d) $(3, \frac{5}{2})$, $(5, \frac{5}{2})$ **23.** (a) $(2, -2)$; (b) $x = 2$;
(c) $(2, 0)$; (d) $(-2, 0)$, $(6, 0)$
25. (b)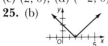

27. (b)

29. (b)

31. (b)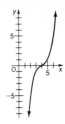

33. (b)

35. (b)

37. (b)

39. (b)

41. (b)

Exercises 3.6 (Page 183)

1. (a) $y = 0$;
   (b) $(-5, 0)$, $(5, 0)$;
   (c) $(0, -3)$, $(0, 3)$;
   (d) $(-4, 0)$, $(4, 0)$
   (e)

3. (a) $x = 0$;
   (b) $(0, -4)$, $(0, 4)$;
   (c) $(-2, 0)$, $(2, 0)$;
   (d) $(0, -2\sqrt{3})$, $(0, 2\sqrt{3})$
   (e)

5. (a) $y = 0$;
   (b) $(-10, 0)$, $(10, 0)$;
   (c) $(0, -6)$, $(0, 6)$;
   (d) $(-8, 0)$, $(8, 0)$
   (e)

7. (a) $x = 0$;
   (b) $(0, -3)$, $(0, 3)$;
   (c) $(-1, 0)$, $(1, 0)$;
   (d) $(0, -2\sqrt{2})$, $(0, 2\sqrt{2})$
   (e)

9. (a) $y = 0$;
(b) $(-8, 0)$, $(8, 0)$;
(c) $(-10, 0)$, $(10, 0)$;
(d)

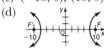

11. (a) $x = 0$;
(b) $(0, -5)$, $(0, 5)$;
(c) $(0, -13)$, $(0, 13)$
(d)

35. Hyperbola;
(a) $(-1, 4)$;
(b) $x = -1$;
(c) $(-1, -1)$, $(-1, 9)$

37. Hyperbola;
(a) $(1, -2)$;
(b) $x = 1$;
(c) $(1, -5)$, $(1, 1)$

13. (a) $y = 0$;
(b) $(-2, 0)$, $(2, 0)$;
(c) $(-\sqrt{13}, 0)$, $(\sqrt{13}, 0)$
(d)

15. (a) $y = 0$;
(b) $(-5, 0)$, $(5, 0)$
(c)

39. Ellipse;
(a) $(2, 0)$
(b) $y = 0$;
(c) $(2 - 3\sqrt{3}, 0)$, $(2 + 3\sqrt{3}, 0)$;
(d) $(2, -3)$, $(2, 3)$

17. (a) $x = 0$;
(b) $(0, -2)$, $(0, 2)$
(c)

19. (a) $x = 0$;
(b) $(0, -6)$, $(0, 6)$
(c)

21. 8.4 m   **23.** $9x^2 + 25y^2 = 5.625(10)^{17}$
25. The right branch of the hyperbola
$16x^2 - 9y^2 = 14{,}400$
29. Ellipse;
(a) $(2, 1)$;
(b) $y = 1$;
(c) $(-1, 1)$, $(5, 1)$;
(d) $(2, -1)$, $(2, 3)$

31. Hyperbola;
(a) $(-3, -2)$;
(b) $y = -2$;
(c) $(-6, -2)$, $(0, -2)$

33. Ellipse;
(a) $(-1, 2)$;
(b) $x = -1$;
(c) $(-1, -8)$, $(-1, 12)$
(d) $(-6, 2)$, $(4, 2)$

Review Exercises for Chapter 3 (Page 185)

1. (b) 5; (c) $(5, \frac{1}{2})$ **3.** (a) $(1, -2)$
5. $|AB| = \sqrt{37}$; $|BC| = 2\sqrt{5}$; $|CA| = 5$

9.   **11.**

13. **15.** **17.**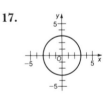

19. $(x - 3)^2 + (y + 5)^2 = 4$; $x^2 + y^2 - 6x + 10y + 30 = 0$
21. $(-2, 3)$; 4   **23.** (b) Slope is $\frac{8}{3}$; (c) $8x - 3y - 17 = 0$
25. (b) $3x + 2y - 11 = 0$ **27.** (a) $m = \frac{2}{3}$, $b = -2$
29. $7x - 3y + 15 = 0$ **33.** $-\frac{34}{3}$
37. (a) $(0, 0)$;   (f)
(b) $x = 0$;
(c) $(0, 4)$;
(d) $y = -4$;
(e) $(-8, 4)$, $(8, 4)$

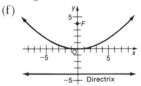

39. (a) $(0, 0)$ (f)

(b) $y = 0$

(c) $(-\frac{5}{2}, 0)$

(d) $x = \frac{5}{2}$

(e) $(-\frac{5}{2}, -5)$, $(-\frac{5}{2}, 5)$

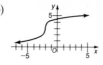

41. $x^2 = 8y$

43. $y^2 = -12x$

45. (a) $(0, -3)$;

(b) $x = 0$;

(c) $(0, -2\frac{3}{4})$ $(0, -\frac{11}{4})$;

(d) $(-\frac{1}{2}, -\frac{11}{4})$, $(\frac{1}{2}, -\frac{11}{4})$

(e)

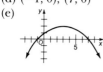

47. (a) $(-16, 4)$;

(b) $y = 4$;

(c) $(-\frac{63}{4}, 4)$;

(d) $(-\frac{63}{4}, \frac{7}{2})$, $(-\frac{63}{4}, \frac{9}{2})$

(e)

49. (a) $(3, 2)$

(b) $x = 3$;

(c) $(3, 0)$;

(d) $(-1, 0)$, $(7, 0)$

(e)

51. (b)

53. (b)

55. (b)

57. (b)

59. Parallelogram   **61.** Rectangle

63. Ellipse,

(a) $(0, 0)$;

(b) $y = 0$;

(c) $(-10, 0)$, $(10, 0)$;

(d) $(0, -6)$, $(0, 6)$;

(e) $(-8, 0)$, $(8, 0)$

65. Hyperbola;

(a) $(0, 0)$;

(b) $x = 0$

(c) $(0, -3)$, $(0, 3)$

67. Hyperbola;

(a) $(0, 0)$;

(b) $y = 0$;

(c) $(-5, 0)$, $(5, 0)$

69. Ellipse;

(a) $(0, 0)$;

(b) $x = 0$;

(c) $(0, -5)$, $(0, 5)$;

(d) $(-3, 0)$, $(3, 0)$;

(e) $(0, -4)$, $(0, 4)$

71. Ellipse; (a) $(2, -4)$; (b) $x = 2$; (c) $(2, -9)$, $(2, 1)$;

(d) $(1, -4)$, $(3, -4)$; (e) $(2, -4 - 2\sqrt{6})$,

$(2, -4 + 2\sqrt{6})$.

73. Hyperbola; (a) $(4, 2)$; (b) $y = 2$; (c) $(\frac{5}{2}, 2)$, $(\frac{11}{2}, 2)$

75. 18.86 m

Chapter 4

Exercises 4.1 (Page 194)

1. Domain: $(-\infty, +\infty)$;

Range: $(-\infty, +\infty)$

3. Domain: $(-\infty, +\infty)$;

Range: $[0, +\infty)$

5. Domain: $(-\infty, +\infty)$;

Range: $(-\infty, 5]$

7. Domain: $[1, +\infty)$;

Range: $[0, +\infty)$

9. Domain = $(-\infty, -2] \cup [2, +\infty)$;
Range = $[0, +\infty)$

11. Domain: $[-3, 3]$;
Range: $[0, 3]$

13. Domain: $(-\infty, +\infty)$;
Range: $[0, +\infty)$

15. Domain: $(-\infty, +\infty)$;
Range: $[0, +\infty)$

17. Domain: $\{x | x \neq -5\}$;
Range: $\{y | y \neq -10\}$

19. Domain: $\{x | x \neq 1\}$;
Range: $\{y | y \neq -2\}$

21. Domain: $(-\infty, +\infty)$;
Range: $\{-2, 2\}$

23. Domain: $(-\infty, +\infty)$
Range: $\{y | y \neq 3\}$

25. Domain: $(-\infty, +\infty)$
Range: $[-4, 5) \cup (5, +\infty)$

27. Domain: $(-\infty, +\infty)$;
Range: $(-\infty, +\infty)$

29. Domain: $(-\infty, +\infty)$;
Range: $(-\infty, 6)$

31. Domain: $(-\infty, +\infty)$;
Range: $(-\infty, -2) \cup [0, 5]$

33. Domain: $\{x | x \neq 2\}$;
Range: $[0, +\infty)$

Exercises 4.2 (Page 201)

1. (a) 5; (b) -5; (c) -1; (d) $2a + 1$; (e) $2x + 1$;
(f) $4x - 1$; (g) $4x - 2$; (h) $2x + 2h - 1$; (i) $2x + 2h - 2$;
(j) 2 **3.** (a) -5; (b) -6; (c) -3; (d) 30;
(e) $2h^2 + 9h + 4$; (f) $8x^4 + 10x^2 - 3$; (g) $2x^4 - 7x^2$;
(h) $2x^2 + (4h + 5)x + (2h^2 + 5h - 3)$;
(i) $2x^2 + 5x + (2h^2 + 5h - 6)$; (j) $4x + 2h + 5$
5. (a) 1; (b) $\sqrt{11}$; (c) 2; (d) 5; (e) $\sqrt{4x + 9}$
7. (a) $x^2 + x - 6$; domain: $(-\infty, +\infty)$; (b) $-x^2 + x - 4$,
domain: $(-\infty, +\infty)$; (c) $x^3 - 5x^2 - x + 5$, domain:
$(-\infty, +\infty)$; (d) $\dfrac{x - 5}{x^2 - 1}$, domain: $\{x | x \neq -1, x \neq 1\}$;
(e) $\dfrac{x^2 - 1}{x - 5}$, domain: $\{x | x \neq 5\}$ **9.** (a) $\dfrac{x^2 + 2x - 1}{x^2 - x}$, domain:
$\{x | x \neq 0, x \neq 1\}$; (b) $\dfrac{x^2 + 1}{x^2 - x}$, domain: $\{x | x \neq 0, x \neq 1\}$;
(c) $\dfrac{x + 1}{x^2 - x}$, domain: $\{x | x \neq 0, x \neq 1\}$;
(d) $\dfrac{x^2 + x}{x - 1}$, domain: $\{x | x \neq 0, x \neq 1\}$;
(e) $\dfrac{x - 1}{x^2 + x}$, domain: $\{x | x \neq -1, x \neq 0, x \neq 1\}$
11. (a) $\sqrt{x} + x^2 - 1$, domain: $[0, +\infty)$; (b) $\sqrt{x} - x^2 + 1$,
domain: $[0, +\infty)$; (c) $\sqrt{x}(x^2 - 1)$, domain: $[0, +\infty)$;
(d) $\dfrac{\sqrt{x}}{x^2 - 1}$, domain: $[0, 1) \cup (1, +\infty)$; (e) $\dfrac{x^2 - 1}{\sqrt{x}}$,
domain: $(0, +\infty)$ **13.** (a) $x^2 + 3x - 1$, domain:
$(-\infty, +\infty)$; (b) $x^2 - 3x + 3$, domain: $(-\infty, +\infty)$;
(c) $3x^3 - 2x^2 + 3x - 2$, domain: $(-\infty, +\infty)$; (d) $\dfrac{x^2 + 1}{3x - 2}$,
domain: $\{x | x \neq \frac{2}{3}\}$; (e) $\dfrac{3x - 2}{x^2 + 1}$, domain: $(-\infty, +\infty)$
15. (a) $\dfrac{x^2 + 2x - 2}{x^2 - x - 2}$, domain: $\{x | x \neq -1, x \neq 2\}$;
(b) $\dfrac{-x^2 - 2}{x^2 - x - 2}$, domain: $\{x | x \neq -1, x \neq 2\}$;
(c) $\dfrac{x}{x^2 - x - 2}$, domain: $\{x | x \neq -1, x \neq 2\}$; (d) $\dfrac{x - 2}{x^2 + x}$,
domain: $\{x | x \neq -1, x \neq 0, x \neq 2\}$; (e) $\dfrac{x^2 + x}{x - 2}$, domain:
$\{x | x \neq -1, x \neq 2\}$ **17.** (a) $x + 5$, domain: $(-\infty, +\infty)$;
(b) $x + 5$, domain: $(-\infty, +\infty)$; (c) $x - 14$, domain:
$(-\infty, +\infty)$; (d) $x + 14$, domain: $(-\infty, +\infty)$
19. (a) $x^2 - 6$, domain: $(-\infty, +\infty)$; (b) $x^2 - 10x + 24$,
domain: $(-\infty, +\infty)$; (c) $x - 10$, domain: $(-\infty, +\infty)$;
(d) $x^4 - 2x^2$, domain: $(-\infty, +\infty)$ **21.** (a) $\sqrt{x^2 - 4}$,

domain: $(-\infty, -2] \cup [2, +\infty)$; (b) $x - 4$, domain: $[2, +\infty)$; (c) $\sqrt{\sqrt{x-2}-2}$, domain: $[6, +\infty)$; (d) $x^4 - 4x^2 + 2$, domain: $(-\infty, +\infty)$ **23.** (a) $\dfrac{1}{\sqrt{x}}$, domain: $(0, +\infty)$; (b) $\dfrac{1}{\sqrt{x}}$, domain: $(0, +\infty)$; (c) x, domain: $\{x | x \neq 0\}$; (d) $\sqrt[4]{x}$, domain: $[0, +\infty)$ **25.** (a) $|x + 2|$, domain: $(-\infty, +\infty)$; (b) $||x| + 2|$, domain: $(-\infty, +\infty)$; (c) $|x|$, domain: $(-\infty, +\infty)$; (d) $||x + 2| + 2|$, domain: $(-\infty, +\infty)$ **27.** (a) $2x^2 - 3$, domain: $(-\infty, +\infty)$; (b) $4x^2 - 12x + 9$, domain: $(-\infty, +\infty)$; (c) $4x - 9$, domain: $(-\infty, +\infty)$ **33.** (a) 0; (b) 4; (c) $4 - 2x$ **35.** (a) 1; (b) -1; (c) 1; (d) -1; (e) 1 if $x \leq 0$, -1 if $x > 0$; (f) 1 if $x \geq -1$, -1 if $x < -1$; (g) 1; (h) -1 if $x \neq 0$, 1 if $x = 0$ **37.** (a) even; (b) neither; (c) odd; (d) even; (e) neither; (f) odd; (g) neither; (h) even; (i) even; (j) odd **39.** (a) even; (b) odd; (c) even; (d) even

Exercises 4.3 (Page 208)

1. (a) $P(x) = -2x^2 + 380x - 12{,}000$; (b) \$5600
3. (a) $V(x) = 4x^3 - 46x^2 + 120x$; (b) $[0, 4]$
5. (a) $A(x) = 120x - x^2$; (b) $[0, 120]$
7. (a) $A(x) = 120x - \frac{1}{2}x^2$; (b) $[0, 240]$
9. (a) $A(x) = 96x - \frac{6}{5}x^2$; (b) $[0, 80]$
11. (a) $C(x) = \dfrac{5x - 160}{9}$; (b) $35°$

13. (a) $P(x) = 45x$; (b) \$675 **15.** (a) $P(x) = \sqrt{\dfrac{x}{2}}$;
(b) 1 sec **17.** (a) $W(x) = \dfrac{3.2(10)^9}{x^2}$; (b) 165 lb

19. (a) $f(x) = \dfrac{9x}{490{,}000}(5000 - x)$; (b) 17.6 people per
day **21.** (a) $f(x) = kx(900{,}000 - x)$; (b) $[0, 900{,}000]$
23. (a) $A(x) = 3x + \dfrac{48}{x} + 30$; $(0, +\infty)$

Exercises 4.4 (Page 217)

1. $-1, 3$ **3.** $\dfrac{1 \pm \sqrt{3}}{2}$ **5.** (a) **7.** (a) **9.** (c)

11. (b) **13.** (c) **15.** 3 is a minimum value **17.** 2 is a maximum value **19.** $\frac{10}{3}$ is a maximum value **21.** $-\frac{9}{4}$ is a minimum value **23.** 5 and 5 **25.** \$95 **27.** 60 m by 60 m **29.** 60 m by 120 m **31.** 2500 **33.** $450{,}000$ **35.** 25 and 25 **37.** 5.5 sec; 499 ft **39.** 130

Exercises 4.5 (Page 224)

1.

3.

5.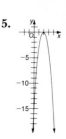

7.

9.

11.

13.

15.

17.

19.

21.

23.

25.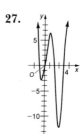

27.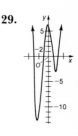

29.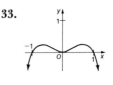

31.

33.

Exercises 4.6 (Page 234)

1. (a) $\{x|x \neq 0\}$;
  (b) no intercepts;
  (c) symmetry with respect to the origin;
  (d) $x = 0$, $y = 0$

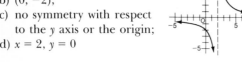

3. (a) $\{x|x \neq 2\}$;
  (b) $(0, -2)$;
  (c) no symmetry with respect to the y axis or the origin;
  (d) $x = 2$, $y = 0$

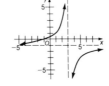

5. (a) $\{x|x \neq 3\}$;
  (b) $(-1, 0)$, $(0, \frac{1}{3})$;
  (c) no symmetry with respect to the y axis or the origin;
  (d) $x = 3$, $y = -1$

7. (a) $\{x|x \neq -4\}$;
  (b) $(2, 0)$, $(0, -1)$;
  (c) no symmetry with respect to the y axis or the origin;
  (d) $x = -4$, $y = 2$

9. (a) $\{x \neq 0\}$;
  (b) no intercepts;
  (c) symmetry with respect to the y axis;
  (d) $x = 0$, $y = 0$

11. (a) $\{x|x \neq 0\}$;
  (b) no intercepts;
  (c) symmetry with respect to the origin;
  (d) $x = 0$, $y = 0$

13. (a) $\{x|x \neq -2\}$;
  (b) $(0, -\frac{1}{4})$;
  (c) no symmetry with respect to the y axis or the origin;
  (d) $x = -2$, $y = 0$

15. (a) $\{x|x \neq \pm 2\}$;
  (b) $(0, 0)$;
  (c) symmetry with respect to the origin;
  (d) $x = -2$, $x = 2$, $y = 0$

17. (a) $\{x|x \neq \pm 4\}$;
  (b) $(0, 0)$;
  (c) symmetry with respect to the origin;
  (d) $x = -4$, $x = 4$, $y = 0$

19. (a) $\{x|x \neq \pm 3\}$;
  (b) $(0, 0)$;
  (c) symmetry with respect to the y axis;
  (d) $x = -3$, $x = 3$, $y = 2$

21. (a) $\{x|x \neq \pm 1\}$;
  (b) $(0, -1)$;
  (c) symmetry with respect to the y axis;
  (d) $x = -1$, $x = 1$, $y = 1$

23. (a) $\{x|x \neq -3, x \neq 2\}$;
  (b) $(-1, 0)$, $(0, -\frac{1}{6})$;
  (c) no symmetry with respect to the y axis or the origin;
  (d) $x = -3$, $x = 2$, $y = 0$

25. (a) $\{x|x \neq 2\}$;
  (b) $(-3, 0)$, $(3, 0)$, $(0, \frac{9}{2})$;
  (c) no symmetry with respect to the y axis or the origin;
  (d) $y = x + 2$

27. (a) $\{x|x \neq 0\}$;
  (b) no intercepts;
  (c) symmetry with respect to the origin;
  (d) $x = 0$, $y = x$

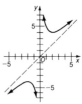

29. (a) $(-\infty, +\infty)$;
  (b) $(0, 1)$;
  (c) symmetry with respect to the y axis;
  (d) $y = 0$

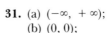

31. (a) $(-\infty, +\infty)$;
  (b) $(0, 0)$;
  (c) symmetry with respect to the y axis;
  (d) $y = 2$

Exercises 4.7 (Page 241)

1. One-to-one   **3.** Not one-to-one   **5.** One-to-one
7. Not one-to-one   **9.** One-to-one   **11.** One-to-one

13. Not one-to-one   **15.** $f^{-1}(x) = \dfrac{3 - x}{4}$

17. $f^{-1}(x) = \sqrt[3]{x - 2}$ **19.** $f^{-1}(x) = \dfrac{1 - x}{x}$

21. (a) $[-5, +\infty)$; (b) $f^{-1}(x) = \sqrt{x + 5}, [-5, +\infty)$
23. (a) $[0, +\infty)$; (b) $f^{-1}(x) = \sqrt{x^2 + 9}, [0, +\infty)$
25. (a) $[0, +\infty)$; (b) $f^{-1}(x) = -\sqrt{x^2 + 9}, [0, +\infty)$
27. (a) $[-\frac{1}{8}, \frac{1}{8}]$; (b) $f^{-1}(x) = 2\sqrt[3]{x}, [-\frac{1}{8}, \frac{1}{8}]$
29. (b) $f^{-1}(x) = \frac{1}{2}(x - 5)$ **31.** $f^{-1}(x) = \sqrt[3]{x} - 1$
33. (b) $f^{-1}(x) = \sqrt{x} + 2$ **35.** $f^{-1}(x) = -\sqrt{4 - x}$
37. $f^{-1}(x) = \frac{5}{9}(x - 32)$

Review Exercises for Chapter 4 (Page 242)

1. Domain: $(-\infty, +\infty)$;
Range: $(-\infty, +\infty)$

3. Domain: $(-\infty, +\infty)$;
Range: $[-4, +\infty)$

5. Domain: $(-\infty, -4]$
$[4, +\infty)$;
Range: $[0, +\infty)$

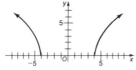

7. Domain: $[-4, 4]$;
Range: $[0, 4]$

9. Domain: $(-\infty, +\infty)$;
Range: $[0, +\infty)$

11. Domain: $\{x | x \neq -4\}$;
Range: $\{y | y \neq -8\}$

13. Domain: $(-\infty, +\infty)$;
Range: $\{y | y \neq -8\}$

15. Domain: $(-\infty, +\infty)$;
Range: $[3, +\infty)$

17. Domain: $(-\infty, +\infty)$;
Range: $[-1, +\infty)$

19. Domain: $(-\infty, +\infty)$;
Range: $(-\infty, 4]$

21. (a) 35; (b) $3x^4 + x^2 + 5$; (c) $-9x^4 + 6x^3 - 31x^2$
$+10x - 25$; (d) $6x + 3h - 1$ **23.** $\dfrac{-1}{\sqrt{1 - x - h} + \sqrt{1 - x}}$

25. (a) $x^2 + 4x - 7$, domain: $(-\infty, +\infty)$; (b) $x^2 - 4x - 1$,
domain: $(-\infty, +\infty)$; (c) $4x^3 - 3x^2 - 16x + 12$,
domain: $(-\infty, +\infty)$; (d) $\dfrac{x^2 - 4}{4x - 3}$, domain: $\{x | x \neq \frac{3}{4}\}$;

(e) $\dfrac{4x - 3}{x^2 - 4}$, domain: $\{x | x \neq \pm 2\}$; (f) $16x^2 - 24x + 5$,

domain: $(-\infty, +\infty)$; (g) $4x^2 - 19$, domain: $(-\infty, +\infty)$;
27. (a) $\sqrt{x + 2} + x^2 - 4$, domain: $[-2, +\infty)$;
(b) $\sqrt{x + 2} - x^2 + 4$, domain: $[-2, +\infty)$;
(c) $\sqrt{x + 2}(x^2 - 4)$, domain: $[-2, +\infty)$;
(d) $\dfrac{\sqrt{x + 2}}{x^2 - 4}$, domain $(-2, 2) \cup (2, +\infty)$; (e) $\dfrac{x^2 - 4}{\sqrt{x + 2}}$,

domain: $(-2, +\infty)$; (f) $\sqrt{x^2 - 2}$,
domain: $(-\infty, -\sqrt{2}] \cup [\sqrt{2}, +\infty)$;
(g) $x - 2$, domain: $[-2, +\infty)$
29. (a) $\dfrac{x^2 - 2x + 1}{x^2 - 2x - 3}$, domain: $\{x | x \neq -1, x \neq 3\}$;

(b) $\dfrac{-x^2 + 4x + 1}{x^2 - 2x - 3}$, domain: $\{x | x \neq -1, x \neq 3\}$;

(c) $\dfrac{x}{x^2 - 2x - 3}$, domain: $\{x | x \neq -1, x \neq 3\}$;

(d) $\dfrac{x + 1}{x^2 - 3x}$, domain: $\{x | x \neq -1, x \neq 0, x \neq 3\}$;

(e) $\dfrac{x^2 - 3x}{x + 1}$, domain: $\{x | x \neq -1, x \neq 3\}$;

(f) $-\dfrac{x + 1}{2x + 3}$, domain: $\{x | x \neq -\frac{3}{2}, x \neq -1\}$;

(g) $\dfrac{1}{x - 2}$, domain: $\{x | x \neq 2, x \neq 3\}$ **31.** (a)   **33.** (b)

35. 6 is a maximum value **37.** -3 is a minimum value

39. **41.** 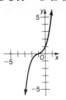 **43.**

45. **47.**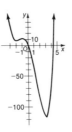

49. (a) $\{x|x \neq 5\}$;
(b) $(0, -\frac{2}{5})$;
(c) no symmetry with respect to the y axis or the origin;
(d) $x = 5$, $y = 0$

51. (a) $\{x|x \neq 0\}$;
(b) no intercepts;
(c) symmetry with respect to the y axis;
(d) $x = 0$, $y = 0$

53. (a) domain: $\{x|x \neq -2\}$;
(b) $(0, 2)$;
(c) no symmetry with respect to the y axis or the origin;
(d) $x = -2$, $y = 0$

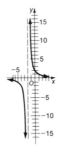

55. (a) $\{x|x \neq \pm 1\}$;
(b) $(0, 0)$;
(c) symmetry with respect to the origin;
(d) $x = -1$, $x = 1$, $y = 0$

57. (a) $\{x|x \neq \pm 3\}$;
(b) $(0, 0)$;
(c) symmetry with respect to the y axis;
(d) $x = -3$, $x = 3$, $y = -4$

59. (a) $\{x|x \neq 1\}$;
(b) $(-2, 0)$, $(2, 0)$, $(0, 4)$;
(c) no symmetry with respect to the y axis or the origin;
(d) $x = 1$, $y = x + 1$

61. (b) $f^{-1}(x) = \frac{1}{5}(x + 2)$ **63.** (b) $f^{-1}(x) = \sqrt[3]{8 - x}$

65. (b) $f^{-1}(x) = \frac{2x + 1}{1 - x}$ **67.** (b) $f^{-1}(x) = \frac{1}{4}(5 - x)$

69. (b) $f^{-1}(x) = 8 - \sqrt[3]{x}$ **71.** (b) $f^{-1}(x) = \sqrt{x + 4}$

73. (a) $f(t) = 16t^2$; (b) 100 ft

75. (a) $f(x) = \frac{x(10,000 - x)}{100,000}$;
(b) 160 fish per week **77.** 6 and 6

79. (a) $f(x) = \begin{cases} 15x & \text{if } 0 \leq x \leq 150 \\ 22.5x - 0.05x^2 & \text{if } 150 > x \leq 250 \end{cases}$;
(b) $[0, 250]$; (c) 225

81. (a) $f(x) = kx(11,000 - x)$; (b) $[0, 11,000]$; (c) 5500

83. (a) $L(r) = kr[200 - r(2 + \frac{1}{2}\pi)]$; (b) $\left[0, \frac{200}{2 + \pi}\right]$

85. $\frac{200}{4 + \pi}$ in. ≈ 28 in.

Chapter 5

Exercises 5.1 (Page 253)

1. (a) $3^{6\sqrt{2}}$; (b) e^{10} **3.** (a) $5^{3\sqrt{10}}$; (b) $5^{\sqrt[3]{x} + \sqrt[3]{x^2}}$
5. (a) $2^{5\sqrt{2}}$ (b) $(\frac{5}{2})^{\sqrt{5}}$ **7.** $e^{2+\sqrt{6}}$ **9.** (a) \$2060;
(b) \$2120; (c) \$2240 **11.** \$10,938.07
13. (a) \$1320; (b) \$1360.49; (c) \$1368.57;
(d) \$1372.79; (e) \$1377.13 **15.** (a) \$849.29;
(b) \$867.55 **17.** (a) \$1094.17; (b) \$1094.16
19. \$3364.86
21. \$3351.60
23. 2.718

25.

Exercises 5.2 (Page 260)

1. **3.**

5. **7.** **9.**

11. **13.** **15.**

17. **19.** **21.**

23. \$907.18 **25.** 1568 lb/ft^2 **27.** (a) 18,221;
(b) 33,201 **29.** (a) $f(t) = 200 \cdot 2^{t/10}$; (b) \$12,800
31. (a) (b) 40; (c) 94; (d) 100

Exercises 5.3 (Page 268)

1. (a) $\log_3 81 = 4$; (b) $\log_5 125 = 3$; (c) $\log_{10}(0.001) = -3$
3. (a) $\log_8 4 = \frac{2}{3}$; (b) $\log_{625}(\frac{1}{125}) = -\frac{3}{4}$; (c) $\log_2 1 = 0$
5. (a) $8^2 = 64$; (b) $3^4 = 81$; (c) $2^0 = 1$
7. (a) $8^{1/3} = 2$; (b) $(\frac{1}{3})^{-2} = 9$; (c) $9^{-1/2} = \frac{1}{3}$ **9.** (a) 2;
(b) $\frac{2}{3}$; (c) -3 **11.** (a) $-\frac{1}{3}$; (b) $-\frac{4}{3}$; (c) $\frac{1}{2}$ **13.** (a) 343;
(b) 81 **15.** (a) $2\sqrt{2}$; (b) $\frac{1}{128}$ **17.** (a) 12; (b) 216
19. (a) 10; (b) 8 **21.** (a) 0; (b) 1 **23.** (a) 3; (b) 0

25. **27.**

29. **31.**

33.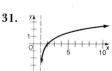

35. 2 years from now **37.** 25 years from now
39. 6.53 years $\approx$ 6 years, 6 months, 11 days
41. 8.66 years $\approx$ 8 years, 7 months, 28 days **43.** 3.5

Exercises 5.4 (Page 276)

1. (a) $\log_b 5 + \log_b x + \log_b y$; (b) $\log_b y - \log_b z$;
(c) $\log_b x - \log_b y - \log_b z$ **3.** (a) $\log_b x + 5\log_b y$;
(b) $\frac{1}{2}(\log_b x + \log_b y)$ **5.** (a) $\frac{1}{3}\log_b x + 3\log_b z$;
(b) $\log_b x + \frac{1}{2}\log_b y - 4\log_b z$
7. (a) $\frac{1}{3}(2\log_b x - \log_b y - 2\log_b z)$;
(b) $\frac{2}{3}\log_b x + \frac{1}{2}(\log_b y + \log_b z)$

9. (a) $\log_{10} x^4\sqrt{y}$; (b) $\log_b \dfrac{\sqrt[4]{x^3}}{y^6\sqrt[5]{z^4}}$ **11.** (a) $\log_{10} \frac{1}{2}gt^2$;

(b) $\ln \frac{1}{3}\pi hr^2$ **13.** (a) 1.1461; (b) 1.1761
15. (a) 1.7993; (b) 2.1461 **17.** (a) 0.3404;
(b) -2.7744 **19.** (a) -0.1761; (b) 0.6309
21. (a) 5.7036; (b) 2.0149 **23.** (a) 0.7851; (b) 2.5108

Exercises 5.5 (Page 284)

1. {1.404} **3.** {0.4307} **5.** {4.301} **7.** {8.638}
9. {32.20} **11.** {1.015} **13.** 2.262 **15.** 4.170
17. 3.638 **19.** 0.9375 **21.** {7} **23.** {125} **25.** {3}
27. {7} **29.** 3.42 years from January 1, 1988, or
June 2, 1991 **31.** 18.6 years **33.** 48,089 ft
35. 1993 **37.** 68.38 years from now **39.** 5.5
41. 55 percent **43.** {0.3828} **45.** {0.8618}
47. {$-2.063, 2.063$}

49. $x = \frac{1}{2}\log \dfrac{1+y}{1-y}$

Review Exercises for Chapter 5 (Page 285)

1. **3.** **5.**

7. **9.** **11.**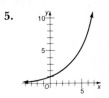

13. 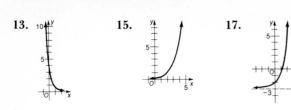 **15.** **17.**

27. (a) 625; (b) $\frac{4}{3}$ **29.** (a) $\frac{1}{64}$; (b) $\dfrac{1}{e^2}$

31. $3 \log_b x + 2 \log_b y + \frac{1}{2} \log_b z$

33. $\log_b x + \frac{1}{3} \log_b y - 4 \log_b z$

35. $\log_{10} \left(\dfrac{\sqrt[3]{y^2}}{x^4 \sqrt[3]{z}} \right)$ **37.** $\ln(\frac{4}{3}\pi r^2 h)$ **39.** $\{2.02\}$

19. 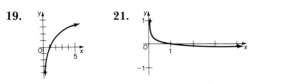 **21.**

41. $\{1.51\}$ **43.** $\{8.08\}$ **45.** $\{-2.54, 2.54\}$

47. 0.9358 **49.** 5.248 **51.** $\{\frac{1}{2}\}$ **53.** $\{400\}$

57. (a) \$960 (b) \$988.80 (c) \$1004.07 **59.** \$1019.97

61. \$14,000 **63.** (a) 5.86 years (b) 5.78 years

65. (a) 57.65 mg (b) 455.8 years **67.** (a) 0; (b) 1.543;

(c) -1.175; (d) 16.543; (e) 3.762

69. (b)

23. 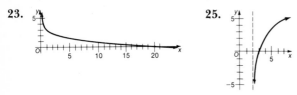 **25.**

71. $S^{-1}(x) = \ln(x + \sqrt{x^2 + 1})$

Chapter 6

Exercises 6.1 (Page 298)

1. (a) first quadrant; (b) second quadrant **3.** (a) first quadrant; (b) fourth quadrant **5.** (a) second quadrant; (b) second quadrant **7.** (a) third quadrant; (b) fourth quadrant **9.** (a) in first quadrant; (b) in second quadrant; (c) quadrantal angle; (d) in third quadrant; (e) in fourth quadrant **11.** (a) in fourth quadrant; (b) quadrantal angle; (c) in third quadrant; (d) in second quadrant; (e) in first quadrant **13.** (a) in first quadrant; (b) in second quadrant; (c) in third quadrant; (d) in fourth quadrant; (e) in second quadrant **15.** (a) $\frac{5}{4}\pi$; (b) $\frac{7}{6}\pi$; (c) $\frac{3}{2}\pi$; (d) $\frac{5}{3}\pi$ **17.** (a) 1; (b) 2.72; (c) 2.03; (d) 1.57 **19.** (a) $\frac{1}{3}\pi$; (b) $\frac{3}{4}\pi$; (c) $\frac{7}{6}\pi$; (d) $-\frac{5}{6}\pi$ **21.** (a) $\frac{1}{4}\pi$; (b) $\frac{5}{2}\pi$; (c) $-\frac{5}{12}\pi$; (d) $\frac{5}{3}\pi$ **23.** (a) 45°; (b) 120°; (c) 330°; (d) $-90°$ **25.** (a) 28.65°; (b) $-114.59°$; (c) 273.87°; (d) 13.18° **27.** (a) 35.37°, 0.62; (b) 102.52°, 1.79 **29.** (a) 315°; (b) 150°; (c) 180°; (d) 240° **31.** (a) 22.56°; (b) 241.76°; (c) 106.15°; (d) 302.36° **33.** (a) 6π in.; (b) 27π in.^{2} **35.** (a) $\frac{9}{2}\pi$ cm; (b) $\frac{27}{2}\pi$ cm^2 **37.** (a) 1.82 in.; (b) 4.30 in.^{2} **39.** 21.70°N **41.** 1009 mi **43.** 560 **45.** 2.79 in. **47.** (a) 611.56; (b) 2342.76

Exercises 6.2 (Page 307)

1. $\sin \theta = \frac{4}{5}$, $\cos \theta = \frac{3}{5}$, $\tan \theta = \frac{4}{3}$, $\csc \theta = \frac{5}{4}$, $\sec \theta = \frac{5}{3}$, $\cot \theta = \frac{3}{4}$ **3.** $\sin \theta = \frac{12}{13}$, $\cos \theta = -\frac{5}{13}$, $\tan \theta = -\frac{12}{5}$, $\csc \theta = \frac{13}{12}$, $\sec \theta = -\frac{13}{5}$, $\cot \theta = -\frac{5}{12}$ **5.** $\sin \theta = -\frac{15}{17}$, $\cos \theta = -\frac{8}{17}$, $\tan \theta = \frac{15}{8}$, $\csc \theta = -\frac{17}{15}$, $\sec \theta = -\frac{17}{8}$, $\cot \theta = \frac{8}{15}$ **7.** $\sin \theta = -\frac{3}{5}$, $\cos \theta = \frac{4}{5}$, $\tan \theta = -\frac{3}{4}$, $\csc \theta = -\frac{5}{3}$, $\sec \theta = \frac{5}{4}$, $\cot \theta = -\frac{4}{3}$ **9.** $\sin \theta = -\dfrac{1}{\sqrt{5}}$, $\cos \theta = \dfrac{2}{\sqrt{5}}$, $\tan \theta = -\frac{1}{2}$, $\csc \theta = -\sqrt{5}$, $\sec \theta = \dfrac{\sqrt{5}}{2}$, $\cot \theta = -2$ **11.** $\sin \theta = -1$, $\cos \theta = 0$, $\csc \theta = -1$, $\cot \theta = 0$, neither $\tan \theta$ nor $\sec \theta$ is defined **13.** $\sin \theta = 0$, $\cos \theta = 1$, $\tan \theta = 0$, $\sec \theta = 1$, neither $\csc \theta$ nor $\cot \theta$ is defined **15.** $\sin \theta = \dfrac{1}{\sqrt{2}}$, $\cos \theta = -\dfrac{1}{\sqrt{2}}$, $\tan \theta = -1$, $\csc \theta = \sqrt{2}$, $\sec \theta = -\sqrt{2}$, $\cot \theta = -1$ **17.** $\sin \theta = -\frac{1}{2}$, $\cos \theta = \dfrac{\sqrt{3}}{2}$, $\tan \theta = -\dfrac{1}{\sqrt{3}}$, $\csc \theta = -2$, $\sec \theta = \dfrac{2}{\sqrt{3}}$, $\cot \theta = -\sqrt{3}$ **19.** (a) fourth quadrant; (b) third quadrant **21.** (a) second quadrant; (b) third quadrant **23.** (a) second quadrant; (b) fourth quadrant **25.** $\cos \theta = \frac{12}{13}$, $\tan \theta = \frac{5}{12}$, $\csc \theta = \frac{13}{5}$, $\sec \theta = \frac{13}{12}$, $\cot \theta = \frac{12}{5}$ **27.** $\sin \theta = \frac{3}{5}$, $\tan \theta = -\frac{3}{4}$, $\csc \theta = \frac{5}{3}$, $\sec \theta = -\frac{5}{4}$, $\cot \theta = -\frac{4}{3}$ **29.** $\sin \theta = -\frac{15}{17}$,

$\cos \theta = -\frac{8}{17}$, $\csc \theta = -\frac{17}{15}$, $\sec \theta = -\frac{17}{8}$, $\cot \theta = \frac{8}{15}$
31. $\sin \theta = -\frac{12}{13}$, $\cos \theta = \frac{5}{13}$, $\tan \theta = -\frac{12}{5}$, $\csc \theta = -\frac{13}{12}$,

$\sec \theta = \frac{13}{5}$ **33.** $\sin \theta = -\frac{1}{2}$, $\cos \theta = -\dfrac{\sqrt{3}}{2}$,

$\tan \theta = \dfrac{1}{\sqrt{3}}$, $\sec \theta = -\dfrac{2}{\sqrt{3}}$, $\cot \theta = \sqrt{3}$ **35.** $\sin \theta = \dfrac{1}{\sqrt{2}}$,

$\cos \theta = -\dfrac{1}{\sqrt{2}}$, $\csc \theta = \sqrt{2}$, $\sec \theta = -\sqrt{2}$, $\cot \theta = -1$

37. $\sin \theta \approx 0.6238$, $\tan \theta \approx 0.7981$, $\csc \theta \approx 1.6031$,
$\sec \theta \approx 1.2794$, $\cot \theta \approx 1.2530$ **39.** $\cos \theta \approx -0.9085$,
$\tan \theta \approx 0.4599$, $\csc \theta \approx -2.3935$, $\sec \theta \approx -1.1007$,
$\cot \theta \approx 2.1744$ **41.** $\sin \theta \approx -0.7639$, $\tan \theta \approx -1.1838$,
$\csc \theta \approx -1.3091$, $\sec \theta \approx 1.5497$, $\cot \theta \approx -0.8447$
43. $\cos \theta \approx -0.9543$, $\tan \theta \approx -0.3133$,
$\csc \theta \approx 3.3445$, $\sec \theta \approx -1.0479$, $\cot \theta \approx -3.1915$

Exercises 6.3 (Page 319)

1. (a) $\dfrac{\sqrt{3}}{2}$; (b) $\dfrac{\sqrt{3}}{2}$; (c) 1 **3.** (a) $\dfrac{1}{\sqrt{2}}$; (b) $\sqrt{3}$; (c) $\sqrt{3}$

5. $\sin 135° = \dfrac{1}{\sqrt{2}}$, $\cos 135° = -\dfrac{1}{\sqrt{2}}$, $\tan 135° = -1$,
$\csc 135° = \sqrt{2}$, $\sec 135° = -\sqrt{2}$, $\cot 135° = -1$

7. $\sin 210° = -\frac{1}{2}$, $\cos 210° = -\dfrac{\sqrt{3}}{2}$, $\tan 210° = \dfrac{1}{\sqrt{3}}$,
$\csc 210° = -2$, $\sec 210° = -\dfrac{2}{\sqrt{3}}$, $\cot 210° = \sqrt{3}$

9. $\sin \frac{5}{3}\pi = -\dfrac{\sqrt{3}}{2}$, $\cos \frac{5}{3}\pi = \frac{1}{2}$, $\tan \frac{5}{3}\pi = -\sqrt{3}$,
$\csc \frac{5}{3}\pi = -\dfrac{2}{\sqrt{3}}$, $\sec \frac{5}{3}\pi = 2$, $\cot \frac{5}{3}\pi = -\dfrac{1}{\sqrt{3}}$

11. $\sin(-\frac{1}{6}\pi) = -\frac{1}{2}$, $\cos(-\frac{1}{6}\pi) = \dfrac{\sqrt{3}}{2}$, $\tan(-\frac{1}{6}\pi) = -\dfrac{1}{\sqrt{3}}$,
$\csc(-\frac{1}{6}\pi) = -2$, $\sec(-\frac{1}{6}\pi) = \dfrac{2}{\sqrt{3}}$, $\cot(-\frac{1}{6}\pi) = -\sqrt{3}$

13. $\sin(-120°) = -\dfrac{\sqrt{3}}{2}$, $\cos(-120°) = -\frac{1}{2}$,
$\tan(-120°) = \sqrt{3}$, $\csc(-120°) = -\dfrac{2}{\sqrt{3}}$,
$\sec(-120°) = -2$, $\cot(-120°) = \dfrac{1}{\sqrt{3}}$

15. $\sin(-\frac{5}{4}\pi) = \dfrac{1}{\sqrt{2}}$, $\cos(-\frac{5}{4}\pi) = -\dfrac{1}{\sqrt{2}}$, $\tan(-\frac{5}{4}\pi) = -1$,
$\csc(-\frac{5}{4}\pi) = \sqrt{2}$, $\sec(-\frac{5}{4}\pi) = -\sqrt{2}$, $\cot(-\frac{5}{4}\pi) = -1$
17. $\sin(-210°) = \frac{1}{2}$, $\cos(-210°) = -\dfrac{\sqrt{3}}{2}$,

$\tan(-210°) = -\dfrac{1}{\sqrt{3}}$, $\csc(-210°) = 2$,

$\sec(-210°) = -\dfrac{2}{\sqrt{3}}$, $\cot(-210°) = -\sqrt{3}$

19. $\sin 480° = \dfrac{\sqrt{3}}{2}$, $\cos 480° = -\frac{1}{2}$, $\tan 480° = -\sqrt{3}$,

$\csc 480° = \dfrac{2}{\sqrt{3}}$, $\sec 480° = -2$, $\cot 480° = -\dfrac{1}{\sqrt{3}}$

21. $\sin 360° = 0$, $\cos 360° = 1$, $\tan 360° = 0$,
$\sec 360° = 1$, $\csc 360°$ and $\cot 360°$ are not defined
23. $\sin(-90°) = -1$, $\cos(-90°) = 0$, $\cot(-90°) = 0$,
$\csc(-90°) = -1$, $\tan(-90°)$ and $\sec(-90°)$ are not
defined **25.** $\sin 3\pi = 0$, $\cos 3\pi = -1$, $\tan 3\pi = 0$,
$\sec 3\pi = -1$, $\cot 3\pi$ and $\csc 3\pi$ are not defined
27. (a) 0.5358; (b) 0.5358; (c) 3.420; (d) 3.420;
(e) 1.402; (f) 1.402 **29.** (a) 0.9677; (b) 0.9252;
(c) 0.6950; (d) 0.0332; (e) 1.646; (f) 21.23
31. (a) 0.3335; (b) 0.9428; (c) 3.113; (d) 0.3212;
(e) 2.010; (f) 4.797 **33.** (a) 0.9898; (b) 0.9239;
(c) 0.8391; (d) 0.2282; (e) 1.051; (f) 1.006
35. (a) 60°; (b) 55°; (c) $\frac{1}{8}\pi$; (d) $\frac{1}{4}\pi$; (e) 0.60; (f) 0.31
37. (a) 62°36′; (b) 52.7°; (c) $\frac{2}{9}\pi$; (d) $\frac{3}{8}\pi$; (e) 0.23;

(f) 1.42 **39.** (a) $\sin 45° = \dfrac{1}{\sqrt{2}}$; (b) $-\cos 30° = -\dfrac{\sqrt{3}}{2}$;

(c) $-\tan 60° = -\sqrt{3}$; (d) $-\cot 30° = -\sqrt{3}$;

(e) $-\csc 60° = -\dfrac{2}{\sqrt{3}}$ **41.** (a) $-\sin \frac{1}{3}\pi = -\dfrac{\sqrt{3}}{2}$;

(b) $\cos \frac{1}{4}\pi = \dfrac{1}{\sqrt{2}}$; (c) $\tan \frac{1}{4}\pi = 1$; (d) $-\sec \frac{1}{6}\pi = -\dfrac{2}{\sqrt{3}}$;

(e) $-\csc \frac{1}{6}\pi = -2$ **43.** (a) $\sin 55°42′ = 0.8261$;
(b) $-\cos 63°36′ = -0.4446$; (c) $-\tan 15°6′ = -0.2698$
45. (a) $-\cos 7.6° = -0.9912$; (b) $\sin 83.8° = 0.9942$;
(c) $\tan 20.2° = 0.3679$ **47.** (a) $\cot 10°30′ = 5.3955$;
(b) $\sec 67.4° = 2.6022$; (c) $\csc 4.5° = 12.7455$
49. (a) $\sin 0.612 = 0.574$; (b) $\cos 0.823 = 0.680$;
(c) $\tan 0.142 = 0.143$ **51.** (a) $\sin 0.183 = 0.182$;
(b) $\cos 0.936 = 0.593$; (c) $-\tan 1.028 = -1.66$
53. (a) $-\sin 1.442 = -0.992$; (b) $-\cos 0.991 = -0.548$;
(c) $\tan 1.274 = 3.27$ **55.** (a) 0.7385; (b) 1.354;
(c) -0.8877; (d) -1.126 **57.** (a) -0.9143;
(b) -1.094; (c) 0.9240; (d) 1.082 **59.** (a) 1.775;
(b) 0.5635; (c) -0.2910; (d) -3.436
61. (a) -0.9511; (b) -1.051; (c) 3.078; (d) 0.3249

Exercises 6.4 (Page 327)

1. $\sin 2 \approx 0.91$, $\cos 2 \approx -0.42$ **3.** $\sin 5.2 \approx -0.88$,
$\cos 5.2 \approx 0.47$ **5.** $\sin(-3) \approx -0.14$, $\cos(-3) \approx -0.99$
7. $\sin(-6.1) \approx 0.18$, $\cos(-6.1) \approx 0.98$ **9.** $x = \frac{1}{2}\sqrt{3}$,
$y = \frac{1}{2}$ **11.** (a) $\sin(-\frac{1}{2}\pi) = -1$, $\cos(-\frac{1}{2}\pi) = 0$;

(b) $\sin \frac{5}{4}\pi = -\dfrac{1}{\sqrt{2}}$, $\cos \frac{5}{4}\pi = -\dfrac{1}{\sqrt{2}}$ **13.** (a) $\sin \frac{7}{6}\pi = -\frac{1}{2}$,

$\cos \frac{7}{6}\pi = -\dfrac{\sqrt{3}}{2}$; (b) $\sin(-\frac{1}{3}\pi) = -\dfrac{\sqrt{3}}{2}$, $\cos(-\frac{1}{3}\pi) = \frac{1}{2}$

15. (a) $\sin \frac{2}{3}\pi = \dfrac{\sqrt{3}}{2}$, $\cos \frac{2}{3}\pi = -\frac{1}{2}$; (b) $\sin(-\frac{7}{6}\pi) = \frac{1}{2}$,

$\cos(-\frac{7}{6}\pi) = -\dfrac{\sqrt{3}}{2}$ **17.** (a) $\sin 4\pi = 0$, $\cos 4\pi = 1$;

(b) $\sin \frac{5}{2}\pi = 1$, $\cos \frac{5}{2}\pi = 0$ **19.** (a) $\sin 5\pi = 0$,

$\cos 5\pi = -1$; (b) $\sin \frac{9}{4}\pi = \dfrac{1}{\sqrt{2}}$, $\cos \frac{9}{4}\pi = \dfrac{1}{\sqrt{2}}$

21. (a) $\sin \frac{11}{3}\pi = -\dfrac{\sqrt{3}}{2}$, $\cos \frac{11}{3}\pi = \frac{1}{2}$;

(b) $\sin(-\frac{11}{4}\pi) = -\dfrac{1}{\sqrt{2}}$, $\cos(-\frac{11}{4}\pi) = -\dfrac{1}{\sqrt{2}}$

23. (a) $\sin \frac{23}{6}\pi = -\frac{1}{2}$, $\cos \frac{23}{6}\pi = \dfrac{\sqrt{3}}{2}$;

(b) $\sin(-\frac{14}{3}\pi) = -\dfrac{\sqrt{3}}{2}$, $\cos(-\frac{14}{3}\pi) = -\frac{1}{2}$

25. (a) $\frac{1}{2}$; (b) $-\frac{1}{2}$; (c) $-\dfrac{1}{\sqrt{2}}$ **27.** (a) 1; (b) -1;

(c) $-\dfrac{\sqrt{3}}{2}$ **29.** (a) $-\dfrac{1}{\sqrt{2}}$; (b) 1; (c) -1 **31.** (a) $\dfrac{1}{\sqrt{2}}$;

(b) $\frac{1}{2}$ **33.** (a) 1; (b) $-\frac{1}{2}$ **35.** (a) 0; (b) $\dfrac{1}{\sqrt{2}}$

37. (a) $-\dfrac{1}{\sqrt{2}}$; (b) $-\frac{1}{2}$ **39.** (a) 0.99833; (b) 0.99893;

(c) 0.99940; (d) 0.99973; (e) 0.99993; (f) 0.99998;

(g) 1.00000; (h) 1

Exercises 6.5 (Page 340)

1. (a) $\frac{2}{3}\pi$; (b) $\frac{1}{3}\pi$; (c) 8π; (d) 10π **3.** (a) $\frac{1}{2}$; (b) 6

5. (a) 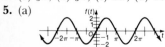   (b)

7. (a)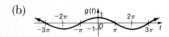

(b)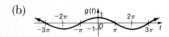

9. (a) 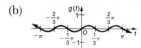   (b)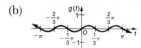

11. (a)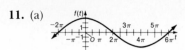

(b)

13.   **15.**

17. 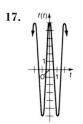   **19.**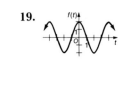

21.   **23.**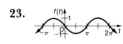

25. 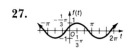   **27.**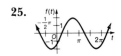

29. 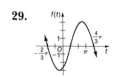   **31.**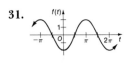

33. 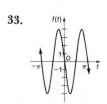   **35.**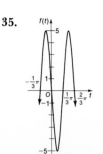

37.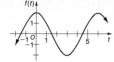

Exercises 6.6 (Page 352)

1. (a) -1; (b) -1; (c) $-\sqrt{3}$; (d) $-\dfrac{1}{\sqrt{3}}$; (e) 0; (f) 0

3. (a) $\frac{1}{4}\pi$; (b) 2π; (c) 3; (d) $\frac{1}{2}$ **5.** (a) $\sqrt{2}$; (b) $-\dfrac{2}{\sqrt{3}}$;

(c) $\dfrac{2}{\sqrt{3}}$; (d) $\sqrt{2}$; (e) -1; (f) -1 **7.** (a) $\frac{2}{5}\pi$; (b) $\frac{8}{3}\pi$;

(c) 1; (d) 3

9. (a) 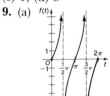   (b) $t = \frac{1}{2}\pi$, $t = \frac{3}{2}\pi$

11. (a) 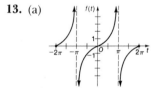   (b) $t = 0$, $t = \frac{1}{2}\pi$, $t = \pi$

13. (a) 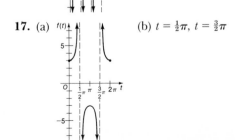   (b) $t = -\pi$, $t = \pi$

15. (a) (b) $t = 0$, $t = \frac{1}{3}\pi$, $t = \frac{2}{3}\pi$, $t = \pi$

17. (a) (b) $t = \frac{1}{2}\pi$, $t = \frac{3}{2}\pi$

19. (a) 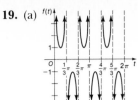   (b) $t = 0$, $t = \frac{1}{3}\pi$, $t = \frac{2}{3}\pi$, $t = \pi$, $t = \frac{4}{3}\pi$, $t = \frac{5}{3}\pi$, $t = \pi$

21. (a)    (b) $t = 0$, $t = 1$, $t = 2$

23. (a) 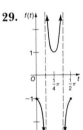   (b) $t = 1$ **25.**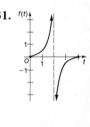

27. 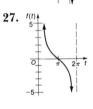   **29.** 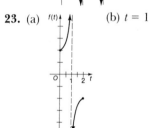   **31.**

33. 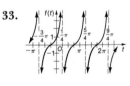   **35.**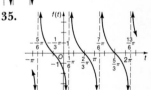

37. 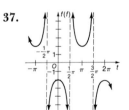   **39.**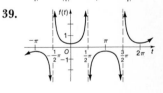

41.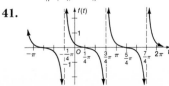

Exercises 6.7 (Page 362)

1. $\sin \theta = \dfrac{4}{\sqrt{65}}$; $\cos \theta = \dfrac{7}{\sqrt{65}}$; $\tan \theta = \frac{4}{7}$; $\csc \theta = \dfrac{\sqrt{65}}{4}$;

$\sec \theta = \dfrac{\sqrt{65}}{7}$; $\cot \theta = \frac{7}{4}$ 3. $\sin \theta = \frac{2}{3}$; $\cos \theta = \dfrac{\sqrt{5}}{3}$;

$\tan \theta = \dfrac{2}{\sqrt{5}}$; $\csc \theta = \frac{3}{2}$; $\sec \theta = \dfrac{3}{\sqrt{5}}$; $\cot \theta = \dfrac{\sqrt{5}}{2}$

5. (a) $\cos 30°$; (b) $\sin 5.7°$; (c) $\cot 40.2°$; (d) $\tan 37.9°$;
(e) $\sec 22.5°$; 7. (a) $\cos 42°42'$; (b) $\sin 18°18'$;
(c) $\cot 44°$; (d) $\tan 34°54'$; (e) $\csc 25°30'$ 9. $a = 2.3$;
$b = 1.4$; $\beta = 30°$ 11. $b = 56$; $c = 79$; $\alpha = 45°$
13. $b = 36$; $c = 39$; $\beta = 66°$ 15. $a = 14$; $b = 42$;
$\alpha = 19°$ 17. $a = 28$; $\alpha = 47°$; $\beta = 43°$ 19. $b = 3.20$;
$c = 5.26$; $\alpha = 52.6°$ 21. $b = 448$; $c = 468$; $\beta = 73.1°$
23. $c = 86.1$; $\alpha = 47.6°$; $\beta = 42.4°$ 25. $a = 532.8$;
$b = 327.4$; $\beta = 31.57°$ 27. $a = 42.36$; $c = 56.06$;
$\alpha = 49.08°$ 29. $b = 746.1$; $\alpha = 22.74°$; $\beta = 67.26°$
31. $a = 16.4$; $b = 35.6$; $\alpha = 24°42'$ 33. $a = 163$;
sq. units $= 330$; $\beta = 60°24'$ 35. $(2.9)10^2$ sq. units
37. $(1.167)10^5$ 39. 333 m   41. $13.32°$ 43. (a) 50
nautical miles; (b) $162°$; (c) $342°$ 45. 108 m   47. 367 m

Exercises 6.8 (Page 370)

1. (a) 3; (b) 12; (c) $\frac{1}{12}$ of a vibration per second; (d) at
0 sec, 3 cm above the central position; at 2 sec, $\frac{3}{2}$ cm
above the central position; at 4 sec, $\frac{3}{2}$ cm below the
central position; at 6 sec, 3 cm below the central

position.   3. (a) 5; (b) π; (c) $\dfrac{1}{\pi}$ of a vibration per

second; (d) at 0 sec, at the central position; at $\frac{1}{4}\pi$ sec,
5 cm above the central position; at $\frac{1}{2}\pi$ sec, at the
central position; at $\frac{3}{4}\pi$ sec, 5 cm below the central
position   5. (a) 8; (b) 1; (c) 1 vibration per second;
(d) at 0 sec, 4 cm above the central position; at $\frac{1}{8}$ sec,
8 cm above the central position; at $\frac{1}{4}$ sec, 4 cm above
the central position; at $\frac{1}{2}$ sec, 4 cm below the central
position   7. Initially the weight is 3 cm above the
central position. In the first 2 seconds the weight
moves downward a distance of $\frac{3}{2}$ cm to a point $\frac{3}{2}$ cm
above the central position. In the next 1 second the
weight moves down to the central position. The speed
is increasing in the first 3 seconds. In the next second
the weight continues downward while its speed begins
to decrease. After a total of 6 seconds the weight is at
its lowest point, 3 cm below its central position with
no speed. The weight then reverses direction and its

speed increases upward until it attains the central
position again at 9 seconds. The weight continues to
move upward with decreasing speed until it has
returned to its initial position after a total of 12
seconds. This completes one cycle of motion.

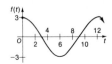

9. Initially, the weight is 4 cm above its central
position. It moves upward with decreasing
speed until $\frac{1}{6}$ second. At $\frac{1}{6}$ second the weight
has reached it highest position of 8 cm and
reverses direction, accelerating downward until
$\frac{5}{12}$ second, when its reaches is central position.
From there it continues downward with decreasing speed until $\frac{2}{3}$ second when it reaches its
lowest point, 8 cm below the central position,
where it again reverses direction and starts
upward again. It continues accelerating up-
ward until $\frac{11}{12}$ second when it passes the central
position again. It continues to move upward
with decreasing speed until at 1 second it
returns to its initial position 4 cm above
the central position.

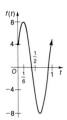

11. (a) $f(t) = 2 \sin 4\pi (t + \frac{1}{8})$ or, equivalently,
$f(t) = 2 \cos 4\pi t$; (b) 1 cm above the central position;
(c) 2 cm below the central position
13. $f(t) = 9 \cos \frac{5}{8}\pi(t - \frac{1}{5})$ 15. (a) $T(t) = 5 \sin (\frac{3}{2}\pi - \frac{1}{12}\pi t)$;
(b) $-2.5°$ Celsius; (c) $4.33°$ Celsius; (d) $4.33°$ Celsius;
(e) $-2.5°$ Celsius   17. (a) $-\sqrt{3}$ cm from the origin
with a velocity of $\frac{1}{3}\pi$ cm/sec and an acceleration of
$\dfrac{\sqrt{3}}{9}\pi^2$ cm/sec^2; (b) at the origin with a velocity of
$\frac{2}{3}\pi$ cm/sec and an acceleration of 0; (c) $\sqrt{3}$ cm from
the origin with a velocity of $\frac{1}{3}\pi$ cm/sec and an

acceleration of $-\dfrac{\sqrt{3}}{9}\pi^2$ cm/sec²; (d) $\sqrt{3}$ cm from the

origin with a velocity of $-\frac{1}{3}\pi$ cm/sec and an

acceleration of $-\dfrac{\sqrt{3}}{9}\pi^2$ cm/sec²; (e) at the origin with

a velocity of $-\frac{2}{3}\pi$ cm/sec and an acceleration of 0.

19. (a) $\frac{1}{25}$; (b) -2 volts; (c) 0; (d) 2 volts; (e) -2 volts
21. (a) $\frac{1}{1400}\pi$; (b) 3.350 amperes; (c) 8.546 amperes;
(d) 9.906 amperes; (e) 2.709 amperes   **23.** (a) $\frac{1}{750}$;
(b) 0.01732 dynes/cm²; (c) -0.02 dynes/cm²;
(d) 0.01564 dynes/cm²; (e) 0

Exercises 6.9 (Page 377)

1.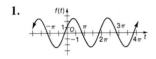

3.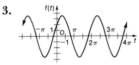

5.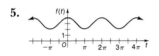

7.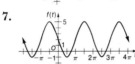

9.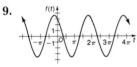

11.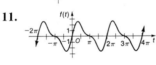

13.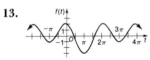

15.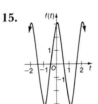

17.

19.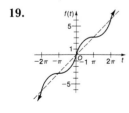

21.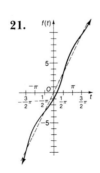

23.

25.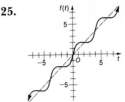

27.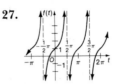

29.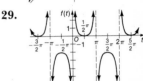

31.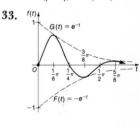

33.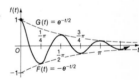

35.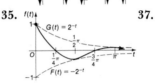

37.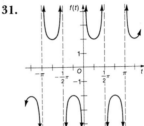

39.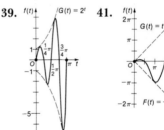
41.
43.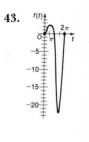

45.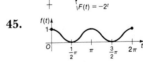

Review Exercises for Chapter 6 (Page 378)

1. (a) $\frac{1}{6}\pi$; (b) $\frac{5}{4}\pi$ **3.** (a) $-\frac{2}{3}\pi$; (b) 1.75   **5.** (a) 60°;
(b) 315°   **7.** (a) $-19.1°$; (b) 135.2°   **9.** (a) 10π cm;
(b) 60π cm²   **11.** $\sin\theta = \frac{15}{17}$, $\cos\theta = -\frac{8}{17}$, $\tan\theta = -\frac{15}{8}$,
$\csc\theta = \frac{17}{15}$, $\sec\theta = -\frac{17}{8}$, $\cot\theta = -\frac{8}{15}$ **13.** $\sin\theta = -\frac{12}{13}$,
$\cos\theta = -\frac{5}{13}$, $\tan\theta = \frac{12}{5}$, $\csc\theta = -\frac{13}{12}$, $\sec\theta = -\frac{13}{5}$,
$\csc\theta = -\frac{13}{12}$ **15.** $\cos\theta = -\frac{4}{5}$, $\tan\theta = -\frac{3}{4}$, $\csc\theta = \frac{5}{3}$,
$\sec\theta = -\frac{5}{4}$, $\cot\theta = -\frac{4}{3}$ **17.** $\sin\theta = -\frac{15}{17}$, $\cos\theta = \frac{8}{17}$,
$\tan\theta = -\frac{15}{8}$, $\csc\theta = -\frac{17}{15}$, $\sec\theta = \frac{17}{8}$ **19.** $\sin 45° = \dfrac{1}{\sqrt{2}}$,

$\cos 45° = \dfrac{1}{\sqrt{2}}$, $\tan 45° = 1$, $\csc 45° = \sqrt{2}$, $\sec 45° = \sqrt{2}$,

$\cot 45° = 1$ **21.** $\sin \frac{2}{3}\pi = \dfrac{\sqrt{3}}{2}$, $\cos \frac{2}{3}\pi = -\frac{1}{2}$,

$\tan \frac{2}{3}\pi = -\sqrt{3}$, $\csc \frac{2}{3}\pi = \dfrac{2}{\sqrt{3}}$, $\sec \frac{2}{3}\pi = -2$,

$\cot \frac{2}{3}\pi = -\dfrac{1}{\sqrt{2}}$ **23.** $\sin(-150°) = -\frac{1}{2}$,

$\cos(-150°) = -\dfrac{\sqrt{3}}{2}$, $\tan(-150°) = \dfrac{1}{\sqrt{3}}$,

$\csc(-150°) = -2$, $\sec(-150°) = -\dfrac{2}{\sqrt{3}}$, $\cot(-150°) = \sqrt{3}$

25. $\sin(-\frac{9}{4}\pi) = -\dfrac{1}{\sqrt{2}}$, $\cos(-\frac{9}{4}\pi) = \dfrac{1}{\sqrt{2}}$, $\tan(-\frac{9}{4}\pi) = -1$,

$\csc(-\frac{9}{4}\pi) = -\sqrt{2}$, $\sec(-\frac{9}{4}\pi) = \sqrt{2}$, $\cot(-\frac{9}{4}\pi = -1)$
27. $\sin 270° = -1$, $\cos 270° = 0$, $\csc 270° = -1$,
$\cot 270° = 0$, $\tan 270°$ and $\sec 270°$ are not defined
29. $\sin(-\pi) = 0$, $\cos(-\pi) = -1$, $\tan(-\pi) = 0$,
$\sec(-\pi) = -1$, $\cot(-\pi)$ and $\csc(-\pi)$ are not defined
31. (a) 45°; (b) 38°; (c) $\frac{1}{6}\pi$; (d) 1.13; (e) $\frac{4}{9}\pi$; (f) 1.33

33. (a) $\sin 30° = \frac{1}{2}$; (b) $-\cos 45° = -\dfrac{1}{\sqrt{2}}$;
(c) $-\tan 60° = -\sqrt{3}$; (d) $-\cot \frac{1}{4}\pi = -1$;
(e) $-\sec \frac{1}{3}\pi = -2$; (f) $-\csc \frac{1}{6}\pi = -2$
35. (a) $\sin 43.6° = 0.6896$; (b) $-\cos 75.2° = -0.2554$;
(c) $-\tan 32.9° = -0.6469$ **37.** (a) $-\sin 0.428 = -0.415$; (b) $-\cos 0.722 = 0.751$; (c) $\tan 0.848 = 1.134$

39. (a) $\sin \frac{1}{4}\pi = \dfrac{1}{\sqrt{2}}$, $\cos \frac{1}{4}\pi = \dfrac{1}{\sqrt{2}}$, $\tan \frac{1}{4}\pi = 1$,
$\csc \frac{1}{4}\pi = \sqrt{2}$, $\sec \frac{1}{4}\pi = \sqrt{2}$, $\cot \frac{1}{4}\pi = 1$; (b) $\sin \frac{1}{6}\pi = \frac{1}{2}$,
$\cos \frac{1}{6}\pi = \dfrac{\sqrt{3}}{2}$, $\tan \frac{1}{6}\pi = \dfrac{1}{\sqrt{3}}$, $\csc \frac{1}{6}\pi = 2$, $\sec \frac{1}{6}\pi = \dfrac{2}{\sqrt{3}}$,
$\cot \frac{1}{6}\pi = \sqrt{3}$; (c) $\sin 0 = 0$, $\cos 0 = 1$, $\tan 0 = 0$,
$\sec 0 = 1$, $\csc 0$ and $\cot 0$ are not defined

41. (a) $\sin \frac{3}{4}\pi = \dfrac{1}{\sqrt{2}}$, $\cos \frac{3}{4}\pi = -\dfrac{1}{\sqrt{2}}$, $\tan \frac{3}{4}\pi = -1$,
$\csc \frac{3}{4}\pi = \sqrt{2}$, $\sec \frac{3}{4}\pi = -\sqrt{2}$, $\cot \frac{3}{4}\pi = -1$;
(b) $\sin(-\frac{5}{6}\pi) = -\frac{1}{2}$, $\cos(-\frac{5}{6}\pi) = -\dfrac{\sqrt{3}}{2}$, $\tan(-\frac{5}{6}\pi) = \dfrac{1}{\sqrt{3}}$,
$\csc(-\frac{5}{6}\pi) = -2$, $\sec(-\frac{5}{6}\pi) = -\dfrac{2}{\sqrt{3}}$, $\cot(-\frac{5}{6}\pi) = \sqrt{3}$;
(c) $\sin 3\pi = 0$, $\cos 3\pi = -1$, $\tan 3\pi = 0$, $\sec 3\pi = -1$,
$\csc 3\pi$ and $\cot 3\pi$ are not defined
43. (a) $\sin(-\frac{4}{3}\pi) = \dfrac{\sqrt{3}}{2}$, $\cos(-\frac{4}{3}\pi) = -\frac{1}{2}$, $\tan(-\frac{4}{3}\pi) = -\sqrt{3}$,
$\csc(-\frac{4}{3}\pi) = \dfrac{2}{\sqrt{3}}$, $\sec(-\frac{4}{3}\pi) = -2$, $\cot(-\frac{4}{3}\pi) = -\dfrac{1}{\sqrt{3}}$;

(b) $\sin \frac{17}{4}\pi = \dfrac{1}{\sqrt{2}}$, $\cos \frac{17}{4}\pi = \dfrac{1}{\sqrt{2}}$, $\tan \frac{17}{4}\pi = 1$,
$\csc \frac{17}{4}\pi = \sqrt{2}$, $\sec \frac{17}{4}\pi = \sqrt{2}$, $\cot \frac{17}{4}\pi = 1$;
(c) $\sin(-\frac{9}{2}\pi) = -1$, $\cos(-\frac{9}{2}\pi) = 0$, $\csc(-\frac{9}{2}\pi) = -1$,
$\cot(-\frac{9}{2}\pi) = 0$, $\tan(-\frac{9}{2}\pi)$ and $\sec(-\frac{9}{2}\pi)$ are not defined
45. (a) $-\dfrac{1}{\sqrt{2}}$; (b) $-\frac{1}{2}$; (c) $\dfrac{2}{\sqrt{3}}$; (d) 0; (e) 0; (f) 1

47. (a) 1; (b) $-\dfrac{1}{\sqrt{3}}$; (c) not defined;
(d) $-\sqrt{3}$; (e) $\sqrt{3}$; (f) not defined
49. (a) $\frac{2}{5}\pi$; (b) 10π; (c) $\frac{4}{5}\pi$; (d) $\frac{8}{3}\pi$
51. (a) $\frac{1}{3}$; (b) $\frac{10}{3}$; (c) $\frac{1}{2}$; (d) 3

53.

55.

57.

59.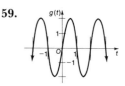

61.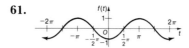

63.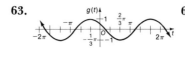

65.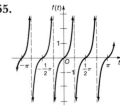

67.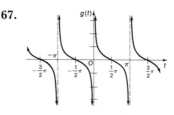

69.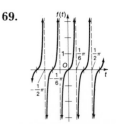

71.

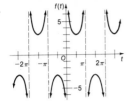

73.

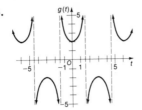

75.

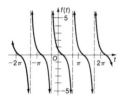

77. (a) cos 15°; (b) sin 27.7°; (c) cot 42.8°; (d) tan 2.4°; (e) csc 34.9°; (f) sec 11.1°   **79.** $a = 9.2$; $c = 18$; $\beta = 60°$
81. $c = 5.8$; $\alpha = 56°$; $\beta = 34°$ **83.** $a = 520.2$; $b = 650.3$; $\alpha = 38.66°$ **85.** 7.7 sq units   **87.** 336
89. (a) 100 ft; (b) 152 ft   **91.** (a) $\frac{2}{3}\pi$; (b) at the central position; (c) 0.28 cm above the central position; (d) 0.56 cm below the central position; (e) 1.30 cm

above the central position   **93.** (a) $f(t) = 6 \sin \frac{10}{9}\pi(t - \frac{9}{20})$ or, equivalently, $f(t) = -6 \cos \frac{10}{9}\pi t$; (b) 1.04 cm above the central position; (c) 5.64 cm above the central position; (d) 4.60 cm below the central position; (e) 1.04 cm below the central position
95. (a) $\frac{1}{50}$; (b) -4 volts; (c) 4 volts; (d) -4 volts; (e) 4 volts
97.

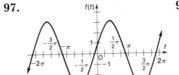

99.

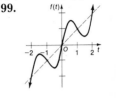

101.

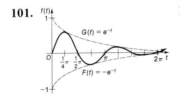

103.

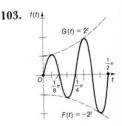

Chapter 7

Exercises 7.1 (Page 386)

1. $\cos\theta = \frac{12}{13}$, $\tan\theta = \frac{5}{12}$, $\csc\theta = \frac{13}{5}$, $\sec\theta = \frac{13}{12}$, $\cot\theta = \frac{12}{5}$
3. $\sin t = \frac{3}{5}$, $\tan t = \frac{3}{4}$, $\csc t = \frac{5}{3}$, $\sec t = \frac{5}{4}$, $\cot t = \frac{4}{3}$
5. $\sin x = -\frac{15}{17}$, $\tan x = -\frac{15}{8}$, $\csc x = -\frac{17}{15}$, $\sec x = \frac{17}{8}$, $\cot x = -\frac{8}{15}$ **7.** $\cos u = \frac{12}{13}$, $\tan u = -\frac{5}{12}$, $\csc u = -\frac{13}{5}$, $\sec u = \frac{13}{12}$, $\cot u = -\frac{12}{5}$ **9.** $\sin\beta = -\frac{5}{6}$, $\cos\beta = -\frac{\sqrt{11}}{6}$, $\tan\beta = \frac{5}{\sqrt{11}}$, $\sec\beta = -\frac{6}{\sqrt{11}}$, $\cot\beta = \frac{\sqrt{11}}{5}$
11. $\sin y = \frac{4}{5}$, $\cos y = -\frac{3}{5}$, $\tan y = -\frac{4}{3}$, $\csc y = \frac{5}{4}$, $\sec y = -\frac{5}{3}$
13. $\sin\theta = \frac{12}{13}$, $\cos\theta = -\frac{5}{13}$, $\csc\theta = \frac{13}{12}$, $\sec\theta = -\frac{13}{5}$, $\cot\theta = -\frac{5}{12}$ **15.** $\sin x = -\frac{3}{\sqrt{34}}$, $\cos x = -\frac{5}{\sqrt{34}}$, $\tan x = \frac{3}{5}$, $\sec x = -\frac{\sqrt{34}}{5}$, $\cot x = \frac{5}{3}$ **17.** 1.302
19. -1.116 **21.** 0.9010   **23.** -2.588 **25.** -1.019
27. 0.8747   **29.** $\frac{1}{\cos t}$ **31.** $\frac{\cos x}{1 - \cos^2 x}$
33. $\sin\theta(1 - \sin^2\theta)$ **35.** $-\sin^2 z$ **37.** $\frac{1}{\sin\gamma}$
39. $\frac{\pm 1}{(1 - \cos^2 y)^{3/2}}$ **41.** $\sin^2\beta - \frac{1}{\sin^2\beta} + 1$

Exercises 7.2 (Page 391)

1. (a) $\sin x$; (b) $\cos^2 x$ **3.** (a) $\frac{1}{\sin x}$; (b) $\cos^2 x$
5. (a) $\frac{1}{\sin x \cos x}$; (b) $\frac{1}{\sin x}$ **43.** Let $t = \frac{1}{4}\pi$ **45.** Let $y = \frac{3}{4}\pi$ **47.** Not an identity; let $\theta = \frac{1}{3}\pi$ **49.** Identity
51. Not an identity; let $x = \frac{3}{4}\pi$ **53.** Identity

Exercises 7.3 (Page 402)

1. (a) $\frac{\sqrt{3} - 1}{2\sqrt{2}}$; (b) $\frac{-\sqrt{3} - 1}{2\sqrt{2}}$ **3.** (a) $\frac{-\sqrt{3} - 1}{2\sqrt{2}}$; (b) $\frac{\sqrt{3} + 1}{2\sqrt{2}}$ **5.** (a) $-2 - \sqrt{3}$; (b) $2 - \sqrt{3}$ **7.** (a) 1; (b) -1 **9.** (a) $\frac{1}{2}$; (b) $-\frac{1}{\sqrt{2}}$ **11.** (a) 1; (b) $-\sqrt{3}$
13. (a) $\frac{323}{325}$; (b) $\frac{36}{325}$; (c) $\frac{253}{325}$; (d) $\frac{204}{325}$; (e) first; (f) first
15. (a) $\frac{16}{65}$; (b) $-\frac{63}{65}$; (c) $\frac{56}{65}$; (d) $-\frac{33}{65}$; (e) second; (f) second   **17.** (a) $-\frac{37}{9}$; (b) $\frac{19}{33}$; (c) second; (d) first
19. (a) $-\frac{38}{55}$; (b) $\frac{62}{25}$; (c) fourth; (d) first   **31.** (a) $\cos 7x$; (b) $\sin 7x$ **33.** (a) $-\cos 10x$; (b) $-\sin x$

35. (a) $\tan 10x$; (b) $-\tan 2x$

43. (a) $f(t) = 2\sin(t + \frac{1}{6}\pi)$;
(b) amplitude is 2, period is 2π, phase shift is $\frac{1}{6}\pi$
(c)

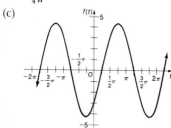

45. (a) $f(t) = 3\sqrt{2}\sin(t - \frac{1}{4}\pi)$;
(b) amplitude is $3\sqrt{2}$, period is 2π, phase shift is $-\frac{1}{4}\pi$
(c)

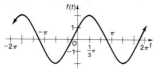

47. (a) $f(t) = 5\sin(6t + 0.93)$; (b) amplitude is 5, period is $\frac{1}{3}\pi$, phase shift is 0.93

49. $f(t) = 0.043\sin(2\pi nt - 0.72)$

51. (a) $s = \dfrac{1 + \sqrt{3}}{\sqrt{2}}\sin(4t + \frac{1}{4}\pi)$;

(b) Amplitude is $\dfrac{1 + \sqrt{3}}{\sqrt{2}}$, frequency is $\dfrac{2}{\pi}$ **53.** $10\sqrt{5}$

Exercises 7.4 (Page 411)

1. (a) $\frac{24}{25}$; (b) $-\frac{7}{25}$; (c) $-\frac{24}{7}$ **3.** (a) $-\frac{120}{169}$; (b) $-\frac{119}{169}$;
(c) $\frac{120}{119}$ **5.** (a) $\frac{240}{289}$; (b) $\frac{161}{289}$; (c) $\frac{240}{161}$ **7.** (a) $-\frac{336}{625}$; (b) $\frac{527}{625}$;
(c) $-\frac{336}{527}$ **9.** $\frac{1}{2}\sqrt{2 - \sqrt{2}}$ **11.** $-\frac{1}{2}\sqrt{2 + \sqrt{3}}$

13. $2 + \sqrt{3}$ **15.** $-1 - \sqrt{2}$ **17.** $\dfrac{1}{\sqrt{3}}$ **19.** $\frac{3}{5}$ **21.** 5

23. (a) $\cos 2x$; (b) $\cos 4x$; (c) $\sin 3x$ **25.** (a) $\sin 2x$;
(b) $2\sin 8x$; (c) $\sin x$ **27.** (a) $\tan 2x$; (b) $-2\tan 4x$;
(c) $3\tan 3x$ **29.** (a) $\tan t$; (b) $\tan 2t$; (c) $-\tan 2t$
31. (a) $\sin 2t$; (b) $\sin 4t$; (c) $2\sin 4t$
33. $\cos 3x = 4\cos^3 x - 3\cos x$
35. $\sin 4x = \pm 4\sin x\sqrt{1 - \sin^2 x}(1 - 2\sin^2 x)$

37. $\tan 3x = \dfrac{3\tan x - \tan^3 x}{1 - 3\tan^2 x}$

39. $\sec 2x = \dfrac{\sec^2 x}{2 - \sec^2 x}$ **51.** (a) $s = 6\sin(8t + \frac{1}{2}\pi)$;

(b) amplitude is 6, frequency is $\dfrac{4}{\pi}$

Exercises 7.5 (Page 420)

1. (a) $\frac{1}{6}\pi$; (b) $-\frac{1}{6}\pi$; (c) $\frac{1}{3}\pi$; (d) $\frac{2}{3}\pi$ **3.** (a) $\frac{1}{6}\pi$; (b) $-\frac{1}{3}\pi$
5. (a) $\frac{1}{2}\pi$; (b) π; (c) $\frac{1}{4}\pi$; (d) 0   **7.** (a) 0.510;
(b) -0.510; (c) 1.061; (d) 2.081   **9.** (a) 0.410
(b) -0.410 **11.** (a) 0.51; (b) -0.51; (c) 1.06; (d) 2.08
13. (a) 0.41; (b) -0.41

15. (a) $\dfrac{2\sqrt{2}}{3}$; (b) $\dfrac{1}{2\sqrt{2}}$; (c) $2\sqrt{2}$; (d) $\dfrac{3}{2\sqrt{2}}$; (e) 3

17. (a) $\dfrac{2\sqrt{2}}{3}$; (b) $-\dfrac{1}{2\sqrt{2}}$; (c) $-2\sqrt{2}$; (d) $\dfrac{3}{2\sqrt{2}}$; (e) -3

19. (a) $-\dfrac{2}{\sqrt{5}}$; (b) $\dfrac{1}{\sqrt{5}}$; (c) $-\dfrac{1}{2}$; (d) $\sqrt{5}$; (e) $-\dfrac{\sqrt{5}}{2}$

21. (a) $\frac{1}{6}\pi$; (b) $-\frac{1}{6}\pi$; (c) $\frac{1}{6}\pi$; (d) $-\frac{1}{6}\pi$
23. (a) $\frac{1}{3}\pi$; (b) $\frac{1}{3}\pi$; (c) $\frac{2}{3}\pi$; (d) $\frac{2}{3}\pi$
25. (a) $\frac{1}{6}\pi$; (b) $-\frac{1}{3}\pi$; (c) $\frac{1}{6}\pi$; (d) $-\frac{1}{3}\pi$

27. (a) $\frac{1}{2}\pi$; (b) $-\frac{1}{4}\pi$; (c) $\sqrt{3}$; (d) $\dfrac{\sqrt{3}}{\sqrt{7}}$

29. (a) $\frac{1}{6}\pi$; (b) $\frac{1}{6}\pi$; (c) $\dfrac{\sqrt{3}}{2}$; (d) $\dfrac{\sqrt{3}}{2}$

31. (a) 0; (b) 0; (c) 0; (d) $\dfrac{1}{\sqrt{2}}$ **33.** $\frac{119}{169}$

35. $\dfrac{2 + 2\sqrt{10}}{9}$ **37.** $\dfrac{7\sqrt{5} + 8\sqrt{2}}{27}$ **39.** $\dfrac{3\sqrt{3} - 4}{3\sqrt{3} + 4}$

41. $\dfrac{4\sqrt{2} + 1}{3\sqrt{5}}$

43.   **45.**   **47.**

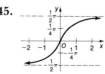 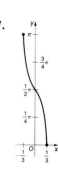

49.

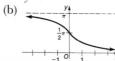

53. (a) $y = \cot^{-1}x$ if and only if $x = \cot y$ and $0 < y < \pi$
(b)

55. (a) $y = \csc^{-1}x$ if and only if $x = \csc y$ and $-\pi < y \le -\frac{1}{2}\pi$ or $0 < y \le \frac{1}{2}\pi$

(b)

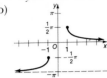

57. (b) 0.279; (c) 0.284; (d) 0.277

Exercises 7.6 (Page 428)

1. (a) $\{\frac{1}{2}\pi\}$; (b) $\{\frac{1}{3}\pi\}$ **3.** (a) $\{\frac{1}{4}\pi\}$; (b) $\{\frac{1}{3}\pi\}$ **5.** (a) $\{0\}$;
(b) $\{\frac{1}{4}\pi\}$ **7.** (a) $\{0, \frac{1}{2}\pi\}$; (b) $\{\frac{1}{3}\pi\}$ **9.** (a) $\{0, \frac{1}{3}\pi\}$;
(b) $\{0, \frac{1}{4}\pi\}$ **11.** (a) $\{\frac{1}{6}\pi, \frac{5}{6}\pi, \frac{7}{6}\pi, \frac{11}{6}\pi\}$; (b) $\{\frac{1}{4}\pi, \frac{3}{4}\pi, \frac{5}{4}\pi, \frac{7}{4}\pi\}$
13. (a) $\{\frac{2}{3}\pi, \pi, \frac{4}{3}\pi\}$; (b) $\{\frac{7}{6}\pi, \frac{11}{6}\pi\}$ **15.** (a) $\{\frac{1}{4}\pi, \frac{5}{4}\pi\}$;
(b) $\{\frac{1}{3}\pi, \pi, \frac{5}{3}\pi\}$ **17.** (a) $\{\frac{3}{4}\pi, \frac{7}{4}\pi\}$; (b) $\{\frac{3}{4}\pi, \frac{7}{4}\pi\}$
19. (a) $\{t | t = \frac{1}{6}\pi + k\pi\} \cup \{t | t = \frac{5}{6}\pi + k\pi\}, k \in J$;
(b) $\{t | t = \frac{1}{4}\pi + k \cdot \frac{1}{2}\pi\}, k \in J$ **21.** (a) $\{t | t = \frac{1}{4}\pi + k\pi\}, k \in J$;
(b) $\{t | t = \frac{1}{3}\pi + k \cdot \frac{2}{3}\pi\}, k \in J$ **23.** $\{48.2°, 311.8°\}$
25. $\{0°, 41.4°, 180°, 318.6°\}$
27. $\{71.6°, 135°, 251.6°, 315°\}$
29. (a) $\{\theta | \theta = 48.2° + k \cdot 360°\} \cup \{\theta | \theta = 311.8° + k \cdot 360°\}$,
$k \in J$ **31.** $\{\theta | \theta = k \cdot 180°\} \cup \{\theta | \theta = 41.4° + k \cdot 360°\} \cup$
$\{\theta | \theta = 318.6° + k \cdot 360°\}, k \in J$
33. (a) $\{\theta | \theta = 71.6° + k \cdot 180°\} \cup \{\theta | \theta = 135° + k \cdot 180°\}$,
$k \in J$ **35.** $\{2.14, 4.14\}$ **37.** $\{0.87, 2.44, 4.01, 5.58\}$
39. (a) $\{\frac{1}{4}\pi, \frac{7}{12}\pi, \frac{11}{12}\pi, \frac{5}{4}\pi, \frac{19}{12}\pi, \frac{23}{12}\pi\}$;
(b) $\{\frac{1}{18}\pi, \frac{5}{18}\pi, \frac{13}{18}\pi, \frac{17}{18}\pi, \frac{25}{18}\pi, \frac{29}{18}\pi\}$
41. (a) $\{\frac{5}{24}\pi, \frac{11}{24}\pi, \frac{17}{24}\pi, \frac{23}{24}\pi, \frac{29}{24}\pi, \frac{35}{24}\pi, \frac{41}{24}\pi, \frac{47}{24}\pi\}$;
(b) $\{\frac{1}{15}\pi, \frac{1}{3}\pi, \frac{7}{15}\pi, \frac{11}{15}\pi, \frac{13}{15}\pi, \frac{17}{15}\pi, \frac{19}{15}\pi, \frac{23}{15}\pi, \frac{5}{3}\pi, \frac{29}{15}\pi\}$
43. (a) $\{\frac{1}{4}\pi, \frac{3}{4}\pi, \frac{5}{4}\pi, \frac{7}{4}\pi\}$; (b) $\{\frac{4}{3}\pi\}$ **45.** (a) $\{\frac{2}{3}\pi, \frac{4}{3}\pi\}$; (b) $\{\frac{1}{2}\pi\}$
47. (a) $\{x | x = \frac{1}{4}\pi + k \cdot \frac{1}{3}\pi\}, k \in J$;
(b) $\{x | x = \frac{1}{18}\pi + k \cdot \frac{2}{3}\pi\} \cup \{x | x = \frac{5}{18}\pi + k \cdot \frac{2}{3}\pi\}, k \in J$
49. (a) $\{x | x = \frac{1}{4}\pi + k \cdot \frac{1}{2}\pi\}, k \in J$; (b) $\{x | x = \frac{4}{3}\pi + k \cdot 2\pi\}, k \in J$
51. $\{0°, 20°, 90°, 100°, 140°, 180°, 220°, 260°, 270°, 340°\}$
53. $\{22.5°, 73.2°, 112.5°, 163.2°,$
$\qquad\qquad 202.5°, 253.2°, 292.5°, 343.2°\}$
55. identity **57.** $\{\frac{1}{4}\pi, \frac{1}{6}\pi, \frac{3}{4}\pi, \frac{5}{6}\pi, \frac{5}{4}\pi, \frac{7}{4}\pi\}$
59. $\{0, \frac{1}{6}\pi, \frac{1}{3}\pi, \frac{1}{2}\pi, \frac{2}{3}\pi, \frac{5}{6}\pi, \pi, \frac{7}{6}\pi, \frac{4}{3}\pi, \frac{3}{2}\pi, \frac{5}{3}\pi, \frac{11}{6}\pi\}$
61. (a) $\frac{1}{25}$; (b) $\frac{1}{150}$; (c) $\frac{1}{50}$; (d) $\frac{1}{75}$ **63.** (a) $\frac{5}{9}$; (b) 0.61
65. (a) $\left\{t \,\middle|\, t = \frac{1}{4\pi} \sin^{-1}\frac{y}{2} - \frac{1}{8} + k \cdot \frac{1}{2}\right\} \cup$

$\left\{t \,\middle|\, t = \frac{1}{8} - \frac{1}{4\pi}\sin^{-1}\frac{y}{2} + k \cdot \frac{1}{2}\right\}, k \in J$; (b) $\frac{1}{12}, \frac{5}{12}, \frac{7}{12}$

Exercises 7.7 (Page 434)

1. (a) $\frac{1}{2}(\sin 5x + \sin 3x)$; (b) $\frac{1}{2}(\cos 3x - \cos 5x)$;
(c) $\frac{1}{2}(\cos 5x + \cos 3x)$ **3.** (a) $\frac{1}{2}(\sin 8x - \sin 4x)$;
(b) $\frac{1}{2}(\cos 4x - \cos 8x)$; (c) $\frac{1}{2}(\cos 8x + \cos 4x)$

5. (a) $\frac{1}{4}(\sqrt{3} + \sqrt{2})$; (b) $\dfrac{-2 - \sqrt{2}}{4}$ **7.** (a) $\dfrac{-2 - \sqrt{2}}{4}$;

(b) $\dfrac{\sqrt{2} - \sqrt{3}}{4}$

9. (a) $2 \sin 4t \cos 2t$; (b) $2 \cos 4t \sin 2t$;
(c) $2 \cos 4t \cos 2t$; (d) $-2 \sin 4t \sin 2t$
11. (a) $2 \sin 5\theta \cos 2\theta$; (b) $-2 \cos 5\theta \sin 2\theta$;

(c) $2 \cos 5\theta \cos 2\theta$; (d) $2 \sin 5\theta \sin 2\theta$ **13.** (a) $\dfrac{\sqrt{6}}{2}$;

(b) $-\dfrac{1}{\sqrt{2}}$ **15.** (a) $-\dfrac{1}{\sqrt{2}}$; (b) $\dfrac{\sqrt{6}}{2}$

39. $\sin 21\pi t + \sin 19\pi t = 2 \sin 20\pi t \cos \pi t$

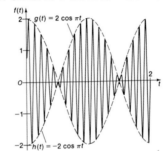

41. $\cos 80\pi t + \cos 100\pi t = 2 \cos 90\pi t \cos 10\pi t$

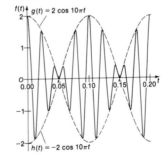

Review Exercises for Chapter 7 (Page 435)

1. $\cos \theta = -\frac{15}{17}$, $\tan \theta = -\frac{8}{15}$, $\csc \theta = \frac{17}{8}$, $\sec \theta = -\frac{17}{15}$,

$\cot \theta = -\frac{15}{8}$ **3.** $\sin x = -\dfrac{\sqrt{21}}{5}$, $\cos x = \frac{2}{5}$,

$\csc x = -\dfrac{5}{\sqrt{21}}$, $\sec x = \frac{5}{2}$, $\cot x = -\dfrac{2}{\sqrt{21}}$

5. (a) 0.7881; (b) 0.7812
7. (a) 3.7624; (b) −0.2569

9. (a) −1.2278; (b) 1.7235 **11.** $\dfrac{1}{\cos^3 t}$ **13.** $\dfrac{\sin \theta}{1 - \sin^2\theta}$

39. (a) $\dfrac{\sqrt{6} - \sqrt{2}}{4}$; (b) $\dfrac{\sqrt{2} - \sqrt{3}}{2}$ **41.** (a) $\dfrac{1 + \sqrt{3}}{1 - \sqrt{3}}$;

(b) $-2 - \sqrt{3}$ **43.** (a) $\frac{253}{325}$; (b) $\frac{204}{325}$; (c) $-\frac{323}{325}$ (d) $-\frac{36}{325}$;

(e) first; (f) third   **45.** -2 **33.** $\dfrac{6 - 4z}{1 + z^2}$

47. (a) $\{\frac{1}{3}\pi, \frac{5}{3}\pi\}$; (b) $\{\frac{1}{4}\pi, \frac{3}{4}\pi, \frac{5}{4}\pi, \frac{7}{4}\pi\}$ **49.** $\{\frac{1}{6}\pi, \frac{5}{6}\pi\}$

51. $\{1.25, \frac{3}{4}\pi, 4.39, \frac{7}{4}\pi\}$

53. (a) $\{\frac{1}{18}\pi, \frac{5}{18}\pi, \frac{7}{18}\pi, \frac{11}{18}\pi, \frac{13}{18}\pi, \frac{17}{18}\pi,$
$\qquad \frac{19}{18}\pi, \frac{23}{18}\pi, \frac{25}{18}\pi, \frac{29}{18}\pi, \frac{31}{18}\pi, \frac{35}{18}\pi\}$;

(b) $\{0.23, 1.34, 1.80, 2.91, 3.37, 4.48, 4.94, 6.05\}$

55. (a) $\{t \mid t = \frac{1}{3}\pi + k \cdot 2\pi\} \cup \{t \mid t = \frac{5}{3}\pi + k \cdot 2\pi\}$, $k \in J$;
(b) $\{t \mid t = \frac{1}{4}\pi + k \cdot \frac{1}{2}\pi\}$, $k \in J$

57. (a) $\{45°, 135°, 225°, 315°\}$; (b) $\{90°, 270°\}$

59. $\{120°, 180°, 240°\}$ **61.** (a) $\{\theta \mid \theta = 45° + k \cdot 90°\}$,
$k \in J$; (b) $\{\theta \mid \theta = 90° + k \cdot 180°\}$, $k \in J$ **63.** identity

65. $\{\frac{1}{4}\pi, \frac{5}{4}\pi\}$ **67.** (a) 0.6475; (b) -0.6475; (c) 0.9233;

(d) 2.2183; (e) 0.5428; (f) -0.5428 **69.** (a) 0.65;
(b) -0.65; (c) 0.92; (d) 2.22; (e) 0.54; (f) -0.54

71. (a) $\frac{4}{5}$; (b) $\frac{3}{5}$; (c) $\frac{3}{4}$; (d) $-\frac{3}{4}$ **73.** (a) $-\frac{120}{169}$; (b) $\frac{3}{4}$

75. $f(t) = 0.02 \sin(200\pi t - \frac{1}{3}\pi)$

77. (a) $s = \sqrt{2} \sin(6t - 0.27)$; (b) amplitude is $\sqrt{2}$,

frequency is $\dfrac{3}{\pi}$ **79.** (a) $\left\{t \,\middle|\, t = \dfrac{1}{2\pi} \sin^{-1}\dfrac{y}{4} - \dfrac{1}{6} + k\right\} \cup$

$\left\{t \,\middle|\, t = \dfrac{1}{3} - \dfrac{1}{2\pi} \sin^{-1}\dfrac{y}{4} + k\right\}$, $k \in J$; (b) $\frac{1}{4}$; (c) 0.20

81. (a) $\theta = \tan^{-1}\left(\dfrac{5x}{x^2 + 36}\right)$; (b) 0.3890; (c) 0.3948;

(d) 0.3906

Chapter 8

Exercises 8.1 (Page 449)

1. $b = 5.6$ **3.** $a = 51$ **5.** $\gamma = 75°$; $b = 41$; $c = 41$
7. $\alpha = 70.0°$; $a = 56.9$; $b = 45.5$ **9.** $\beta = 83.0°$;
$a = 2.22$; $c = 0.935$ **11.** $\gamma = 51°42'$; $a = 335$; $c = 331$
13. One triangle; $\beta = 29°$; $\gamma = 109°$; $c = 9.0$
15. One triangle; $\alpha = 60°$; $\gamma = 90°$; $a = 29$
17. Two triangles; $\alpha_1 = 67.3°$; $\beta_1 = 77.5°$; $b_1 = 72.0$;
$\alpha_2 = 112.7°$; $\beta_2 = 32.1°$; $b_2 = 39.2$ **19.** One triangle;
$\alpha = 27.2°$, $\gamma = 36.4°$; $c = 560$ **21.** No triangle
23. Two triangles; $\alpha_1 = 64.48°$; $\gamma_1 = 65.29°$;
$a_1 = 4.182$; $\alpha_2 = 15.06°$; $\gamma_2 = 114.71°$; $a_2 = 1.204$
25. $(4.7)10^2$ **27.** 1.03 **29.** 14 **31.** $(1.08)10^5$
33. 61.8 m   **35.** 296 m   **37.** 9.80 ft

Exercises 8.2 (Page 456)

1. $c = 5.6$ **3.** $b = 34$ **5.** $\beta = 55°$; $\gamma = 43°$; $a = 4.1$
7. $\alpha = 26.4°$; $\beta = 37.2°$; $c = 22.6$ **9.** $\alpha = 18.51°$;
$\gamma = 150.97°$; $b = 1176$ **11.** $\alpha = 26°23'$; $\beta = 18°25'$;
$c = 8.34$ **13.** 12.3 **15.** 76.7 **17.** $(5.836)10^5$
19. $\alpha = 44°$; $\beta = 108°$; $\gamma = 28°$ **21.** $\alpha = 57.5°$;
$\beta = 24.6°$; $\gamma = 97.9°$ **23.** $\alpha = 98.3°$; $\beta = 38.4°$;
$\gamma = 43.3°$ **25.** $\alpha = 144.36°$; $\beta = 27.74°$; $\gamma = 7.90°$
27. (a) $33°$; (b) $(3.9)10^2$ **29.** (a) 30.4 cm; (b) 194 cm^2
31. (a) 17.3 nautical miles; (b) $188.1°$ **33.** 58.7 m
35. $31.2°$ **37.** $56°$ **39.** (a) 39.2 mi; (b) $212.5°$
45. $(3.83)10^4$ **47.** 20 **49.** 27.5 **51.** 78.8

Exercises 8.3 (Page 467)

1. (a) $|\mathbf{A}| = 5$; (b) $|\mathbf{A}| = 2$ **3.** (a) $\langle 2, -3 \rangle$; (b) $\langle -2, -7 \rangle$
5. (a) $\langle -4, -3 \rangle$; (b) $\langle 12, -5 \rangle$ **7.** (a) $\langle -1, 9 \rangle$; (b) $\langle 1, -5 \rangle$
9. (a) $\langle 7, 3 \rangle$; (b) $\langle -9, -4 \rangle$ **11.** (a) $\langle 6, 1 \rangle$; (b) $\sqrt{13}$
13. $\sqrt{61}$ **15.** $\langle 14.6, 10.6 \rangle$ **17.** $\langle -12.0, -32.9 \rangle$
19. (a) 135 lb; (b) $17°$ **21.** (a) 135 lb; (b) $17°$
23. $29.0°$ **25.** (a) $28.1°$; (b) 1.7 mi/hr; (c) 0.53 mi
27. (a) $346.1°$; (b) 243 mi/hr   **29.** 15.3 knots, $191.3°$

Exercises 8.4 (Page 474)

5. (a) $(-4, \frac{5}{4}\pi)$; (b) $(4, -\frac{7}{4}\pi)$; (c) $(-4, -\frac{3}{4}\pi)$
7. (a) $(-2, \frac{3}{2}\pi)$; (b) $(2, -\frac{3}{2}\pi)$; (c) $(-2, -\frac{1}{2}\pi)$
9. (a) $(-\sqrt{2}, \frac{3}{4}\pi)$; (b) $(\sqrt{2}, -\frac{1}{4}\pi)$; (c) $(-\sqrt{2}, -\frac{5}{4}\pi)$
11. (a) $(-2, \frac{3}{4}\pi)$; (b) $(-2, -\frac{5}{4}\pi)$; (c) $(2, \frac{15}{4}\pi)$
13. $(3, \frac{4}{3}\pi)$; $(-3, \frac{1}{3}\pi)$ **15.** $(-4, -\frac{7}{6}\pi)$; $(4, -\frac{1}{6}\pi)$
17. $(-2, \frac{3}{4}\pi)$; $(2, \frac{7}{4}\pi)$ **19.** $(2, 2\pi + 6)$; $(-2, 6 - \pi)$
21. (a) $(-3, 0)$; (b) $(-1, -1)$; (c) $(2, -2\sqrt{3})$;
(d) $(\frac{1}{2}\sqrt{3}, -\frac{1}{2})$ **23.** (a) $(\sqrt{2}, \frac{7}{4}\pi)$; (b) $(2, \frac{5}{6}\pi)$;
(c) $(2\sqrt{2}, \frac{1}{4}\pi)$; (d) $(5, \pi)$ **25.** $r = |a|$

27. $r = \dfrac{2}{1 - \cos\theta}$ **29.** $r = 6\sin\theta$ **31.** $r^2 = 4\cos 2\theta$

33. $r = \dfrac{3a\sin 2\theta}{2(\sin^3\theta + \cos^3\theta)}$ **35.** $(x^2 + y^2)^2 = 4xy$

37. $(x^2 + y^2)^3 = x^2$ **39.** $y = x\tan(x^2 + y^2)$
41. $x = -1$ **43.** $4x^2 - 5y^2 - 36y - 36 = 0$

Exercises 8.5 (Page 481)

9. (a) $\sqrt{29}$; (b) $\sqrt{5}$ **11.** (a) 3; (b) 3 **13.** (a) 2; (b) 2
15. (a) 10; (b) 10 **17.** (a) $\frac{3}{2} + \frac{3}{2}\sqrt{3}i$; (b) $-\frac{3}{2} - \frac{3}{2}\sqrt{3}i$
19. (a) $-3\sqrt{3} + 3i$; (b) $3\sqrt{3} - 3i$ **21.** (a) $2i$; (b) $-\frac{2}{3}$
23. (a) $4\sqrt{2}(\cos\frac{7}{4}\pi + i\sin\frac{7}{4}\pi)$; (b) $6(\cos 0 + i\sin 0)$;
(c) $\cos\frac{1}{2}\pi + i\sin\frac{1}{2}\pi$ **25.** (a) $6(\cos\frac{5}{8}\pi + i\sin\frac{5}{8}\pi)$;
(b) $7(\cos\pi + i\sin\pi)$; (c) $7(\cos\frac{3}{2}\pi + i\sin\frac{3}{2}\pi)$
27. (a) $5[\cos(\arctan\frac{4}{3}) + i\sin(\arctan\frac{4}{3})]$;
(b) $5(\cos 0.93 + i\sin 0.93)$; (c) $5(\cos 53.1° + i\sin 53.1°)$
29. (a) $\sqrt{5}[\cos(\arctan(-2) + 2\pi) + i\sin(\arctan(-2) + 2\pi)]$;
(b) $\sqrt{5}(\cos 5.18 + i\sin 5.18)$;
(c) $\sqrt{5}(\cos 296.6° + i\sin 296.6°)$ **31.** $0 + 10i$
33. $1 + \sqrt{3}i$ **35.** $-2\sqrt{2} + 2\sqrt{2}i$ **37.** $2\sqrt{3} - 2i$
39. $\sqrt{3} + i$ **41.** $-2\sqrt{2} + 2\sqrt{2}i$ **43.** $\frac{5}{2} - \frac{5}{2}\sqrt{3}i$
45. $0 - \frac{2}{3}i$

Exercises 8.6 (Page 492)

1. $2 + 2\sqrt{3}i$ **3.** 64 **5.** -1 **7.** $-16 - 16\sqrt{3}i$
9. $-2 + 2i$ **11.** 64 **13.** $4096 - 4096i$ **15.** -1
17. 1 **19.** $-6^{30} \approx (-2.2107)10^{23}$
21. $-\sqrt{2} + \sqrt{2}i, \sqrt{2} - \sqrt{2}i$ **23.** $5, -\frac{5}{2} \pm \frac{5}{2}\sqrt{3}i$
25. $2(\cos\frac{5}{9}\pi + i\sin\frac{5}{9}\pi) \approx -0.35 + 1.97i$;
$2(\cos\frac{11}{9}\pi + i\sin\frac{11}{9}\pi) \approx -1.53 - 1.29i$;
$2(\cos\frac{17}{9}\pi + i\sin\frac{17}{9}\pi) \approx 1.88 - 0.68i$
27. $\frac{1}{2}\sqrt{3} + \frac{1}{2}i; -\frac{1}{2} + \frac{1}{2}\sqrt{3}i; -\frac{1}{2}\sqrt{3} - \frac{1}{2}i; \frac{1}{2} - \frac{1}{2}\sqrt{3}i$
29. $2(\cos\frac{7}{30}\pi + i\sin\frac{7}{30}\pi) \approx 1.49 + 1.34i$;
$2(\cos\frac{19}{30}\pi + i\sin\frac{19}{30}\pi) \approx -0.81 + 1.83i$;
$2(\cos\frac{31}{30}\pi + i\sin\frac{31}{30}\pi) \approx -1.99 - 0.21i$;
$2(\cos\frac{43}{30}\pi + i\sin\frac{43}{30}\pi) \approx -0.42 - 1.96i; \sqrt{3} - i$
31. (a) $1, -\frac{1}{2} \pm \frac{1}{2}\sqrt{3}i$; (b) $3, -\frac{3}{2} \pm \frac{3}{2}\sqrt{3}i$
33. (a) $\pm 1, \pm i, \frac{1}{2}\sqrt{2} \pm \frac{1}{2}\sqrt{2}i, -\frac{1}{2}\sqrt{2} \pm \frac{1}{2}\sqrt{2}i$;
(b) $\pm 2, \pm 2i; \sqrt{2} \pm \sqrt{2}i, -\sqrt{2} \pm \sqrt{2}i$ **35.** $\pm 3, \pm 3i$
37. $i, \frac{1}{2}\sqrt{3} - \frac{1}{2}i, -\frac{1}{2}\sqrt{3} - \frac{1}{2}i$ **39.** $-2i, \sqrt{3} + i,$
$-\sqrt{3} + i$ **41.** $\cos\frac{1}{5}\pi + i\sin\frac{1}{5}\pi, \cos\frac{3}{5}\pi + i\sin\frac{3}{5}\pi,$
$-1, \cos\frac{7}{5}\pi + i\sin\frac{7}{5}\pi, \cos\frac{9}{5}\pi + i\sin\frac{9}{5}\pi$
43. $1.74 + 0.22i, -1.06 + 1.39i, -0.68 - 1.62i$
45. $1.21 + 0.66i, -0.25 + 1.36i, -1.37 + 0.18i,$
$-0.59 - 1.25i, 1.00 - 0.95i$

Exercises 8.7 (Page 504)

1. (a) line through the pole with slope $\sqrt{3}$; (b) circle

with center at the pole and radius $\frac{1}{3}\pi$ **3.** (a) line through the pole with slope $\tan^{-1} 2$; circle with center at the pole and radius 2 **5.** (a) line parallel to the $\frac{1}{2}\pi$ axis and 4 units to the right of it; (b) circle tangent to the $\frac{1}{2}\pi$ axis with center on the polar axis and radius 2 **7.** (a) line parallel to the polar axis and 4 units below it; (b) circle tangent to the polar axis with center on the extension of the $\frac{1}{2}\pi$ axis and radius 2

9. Cardioid

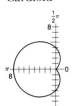

11. Cardioid

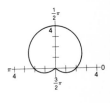

13. Limaçon with a loop

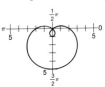

15. Limaçon with a dent

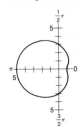

17. Convex limaçon

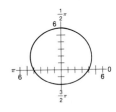

19. Three-leafed rose

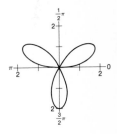

21. Eight-leafed rose

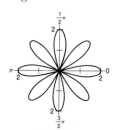

23. Four-leafed rose

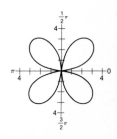

25. Logarithmic spiral, containing the points (r, θ) given in the following table.

r	1	$e^{\pi/2} \approx 5$	$e^{\pi} \approx 23$	$e^{3\pi/2} \approx 111$	$e^{2\pi} \approx 535$	$e^{5\pi/2} \approx 2576$	$e^{3\pi} \approx 12{,}392$
θ	0	$\frac{1}{2}\pi$	π	$\frac{3}{2}\pi$	2π	$\frac{5}{2}\pi$	3π

27. Reciprocal spiral, containing the points (r, θ) given in the following table.

r	$\dfrac{6}{\pi} \approx 1.9$	$\dfrac{3}{\pi} \approx 0.95$	$\dfrac{2}{\pi} \approx 0.63$	$\dfrac{1}{\pi} \approx 0.32$	$\dfrac{1}{2\pi} \approx 0.16$	$\dfrac{1}{3\pi} \approx 0.12$	$\dfrac{1}{4\pi} \approx 0.08$	$\dfrac{1}{6\pi} \approx 0.05$
θ	$\dfrac{1}{6}\pi$	$\dfrac{1}{3}\pi$	$\dfrac{1}{2}\pi$	π	2π	3π	4π	6π

29. Lemniscate

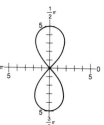

31. Lemniscate

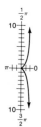

33. Cissoid

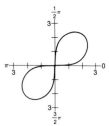

35. Cissoid

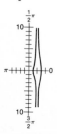

37. Four-leafed rose

Review Exercises for Chapter 8 (Page 504)

1. 3.7 **3.** 70 **5.** $\gamma = 75.4°$; $a = 20.9$; $c = 29.5$
7. $\alpha = 28.7°$; $\beta = 46.0°$; $c = 43.4$ **9.** $\alpha = 54.46°$;
$\beta = 43.67°$, $\gamma = 81.87°$ **11.** 271 **13.** $(1.128)10^5$
15. two triangles; $\beta_1 = 72.0°$; $\gamma_1 = 53.6°$; $c_1 = 111$;
$\beta_2 = 108.0°$; $\gamma_2 = 17.6°$; $c_2 = 41.6$
17. one triangle; $\alpha = 107.7°$; $\gamma = 32.6°$; $a = 19.1$
19. (a) $\langle 5, 1 \rangle$; (b) $\langle -5, -1 \rangle$
21. (a) $2\sqrt{13}$; (b) $2\sqrt{26}$; (c) $5\sqrt{2} + \sqrt{106}$
23. (a) $\langle 6\sqrt{3}, 6 \rangle$; (b) $\langle -13.5, 33.4 \rangle$
25. (a) $(2, \frac{11}{4}\pi)$, $(-2, \frac{7}{4}\pi)$; (b) $(-3, -\frac{5}{6}\pi)$, $(3, \frac{1}{6}\pi)$
27. (a) $(0, 1)$; (b) $(1, -\sqrt{3})$;
(c) $(-2\sqrt{2}, -2\sqrt{2})$, $(-\frac{3}{2}\sqrt{3}, -\frac{3}{2})$
29. (a) $(4\sqrt{2}, \frac{3}{4}\pi)$; (b) $(2, \frac{5}{3}\pi)$; (c) $(6, \frac{1}{2}\pi)$; (d) $(4, \frac{7}{6}\pi)$
31. $r^2(4 \cos^2 \theta - 9 \sin^2 \theta) = 36$
33. $r = 9 \cos \theta - 8 \sin \theta$ **35.** $xy = 2$
37. $x^4 + 2x^2y^2 + y^4 - y^2 = 0$
41. (a) 5; (b) $2\sqrt{10}$
43. (a) $2\sqrt{3} (\cos \frac{5}{6}\pi + i \sin \frac{5}{6}\pi)$;
(b) $4(\cos \frac{1}{6}\pi + i \sin \frac{1}{6}\pi)$; (c) $4(\cos \frac{3}{2}\pi + i \sin \frac{3}{2}\pi)$;
(d) $\sqrt{2}(\cos \frac{5}{4}\pi + i \sin \frac{5}{4}\pi)$ **45.** $\sqrt{3} + i$

47. $-2\sqrt{2} + 2\sqrt{2}i$ **49.** $-\frac{3}{2} + \frac{3}{2}\sqrt{3}i$ **51.** $-\sqrt{2} - \sqrt{2}i$
53. $0 + 8i$ **55.** $-\frac{1}{512}i$ **57.** $1 \pm \sqrt{3}i$ **59.** $1.21 +$
$0.14i, -0.14 + 1.21i, -1.21 - 0.14i, 0.14 - 1.21i$
61. $0.73 + 1.19i, -0.91 + 1.07i, -1.29 - 0.53i,$
$0.11 - 1.40i, 1.36 - 0.33i$ **63.** $60°$ **65.** (a) 241 mi;
(b) $73.0°$ **67.** 40.6 ft **69.** $\langle 6\sqrt{3}, 6 \rangle$ **71.** $95.3°$
73. $5\sqrt{3} + 5i, 10i, -5\sqrt{3} + 5i, -5\sqrt{3} - 5i, -10i,$
$5\sqrt{3} - 5i$ **75.** (a) line through the pole with slope 1;
(b) circle with center at the pole and radius 4
77. Four-leafed rose   **79.** Cardioid

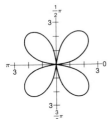

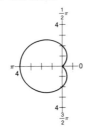

81.

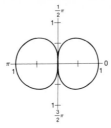

83. Lemniscate

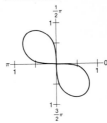

85. (a) Reciprocal spiral

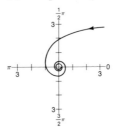

(b) spiral of
Archimedes

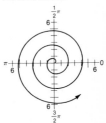

Chapter 9

Exercises 9.1 (Page 516)

1. (i); $\{(3, -5)\}$ **3.** (i); $\{(\frac{9}{4}, \frac{3}{2})\}$ **5.** (iii) **7.** (ii)
9. (i); $\{(-1.5, -1.3)\}$ **11.** $\{(1, 0)\}$ **13.** $\{(0, -2)\}$
15. $\{(\frac{1}{2}, \frac{1}{6})\}$ **17.** $\{(\frac{2}{3}, \frac{1}{2})\}$ **19.** $\{(-4, 7)\}$ **21.** $\{(-\frac{12}{17}, \frac{42}{17})\}$
23. $\{(-\frac{15}{4}, -\frac{15}{2})\}$ **25.** $\{(\frac{3}{4}, -\frac{1}{5})\}$ **27.** Inconsistent
29. Consistent; $\{(-3, -2)\}$ **31.** \$6.30 per pound of
tea; \$2.60 per pound of coffee **33.** 20; \$120
35. $\frac{400}{3}$ cm^3 of the 15 percent acid solution;
$\frac{800}{3}$ cm^3 of the 6 percent acid solution **37.** $17\frac{1}{2}$ mi/hr,
$2\frac{1}{2}$ mi/hr **39.** $6\frac{1}{4}$ mi/hr, 12 mi/hr **41.** Girl, 20 min;
brother, 25 min **43.** Any fraction of the form
$\frac{t}{2t - 4}$, where t is an integer, $t \neq 0, t \neq 2$

Exercises 9.2 (Page 526)

1. $\{(3, 1, 0)\}$ **3.** $\{(1, 2, 1)\}$ **5.** $\{(-3, 2, -4)\}$
7. $\emptyset$; inconsistent **9.** $\{(3, 1, -2)\}$ **11.** $\{(-3, -5, 0)\}$
13. $\{(\frac{1}{4}, \frac{1}{2}, 1)\}$ **15.** $\{(t - 3, 2t - 4, t)\}$; $(-3, -4, 0)$,
$(-2, -2, 1), (-1, 0, 2), (-4, -6, -1), (-5, -8, -2)$
17. $\{(t + 1, 3, t)\}$; $(1, 3, 0), (2, 3, 1), (3, 3, 2), (0, 3, -1)$,
$(-1, 3, -2)$ **19.** Consistent; $\{(3, 2, -4)\}$

21. Inconsistent **23.** $x^2 + y^2 + 4x - 6y - 12 = 0$
25. $y = 4x^2 - 8x + 3$ **27.** \$10,000 at 10 percent;
\$5000 at 12 percent; \$10,000 at 16 percent
29. Twenty oz of first alloy; forty oz of second alloy;
ten oz of third alloy **31.** Two large-size cards, seven
medium-size cards, and one small-size card; three
large-size cards, five medium-size cards, and two small-
size cards; four large-size cards, three medium-size
cards, and three small-size cards; or five large-size
cards, one medium-size card, and four small-size cards
33. (b) $0 \leq x \leq \frac{5}{2}; \frac{5}{3} \leq y \leq \frac{5}{2}; 0 \leq z \leq \frac{10}{3}$

Exercises 9.3 (Page 534)

1. $\{(3, 4), (-4, -3)\}$

3. $\{(-1, 0), (\frac{3}{2}, \frac{5}{4})\}$

5. $\{(\frac{17}{5}, \frac{8}{5})\}$

7. $\left\{\left(\dfrac{1 + \sqrt{5}}{2}, \dfrac{-5 + \sqrt{5}}{2}\right), \left(\dfrac{1 - \sqrt{5}}{2}, \dfrac{-5 - \sqrt{5}}{2}\right)\right\}$

9. $\{(-2 + 2\sqrt{3}, 2 - \sqrt{3}), (-2 - 2\sqrt{3}, 2 + \sqrt{3})\}$

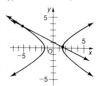

11. $\{(i, -4i), (-i, 4i), (-4, -1), (4, 1)\}$

13. $\{(2, 0), (-2, 0), (0, 2)\}$; the graph of the first equation is a circle and the graph of the second is a parabola; the circle and parabola are tangent at the point $(0, 2)$

15. $\{(2\sqrt{3}, \sqrt{13}), (-2\sqrt{3}, \sqrt{13}), (2\sqrt{3}, -\sqrt{13}), (-2\sqrt{3}, -\sqrt{13})\}$

17. $\{(-3, 0), (5, 4), (5, -4)\}$

19. $\{(2, 4), (-2, -4)\}$

21. $\{(-1, 2), (1, -2), (\frac{2}{3}\sqrt{6}, \frac{1}{3}\sqrt{6}), (-\frac{2}{3}\sqrt{6}, -\frac{1}{3}\sqrt{6})\}$

23. $\{(1, 2), (-1, 2), (1, -2), (-1, -2)\}$

25. 6, 10

27. 12 cm, 35 cm

29. 8 m, 12 m

31. 25, $7.50

33. 9.9 cm, 49.1 cm

35. 4 ft by 4 ft by 7 ft

Exercises 9.4 (Page 544)

1. (a) (b)

3. (a) (b)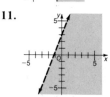

5. 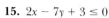 **7.**

9. 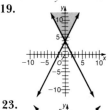 **11.**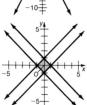

13. $3x + 2y - 12 < 0$ **15.** $2x - 7y + 3 \le 0$

17. **19.**

21. No region **23.**

25.

27. $\frac{219}{16}$ at $(\frac{25}{8}, \frac{17}{8})$ **29.** 64.8 at $(\frac{44}{5}, \frac{12}{5})$

31. 19 at $(2, 3)$

33. (a) $\begin{cases} x > 0 \\ y > 0 \\ x - y > 0 \\ x - y \le 10 \\ x + y \le 36 \end{cases}$ (b)

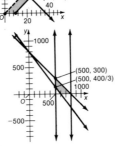

35. (a) $\begin{cases} x \ge 500 \\ x \le 800 \\ y \ge 0 \\ x + y \le 800 \\ 4x + 3y \ge 2400 \end{cases}$ (b)

37. 8 units of A and 16 units of B
39. 30 lb of A and 10 lb of B

Exercises 9.5 (Page 552)

1. $\{(1, 0)\}$ **3.** $\{(\frac{1}{2}, \frac{1}{6})\}$ **5.** $\{(-\frac{12}{17}, \frac{42}{17})\}$ **7.** $\{(\frac{1}{2}, -1)\}$
9. $\{(4, 6)\}$ **11.** $\{(3, 1, 0)\}$ **13.** $\{(-3, 2, -4)\}$
15. $\varnothing$; inconsistent   **17.** $\{(3, -1, 2)\}$ **19.** $\{(2, -3, 8)\}$
21. $\{(-7t - 10, 6t + 8, t)\}$ **23.** $\{(-2, 2, -3, 1)\}$
25. $\varnothing$; inconsistent

Exercises 9.6 (Page 564)

1. (a) 17; (b) 16   **3.** (a) -4; (b) -27 **5.** -1 **7.** 11
9. 1   **11.** ± 2 **13.** -2 **15.** $-\frac{3}{2}, 1$ **17.** $\{(1, 0)\}$
19. $\{(\frac{1}{2}, \frac{1}{6})\}$ **21.** $\{(-\frac{12}{17}, \frac{42}{17})\}$ **23.** $\{(-\frac{1}{7}, -\frac{10}{7})\}$
25. $\{(-\frac{1}{2}, \frac{2}{3})\}$ **27.** $\{(3, 1, 0)\}$ **29.** $\{(-3, 2, -4)\}$
31. $\varnothing$; inconsistent   **33.** $\{(2, 3, 2)\}$ **35.** $\{(-\frac{1}{2}t, -t, t)\}$;
dependent   **37.** $\{(-\frac{5}{12}, -\frac{1}{6}, -\frac{4}{3})\}$ **39.** $\varnothing$; inconsistent
41. $\{(3, -1, 2)\}$ **45.** -67

Exercises 9.7 (Page 574)

1. (a) 2×4; (b) 3×1; (c) -2;

(d) $\begin{bmatrix} -3 & 2 & -7 & 0 \\ -4 & -5 & 1 & -8 \end{bmatrix}$; (e) $\begin{bmatrix} 6 \\ -2 \\ -5 \end{bmatrix}$;

(f) $\begin{bmatrix} 9 & -6 & 21 & 0 \\ 12 & 15 & -3 & 24 \end{bmatrix}$; (g) $\begin{bmatrix} 12 \\ -4 \\ -10 \end{bmatrix}$

3. (a) 4×3; (b) 2×2; (c) 0; (d) $\begin{bmatrix} 0 & 0 & 0 \\ 0 & 0 & 0 \\ 0 & 0 & 0 \\ 0 & 0 & 0 \end{bmatrix}$;

(e) $\begin{bmatrix} 0 & 0 \\ 0 & 0 \end{bmatrix}$; (f) $\begin{bmatrix} -28 & -8 & 4 \\ 12 & 16 & 0 \\ 8 & -4 & 20 \\ 0 & 16 & 4 \end{bmatrix}$; (g) $\begin{bmatrix} -18 & 30 \\ 6 & -12 \end{bmatrix}$

5. $\begin{bmatrix} -1 & -4 \\ -9 & 8 \end{bmatrix}$ **7.** $\begin{bmatrix} -11 & 5 & -6 \\ 13 & 5 & -8 \end{bmatrix}$

9. $\begin{bmatrix} -5 & 0 & 3 \\ 5 & -9 & 0 \\ -5 & -5 & -1 \end{bmatrix}$

11. $a = 13, b = -7, c = -3, d = -1$ **13.** $\begin{bmatrix} -4 & 8 \\ 12 & 20 \\ -24 & 4 \end{bmatrix}$

15. $\begin{bmatrix} 6 & -12 & 2 \\ -8 & 4 & 0 \\ 10 & -2 & 16 \end{bmatrix}$ **17.** $\begin{bmatrix} 4 & 7 \\ -2 & 16 \end{bmatrix}$

19. $\begin{bmatrix} 2 & 2 \\ 7 & 3 \end{bmatrix}$ **21.** $\begin{bmatrix} 7 \end{bmatrix}$ **23.** $\begin{bmatrix} 2 & 4 & -6 \\ 1 & 2 & -3 \\ -1 & -2 & 3 \end{bmatrix}$

25. $\begin{bmatrix} -8 & 15 \\ 0 & -6 \\ -10 & 12 \end{bmatrix}$ **27.** $\begin{bmatrix} 5 & -26 & 2 \\ 8 & -1 & 13 \end{bmatrix}$

41. $\begin{bmatrix} 4 & 2 \\ 10 & 7 \end{bmatrix}$ **43.** $\begin{bmatrix} -2 & -1 \\ -4 & -1 \\ 1 & 2 \end{bmatrix}$

Exercises 9.8 (Page 582)

1. $\begin{bmatrix} 2 & -3 \\ -4 & 5 \\ 1 & 0 \end{bmatrix}$ **3.** $\begin{bmatrix} -5 & 3 & 0 \\ 0 & -6 & -1 \\ 2 & 8 & 4 \end{bmatrix}$ **5.** $\begin{bmatrix} 1 & -2 \\ -2 & 5 \end{bmatrix}$

7. Singular   **9.** $\begin{bmatrix} -\frac{2}{3} & \frac{1}{3} \\ -\frac{5}{3} & \frac{4}{3} \end{bmatrix}$ **11.** $\begin{bmatrix} 1 & 0 & -1 \\ 1 & 1 & -2 \\ -5 & -3 & 9 \end{bmatrix}$

13. $\begin{bmatrix} \frac{1}{3} & \frac{1}{3} & 0 \\ -\frac{1}{3} & -\frac{1}{3} & 1 \\ 0 & -1 & 2 \end{bmatrix}$ **15.** $\begin{bmatrix} 3 & 4 \\ 2 & 3 \end{bmatrix}$ **27.** $\{(1, 4)\}$

29. $\{(-3, 4)\}$ **31.** $\{(-3, 1, 4)\}$ **33.** $\{(-\frac{58}{7}, -\frac{18}{7}, -8)\}$

Review Exercises for Chapter 9 (Page 583)

1. (i); $\{(3, -2)\}$ **3.** (iii)   **5.** $\{(2, -5)\}$
7. Inconsistent   **9.** $\{(\frac{12}{5}, -3)\}$ **11.** $\{(3, 1, -2)\}$
13. $\{(3t + 1, 2t - 1, t)\}$; dependent
15. $\{(0, -3), (\frac{36}{13}, \frac{15}{13})\}$ **17.** $\{(-3, 2), (-\frac{3}{2}, -\frac{1}{4})\}$
19. $\{(1, \sqrt{2}), (1, -\sqrt{2}), (-1, \sqrt{2}), (-1, -\sqrt{2})\}$
21. $\{(1, -2), (-1, 2), (\frac{1}{2}, -\frac{5}{2}), (-\frac{1}{2}, \frac{5}{2})\}$

23. (a) 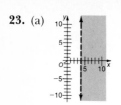 (b)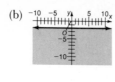

25. 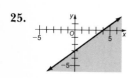 **27.**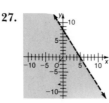

29. 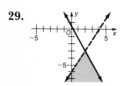 **31.**

33. -6 at $(2, -2)$ **35.** $\{(5, 3)\}$ **37.** $\{(-4, -2, 3)\}$
39. $\{(1, -1, 2, \frac{1}{3})\}$ **41.** 1 **43.** -35 **45.** 10
47. $\frac{1}{2}, 1$ **49.** $\{(\frac{2}{3}, -1)\}$ **51.** $\{(\frac{2}{5}, \frac{7}{5})\}$ **53.** $\{(-1, 2, 1)\}$
55. $\{(\frac{3}{4}t - \frac{1}{2}, 1 - \frac{3}{2}t, t)\}$, dependent **57.** \$34,000
59. (a) 2800; (b) 4000 **61.** 28 days for A and
21 days for B **63.** $x^2 + y^2 - 4x + 8y - 5 = 0$
65. Thirty 20-cent stamps; forty-five 30-cent stamps;
twenty-five 42-cent stamps **67.** 400 type A and 400
type B **69.** 12 in. by 10 in.

71. $\begin{bmatrix} -2 & -6 \\ -7 & -21 \\ 19 & 5 \end{bmatrix}$ **73.** $\begin{bmatrix} 0 & 2 \\ -1 & 1 \end{bmatrix}$ **75.** $\begin{bmatrix} 2 & -1 & 3 \\ -2 & -4 & -3 \end{bmatrix}$

77. $\begin{bmatrix} -1 & 3 \\ 0 & 2 \end{bmatrix}$ **79.** $\begin{bmatrix} 2 & 3 \\ -1 & 4 \end{bmatrix}$ **81.** $\begin{bmatrix} 0 & 0 & \frac{1}{2} \\ \frac{1}{3} & 0 & -\frac{1}{6} \\ -\frac{1}{3} & 1 & \frac{1}{6} \end{bmatrix}$

83. $\{(-1, 2, 1)\}$

Chapter 10

Exercises 10.1 (Page 596)

1. 20 **3.** 5 **5.** 1 **7.** -8 **9.** Yes **11.** No
13. Yes **15.** -7 **17.** -3 and 1
19. $2x^2 + 7x + 31, 136$ **21.** $2x^3 - 3x^2 + 12x - 50, 199$
23. $3z^4 + 7z^3 + 14z^2 + 24z + 48, 103$
25. $6x^2 - 3x + 3, 1$
27. $x^6 + x^5 + x^4 + x^3 + x^2 + x + 1, 0$ **29.** $8, -157$
31. $-7, 126$ **33.** $4, \frac{19}{2}$ **39.** No **41.** Yes
43. (a) Yes; (b) No; (c) No; (d) Yes **45.** All positive
integers **47.** All positive odd integers

Exercises 10.2 (Page 609)

1. 4, multiplicity two; -2; 2 **3.** 0, multiplicity two;
-1, multiplicity two; $-\sqrt{3}$; $\sqrt{3}$ **5.** -7, multiplicity
three; $-\sqrt{7}$, multiplicity two; $\sqrt{7}$, multiplicity two
7. $-\frac{4}{3}$, multiplicity three; $-\frac{3}{2}$, multiplicity four; $\frac{3}{2}$,
multiplicity two **9.** $-1, 4$ **11.** $\dfrac{-1 \pm \sqrt{5}}{2}$ **13.** $1, \frac{2}{5}$
15. $-2 \pm \sqrt{3}$ **17.** $-1, 1, 3$ **19.** $-2, -1, 3$
21. $-5, 3, \frac{1}{2}(-1 \pm i\sqrt{3})$ **23.** $\frac{1}{3}, \dfrac{-3 \pm \sqrt{5}}{2}$
25. $\frac{1}{2}, \frac{5}{3}, 2 \pm \sqrt{3}$ **27.** 2 is the only rational zero
29. 1, multiplicity two; -4 **31.** $\frac{3}{2}, 2, 3$

33. $-4, -2, 0, 1, 3$ **35.** $-\frac{1}{4}, \frac{2}{3}, \pm\sqrt{3}$
37. $-6, \frac{1}{6}(1 \pm i\sqrt{11})$ **39.** -2 is the only rational root
45. 2 cm or $(4 - \sqrt{11})$ cm ≈ 0.68 cm
47. 3 ft or $\frac{1}{2}(1 + \sqrt{13})$ ft ≈ 2.3 ft

Exercises 10.3 (Page 615)

1. 1 positive, 0 negative, 2 imaginary **3.** 2 positive,
1 negative, and 0 imaginary; or 0 positive, 1 negative,
and 2 imaginary **5.** 3 positive, 0 negative, and 0
imaginary; or 1 positive, 0 negative, and 2 imaginary
7. 1 positive, 1 negative, and 2 imaginary
9. 0 positive, 1 negative, and 2 imaginary; one root is 0
11. 3 positive, 0 negative, and 2 imaginary;
or 1 positive, 0 negative, and 4 imaginary **13.** One
negative irrational root between -3 and -2; one
positive irrational root between 0 and 1; one positive
irrational root between 2 and 3 **15.** One negative
irrational root between -1 and 0; one positive
irrational root between 0 and 1; two imaginary roots
17. -1 is a root; one positive irrational root between
1 and 2; two imaginary roots **19.** 4 is a root; one
negative irrational root between -1 and 0; one
positive irrational root between 0 and 1; one positive
irrational root between 2 and 3 **21.** 2.65 **23.** 0.51
25. -1.12 **33.** 3.9 in. **35.** 1.33 in., 4.13 in.

Exercises 10.4 (Page 623)

1. $3i, 1, 2$ **3.** $2 - i, 3$ **5.** $2, 1 \pm i$ **7.** $-1, \frac{1}{2}, 2 \pm 3i$
9. $2, 2i$ (multiplicity two), $-2i$ (multiplicity two)
11. $\frac{1}{3}$ (multiplicity two), $2 \pm \sqrt{2}i$ **13.** $-4, \dfrac{-1 \pm i\sqrt{3}}{2}$
15. $1, 2, \pm\sqrt{2}i$ **17.** $-\frac{1}{2}$ (multiplicity two), $-1 \pm i$
19. $x^2 - 8x + 25$ **21.** $x^3 - 7x + 36$
23. $x^4 + 18x^2 + 81$
25. $x^5 - 9x^4 + 31x^3 - 51x^2 + 58x - 66$
27. $\left\{ \dfrac{1 \pm \sqrt{31}}{3}, -2i, 2i \right\}$
29. $\left\{ \dfrac{1 \pm i}{2}, -2 - 4i, -2 + 4i \right\}$
31. $(x - 2)(x^2 - 2x + 2)$
33. $(x + 1)(2x - 1)(x^2 - 4x + 13)$ **35.** $(x - 2)(x^2 + 4)^2$

Exercises 10.5 (Page 630)

1. $\dfrac{3}{x - 2} - \dfrac{3}{x + 2}$ **3.** $-\dfrac{1}{x} + \dfrac{2}{x + 1}$
5. $\dfrac{4}{x - 3} - \dfrac{3}{x - 1}$ **7.** $\dfrac{-5}{3x + 1} + \dfrac{2}{x - 2}$
9. $\dfrac{2}{x} - \dfrac{3}{3x - 2} - \dfrac{1}{2x + 3}$ **11.** $2x + \dfrac{5}{x - 2} + \dfrac{3}{x + 2}$
13. $2x - 5 + \dfrac{3}{2x - 3} + \dfrac{2}{x + 2}$ **15.** $\dfrac{5}{x^2} - \dfrac{4}{x} + \dfrac{7}{x - 2}$
17. $\dfrac{2}{x + 2} - \dfrac{3}{(x - 2)^2} - \dfrac{1}{x - 2}$
19. $\dfrac{58}{9(2x + 1)^2} + \dfrac{134}{27(2x + 1)} + \dfrac{19}{9(x - 1)^2} - \dfrac{67}{27(x - 1)}$
21. $1 + \dfrac{16}{(x - 2)^3} + \dfrac{10}{(x - 2)^2} + \dfrac{2}{x - 2}$
23. $\dfrac{4}{x} - \dfrac{x + 5}{x^2 + x + 1}$ **25.** $\dfrac{1}{x - 2} + \dfrac{2x + 4}{x^2 + 2x + 4}$
27. $-\dfrac{2}{x - 1} + \dfrac{4x - 3}{x^2 + 1}$ **29.** $\dfrac{4}{x + 4} + \dfrac{3x - 1}{2x^2 + 3}$
31. $\dfrac{x + 2}{x^2 + 1} - \dfrac{x - 2}{x^2 - x - 1}$ **33.** $\dfrac{2}{x^2} - \dfrac{1}{x} + \dfrac{x}{x^2 + 2x + 3}$

35. $\dfrac{x - 1}{x^2 + 1} - \dfrac{x - 1}{(x^2 + 1)^2}$
37. $\dfrac{x - 15}{(x^2 + 2)^2} + \dfrac{3x + 4}{x^2 + 2} - \dfrac{2}{x - 1}$

Review Exercises for Chapter 10 (Page 631)

1. -7 **3.** Yes **5.** $\frac{5}{7}$ **7.** $2x^3 + x^2 - 3x + 5, -10$
9. $x^5 + 2x^4 + 4x^3 + 8x^2 + 16x + 32, 0$ **11.** (a) 17;
(b) 57 **13.** -3, multiplicity two; 1; $\frac{5}{2}$
15. -2, multiplicity two; 2, multiplicity two; -3; $-\frac{5}{3}$; $\frac{3}{2}$; 3
17. $-1 \pm \sqrt{5}$ **19.** $-3, 2, 2 \pm \sqrt{2}$ **21.** $\frac{2}{3}, \dfrac{3 \pm \sqrt{5}}{2}$
23. $3, -2 \pm \sqrt{7}$ **25.** $-\frac{1}{3}, \frac{1}{2}, 2 \pm \sqrt{5}$
27. $\{\pm\sqrt{2}, -1 \pm \sqrt{3}\}$ **29.** 2 positive, 1 negative, and
0 imaginary; or 0 positive, 1 negative, and
2 imaginary **31.** 1 positive; 1 negative; and
2 imaginary **33.** One positive irrational root
between 3 and 4; two imaginary roots **35.** 2 is a root;
one negative irrational root between -4 and -3; one
positive irrational root between 0 and 1; one positive
irrational root between 3 and 4 **37.** -1.58
39. $\{\frac{3}{4}, 1 \pm 2i\}$ **41.** $\{-\frac{1}{3}, \frac{1}{2}, 2 \pm \sqrt{5}\}$
43. (a) $\frac{1}{2}, -1 \pm \sqrt{3}i$; (b) $(2x - 1)(x^2 + 2x + 4)$
45. (a) $\pm 2i, -1 \pm i$; (b) $(x^2 + 4)(x^2 + 2x + 2)$
47. $\left\{ -1 \pm i, \dfrac{-1 \pm \sqrt{5}}{2} \right\}$
49. $x^4 - 4x^3 + 10x^2 - 12x + 8$
51. $x^6 - 4x^5 + 24x^4 - 72x^3 + 189x^2 - 324x + 486$
53. 1 **55.** 3 cm, 4 cm, 5 cm **57.** 0.98 in.
59. $-\dfrac{3}{x - 1} - \dfrac{1}{x + 2} + \dfrac{6}{x + 4}$
61. $\dfrac{6}{7(x - 2)} + \dfrac{x + 10}{7(x^2 + x + 1)}$
63. $1 + \dfrac{x - 2}{x^2 + 1} - \dfrac{3}{(x^2 + 1)^2}$

Chapter 11

Exercises 11.1 (Page 641)

1. $5, 7, 9, 11, 13, 15, 17, 19$
3. $\frac{2}{1}, \frac{5}{2}, \frac{10}{3}, \frac{17}{4}, \frac{26}{5}, \frac{37}{6}, \frac{50}{7}, \frac{65}{8}$
5. $\frac{2}{1}, -1, \frac{4}{5}, -\frac{5}{7}, \frac{2}{3}, -\frac{7}{11}, \frac{8}{13}, -\frac{3}{5}$

7. $-\frac{2}{3}, \frac{4}{5}, -\frac{8}{9}, \frac{16}{17}, -\frac{32}{33}, \frac{64}{65}, -\frac{128}{129}, \frac{256}{257}$
9. $\frac{1}{3}x, -\frac{1}{4}x^2, \frac{1}{5}x^3, -\frac{1}{6}x^4, \frac{1}{7}x^5, -\frac{1}{8}x^6, \frac{1}{9}x^7, -\frac{1}{10}x^8$
11. $0, 4, 0, 8, 0, 12, 0, 16$ **13.** $1, 2, \frac{1}{2}, 2, \frac{1}{3}, 2, \frac{1}{4}, 2$
15. $1, 1, 2, 2, 3, 3, 4, 4$ **17.** $1 + 5 + 9 + 13 + 17; 45$

19. $\frac{2}{1} + \frac{3}{2} + \frac{4}{3} + \frac{5}{4} + \frac{6}{5}$; $\frac{437}{60}$

21. $5 + 5 + 5 + \ldots + 5$, (100 terms); 500

23. $1 + \frac{1}{2} + \frac{1}{4} + \frac{1}{8} + \frac{1}{16} + \frac{1}{32}$; $\frac{63}{32}$

25. $x - x^3 + x^5 - x^7 + x^9 - x^{11} + x^{13} - x^{15}$

27. $a_0 - 2a_1 + 3a_2 - 4a_3 + 5a_4 - 6a_5 + 7a_6 - 8a_7$

29. $f(x_0) + f(x_1) + f(x_2) + \ldots + f(x_{n-1})$ **31.** $\displaystyle\sum_{i=1}^{6} (2i - 1)$

33. $\displaystyle\sum_{i=1}^{5} (-1)^{i-1}(3i + 1)$ **35.** $\displaystyle\sum_{i=1}^{6} \frac{2i - 1}{i^2}$

37. $\displaystyle\sum_{i=1}^{5} (-1)^{i-1} \frac{x^{2i-2}}{2i}$

39. $1, 3, 5, 7, 9, 11, \ldots, 2n - 1, \ldots$;

$1, 3, 5, 5, 9, 7, \ldots, a_n, \ldots$, where

$$a_n = \begin{cases} 2n - 1, & \text{if } n \text{ is odd} \\ n + 1, & \text{if } n \text{ is even} \end{cases}$$

$1, 3, 5, 13, 33, 71, \ldots, a_n, \ldots$, where

$$a_n = 2n - 1 + (n - 1)(n - 2)(n - 3)$$

Exercises 11.3 (Page 658)

1. $5, 8, 11, 14, 17$ **3.** $10, 6, 2, -2, -6$

5. $-5, -12, -19, -26, -33$

7. $x, x + 2y, x + 4y, x + 6y, x + 8y$

9. An arithmetic sequence; $-13, -17$

11. Not an arithmetic sequence   **13.** An arithmetic sequence; $0, -\frac{1}{12}$ **15.** An arithmetic sequence; $4x + 3y, 5x + 4y$ **17.** 35   **19.** 16   **21.** Thirtieth   **23.** $\frac{26}{5}, \frac{27}{5}, \frac{28}{5}, \frac{29}{5}$ **25.** $\frac{143}{2}$ **27.** 108   **29.** 2500

31. $-\frac{25}{2}$ **33.** 2430   **35.** $280, $260, $240, $220

37. 120   **39.** 18

Exercises 11.4 (Page 667)

1. $5, 15, 45, 135, 405$ **3.** $8, -4, 2, -1, \frac{1}{2}$

5. $-\frac{9}{16}, \frac{3}{8}, -\frac{1}{4}, \frac{1}{6}, -\frac{1}{9}$ **7.** $\frac{x}{y}, -1, \frac{y}{x}, -\frac{y^2}{x^2}, \frac{y^3}{x^3}$

9. A geometric sequence; $27, 81$ **11.** A geometric sequence; $3\sqrt{6}, 9\sqrt{2}$ **13.** Not a geometric sequence

15. A geometric sequence; $\frac{2}{9}, -\frac{2}{27}$ **17.** $\frac{729}{4}$ **19.** -3

21. Eleventh   **23.** $2, 4, 8, 16, 32$ **25.** $\sqrt[3]{9}, \sqrt[6]{243}$

27. -2 **29.** 21   **31.** 510   **33.** $\frac{5465}{729}$ **35.** 11.17

37. $6, 9, 12, 16$; or, $6, 1, -4, 16$ **39.** $\frac{117,649}{262,144}$

41. $33,066

Exercises 11.5 (Page 673)

1. $\frac{64}{3}$ **3.** $\frac{600}{11}$ **5.** $\frac{4}{5}$ **7.** $\frac{16}{21}$ **9.** $\dfrac{9 + 3\sqrt{3}}{2}$ **11.** $\frac{2}{3}$

13. $\frac{9}{11}$ **15.** 3   **17.** $\frac{137}{111}$ **19.** $\frac{47}{101}$ **21.** $\frac{29,029}{9,000}$

23. $\frac{244}{99}$ **25.** 84 m   **27.** 400 cm   **29.** 1500

Exercises 11.6 (Page 685)

1. (a) $\frac{1}{504}$; (b) $\frac{33}{2}$; (c) $\frac{1}{168}$ **3.** (a) 60; (b) 720; (c) 7

5. (a) 35; (b) 1225; (c) 1   **7.** (a) 60; (b) 125

9. (a) 180; (b) 240   **11.** (a) 36; (b) 216   **13.** 360

15. 17,280   **17.** (a) 362,880; (b) 79,833,600

19. 220   **21.** 84   **23.** 21   **25.** 210   **27.** 540

29. (a) 2880; (b) 5040   **31.** (a) 6720; (b) 3780

33. 2240   **35.** 12,600   **37.** $\dfrac{52!}{13! \, 39!}$ **39.** 720

41. 5040   **43.** 240   **45.** 1,201,200

Exercises 11.7 (Page 692)

1. (a) $\frac{1}{6}$; (b) $\frac{1}{2}$ **3.** (a) $\frac{1}{18}$; (b) $\frac{1}{6}$ **5.** $\frac{1}{6}$ **7.** (a) $\frac{1}{4}$; (b) $\frac{3}{4}$

9. (a) $\frac{1}{13}$; (b) $\frac{1}{4}$; (c) $\frac{4}{13}$ **11.** (a) $\frac{1}{2}$; (b) $\frac{3}{10}$; (c) $\frac{1}{2}$

13. (a) $\frac{2}{9}$; (b) $\frac{2}{9}$; (c) $\frac{24}{45}$ **15.** (a) $\frac{2}{5}$; (b) $\frac{2}{5}$; (c) $\frac{9}{20}$

17. $\frac{29}{60}$ **19.** $\frac{1}{462}$ **21.** $\frac{1}{5}$ **23.** $\frac{10}{21}$ **25.** $\frac{1}{126}$ **27.** $\frac{7}{9}$

29. $(4)\dfrac{_{13}C_5}{_{52}C_5} \approx 0.002$ **31.** (a) $\frac{1}{8}$; (b) $\frac{3}{8}$; (c) $\frac{1}{2}$; (d) $\frac{1}{2}$

33. $\frac{15}{16}$

Exercises 11.8 (Page 700)

1. $a^8 + 8a^7b + 28a^6b^2 + 56a^5b^3 + 70a^4b^4 + 56a^3b^5 + 28a^2b^6 + 8ab^7 + b^8$ **3.** $x^9 - 9x^8y + 36x^7y^2 - 84x^6y^3 + 126x^5y^4 - 126x^4y^5 + 84x^3y^6 - 36x^2y^7 + 9xy^8 - y^9$

5. $x^5 + 15x^4y + 90x^3y^2 + 270x^2y^3 + 405xy^4 + 243y^5$

7. $4096 - 6144ab + 3840a^2b^2 - 1280a^3b^3 + 240a^4b^4 - 24a^5b^5 + a^6b^6$ **9.** $243e^{5x} - 810e^{3x} + 1080e^x - 720e^{-x} + 240e^{-3x} - 32e^{-5x}$ **11.** $a^4 + 8a^{7/2}b^{1/2} + 28a^3b + 56a^{5/2}b^{3/2} + 70a^2b^2 + 56a^{3/2}b^{5/2} + 28ab^3 + 8a^{1/2}b^{7/2} + b^4$

13. $4096x^{24} + 24,576x^{22}y^2 + 67,584x^{20}y^4 + 112,640x^{18}y^6$ **15.** $\dfrac{1}{u^{11}} - 33\dfrac{v^2}{u^{10}} + 495\dfrac{v^4}{u^9} - 4455\dfrac{v^6}{u^8}$

17. $a^3 - 9a^{8/3}b^{1/3} + 36a^{7/3}b^{2/3} - 84a^2b$ **19.** $3003a^8b^6$

21. $-489,888x^4$ **23.** $\dfrac{924x^{18}}{y^{12}}$ **25.** $\frac{231}{16}a^6$ **27.** -8064

29. $a^9 + 9a^8b + 36a^7b^2 + 84a^6b^3 + 126a^5b^4 + 126a^4b^5 + 84a^3b^6 + 36a^2b^7 + 9ab^8 + b^9$

31. $r^{12} - 12r^{11}t + 66r^{10}t^2 - 220r^9t^3 + 495r^8t^4 -$

$792r^7t^5 + 924r^6t^6 - 792r^5t^7 + 495r^4t^8 - 220r^3t^9 + 66r^2t^{10} - 12rt^{11} + t^{12}$ **33.** 1.172 **35.** $(9.510)10^9$

Review Exercises for Chapter 11 (Page 700)

1. $\frac{1}{3}, -\frac{1}{3}, \frac{5}{27}, -\frac{7}{81}, \frac{1}{27}, -\frac{11}{729}$

3. $3 + 2 + \frac{5}{3} + \frac{3}{2} + \frac{7}{5} + \frac{4}{3} = \frac{109}{10}$

5. $\displaystyle\sum_{i=1}^{4} (-1)^{i-1} \frac{x^{2i}}{2^i}$ **7.** $\frac{3069}{1024}$ **9.** 130 **11.** (a) 55; (b) $\frac{8}{7}$

13. $\frac{7}{16}$ **15.** $\frac{1}{9}$ or $-\frac{1}{9}$ **17.** Twenty-first

19. $96, 48, 24, 12, 6$ **21.** $\frac{8}{19}$ **23.** (a) 50; (b) $15,000$
25. $\frac{8}{11}$ **31.** (a) 5040; (b) 35; (c) 120
33. $128x^7 - 448x^6y + 672x^5y^2 - 560x^4y^3 + 280x^3y^4 - 84x^2y^5 + 14xy^6 - y^7$ **35.** $x^{20} + 40x^{18} + 760x^{16} + 9120x^{14}$
37. $-5005t^{-3/2}$ **39.** $-\frac{231}{16}z^{12}$ **41.** $x^9 - 9x^8y + 36x^7y^2 - 84x^6y^3 + 126x^5y^4 - 126x^4y^5 + 84x^3y^6 - 36x^2y^7 + 9xy^8 - y^9$ **43.** 455 **45.** (a) $1,049,000$;
(b) $2,097,000$ **47.** $20, 10, 5,$ or $5, 10, 20$
49. 107 km **51.** $362,880$ **53.** 252 **55.** 900
57. 450 **59.** (a) 180; (b) 36 **61.** (a) $\frac{1}{3}$; (b) $\frac{1}{2}$
63. (a) $\frac{3}{4}$; (b) $\frac{3}{4}$ **65.** (a) $\frac{1}{8}$; (b) $\frac{1}{2}$ **67.** $\frac{2}{35}$

Appendix

Exercises A.1 (Page A-5)

1. Commutative law for multiplication (Axiom 2)
3. Existence of additive identity element (Axiom 5)
5. Commutative law for addition (Axiom 2)
7. Associative law for multiplication (Axiom 3)
9. Existence of additive inverse (Axiom 6)
11. Commutative law for addition (Axiom 2)
13. Existence of multiplicative inverse (Axiom 7)
15. Commutative law for multiplication (Axiom 2)
17. Existence of additive identity element (Axiom 5)
19. Existence of multiplicative identity element
(Axiom 5) **21.** Distributive law (Axiom 4)

23. Associative law for addition (Axiom 3)
25. Closed; $2 + 3 = 5$ **27.** Closed; $3 \cdot 5 = 15$
29. Closed; $0 + 0 = 0$ **31.** Closed; $1 \cdot 1 = 1$
33. Not closed; $1 + 2 = 3$ **35.** Closed; $0 \cdot 1 = 0$,
$0 \cdot 0 = 0, 1 \cdot 1 = 1$

Exercises A.2 (Page A-10)

1. 0.4582 **3.** 0.9772 **5.** 1.145 **7.** 1.751
9. 4.143 **11.** 0.7930 **13.** 1.5229 **15.** 0.7292
17. 1.0153 **19.** 4.5182 **21.** 0.454 **23.** 0.972
25. 0.715 **27.** 0.369 **29.** 1.007 **31.** $59.33°$
33. $37.42°$ **35.** $68.04°$ **37.** $28.93°$ **39.** $75.48°$

Index